# Measure and Integral

This accessible introduction to the topic covers the theory of measure and integral, as introduced by Lebesgue and developed in the first half of the 20th century. It leads naturally to Banach spaces of functions and linear operators acting on them.

The material in *Measure and Integral: Theory and Practice* is typically covered in a graduate course and is almost always treated in an abstract way, with little or no motivation. The author employs a plethora of examples and exercises and strives to motivate every concept with its historical background. This textbook is accessible to a wider range of students, including at the undergraduate level.

A major problem facing anyone teaching measure theory is how to combine the elementary approach (measure on the real line or in the plane) and the abstract measure theory. The author develops a theory of measure in the plane, then shows how to generalize these ideas to an abstract setting.

The result is a textbook accessible to a wider range of students.

The material requires a good understanding of topics often referred to as advanced calculus, such as Riemann integration on Euclidean spaces and series of functions. Also, a reader is expected to be proficient in the basics of set theory and point-set topology, preferably including metric spaces.

**John Srdjan Petrovic** was born in Belgrade, Yugoslavia. He earned a PhD from the University of Michigan under the direction of Dr. Carl Pearcy. His research area is the theory of operators on Hilbert space, and he has published more than 30 articles in prestigious journals. He is a professor of mathematics at Western Michigan University and his visiting positions include Texas A&M University, Indiana University, and University of North Carolina Charlotte. His text, *Advanced Calculus: Theory and Practice,* is in its second edition (CRC Press).

**Textbooks in Mathematics**
Series editors:
Al Boggess, Kenneth H. Rosen

**An Introduction to Optimization with Applications in Data Analytics and Machine Learning**
*Jeffrey Paul Wheeler*

**Encounters with Chaos and Fractals, Third Edition**
*Denny Gulick and Jeff Ford*

**Differential Calculus in Several Variables**
A Learning-by-Doing Approach
*Marius Ghergu*

**Taking the "Oof!" Out of Proofs**
*Alexandr Draganov*

**Vector Calculus**
*Steven G. Krantz and Harold Parks*

**Intuitive Axiomatic Set Theory**
José Luis García

**Fundamentals of Abstract Algebra**
*Mark J. DeBonis*

**A Bridge to Higher Mathematics**
*James R. Kirkwood and Raina S. Robeva*

**Advanced Linear Algebra, Second Edition**
*Nicholas Loehr*

**Mathematical Biology: Discrete and Differential Equations**
*Christina Alvey and Daniel Alvey*

**Numerical Methods and Analysis with Mathematical Modelling**
*William P. Fox and Richard D. West*

**Business Process Analytics**
Modeling, Simulation, and Design
*Manuel Laguna and Johan Marklund*

**Quantitative Literacy Through Games and Gambling**
*Mark Hunacek*

**Measure and Integral**
Theory and Practice
*John Srdjan Petrovic*

https://www.routledge.com/Textbooks-in-Mathematics/book-series/CANDHTEX-BOOMTH

# Measure and Integral
## Theory and Practice

John Srdjan Petrovic

CRC Press
Taylor & Francis Group
Boca Raton London New York

CRC Press is an imprint of the
Taylor & Francis Group, an **informa** business
A CHAPMAN & HALL BOOK

Seventh edition published 2025
by CRC Press
2385 NW Executive Center Drive, Suite 320, Boca Raton FL 33431

and by CRC Press
4 Park Square, Milton Park, Abingdon, Oxon, OX14 4RN

*CRC Press is an imprint of Taylor & Francis Group, LLC*



*Library of Congress Cataloging-in-Publication Data*
Names: Petrovic, John Srdjan, author.
Title: Measure and integral : theory and practice / John Srdjan
Petrovic.
Description: First edition. | Boca Raton : CRC Press, 2025. | Series:
Textbooks in mathematics series | Includes bibliographical references
and index.
Identifiers: LCCN 2024036271 (print) | LCCN 2024036272 (ebook) | ISBN
9781032712420 (hbk) | ISBN 9781032716534 (pbk) | ISBN 9781032716497
(ebk)
Subjects: LCSH: Measure theory. | Lebesgue integral.
Classification: LCC QA312 .P453 2025 (print) | LCC QA312 (ebook) | DDC
515/.42--dc23/eng/20240929
LC record available at https://lccn.loc.gov/2024036271
LC ebook record available at https://lccn.loc.gov/2024036272

ISBN: 978-1-032-71242-0 (hbk)
ISBN: 978-1-032-71653-4 (pbk)
ISBN: 978-1-032-71649-7 (ebk)

DOI: 10.1201/9781032716497

Typeset in Latin Modern font
by KnowledgeWorks Global Ltd.

Publisher's note: This book has been prepared from camera-ready copy provided by the authors.

# Contents

# Preface

This text was written for a one-semester or a two-semester course that encompasses the theory of measure and integration, and an introduction to functional analysis. It has become standard practice to merge them all under the label of real analysis, and I will follow this practice in the book. The monograph has several goals: to present measure theory, both from theoretical and practical standpoint; to help students become comfortable with abstract and axiomatic systems; to introduce Lebesgue integration theory and point out at its advantages compared with Riemann integration theory; to make the reader aware of the rapid growth of real analysis during the early years of the 20th century and the reasons behind it; and to point out the connection between various topics and how their development inspired each other. In writing this book, I was guided by my long-standing experience and interest in teaching analysis at a variety of levels. In addition, in the functional analysis sections, I relied on my research experience in operator theory.

For the student, my purpose was to lay out material in a precise, readable manner, with the concepts and techniques clearly presented. I wanted to give students an understanding of abstract mathematical theories together with a sense of why these concepts matter and where they come from. My goal was to ensure that students acquire good mathematical skills, both in understanding why a theorem is true and how to prove it. Finally, my intention was that a student become proficient in using these wonderful tools acquired throughout the material.

For the instructor, my purpose was to create a flexible, comprehensive teaching tool using proven pedagogical techniques in mathematics. I wanted to provide instructors with a text that they could use to teach real analysis effectively and efficiently in the most appropriate manner for their particular set of students.

This book is designed to meet the needs of a one- or two-term real analysis course. It is very flexible and highly comprehensive. It is intended not only as a successful textbook but also as a valuable resource students can consult throughout their studies and professional life.

## Goals of a Real Analysis Course

A real analysis course has several purposes.

- Students should learn a specific set of mathematical objects, ideas, and techniques, and the role they play in the theory of measure and integration and in functional analysis.
- They should become proficient in establishing further properties of these objects and their relationship.
- They should understand the origins of particular concepts.
- They should learn to find analogies and similarities between different branches of real analysis.

## Features of the Book

- **Motivation of Theorems and Definitions.** An effort is made to explain the reason for a large number of definitions and to motivate theorems by offering examples when the asserted property is on display. On a large scale, measure on $\mathbb{R}^2$ is developed gradually in Chapter 4, from rectangles through elementary and open sets to those that can be approximated by the latter. In a sense, the purpose of the whole chapter is to explain why one needs a tool to extend a measure from a ring to a $\sigma$-algebra and to give sense to a rather obscure Carathéodory construction. On a smaller scale, the standard definition of a measurable function is preceded by a description of properties that the class of such functions should possess, and this is then articulated in Exercise 6.2.25. See also the passage following Remark 12.1.12 that sets the stage for the Hahn–Banach Theorem. To a student, such insights can help make the theory more natural.
- **Selection of Proofs.** Whenever possible, the proofs followed the same philosophy: prove the result for the simplest object (a rectangle, or a characteristic function of a measurable set) then use the established fact to gradually increase the level of generality. As an example, Fubini's Theorem is first considered for characteristic functions (Theorem 8.2.1), then for simple functions (by the linearity of integral), nonnegative functions (Tonelli's Theorem), and finally for integrable functions. For a student, embracing such an approach is an important part of conquering the subject.
- **Historical Information.** A significant effort was made for the book to include a historical prospective. A reader will get a glimpse into the development of real analysis and its ideas from the late 19th century, all the way into the 1930s. Some of the benefits of such an approach include deeper understanding of

- *A concept.* Sets of measure zero were originally confusing but gained importance once it was understood that many results do not hold for all members of a set, but the exceptional set is frequently of measure zero. The definition itself morphed after understanding that one can use infinite covers.
- *The development of some concepts.* Abstract ideas are not easy to digest, so it is comforting to find out that they were not an overnight success. For example, Fréchet generalized Lebesgue integration theory to abstract measure spaces in 1915. Yet, as late as 1933, in the first edition of his classic book *Theory of the Integral* Saks relegated abstract integration to an appendix. Only in the second edition in 1937, the abstract Lebesgue integral comes at the beginning, as the topic of Chapter I.
- *The unity of real analysis.* While the dependence of integration theory on measure theory will be glaringly obvious to a student, the connection of these two with functional analysis may not be. At least, not without understanding the time in which the latter was developing and the concurrent achievements in the former two. Even though the idea of an abstract linear space had been brewing since Grassmann (perhaps even Bolzano) and by the end of the 19th century it appeared in the works of several mathematicians, it was too early to come up with axioms of a Banach space. When Banach eventually did it, he had an abundance of concrete examples to look at, many of which formulated some convergence condition using integrals.
- *Mathematics as a developing science.* It is helpful to know that some of the "natural" ideas were accepted until they proved to be flawed, and that it took a long time and several tries to come up with a proper formulation. It also helps students understand that mathematics has had (and still has) its share of disagreements and controversies, and that it is a lively area full of opportunities to contribute to its advancement. There is a long list of references which will allow the instructor to direct students to further research the history of a concept or a result.

- **Worked Examples.** The book contains close to 200 worked-out examples. In most cases a question is first asked, then its solution is presented. To signal the end of the example, the symbol ♦ is used. Examples have several different aims: to introduce a new concept; to help a student develop their problem-solving skills; to motivate and lead students to important theorems and definitions; and to help students acquire a habit of scrutinizing theorems. Does the converse hold? Will it remain true if one of the hypotheses is relaxed? If not, what is the counterexample?

- **Exercises.** The book contains more than 1,600 exercises that will help students develop their problem-solving skills. Whenever appropriate, exercises are grouped and solutions or hints are provided for at least one of the questions. An effort was made to include exercises that lead the student through the original proofs of important results. Harder problems are marked by an asterisk, sometimes with two if the solution is really hard to come by.

- **Key Terms and Authors.** To facilitate the use of the book, there is a comprehensive subject index and a separate author index.

- **The Philosophy.** The exposition follows the idea that learning goes from specific to general. Perhaps it was best formulated by Ralph Boas in [11]:

  > *Suppose that you want to teach the 'cat' concept to a very young child. Do you explain that a cat is a relatively small, primarily carnivorous mammal with retractible claws, a distinctive sonic output, etc.? I'll bet not. You probably show the kid a lot of different cats, saying 'kitty' each time, until it gets the idea. To put it more generally, generalizations are best made by abstraction from experience.*

  John B. Conway echoes the same idea in [25]:

  > *To many, mathematics is a collection of theorems. For me, mathematics is a collection of examples; a theorem is a statement about a collection of examples and the purpose of proving theorems is to classify and explain the examples...*

- **Illustrations.** The book contains 34 illustrations that will help the student follow the reasoning in examples or proofs.

- **Accessibility.** The prerequisites for a course based on this book are all covered in a standard mathematics major, including what is known as "Advanced Calculus". In particular, a solid foundation in the theory of Riemann integrals is indispensable, both in one and several variables. A reader should also feel comfortable with topics such as sequences of functions and their convergence (uniform or pointwise) and topology in Euclidean spaces. Some knowledge of the Fourier series and problems arising in connection with them would be very useful because parts of functional analysis come to life when applied to classical questions from the theory of functions. A student should be familiar with the fundamental role that the Fourier series had in the development of modern analysis.

- **Homogeneity.** The philosophy followed in writing the book was to define the starting point (the "Epilogue") and the ending (the "Prologue"), with the material logically flowing from the former to the latter. That way the reader can appreciate the reasons for the development of each topic covered in the text and have an understanding of where they evolved beyond the content of the monograph.

- **Focus.** Each topic in the book is examined from various viewpoints. One that is frequently ignored or brushed aside in textbooks concerns the word *practice* in the title of the book. For example: how does one calculate the Lebesgue integral of a function? (Section 7.4.) What are some applications of functional analysis? (Theorems 10.7.11 and 11.3.5.)

- **Mathematical Rigor and Precision.** All definitions and theorems in this text are stated extremely carefully so that students will appreciate the precision of language and rigor needed in calculus. Every step of a proof is either carefully justified or stated as an exercise.

- **Flexibility.** An effort was made to present the material in a unified way. Most sections allow the instructor to cover the material in a 50-minute class, assuming that some of the proofs are left for students to read on their own. Nevertheless, there are parts that can be omitted. If students are already familiar with the content of the part dealing with preliminaries, the first two chapters can be omitted. The only standout is the Baire Category Theorem which can be covered independently. Also, it is possible to skip Chapter 4 and go directly to the abstract treatment of the measure, but that requires a higher level of preparedness on the part of students. In that scenario, Section 5.3 would become unnecessary, except for a few results specific to $\mathbb{R}^n$, such as Theorems 5.3.7–5.3.10.

- **Writing style.** The writing style in this book is direct and pragmatic. Precise mathematical language is used without rhetorical flourishes. Care has been taken to make things simple and clear.

- **Writing projects.** Throughout the text, efforts were made to include the historical development of important theorems. These can be expanded by students by supplying either the proofs of original (frequently weaker) results or finding in the literature examples that contradicted them. The book contains numerous references for students to conduct this research.

- **Suggested reading.** While the book is trying to be comprehensive, it is impossible to cover everything. Whenever there is more to read about a topic, a reference is given. These include classical textbooks, expository articles, and research publications.

## How to Use This Book

The book has been carefully written and constructed to support real analysis courses at different levels and with differing goals. It can be used either for a year-long course or for the study of measure and integration in a one-semester course. For schools that operate on a quarter system, each part of the book can be covered in one quarter.

The first part, titled "Preliminaries" is likely known to students and can be covered at the pace that is appropriate to their background or even completely omitted. In particular, Sections 1.1 and 1.2 contain set-theoretic formulas that are used later in the text, and their verification can be postponed.

Part IV, titled "An Introduction to Functional Analysis", need not be completely covered. Students will benefit from seeing the integration theory in action and getting to know the direction in which the theory evolved. Yet, if pressed with time, this part can be reduced to Chapter 10 with Section 10.7 dispensable.

Parts II and III are the essence of the book. Yet, even there some sections need not be covered. While we feel that there is a strong pedagogical reason to include the Jordan measure (Section 4.1) and signed measures (Section 5.5), the remainder of the book is independent of them.

To help the instructor design a course, the table below identifies the optional sections. The symbol $R$ denotes a review of material that should be familiar to students from classes they have almost certainly taken.

| Chapter | Optional |
|---|---|
| 1 | All$^R$ |
| 2 | All$^R$ |
| 3 | All$^R$ |
| 4 | 4.1 |
| 5 | 5.5 |
| 7 | 7.1$^R$, 7.4 |
| 10 | 10.7 |
| 12 | 12.4-12.5 |

The optional and review sections

## Acknowledgments

It is my pleasant duty to thank several people who have made significant contributions to the quality of the book. Shelley Speiss is a rare combination of an artist and a computer whiz. She has made all the illustrations in the book, on some occasions making an immeasurable improvement of my original sketches. Daniel Sievewright has read the whole text and even checked the calculations in many examples and exercises (I know that for a fact because he discovered numerous errors, both of a typographical and computational nature). Darryl Jent also read the whole text. He pointed at sections that needed polishing and provided direction in clarifying parts of the exposition.

I am grateful to Mr. Robert E. Ross, senior editor at CRC Press, who has shown a lot of patience and willingness to help throughout the publishing process. Several anonymous reviewers have made valuable suggestions, and my thanks goes to them.

My family has been, as always, extremely supportive and helped me sustain the energy level needed to finish the project.

This book is dedicated to my Real Analysis teachers, Slobodan Aljančić and Branislav Mirković, whose beautiful lectures and contagious enthusiasm contributed to my affinity for the subject.

John Srdjan Petrovic
Western Michigan
University

# Prologue

The concept of the integral has an interesting past. Newton defined it using infinitesimals which did not always inspire confidence, so until the beginning of the 19th century, mathematicians preferred to think of it as the antiderivative. Cauchy was the first to define it rigorously in 1821 for the case when the integrand is a continuous function. The evolution of the concept of a function dictated that the class of integrable functions be expanded, and much of the work during the following few decades was aimed at decreasing the role of continuity. After some unsuccessful attempts by Dirichlet, Lipschitz, and Cauchy, himself, Riemann accomplished this in 1854. In effect, instead of the continuity of the integrand $f$, he required only the convergence of the sequence of sums

$$\sum_{k=1}^{n} f(\xi_k)[x_k - x_{k-1}],$$

regardless of the choice of $\xi_k \in [x_{k-1}, x_k]$, as long as the norm of the partition $\{a = x_0, x_1, \ldots, x_n = b\}$ went to 0. While we call them Riemann sums, they were actually introduced by Cauchy (although Cauchy used them for continuous functions only). Within the framework of such sums, Riemann expanded the class of integrable functions as far as it was possible. Still, there was a level of dissatisfaction among mathematicians because the feeling was that the class was still not large enough to encompass many functions that appeared in applications.

Actually, this integral was inadequate in more than one way, but we will return to this shortly. At this point we mention that the transition to the multivariable case revealed a naturally arising problem. When the integrand is one, the value of the integral should give a "measure" of the domain of integration, and so it was necessary to develop some kind of theory that would associate to each set in the plane a nonnegative number—its measure. The foundation was laid by Jordan who developed the theory of content around 1890. At the same time, Peano did similar work.

The development of measure theory provided a different view at the concept of the integral. If $f$ is nonnegative, the value of the integral

$$\int_a^b f(x)\,dx$$

represents the area under the graph. More precisely, let $E$ denote the set of points in the plane bounded by the graph of $f$, the lines $x = a$ and $x = b$, and the $x$-axis. Then we can say that $f$ is Riemann integrable if and only if the set $E$ is Jordan measurable with measure $c(E)$, and

$$\int_a^b |f| = c(E), \quad \int_a^b f = c(E^+) - c(E^-),$$

where the sets $E^+$ and $E^-$ are the parts of $E$ above and below the $x$-axis, respectively. Thus, generalizing the idea of the content would lead to more integrable functions.

Another measure-theoretic-based approach to the integral is related to the so-called lower and upper Darboux sums. (They were introduced by Riemann in 1854.) Assuming that $f$ is a bounded function, we define the numbers $M_i = \sup\{f(x) : x \in [x_{i-1}, x_i]\}$ and $m_i = \inf\{f(x) : x \in [x_{i-1}, x_i]\}$, and the sums

$$L(f,P) = \sum_{i=1}^n m_i[x_k - x_{k-1}] \text{ and } U(f,P) = \sum_{i=1}^n M_i[x_k - x_{k-1}],$$

are the **lower** and the **upper** Darboux sums corresponding to a partition $P = \{a = x_0, x_1, \dots, x_n = b\}$. If, instead, the interval $[a,b]$ is partitioned into a disjoint union of measurable sets $\{E_i : 1 \leq i \leq n\}$, then we can consider the sums

$$\tilde{L}(f,\tilde{P}) = \sum_{i=1}^n m_i c(E_i), \quad \tilde{U}(f,\tilde{P}) = \sum_{i=1}^n M_i c(E_i).$$

Since the classical partitions $\mathcal{P}$ form a subset of the more general partitions $\tilde{\mathcal{P}}$, we have that

$$\begin{aligned}\underline{\int_a^b} f(x)\,dx &= \sup\{L(f,P) : P \in \mathcal{P}\} \leq \sup\{\tilde{L}(f,\tilde{P}) : \tilde{P} \in \tilde{\mathcal{P}}\} \\ &\leq \inf\{\tilde{U}(f,\tilde{P}) : \tilde{P} \in \tilde{\mathcal{P}}\} \leq \inf\{U(f,P) : P \in \mathcal{P}\} = \overline{\int_a^b} f(x)\,dx.\end{aligned}$$

Once again, generalizing the idea of the content would lead to more integrable functions.

The question as to how to extend the Jordan measure was brewing for a while until Lebesgue (and independently W. H. Young) obtained their generalizations of the Jordan measure and Riemann integral. It should be said that Vitali also developed an equivalent measure theory but failed to apply it to integration until he became acquainted with Lebesgue's results. Not only did this new theory bring forward a new integral and expand the class of integrable functions, but it rectified many of the shortcomings of the Riemann integral, such as:

- It does not commute with the limit process (including derivatives and integrals).
- The derivative of a differentiable function need not be integrable.
- The limit of a sequence of integrable functions need not be integrable.

Lebesgue presented his work in a series of papers between 1899 and 1902. The first decade of the 20th century saw an increase in activity in developing this theory further. In addition to Lebesgue, Guido Fubini, Pierre Fatou, Ernst Fischer, and Frigyes Riesz contributed substantially to the field. In 1913, Johann Radon created the so-called Lebesgue-Stieltjes integral and opened the door to the abstract treatment of the theory of measure and integration.

# Part I

# Preliminaries

# 1

# *Set Theory*

One of the goals of this monograph is to develop measure theory. By measure, we mean a function that assigns a number to a set. In $\mathbb{R}^2$ we call it the *area*, and we know how to associate such a number to familiar sets like a triangle or a disk. Our goal will be more ambitious: assign such a number to all subsets of $\mathbb{R}^2$. This brings up some difficulties. Suppose you only know how to calculate the area of a rectangle with sides parallel to coordinate axes. Now, if two rectangles are disjoint, we have a good idea of how to calculate the area of their union. What if they are not disjoint? That will require us to break up their union into disjoint rectangles. What if we are looking for the area of a disk? That will require some approximation. The moral is that we have to be able to manipulate sets, and once the underlying space becomes more complicated than the real plane, we will have to rely on set theory and operations like union, intersection, etc. At the very least, we should be comfortable with these operations and the rules they follow. In this chapter you will find many identities and inclusions, starting from the most elementary up to some that are less frequently used, involving, e.g., the limit superior of a sequence of sets (Definition 1.1.18).

Let us start with the basics. Sets belong to the list of mathematical objects that are not defined. Each set is characterized by its elements (or members). If $x$ is an element of a set $S$, we write $x \in S$.

A set $A$ is a **subset** of $B$ if every element of $A$ is also an element of $B$. We write $A \subset B$.

The set that has no elements is called the empty set and is denoted by $\emptyset$.

## 1.1 Operations with Sets

We review the basic (and not so basic) set-theoretic operations.

- The **union** of the sets $A$ and $B$, denoted $A \cup B$, is the set of all objects that are members of $A$, or $B$, or both.

- The **intersection** of the sets $A$ and $B$, denoted $A \cap B$, is the set of all objects that are members of both $A$ and $B$.

DOI: 10.1201/9781032716497-1

• The **set difference** of the sets $A$ and $B$, denoted $A \setminus B$, is the set of all members of $A$ that are not members of $B$.

• When $A$ is a subset of $X$, the set difference $X \setminus A$ is also called the **complement** of $A$ in $X$. If the choice of $X$ is clear from the context, the notation $A^C$ is used instead. In that case $X$ is called the **universal set**.

• The **symmetric difference** of sets $A$ and $B$, denoted $A \triangle B$, is the set of all objects that are a member of exactly one of $A$ and $B$ (elements which are in one of the sets, but not in both). In other words,

$$A \triangle B = (A \setminus B) \cup (B \setminus A) = (A \cup B) \setminus (A \cap B). \tag{1.1}$$

• The **Cartesian product** of the sets $A$ and $B$, denoted $A \times B$, is the set of all ordered pairs $(a, b)$ where $a \in A$ and $b \in B$.

• The **limit superior** of $\{E_n : n \in \mathbb{N}\}$, denoted $\limsup\limits_{n \to \infty} E_n$ or simply $\limsup E_n$, is defined as:

$$\bigcap_{i=1}^{\infty} \bigcup_{n=i}^{\infty} E_n.$$

• The **limit inferior**, denoted $\liminf\limits_{n \to \infty} E_n$ or simply $\liminf E_n$, is defined as:

$$\bigcup_{i=1}^{\infty} \bigcap_{n=i}^{\infty} E_n.$$

We list some fundamental set-theoretic identities that are very easy to prove. More complicated identities and inclusions can be found among exercises. They are all used throughout the text. Here, $X$ is the universal set and $A, B, C \subset X$.

**Commutative property:**
$A \cup B = B \cup A, \quad A \cap B = B \cap A, \quad A \triangle B = B \triangle A.$

**Associative property:**
$(A \cup B) \cup C = A \cup (B \cup C), \quad (A \cap B) \cap C = A \cap (B \cap C).$

**Distributive property:**
$A \cup (B \cap C) = (A \cup B) \cap (A \cup C), \quad A \cap (B \cup C) = (A \cap B) \cup (A \cap C).$

**Identity:** $A \cup \emptyset = A, \quad A \cap X = A.$

**Complement:** $A \cup A^C = X, \quad A \cap A^C = \emptyset.$

**Idempotent laws:** $A \cup A = A, \quad A \cap A = A.$

**Domination laws:** $A \cup X = X, \quad A \cap \emptyset = \emptyset.$

**Absorption laws:** $A \cup (A \cap B) = A, \qquad A \cap (A \cup B) = A.$

**De Morgan's laws:** $(A \cup B)^C = A^C \cap B^C, \qquad (A \cap B)^C = A^C \cup B^C.$

**Double complement or involution law:** $\left(A^C\right)^C = A.$

**Complement laws for the universal set and the empty set:**
$\emptyset^C = X, \qquad X^C = \emptyset.$

**Alternative definition of difference:** $A \setminus B = A \cap B^C.$

Some identities are traditionally stated for two sets even though they are valid for a larger, possibly infinite collection. When the collection is countable, it is customary to think about it as a *sequence* and use the notation $\{A_n\}_{n=1}^{\infty}$ or just $\{A_n\}$. However, we write

$$\{A_i : i \in I\}, \qquad \text{or} \qquad \{A_i\}_{i \in I},$$

when $I$ is a set of any cardinality, finite or infinite. In these situations we call $I$ the *index set.*

Many set-theoretic identities are based on a *tautology*, i.e., a logical statement that is true in all interpretations. We will assume that the reader is proficient with such techniques and provide an example for illustration, not as a teaching tool.

**Example 1.1.1. Proving one of the Absorption laws.** $A \cup (A \cap B) = A$.

**Solution:** Denote by $p$ the statement $x \in A$ and by $q$ the statement $x \in B$. Since $x \in A \cup (A \cap B)$ if and only if $x \in A$ or $x \in A \cap B$, and $x \in A \cap B$ if and only if $x \in A$ and $x \in B$, we see that $x \in A \cup (A \cap B)$ can be stated as $p$ or ($p$ and $q$), i.e., $p \vee (p \wedge q)$. On the other hand, $x \in A$ is $p$, so proving the Absorption law under scrutiny amounts to verifying that $p \vee (p \wedge q) \Leftrightarrow p$ for all possible truth values of $p$ and $q$. ♦

Unfortunately, verifying a tautology is relatively easy with up to three statements, so the method of Example 1.1.1 is useful when there are no more than 3 sets. In many cases it is more efficient to use the set algebra and try to use listed identities.

**Example 1.1.2. Proving a set-theoretic identity.**

Prove that $(A \triangle B)^C = A^C \triangle B$.

**Solution:**

$$\begin{aligned}
(A \triangle B)^C &= ((A \setminus B) \cup (B \setminus A))^C && \text{by definition of symmetric difference} \\
&= ((A \cap B^C) \cup (B \cap A^C))^C && \text{by definition of set difference} \\
&= (A \cap B^C)^C \cap (B \cap A^C)^C && \text{by De Morgan's laws} \\
&= (A^C \cup B) \cap (B^C \cup A) && \text{by De Morgan's laws}
\end{aligned}$$

$= (A^C \cap B^C) \cup (A^C \cap A) \cup (B \cap B^C) \cup (B \cap A)$ by Distributive property
$= (A^C \cap B^C) \cup \emptyset \cup \emptyset \cup (B \cap A)$ by Complement
$= (A^C \cap B^C) \cup (B \cap A)$ by Identity
$= (A^C \setminus B) \cup (B \setminus A^C)$ by definition of set difference
$= A^C \triangle B$ by definition of symmetric difference. ♦

The exercises below were selected to reflect the needs of the course, and many will be of use in subsequent portions of the book.

## Exercises

1.1.1. Prove the identities:

(a) $(A_1 \times B_1) \cap (A_2 \times B_2) = (A_1 \cap A_2) \times (B_1 \cap B_2)$

(b) $(A_1 \times B_1) \setminus (A_2 \times B_2) = [(A_1 \setminus A_2) \times B_1] \cup [(A_1 \cap A_2) \times (B_1 \setminus B_2)]$

(c) $A \times (B \cup C) = (A \times B) \cup (A \times C)$

1.1.2. Prove that the following identities hold for the universal set $X$ and $A, B, C \subset X$.

(a) $C \setminus (A \cap B) = (C \setminus A) \cup (C \setminus B)$.

(b) $C \setminus (A \cup B) = (C \setminus A) \cap (C \setminus B)$.

(c) $(A \cup B) \setminus C = (A \setminus C) \cup (B \setminus C)$.

(d) $(A \cap B) \setminus C = (A \setminus C) \cap (B \setminus C)$.

1.1.3. Prove that the following identities hold for the universal set $X$ and $A, B, C \subset X$.

(a) $C \setminus (B \setminus A) = (A \cap C) \cup (C \setminus B)$.

(b) $(B \setminus A) \cap C = (B \cap C) \setminus A = B \cap (C \setminus A)$.

(c) $(B \setminus A) \cup C = (B \cup C) \setminus (A \setminus C)$.

(d) $(B \setminus A) \setminus C = B \setminus (A \cup C)$.

1.1.4.* Prove that $A \setminus B \subset (A \setminus C) \cup (C \setminus B)$.

1.1.5. Prove that $A \triangle B \subset (A \triangle C) \cup (C \triangle B)$.

1.1.6. Prove that $A \setminus (B \cap C) \subset (A \setminus B) \cup ((A \cap B) \setminus C)$.

1.1.7. Prove that $A \setminus B = (A^C \cup B)^C$.

1.1.8. Prove that $A \cap (E_1 \cup E_2) = (A \cap E_1) \cup (A \cap E_2 \cap E_1^C)$.

1.1.9. Prove that $(A \cup B) \triangle (C \cup D) \subset (A \triangle C) \cup (B \triangle D)$.

1.1.10. Prove that $(A \triangle B) = A^C \triangle B^C$.

1.1.11. Prove that $(A \cap B) \triangle (C \cap D) \subset (A \triangle C) \cup (B \triangle D)$.

1.1.12. Suppose that $E \subset U \cup V$. Prove that $E \setminus (A \cup B) \subset (U \setminus A) \cup (V \setminus B)$.

1.1.13. Suppose that $A \cap B \subset C$. Prove that $A \setminus C \subset A \setminus B$.

1.1.14. Prove that $(A \triangle B) \triangle B = A$.

1.1.15. Prove that $(A \cap C) \setminus (B \cap C) = (A \setminus B) \cap C = (A \cap C) \setminus B$.

1.1.16. Prove that $(A \cap C) \triangle (B \cap C) = (A \triangle B) \cap C$.

1.1.17. Suppose that $A \subset C$. Prove that $A^C \cap B^C = (B \cup C)^C \cup (C \setminus (A \cup B))$.

1.1.18. Prove that $[(A \cap S) \cup (B \cap S^C)]^C = (A^C \cap S) \cup (B^C \cap S^C)$.

1.1.19.* Prove that $(A \setminus B) \triangle (C \setminus D) \subset (A \triangle C) \cup (B \triangle D)$.

1.1.20. Let $\{A_n\}_{n=1}^{\infty}$ and $\{B_n\}_{n=1}^{\infty}$ be arbitrary collections of sets and $A = \cup_{n=1}^{\infty} A_n$, $B = \cup_{n=1}^{\infty} B_n$. Prove that

$$A \triangle B \subset \bigcup_{n=1}^{\infty} A_n \triangle B_n.$$

1.1.21.* Let $\{A_n\}_{n=1}^{\infty}$ be an arbitrary collection of sets. Prove that

$$\bigcap_{n=1}^{\infty} A_n = A_1 \setminus \bigcup_{n=1}^{\infty} (A_1 \setminus A_n).$$

1.1.22.* Let $\{A_n\}_{n=1}^{\infty}$ and $\{B_n\}_{n=1}^{\infty}$ be decreasing collections of sets (meaning that $A_n \supset A_{n+1}$ and $B_n \supset B_{n+1}$ for all $n \in \mathbb{N}$).

(a) Prove that

$$\bigcap_{n=1}^{\infty} (A_n \cup B_n) = \left( \bigcap_{n=1}^{\infty} A_n \right) \cup \left( \bigcap_{n=1}^{\infty} B_n \right).$$

(b) Give an example that equality in (a) need not hold if the sets are not decreasing.

1.1.23. Let $A_0$ be a set and let $A_n \subset A_0$ for all $n \in \mathbb{N}$. Prove that

$$A_0 = (A_0 \setminus A_1) \cup (A_1 \setminus A_2) \cup \cdots \cup (\cap_{n=1}^{\infty} A_n).$$

1.1.24.* Let $X$ be a set and $\{E_n\}_{n \in \mathbb{N}}$ subsets of $X$. Let $A_1 = E_1$ and, for $k \geq 2$, $A_k = E_k \setminus (E_1 \cup E_2 \cup \cdots \cup E_{k-1})$. Prove that $\{A_k\}_{k=1}^{\infty}$ is a sequence of pairwise disjoint sets and that $\cup_{k=1}^{\infty} A_k = \cup_{k=1}^{\infty} E_k$.

1.1.25. Let $\{E_n\}$ be subsets of $X$. Prove that

$$(\limsup E_n)^C = \liminf E_n^C.$$

1.1.26.* Let $\{A_n\}_{n=1}^{\infty}$ and $\{B_n\}_{n=1}^{\infty}$ be arbitrary collections of sets. Prove that:

(a) $\liminf(A_n \cap B_n) = (\liminf A_n) \cap (\liminf B_n)$.

(b) $\limsup(A_n \cup B_n) = (\limsup A_n) \cup (\limsup B_n)$.

(c) $\liminf(A_n \cup B_n) \supset (\liminf A_n) \cup (\liminf B_n)$.

(d) inclusion in (c) can be strict.

(e) $\limsup(A_n \cap B_n) \subset (\limsup A_n) \cap (\limsup B_n)$.

(f) inclusion in (e) can be strict.

1.1.27.* Let $\{A_n\}_{n=1}^{\infty}$ be an arbitrary collection of sets. Prove that:

(a) $\cap_{n=1}^{\infty} A_n \subset \liminf A_n \subset \limsup A_n \subset \cup_{n=1}^{\infty} A_n$.

(b) each inclusion in (a) can be strict.

(c) $A \triangle (\liminf A_n) \subset \limsup(A \triangle A_n)$.

(d) inclusion in (c) can be strict.

(e) $A \triangle (\limsup A_n) \subset \limsup(A \triangle A_n)$.

(f) inclusion in (e) can be strict.

1.1.28. Prove that $\limsup\limits_{n\to\infty} E_n = \{x : x \in E_i \text{ for infinitely many } i\}$.

1.1.29. Prove that $\liminf\limits_{n\to\infty} E_n = \{x : x \in E_i \text{ for all but finitely many } i\}$.

1.1.30. Calculate $\limsup E_n$ and $\liminf E_n$ if $E_n = (-1/n, 1 - 1/n]$.

1.1.31. Calculate $\limsup E_n$ and $\liminf E_n$ if $E_n = ((-1)^n/n, 1 - (-1)^n/n]$.

1.1.32. Calculate $\limsup E_n$ and $\liminf E_n$ if $E_n = \{0, 1/n, 2/n, \ldots, (n-1)/n, 1\}$.

1.1.33. Let $\{E_n\}$ be subsets of $X$ and let $\{A_n\}$ be a subsequence of $\{E_n\}$ obtained by discarding the first $N$ members, i.e., $A_1 = E_{N+1}$, $A_2 = E_{N+2}$, etc. Prove that

$$\limsup E_n = \limsup A_n.$$

## 1.2 Functions

It is customary to think of a function as an assignment; i.e., if $f : X \to Y$, then $f$ assigns to each element of $X$ an element of $Y$. While this definition is intuitive it is rather informal and the movement to make mathematics more formal required that the definition of a function be revised.

**Definition 1.2.1.** A **function** with domain $X$ and codomain $Y$ is a subset $R$ of $X \times Y$ such that for each $x \in X$ there is a unique $y \in Y$ for which $(x, y) \in R$. If this **mapping** or a function is denoted by $f$ we write

$$y = f(x) \quad \text{or} \quad f : X \to Y.$$

If $A \subset X$ then the **image** of $A$ is

$$f(A) = \{f(x) : x \in A\}.$$

The **inverse image** of a set $B \subset Y$ is

$$f^{-1}(B) = \{x \in X : f(x) \in B\}.$$

In the special case when $X = \mathbb{N}$, we use the word **sequence**. As usual, when $a : \mathbb{N} \to Y$, instead of $a(n)$ we write $a_n$. The word sequence will also be used to denote the set of values of $a$, i.e., $\{a_1, a_2, a_3, \dots\}$ or $\{a_n\}_{n=1}^{\infty}$.

A function $f : X \to Y$ is **injective** if whenever $(x_1, y)$ and $(x_2, y)$ are in $R$ then $x_1 = x_2$. Alternatively, $f(x_1) = f(x_2)$ implies $x_1 = x_2$.

A function $f : X \to Y$ is **surjective** if for every $y \in Y$ there is $x \in X$ for which $(x, y) \in R$.

If $W \subset X$, a function $f : X \to Y$ can be viewed as a function $g : W \to Y$, where $f(x) = g(x)$ for $x \in W$. The function $g$ is then called the **restriction** of $f$ to $W$. This relationship can be considered in the other direction as well. If we start with $g : W \to Y$ then $f$ is an **extension** of $g$.

In this section we will take a quick look at the interplay of functions and set-theoretic operations. Here are some examples.

**Example 1.2.2. Proving a set-theoretic inclusion.**

Let $f : X \to Y$ and $A, B \subset X$. Then $A \subset B$ implies that $f(A) \subset f(B)$.

**Solution:** Let $y \in f(A)$. Then there is $x \in A$ such that $y = f(x)$. But $x \in B$ as well, so $y \in f(B)$. ♦

**Example 1.2.3. Proving a set-theoretic identity.**

Let $f : X \to Y$ and $A, B \subset X$. Then $f(A \cup B) = f(A) \cup f(B)$.

**Solution:** First, we prove that $f(A) \cup f(B) \subset f(A \cup B)$. By Example 1.2.2, $A \subset A \cup B$ implies $f(A) \subset f(A \cup B)$. Similarly, $f(B) \subset f(A \cup B)$, whence $f(A) \cup f(B) \subset f(A \cup B)$.

To prove the opposite inclusion, let $y \in f(A \cup B)$. Then there is $x \in A \cup B$ such that $y = f(x)$. If $x \in A$ then $y \in f(A)$. If $x \in B$ then $y \in f(B)$. Thus, $y \in f(A) \cup f(B)$. ♦

It is possible to extend the identity in Example 1.2.3 to more than two sets. If the union is finite,

$$f\left(\bigcup_{i=1}^{n} A_i\right) = \bigcup_{i=1}^{n} f(A_i)$$

can be proved using induction. What if the union is infinite?

Among the exercises provided at the end of the section, there are many identities that are stated for two sets even though they are valid for an arbitrary collection of sets, possibly an infinite one. In the latter case we denote such a collection by

$$\{A_i : i \in I\}, \qquad \text{or} \qquad \{A_i\}_{i \in I},$$

and we call $I$ the *index set*. In a special case when the index set is countable it is customary to use notation $\{A_n\}_{n=1}^{\infty}$ or just $\{A_n\}$. Here is an infinite version of Example 1.2.3.

**Example 1.2.4. Proving a set-theoretic identity.**

Let $f : X \to Y$ and let $\{A_i\}_{i \in I}$ be a collection of subsets of $X$. Then $f\left(\cup_{i \in I} A_i\right) = \cup_{i \in I} f(A_i)$.

**Solution:**

$$\begin{aligned} y \in f\left(\cup_{i \in I} A_i\right) &\Leftrightarrow \exists x \in \cup_{i \in I} A_i \text{ such that } y = f(x) \\ &\Leftrightarrow \exists x \exists i \text{ such that } x \in A_i \text{ and } y = f(x) \\ &\Leftrightarrow \exists i \exists x \text{ such that } x \in A_i \text{ and } y = f(x) \\ &\Leftrightarrow \exists i \text{ such that } y \in f(A_i) \\ &\Leftrightarrow y \in \cup_{i \in I} f(A_i). \end{aligned}$$

◆

We remark here that if the identity holds for an arbitrary index set $I$, then it includes both the finite and the countable case. Thus, the proof in Example 1.2.4 makes the one in Example 1.2.3 unnecessary. However, it is sometimes easier to prove an identity for a finite collection of sets, and it may be all we need at the moment. Also, as examples get more complicated, we will see that some identities hold only in a finite case.

Example 1.2.4 exhibits a nice relationship between functions and the union of sets. When it comes to the intersection of sets, the symmetry is broken.

**Example 1.2.5. Proving a set-theoretic inclusion.**

Let $f : X \to Y$ and $A, B \subset X$. Then $f(A \cap B) \subset f(A) \cap f(B)$ and the inclusion can be proper.

**Solution:** First, $A \cap B \subset A$ so, by Example 1.2.2, $f(A \cap B) \subset f(A)$. Similarly, $f(A \cap B) \subset f(B)$, hence $f(A \cap B) \subset f(A) \cap f(B)$.

As an example when the inclusion is strict, consider $f(x) = x^2$, $A = [-2, -1]$, and $B = [1, 2]$. Then $f(A \cap B) = f(\emptyset) = \emptyset$, but $f(A) = f(B) = [1, 4]$, so $f(A) \cap f(B) = [1, 4]$. ◆

Strangely, the equality holds if instead of the images, one considers the inverse images of sets.

**Example 1.2.6. Proving a set-theoretic inclusion.**

Let $f : X \to Y$ and $A, B \subset Y$. Then $f^{-1}(A \cap B) = f^{-1}(A) \cap f^{-1}(B)$.

**Solution:** First, we prove that $f^{-1}(A \cap B) \subset f^{-1}(A) \cap f^{-1}(B)$. We will rely on Exercise 1.2.4 (a) to conclude that, since $A \cap B \subset A$, $f^{-1}(A \cap B) \subset f^{-1}(A)$. Similarly, $f^{-1}(A \cap B) \subset f^{-1}(B)$, hence $f^{-1}(A \cap B) \subset f^{-1}(A) \cap f^{-1}(B)$.

To prove the opposite inclusion, let $x \in f^{-1}(A) \cap f^{-1}(B)$. Then $x \in f^{-1}(A)$ so $f(x) \in A$. Also, $x \in f^{-1}(B)$ so $f(x) \in B$. Thus, $f(x) \in A \cap B$ whence $x \in f^{-1}(A \cap B)$. ♦

Next, we introduce a class of functions that will play a significant role throughout the text.

**Definition 1.2.7.** Let $A \subset X$. The **characteristic function** of $A$, denoted by $\chi_A$ is a function defined by

$$\chi_A(x) = \begin{cases} 1 & \text{if } x \in A \\ 0 & \text{if } x \notin A. \end{cases}$$

Notice that $A$ can be any subset of $X$. The function is sometimes called the **indicator function** of a set $A$. In addition to the notation $\chi_A$, some authors use $\mathbb{1}_A$.

**Example 1.2.8.** The characteristic function of the set $\mathbb{Q}$ is the famous Dirichlet function. ♦

Characteristic functions satisfy some identites involving set-theoretic operations.

**Example 1.2.9. Proving a set-theoretic identity.**

Let $A, B$ be disjoint sets. Then $\chi_{A \cup B} = \chi_A + \chi_B$.

**Solution:** If $x \in A$ then $x \in A \cup B$ but $x \notin B$. Thus, the equality becomes $1 = 1 + 0$. A similar situation occurs if $x \in B$. Finally, if $x$ belongs to neither $A$ nor $B$, then $x \notin A \cup B$ so the equality becomes $0 = 0 + 0$. ♦

> The concept of the characteristic function of a set was introduced by a Belgian mathematician Charles de la Vallée Poussin (1866–1962) in 1915 [28]. He produced several significant results in the theory of Fourier series and integrals depending on a parameter, but he is best known for the proof of the *Prime Number Theorem* in 1896. (The number of primes $\leq x$ is asymptotically equivalent to $x/\ln x$, as $x \to \infty$.) His major work was *Cours d'Analyse* whose second edition (1909–1912) brought forward a number of modern topics, such as set theory and the Lebesgue integral. He had a doctoral student Georges Lemaître, who was later the first to propose a Big Bang theory of the history of the Universe.

Further properties of functions require that their domain and codomain have additional structure. We will follow this path in the next chapter.

## Exercises

1.2.1. Let $f : X \to Y$ and $A, B \subset X$. Prove that:

(a) $f(A) \setminus f(B) \subset f(A \setminus B)$.

(b) inclusion in (a) can be strict.

1.2.2. Let $f : X \to Y$, let $A \subset X$, and $B \subset Y$. Prove that:

(a) If $f$ is surjective then $f(f^{-1}(B)) = B$.

(b) If $f$ is injective then $f^{-1}(f(A)) = A$.

1.2.3. Let $f : X \to Y$ and $A, B \subset X$. Prove that:

(a) $f$ is injective if and only if $f(A \cap B) = f(A) \cap f(B)$ for all $A, B$.

(b) $f$ is surjective if and only if $f(A)^C \subset f(A^C)$ for all $A$.

(c) $f(A)^C = f(A^C)$ for all $A$ if and only $f$ is bijective.

1.2.4. Let $f : X \to Y$ and $A, B \subset Y$. Prove that

(a) $A \subset B$ implies that $f^{-1}(A) \subset f^{-1}(B)$.

(b) $f^{-1}(A \cup B) = f^{-1}(A) \cup f^{-1}(B)$.

(c) $f^{-1}(B^C) = (f^{-1}(B))^C$.

1.2.5. Let $f : X \to Y$, $A \subset X$, and $B \subset Y$. Prove that $f(A \cap f^{-1}(B)) = f(A) \cap B$.

1.2.6. Let $f : X \to Y$, let $X_0 \subset X$, let $f_0$ be the restriction of $f$ to $X_0$, and let $E \subset Y$. Prove that:

(a) $f_0^{-1}(E) = f^{-1}(E) \cap X_0$.

(b) $f^{-1}(E) = f_0^{-1}(E) \cup (f^{-1}(E) \cap (X \setminus X_0))$.

1.2.7. Let $f : X \to Y$, let $\{A_i\}_{i \in I}$ be a collection of subsets of $X$, and let $\{B_i\}_{i \in I}$ be a collection of subsets of $Y$. Prove that:

(a) $f^{-1}(\cap_{i \in I} B_i) = \cap_{i \in I} f^{-1}(B_i)$.

(b) $f(\cap_{i \in I} A_i) \subset \cap_{i \in I} f(A_i)$.

(c) The equality holds in (b) if and only if $f$ is injective.

(d) $f^{-1}(\cup_{i \in I} B_i) = \cup_{i \in I} f^{-1}(B_i)$.

1.2.8. Let $\{A_i\}_{i \in I}$ and $\{B_i\}_{i \in I}$ be subsets of $X$. Prove that

(a) $(\cup_{i \in I} A_i) \setminus (\cup_{i \in I} B_i) \subset \cup_{i \in I}(A_i \setminus B_i)$.

(b) Inclusion in (a) can be strict.

1.2.9. Let $f : X \to \mathbb{R}$ and let $A, B \subset X$. Prove that:

(a) $\chi_{A\cap B} = \chi_A \chi_B$.

(b) $\chi_{A\cup B} = \chi_A + \chi_B - \chi_A \chi_B$.

(c) $\chi_{A\setminus B} = \chi_A - \chi_A \chi_B$.

(d) $\chi_A \circ f = \chi_{f^{-1}(A)}$.

(e) $|\chi_A - \chi_B| = \chi_{A\triangle B}$.

## 1.3 Cardinal and Ordinal Numbers

In this section, we are interested in classifying sets based on the size or of the structure. We will start with the former. The goal is to give a reader an understanding of the concepts without going too deep into the theory. To borrow the expression from Paul Halmos and his book [52], we will dwell in the "naive set theory".

When considering the number of objects in a finite set, it is customary to point at one of them saying "one", then point at another saying "two", and so on until all objects are accounted for. Mathematically speaking, we construct a bijection between the elements of the set in question and an initial segment of the set $\mathbb{N}$. It makes sense to apply the same principle to sets with infinitely many elements. Namely, we will be looking for a bijection between two infinite sets and, if there is one, we say that they have the same **cardinal number**. We use the notation $\operatorname{card} A$ for the cardinal number of a set $A$. Instead of *cardinal number* we frequently say *cardinality*. When a set is finite its cardinal number is the *number* of its elements. For an infinite set, we will explain how to associate a cardinal number with it.

In our study, we start with the smallest (infinite) cardinal number. When there is a bijection between the given infinite set $A$ and $\mathbb{N}$ we say that $A$ is **countable**. Our first theorem shows that, in a sense, these are the "smallest" infinite sets.

**Theorem 1.3.1.** *Every infinite subset of* $\mathbb{N}$ *is countable.*

*Proof.* Let $A$ be an infinite subset of $\mathbb{N}$. We will construct a bijection $f : \mathbb{N} \to A$ by induction. We identify the smallest element of $A$, $a_1 = \min A$, and we define $f(1) = a_1$. Suppose that we have determined $a_k := f(k)$, for $k \leq n$. Since $A$ is infinite, the set $A\setminus\{a_1, \dots, a_n\}$ is nonempty. Let $a_{n+1} = \min(A\setminus\{a_1, \dots, a_n\})$ and $f(n+1) = a_{n+1}$. It is easy to verify that $f : \mathbb{N} \to A$ is a bijection. $\square$

The same conclusion holds for infinite subsets of any countable set. (See Exercise 1.3.1.)

Following Cantor, we say that every countable set has the cardinal number $\aleph_0$. Theorem 1.3.1 shows that $\aleph_0$ is the smallest infinite cardinal number. This leads to a natural question of whether there are sets "bigger" than $\mathbb{N}$. The following result which is due to Cantor shows that the answer is in the affirmative.

**Theorem 1.3.2.** *The set* $\mathbb{R}$ *is not countable.*

*Proof.* Suppose to the contrary that there is a bijection $f : \mathbb{R} \to \mathbb{N}$. Then the restriction of $f$ to the interval $[0, 1)$ is a bijection between $[0, 1)$ and an infinite subset of $\mathbb{N}$. By Theorem 1.3.1, this set is also countable and so is $[0, 1)$. Then, real numbers in $[0, 1)$ can be arranged into a sequence $\{a_n\}$. Select an integer $d_1 \in \{0, 1\}$ that is different from the first decimal of $a_1$, an integer $d_2 \in \{0, 1\}$ that is different from the second decimal of $a_2$, and so on. We obtain a real number $0.d_1 d_2 \dots$, which is in $[0, 1)$, but differs from each $a_n$ on at least one decimal. The obtained contradiction shows that there can be no bijection between $\mathbb{R}$ and $\mathbb{N}$, so $\mathbb{R}$ is not countable. □

In fact, for every cardinal number, there is a bigger one. (See Exercise 1.3.9.) This brings forward the idea to introduce an *order* between them.

**Definition 1.3.3.** A **partial order** is a binary relation $\leq$ on a set $\mathcal{F}$ that is reflexive, antisymmetric, and transitive. That is, for all $a, b, c \in \mathcal{F}$,

(i) $a \leq a$ (**reflexivity**);

(ii) if $a \leq b$ and $b \leq a$, then $a = b$ (**antisymmetry**);

(iii) if $a \leq b$ and $b \leq c$, then $a \leq c$ (**transitivity**).

A set with a partial order is called a **partially ordered set**. For $a, b$, elements of a partially ordered set $A$, if $a \leq b$ or $b \leq a$, then $a$ and $b$ are **comparable**. A partially ordered set in which every pair of elements is comparable is called a **totally ordered set** or a **chain**.

When considering cardinal numbers, we define $\operatorname{card} A \leq \operatorname{card} B$ if there is an injective map $f : A \to B$. (Of course, if it is surjective as well, then $\operatorname{card} A = \operatorname{card} B$.) The only nontrivial part of verifying that this defines a total order on the set of cardinal numbers, concerns property (ii). This is the famous Cantor–Bernstein–Schröder Theorem.

**Theorem 1.3.4.** ***Cantor–Bernstein–Schröder Theorem.*** *If* $A$ *and* $B$ *are sets such that* $\operatorname{card} A \leq \operatorname{card} B$ *and* $\operatorname{card} B \leq \operatorname{card} A$ *then* $\operatorname{card} A = \operatorname{card} B$.

*Proof.* By assumption, there are bijections $f : A \to B_1 \subset B$ and $g : B \to A_1 \subset A$. It suffices to exhibit a bijection $h : A \to A_1$ because then $g^{-1} \circ h : A \to B$. We start by inductively defining two decreasing sequences of sets $\{A_n\}_{n=1}^{\infty}$ and $\{B_n\}_{n=1}^{\infty}$. Denoting $A = A_0$ and $B = B_0$,

$$A_{n+1} := g(B_n), \qquad B_{n+1} := f(A_n), \qquad \text{for } n = 0, 1, 2, \dots.$$

By assumption, $A_1 \subset A_0$ and $B_1 \subset B_0$. Suppose now that $A_0 \supset A_1 \supset \cdots \supset A_n$ and $B_0 \supset B_1 \supset \cdots \supset B_n$. Then $A_{n+1} = g(B_n) \subset g(B_{n-1}) = A_n$ and, similarly, $B_{n+1} \subset B_n$ so both $\{A_n\}$ and $\{B_n\}$ are decreasing sequences. By Exercise 1.1.23,

$$A_0 = (A_0 \setminus A_1) \cup (A_1 \setminus A_2) \cup \cdots \cup (\cap_{n=1}^{\infty} A_n), \quad \text{and}$$
$$A_1 = (A_1 \setminus A_2) \cup (A_2 \setminus A_3) \cup \cdots \cup (\cap_{n=1}^{\infty} A_n).$$

Clearly, it suffices to establish the existence of a bijection between $(A_0 \setminus A_1) \cup (A_2 \setminus A_3) \cup \ldots$ and $(A_2 \setminus A_3) \cup (A_4 \setminus A_5) \cup \ldots$. Furthermore, both infinite unions consist of pairwise disjoint sets so we only need to prove that

$$g \circ f(A_{2k} \setminus A_{2k+1}) = A_{2k+2} \setminus A_{2k+4}, \qquad \text{for } k = 0, 1, 2, \ldots.$$

By Exercise 1.2.3, using the fact that $g \circ f$ is a bijection (since $f$ and $g$ are),

$$\begin{aligned} g \circ f(A_{2k} \setminus A_{2k+1}) &= g \circ f(A_{2k} \cap A_{2k+1}^C) = g \circ f(A_{2k}) \cap g \circ f(A_{2k+1}^C) \\ &= g \circ f(A_{2k}) \cap [g \circ f(A_{2k+1})]^C = g(B_{2k+1}) \setminus g(B_{2k+2}) \\ &= A_{2k+2} \setminus A_{2k+4}, \end{aligned}$$

and the proof is complete. □

In the case when there is no bijection between sets $A$ and $B$, there may be a bijection between $A$ and a proper subset of $B$ in which case we write $\operatorname{card} A < \operatorname{card} B$. By Theorem 1.3.2, $\aleph_0 < c$ where $c$ is the cardinality of $\mathbb{R}$.

We now turn our attention to ordinal numbers. In this direction, we start with a definition.

**Definition 1.3.5.** A totally ordered set $A$ is **well-ordered** if every nonempty subset of $A$ has the least element.

We will be interested in well-ordered sets and, for the moment, we will ignore those that are not. (More about it in the next section.) When classifying sets based on their cardinality, we used bijections between sets. For ordinal numbers, we use *order isomorphisms*.

**Definition 1.3.6.** A mapping $f : A \to B$ is an **order isomorphism** if it is a bijection and preserves order. More precisely, if $x, y \in A$ and $x < y$ then $f(x) < f(y)$. We will use notation $A \sim B$ when $A$ and $B$ are order isomorphic.

**Example 1.3.7. An order isomorphism.**
Let $2\mathbb{N} = \{2, 4, 6, \ldots\}$ be the set of all even positive integers and let $2\mathbb{N} - 1 = \{1, 3, 5, \ldots\}$ be the set of odd ones. The function $f : 2\mathbb{N} \to 2\mathbb{N} - 1$ defined by $f(2n) = 2n - 1$ is an order isomorphism. ♦

It is easy to see that any two finite sets with the same number of elements are order isomorphic. This is because if $A$ and $B$ are ordered sets, $A = \{a_1, a_2, \ldots, a_n\}$, $B = \{b_1, b_2, \ldots, b_n\}$, then $f(a_k) = b_k$ defines an order isomorphism. The situation is much more complicated in the infinite-dimensional case.

**Example 1.3.8. Two infinite sets that are not order isomorphic.**
Let $A = 2\mathbb{N} = \{2, 4, 6, \dots\}$, $B = 2\mathbb{N} - 1 = \{1, 3, 5, \dots\}$, and define $C = A \cup B$ ordered as $\{2, 4, 6, \dots; 1, 3, 5, \dots\}$. Then $C$ is not order isomorphic to $\mathbb{N}$.

Indeed, if $f : C \to \mathbb{N}$ is an order isomorphism, then there is $n \in \mathbb{N}$ such that $f(1) = n$. Thus, $f$ would map the set $A$ to $\{1, 2, \dots, n-1\}$ so it could not be a bijection. ♦

Example 1.3.8 shows that even though sets $C$ and $\mathbb{N}$ are both countable, so that they have the same cardinal number, they are not order isomorphic. We say that they have different **ordinal numbers**. In fact, $\mathbb{N}$ is order isomorphic to the set $A$ which represents an initial segment of $C$. In such a situation we say that the ordinal number of $A$ is smaller than the ordinal number of $C$ and we write $\operatorname{ord} A < \operatorname{ord} C$. The initial segment of $C$ that is order isomorphic to $A$ is $s(1)$. In general, if $x$ is an element of a totally ordered set $X$, $s(x) := \{y \in X : y < x\}$.

When studying the linear structure of sets, it is helpful to identify the ordinal number of a set with a set of all smaller ordinals. This is not hard to understand when a set is finite. The ordinal 5 can be identified with the set $\{0, 1, 2, 3, 4\}$. (The ordinal 0 is necessary because we ascribe it to the empty set.) Of course, $\{0, 1, 2, 3, 4\}$ is not the only set with five elements, but they are all order isomorphic to it. The first infinite ordinal can be identified with the set of all finite ordinals, hence with $\{0, 1, 2, 3, \dots\} = \mathbb{N}_0$. Traditionally, this infinite ordinal is denoted by $\omega$. The next one is $\omega + 1$.

**Example 1.3.9. A set with ordinal number $\omega + 1$.**
The set $\{1, 3, 5, 7, \dots; 2\}$ has ordinal number $\omega + 1$. ♦

Notice that $\omega + 1 \neq 1 + \omega$. Indeed, $1 + \omega$ would be the ordinal number of a set $\{2; 1, 3, 5, 7, \dots\}$ which is order isomorphic to $\mathbb{N}$, so its ordinal number is $\omega$. In other words, $1 + \omega = \omega$.

The progression of ordinals continues with $\omega+2, \omega+3, \dots$. After all ordinal numbers of the form $\omega + n$ comes $\omega + \omega$, which is written as $2\omega$. The same reasoning leads to $3\omega$, $4\omega$, etc., culminating with $\omega \cdot \omega = \omega^2$. Making the linear structure of a set ever more complicated, we get $\omega^3$, $\omega^4$, etc., leading to $\omega^\omega$. This procedure can continue, producing sets with ordinal numbers $\omega^{\omega^\omega}$, $\omega^{\omega^{\omega^\omega}}$, etc. Notice that all these ordinals correspond to sets that are countable. The first *uncountable* ordinal is denoted by $\omega_1$.

An astute reader may have noticed that we avoided defining what an ordinal number is. Since order isomorphisms define an equivalence relation (two sets are equivalent if they are order isomorphic), a naive approach would be to talk about ordinal numbers as equivalence classes relative to this relation. However, considering sets whose elements are sets is a perilous path, frequently leading to paradoxes. In order to avoid such a predicament, the standard practice is to select a representative from each class and call it the ordinal number, with the understanding that each set in the same class is assigned the same ordinal number. Going back to our example where a set had exactly five elements, we selected $\{0, 1, 2, 3, 4\}$ as the representative. We will leave aside the question of how to make the selection in general and focus on showing that the set of representatives is well-ordered.

When comparing well-ordered sets we use the following terminology. We say that $A$ is **shorter** than $B$ if $A$ is order isomorphic to an initial segment of $B$. The next theorem shows that if $A$ is order isomorphic to a proper initial segment of $B$ then $A$ cannot be ordered isomorphic to $B$, and an order isomorphism maps initial segments onto initial segments..

**Theorem 1.3.10.** *Let $A, B$ be well-ordered sets, let $a \in A$, let $b \in B$, and suppose that $s(a) \sim s(b)$.*

> *(a) $s(a)$ is order isomorphic to a unique initial segment of $B$.*
>
> *(b) If $a' < a$ then there is $b' < b$ for which $s(a') \sim s(b')$. Furthermore, if $f : s(a) \to s(b)$ is the order isomorphism then $f|s(a')$ is an order isomorphism onto $s(b')$ and $b' = f(a')$.*

*Proof.* (a) Suppose to the contrary that $b' \in B$ and that $s(a) \sim s(b')$. Then $s(b) \sim s(b')$ so there is an order isomorphism $f : s(b') \to s(b)$. Without loss of generality, we may assume that $b < b'$. This implies that $f(b) \in s(b)$, so $f(b) < b$. It follows that the set $C := \{x \in A : f(x) < x\}$ is nonempty and the well-ordering of $A$ ensures that there is $c = \min C$. Then $f(c) < c$ hence $f(f(c)) < f(c)$, and we conclude that $f(c) \in C$. On the other hand, $f(c) < c$ contradicting the minimality of c, whence $b = b'$.

(b) Define $b' := f(a')$. Clearly, $f|s(a')$ is injective and preserves order, so $s(a') \sim f(s(a'))$. Thus, $c < a'$ if and only if $f(c) < b'$,so $f(s(a')) = s(b')$, i.e., $s(a') \sim s(b')$. □

When well-ordered sets $A, B$ are finite, say $A = \{a, b, c\}$ and $B = \{u, v\}$, then $s(u) \sim s(a)$ and $s(v) \sim s(b)$ but there is no $x \in B$ for which $s(x) \sim s(c)$. If we denote

$$S = \{x \in A : \exists y \in B \text{ such that } s(x) \sim s(y)\}, \quad \text{and}$$
$$T = \{y \in B : \exists x \in A \text{ such that } s(y) \sim s(x)\},$$

then $S = s(c)$ and $T = B$. This type of dichotomy persists for all well-ordered sets.

**Theorem 1.3.11.** *Let $A, B$ be well-ordered sets, and let $S = \{x \in A : \exists y \in B \text{ such that } s(x) \sim s(y)\}$.*

> *(a) $S = A$ or $S$ is an initial segment of $A$.*
>
> *(b) If $T = \{y \in B : \exists x \in A \text{ such that } s(y) \sim s(x)\}$, then $S \sim T$.*

*Proof.* (a) Suppose $S \neq A$. Since $A$ is well-ordered there is $a := \min\{A \setminus S\}$. We will show that $S = s(a)$.

($\supset$) Let $x \in s(a)$. Then $x < a$ so $x \notin A \setminus S$, whence $x \in S$.

($\subset$) Let $x \in S$ and suppose that $x \notin s(a)$. Then $x \geq a$. By Theorem 1.3.10 (b), $s(a)$ is order isomorphic to an initial segment of $B$, hence $a \in S$ which is a contradiction because $a \in A \setminus S$.

(b) Let $x \in S$. By definition of $S$, there is $y \in B$ such that $s(x) \sim s(y)$. By Theorem 1.3.10 (a), the element $y$ is unique, and it clearly belongs to $T$, so we can define $f : S \to T$ by $f(x) = y$. It is easy to see that $f$ is a bijection, so it remains to establish that it preserves order. Let $x', x \in S$, $x' < x$, $y = f(x)$, and $y' = f(x')$. Since $s(x) \sim s(y)$, there is an order isomorphism $g : s(x) \to s(y)$. By Theorem 1.3.10 (b), $g|s(x')$ is an order isomorphism between $s(x')$ and $s(g(x'))$. On the other hand, by definition of $S$, $s(x') \sim s(f(x')) = s(y')$. By Theorem 1.3.10 (a), $s(g(x')) = s(y')$, whence $y' = g(x') \in s(y)$ and it follows that $y' < y$. □

Now we can prove that any two well-ordered sets are comparable.

**Theorem 1.3.12.** *Let $A, B$ be well-ordered sets. Then $A$ is shorter than $B$, $A \sim B$, or $B$ is shorter than $A$.*

*Proof.* Let $S$ and $T$ be as in Theorem 1.3.11. It was proved there that $S \sim T$ and that either $S = A$ or $S = s(a)$, for some $a \in A$, and either $T = B$ or $T = s(b)$, for some $b \in B$. Thus, there are four cases to consider.

(i) $S = A$ and $T = B$. Then $A \sim B$.

(ii) $S = A$ and $T = s(b)$. Then $A$ is shorter than $B$.

(iii) $S = s(a)$ and $T = B$. Then $B$ is shorter than $A$.

(iv) $S = s(a)$ and $T = s(b)$. Then $s(a) \sim s(b)$ so, by definition of $S$, $a \in S = s(a)$, which is impossible. □

Consider the relation $\leq$ on the collection of well-ordered sets, defined by $A \leq B$ if $A$ is shorter than $B$ or order isomorphic to $B$. By Theorem 1.3.12, the collection of ordinal numbers is a totally ordered set. In fact, it is well-ordered. (See Exercise 1.3.10.)

We have defined the relation of order among well-ordered sets by appealing to order isomorphisms between initial segments of such sets, including ordinals. It would be much easier to have a concrete choice of ordinals, i.e., representatives from equivalence classes modulo order isomorphisms. Such a construction is presented here:

(i) $0 = \emptyset$;

(ii) $\alpha + 1 = \alpha \cup \{\alpha\}$;

(iii) If $\beta$ is defined whenever $\beta < \alpha$, then $\alpha = \cup_{\beta<\alpha}\beta$.

The meaning is that the ordinal number 0 is associated with the empty set. Using (ii), we have that the ordinal number 1 corresponds to the set $\emptyset \cup \{\emptyset\} = \{\emptyset\}$. The ordinal number 2 corresponds to the set $\{\emptyset\} \cup \{\{\emptyset\}\} = \{\emptyset, \{\emptyset\}\}$. The ordinal number 3 corresponds to the set $\{\emptyset, \{\emptyset\}\} \cup \{\{\emptyset, \{\emptyset\}\}\}$ which is $\{\emptyset, \{\emptyset\}, \{\emptyset, \{\emptyset\}\}$. And so on. This is a rather cumbersome way of writing, so we use numerals $0, 1, 2, \ldots$ with the understanding that, e.g., 2 stands for the set $\{\emptyset, \{\emptyset\}\}$. Finally, part (iii) allows us to define ordinal numbers of the so-called *limit elements*, i.e., ordinal numbers that do not have an immediate predecessor. Thus, $\omega = \cup_{\beta<\omega} \beta$, the union of all finite ordinals. Formally,

$$\omega = \{\emptyset, \{\emptyset\}, \{\emptyset, \{\emptyset\}\}, \{\emptyset, \{\emptyset\}, \{\emptyset, \{\emptyset\}\}, \ldots\}.$$

It is not hard to see that the set of ordinals, obtained in this manner, i.e., using (i)–(iii) is well-ordered with respect to inclusion.

Cantor proved that for every cardinal number there exists one that is strictly bigger. He labeled all possible cardinal numbers in the increasing order: $\aleph_0, \aleph_1, \aleph_2, \ldots$. He conjectured that the cardinality of $\mathbb{R}$ is $\aleph_1$ but was unable to prove it. This question, known as the *continuum hypothesis* became the first on David Hilbert's list of important open questions that was presented at the International Congress of Mathematicians in the year 1900 in Paris. Kurt Gödel proved in 1940 that the negation of the continuum hypothesis, i.e., the existence of a set with intermediate cardinality, could not be proved in standard set theory. The second half of the independence of the continuum hypothesis—i.e., unprovability of the nonexistence of an intermediate-sized set—was proved in 1963 by Paul Cohen. Cantor also introduced ordinal numbers in [19]. When various paradoxes showed that it was not prudent to consider the set of equivalence classes under order isomorphism, ordinal numbers were redefined as specific abstract representatives from these classes. John von Neumann improved on this by defining each ordinal to be "the well-ordered set of all smaller ordinals".

## Exercises

1.3.1. Prove that every infinite subset of a countable set is countable.

1.3.2. Prove that $\text{card}\,[0,1] = \text{card}\,[0,1) = \text{card}\,(0,1) = \text{card}\,[a,b]$, if $a < b$.

1.3.3. Suppose that $A$ is an infinite set and $F$ is a finite subset of $A$. Prove that $\text{card}\,A \setminus F = \text{card}\,A$.

1.3.4. Prove that a countable union of finite sets is countable.

1.3.5. Prove that a countable union of pairwise disjoint, countable sets is countable.

1.3.6. Prove that the plane $\mathbb{R}^2$ is not the union of a countable number of lines.

1.3.7. Prove that $\text{card}\,\mathbb{R} = \text{card}\,\mathbb{R}^n$, for all $n \in \mathbb{N}$.

1.3.8. Let $A$ be the set of all real numbers in $(0,1)$ that have infinitely many threes in their decimal expansion. Prove that $\text{card}\,A = \text{card}\,(0,1)$.

1.3.9. Prove that for any set $A$, $\text{card}\,A < \text{card}\,\mathbb{P}(A)$, where $\mathbb{P}(A)$ is the collection of all subsets of $A$, including the empty set and $A$ itself.

1.3.10. Prove that the collection of ordinals is well-ordered.

1.3.11. Let $S$ be a subset of a well-ordered set $A$ and suppose that:

(i) $S$ contains $a_0 := \min A$.

(ii) If $a \in A$ and $s(a) \subset S$ then $a \in S$.

Prove that $S = A$.

1.3.12. Let $A$ be a well-ordered set and $S = \{s(a) : a \in A\}$, ordered with inclusion. Prove that $f : A \to S$, defined by $f(x) = s(x)$ is an order isomorphism.

1.3.13. Let $A, B$ be well-ordered sets that are order isomorphic. Prove that there is only one order isomorphism $f : A \to B$.

1.3.14. Prove that a well-ordered set cannot be ordered isomorphic to one of its initial segments.

1.3.15. Prove that two different initial segments of a well-ordered set cannot be order isomorphic.

1.3.16. Let $\lambda, \mu$ be ordinal numbers, $\lambda = \operatorname{ord} A$, and let $\mu < \lambda$. Prove that there is a unique $a \in A$ such that $\mu = \operatorname{ord} s(a)$.

1.3.17. Let $s(\lambda)$ be the set of ordinal numbers smaller than the ordinal $\lambda$. Prove that $\lambda = \operatorname{ord} s(\lambda)$.

## 1.4 The Axiom of Choice

It would take us too far to discuss the axioms of set theory. We will, though, look closely at one that stands out. The reason is that it postulates the existence of a function without actually constructing it. We will denote by $\mathbb{P}(X)$ the collection of all subsets of the set $X$, including the empty set and $X$ itself. We call it the **power set** of $X$. The name comes from the fact that if $X$ is a finite set with $n$ elements then $\mathbb{P}(X)$ has $2^n$ elements.

**The Axiom of Choice.** *There exists a function $f : \mathbb{P}(X) \to X$ such that, for every $A \subset X$, $f(A) \in A$.*

The statement seems reasonable. To explain it in a more colorful way, we reach for an example due to English mathematician and philosopher Bertrand Russel (1872–1970). Suppose that you have infinitely many pairs of shoes and you have to select one shoe out of each pair. You could, for example, settle for the left shoe in each pair. Suppose now that you have infinitely many pairs of socks. There does not seem an obvious way to select one out of each pair, and you would have to go pair by pair, an impossible task because there are infinitely many pairs. Yet, "there should be a way." This is what the Axiom of Choice allows us to do.

In order to convince the reader of its usefulness we will present a few examples here. More will come throughout the text.

**Example 1.4.1. An application of Axiom of Choice.**

Every infinite set has a countably infinite subset.

**Proof:** Suppose that $X$ is infinite. Let $f : \mathbb{P}(X) \to X$ be a function supplied by the Axiom of Choice. Let $g : \mathbb{N} \to X$ be defined by $g(1) = f(X)$ and $g(n) = f(X \setminus \{g(1), g(2), \ldots, g(n-1)\})$ for $n > 1$. Since $X$ is infinite, $X \setminus \{g(1), g(2), \ldots, g(n-1)\}$ is nonempty for each $n \in \mathbb{N}$. Moreover, for every $n \in \mathbb{N}$, $g(n) \notin \{g(0), \ldots, g(n-1)\}$, so $g$ is injective. Also, this shows that $g(\mathbb{N})$ is infinite. Finally, $g$ is a bijection from $\mathbb{N}$ onto $g(\mathbb{N})$, whence $g(\mathbb{N})$ is a countable subset of $X$. ♦

Natural as the Axiom of Choice may seem, it is impossible to prove it using other axioms of set theory. In other words, we can either take it as an axiom or dismiss it. There is a long line of dissenters that include some immortals like Baire, Borel, and Lebesgue, for whom a mathematical object could be asserted to exist only if it can be defined in such a way as to characterize it uniquely.

In fact, there is a movement called *intuitionism* that rejects all mathematics that is not constructive. For example, within such mathematics, proofs of the type "since statement $A$ is false its negation must be true" are not acceptable. Proponents of this philosophy question the existence of objects that are obtained using the argument above and accept only those that can be constructed. Since the Axiom of Choice does not offer an algorithm how to select a member of the given set, it is not a part of the intuitionist's mathematics. We will nevertheless accept it, as well as those statements that are equivalent to it. Not only because it looks legitimate but also because some important results are unprovable without it. That is, in the axiomatic system where the Axiom of Choice has been removed, such statements could be true, or they could be false, and neither would contradict the remaining axioms.

On the flip side, the Axiom of Choice also allows us to prove some very counterintuitive theorems such as the famous Banach–Tarski paradox in Section 5.4. Even worse, the axiom is equivalent to some assertions that seem impossible. On of them concerns the notion of *order* introduced in Section 1.3. We first look at a few examples.

**Example 1.4.2. A partially ordered set.**

The set of subsets of a given set $X$ (its power set) is a partially ordered set with respect to inclusion. ♦

**Example 1.4.3. A partially ordered set.**

The set $\mathbb{N}$ with respect to the relation of divisibility: $a \leq b$ if and only if $b$ is divisible by $a$. ♦

**Example 1.4.4. A totally ordered subset of a partially ordered set.**

Let $\mathcal{F}$ be the set of all intervals $[a, b]$. Then $A$ is a partially ordered set with respect to inclusion. The set $\mathcal{F}_0$ of all intervals $[0, b]$ is a totally ordered subset of $\mathcal{F}$. ♦

Now we can present one of the statements equivalent to the Axiom of Choice.

**Hausdorff maximal principle.** *In every partially ordered set there exists a maximal totally ordered subset.*

By *maximal*, we mean one that is not properly contained in another totally ordered subset.

**Example 1.4.5. A maximal totally ordered subset of a partially ordered set.**

The set $\mathcal{F} = \{1, 2, 4, 8, 16, \dots\}$ is a maximal totally ordered subset of $\mathbb{N}$ with respect to the relation of divisibility. ♦

Just like the Axiom of Choice, Hausdorff's maximal principle can be a powerful tool. Here is one application.

**Example 1.4.6. An application of Hausdorff maximal principle.**

Every vector space has a basis.

**Proof:** Let $V$ be a vector space and let $\mathcal{F}$ be the family of all linearly independent subsets of $V$. Assuming that $V \neq \{0\}$ every singleton $\{x\} \subset V$ is a linearly independent set, so $\mathcal{F}$ is nonempty. It is a partially ordered set by inclusion so, by Hausdorff maximal principle, there is a maximal totally ordered subset $\mathcal{C} = \{A_i\}_{i \in I}$. Let $W = \cup_{i \in I} A_i$. We show that $W$ is a basis for $V$. Let $u_1, u_2, \dots, u_n$ be vectors in $W$. For each $k$, $1 \leq k \leq n$, there is $A_k \in \mathcal{C}$ such that $u_k \in A_k$. Since $\mathcal{C}$ is totally ordered, one of $A_k$ contains the others, so all vectors $u_i$ are in that set whence they are linearly independent because $A_k \in \mathcal{F}$. Also, $W$ spans $V$. Otherwise, there would be $v \in V$ that is not in the span of $W$ which would imply that $W \cup \{v\}$ is a linearly independent subset of $V$, hence in $\mathcal{F}$. Also, it would contain $W$, contradicting the maximality of $\mathcal{C}$. ♦

> This was proved by a German mathematician Georg Hamel (1877–1954) in the special case of $\mathbb{R}$ as a vector space over $\mathbb{Q}$ in 1905. The result in full generality is due to Hausdorff in 1932.

For the third statement equivalent to the Axiom of Choice (and Hausdorff maximal principle) we need some additional concepts.

**Definition 1.4.7.** For a subset $\mathcal{F}_0$ of a partially ordered set $\mathcal{F}$, an element $x \in \mathcal{F}$ is an **upper bound** of $\mathcal{F}_0$ if $b \leq x$, for each element $b \in \mathcal{F}_0$. An element $y \in \mathcal{F}$ is a **lower bound** of $\mathcal{F}_0$ if $y \leq b$, for each element $b \in \mathcal{F}_0$.

**Example 1.4.8. A partially ordered set with a lower bound but no upper bound.**

The set of positive integers divisible by 5 has 1 as a lower bound, but no upper bound with respect to divisibility in $\mathbb{N}$. Notice that 5 is another lower bound. ♦

We also need the notion of a *maximal element.*

**Definition 1.4.9.** An element $x$ in a partially ordered set $\mathcal{F}$ is a **maximal element** if there is no element $a$ in $\mathcal{F}$ such that $x < a$ (meaning $x \leq a$ and $x \neq a$). Similarly, an element $y$ in $\mathcal{F}$ is a **minimal element** if there is no element $a \in \mathcal{F}$ such that $a < y$.

**Example 1.4.10. A maximal and a minimal element in a partially ordered set.**

Let $\mathcal{F}$ be the set of all intervals of length no more than 1. It is a partially ordered set with respect to inclusion. The interval $[0, 1]$ is a maximal element. If we allow degenerate intervals, then $\{2\}$ is a minimal element. Otherwise, $\mathcal{F}$ does not have a minimal element. ♦

The following statement is equivalent to the Axiom of Choice. It is known as Zorn's lemma.

**Zorn's Lemma.** *Suppose a partially ordered set $\mathcal{F}$ has the property that every chain (i.e., totally ordered subset) in $\mathcal{F}$ has an upper bound in $\mathcal{F}$. Then the set $\mathcal{F}$ contains at least one maximal element.*

We remark that Zorn's Lemma and Hausdorff's maximal principle are very similar to each other. Nevertheless, here is one application of Zorn's Lemma.

**Example 1.4.11. Cardinal numbers arithmetic.**

If $\zeta$ is an infinite cardinal number then $\zeta\aleph_0 = \zeta$.

**Proof:** Let $\operatorname{card} X = \zeta$. Consider $\mathcal{F} = \{A_i\}_{i \in I}$, where $\operatorname{card} A_i = \aleph_0$ and $\{A_i\}$ are pairwise disjoint subsets of $X$. Now, let $\mathcal{A}$ be the collection of all such $\mathcal{F}$. Clearly, $\mathcal{A}$ is partially ordered with inclusion. Moreover, if $\{\mathcal{F}_\alpha\}_{\alpha \in A}$ is a chain in $\mathcal{A}$ then $\cup_{\alpha \in A} \mathcal{F}_\alpha \in \mathcal{A}$ and it is an upper bound for $\{\mathcal{F}_\alpha\}_{\alpha \in A}$. By Zorn's Lemma, $\mathcal{A}$ contains a maximal element $\mathcal{M} = \{B_j\}_{j \in J}$. Let $Y = \cup_{j \in J} B_j$. Notice that $X \setminus Y$ must be a finite set. Indeed, if it contained a countably infinite set $C$, then $\mathcal{M} \cup \{C\}$ would properly contain $\mathcal{M}$. Let $B_{j_0}$ be a nonempty element of $\mathcal{M}$ and define $B'_{j_0} = B_{j_0} \cup (X \setminus Y)$. If we replace $B_{j_0}$ with $B'_{j_0}$ and keep the remaining elements of $\mathcal{M}$ we get a collection $\mathcal{M}'$ of disjoint, countably infinite sets, whose union is $X$. Therefore, $\zeta = \operatorname{card} \mathcal{M}' \cdot \aleph_0$ which implies that $\zeta\aleph_0 = \operatorname{card} \mathcal{M}' \cdot \aleph_0 \cdot \aleph_0 = \operatorname{card} \mathcal{M}' \cdot \aleph_0 = \zeta$. ♦

We finish with one more statement equivalent to the Axiom of Choice, the so-called *Well Ordering Principle.*

**The Well Ordering Principle.** *Every set can be well-ordered.*

This is puzzling because we know that $\mathbb{R}$ is not well-ordered. The Well Ordering Principle does not contradict that. It simply states that there is another relation of order that can be defined on $\mathbb{R}$ (probably very different from $\leq$) and that $\mathbb{R}$ is well-ordered relative to this new order. How can that be useful when we don't know what the relation looks like?

**Example 1.4.12. Cardinal numbers comparison.**

If $A$ and $B$ are any sets then either $\operatorname{card} A \leq \operatorname{card} B$ or $\operatorname{card} B \leq \operatorname{card} A$.

**Proof:** By the Well Ordering Principle, every set can be well-ordered, so there is a bijection between it and an ordinal. Let $\alpha = \operatorname{ord} A$ and $\beta = \operatorname{ord} B$. Since $\alpha$ and $\beta$ are ordinals, either $\alpha \leq \beta$ or $\beta \leq \alpha$. ♦

Another consequence of the Well Ordering Principle will be useful later.

**Example 1.4.13. A well ordering of $[0,1]$.**

There is a a well-ordered set $W$ and a bijection $j : [0,1] \to W$ such that for any $x \in [0,1]$, $j(x)$ has at most a countable number of predecessors.

**Proof:** By the Well Ordering Principle, every set can be well-ordered, so there is a well-ordered set $X$ and a bijection $f : [0,1] \to X$. Let $Z = \{x \in X : x \text{ has an uncountable number of predecessors}\}$. If $Z = \emptyset$, then we take $W := X$ and we are done. Otherwise, the set $Z$ has a minimal element $z$. Let $W := \{x \in X : x < z\}$. Then every element of $W$ has at most a countable number of predecessors, otherwise it would belong to $Z$, contradicting the minimality of $z$. On the other hand, $W$ is uncountable, otherwise $z$ would have at most a countable number of predecessors, hence it would not belong to $Z$. Finally, $\operatorname{card} W = \operatorname{card}[0,1]$, so there is a bijection $j : [0,1] \to W$. ♦

Although the 4 statements are all equivalent, there is a different level of disbelief with which they are met. For example, as we mentioned, it is not easy to understand how the set $\mathbb{R}$ can be well-ordered. Jerry Bona is credited with the humorous remark: *The Axiom of Choice is obviously true, the Well-ordering principle obviously false, and who can tell about Zorn's lemma?*

## Exercises

1.4.1. Prove that an arbitrary cartesian product of nonempty sets is nonempty.

1.4.2. Prove that the union of a countable collection of countable sets is countable.

1.4.3. Prove that every surjection has a right inverse.

1.4.4. Let $R$ be a subset of $A \times B$ such that for every $a \in A$ there is $b \in B$ for which $(a,b) \in R$. Prove that there is $F \subset R$ that defines a function from $A$ to $B$, i.e., for every $a \in A$ there is exactly one $b \in B$ for which $(a,b) \in F$.

# 2

# *Metric Spaces*

Nineteenth century brought many new mathematical ideas and concepts. Towards the end of it, it became apparent that there were analogies between various topics and that it was beneficial to create theories that would encompass more than one line of research. In analysis, a crucial idea was to consider sets whose elements are functions. We will study specific sets of this type in a later part of this monograph. In this chapter we will consider even more general structures.

## 2.1 Elementary Theory of Metric Spaces

We start with the definition.

**Definition 2.1.1.** A **metric space** is an ordered pair $(M, d)$, where $M$ is a set and $d$ is a **metric** on $M$, i.e., a function $d\colon M \times M \to \mathbb{R}$ such that for any $x, y, z \in M$, the following holds:

1. $d(x, y) \geq 0$ and $d(x, y) = 0$ if and only if $x = y$;
2. $d(x, y) = d(y, x)$ (symmetry);
3. $d(x, z) \leq d(x, y) + d(y, z)$ (subadditivity or triangle inequality).

When it is clear what metric we have in mind, we will call $M$ a metric space.

Notice that in this definition there is no mention of the nature of elements of the metric space. However, when Fréchet wrote down the axioms of a metric space he had some well-known examples in mind.

**Example 2.1.2. A metric space.**
The set $\mathbb{R}$ with the distance function $d(x, y) = |y - x|$ is a metric space. (Exercise 2.1.1.) ♦

**Example 2.1.3. A metric space.**
Let $M = C[a, b]$, the set of continuous functions defined on $[a, b]$, and let the distance be defined by $d(f, g) := \max\{|f(x) - g(x)| : a \leq x \leq b\}$. Then $(M, d)$ is a metric space. (Exercise 2.1.2.) ♦

DOI: 10.1201/9781032716497-2

*Remark* 2.1.4. It is possible to define more than one metric on a set $M$. However, in many examples there is a "usual" one. Such are the metrics in Examples 2.1.2–2.1.3. From now on, unless we say differently, the metric on $\mathbb{R}$ and $C[a,b]$ will be assumed to be these two.

If $(M,d)$ is a metric space and $E \subset M$, then $(E, d|E)$ is a metric space and we refer to $E$ as a **subspace** of $M$. The name is somewhat misleading because there is usually a difference between a *subset* and a *subspace*, in the sense that the latter typically has to be closed under some operations. There is no such restriction here so it would make sense to call $E$ merely a subset of $M$. On the other hand, $(E, d|E)$ is a legitimate metric space so there is no harm to call it a subspace. We will often take advantage of this freedom and formulate properties to hold on the whole metric space $M$, fully aware that if they hold only on a subset $E$, we can still claim that the property under scrutiny holds on a metric space. This is going to be especially useful when talking about functions where we will always take their domain to be a metric space.

Example 2.1.3 brings forward a metric space whose elements are functions. We will see many more such metric spaces later on. Once we verify that they satisfy the conditions of Definition 2.1.1, everything presented in this chapter will hold in those more special circumstances.

One type of metric space deserves attention. Recall that the set of all ordered $n$-tuples $(x_1, x_2, \dots, x_n)$ is the $n$-dimensional Euclidean space $\mathbb{R}^n$. It is a vector space with respect to addition and scalar multiplication. These operations in $\mathbb{R}^n$ are defined coordinatewise. This means that, if $\mathbf{a}, \mathbf{b} \in \mathbb{R}^n$ and $\alpha \in \mathbb{R}$, then

$$\mathbf{a} + \mathbf{b} = (a_1 + b_1, a_2 + b_2, \dots, a_n + b_n), \quad \text{and} \quad \alpha\mathbf{a} = (\alpha a_1, \alpha a_2, \dots, \alpha a_n).$$

We also mention that $\mathbf{0}$ stands for a point with all coordinates equal to 0.

Further, $\mathbb{R}^n$ is a normed space, meaning that there is a norm on $\mathbb{R}^n$.

**Definition 2.1.5.** If $\mathbf{a} \in \mathbb{R}^n$, we define the **Euclidean norm** of $\mathbf{a}$ by

$$\|\mathbf{a}\| = \sqrt{a_1^2 + a_2^2 + \cdots + a_n^2}. \tag{2.1}$$

The following theorem states that the norm shares some of the most useful features of the distance in $\mathbb{R}^2$.

**Theorem 2.1.6.** *The Euclidean norm has the following properties:*

*(a) for any* $\mathbf{a} \in \mathbb{R}^n$, $\|\mathbf{a}\| \geq 0$;

*(b)* $\|\mathbf{a}\| = 0$ *if and only if* $\mathbf{a} = \mathbf{0}$;

*(c) if* $\alpha \in \mathbb{R}$ *and* $\mathbf{a} \in \mathbb{R}^n$, $\|\alpha\mathbf{a}\| = |\alpha|\|\mathbf{a}\|$;

*(d) if* $\mathbf{a}, \mathbf{b} \in \mathbb{R}^n$, $\|\mathbf{a} + \mathbf{b}\| \leq \|\mathbf{a}\| + \|\mathbf{b}\|$.

We leave the proof to the reader (Exercise 2.1.4). As an immediate consequence we obtain the following result (Exercise 2.1.5).

**Theorem 2.1.7.** *The set* $\mathbb{R}^n$ *with the distance function* $d(x, y) = \|y - x\|$ *is a metric space.*

Theorem 2.1.7 shows that every Euclidean space is a metric space with respect to the Euclidean norm, and this is the "standard" norm on it. However, it has the additional structure of a vector space, which allows us to define a special type of metric, namely the norm. In Chapter 10, the main focus of the investigation will be vector spaces with a norm. At that time we will establish that they are all metric spaces so all results of this chapter will apply to them.

One should bear in mind that metric spaces are abstract, much like topological spaces. We do not talk about the nature of elements of such a space, nor of its main feature *the distance* (a.k.a., the metric) defined only through a set of axioms in Definition 1.2.1. Nevertheless, Euclidean spaces are all metric spaces and many proofs in general metric spaces are verbatim the same as in $\mathbb{R}^n$ (and will be in other concrete spaces, which is the reason to go for the general theory). Thinking of a concrete metric space (such as the real line) can also lead to a better insight where some ideas may be coming from. However, the reader should be warned that the real line can be a very confusing place.

> In 1906, Fréchet introduced metric spaces in his dissertation [35]. However, the name is due to Hausdorff and appears in 1914 in his book [56].

The most important consequence of the presence of a metric is that it is possible to introduce topology. The first step is to define open and closed balls.

**Definition 2.1.8.** An **open ball** with center $c \in M$ and radius $r > 0$ is the set

$$B(c, r) = \{x \in M : d(c, x) < r\}.$$

If $<$ is replaced by $\leq$, then it is a **closed ball** with center $c$ and radius $r$

$$B[c, r] = \{x \in M : d(c, x) \leq r\}.$$

On the other hand, if $<$ is replaced by $=$, then it is a **sphere** with center $c$ and radius $r$

$$S(c, r) = \{x \in M : d(c, x) = r\}.$$

Euclidean plane is a familiar set so our first examples will rely on its geometry. Still, the choice of a metric may lead to balls of surprising shapes. For more examples see Exercises 2.1.15–2.1.18.

**Example 2.1.9. A ball in a metric space.**
Let $M = \mathbb{R}^2$ and let

$$d((x_1, x_2), (y_1, y_2)) := |x_1 - y_1| + |x_2 - y_2|.$$

Then $(\mathbb{R}^2, d)$ is a metric space (Exercise 2.1.6). The open ball with the center at $(0,0)$ and radius 1 is the region inside a square with vertices at $(1,0)$, $(0,1)$, $(-1,0)$, $(0,-1)$ (Figure 2.1). ♦

**Example 2.1.10. A ball in a metric space.**
Let $M = \mathbb{R}^2$ and let

$$d((x_1, x_2), (y_1, y_2)) := \max\{|x_1 - y_1|, |x_2 - y_2|\}.$$

Then $(\mathbb{R}^2, d)$ is a metric space (Exercise 2.1.7). The closed ball with center at $(0,0)$ and radius 1 is the region inside a square with vertices at $(1,1)$, $(-1,1)$, $(-1,-1)$, $(1,-1)$ (Figure 2.2). ♦

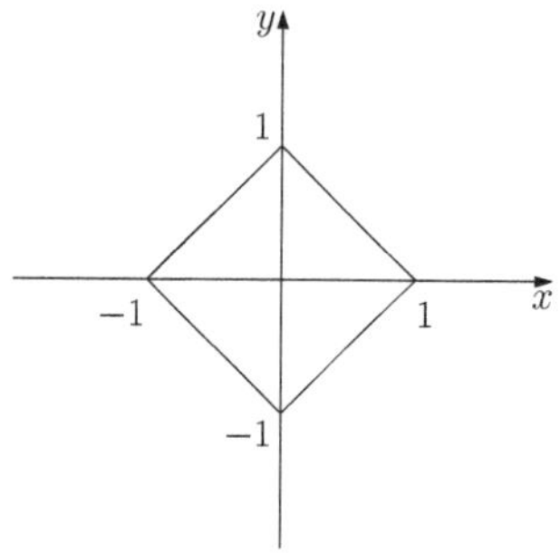


Figure 2.1: A ball in $\mathbb{R}^2$

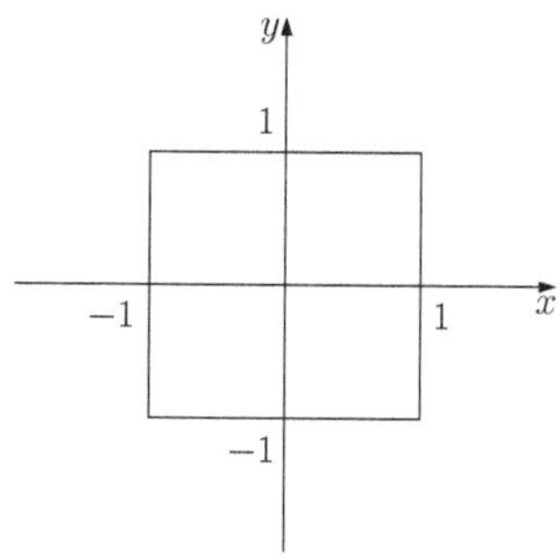


Figure 2.2: A ball in $\mathbb{R}^2$

Every metric space is a topological space because open balls form the base for a topology. This is specified by the following definition.

**Definition 2.1.11.** A subset $G$ of a metric space $M$ is called **open** if for every $x$ in $G$ there exists an $r > 0$ such that $B(x, r)$ is contained in $G$. A set $F$ is called **closed** if its complement is an open set.

We present two examples of open sets in metric spaces and leave the verification to the reader (Exercises 2.1.19 and 2.1.28).

**Example 2.1.12. An open set.**
The set $G = \{(x, y) \in \mathbb{R}^2 : x^2 + y^2 \neq 1\}$ is an open set in the Euclidean space $\mathbb{R}^2$. ♦

**Example 2.1.13. A closed set.**
The set of all rational numbers between $\sqrt{2}$ and $\pi$ is a closed set in the metric space $\mathbb{Q}$ with the usual distance as in Example 2.1.2. It is also an open set. ♦

It is well known that there are subsets of $\mathbb{R}^2$, such as the rectangle $[0, 1] \times [0, 1]$ which are not open. The culprits are sides of the rectangle. Indeed, if $x$ lies on one of them then one cannot find an open ball $B(x, r)$ that would be contained in the rectangle (Figure 2.3). This motivates the following definition.

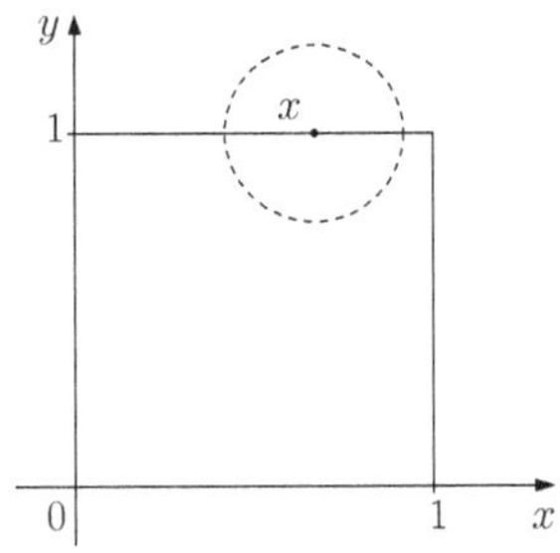


Figure 2.3: A point on the side of the rectangle is not an interior point

**Definition 2.1.14.** A point $x$ in a set $A \subset M$ is an **interior point** of $A$ if there is a ball $B(x, r)$ contained in $A$. The set of all interior points of a set $A$ is called the **interior** of $A$ and is denoted by $\operatorname{Int} A$.

We will now establish an important property of the interior of a set.

**Theorem 2.1.15.** *If $A$ is a set in a metric space then* $\operatorname{Int} A$ *is the largest open set contained in $A$.*

*Proof.* First, $\operatorname{Int} A$ is a subset of $A$. Second, it is open. Indeed, let $x \in \operatorname{Int} A$. By definition, there is $B(x, r) \subset A$ but we need it to be contained in $\operatorname{Int} A$. It is, because $B(x, r)$ is an open set (Exercise 2.1.20). Thus, if $y \in B(x, r)$ there is $B(y, r') \subset B(x, r) \subset A$, so $y \in \operatorname{Int} A$. Finally, $\operatorname{Int} A$ is the largest open set contained in $A$. If $G$ is open and $\operatorname{Int} A \subset G \subset A$ then every $x \in G$ is an interior point of $G$, so there is $B(x, r) \subset G \subset A$, hence $x \in \operatorname{Int} A$. □

The next example presents the interior of a familiar set. We leave the proof to the reader (Exercise 2.1.23).

**Example 2.1.16. Interior of a set.**
The interior of $[a, b)$ is the open interval $(a, b)$. ♦

In the previous exercise, almost all points of $[a, b)$ were interior points. In the next one we will go to the other extreme.

**Example 2.1.17. Interior of a set.**
The interior of $\mathbb{Q}$ is $\emptyset$. We leave the proof as Exercise 2.1.23. ♦

By Exercise 2.1.25, the union of any number of open sets is open, but when it comes to their intersection we can be sure only if the collection is finite. By taking complements, we can deduce a similar result for closed sets. Yet, infinite intersections of open sets and infinite unions of closed sets will show up frequently throughout the text.

**Definition 2.1.18.** A set is said to be a $\boldsymbol{G_\delta}$ **set** if it is a countable intersection of open sets. A set is said to be an $\boldsymbol{F_\sigma}$ **set** if it is a countable union of closed sets.

The following examples show that $F_\sigma$ as well as $G_\delta$ sets may be neither open nor closed.

**Example 2.1.19. An $F_\sigma$ set.**
The set $\mathbb{Q}$ of rational numbers is an $F_\sigma$ set.
Reason: if $\mathbb{Q} = \{r_n\}_{n=1}^\infty$, then $\mathbb{Q} = \cup_{n=1}^\infty \{r_n\}$. ♦

**Example 2.1.20. A $G_\delta$ set.** The set of irrational numbers is a $G_\delta$ set. We leave the verification to the reader (Exercise 2.1.32). ♦

While interior points are used to describe open sets (see Exercise 2.1.21) we need another type of points for closed sets.

**Definition 2.1.21.** A point $x$ in a metric space $M$ is a **cluster point** of $A$ if every ball $B(x, r)$ contains at least one point of $A$ other than itself. A point $x$ in a set $A \subset M$ is an **isolated point** of $A$ if there is a ball $B(x, r)$ that contains no other points of $A$. The set of all points of a set $A$ together with all of its cluster points is called the **closure** of $A$ and is denoted by $\overline{A}$.

An alternate description of the closure is very useful.

**Theorem 2.1.22.** *The closure $\overline{A}$ is the smallest closed set that contains $A$.*

*Proof.* First, $\overline{A}$ is closed because $(\overline{A})^C$ is open. Indeed, if $x \in (\overline{A})^C$ then $x \notin A$ and $x$ is not a cluster point of $A$, so there is $B(x, r)$ that has empty intersection with $\overline{A}$, i.e., $B(x, r) \subset (\overline{A})^C$, whence $(\overline{A})^C$ is open. By definition, $A \subset \overline{A}$. Finally, let $F$ be a closed set that contains $A$ and let $x$ be a cluster point of $A$. We show that $x \in F$ (hence $\overline{A} \subset F$). Otherwise, $x \in F^C$ which is open so there is $B(x, r)$ such that $B(x, r) \cap F = \emptyset$ and all the more $B(x, r) \cap A = \emptyset$, contradicting the choice of $x$. □

Next, we look at closures of the sets from Examples 2.1.16 and 2.1.17. We leave the proofs to the reader (Exercises 2.1.33 and 2.1.34).

**Example 2.1.23. Closure of a set.**
The closure of $[a, b)$ is $[a, b]$. ♦

**Example 2.1.24. Closure of a set.**
The closure of $\mathbb{Q}$ is $\mathbb{R}$. ♦

> The closure was introduced by a Polish mathematician Kazimierz Kuratowski (1896–1980) in his 1922 paper [67]. *Analysis situs* in the title of the article is the old name for topology. This article was actually a part of his doctoral dissertation.

We close this subsection with a very important concept.

**Definition 2.1.25.** The **boundary** of a set $A \subset M$ is $\overline{A} \setminus \operatorname{Int} A$. We denote it by $\partial A$.

Using this concept we have another definition of closure.

**Theorem 2.1.26.** *If $A$ is a set in a metric space then* $\overline{A} = A \cup \partial A$.

*Proof.* By definition,

$$A \cup \partial A = A \cup (\overline{A} \setminus \operatorname{Int} A) = A \cup (\overline{A} \cap (\operatorname{Int} A)^C)) = (A \cup \overline{A}) \cap (A \cup (\operatorname{Int} A)^C) = \overline{A},$$

because $A \subset \overline{A}$ (Theorem 2.1.22) and $\operatorname{Int} A \subset A$ (Theorem 2.1.15) whence $A \cup (\operatorname{Int} A)^C \supset A \cup A^C = X$. □

We leave the verification in the following examples to the reader (Exercises 2.1.40 and 2.1.41).

**Example 2.1.27. Boundary of a set.**
The boundary of $[a, b)$ is $\{a, b\}$. ♦

**Example 2.1.28. Boundary of a set.**
The boundary of $\mathbb{Q}$ is $\mathbb{R}$. ♦

FELIX HAUSDORFF (1868–1942) was a German mathematician. He came from a wealthy Jewish family in Leipzig and never needed to work to support himself. He studied at Leipzig University and received a doctorate in applications of mathematics to astronomy. His first teaching position was at Leipzig University. He wrote several papers in astronomy and optics, but he also published a literary work of 378 pages, a philosophy book, a book of poems, and a number of articles in philosophy and literature. His 1904 comedy was performed more than 300 times between 1904 and 1912. Starting with 1904, he became interested in topology and set theory. He introduced the concept of a partially ordered set and proved many results about them. In 1910 he moved to Bonn, then three years later to Greifswalf. Hausdorff returned to Bonn in 1921 and worked there until 1935 when the Nazi regime forced him to retire. Faced with the prospect of being sent to a concentration camp, he and his wife committed suicide. His masterpiece on set theory and topology [56] is considered to be one of the most important mathematical books of the twentieth century. It created the theory of topological and metric spaces. In addition, he worked in measure theory (Hausdorff dimension), probability (the first correct proof of the strong law of large numbers), and functional analysis (Hausdorff-Young inequality).

## Exercises

2.1.1. Prove that the set $\mathbb{R}$ is a metric space with

$$d(x, y) = |y - x|.$$

2.1.2. Prove that the set $C[a, b]$ is a metric space with

$$d(f, g) := \max\{|f(x) - g(x)| : a \le x \le b\}.$$

2.1.3. Prove that the set $C[a, b]$ is a metric space with

$$d(f, g) := \int_a^b |f(x) - g(x)|\, dx$$

2.1.4. Prove Theorem 2.1.6.

2.1.5. Prove Theorem 2.1.7.

2.1.6. Prove that $\mathbb{R}^2$ with the distance $d((x_1, x_2), (y_1, y_2)) := |x_1 - y_1| + |x_2 - y_2|$ is a metric space.

2.1.7. Prove that $\mathbb{R}^2$ with the distance $d((x_1, x_2), (y_1, y_2)) := \max\{|x_1 - y_1|, |x_2 - y_2|\}$ is a metric space.

2.1.8.* Prove that the set $\mathbb{R}$ is a metric space with

$$d(x, y) := \frac{|x - y|}{1 + |x - y|}.$$

2.1.9. Prove that the set $C^1[a, b]$ of differentiable functions on $[a, b]$ with continuous derivatives, is a metric space with

$$d(f, g) := \max\{|f(x) - g(x)| : a \le x \le b\} + \max\{|f'(x) - g'(x)| : a \le x \le b\}.$$

2.1.10.* Prove that the set $C(-\infty, \infty)$ is a metric space with

$$d(f, g) := \sum_{n=1}^{\infty} \frac{1}{2^n} \frac{\max\{|f(x) - g(x)| : |x| \le n\}}{1 + \max\{|f(x) - g(x)| : |x| \le n\}}.$$

2.1.11.* Determine the values of $p$ for which $|x - y|^p$ is a metric on $\mathbb{R}$.

2.1.12. Let $M$ be any nonempty set and define $d : M \to \mathbb{R}$ by $d(x, y) = 0$ if and only if $x = y$ and $d(x, y) = 1$ otherwise. Prove that $(M, d)$ is a metric space

2.1.13. Let $c$ be the set of all bounded sequences and, for $\{a_n\}, \{b_n\} \in c$, let $d(\{a_n\}, \{b_n\}) = \sup\{|a_n - b_n| : n \in \mathbb{N}\}$. Prove that $(c, d)$ is a metric space.

2.1.14. Suppose that, for $1 \le i \le n$, $(M_i, d_i)$ are metric spaces. Let $M = M_1 \times M_2 \times \cdots \times M_n$ and for $x = (x_1, x_2, \ldots, x_n), y = (y_1, y_2, \ldots, y_n) \in M$ let $d(x, y) = \sum_{k=1}^{n} d_k(x_k, y_k)$. Prove that $(M, d)$ is a metric space.

2.1.15. Describe open and closed balls in the metric space $(\mathbb{R}^2, d)$, where the distance $d$ is defined by $d((x_1, x_2), (y_1, y_2)) := \left(|x_1 - y_1|^3 + |x_2 - y_2|^3\right)^{1/3}$.

2.1.16. Describe open and closed balls in the metric space $(C[0, 1], d)$, where $d(f, g) := \sup\{|f(x) - g(x)| : x \in [0, 1]$.

2.1.17. Let $u(x) = \sin x$ and $v(x) = 0$ and let $C[0, 2\pi]$ denote the metric space of continuous functions on $[0, 2\pi]$ with the distance $d(f, g) := \sup\{|f(x) - g(x)| : x \in [0, 2\pi]\}$. Give an example of a function that belongs to both $B(u, 2/3)$ and $B(v, 2/3)$.

2.1.18. In the metric space $(\mathbb{R}, d)$ where

$$d(x, y) := \frac{|x - y|}{1 + |x - y|}$$

describe open balls $B(0, r)$ for: (a) $r = 1/2$; (b) $r = 1$. (See Exercise 2.1.8.)

2.1.19. Prove that $G := \{(x, y) \in \mathbb{R}^2 : x^2 + y^2 \neq 1\}$ is an open set in $\mathbb{R}^2$.

2.1.20. Prove that an open ball in a metric space is an open set.

2.1.21. Prove that the following are equivalent for a set $A$ in a metric space:

(a) $A$ is open.

(b) Each point of $A$ is an interior point.

(c) $A$ is equal to its interior.

2.1.22. Prove that in a metric space:

(a) $\text{Int}\,(\text{Int}\,A) = \text{Int}\,A$.

(b) If $A \subset B$ then $\text{Int}\,A \subset \text{Int}\,B$.

(c) $\text{Int}\,(A \cap B) = \text{Int}\,A \cap \text{Int}\,B$.

(d) $\text{Int}\,(\bigcap_{i \in I} A_i) \subset \bigcap_{i \in I} \text{Int}\,A_i$.

(e) Give an example to show that the inclusion in (d) can be proper.

(f) $\text{Int}\,A \cup \text{Int}\,B \subset \text{Int}\,(A \cup B)$.

(g) Give an example to show that the inclusion in (f) can be proper.

2.1.23. Prove that the interior of the interval $[a, b)$ is the open interval $(a, b)$.

2.1.24. Prove that the set $\mathbb{Q}$ has empty interior.

2.1.25. Let $(M, d)$ be a metric space. Prove that:

(a) The empty set and the whole space $M$ are open sets.

(b) If $\{G_i\}_{i \in I}$ is an arbitrary collection of open sets then their union is an open set.

(c) If $\{G_i\}_{i=1}^n$ is a finite collection of open sets then their intersection is an open set.

2.1.26. Let $A, B$ be subsets of $\mathbb{R}$. Prove that $\text{Int}\,(A \times B) = \text{Int}\,A \times \text{Int}\,B$.

2.1.27. Let $(M, d)$ be a metric space. Prove that:

(a) The empty set and the whole space $M$ are closed sets.

(b) If $\{F_i\}_{i \in I}$ is an arbitrary collection of closed sets then their intersection is a closed set.

(c) If $\{F_i\}_{i=1}^n$ is an arbitrary collection of closed sets then their union is a closed set.

2.1.28. Prove that the set of all rational numbers between $\sqrt{2}$ and $\pi$ is a closed set in the metric space $\mathbb{Q}$ with the distance $d(x, y) = |x - y|$.

2.1.29. Prove that a closed ball in a metric space is a closed set.

2.1.30. Prove that in a metric space $(M,d)$ if $a \in M$ and $r > 0$ then the set $\{x : d(x,a) = r\}$ is closed.

2.1.31.* Give an example of two closed sets $A, B$ such that $A \cap B = \emptyset$ but $\operatorname{dist}(A,B) = 0$.

2.1.32. Prove that the set of irrational numbers is a $G_\delta$ set.

2.1.33. Prove that the closure of the interval $[a,b)$ is the closed interval $[a,b]$.

2.1.34. Prove that the closure of the set $\mathbb{Q}$ is $\mathbb{R}$.

2.1.35. Prove that in a metric space:

(a) $\overline{(\overline{A})} = \overline{A}$.

(b) If $A \subset B$ then $\overline{A} \subset \overline{B}$.

(c) $\overline{A \cup B} = \overline{A} \cup \overline{B}$.

(d) $\bigcup_{i\in I} \overline{A_i} \subset \overline{\bigcup_{i\in I} A_i}$.

(e) Give an example to show that the inclusion in (d) can be proper.

(f) $\overline{\bigcap_{i\in I} A_i} \subset \bigcap_{i\in I} \overline{A_i}$.

(g) Give an example to show that the inclusion in (f) can be proper.

2.1.36. Prove that the following are equivalent for a set $A$ in a metric space:

(a) $A$ is closed.

(b) $A$ contains all of its cluster points.

(c) $A$ is equal to its closure.

2.1.37. Prove that the closure of a set $A$ equals the intersection of all closed sets that contain $A$.

2.1.38. Prove that in a metric space:

(a) $(\overline{A})^C = \operatorname{Int}(A^C)$.

(b) $(\operatorname{Int} A)^C = \overline{A^C}$.

2.1.39.* Prove or disprove: in a metric space $(M,d)$ the closure of an open ball $B(c,r)$ is the closed ball $B[c,r]$.

2.1.40. Prove that the boundary of the interval $[a,b)$ is the set $\{a,b\}$.

2.1.41. Prove that the boundary of $\mathbb{Q}$ is the set $\mathbb{R}$.

2.1.42. Prove that the boundary of any set is closed.

2.1.43. Prove that a point $x$ in a metric space belongs to $\partial A$ if and only if every ball $B(x,r)$ has a nonempty intersection with both $A$ and $A^C$.

2.1.44. Prove that a set is closed if and only if it contains its boundary.

2.1.45.* Prove that the interior of the boundary of a closed set is the empty set.

2.1.46.* Give an example of a metric space in which the boundary of the open unit ball is a proper subspace of the unit sphere with the same center.

## 2.2 Completeness

In the previous section we saw that a metric space has a natural topology. This topology comes with some nice features that will allow us to establish more easily some fundamental properties that a metric space may have. More concretely, a study of a metric space can be conducted by relying on the behaviour of a sequence of points in it. (The technical term is that a metric space is a first-countable Hausdorff topological space.) Therefore, the natural place to begin is with a definition of a sequence.

**Definition 2.2.1.** A **sequence** $\{x_n\}$ in a metric space $M$ is a function $x : \mathbb{N} \to M$. Instead of the functional notation $x(n)$ we always write $x_n$.

*Remark* 2.2.2. For emphasis, or when the domain is a subset of $\mathbb{N}$, we may include it as in $\{x_n\}_{n=k}^{m}$, where $k \in \mathbb{N}$ and $m$ is either a positive integer no less than $k$, or $+\infty$. Even though we use notation $\{x_n\}$ for a sequence, when we want to list its members we write $(x_1, x_2, \dots)$.

**Example 2.2.3. A sequence in a metric space.**
Let $M$ be the metric space $\mathbb{R}^2$ with Euclidean distance, and let $x_n = (n, 2n)$. ♦

Next, we define a convergent sequence.

**Definition 2.2.4.** We say that a sequence $\{x_n\}$ is **convergent** in $M$ and that $x \in M$ is its **limit**, denoted $\lim x_n = x$, if for any $\varepsilon > 0$ there exists a positive integer $N$ such that $d(x_n, x) < \varepsilon$, for all $n \geq N$.

We have seen that the real line with the distance $d(x, y) = |x-y|$ is a metric space. Many properties of $\mathbb{R}$ carry over to general metric spaces. Among them are many that concern sequences. Here is an illustration.

**Theorem 2.2.5.** *A convergent sequence in a metric space can have only one limit.*

*Proof.* Suppose to the contrary that $\{x_n\}$ has two distinct limits $y$ and $z$. Then $d(y, z) > 0$. By definition, since $\lim x_n = y$ there is $N_y \in \mathbb{N}$ such that $d(x_n, y) < d(y, z)/2$, for all $n \geq N_y$. Similarly, there is $N_z \in \mathbb{N}$ such that $d(x_n, z) < d(y, z)/2$, for all $n \geq N_z$. If now $n \geq \max\{N_y, N_z\}$ then

$d(y,z) \le d(y,x_n) + d(x_n,z) < d(y,z)/2 + d(y,z)/2 = d(y,z)$. The obtained contradiction shows that there can be no two distinct limits of a sequence. □

We mentioned that using sequences often makes proofs easier. The following characterization of a closure of a set is a good example.

**Theorem 2.2.6.** *In a metric space, $x$ belongs to the closure of a set $A$ if and only if there exists a sequence $\{a_n\} \subset A$ that converges to $x$.*

*Proof.* ($\Rightarrow$) By definition, either $x \in A$ or $x$ is a cluster point of $A$. If $x \in A$ we can take $\{a_n\}$ to be the constant sequence $(x, x, x, \dots)$. If $x$ is a cluster point of $A$, by definition, every ball $B(x, r)$ contains a member of $A$. Taking $r$ to be $1, 1/2, 1/3, \dots$ we obtain that, for every $n \in \mathbb{N}$, there is $a_n \in A$ such that $a_n \in B(x, 1/n)$. Since $d(x, a_n) < 1/n$, Definition 2.2.4 implies that $\lim a_n = x$.

($\Leftarrow$) Suppose that $\lim a_n = x$ but $x \notin \overline{A}$. Then $x$ belongs to the open set $(\overline{A})^C$, so there is $r > 0$ such that $B(x, r) \subset (\overline{A})^C$. Since $(\overline{A})^C \subset A^C$, the ball $B(x, r)$ contains no points of $A$. In particular, it does not contain any members of $\{a_n\}$, contradicting the assumption that $\lim a_n = x$. □

Numerous applications of Theorem 2.2.6 can be found in among exercises at the end of the section. Here is a sample.

**Theorem 2.2.7.** *If $A, B$ are subsets of $\mathbb{R}$ then $\overline{A \times B} = \overline{A} \times \overline{B}$.*

*Proof.* By Theorem 2.2.6, $(x, y) \in \overline{A \times B}$ if and only if there is a sequence $\{(x_n, y_n)\} \subset A \times B$ that converges to $(x, y)$. This is further equivalent to the existence of sequences $\{x_n\} \subset A$ and $\{y_n\} \subset B$ such that $x = \lim x_n$ and $y = \lim y_n$. Applying Theorem 2.2.6 again, we see that this is equivalent to $x \in \overline{A}$ and $y \in \overline{B}$, hence to $(x, y) \in \overline{A} \times \overline{B}$. □

Completeness is one of the very much desired properties of a metric space. Recall that the set of rational numbers is not complete and that the so-called *The Completeness Axiom* ensures that the set of real numbers is complete. In this section we will talk about the completeness of metric spaces. Before that, we need the definition of a Cauchy sequence.

**Definition 2.2.8.** A sequence $\{x_n\}$ is a **Cauchy sequence** in $M$ if, for any $\varepsilon > 0$, there exists a positive integer $N$ such that, if $m \ge n \ge N$, then $d(x_m, x_n) < \varepsilon$.

**Example 2.2.9. A Cauchy sequence**

Let $M$ be the interval $(0, 1)$ with the distance $d(x, y) = |x - y|$. The sequence $\{1/n\}$ is a Cauchy sequence. We leave the proof to the reader. ◆

For more examples see Exercises 2.2.25–2.2.32.

Just like in Euclidean spaces every convergent sequence is a Cauchy sequence (Exercise 2.2.1). As Example 2.2.9 shows, a Cauchy sequence need not converge. (The limit would be 0 which is not in $(0, 1)$.) The issue at stake is whether the metric space is complete.

**Definition 2.2.10.** A metric space $M$ is said to be **complete** if every Cauchy the sequence in $M$ converges.

It is not hard to determine whether metric spaces in the following examples are complete or not. We leave the task to the reader (Exercises 2.2.13–2.2.15).

**Example 2.2.11. A complete metric space.**
Interval $[0, 1]$ with the usual distance is complete. ♦

**Example 2.2.12. A complete metric space.**
Euclidean space $\mathbb{R}^n$ is complete. ♦

**Example 2.2.13. A metric space that is not complete.**
The set $\mathbb{Q}$ with the usual distance is not complete. ♦

An important and valuable property of complete metric spaces is that their defining feature is inherited by their closed subspaces.

**Theorem 2.2.14.** *A closed subset of a complete metric space is complete.*

*Proof.* Let $A$ be a closed set in a complete metric space $M$ and let $\{x_n\}$ be a Cauchy sequence in $A$. Then $\{x_n\}$ is a Cauchy sequence in $M$, hence it converges to $x \in M$. By definition, every ball $B(x, r)$ contains a member of $\{x_n\}$, hence an element of $A$. So either $x \in A$ (in which case $A$ is complete) or $x$ is a cluster point of $A$ so it belongs to $A$ because $A$ is closed (Exercise 2.1.36). □

It is not always easy to determine whether in a specific metric space every Cauchy sequence converges. We will now state a few results that can be useful when trying to answer this question. First, we need to define a *subsequence.*

**Definition 2.2.15.** Let $\{x_n\}_{n=1}^{\infty}$ be a sequence in a metric space $M$ and let $\{n_k\}_{k=1}^{\infty}$ be an infinite, strictly increasing sequence of positive integers. We say that a sequence $(x_{n_1}, x_{n_2}, \dots)$ is a **subsequence** of $\{x_n\}_{n=1}^{\infty}$.

**Example 2.2.16. A subsequence.**
The sequence $\{1/n\}_{n=1}^{\infty}$ has a subsequence $\{1/2^k\}_{k=1}^{\infty}$. ♦

The following proposition gives us a sufficient condition for a Cauchy sequence to converge.

**Proposition 2.2.17.** *A Cauchy sequence in a metric space converges if it has a convergent subsequence.*

*Proof.* Let $\{x_n\}$ be a Cauchy sequence in a metric space $(M, d)$, let $\{x_{n_k}\}$ be a convergent subsequence of $\{x_n\}$, and denote $\lim x_{n_k} = x \in M$. Let $\varepsilon > 0$. Since $\{x_n\}$ is a Cauchy sequence, there exists $N \in \mathbb{N}$ such that

$$m, n \geq N \quad \Rightarrow \quad d(x_n, x_m) < \frac{\varepsilon}{2}.$$

Also, $x_{n_k} \to x$ so there exists $k \in \mathbb{N}$ such that $n_k > N$ and

$$d(x_{n_k}, x) < \frac{\varepsilon}{2}.$$

Let $n \geq N$. Combining the last two estimates,

$$d(x_n, x) \leq d(x_n, x_{n_k}) + d(x_{n_k}, x) < \varepsilon.$$

□

Proposition 2.2.17 points out the utility of finding a convergent subsequence. In order to do so, we will define a special type of Cauchy sequence.

**Definition 2.2.18.** A sequence $\{x_n\}_{n=1}^{\infty}$ in a metric space $(M, d)$ is **rapidly Cauchy** if there is a convergent series of positive numbers $\sum_{k=1}^{\infty} \varepsilon_k$ such that $d(x_{k+1}, x_k) \leq \varepsilon_k^2$, for all $k$.

**Example 2.2.19. A rapidly Cauchy sequence.**
The sequence $\{1/n^2\}_{n=1}^{\infty}$ is a rapidly Cauchy sequence.
**Proof:** For any $k \in \mathbb{N}$,

$$\begin{aligned} d(x_{k+1}, x_k) &= \left| \frac{1}{(k+1)^2} - \frac{1}{k^2} \right| = \frac{2k+1}{k^2(k+1)^2} \\ &< \frac{2k+2}{k^2(k+1)^2} = \frac{2}{k^2(k+1)} < \frac{4}{k^3}. \end{aligned}$$

If we denote $\varepsilon_k := 2/k^{3/2}$ then $d(x_{k+1}, x_k) < \varepsilon_k^2$ for all $k$ and $\sum_{k=1}^{\infty} \varepsilon_k = \sum_{k=1}^{\infty} 2/k^{3/2}$ converges. ♦

**Example 2.2.20. A sequence that is not a rapidly Cauchy sequence.**
Let $s_n = \sum_{k=1}^{n} 1/k^2$. Then $\{s_n\}$ is not a rapidly Cauchy sequence.
**Proof:** Members of the sequence $\{s_n\}$ are partial sums of a convergent series, so $\{s_n\}$ is convergent itself. By Exercise 2.2.1, it is a Cauchy sequence. However, it is not a rapidly Cauchy sequence, because if $|s_{n+1} - s_n| \leq \varepsilon_n^2$ then $1/(n+1)^2 \leq \varepsilon_n^2$ so $\varepsilon_n \geq 1/(n+1)$, whence the series $\sum \varepsilon_n$ diverges. ♦

The following result justifies the name given to rapidly Cauchy sequences.

**Proposition 2.2.21.** *Every rapidly Cauchy sequence in a metric space is a Cauchy sequence.*

*Proof.* Let $\{x_n\}$ be a rapidly Cauchy sequence in a metric space $(M, d)$. By definition, there is a convergent series of positive numbers $\sum_{k=1}^{\infty} \varepsilon_k$ such that $d(x_{k+1}, x_k) \leq \varepsilon_k^2$, for all $k$. Let $\varepsilon > 0$. The convergence of $\sum_{k=1}^{\infty} \varepsilon_k$ implies that $\sum_{k=1}^{\infty} \varepsilon_k^2$ is also convergent. Indeed, $\{\varepsilon_k\}$ must converge to 0, so for sufficiently large $k$, $\varepsilon_k < 1$ hence $\varepsilon_k^2 < \varepsilon_k$. Thus, there is $N \in \mathbb{N}$ such that $\sum_{k=N}^{\infty} \varepsilon_k^2 < \varepsilon$.

For $m \geq n \geq N$, we have

$$\begin{aligned} d(x_m, x_n) &\leq d(x_n, x_{n+1}) + d(x_{n+1}, x_{n+2}) + \cdots + d(x_{m-1}, x_m) \\ &\leq \sum_{k=n}^{m-1} \varepsilon_k^2 \leq \sum_{k=N}^{\infty} \varepsilon_k^2 < \varepsilon, \end{aligned}$$

so $\{x_n\}$ is a Cauchy sequence. □

Proposition 2.2.17 shows that one should be on the lookout for a suitable subsequence of a given Cauchy sequence. The following result asserts that one can always find a rapidly Cauchy subsequence.

**Proposition 2.2.22.** *Every Cauchy sequence in a metric space has a rapidly Cauchy subsequence. Moreover, if $\{x_n\}_{n=1}^{\infty}$ is a Cauchy sequence in a metric space $(M, d)$ and $\{\varepsilon_n\}$ is a sequence of positive numbers that converges to 0 then there is a subsequence $\{y_n\} \subset \{x_n\}$ such that $d(y_n, y_m) < \varepsilon_n$, for $m \geq n$.*

*Proof.* By definition, there is $N_1 \in \mathbb{N}$ such that $m \geq n \geq N_1 \Rightarrow d(x_m, x_n) < \varepsilon_1$. Next, there is $N_2 \geq N_1$ such that $m \geq n \geq N_2 \Rightarrow d(x_m, x_n) < \varepsilon_2$. Continuing, we obtain an increasing sequence $\{N_k\}$ such that $m \geq n \geq N_k \Rightarrow d(x_m, x_n) < \varepsilon_k$. Now define $y_k = x_{N_k}$. Then $\{y_n\} \subset \{x_n\}$ and $d(y_n, y_m) < \varepsilon_n$, for $m \geq n$. □

This brings us to the following theorem. It will be a powerful tool to establish the completeness of some important metric spaces. Since it follows directly from Propositions 2.2.17 and 2.2.21 we leave the proof to the reader (Exercise 2.2.33).

**Theorem 2.2.23.** *A metric space is complete if and only if every rapidly Cauchy sequence converges.*

Another very important tool is a characterization of completeness contained in the following result. The reader is likely familiar with its special case when $M$ is either $\mathbb{R}$ or any Euclidean space $\mathbb{R}^n$. Here, the word diameter means the smallest possible $K$ such that $d(x, y) \leq K$ for all $x$ and $y$ in $M$. (See Definition 2.3.4.)

**Theorem 2.2.24. Cantor's Intersection Theorem.** *A metric space $M$ is complete if and only if every sequence of nested, nonempty, closed sets whose diameters tend to be zero has a nonempty intersection consisting of exactly one point.*

*Proof.* ($\Rightarrow$) Let $\{F_k\}_{k=1}^{\infty}$ be a sequence of nested, nonempty, closed sets in a metric space $(M, d)$ and suppose that their diameters tend to zero. Without loss of generality we may assume that for every $k \in \mathbb{N}$, $F_{k+1}$ is a proper subset of $F_k$. This allows us to select $x_k \in F_k \setminus F_{k+1}$ and we obtain a sequence $\{x_k\}_{k=1}^{\infty}$. We will show that it is a Cauchy sequence. Let $\varepsilon > 0$. Since the diameters of the sets $F_n$ go to 0, there is $N \in \mathbb{N}$ such that, if $n \geq N$ then

$\operatorname{diam} F_n < \varepsilon$. For that $N$ and $m \geq n$ we have that $d(x_m, x_n) \leq \operatorname{diam} F_N < \varepsilon$. Thus, $\{x_k\}_{k=1}^{\infty}$ is a Cauchy sequence and it follows that it is convergent because $M$ is complete. Let $x$ denote the limit of $\{x_k\}_{k=1}^{\infty}$. It is clear that, for every $n \in \mathbb{N}$, $\{x_k\}_{k=n}^{\infty} \subset F_n$ so $x$ is a cluster point of a closed set $F_n$. By Exercise 2.1.36, $x \in F_n$. Since this holds for all $n$, $x \in \cap_{n=1}^{\infty} F_n$, so the intersection under scrutiny is not empty. Finally, there can be no 2 points $x, y \in \cap_{n=1}^{\infty} F_n$. Indeed, if $x, y \in F_n$ then $d(x, y) < \operatorname{diam} F_n$, so letting $n \to \infty$ we would get $d(x, y) = 0$, whence $x = y$.

($\Leftarrow$) Let $\{x_n\}_{n=1}^{\infty}$ be a Cauchy sequence in a metric space $(M, d)$. For every $k \in \mathbb{N}$, we define $F_k$ to be the closure of the set $\{x_k, x_{k+1}, \dots\}$. Then $\{F_k\}$ is obviously a sequence of closed, nonempty, nested sets. Moreover, their diameters tend to 0. Indeed, let $\varepsilon > 0$. Since $\{x_n\}_{n=1}^{\infty}$ is a Cauchy sequence there is $N \in \mathbb{N}$ such that if $m \geq n \geq N$ then $d(x_m, x_n) < \varepsilon/3$. Further, if $n \geq N$ and $y, z \in F_n$ then there are $x_i, x_j$, with $i, j \geq n$ such that $d(y, x_i) < \varepsilon/3$ and $d(z, x_j) < \varepsilon/3$. Now,

$$d(y, z) \leq d(y, x_i) + d(x_i, x_j) + d(x_j, z) < \frac{\varepsilon}{3} + \frac{\varepsilon}{3} + \frac{\varepsilon}{3} = \varepsilon,$$

so $\operatorname{diam} F_n < \varepsilon$ which shows that diameters of $\{F_n\}$ go to 0. By assumption, there is a unique point $x \in \cap_{n=1}^{\infty} F_n$. It remains to establish that $\lim x_n = x$. With $N$ as above, we have that $\operatorname{diam} F_N < \varepsilon$. If $n \geq N$ then both $x_n$ and $x$ are in $F_N$ so $d(x_n, x) < \varepsilon$. □

## Exercises

2.2.1. Prove that in a metric space every convergent sequence is a Cauchy sequence.

2.2.2. Prove that if a sequence $\{x_n\}_{n=1}^{\infty}$ does not converge to $x$ then there is an open ball $B(x, r)$ that contains no members of the sequence.

2.2.3. Let $G$ be an open set in $\mathbb{R}^n$, let $x_0 \in \mathbb{R}_n$, and let $\{x_n\}$ be a sequence in $\mathbb{R}^n$ that converges to $x_0$. Prove that $G + x_0 \subset \liminf(G + x_n)$.

2.2.4. Let $F$ be a closed set in $\mathbb{R}^n$, let $x_0 \in \mathbb{R}_n$, and let $\{x_n\}$ be a sequence in $\mathbb{R}^n$ that converges to $x_0$. Prove that $F + x_0 \supset \limsup(F + x_n)$.

2.2.5. Let $A$ be a set in a metric space $M$ and let $a$ be a cluster point of $A$. Prove that there is a sequence $\{a_n\} \subset A$ that converges to $a$.

2.2.6. Let $(M, d)$ be a metric space, and let $\lim x_n = x$, $\lim y_n = y$. Prove that $\lim d(x_n, y_n) = d(x, y)$.

In Exercises 2.2.7–2.2.9, $\{a_{ij}\}_{i,j=1}^{\infty}$ is a sequence of nonnegative real numbers.

2.2.7. Prove that $\sum_{i=1}^{\infty} \sum_{j=1}^{\infty} a_{ij} = \sum_{j=1}^{\infty} \sum_{i=1}^{\infty} a_{ij}$.

2.2.8. Prove that $\sup_{i=1}^{\infty} \sup_{j=1}^{\infty} a_{ij} = \sup_{j=1}^{\infty} \sup_{i=1}^{\infty} a_{ij}$.

2.2.9. Prove that $\sup_{i=1}^{\infty} \sum_{j=1}^{\infty} a_{ij} \leq \sum_{j=1}^{\infty} \sup_{i=1}^{\infty} a_{ij}$. Give an example to show that the inequality may be strict.

2.2.10. Let $F \subset \mathbb{R}$ be a closed set that is contained in the disjoint union of intervals $(a, b)$ and $(c, d)$. Prove that $F \cap (a, b)$ is a closed set.

2.2.11. Let $A, B$ be closed subsets of $\mathbb{R}$. Prove that $A \times B$ is a closed subset of $\mathbb{R}^2$.

2.2.12. Prove that in a metric space every closed set is a $G_\delta$ set and that every open set is an $F_\sigma$ set.

2.2.13. Prove that the metric space $([0, 1], d)$, where $d(x, y) = |x - y|$, is complete.

2.2.14. Prove that Euclidean space $\mathbb{R}^n$ is complete.

2.2.15. Prove that the metric space $(\mathbb{Q}, d)$, where $d(x, y) = |x - y|$, is not complete.

2.2.16. Prove that the set of all polynomials on $[0, 1]$ with the metric $d(p, q) := \max\{|p(x) - q(x)| : 0 \leq x \leq 1\}$ is not a complete metric space.

2.2.17. Prove that the set $C[0, 1]$ with the metric

$$d(f, g) := \int_0^1 |f(x) - g(x)| \, dx$$

is not complete.

2.2.18. Prove that if a subspace of a metric space is complete then it is closed.

2.2.19. Give an example of a metric space $(M, d)$ and a closed set $A \subset M$ that is not complete.

2.2.20. Let $\{x_n\}$ be a convergent sequence in a metric space. Prove that every subsequence of $\{x_n\}$ converges to the same limit.

2.2.21. Let $\{x_n\}$ be a sequence in a metric space and suppose that each of its three subsequences $\{x_{2n}\}$, $\{x_{2n+1}\}$, and $\{x_{3n}\}$ is convergent. Prove that $\{x_n\}$ is a convergent sequence.

2.2.22.* Let $\{x_n\}$ be a sequence in a metric space and suppose that each of its infinitely many subsequences $\{x_{sn}\}$, where $s = 2, 3, \ldots$, is convergent. Prove or disprove: $\{x_n\}$ is a convergent sequence.

2.2.23.* Let $\{x_n\}$ be a sequence in a metric space $(M, d)$, let $x \in M$, and suppose that each subsequence $\{x_{n_k}\} \subset \{x_n\}$ has a further subsequence $\{y_n\}$ which converges to $x$. Prove that $\{x_n\}$ converges to $x$.

2.2.24. Let $(X, d)$ be a metric space that is not complete and let $\{x_n\}$ be a Cauchy sequence in $X$ that is not convergent. Prove that the set $E = \{x_n\}$ is closed.

In Exercises 2.2.25–2.2.28, determine which of the following sequences is (a) a Cauchy sequence, (b) a rapidly Cauchy sequence in $(0, 1)$.

2.2.25. $x_n = 1/n$.

2.2.26. $1, 0, 1/2, 0, 1/3, 0, 1/4, 0, \ldots$.

2.2.27. $x_n = e^n/n!$.

2.2.28. $1, 1, 1/4, 1/4, 1/9, 1/9, 1/16, 1/16, \ldots$.

In Exercises 2.2.29–2.2.32, determine which of the following sequences is (a) a Cauchy sequence, (b) a rapidly Cauchy sequence in the metric space $C[0, 1]$ (see Exercise 2.1.2).

2.2.29. $f_n(x) = x^n$

2.2.30. $f_n(x) = x^n/n$

2.2.31. $f_n(x) = x^n/\sqrt{n}$

2.2.32. $f_n(x) = \sin nx$

2.2.33. Prove Theorem 2.2.23.

## 2.3 Compactness

There are several types of compactness, and we will show that in a metric space they are all equivalent. We start with the so-called *sequential* compactness.

**Definition 2.3.1.** A set $A$ in a metric space $M$ is **sequentially compact** if every sequence of points in $A$ has a convergent subsequence that converges to a point in $A$.

**Example 2.3.2. A sequentially compact metric space.**
Interval $[0, 1]$ with the usual distance is sequentially compact (by Bolzano–Weierstrass Theorem). ♦

**Example 2.3.3. A metric space that is not sequentially compact.**
Interval $(0, 1)$ with the usual distance is not sequentially compact because the sequence $\{1/n\}$ does not have a convergent subsequence. ♦

These examples are a reminder that in Euclidean spaces a set is sequentially compact if and only if it is closed and bounded. We will see that a modification of this result holds in metric spaces. First, we need to define boundedness.

**Definition 2.3.4.** A metric space $M$ is called **bounded** if there exists some number $K$, such that $d(x, y) \leq K$ for all $x$ and $y$ in $M$. The smallest possible such $K$ is called the **diameter** of $M$ (notation $\text{diam}(M)$). The space $M$ is called **totally bounded** if for every $r > 0$ there exist finitely many open balls of radius $r$ whose union covers $M$.

It follows from the definition that every totally bounded space is bounded (Exercise 2.3.2). In the metric space $(\mathbb{R}^n, d)$, where $d$ is the Euclidean distance, the converse holds as well (Exercise 2.3.3). However, the last implication is not universally true.

**Example 2.3.5. A bounded set that is not totally bounded.**
Let $M = \mathbb{R}$ and let

$$d(x, y) = \begin{cases} 1 & \text{if } x \neq y, \\ 0 & \text{if } x = y. \end{cases}$$

It is easy to verify that $(M, d)$ is a metric space, and it is bounded because the distance between any two real numbers cannot be more than 1. In other words, $M$ coincides with the closed ball with center zero and radius 1. However, for any $r < 1$, a ball $B(c, r)$ contains only $c$, i.e., $B(c, r) = \{c\}$. It follows that a finite number of such balls cannot cover $M$, so it is not totally bounded. ♦

It is possible to define compactness in a metric space by relying on open sets rather than sequences.

**Definition 2.3.6.** A set $A$ in a metric space $M$ is **compact** if whenever $\mathcal{G}$ is a collection of open sets such that $A \subset \cup_{G \in \mathcal{G}} G$, then there exists a finite collection of sets $G_i \in \mathcal{G}$, $1 \leq i \leq m$, such that $A \subset \cup_{i=1}^{m} G_i$. In short, every open cover of $A$ has a finite subcover.

Now we come to the announced equivalent versions of compactness in a metric space. For additional characterizations see Exercises 2.3.18 and 2.3.19.

**Theorem 2.3.7.** *Let $(M, d)$ be a metric space. The following are equivalent:*

*(a) $M$ is compact.*

*(b) $M$ is sequentially compact.*

*(c) $M$ is complete and totally bounded.*

*Proof.* (a) $\Rightarrow$ (b) Suppose that $M$ is not sequentially compact and let $\{x_n\}_{n=1}^{\infty}$ be a sequence without a convergent subsequence. Let $a \in M$. Since it is not a limit of a subsequence of $\{x_n\}_{n=1}^{\infty}$, there exists a ball $B(a, r_a)$ that contains only finitely many members of $\{x_n\}_{n=1}^{\infty}$. The collection $\{B(a, r_a) : a \in M\}$ is an open cover of $M$ (which is compact), so there is a finite subcover $\{B(a_n, r_{a_n})\}_{n=1}^{N}$. Each of these balls contains a finite number of members of $\{x_n\}_{n=1}^{\infty}$ which leads to a contradiction.

(b) $\Rightarrow$ (c) Let $\{x_n\}$ be a Cauchy sequence in $M$. By (b), it contains a subsequence converging to some point $x \in M$. By Proposition 2.2.17, the whole sequence $\{x_n\}$ converges to $x$, whence $M$ is complete.

If $M$ were not totally bounded, there would exist $r > 0$ such that $M$ is not a union of finitely many balls of radius $r$. Let $x_1 \in M$ be arbitrary. We proceed by induction. Suppose that $x_1, x_2, \ldots, x_n$ have been selected. The set $\cup_{k=1}^{n} B(x_k, r)$ does not cover $M$, so there exists

$$x_{n+1} \in M \setminus \left( \bigcup_{k=1}^{n} B(x_k, r) \right).$$

It follows that for any $m, n \in \mathbb{N}$, $d(x_m, x_n) \geq r$, so $\{x_n\}$ cannot have a Cauchy subsequence, let alone a convergent subsequence. Consequently, it cannot be sequentially compact.

(c) $\Rightarrow$ (a) We will argue by contradiction. Suppose that $M$ is not compact and let $\{G_i\}_{i \in I}$ be an open cover of $M$ without a finite subcover. Since $M$ is totally bounded, it may be covered by finitely many balls of radius $1/2$. Then at least one of these balls cannot be covered by a finite subcollection of $\{G_i\}_{i \in I}$. Let $B(x_1, 1/2)$ be one such ball. Using total boundedness of $M$, we obtain that $B(x_1, 1/2)$ can be covered by finitely many balls of radius $1/4$ and at least one of them cannot be covered by a finite subcollection of $\{G_i\}_{i \in I}$. Let $B(x_2, 1/4)$ be one such ball and notice that $d(x_1, x_2) < 1/2 + 1/4$ (Figure 2.4). Continuing we obtain a sequence of balls $B(x_n, 1/2^n)$ such that

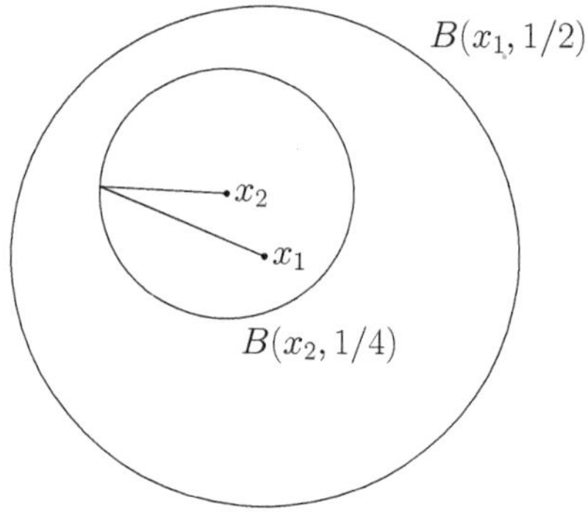


Figure 2.4: $d(x_1, x_2) < 1/2 + 1/4$

$$d(x_n, x_{n+1}) < \frac{1}{2^n} + \frac{1}{2^{n+1}} = \frac{3}{2^{n+1}} < \frac{1}{2^{n-1}}.$$

The series $\sum_{n=1}^{\infty} 1/(\sqrt{2})^{n-1}$ converges and it follows that $\{x_n\}$ is a rapidly Cauchy sequence in $M$. By Proposition 2.2.21, $\{x_n\}$ is a Cauchy sequence. Since $M$ is complete, this sequence converges to some $x \in M$. Now, there exists $i \in I$ for which $x \in G_i$. By assumption, $G_i$ is open so there exists $r > 0$ such that $B(x, r)$ is contained in $G_i$. The fact that $\{x_n\}$ converges to $x$, implies that there is $N \in \mathbb{N}$ for which

$$B(x_N, 1/2^N) \subset B(x, r) \subset G_i.$$

This is a contradiction because the ball $B(x_N, 1/2^N)$ was selected as one that could not be covered by finitely many sets $G_i$, and the theorem is proved. □

We see that if $K$ is a compact set then every open cover of $K$ can be reduced to a finite subcover. In fact, more is true.

**Theorem 2.3.8.** ***Lebesgue Covering Lemma.*** *If $\{G_i : i \in I\}$ is an open cover of a compact metric space $(M, d)$, then there exists $\delta > 0$ such that for every $x \in M$ there is $i \in I$ with $B(x, \delta) \subset G_i$.*

*Proof.* Suppose that no such $\delta > 0$ exists. Then for every $n \in \mathbb{N}$ there exists $x_n \in M$ such that $B(x_n, 1/n)$ is not contained in $G_i$ for any $i \in I$. Since $M$ is sequentially compact, there is $\{x_{n_k}\} \subset \{x_n\}$ that converges to $x \in M$. Therefore, there is $i \in I$ such that $x \in G_i$. Moreover, $G_i$ is open so there is $\varepsilon > 0$ such that $B(x, \varepsilon) \subset G_i$. Now $x_{n_k} \to x$ and $1/n_k \to 0$ as $k \to \infty$, so there exists $k \in \mathbb{N}$ such that $d(x, x_{n_k}) < \varepsilon/2$ and $1/n_k < \varepsilon/2$. It follows that $B(x_{n_k}, 1/n_k) \subset B(x, \varepsilon)$, so $B(x_{n_k}, 1/n_k)$ is contained in $G_i$, contradicting the choice of $x_{n_k}$. □

Another nice trait of compactness concerns separation between sets. We make a first step here.

**Proposition 2.3.9.** *If $K$ is a compact subset of a metric space $(M, d)$ and $x \in M \setminus K$ then there exist disjoint open sets $G_1, G_2$ such that $K \subset G_1$ and $x \in G_2$.*

*Proof.* For $y \in M$ define $B_y = B(y, d(x, y)/2)$. Then $\{B_y : y \in K\}$ is an open cover of $K$ so there are $\{y_k\}_{k=1}^n$ such that $K \subset G_1 := \cup_{k=1}^n B_{y_k}$. Let $G_2 = B(x, r)$, where $r = \min\{d(x, y_k)/2 : 1 \le k \le n\}$. The sets $G_1$ and $G_2$ are open by Exercises 2.1.20 and 2.1.25, and they contain $K$ and $x$, respectively. Finally, they are disjoint because if $z \in G_1 \cap G_2$ then $z \in B_{y_k}$ for some $k$, hence $d(z, y_k) < d(x, y_k)/2$. Also, $z \in G_2$ implies that $d(x, z) < r$ so

$$\begin{aligned} d(x, y_k) &\le d(x, z) + d(z, y_k) < r + \frac{d(x, y_k)}{2} \\ &= \min_{1 \le k \le n} \left\{ \frac{d(x, y_k)}{2} \right\} + \frac{d(x, y_k)}{2} \le d(x, y_k), \end{aligned}$$

a contradiction. □

Intrinsic traits of metric spaces are of interest to topologists. Analysts are interested in functions defined on a metric space and their properties. Let us start with continuity. The reader is encouraged to compare the definition below with the special case when $M_1 = M_2 = \mathbb{R}$.

**Definition 2.3.10.** Let $(M_1, d_1)$ and $(M_2, d_2)$ be metric spaces. The map $f : M_1 \to M_2$ is **continuous** at $x \in M_1$ if for every $\varepsilon > 0$ there exists $\delta > 0$ such that for all $y$ in $M_1$ we have

$$d_1(x, y) < \delta \Rightarrow d_2(f(x), f(y)) < \varepsilon.$$

The map $f$ is continuous on $M_1$ if it is continuous at every point of $M_1$.

The following characterization of continuity is well known when $f$ is a map between Euclidean spaces and the proof goes along the same lines.

**Theorem 2.3.11.** *Let $(M_1, d_1)$ and $(M_2, d_2)$ be metric spaces and let $f : M_1 \to M_2$. The following are equivalent:*

*(a) $f$ is continuous;*

*(b) for every open set $G \subset M_2$, the preimage $f^{-1}(G)$ is open in $M_1$;*

*(c) if $\{x_n\}$ is a sequence in $M_1$ that converges to $x$ in $M_1$ then the sequence $\{f(x_n)\}$ converges to $f(x)$ in $M_2$.*

*Proof.* (a) ⇒ (b) Let $G$ be an open set and let $x \in f^{-1}(G)$. Then $f(x) \in G$ so there is a ball

$$B(f(x), \varepsilon) \subset G.$$

Since $f$ is continuous at $x$ there is $\delta > 0$ such that $d_1(x, y) < \delta$ implies $d_2(f(x), f(y)) < \varepsilon$. In other words, if $y \in B(x, \delta)$ then $f(y) \in B(f(x), \varepsilon) \subset G$, hence

$$B(x, \delta) \subset f^{-1}(B(f(x), \varepsilon)) \subset f^{-1}(G),$$

whence $f^{-1}(G)$ is open.

(b) ⇒ (c) Let $\varepsilon > 0$ and suppose that $\lim x_n = x$. Since $B(f(x), \varepsilon)$ is open then so is $f^{-1}(B(f(x), \varepsilon))$ and it contains $x$. Therefore, there is $\delta > 0$ such that

$$B(x, \delta) \subset f^{-1}(B(f(x), \varepsilon)).$$

By definition of the limit, there is $N \in \mathbb{N}$ such that $n \geq N$ implies $d_1(x_n, x) < \delta$. For all such $n$,

$$x_n \in B(x, \delta) \subset f^{-1}(B(f(x), \varepsilon)),$$

so $f(x_n) \in B(f(x), \varepsilon)$, hence $d_2(f(x_n), f(x)) < \varepsilon$. Thus, $\lim f(x_n) = f(x)$.

(c) ⇒ (a) We will prove the contrapositive. Assuming that $f$ is not continuous at $x$, there is $\varepsilon_0 > 0$ such that for every $\delta > 0$ there is $y \in M_1$ satisfying $d_1(x, y) < \delta$ and $d_2(f(x), f(y)) \geq \varepsilon_0$. In symbolic notation,

$$\exists \varepsilon_0 > 0 \quad \forall \delta > 0 \quad \exists y \in M_1 \quad d_1(x, y) < \delta \text{ and } d_2(f(x), f(y)) \geq \varepsilon_0.$$

This implies that for every $n \in \mathbb{N}$ there is $y_n \in M_1$ satisfying $d_1(x, y_n) < 1/n$ and $d_2(f(x), f(y_n)) \geq \varepsilon_0$. Applying the Axiom of Choice, we get a sequence $\{y_n\}$ and, by definition, $\lim y_n = x$, but it is not true that $\lim f(y_n) = f(x)$. □

**Example 2.3.12. A continuous function.** Let $a \in [0, 1]$. The function $F : C[0, 1] \to \mathbb{R}$, defined by $F(f) = f(a)$, is continuous on the metric space $C[0, 1]$.

**Proof:** Let $\varepsilon > 0$ and set $\delta = \epsilon$. If $d(f, g) < \delta$ then $\sup\{|f(x) - g(x)| : x \in [0, 1]\} < \delta = \epsilon$, so $|F(f) - F(g)| = |f(a) - g(a)| < \epsilon$. ♦

**Example 2.3.13. A discontinuous function**

The famous Dirichlet function is not continuous on the metric space $\mathbb{R}$ with the usual distance. ♦

We continue with a concept familiar with advanced calculus.

**Definition 2.3.14.** Let $(M_1, d_1)$ and $(M_2, d_2)$ be metric spaces. The map $f : M_1 \to M_2$ is **uniformly continuous** if for every $\varepsilon > 0$ there exists $\delta > 0$ such that

$$d_1(x, y) < \delta \Rightarrow d_2(f(x), f(y)) < \varepsilon \quad \text{for all} \quad x, y \in M_1.$$

**Example 2.3.15. A uniformly continuous function.** Let $d : \mathbb{R} \to \mathbb{R}$ be defined by $d(x, y) = 0$ if and only if $x = y$ and $d(x, y) = 1$ otherwise. (See Exercise 2.1.12.) The function $f(x) = x^2$ is uniformly continuous on $(\mathbb{R}, d)$.

**Proof:** Let $\varepsilon > 0$. Choose $\delta = 1/2$. If $d(x, y) < \delta$ then $x = y$, so $|f(x) - f(y)| = 0 < \varepsilon$. ♦

**Example 2.3.16. A function that is not uniformly continuous.**

Let $d : \mathbb{R} \to \mathbb{R}$ be defined by

$$d(x, y) := \frac{|x - y|}{1 + |x - y|}.$$

(See Exercise 2.1.8.) The function $f : (\mathbb{R}, d) \to (\mathbb{R}, d)$, defined by $f(x) = x^2$, is not uniformly continuous.

**Proof:** Let $\varepsilon_0 = 4/9$. If $\delta > 0$ choose $x = 1/\delta^2$ and $y = 1/\delta^2 + \delta/2$. Then

$$d(x, y) = \frac{\delta/2}{1 + (\delta/2)} \leq \frac{\delta}{2} < \delta.$$

On the other hand, $|x^2 - y^2| = \delta^2/4 + 1/\delta$ so

$$d(f(x), f(y)) = \frac{\delta^2/4 + 1/\delta}{1 + \delta^2/4 + 1/\delta} = \frac{\delta^3 + 4}{4\delta + \delta^3 + 4}.$$

Assuming that $\delta < 1$ we get $d(f(x), f(y)) > 4/9$. ♦

The next two theorems bring forward some nice ties between the continuity of a function and the compactness of the metric space on which it is defined. We leave their proofs to the reader (Exercises 2.3.28 and 2.3.40) because they can be argued in the similar way as in the case of Euclidean spaces.

**Theorem 2.3.17.** *Let $(M, d)$ be a metric space. Every uniformly continuous map $f : M_1 \to M_2$ is continuous. If, in addition, $M_1$ is compact then every continuous map $f : M_1 \to M_2$ is uniformly continuous.*

**Theorem 2.3.18.** *Let $(M, d)$ be a compact metric space and let $f : M \to \mathbb{R}$ be continuous. Then $f$ is bounded and it attains its extreme values on $M$.*

Since many functions are not continuous it is natural to relax this requirement.

**Definition 2.3.19.** Let $(M, d)$ be a metric space. The map $f : M \to \mathbb{R}$ is **upper semicontinuous** at $x_0 \in M$ if for every $y > f(x_0)$ there exists $\delta > 0$ such that for all $x$ in $M$ we have

$$d(x, x_0) < \delta \Rightarrow f(x) < y,$$

and **lower semicontinuous** at $x_0 \in M$ if for every $y < f(x_0)$ there exists $\delta > 0$ such that for all $x$ in $M$ we have

$$d(x, x_0) < \delta \Rightarrow f(x) > y.$$

**Example 2.3.20.** Let $f = \chi_{(0,1]}$. Then $f$ is upper semicontinuous at every point except at $x = 0$ and it is lower semicontinuous at every point except at $x = 1$. (See Exercise 2.3.45.) ♦

Finally, there is a very important theorem of Banach that is often referred to as the *Banach fixed-point theorem.* A function $f$ mapping a metric space $(M, d)$ to itself is a *contraction* if there is $0 < q < 1$ such that $d(f(x_1), f(x_2)) \leq qd(x_1, x_2)$.

**Theorem 2.3.21.** ***Banach Fixed-point Theorem.*** *A contraction on a metric space has exactly one fixed point, i.e., the point $x$ such that $f(x) = x$.*

The proof is left to the reader (Exercise 2.3.47).

## Exercises

2.3.1. Prove that a Cauchy sequence in a metric space is bounded.

2.3.2. Prove that every totally bounded metric space is bounded.

2.3.3. Prove that every bounded subset of $\mathbb{R}^n$ is totally bounded.

2.3.4. Prove that the closure of a totally bounded set in a metric space is totally bounded.

2.3.5.* Prove that a metric space $(M, d)$ is totally bounded if and only if every sequence in $M$ has a Cauchy subsequence.

2.3.6. Prove that the closure of a bounded set in a metric space is bounded.

2.3.7. Prove that a compact set in a metric space is closed.

2.3.8. Prove that a closed subset of a compact metric space is compact.

2.3.9. Prove that a metric space $(M, d)$ is compact if and only if every infinite subset of $M$ has a cluster point in $M$.

2.3.10. Prove that a subset of a complete metric space $M$ is compact if and only if it is totally bounded and closed.

2.3.11. In the metric space $(\mathbb{Q}, d)$, where $d(x, y) = |x - y|$, determine all subspaces that are:

(a) complete.

(b) compact.

2.3.12. Prove or disprove: the set $\mathbb{N}$ is a compact subset of $(\mathbb{R}, d)$, where $d(x, y) = |x - y|$.

2.3.13. Let $(M, d)$ be a complete metric space. Prove that a set $A \subset M$ is totally bounded if and only if $\overline{A}$ is compact.

2.3.14. Let $(M, d)$ be the metric space of all bounded sequences, with metric $d(a, b) = \sup\{|a_n - b_n| : n \in \mathbb{N}\}$, where $a = \{a_n\}$ and $b = \{b_n\}$. (See Exercise 2.1.13.) Let $B$ be the closed ball of radius 1 centered at the zero sequence $(0, 0, \dots)$. Prove that $B$ is bounded and closed but not sequentially compact.

2.3.15. Let $(M, d)$ be a metric space and let $A \subset M$ be a set such that $A \cap K$ is compact for every compact set $K \subset M$. Prove that $A$ is closed in $M$.

2.3.16. Let $(M, d)$ be a metric space. Prove that the distance between a closed set and a compact set that are disjoint cannot be 0. Give an example to show that this is false if both sets are only closed.

2.3.17. Let $K_1, K_2$ be disjoint compact subsets of a metric space $(M, d)$. Prove that there exist disjoint, open sets $G_1, G_2$ such that $K_1 \subset G_1$ and $K_2 \subset G_2$.

2.3.18. Prove that a metric space $M$ is compact if and only if every collection of closed subsets of $M$ with the finite intersection property (meaning that every finite subcollection has non-empty intersection) has a non-empty intersection.

2.3.19. Prove that a metric space $M$ is compact if and only if every decreasing sequence of nonempty closed sets in $M$ has nonempty intersection.

2.3.20. Prove that the union of a finite number of compact sets in a metric space is compact.

2.3.21. Let $M$ be a compact metric space and let $\{G_n\}_{n=1}^{\infty}$ be an increasing sequence of open subsets of $M$ such that $M = \cup_{n=1}^{\infty} G_n$. Prove that $M = G_k$ for some $k \in \mathbb{N}$.

2.3.22. Let $\{M_i\}_{i=1}^{n}$ be compact metric spaces. Prove that $M := M_1 \times M_2 \times \cdots \times M_n$ is a compact metric space with respect to metric $d$ as in Exercise 2.1.14.

2.3.23. Determine if $C[0, 1]$ is sequentially compact.

2.3.24. Determine if the metric space $(M, d)$ is sequentially compact, where $d(x, y) = 0$ if $x = y$ and $d(x, y) = 1$ if $x \neq y$.

2.3.25. Determine if the metric space $(\mathbb{R}, d)$ is sequentially compact, where

$$d(x, y) := \frac{|x - y|}{1 + |x - y|}.$$

2.3.26.* Determine if the metric space $(C[0,1], d)$ is sequentially compact, where

$$d(f,g) := \int_0^1 |f(x) - g(x)|\, dx$$

2.3.27.* Determine if the metric space $(C(-\infty,\infty), d)$ is sequentially compact, where

$$d(f,g) := \sum_{n=1}^{\infty} \frac{1}{2^n} \frac{\max\{|f(x) - g(x)| : |x| \le n\}}{1 + \max\{|f(x) - g(x)| : |x| \le n\}}.$$

2.3.28. Prove Theorem 2.3.18.

2.3.29. Let $(M,d)$ be a metric space and let $f, g : M \to \mathbb{R}$ be continuous. Prove that $\max\{f,g\}$ and $\min\{f,g\}$ are continuous.

2.3.30. Let $f : \mathbb{R} \to \mathbb{R}$, let $x_0 \in \mathbb{R}$, and suppose that for every interval $(a,b)$ that contains $x_0$, the function $f$ is unbounded on $(a,b)$. Prove that $f$ is not continuous at $x_0$.

2.3.31. Let $(M,d)$ be a metric space, let $f : M \to \mathbb{R}$ be continuous, and let $E \subset M$. Prove that $\sup\{|f(x)| : x \in E\} = \sup\{|f(x)| : x \in \overline{E}\}$.

2.3.32.* Let $f : \mathbb{R} \to \mathbb{R}$ be a strictly increasing continuous function. Prove that $f^{-1}$ is continuous.

2.3.33. Let $(M,d)$ be a metric space, let $f, g : M \to \mathbb{R}$ be continuous, and let $\alpha \in \mathbb{R}$. Prove that $\alpha f$, $f + g$ and $fg$ are continuous functions. Moreover, if $g(x) \neq 0$ on $M$, prove that $f/g$ is continuous.

2.3.34.** Let $f$ be a real-valued function on a metric space $X$. Show that the set of points at which $f$ is continuous is the intersection of a countable collection of open sets. Conclude that there is no real-valued function on $\mathbb{R}$ that is continuous at the rational numbers only.

2.3.35. Let $(M,d)$ be a metric space and $K$ a compact subset of $M$. Suppose that a function $f : K \to \mathbb{R}$ is continuous. Prove that $f(K)$ is a compact subset of $\mathbb{R}$.

2.3.36. Let $(M,d)$ be a metric space and $A \subset M$. Suppose that functions $f, g : M \to \mathbb{R}$ are continuous and $f(x) = g(x)$ when $x \in A$. Prove that $f(x) = g(x)$ for all $x \in \overline{A}$.

2.3.37. Let $(M,d)$ be a metric space, let $A \subset M$, and let $h(x) = d(x,A) = \inf\{d(x,a) : a \in A\}$. Prove that $d$ is a continuous function on $M$.

2.3.38. Let $(M,d)$ be a metric space and let $A$ and $B$ be disjoint, closed subsets of $M$. Prove that there exists a continuous, real-valued function $f$ on $M$ such that $0 \le f(x) \le 1$, for all $x \in M$, $f(x) = 0$ on $A$, and $f(x) = 1$ on $B$.

2.3.39. Let $\{f_n\}$ be a sequence of continuous functions that converge to $f$ uniformly on $[a,b]$ and let $\{x_n\}$ be a sequence in $[a,b]$ that converges to $x$. Prove that $\{f_n(x_n)\}$ converges to $f(x)$.

2.3.40. Prove Theorem 2.3.17.

2.3.41. Determine whether the function $F$ in Example 2.3.12 is uniformly continuous on $C[0,1]$.

2.3.42. Determine whether the function $F$ in Example 2.3.12 is continuous on $(C[0,1], d)$ with $d$ defined as in Exercise 2.1.3, by

$$d(f,g) = \int_0^1 |f(x) - g(x)|\, dx.$$

Is it uniformly continuous?

2.3.43. Let $(M, d)$ be a metric space and $\{x_n\} \subset A \subset M$ a Cauchy sequence. Suppose that a function $f : M \to \mathbb{R}$ is uniformly continuous on $A$. Prove that $\{f(x_n)\}$ is a Cauchy sequence in $\mathbb{R}$. What if $f$ is assumed to be only continuous?

2.3.44. Let $(M, d)$ be a metric space and let $f : M \to \mathbb{R}$. Prove that the following statements are equivalent:

(a) The function is upper semicontinuous at every point of $M$.

(b) All sets $\{x \in M : f(x) < y\}$ with $y \in \mathbb{R}$ are open in $M$.

(c) All sets $\{x \in M : f(x) \geq y\}$ with $y \in \mathbb{R}$ are closed in $M$.

2.3.45. Prove that the function $f = \chi_{(0,1]}$ is upper semicontinuous at every point except at $x = 0$ and it is lower semicontinuous at every point except at $x = 1$.

2.3.46. Let $(M, d)$ be a metric space. Suppose that, for every $\varepsilon > 0$, there exists $\delta > 0$, a metric space $(W, \rho)$, and a mapping $\Phi : M \to W$ so that $\Phi(M)$ is totally bounded, and whenever $x, y \in M$ are such that $\rho(\Phi(x), \Phi(y)) < \delta$, then $d(x, y) < \varepsilon$. Prove that $M$ is totally bounded.

2.3.47. Prove Banach Fixed-point Theorem (Theorem 2.3.21).

## 2.4 Limits of Functions

In this monograph we will encounter three types of limits. We have already come across limits of sequences and to some extent limits of sets. We now turn our attention to the limits of functions.

**Definition 2.4.1.** Let $(M_1, d_1)$ and $(M_2, d_2)$ be metric spaces, let $f : M_1 \to M_2$, and let $a$ be a cluster point of $M_1$. We say that $L$ is the **limit** of $f$ as $x$ approaches $a$, and we write $\lim_{x \to a} f(x) = L$, if for any $\varepsilon > 0$ there exists $\delta > 0$ such that $d_2(f(x), L) < \varepsilon$, whenever $0 < d(x, a) < \delta$ and $x \in M_1$.

**Example 2.4.2. A limit of a function.** Let $a \in [0,1]$, let $F : C[0,1] \to \mathbb{R}$ be defined by $F(f) = f(a)$, and let $g(x) = x^2$. Then

$$\lim_{f \to g} F(f) = a^2.$$

**Proof:** Let $\varepsilon > 0$ and set $\delta = \epsilon$. If $d(f,g) < \delta$ then $\sup\{|f(x) - g(x)| : x \in [0,1]\} < \delta = \epsilon$, so $|F(f) - a^2| = |f(a) - g(a)| < \epsilon$. ♦

It is well-known that, when $f$ is defined on a subset of $\mathbb{R}^n$, there is a definition equivalent to Definition 2.4.1 that is stated using sequences. In view of the remark made at the beginning of Section 2.2, the following should not come as a surprise.

**Theorem 2.4.3.** *Let $(M,d)$ be a metric space, $A \subset M$, $f : M \to \mathbb{R}$, and $a$ cluster point of $A$. Then $\lim_{x\to a} f(x) = L$ if and only if, for every sequence $\{a_n\} \subset A$ converging to $a$, with $a_n \neq a$, $\lim_{n\to\infty} f(a_n) = L$.*

*Proof.* Suppose first that $\lim_{x\to a} f(x) = L$, and let $\{a_n\}$ be an arbitrary sequence in $A$ converging to $a$. We will show that $\lim_{n\to\infty} f(a_n) = L$. So, let $\varepsilon > 0$. Since $\lim_{x\to a} f(x) = L$, there exists $\delta > 0$ such that

$$0 < d(x,a) < \delta \Rightarrow |f(x) - L| < \varepsilon. \tag{2.2}$$

Further, $\lim a_n = a$ and $a_n \neq a$ so there exists $N \in \mathbb{N}$ such that if $n \geq N$ then $0 < d(a_n, a) < \delta$. By (2.2), if $n \geq N$, we have that $|f(a_n) - L| < \varepsilon$.

Now we will prove the converse. Namely, we will assume that, for every sequence $\{a_n\}$ converging to $a$ and satisfying $a_n \neq a$, $\lim_{n\to\infty} f(a_n) = L$, and we will establish that $\lim_{x\to a} f(x) = L$. Suppose, to the contrary, that $\lim_{x\to a} f(x) \neq L$. By carefully negating the definition of the limit (Definition 2.4.1), we see that this means:

$$(\exists\varepsilon)(\forall\delta)(\exists x)\, 0 < d(x,a) < \delta \text{ and } |f(x) - L| \geq \varepsilon.$$

Let $\varepsilon_0$ be such a number, and let $\delta_n = 1/n$, $n = 1, 2, \dots$. Then, for each $n \in \mathbb{N}$, there exists $x_n$ such that $0 < d(x_n, a) < 1/n$ and $|f(x_n) - L| \geq \varepsilon_0$. Since $d(x_n, a) < 1/n$, for each $n \in \mathbb{N}$, we see that $\lim x_n = a$. Also, $0 < d(x_n, a)$ shows that $x_n \neq a$. Our assumption now allows us to conclude that $\lim_{n\to\infty} f(x_n) = L$ which contradicts the inequality $|f(x_n) - L| \geq \varepsilon_0$. Therefore, $\lim_{x\to a} f(x) = L$. □

In situations when the limit does not exist, it is worth considering a weaker concept. In elementary calculus, one deals with *left* and *right* limits. This cannot be done in a metric space in which there may be no concept of left and right. An interesting situation is created when the codomain $M_2 = \mathbb{R}$ in which case we can take advantage of its linear structure and talk about *limit superior* and *limit inferior*. We start with sequences.

**Definition 2.4.4.** Let $\{a_n\}$ be a sequence of real numbers, let $\{a_{n_k}\}$ be a convergent subsequence of $\{a_n\}$, and let $\lim a_{n_k} = c$. We say that $c$ is an **accumulation point** of the sequence $\{a_n\}$. If $\{a_n\}$ is a bounded sequence, it has a largest and a smallest accumulation point which we call **the limit superior** and **the limit inferior**, respectively. They are denoted by $\limsup a_n$ and $\liminf a_n$. When $\{a_n\}$ is not bounded above (resp., below), we will say that $\limsup a_n = +\infty$ (resp., $\liminf a_n = -\infty$).

The following is one of the well-known equivalent definitions of the limit superior and inferior (for sequences of real numbers), so we leave the proof to the reader (Exercise 2.4.6).

**Theorem 2.4.5.** *Let $\{a_n\}_{n=1}^{\infty}$ be a sequence of real numbers. Then*

$$\limsup a_n = \inf_{k\geq 1}\sup_{n\geq k} a_n \quad \textit{and} \quad \liminf a_n(x) = \sup_{k\geq 1}\inf_{n\geq k} a_n$$

Next, we turn our attention to functions.

**Definition 2.4.6.** We say that $L$ is the **limit superior** of $f$ as $x$ approaches $a$, and we use the notation $\limsup_{x\to a} f(x) = L$, if

$$L = \lim_{h\to 0+}\left[\sup\{f(x) : 0 < d(x,a) < h\}\right].$$

Similarly, $L$ is the **limit inferior** to $f$ as $x$ approaches to $a$, and we write $\liminf_{x\to a} f(x) = L$, if

$$L = \lim_{h\to 0+}\left[\inf\{f(x) : 0 < d(x,a) < h\}\right].$$

**Example 2.4.7. Limit superior and limit inferior of a function.**
Let $f(x) = \sin(1/x)$. Then

$$\limsup_{x\to 0} f(x) = 1 \quad \text{and} \quad \liminf_{x\to 0} f(x) = -1. \tag{2.3}$$

**Proof:** Let $h > 0$. The condition $0 < d(x,a) < h$ means here that $0 < |x| < h$. One such $x$ is $1/[\pi/2 + 2n\pi]$ where $n = \lfloor 1/h \rfloor + 1$. Indeed, $n > 1/h$ implies that $\pi/2 + 2n\pi > 1/h$ whence $x < h$. Further, $f(x) = 1$ so $\sup\{f(x) : 0 < d(x,a) < h\} = 1$. Since $h > 0$ is arbitrary, we see that $\limsup_{x\to 0} f(x) = 1$, and a similar proof can be used to show that $\liminf_{x\to 0} f(x) = -1$. ♦

Calculus is oriented towards continuous functions so the limit superior and the limit inferior are not essential therein. In this monograph, they will be of some use.

Both limit superior and limit inferior can also be expressed using sequences. In the situation where $A$ is a subset of a metric space $(M,d)$, $f : M \to \mathbb{R}$, and $a$ is a cluster point of $A$ we will use notation

$$W(f,a) = \{w : \exists\{a_n\} \subset A \setminus \{a\},\ \lim a_n = a,\ \lim f(a_n) = w\}. \tag{2.4}$$

Our goal is to show that $\limsup_{x\to a} f(x) = \sup W(f,a)$ and $\liminf_{x\to a} f(x) = \inf W(f,a)$. In this direction, the following theorem is instrumental.

**Theorem 2.4.8.** *Let $(M, d)$ be a metric space, $A \subset M$, and $f : M \to \mathbb{R}$. If $a$ is a cluster point of $A$ then:*

> *(a) If $y > \sup W(f, a)$, then there exists $\delta > 0$ such that $f(x) < y$ whenever $0 < d(x, a) < \delta$, $x \in A$.*
>
> *(b) If $y < \sup W(f, a)$, then for every $\delta > 0$, there exists $x \in A$ with $0 < d(x, a) < \delta$, such that $f(x) > y$.*

*Proof.* (a) Suppose to the contrary that for every $\delta > 0$ there is $x \in A$ such that $0 < d(x, a) < \delta$ and $f(x) \geq y$. Taking $\delta = 1/n$, we obtain a sequence $\{x_n\} \in A \setminus \{a\}$ such that $\lim x_n = a$ and $f(x_n) \geq y$. If $b$ is an accumulation point of $\{f(x_n)\}$ then $b \in W(f, a)$, hence $b \leq \sup W(f, a)$. Taking supremum over all accumulation points of $\{f(x_n)\}$, we get $\sup W(f, a) \geq \limsup_{n\to\infty} f(x_n) \geq y$.

(b) Since $y < \sup W(f, a)$, there is $w \in W(f, a)$ such that $w > y$. Thus, there exists a sequence $\{x_n\} \in A \setminus \{a\}$ such that $\lim x_n = a$ and $f(x_n) \to w$. Since $w > y$, there is $N_1 \in \mathbb{N}$ such that $n \geq N_1$ implies $f(x_n) > y$. Further, for any $\delta > 0$, there is $N_2 \geq N_1$ such that $0 < d(a, x_n) < \delta$, for $n \geq N_2$. The proof is completed by taking $x = x_{N_2}$. □

A similar result is valid for limit inferior (Exercise 2.4.18).

From Theorem 2.4.8 we derive an easy corollary. We leave the proof to the reader (Exercise 2.4.19).

**Corollary 2.4.9.** *Let $(M, d)$ be a metric space, let $f : M \to \mathbb{R}$, and let $A \subset M$. If $a$ is a cluster point of $A$ then*

$$\limsup_{x\to a} f(x) = \sup W(f, a).$$

We leave the analogous result for limit inferior as Exercise 2.4.20.

The next theorem brings forward a relationship between the concepts presented in this section.

**Theorem 2.4.10.** *Let $(M, d)$ be a metric space, $f : M \to \mathbb{R}$, and $a \in M$. The limit $\lim_{x\to a} f(x)$ exists if and only if $\limsup_{x\to a} f(x) = \liminf_{x\to a} f(x)$ and in this situation, they are all equal.*

*Proof.* ($\Rightarrow$) By Theorem 2.4.3, for every sequence $\{a_n\} \subset A \setminus \{a\}$ converging to $a$, we have $\lim_n f(a_n) = \lim_{x\to a} f(x)$. Thus, $W(f, a)$ is a singleton $\{\lim_{x\to a} f(x)\}$, so $\sup W(f, a) = \inf W(f, a) = \lim_{x\to a} f(x)$.

($\Leftarrow$) Since $\sup W(f, a) = \inf W(f, a)$ the set $W(f, a)$ is a singleton, which implies that for every sequence $\{a_n\} \subset A\setminus\{a\}$ converging to $a$ we have $\{f(a_n)\}$ converge to the same element. Indeed, if $b$ is an accumulation point of $\{f(a_n)\}$ then $b \in W(f, a)$, hence $\inf W(f, a) \leq b \leq \sup W(f, a)$. By Theorem 2.4.3, that unique accumulation point is precisely $\lim_{x\to a} f(x)$. □

When $M$ is also $\mathbb{R}$, in addition to limit superior and limit inferior of a function it is possible to consider their one-sided versions:

$$\limsup_{x\to a+} f(x) = \lim_{h\to 0+} \left[\sup\{f(x) : a < x < a+h\}\right],$$
$$\limsup_{x\to a-} f(x) = \lim_{h\to 0+} \left[\sup\{f(x) : a-h < x < a\}\right],$$

and similarly for limit inferior. They are useful when $\limsup_{x\to a} f(x)$ or $\liminf_{x\to a} f(x)$ do not provide a complete picture about the behaviour of a function.

**Example 2.4.11. One-sided limit superior and limit inferior of a function.**

Let

$$f(x) = \begin{cases} \sin(1/x) & \text{if } x > 0 \\ 2\sin(1/x) & \text{if } x < 0 \end{cases}.$$

Then

$$\limsup_{x\to 0^+} f(x) = 1 \quad \text{and} \quad \liminf_{x\to 0^+} f(x) = -1 \quad \text{and}$$
$$\limsup_{x\to 0^-} f(x) = 2 \quad \text{and} \quad \liminf_{x\to 0^-} f(x) = -2.$$ ♦

Notice that in Example 2.4.11 $\limsup_{x\to 0} f(x) = 2$ and $\liminf_{x\to 0} f(x) = -2$. Considering one-sided limits gives additional knowledge about the lack of symmetry about the $y$-axis, something that limit superior and limit inferior cannot provide.

When $M_1 = M_2 = \mathbb{R}$ we can also consider the case $a = +\infty$ or $a = -\infty$. Definition of the limit in such a situation is well-known and belongs to advanced calculus. We direct a reader who is in search of a refresher to [86, Sections 3.4–3.5].

**Definition 2.4.12.** We say that $L$ is the **limit superior** of $f$ as $x$ approaches to $+\infty$, and we write $\limsup_{x\to+\infty} f(x) = L$, if

$$L = \lim_{M\to+\infty} \left[\sup\{f(x) : x > M\}\right].$$

Similarly, $L$ is the **limit inferior** of $f$ as $x$ approaches to $+\infty$, and we write $\liminf_{x\to+\infty} f(x) = L$, if

$$L = \lim_{M\to+\infty} \left[\inf\{f(x) : x > M\}\right].$$

*Remark* 2.4.13. We omit the analogous definition for the limit superior and the limit inferior when $x \to -\infty$.

**Example 2.4.14. Limit superior and limit inferior of a function.**
Let $f(x) = \sin^2 x$. Then

$$\limsup_{x\to+\infty} f(x) = 1 \quad \text{and} \quad \liminf_{x\to+\infty} f(x) = 0. \tag{2.5}$$

**Proof:** Let $M > 0$. The condition $x > M$ is satisfied by $\pi/2 + 2n\pi$ where $n = \lfloor M/(2\pi)\rfloor + 1$. Indeed, this implies that $\pi/2 + 2n\pi > M$ and $f(\pi/2 + 2n\pi) = 1$ so $\sup\{f(x) : x > M\} = 1$. Since $M > 0$ is arbitrary, we see that $\limsup_{x\to+\infty} f(x) = 1$, and a similar proof can be used to show that $\liminf_{x\to+\infty} f(x) = 0$. ♦

Formula (2.4) can be extended in a natural way to the case $a = \infty$ (or $a = -\infty$). Corollary 2.4.9 and the analogous result for limit inferior still hold. (See Exercise 2.4.24.)

Since we have no less than three different uses of the symbol lim sup (and lim inf) we should point out that there is a connection between all of them. While Corollary 2.4.9 and Exercise 2.4.20 provide ample evidence in support of such a thesis, some less obvious corroboration can be found in Exercises 2.4.8 and 2.4.12.

Some of the most interesting metric spaces have functions as their elements. We have encountered $C[a, b]$ but there will be many more later in the text. Definition 2.2.4 applied to $M = C[a, b]$ elucidates one type of convergence of a sequence of functions, namely the uniform convergence. There is also another one that does not involve the metric.

**Definition 2.4.15.** Let $(M, d)$ be a metric space and $A \subset M$. We say that a sequence of real-valued functions $\{f_n\}$, defined on $A$, converges **pointwise** to a function $f$ on $A$ if, for any $x \in A$, the sequence of real numbers $\{f_n(x)\}$ converges to a real number $f(x)$.

In a similar way, the definitions of limit superior and limit inferior extend to sequences of functions and their *pointwise* behaviour when $n \to \infty$.

**Example 2.4.16. Limit superior and limit inferior of a sequence of functions.**
Let $\{f_n\}$ be a sequence of functions defined by

$$f_n(x) = \sin nx, \qquad \text{for } x \in \mathbb{R} \text{ and } n \in \mathbb{N}.$$

Then $\limsup f_n(x) = 1$ and $\liminf f_n(x) = -1$, i.e., the constant functions 1 and $-1$, respectively. ♦

When talking about the derivative of a function we will assume that both its domain and codomain are subsets of $\mathbb{R}$. Now that we have the concept of the limit available, we can define differentiability. In reality, this is verbatim taken from a course in calculus.

**Definition 2.4.17.** Let $A$ be a subset of $\mathbb{R}$, let $f : A \to \mathbb{R}$, and let $x \in A$. We say that the function $f$ is **differentiable** at $x$ if the limit

$$\lim_{h \to 0} \frac{f(x+h) - f(x)}{h}$$

exists as a finite number. In that case we call this limit the **derivative** of $f$ at $x$ and we denote it by $f'(x)$.

There is a common trait of continuity and differentiability—they both deal with the relationship between the distance of function values and the distance of the inputs. Another concept that involves these quantities is described in the following definition.

**Definition 2.4.18.** Let $(M_1, d_1)$ and $(M_2, d_2)$ be metric spaces. The map $f : M_1 \to M_2$ is **Lipschitz on** $M_1$ if there exists $K \geq 0$ such that

$$d_2(f(u), f(v)) \leq K\, d_1(u, v), \qquad \text{for all } u, v \in M_1.$$

**Example 2.4.19. A Lipschitz function.**
Let $f(x) = |x|$, for $x \in \mathbb{R}$. Then $f$ is Lipschitz with constant 1, because

$$\big|\,|u| - |v|\,\big| \leq |u - v|.$$ ♦

Notice that $f$ is not differentiable at $x = 0$. When $f$ is differentiable we have an easy test to detect whether $f$ is Lipschitz or not.

**Theorem 2.4.20.** *Let $f$ be a differentiable function on a set $E \subset \mathbb{R}$. Then $f$ is Lipschitz with constant $K = \sup\{|f'(x)| : x \in E\}$ if and only if it has a bounded derivative.*

The proof is straightforward and we leave it as Exercise 2.4.25.

**Example 2.4.21. A Lipschitz function.**
Let $f(x) = \sin x$, for $x \in \mathbb{R}$. Then $f$ is Lipschitz with constant 1. Indeed, $f'(x) = \cos x$ and $|\cos x| \leq 1$. ♦

**Example 2.4.22. A function that is not Lipschitz.**
The function $f : [0, 1] \to \mathbb{R}$ defined by $f(0) = 0$ and $f(x) = x^{3/2} \sin(1/x)$ for $x \neq 0$ is differentiable but not Lipschitz because its derivative is unbounded in the vicinity of 0. ♦

## Exercises

2.4.1. Let $(M, d)$ be a metric space, $A \subset M$, $f, g : M \to \mathbb{R}$, and let $a$ be a cluster point of $A$. Also, let $\alpha$ be a real number. If $\lim_{x \to a} f(x)$ and $\lim_{x \to a} g(x)$ both exist prove that

(a) $\lim_{x\to a}(\alpha f(x)) = \alpha \lim_{x\to a} f(x)$;

(b) $\lim_{x\to a}(f(x)+g(x)) = \lim_{x\to a} f(x) + \lim_{x\to a} g(x)$;

(c) $\lim_{x\to a}(f(x)g(x)) = (\lim_{x\to a} f(x))(\lim_{x\to a} g(x))$;

(d) $\lim_{x\to a}(f(x)/g(x)) = \lim_{x\to a} f(x)/\lim_{x\to a} g(x)$, with the additional assumption that $\lim_{x\to a} g(x) \neq 0$.

2.4.2. Let $(M,d)$ be a metric space, $A \subset M$, and $f, g : M \to \mathbb{R}$ satisfy $f(x) < g(x)$ for all $x \in A$. If $a$ is a cluster point of $A$ and if $\lim_{x\to a} f(x)$ and $\lim_{x\to a} g(x)$ both exist to prove that $\lim_{x\to a} f(x) \leq \lim_{x\to a} g(x)$.

2.4.3. Let $F : C[0,1] \to \mathbb{R}$ be defined by $F(f) = f(0)$, and let $g(x) = x^2$. Find $\lim_{f\to g} F(f)$.

2.4.4.* Let $d$ be a metric on $C[0,1]$ defined by

$$d(f,g) = \int_0^1 |f(x) - g(x)|\, dx,$$

(see Exercise 2.1.3). Let $F : C[0,1] \to \mathbb{R}$ be defined by $F(f) = f(0)$, and let $g(x) = x^2$. Find $\lim_{f\to g} F(f)$.

2.4.5.* Let $d$ be a metric on $\mathbb{R}$ defined by

$$d(x,y) := \frac{|x-y|}{1+|x-y|},$$

(see Exercise 2.1.8), and let $\rho$ denote the usual distance on $\mathbb{R}$, $\rho(x,y) = |x-y|$. Let $f : (\mathbb{R}, d) \to (\mathbb{R}, \rho)$ be defined by $f(x) = x^2$. Find $\lim_{x\to 2} f(x)$.

2.4.6.* Prove Theorem 2.4.5.

2.4.7. Let $(M,d)$ be a metric space, let $f : M \to \mathbb{R}$, and let $a \in M$. Prove that $\limsup_{x\to a} f(x)$ and $\liminf_{x\to a} f(x)$ always exist either as finite numbers or $\pm\infty$.

2.4.8. Let $\{a_n\}$ be a sequence of real numbers. Prove that $\limsup a_n$ equals $\limsup_{x\to 0} f(x)$, where the function $f : \{1, 1/2, 1/3, \ldots\} \to \mathbb{R}$ is defined by $f(1/n) = a_n$.

2.4.9. Prove that $\lim_{x\to a} f(x)$ exists if and only if both limit superiors and both limit inferiors (left and right) are all equal.

2.4.10. If $f$ is defined on $[-c,c]$ and $x_0 \in (-c,c)$, prove the following:

(a) $\limsup_{x\to x_0+}(-f(x)) = -\liminf_{x\to x_0+} f(x)$, $\limsup_{x\to x_0-}(-f(x)) = -\liminf_{x\to x_0-} f(x)$.

(b) $\limsup_{x\to x_0+}(f(-x)) = \limsup_{x\to -x_0-} f(x)$, $\liminf_{x\to x_0+}(f(-x)) = \liminf_{x\to -x_0-} f(x)$.

(c) $\limsup_{x\to x_0-}(f(-x)) = \limsup_{x\to -x_0+} f(x)$, $\liminf_{x\to x_0-}(f(-x)) = \liminf_{x\to -x_0+} f(x)$.

2.4.11.* Let $(M, d)$ be a metric space, let $f : M \to \mathbb{R}$, and let $g : M \to \mathbb{R}$ be defined by $g(x) = \limsup_{t\to x} f(t)$. Prove that $g$ is upper semicontinuous.

2.4.12. Let $\{A_n\}$ be an arbitrary collection of subsets of a set $X$. Prove that

$$\limsup A_n = \{x \in X : \limsup \chi_{A_n}(x) = 1\}.$$

2.4.13. Let $(M, d)$ be a metric space and for every $n \in \mathbb{N}$ let $f_n : M \to [0, +\infty]$. Prove that

$$\sup\{\sum_{k=1}^{\infty} f_k(x_k) : \{x_k\} \subset M\} = \sum_{k=1}^{\infty} \sup\{f_k(x) : x \in M\}.$$

2.4.14. Let $f$ and $g$ be real-valued functions on an interval $(a-\eta, a+\eta)$. Prove that

$$\begin{aligned}\liminf_{x\to a} f(x) + \liminf_{x\to a} g(x) \le \liminf_{x\to a}(f(x)+g(x)) \le \liminf_{x\to a} f(x) + \limsup_{x\to a} g(x)\\ \le \limsup_{x\to a}(f(x)+g(x)) \le \limsup_{x\to a} f(x) + \limsup_{x\to a} g(x).\end{aligned}$$

2.4.15.* Let $f$ and $g$ be strictly positive functions on an interval $(a-\eta, a+\eta)$ and suppose that $\liminf_{x\to a} f(x)$ and $\limsup_{x\to a} f(x)$ exist and are not 0. Prove that:

(a) $\limsup_{x\to a} \frac{1}{f(x)} = \frac{1}{\liminf_{x\to a} f(x)}$.

(b) $\liminf_{x\to a} \frac{1}{f(x)} = \frac{1}{\limsup_{x\to a} f(x)}$.

2.4.16. Let $f$ and $g$ be nonnegative functions on an interval $(a-\eta, a+\eta)$ and suppose that $\liminf_{x\to a} f(x)$, $\limsup_{x\to a} f(x)$, $\liminf_{x\to a} g(x)$, and $\limsup_{x\to a} g(x)$ exist and are not 0. Prove that

$$\begin{aligned}\liminf_{x\to a} f(x) \cdot \liminf_{x\to a} g(x) \le \liminf_{x\to a}(f(x)\cdot g(x)) \le \liminf_{x\to a} f(x) \cdot \limsup_{x\to a} g(x)\\ \le \limsup_{x\to a}(f(x)\cdot g(x)) \le \limsup_{x\to a} f(x) \cdot \limsup_{x\to a} g(x).\end{aligned}$$

2.4.17. Let $f$ and $g$ be nonnegative functions on an interval $(a-\eta, a+\eta)$ and suppose that $\lim_{x\to a} f(x)$ exists. Prove that:

(a) $\limsup_{x\to a}(f(x)+g(x)) = \lim_{x\to a} f(x) + \limsup_{x\to a} g(x)$.

(b) $\liminf_{x\to a}(f(x)+g(x)) = \lim_{x\to a} f(x) + \liminf_{x\to a} g(x)$.

(c) $\limsup_{x\to a}(f(x)\cdot g(x)) = \lim_{x\to a} f(x) \cdot \limsup_{x\to a} g(x)$.

(d) $\liminf_{x\to a}(f(x)\cdot g(x)) = \lim_{x\to a} f(x) \cdot \liminf_{x\to a} g(x)$.

2.4.18. Let $(M, d)$ be a metric space, $A \subset M$, and $f : M \to \mathbb{R}$. If $a$ is a cluster point of $A$ prove that:

(a) If $y < \inf W(f,a)$, then there exists $\delta > 0$ such that $f(x) > y$ whenever $0 < d(x,a) < \delta$, $x \in A$.

(b) If $y > \inf W(f,a)$, then for every $\delta > 0$, there exists $x \in A$ with $0 < d(x,a) < \delta$, such that $f(x) < y$.

2.4.19. Prove Corollary 2.4.9.

2.4.20. Let $(M,d)$ be a metric space, let $f : M \to \mathbb{R}$, and let $A \subset M$. If $a$ is a cluster point of $A$ prove that

$$\liminf_{x \to a} f(x) = \inf W(f,a).$$

2.4.21. Let $a \in \mathbb{R}$ and $f : (a,+\infty) \to \mathbb{R}$. Prove that the limit $\lim_{x\to+\infty} f(x)$ exists if and only if $\limsup_{x\to+\infty} f(x) = \liminf_{x\to+\infty} f(x)$ and in this situation they are all equal.

2.4.22. Let $f : \mathbb{R} \to \mathbb{R}$. Prove that

(a) $\limsup_{x\to+\infty}(-f(x)) = -\liminf_{x\to+\infty} f(x)$, $\limsup_{x\to+\infty}(-f(x)) = -\liminf_{x\to+\infty} f(x)$.

(b) $\limsup_{x\to+\infty}(f(-x)) = \limsup_{x\to-\infty} f(x)$, $\liminf_{x\to+\infty}(f(-x)) = \liminf_{x\to-\infty} f(x)$.

2.4.23. Let $a \in \mathbb{R}$ and $f : (a,+\infty) \to \mathbb{R}$. Prove that $\limsup_{x\to+\infty} f(x) = L$ if and only if, for any sequence $\{a_n\}$ converging to $+\infty$, $\limsup_{x\to+\infty} f(a_n) = L$.

2.4.24. Let $f : (a,+\infty) \to \mathbb{R}$. Prove that

$$\limsup_{x\to\infty} f(x) = \sup W(f,\infty) \quad \text{and} \quad \liminf_{x\to\infty} f(x) = \inf W(f,\infty)$$

2.4.25. Prove Theorem 2.4.20.

2.4.26. Suppose that $f$ is a real-valued Lipschitz function on $(a,b)$. Prove that $f$ is uniformly continuous on $(a,b)$.

2.4.27. Let $(M,d)$ be a metric space and $x_0 \in M$. Let $f : M \to \mathbb{R}$ be defined by $f(x) = d(x,x_0)$. Prove that $f$ is uniformly continuous.

2.4.28.* Give an example of a function that is continuous but not uniformly continuous. Give an example of a function that is uniformly continuous but not Lipschitz.

2.4.29. Give an example of a function that is Lipschitz on a set $E$ but not differentiable at a point $x \in E$.

2.4.30. Let $f$ be a real-valued function defined on a set $A \subset \mathbb{R}^2$ such that it is a continuous function of $x$ (with $y$ fixed), and let $f$ satisfy the Lipschitz condition with respect to $y$, i.e., suppose that there exists $M > 0$ such that

$$|f(x,y_1) - f(x,y_2)| \leq M|y_1 - y_2|$$

for all $(x,y_1),(x,y_2) \in A$. Prove that $f$ is continuous on $A$.

2.4.31.* Let $d(x,y) = \sqrt{|x-y|}$, let $\rho(x,y) = |x-y|$, and consider $f : ([0,1], d) \to ([0,1], \rho)$ defined by $f(x) = x^\alpha$. Determine what values of $\alpha > 0$ is $f$ Lipschitz.

2.4.32. Let $(C[0,1], d)$ be the metric space with $d(f,g) = \sup\{|f(x) - g(x)| : x \in [0,1]\}$, and let $F : C[0,1] \to C[0,1]$ be defined by $F(f) = f^2$. Determine whether $F$ is Lipschitz.

2.4.33. Let $(C[0,1], d)$ be the metric space with $d(f,g) = \sup\{|f(x) - g(x)| : x \in [0,1]\}$, and let $F : C[0,1] \to C[0,1]$ be defined by $F(f) = \sin f$. Determine whether $F$ is Lipschitz.

2.4.34. Let $d : \mathbb{R} \to \mathbb{R}$ be defined by $d(x,y) = 1$ if $x \neq y$ and $d(x,y) = 0$ if $x = y$, let $\rho(x,y) = |x-y|$, and consider $f : (\mathbb{R}, d) \to (\mathbb{R}, \rho)$ defined by $f(x) = x^2$. Determine if $f$ is uniformly continuous. Is it Lipschitz?

## 2.5 Baire's Theorem

In this section we will present a result of a topological nature that is very simple to state and not very hard to prove yet has unexpectedly far-reaching consequences, many of which we will encounter in the later part of the text.

**Theorem 2.5.1.** ***The Baire Category Theorem.*** *Let $X$ be a complete metric space and $\{F_n\}_{n=1}^\infty$ closed subsets of $X$. If $F := \cup_{n=1}^\infty F_n$ has a nonempty interior, then at least one of the sets $F_n$ does. In particular, if $X = \cup_{n=1}^\infty F_n$, at least one of the sets $F_n$ has a nonempty interior.*

*Proof.* Suppose to the contrary that each $F_n$ has an empty interior. Since $F$ has an nonempty interior, the set $F_1$ is not equal to $\operatorname{Int} F$. Therefore, there exists $x_1$ in the open set $(\operatorname{Int} F) \setminus F_1$ and consequently, there exists a closed ball $B_1 = \overline{B(x_1, r_1)} \subset F \setminus F_1$ such that $r_1 < 1/2$ (Figure 2.5). Next, $F_1 \cup F_2$ has empty interior so it cannot be equal to $\operatorname{Int} B_1$, and it follows that there exists a closed ball $B_2 = \overline{B(x_2, r_2)} \subset B_1 \setminus (F_1 \cup F_2)$ such that $r_2 < 1/2^2$. Continuing, we get a sequence of closed nested balls $\{B_n\}_{n=1}^\infty$. By Cantor's Intersection Theorem (Theorem 2.2.24), there is a point $x \in \cap_{n=1}^\infty B_n$. The definition of $B_n$ implies that $x \notin F_n$ and this is true for all $n \in \mathbb{N}$, so $x \notin F$. This is a contradiction, because $B_n \subset F$, for all $n \in \mathbb{N}$. □

There are several equivalent versions of Baire's theorem. Before we can present them we need to define the concept of *density*.

**Definition 2.5.2.** A set $A$ is **dense** in a metric space $(M, d)$ if the closure of $A$ equals $M$.

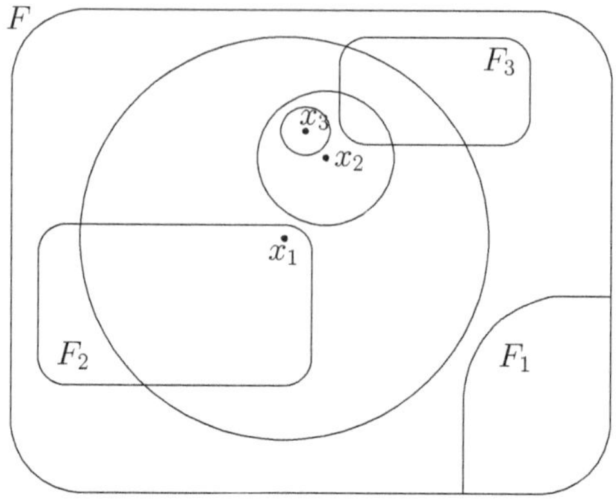


Figure 2.5: The construction in the proof of Baire's Theorem

**Example 2.5.3. A dense set in a metric space.**
Let $M$ be the interval $[0,1]$ with the usual distance. The set of rational numbers in $[0,1]$ is dense in $[0,1]$. The same is true for the set of irrational numbers in $[0,1]$. (See Exercise 2.1.34.) ♦

**Example 2.5.4. A dense set in a metric space.**
By the Weierstrass Theorem, polynomials are dense in $C[a,b]$. ♦

Another characterization of density is contained in the following proposition.

**Proposition 2.5.5.** *A set is dense in a metric space $M$ if and only if it has a nonempty intersection with every nonempty open set.*

*Proof.* ($\Rightarrow$) Let $A$ be a dense set in $M$, let $G$ be open, and $x \in G$. Then there is $B(x,r) \subset G$. Since $A$ is dense, there is $a \in A \cap B(x,r) \subset A \cap G$.
($\Leftarrow$) Suppose $A$ is not dense. By definition, $\overline{A} \neq M$ so $(\overline{A})^C \neq \emptyset$. Now $(\overline{A})^C$ is an open set and $A \cap (\overline{A})^C = \emptyset$. □

Open and closed sets play a fundamental role in understanding a metric space. Unfortunately, not all sets are open so sometimes it helps if the set under scrutiny contains an open set, (i.e., if its interior is nonempty). A set like $\mathbb{Q} \cap [0,1]$ does not but, by Example 2.5.3, its closure does since it is dense in $[0,1]$. At the extreme, there are sets where even taking closure does not help.

**Definition 2.5.6.** A subset $E$ of a metric space $X$ is **nowhere dense** if its closure $\overline{E}$ has empty interior.

**Example 2.5.7. A nowhere dense subset of $\mathbb{R}$.**
The set $\mathbb{N}$ is nowhere dense in $\mathbb{R}$ because it is closed (its complement is open) and it does not contain any intervals. ♦

**Example 2.5.8. A nowhere dense subset of $\mathbb{R}^2$.**

The graph of $y = x$ is nowhere dense in $\mathbb{R}^2$. (Exercise 2.5.9.) ♦

It is sometimes convenient to use a different characterization of the property of being nowhere dense. For subsets $A, B \subset M$ we say that $A$ is dense in $B$ if $B \subset \overline{A}$.

**Proposition 2.5.9.** *If $A$ is a subset of a metric space $(M, d)$ then the following statements are equivalent:*

*(a) $A$ is nowhere dense.*

*(b) $A$ is not dense in any ball in $(M, d)$.*

*(c) $\overline{A}^C$ is dense in $M$.*

*Proof.* (a) $\Leftrightarrow$ (b) $A$ is *not* nowhere dense if and only if there is $B(x, r) \subset \overline{A}$, which is equivalent to $A$ being dense in $B(x, r)$.

(a) $\Leftrightarrow$ (c) $\text{Int}\,(\overline{A}) = \emptyset$ if and only if $[\text{Int}\,(\overline{A})]^C = M$. By Exercise 2.1.38, this is equivalent to $\overline{[(\overline{A})^C]} = M$ which is equivalent to $(\overline{A})^C$ being dense in $M$. □

Now we can state one equivalent formulation of the Baire Category Theorem.

**Theorem 2.5.10.** *A complete metric space cannot be a countable union of nowhere dense sets.*

Rather than talk about countable unions of nowhere dense sets we introduce the following terminology.

**Definition 2.5.11.** A subset of metric space is of the **first category** (or **meager**) if it is a countable union of nowhere dense sets. If it is not of the first category then it is of the **second category**.

Theorem 2.5.10 can now be restated.

**Theorem 2.5.12.** *A complete metric space is of second category.*

*Remark* 2.5.13. The assertion in Theorem 2.5.12 is an implication which cannot be reversed. (See Exercise 2.5.21.)

We present a result that sheds more light on the properties described above.

**Proposition 2.5.14.** *If $A$ is a set of first category in a complete metric space then $A^C$ is dense.*

*Proof.* Suppose $A^C$ is not dense. Then $\overline{A^C} \neq M$ so $[\overline{A^C}]^C$ is a nonempty open set. By Exercise 2.1.38, $[\overline{A^C}]^C = \text{Int}\,A$, so $A$ has a nonempty interior. On the other hand, $A = \cup A_n$ with each $A_n$ nowhere dense, i.e., $\text{Int}\,\overline{A_n} = \emptyset$. By Theorem 2.5.1, $\cup \overline{A_n}$ has an empty interior, and so does its subset $A = \cup A_n$, which is a contradiction. □

Baire's Theorem has numerous applications. Some of the fundamental results like Theorems 11.3.1 and 11.4.3 rely on it. It can also be used to solve some exercises in this chapter.

Baire wrote his dissertation [1] in 1899 about limits of continuous functions. As a part of this study he introduced the concepts of the sets of first and second categories and gave a brief argument that a closed interval on the real line is of the second category. His argument can be easily adapted to other Euclidean spaces. Hausdorff formulated and proved the metric space version in [56] in 1914. Before Baire, an American mathematician William Osgood (1864–1943) proved exactly the same fact in 1897 in [82]. Osgood studied for 2 years in Götingen, then finished his doctoral studies in Erlangen. He returned to Boston and became a professor at Harvard where he was a chair for a while. He stayed at Harvard until 1933 and then taught for 2 years at the National University of Peking. He was an important factor in bringing the European mathematics to the United States and transforming Harvard into a research university.

Although not directly connected to Baire's theorem, *separability* is another very important property that a metric space may have, and it is related to density.

**Definition 2.5.15.** A metric space is **separable** if it has a countable, dense subset.

**Example 2.5.16. A separable metric space.**

The set $\mathbb{R}^n$ with Euclidean distance is separable because the points with all rational coordinates form a dense set. ♦

**Example 2.5.17. A nonseparable metric space.**

Let $M$ be an uncountable set and define $d : M \to \mathbb{R}$ by $d(x, y) = 0$ if and only if $x = y$ and $d(x, y) = 1$ otherwise. The metric space $(M, d)$ is not separable. (See Exercise 2.1.25.)

**Proof:** Since $B(x, 1/2) = \{x\}$, every singleton $\{x\}$ is an open set so, by Exercise 2.1.25, every subset of $M$ is open, hence every subset of $M$ is closed. Thus, the closure of every countable set $S$ equals $S$ and cannot equal $M$ which is uncountable. ♦

RENE-LOUIS BAIRE (1874–1932) was a French mathematician. His father was a tailor, and the family had a hard time making ends meet. At the age of 12, he earned a scholarship to Lycée Lakanal in Paris, where he showcased his academic prowess, winning awards at the prestigious Concours Général competition. Attending Ecole Normale Supérieure, he learned from eminent mathematicians like Charles Hermite (1822–1901), Émile Picard (1856–1941), and Henri Poincaré (1854–1912). Despite excelling in written examinations, his oral exams revealed gaps in understanding of function continuity. This drove him to delve into function theory and limits, culminating in his eponymous classification of functions. Despite completing his doctoral thesis on discontinuous functions in 1899, his novel ideas faced significant opposition, hindering his prospects of securing a university position. He had academic disputes with contemporaries such as Lebesgue and de la Vallée Poussin. A brief respite came in 1905 with a university post in Dijon, yet declining health forced him to take leave in 1914 and retire in 1925, spending his final years around Lake Geneva, where he received the Chevalier de la Legion d'Honneur and was elected to the French Academy of Sciences. He lived modestly on a pension, tragically ending his life by suicide. Despite his struggles, Baire left enduring contributions in analysis, leaving behind a complex legacy remembered by his contemporaries and students.

The next result gives several statements equivalent to separability. We say that a set $M$ is discrete if all its points are isolated, i.e., for every $x \in M$ there is an open ball $B(x,r)$ that contains no other points of $M$ except $x$. Also, a set $\mathcal{B}$ is a base of topology if every open set can be written as a union of elements of $\mathcal{B}$.

**Theorem 2.5.18.** *Let $(M,d)$ be a metric space. The following statements are equivalent:*

*(a) $M$ is separable.*

*(b) $M$ has a countable base of topology.*

*(c) Every discrete subset of $M$ is countable.*

*(d) Every disjoint family of open subsets of $M$ is countable.*

*(e) Every subspace of $M$ is separable.*

*Proof.* (a) $\Rightarrow$ (b) Let $A$ be a countable dense set in $M$ and define $\mathcal{B} = \{B(a,1/n) : a \in A,\, n \in \mathbb{N}\}$. We will show that if $G$ is an open set and $x \in G$ then there is $B \in \mathcal{B}$ such that $x \in B \subset G$. Since $G$ is open there is $n \in \mathbb{N}$ such that $B(x,1/n) \subset G$. The set $A$ is dense so there is $a \in A \cap B(x,1/(2n))$ (Figure 2.6). Then $B := B(a,1/(2n)) \in \mathcal{B}$ and $x \in B$ because $a \in A \cap B(x,1/(2n))$,

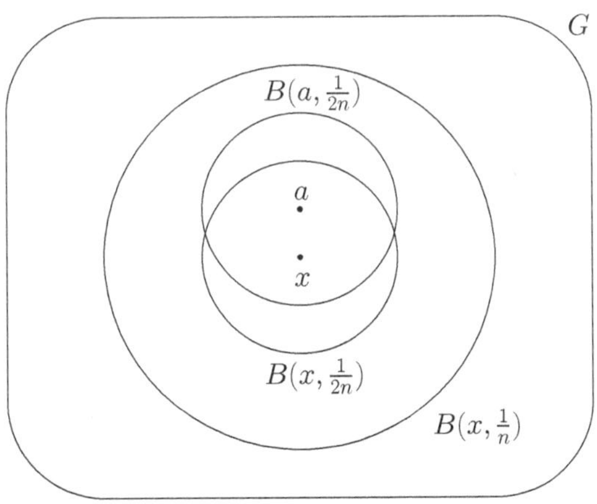


Figure 2.6: Countable base of topology

so $d(a,x) < 1/(2n)$. Finally, $B \subset B(x,1/n) \subset G$. Indeed, if $y \in B$ then $d(y,x) \le d(y,a) + d(a,x) < 1/(2n) + 1/(2n) = 1/n$.

(b) $\Rightarrow$ (e) Let $X \subset M$ and $\mathcal{B} = \{B_n\}$ a countable base of topology. Define $\mathcal{B}' = \{B_n \cap X\}$. Then $\mathcal{B}'$ is a base of topology in $X$ and it is at most countable. Now, for each $n$, by Axiom of Choice, there is $b_n \in B_n \cap X$. The set $\{b_n\}$ is dense in $X$. Indeed, let $x \in X$ and $\varepsilon > 0$. By definition, there is $B_n \cap X \subset B(x,\varepsilon)$ so $b_n \in B(x,\varepsilon)$, i.e., $d(x,b_n) < \varepsilon$.

(e) ⇒ (c) Let $Y = \{y_i : i \in I\}$ be a discrete subset of $M$. Then for each $i \in I$ there is $B(y_i, r_i)$ that contains no other member of $Y$. Let $\mathcal{A} = \{B(y_i, r_i/3)\}_{i\in I}$. These balls are pairwise disjoint because if $y \in B(y_i, r_i/3) \cap B(y_j, r_j/3)$ then $d(y_i, y_j) \le d(y_i, y) + d(y, y_j) \le (r_i + r_j)/3 < \max\{r_i, r_j\}$, a contradiction. By (a) ⇒ (b), $Y$ has a countable base of topology $\mathcal{B} = \{B_n\}_{n=1}^{\infty}$ so every ball $B(y_i, r_i/3)$ contains a ball from $\mathcal{B}$. Thus, $\mathcal{A}$ is countable, hence $Y$ is countable.

(c) ⇒ (d) Let $\mathcal{G} = \{G_i \in I\}$ be a disjoint family of nonempty open subsets of $M$. By Axiom of Choice, we can select $a_i \in G_i$. Let $A = \{a_i : i \in I\}$. Since $A$ is a discrete set it must be countable whence $\mathcal{G}$ is countable.

(d) ⇒ (a) For each $r > 0$ let $\mathcal{A}_r = \{S \subset M : x \neq y,\, x, y \in S \implies d(x, y) > r\}$. Let $S \in \mathcal{A}_r$. Then $S$ is discrete and the balls with centers at the points of $S$ and radius $r/2$ are disjoint. By assumption, there can be only countably many of those, and therefore each $S \in \mathcal{A}_r$ is at most countable. The collection $\mathcal{A}_r$ is partially ordered with respect to inclusion so, by Hausdorff maximal principle, there is a maximal totally ordered subset $\{S_i : i \in I\}$. Let $\mathcal{B}_r = \cup_{i\in I} S_i$. Once again, if $x \neq y$ are elements of $\mathcal{B}_r$ then $x \in S_i$ and $y \in S_j$, and one of these two sets is contained in the other, so $x, y$ are in the same one and thus $d(x, y) > r$. It follows that $\mathcal{B}_r$ is also countable, and so is $\mathcal{S} = \cup\{\mathcal{B}_q : q \in \mathbb{Q},\, q > 0\}$. Finally, $\mathcal{S}$ is dense in $M$. Indeed, for every positive rational number $q$, $d(x, \mathcal{S}) \le d(x, \mathcal{B}_q) \le 2q$, otherwise $x$ could be included in $\mathcal{B}_q$ contradicting its maximality. □

## Exercises

2.5.1. Let $(M, d)$ be a metric space and $A, B, C \subset M$. Prove that if $A$ is dense in $B$ and $B$ is dense in $C$, then $A$ is dense in $C$.

2.5.2. Let $(M, d)$ be a metric space and $A \subset M$. Prove that $A$ is dense in $M$ if and only if the interior of $A^C$ is empty.

2.5.3. Let $(M, d)$ be a metric space, let $A, B$ be dense subsets of $M$, and suppose that $A$ is open.

(a) Prove that: $A \cap B$ is dense in $M$.

(b) Give an example to show that (a) fails if we drop the assumption that $A$ is open.

2.5.4. Let $(M, d)$ be a complete metric space and $\{G_n\}_{n=1}^{\infty}$ open dense subsets of $X$. Prove that the intersection $\cap_{n=1}^{\infty} G_n$ is dense in $M$.

2.5.5.* Let $G$ be an open dense set in a metric space $M$, and let $N$ be a set with empty interior. Prove that $G \setminus N$ is dense in $M$

2.5.6. Let $\mathcal{A}$ be the collection of all nowhere dense subsets of a metric space $(M, d)$. Prove that:

(a) If $A \in \mathcal{A}$ then $\overline{A} \in \mathcal{A}$.

(b) If $\{A_k\}_{k=1}^n \subset \mathcal{A}$ then $\cup_{k=1}^n A_k \in \mathcal{A}$.

(c) If $A \in \mathcal{A}$ and $B \subset A$ then $B \in \mathcal{A}$.

(d) Give an example to show that $\mathcal{A}$ is not closed under countable unions.

2.5.7. Let $A$ be a closed subset of a metric space $(M, d)$. Prove that the boundary of $A$ is nowhere dense. What if $A$ is an open set?

2.5.8. Let $A$ be a dense subset of a metric space. Prove or disprove: the set $A^C$ is nowhere dense.

2.5.9. Prove that the graph of $y = x$ is nowhere dense in $\mathbb{R}^2$.

2.5.10.* Prove that a set is the boundary of some open set if and only if it is closed and nowhere dense.

2.5.11. Let $A$ be a closed subset of a metric space such that every point of $A$ is a cluster point of both $A$ and $A^C$. Prove that $A$ is nowhere dense.

2.5.12. Prove that a set $A$ in a metric space is nowhere dense if and only if for every open set $G$, there is an open set $G_1 \subset G$ such that $G_1 \cap A = \emptyset$.

2.5.13. Let $E$ be a subset of a complete metric space $M$ and suppose that both $E$ and $E^C$ are dense. Prove that at most one of them is the union of a countable collection of closed sets. Conclude that the set of rational numbers in $[0, 1]$ is not the intersection of a countable collection of open sets.

2.5.14. Let $M$ be a complete metric space without isolated points and let $\{F_n\}$ be a countable collection of closed, nowhere dense sets. Prove that $(\cup_{n=1}^{\infty} F_n)^C$ is dense in $M$ and uncountable.

2.5.15. Let $(M, d)$ be a complete metric space without isolated points. Prove that $M$ is uncountable. Conclude that the set of all real numbers is uncountable.

2.5.16. Prove that the set of rational numbers is not a $G_\delta$ set.

2.5.17. Let $(M, d)$ be a metric space. Prove that it is of the second category if and only if, whenever $M = \cup_{n=1}^{\infty} F_n$ where each $F_n$ is closed, at least one of the sets $F_k$ contains an open ball.

2.5.18. Prove that if a subset of a metric space is of second category then it must be uncountable. Conclude that the interval $[0, 1]$ is uncountable.

2.5.19.* Prove that in a complete metric space, a set is of the first category if and only if its complement contains a dense $G_\delta$ set.

2.5.20.* Prove that if $A \subset \mathbb{R}$ is a set is of first category then $A^C + A^C = \mathbb{R}$, if $A^C + A^C = \{x + y : x \in A^C,\, y \in A^C\}$.

2.5.21. Give an example of a metric space that is of second category but not complete.

2.5.22. Let $(M, d)$ be a separable metric space and $E$ a bounded subset of $M$. Prove that $(E, d|E)$ is separable.

2.5.23. Prove that every totally bounded metric space is separable.

2.5.24. Prove that a compact metric space is separable.

2.5.25.** If $P$ is an infinitely differentiable function such that for each $x$, there is an $n$ with $P^{(n)}(x) = 0$, prove that $P$ is a polynomial. (Note $n$ depends on $x$.)

2.5.26.* Suppose that $f : [0, +\infty) \to [0, +\infty)$ is a continuous function with the following property: for all $x \in [0, +\infty)$, the sequence $f(x), f(2x), f(3x), \dots$ tends to 0. Prove that $\lim_{t\to\infty} f(t) = 0$.

2.5.27.* Let $\{f_n\}$ be a sequence of real-valued functions on $[0, 1]$ defined by $f_0 = f \in C[0, 1]$ and $f_n$ is an antiderivative of $f_{n-1}$. Suppose that for each $x \in [0, 1]$ there is $n \in \mathbb{N}$ for which $f_n(x) = 0$. Prove that $f = 0$.

2.5.28.* Let $(M, d)$ be a metric space, let $\{f_n\}$ be a sequence of continuous, real-valued functions on $M$, and suppose that there is a set $A \subset M$ of second category such that $\sup\{|f_n(x)| : n \in \mathbb{N}\}$ is finite for each $x \in A$. Prove that there exists a ball $B(x_0, r)$ and $K > 0$ such that $|f_n(x)| \le K$, for all $n \in \mathbb{N}$ and all $x \in B(x_0, r)$.

2.5.29. Let $(M, d)$ be a metric space and let $G$ be an open subset of $M$. Prove that $G$ is of second category in itself.

2.5.30.* Let $(M, d)$ be a metric space and let $G$ be a $G_\delta$ subset of $M$. Prove that $G$ is of the second category in itself.

2.5.31. Let $\{f_n\} \subset C[0, 1]$ that converges to a real-valued function $f$, for all $x \in [0, 1]$. Prove that for every $\varepsilon > 0$ there is a dense subset $D_\varepsilon \subset [0, 1]$ such that if $x \in D_\varepsilon$, then there is an open interval $I$ containing $x$ and a positive integer $N_x$ such that for all $n > N_x$, $\sup\{|f_n(y) - f(y)| : y \in I\} \le \varepsilon$.

# 3

# *Geometry of the Line and the Plane*

As we stated earlier, the motivation for the development of measure theory was the desire to assign a "correct" number to each subset of $\mathbb{R}^2$, and more generally to do so in all Euclidean spaces. In this chapter we will take a closer look at some of the geometric properties of $\mathbb{R}^n$. We will also introduce some special subsets of $\mathbb{R}$ that will give us an insight into how complicated these can be and how formidable a task is ahead of us.

We expect many sets to have infinite measures. Because of that, in addition to the set $\mathbb{R}$ it will be useful to consider its extension $\mathbb{R}^*$ which is

$$\mathbb{R} \cup \{-\infty, +\infty\}.$$

The **extended real line** has the ordering induced by the order relation in $\mathbb{R}$. Namely, for any $x, y \in \mathbb{R}$ the usual order applies, while for any $x \in \mathbb{R}^*$, $-\infty \leq x \leq +\infty$. Thus, the collection of intervals will now comprise of the sets of the form

$$[a,b] := \{x \in \mathbb{R} : a \leq x \leq b\}, \quad [a,b) := \{x \in \mathbb{R} : a \leq x < b\},$$
$$(a,b] := \{x \in \mathbb{R} : a < x \leq b\}, \quad (a,b) := \{x \in \mathbb{R} : a < x < b\},$$

where $a, b \in \mathbb{R}^*$. For example, $x \geq \alpha$ is the same as $x \in [\alpha, +\infty]$. As usual, $x < y$ is short for $x \leq y$ and $x \neq y$.

Some of the algebraic operations can be extended in an obvious way from $\mathbb{R}$ to $\mathbb{R}^*$. For example, if $x \in \mathbb{R}$ then

$$x + (+\infty) = (+\infty) + x = +\infty, \quad x + (-\infty) = (-\infty) + x = -\infty,$$

$$x \cdot (+\infty) = (+\infty) \cdot x = \begin{cases} +\infty & \text{if } x > 0, \\ -\infty & \text{if } x < 0, \\ 0 & \text{if } x = 0, \end{cases}$$

$$\frac{x}{+\infty} = \frac{x}{-\infty} = 0.$$

Naturally, expressions like $(+\infty) + (-\infty)$ are not defined.

Topologically, $\mathbb{R}^*$ is homeomorphic to the closed interval $[-1, 1]$ and a homeomorphism $F : \mathbb{R}^* \to [-1, 1]$ is given by

$$F(x) = \begin{cases} \frac{x}{1+|x|} & \text{if } x \in \mathbb{R}, \\ 1 & \text{if } x = +\infty, \\ -1 & \text{if } x = -\infty. \end{cases} \tag{3.1}$$

DOI: 10.1201/9781032716497-3

We leave the proof that $F$ is a **homeomorphism**, i.e., a continuous bijection with a continuous inverse to the reader (Exercise 3.1.1). From here, we can conclude that the following sets are open: intervals of the form $(a, b)$ (bounded or unbounded), $[-\infty, b)$, and $(a, +\infty]$, as well as all possible unions of these sets. (See Exercise 3.1.2). It turns out that there are no others, but to prove that we need the following fundamental result.

**Theorem 3.1.1.** *A subset of $\mathbb{R}^*$ is connected if and only if it is an interval.*

*Proof.* Since the homeomorphism (3.1) is an increasing function, it preserves the order, so it suffices to prove the theorem for subsets of $[-1, 1]$. Clearly, every interval is path-connected, hence connected. In the other direction, suppose that a set $A$ is not an interval. If we define $a = \inf A$ and $b = \sup A$ then the assumption that $A$ is not an interval leads to the conclusion that there is $x \in [-1, 1]$ such that $a < x < b$ but $x \notin A$. Then $[-1, x) \cap A$ and $(x, 1] \cap A$ form a separation of $A$, so it is not connected. □

Now we can give a characterization of open sets on the real line.

**Theorem 3.1.2.** *Every nonempty open subset of $\mathbb{R}^*$ is a countable disjoint union of open intervals.*

*Proof.* In view of the homeomorphism (3.1), we may assume that the open set $G$ is a subset of $[-1, 1]$. If $G$ is connected, Theorem 3.1.1 shows that it must be an open interval. Otherwise, $G$ is a union of its components, and each of them must be an open interval. Finally, the density of rational numbers ensures that each component must contain a different rational number, so there can be at most a countable number of them. □

Next, we look for a generalization of Theorem 3.1.2 to open subsets of Euclidean spaces. Recall that a **rectangle** in $\mathbb{R}^2$ is a Cartesian product $R := I_1 \times I_2$ of intervals $I_1, I_2$. It is natural to wonder whether Theorems 3.1.1 and 3.1.2 hold in $\mathbb{R}^2$ if the word interval is replaced by *rectangle.* The answer is that neither one does, not only in $\mathbb{R}^2$, but in $\mathbb{R}^n$, for $n > 2$. To convince ourselves, notice that in $\mathbb{R}^2$ every open disk is a connected set, so a generalization of Theorem 3.1.1 fails. Moreover, a disjoint union of any two open sets cannot be connected. Thus, a connected open set in $\mathbb{R}^n$ (such as an open disk in $\mathbb{R}^2$) is not a disjoint union of open rectangles (unless it is an open rectangle itself). However, in $\mathbb{R}^n$ we have the following result.

**Theorem 3.1.3.** *Every nonempty open set in $\mathbb{R}^n$ is a countable union of closed rectangles with disjoint interiors.*

*Proof.* We will present the proof in the case $n = 2$. The general case follows along the same lines.

Let $r \in \mathbb{N}$ be fixed and denote by $\mathfrak{C}_r$ the grid consisting of lines

$$x = \frac{n}{2^r}, \quad y = \frac{m}{2^r}, \quad m, n \in \mathbb{Z}.$$

Such a grid can be constructed for any $r \in \mathbb{N}$ and during the transition from $\mathfrak{C}_r$ to $\mathfrak{C}_{r+1}$ each square is divided into 4 squares.

Let $G$ be an open set in $\mathbb{R}^2$ and let $x \in G$. For each $r \in \mathbb{N}$ there is a closed square $Q_r$ that contains $x$, so we get a decreasing sequence $\{Q_r\}$ of squares. On the other hand, $x$ lies in the open set $G$, so there is a ball $B$ such that $x \in B \subset G$. For $r$ large enough $Q_r \subset B$, hence $Q_r \subset G$.

We denote by $\mathcal{A}_1$ the set of all squares in $\mathfrak{C}_1$ that are contained in $G$; $\mathcal{A}_2$ is the set of all squares in $\mathfrak{C}_2$ that are contained in $G$ but not in any of the squares from $\mathcal{A}_1$. Inductively, $\mathcal{A}_{n+1}$ is the set of all squares in $\mathfrak{C}_{n+1}$ that are contained in $G$ but not in any of the squares from $\mathcal{A}_1, \mathcal{A}_2, \ldots, \mathcal{A}_n$. Let $\mathcal{A}$ be the union of all sets $\mathcal{A}_n$, $n \in \mathbb{N}$. Since each $\mathcal{A}_n$ is a finite set, $\mathcal{A}$ is countable so we can arrange all the squares in $\mathcal{A}$ in a sequence $\{S_n\}$. It follows from the construction that

$$\bigcup_{n=1}^{\infty} S_n \subset G.$$

We will show the opposite inclusion.

Let $x \in G$. We have seen that there exists $r$ such that $x \in Q_r \subset G$. If we denote by $m$ the smallest value of $r$ with this property, then $Q_m$ belongs to $\mathcal{A}_m$, so it is one of the sets in $\{S_n\}$. It follows that

$$G \subset \bigcup_{n=1}^{\infty} S_n$$

and the theorem is proved. □

Even though we are familiar with Euclidean spaces, and in particular with the real line, their geometry and topology can be quite deceptive. Nineteenth century provides an abundance of examples where the best of minds went astray, deceived by their intuition, and made claims that are false. As an illustration, in advanced calculus you learned that a continuous function need not be differentiable at any point, a fact that shocked (and disgusted) many mathematicians of the era. In fact, as we will see later, differentiable functions present in some sense a small minority among all continuous functions. (They form a set that is nowhere dense in the metric space $C[a, b]$.)

This is not an isolated example. Mathematicians tend to formulate general results based on specific cases, usually the most obvious and natural ones. Unfortunately, it happens frequently that an assertion is false but it is very hard to come up with a counterexample. One such situation will be our segue into the content of this section. Namely, we will ask whether there is a set $A \subset \mathbb{R}$ such that (a) every point of $A$ is a cluster point of both $A$ and $A^C$; (b) the set $A$ is closed. The answer to (a) is yes: take the set of rational numbers. However, it is far from clear whether there is a set satisfying both (a) and (b).

We have seen in Theorem 3.1.2 that every open set on the real line is a countable union of open intervals. This implies that every closed set can be obtained by deleting a countable number of open intervals. For example, closed interval $[a, b]$ can be obtained by removing open rays $(-\infty, a)$ and $(b, +\infty)$.

Therefore, if there is an affirmative answer to the question above, such a set can be obtained by removing a collection of open sets from $\mathbb{R}$ or one of its subsets, such as the interval $I = [0, 1]$.

We will start with $[0, 1]$ and keep deleting the middle thirds (Figure 3.1):

$$F_1 = [0, 1/3] \cup [2/3, 1]$$
$$F_2 = [0, 1/9] \cup [2/9, 1/3] \cup [2/3, 7/9] \cup [8/9, 1]$$
$$\dots$$

We define the so-called Cantor set as $F = \bigcap_{n=1}^{\infty} F_n$.

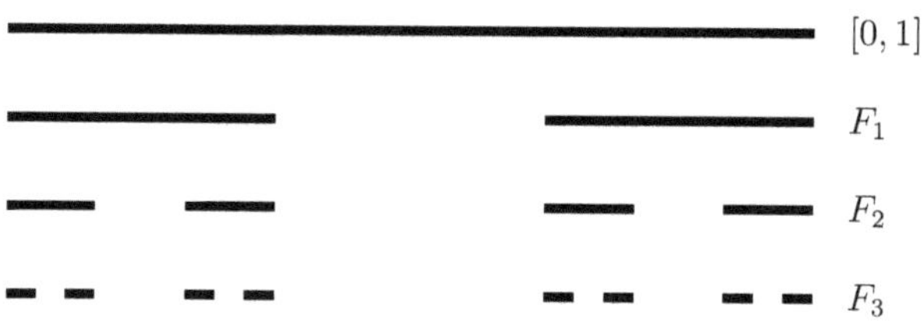


Figure 3.1: Constructing Cantor set

**Theorem 3.1.4.** *The Cantor set $F$ is a closed, uncountable, nowhere dense set whose every point is a cluster point of both $F$ and $F^C$.*

*Proof.* First, $F$ is closed, by Exercise 2.1.27, because each set $F_n$ is closed.

Second, we will show that $F$ is uncountable. Suppose to the contrary that $F$ is countable, $F = \{a_n\}$. The set $F_1$ consists of 2 disjoint intervals. Let $I_1$ denote one that does not contain $a_1$. Now, the set $F_2 \cap I_1$ consists of 2 disjoint intervals. Let $I_2$ denote one that does not contain $a_2$. Continuing, we obtain a sequence $\{I_n\}$ of nested intervals. By Cantor's Intersection Theorem, there exists $x \in \cap_{n \in \mathbb{N}} I_n \subset F$. However, $x$ cannot equal any $a_i$. This contradiction shows that $F$ is uncountable.

Next, we will establish that every point $x_0 \in F$ is a cluster point of $F$. To that end, we will construct a sequence $\{x_n\} \subset F \setminus \{x_0\}$ that converges to $x_0$. Notice that for each $n \in \mathbb{N}$, $x_0$ belongs to one of the $2^n$ closed intervals whose union is $F_n$. Let us denote that interval by $\tilde{I}_n$ and let $x_n$ be one of its endpoints that is not equal to $x_0$. Then, $\{x_n\} \subset F \setminus \{x_0\}$ and $|x_n - x_0| \leq 1/3^n$ so, when $n \to \infty$, we get the desired conclusion.

A similar construction can be used to deduce that $x_0$ is a cluster point of $F^C$. Let $x_n$ and $\tilde{I}_n$ be as in the previous paragraph, and let $J_n$ be a deleted open interval of length $1/3^n$ (i.e., one of the intervals removed when passing from $F_{n-1}$ to $F_n$) that is nearest to $x_n$. Let $y_n \in J_n$. Then, $|x_n - y_n| < 1/3^n + 1/3^n = 2/3^n$, whence $|x_0 - y_n| < 2/3^n + 1/3^n = 1/3^{n-1}$. Since each $y_n \in F^C$ and $\lim y_n = x_0$, we conclude that $x_0$ is a cluster point of $F^C$.

Finally, the fact that $F$ is nowhere dense follows from Exercise 2.5.11. □

*Remark* 3.1.5. A set $A$ is **perfect** if it is closed and has no isolated points. Thus, Cantor set is an example of a perfect set.

The Cantor set was actually discovered in 1874 by an Irish mathematician and amateur astronomer Henry Smith (1826–1883) in [114] and introduced by Cantor in 1883, in [19]. It is often referred to as the "middle thirds set". Cantor was looking for examples of perfect sets that are not dense. There is an indication that a German mathematician du Bois-Reymond (1831–1889) communicated this example to Cantor some years earlier because upon learning that Cantor was going to publish it, he added an announcement of his in a footnote to his paper *Der Beweis des Fundamentalsatzes der Integralrechnung*, (The proof of the fundamental theorem of integral calculus). The example was also discovered in 1881 by Volterra [124]. Smith was a graduate of Oxford and spent his life there as a professor and administrator: he was the Keeper of the Oxford University Museum, a member of the Royal Commission to review scientific education practice, a member of the commission to reform University of Oxford governance, chairman of the committee of scientists overseeing the Meteorological Office, etc. Du Bois-Reymond studied medicine at the University of Zürich but changed to mathematical physics obtaining a PhD in 1853 at the University of Berlin. At the University of Tübingen he was a doctoral adviser of Hölder. He constructed a continuous function whose Fourier series diverges at a dense set of points and a continuous function that has no derivative at any point. He was the first to use, in 1875, the "diagonal argument", which Cantor would later make famous.

We will use a Cantor set and construct a function with some unexpected properties. Let $\{O_n\}$ denote the deleted middle thirds:

$$\begin{aligned} O_1 &= (1/3, 2/3) \\ O_2 &= (1/9, 2/9) \cup (1/3, 2/3) \cup (7/9, 8/9) = O_{21} \cup O_{22} \cup O_{23} \\ O_3 &= (1/27, 2/27) \cup (1/9, 2/9) \cup (7/27, 8/27) \cup \cdots \cup (25/27, 26/27) \\ &= O_{31} \cup O_{32} \cup \cdots \cup O_{37} \\ &\ldots, \end{aligned}$$

and let $O = \cup_{n=1}^{\infty} O_n$.

Next, we define a sequence of functions $\{\varphi_n\}$ as follows. Each function $\varphi_n$ will satisfy $\varphi_n(0) = 0$ and $\varphi_n(1) = 1$. For $n = 1$, we define $\varphi_1(x) = 1/2$ on $O_1$, and we extend it linearly to $[0, 1]$. For $n = 2$, we define $\varphi_2$ on the components of $O_2$ by

$$\varphi_2(x) = \begin{cases} 1/4 & \text{if } x \in (1/9, 2/9), \\ 2/4 & \text{if } x \in (1/3, 2/3), \\ 3/4 & \text{if } x \in (7/9, 8/9), \end{cases}$$

and extend it linearly to $[0, 1]$. The graphs of $\varphi_1$ and $\varphi_2$ are shown in Figure 3.2. On the components of $O_n$, $\varphi_n$ takes values $1/2^n, 2/2^n, \ldots, (2^n - 1)/2^n$, and we extend it linearly to $[0, 1]$. Thus, we get a sequence of continuous functions $\{\varphi_n\}$. We leave the following as Exercise 3.1.24:

**Proposition 3.1.6.** *The sequence* $\{\varphi_n\}$ *defined above satisfies:*

*(a)* $|\varphi_{n+1}(x) - \varphi_n(x)| < 2^{-n}$ *for every* $x \in [0, 1]$.

*(b) It converges uniformly on* $[0, 1]$.

We call $\varphi := \lim \varphi_n$ the **Cantor–Lebesgue function**. The following result presents some curious properties of this function. The proof is a direct consequence of the construction and we leave it to the reader (Exercise 3.1.25).

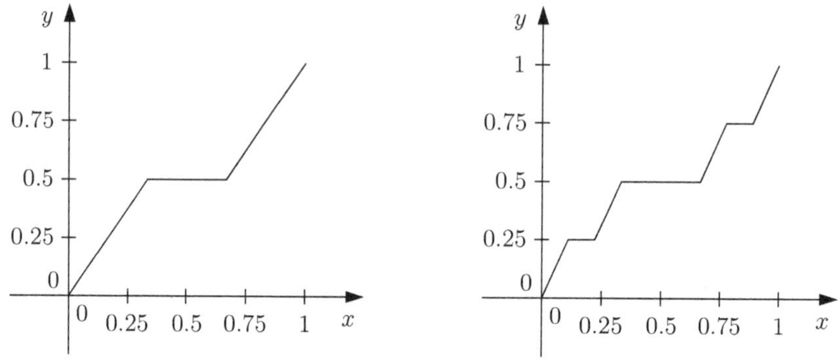


Figure 3.2: Graphs of $\varphi_1$ and $\varphi_2$

**Theorem 3.1.7.** *The Cantor–Lebesgue function $\varphi$ is an increasing, continuous function that maps $[0,1]$ onto $[0,1]$. Its derivative exists in $O$ and $\varphi' = 0$ on $O$.*

> The Cantor–Lebesgue function is also referred to as the Cantor ternary function, the Lebesgue function, Lebesgue's singular function, the Cantor–Vitali function, the Devil's staircase, and the Cantor staircase function. Cantor (1884) introduced the Cantor function [20] and mentioned that a German mathematician and university teacher Ludwig Scheeffer (1859–1885) had pointed out that it was a counterexample to an extension of the fundamental theorem of calculus claimed by Harnack. The Cantor function was discussed and popularized by Scheeffer (1884) [106], Lebesgue (1904) [69], and Vitali (1905) [120]. Scheeffer studied in Berlin and Munich where he started his university career. He died of typhoid fever at the age of 26. Despite the brevity of his life and academic activity, he published a number of important writings and essays.

It is possible to delete different intervals instead of middle thirds. In the following construction we delete the middle quarters, then middle sixteenths, and intervals of length $1/4^n$ in step $n$:

$$F_1 = [0, 3/8] \cup [5/8, 1]$$
$$F_2 = [0, 5/32] \cup [7/32, 3/8] \cup [5/8, 25/32] \cup [27/32, 1]$$
$$\dots$$

We define the "Fat Cantor set" as $F = \cap_{n=1}^{\infty} F_n$. The following proposition lists the most important features of $F$. We leave the proof as Exercise 3.1.21.

**Theorem 3.1.8.** *The Fat Cantor set $F$ is a closed, uncountable, perfect, nowhere dense set whose every point is an accumulation point of $F^C$.*

> This set is known as the Smith–Volterra–Cantor set or Fat Cantor set. In his 1875 paper [114], Smith discussed a nowhere dense set of positive measure on the real line, and Volterra introduced a similar example in 1881 in [124]. Cantor's name seems to be associated because of his ternary set discussed above. (Positive measure means that no matter how tightly it is covered by open intervals, their total length is bigger than some fixed positive number. In that sense the Fat Cantor set is "bigger" than the Cantor set, hence the name. More about this in subsequent chapters.)

GEORG CANTOR (1845–1918) was a German mathematician, best known as the inventor of set theory. He was born in Saint Petersburg, Russia and raised in Germany. He studied mathematics at the Polytechnic of Zürich and the University of Berlin. In 1869 he got a position at the University of Halle, where he remained for the rest of his career. It is there that he became interested in analysis under the influence of German mathematician Eduard Heine (1821–1881), and he proved the uniqueness of the representation of a function by a trigonometric series. He became a lifelong friend with Dedekind and likely inspired "Dedekind cuts". In 1873 he proved that rational numbers were countable, and a year later that real numbers were not. This led him to introduce cardinal numbers which he denoted by the Hebrew letter $\aleph$ (aleph) with a natural number subscript. The Continuum hypothesis, introduced by Cantor, was presented by Hilbert as the first of his twenty-three open problems at the 1900 International Congress of Mathematicians in Paris. It conjectures that there is no cardinal number between those of the natural numbers ($\aleph_0$) and real numbers (continuum). Cantor also proved that there exists a 1-1 correspondence of points on the interval $I = [0, 1]$ and $I^n$ which prompted him to write *I see it, but I don't believe it!* From 1884 until the end of his life, he suffered from depression and mental illness. He continued doing research but his best years were behind him. He retired from teaching in 1913 and died in 1918 from a heart attack.

## Exercises

3.1.1. Prove that the mapping $F$ given by (3.1) is a homeomorphism.

3.1.2. Let $\mathcal{T}$ consist of all sets of the form $(a, b)$, $[-\infty, b)$, $(a, +\infty]$ and any union of such sets. Prove that $\mathcal{T}$ is a topology on $[-\infty, +\infty]$.

3.1.3. Suppose that $A \subset \mathbb{R}$ is a union of nondegenerate closed intervals. Prove that $A$ can be represented as a countable disjoint union of intervals.

3.1.4. Prove that $\mathbb{R}^*$ is a compact set in the topology described in Exercise 3.1.2.

3.1.5. Prove that $\mathbb{R}^*$ is complete in the topology described in Exercise 3.1.2.

3.1.6. Let $f : [0, \infty) \to \mathbb{R}$ be a continuous function such that $\lim_{x\to\infty} f(x) = L$, and $L < f(0)$. Prove that $f$ attains a maximum value in $[0, \infty)$.

3.1.7.* Let $\mathcal{F}$ be the collection of finite unions of intervals of $\mathbb{R}$. Prove that $\mathcal{F}$ is closed under complementation.

3.1.8.* Let $\mathcal{F}$ be the set of all countable subsets of $\mathbb{R}$ and suppose that $E \subset \mathbb{R}$ but $E \notin \mathcal{F}$. Let $A := \{x \in \mathbb{R} : (-\infty, x) \cap E \in \mathcal{F}\}$ and $B := \{x \in \mathbb{R} : (x, \infty) \cap E \in \mathcal{F}\}$. Prove that $A \cap B = \emptyset$ and $A \cup B \neq \mathbb{R}$.

3.1.9.* Let $\{A_i : i \in I\}$ be a collection of closed intervals in $\mathbb{R}$. Prove that the set

$$\bigcup_{i\in I} A_i \setminus \bigcup_{i\in I} \operatorname{Int} A_i$$

is countable.

3.1.10.* Let $\mathcal{F}$ be the collection of all closed subsets of $\mathbb{R}^n$. Prove that $\mathcal{F}$ has the same cardinality as $\mathbb{R}$.

3.1.11. Prove that every open set in $\mathbb{R}^n$ is a countable union of open balls.

3.1.12. Prove that every element of the Cantor set can be written in a unique way as $\sum_{n=1}^{\infty} a_n/3^n$, where each $a_n$ is either 0 or 2.

3.1.13. Let $F$ denote the Cantor ternary set. Prove that $F+F=[0,2]$, if $F+F=\{x+y : x \in F,\, y \in F\}$.

3.1.14. Let $F$ denote the Cantor ternary set. Prove that $F-F=[-1,1]$, if $F-F=\{x-y : x \in F,\, y \in F\}$.

3.1.15. Prove or disprove: if $F$ is an uncountable dense subset of $\mathbb{R}$ then it is of second category.

3.1.16. Give examples of rational and irrational numbers that belong or do not belong to the Cantor ternary set.

3.1.17. Let $S$ denote the set of all real numbers whose base 3 representation requires using digit 1. Prove that $S$ is dense in $\mathbb{R}$.

3.1.18. Let $\mathcal{F}$ be the collection of countable unions of intervals of $\mathbb{R}$. Prove that $\mathcal{F}$ is not closed under complementation.

3.1.19. Prove or disprove: every closed set in $\mathbb{R}^*$ is a countable union of possibly degenerate closed intervals.

3.1.20.* Prove that no countably infinite set can be perfect.

3.1.21. Prove Theorem 3.1.8.

3.1.22. Let $F$ denote the Fat Cantor set obtained by deleting at step $n$ from each of the $2^{n-1}$ intervals the middle interval of length $\alpha^n I$, where $0 < \alpha < 1/2$. (In the classical Cantor set $\alpha = 1/3$.) Prove that $F$ is a closed, uncountable, perfect, nowhere dense set whose every point is an accumulation point of $F^C$.

3.1.23. Prove that the characteristic function of a Fat Cantor Set $F$ is discontinuous at every point of $F$ and continuous at every point of $F^C$.

3.1.24. Prove Proposition 3.1.6.

3.1.25. Prove Theorem 3.1.7.

3.1.26. Let $\varphi$ be the Cantor–Lebesgue function. Prove that $\varphi(1-x) = 1-\varphi(x)$, for all $x \in [0,1]$.

3.1.27.** Let $\varphi$ be the Cantor–Lebesgue function and let $0 < \alpha \le \log_3 2$ be fixed. Prove that there exists $C = C(\alpha) > 0$ such that for all $x, y \in [0,1]$, $|\varphi(x) - \varphi(y)| \le C|x-y|^\alpha$.

3.1.28.** Prove that if $x = \sum_{n=1}^{\infty} a_n/3^n$ is an element of the Cantor set, and $\varphi$ is the Cantor–Lebesgue function, then $\varphi(x) = \sum_{n=1}^{\infty} a_n/2^{n+1}$.

3.1.29.** Let $\varphi$ be the Cantor–Lebesgue function. Prove that $\varphi(x) = 2\varphi(x/3)$, for all $x \in [0,1]$.

3.1.30.** Let $F$ denote the Cantor set and $\varphi$ the Cantor–Lebesgue function. Prove or disprove: there is $c \in F \cap (0,1)$ for which $\varphi'(c)$ exists and it is finite.

3.1.31.* Find the length of the graph of Cantor–Lebesgue function.

3.1.32. Determine whether Cantor–Lebesgue function satisfies Lipschitz condition on $[0, 1]$.

3.1.33. Let $f$ be an increasing function on $[0, 1]$ that agrees with the Cantor–Lebesgue function $\varphi$ on the complement of the Cantor set. Prove that $f = \varphi$ on $(0, 1)$.

# Part II

# Measure Theory

# 4

# *Lebesgue Measure on* $\mathbb{R}^2$

As stated earlier, there were many reasons that mathematicians were not pleased with the state of affairs regarding integration theory in the 19th century and several came down to the lack of a measure theory. Let us explain this.

First, the way the integral was defined by Cauchy, the integrand was supposed to be a continuous function. Riemann made a generalization by replacing the continuity by *integrability*. What that meant was that he considered sums of the form $\sum_{k=1}^{n} \omega_k \Delta_k$ where $\{\Delta_k\}$ are subintervals obtained by partitioning the domain $[a, b]$ and $\omega_k$ is the **oscillation** of the integrand $f$ on $\Delta_k$, i.e., the difference between the maximum and the minimum value of $f$. He took it as granted that, as the norm of the partition (the maximum length of $\Delta_k$) went to 0, the sum would go to 0 as well. It was a great idea because it implies that

$$\sum_{k=1}^{n} \sup\{f(x) : x \in \Delta_k\}\Delta_k \quad \text{and} \quad \sum_{k=1}^{n} \inf\{f(x) : x \in \Delta_k\}\Delta_k$$

converge to the same limit, which he then defined as the integral of $f$. Much of the effort was directed toward understanding what it takes for the equality

$$\lim_{\|P\|\to 0} \sum_{k=1}^{n} \omega_k \Delta_k = 0$$

to hold. One of the discoveries was that $f$ need not be continuous, but the points at which it is discontinuous must be nowhere dense. Looking at examples of such sets, at least those that were known to mathematicians of the period, one might conclude that they are of small size (measure, content), and around 1870 there were first attempts to formulate what it means for a set to have measure 0.

Second, the theory of a double Riemann integral proved to be inadequate. Even when the integrand is a function that takes value 1 on a planar region and 0 elsewhere, in which case the result should be the area of the region, it is not always clear what number to associate with the integral.

In this chapter we will expound the measure theory in $\mathbb{R}^2$, the result of the contributions of many mathematicians. Even though it is customary to start with the one-dimensional case, the measure theory of subsets of $\mathbb{R}$, studying planar sets has many advantages. Namely, it gives a great insight into

DOI: 10.1201/9781032716497-4

the difficulties that one faces when developing the notion of measure, it is more natural, there is a high variety of reasonable regions, and it is easier to generalize because, for the most part, the distinction between various multi-dimensional Euclidean spaces is more in the size of the formulas rather than their content. To emphasize this analogy we will refer to the measure of a rectangle as its volume rather than area, and we will use the abbreviation *vol.* Measure theory in other Euclidean and even more abstract spaces will be treated in Chapter 5.

Our goal in this chapter is to construct a function $m$ defined on the power set $\mathbb{P}(\mathbb{R}^2)$. Such a function should possess certain properties that we would expect from the area. Here is a wish list:

(i) For any rectangle $R$ in $\mathbb{R}^2$, $m(R) = \text{vol}(R)$.

(ii) **Countable additivity:** For any disjoint collection $\{E_n\}_{n\in\mathbb{N}}$,

$$m\left(\bigcup_{n=1}^{\infty} E_n\right) = \sum_{n=1}^{\infty} m(E_n).$$

(iii) **Translation invariance:** For any $y \in \mathbb{R}^2$ and any $E \subset \mathbb{R}^2$,

$$m(E + y) = m(E).$$

This list appears in Lebesgue's 1902 article [68] that contains his doctoral dissertation. He was influenced in that Borel and his 1898 book [14] where he favored the axiomatic approach to function theory. This type of exposition can be further traced to a book that Borel his former fellow student Drach coauthored, based on lectures by their professor Tannery given in 1891/92. Borel's list was different and he made only a preliminary investigation, establishing neither the existence nor the uniqueness of such a measure. Jules Drach (1871–1949) was a French mathematician who worked mostly in mathematical analysis and its applications to mechanics. In his thesis, he developed a Galois theory for differential equations. He spent much of his career as Professor for Analytic Mechanics and Higher Analysis at the Sorbonne. Jules Tannery (1848–1910) was a French mathematician who studied under Charles Hermite and was the PhD advisor of Jacques Hadamard. Tannery's theorem gives sufficient conditions for the interchanging of the limit and infinite summation operations. He was an advocate for the mathematics education of children through synthetic geometry (geometry without coordinates) and mathematical proofs.

## 4.1 Jordan Measure

In this section we will look at Jordan's construction, which was inspired by a similar work of Peano. Following a recurring theme in measure theory, we will start with simpler sets and work towards more complicated. The simplest ones are rectangles, by which we mean those whose sides are parallel to coordinate axes. Condition (i) on the wish list reminds us that for them the measure is just the area. Clearly, this number is the same regardless of whether the rectangle under consideration is closed, open, or contains a part of its boundary. In other words, $\operatorname{vol}(R) = \operatorname{vol}(\operatorname{Int} R) = \operatorname{vol}(\overline{R})$, for any rectangle $R$.

The following class represents the next level of complexity.

**Definition 4.1.1.** An **elementary set** is any subset of $\mathbb{R}^2$ which is a union of a finite number of rectangles. We denote by $\mathcal{E}$ the class of elementary sets.

Although the definition does not require the rectangles to be disjoint, this can be accomplished.

**Proposition 4.1.2.** *Every elementary set can be expressed as a finite union of disjoint rectangles.*

*Proof.* By definition, the given elementary set $E$ is the union of finitely many rectangles $R_1, \dots, R_m$. We will prove that $E$ can be expressed as the finite union of disjoint rectangles using induction on $m$. The case $m = 1$ is obvious so we assume that $E_m := R_1 \cup \dots \cup R_m$ is a union of $M$ disjoint rectangles $R'_k$:

$$E_m = \bigcup_{k=1}^{M} R'_k$$

and we consider $E_{m+1} = E_m \cup R_{m+1}$. Since

$$E_{m+1} = \left(\bigcup_{k=1}^{M} R'_k\right) \cup R_{m+1} = \left(\bigcup_{k=1}^{M} (R'_k \setminus R_{m+1})\right) \cup R_{m+1},$$

(see Figure 4.1), and the $M + 1$ sets on the right-hand side are pairwise

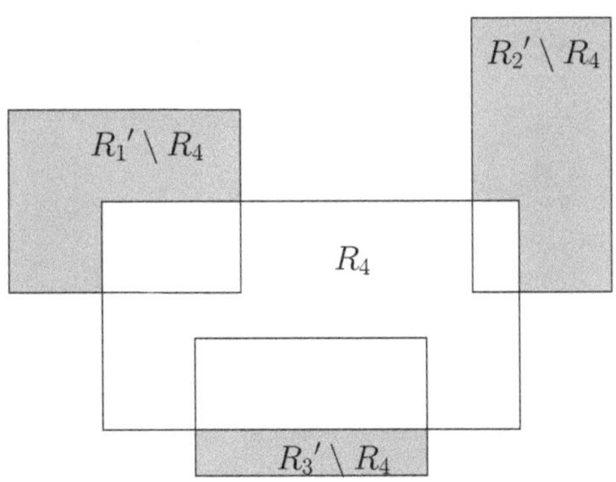


Figure 4.1: $R'_1 \cup R'_2 \cup R'_3 \cup R_4$ is a finite union of disjoint rectangles

disjoint, it suffices to prove that $R'_k \setminus R_{m+1}$ can be represented as a finite union of disjoint rectangles which follows from Exercise 1.1.1. □

Proposition 4.1.2 allows us to define the measure of an elementary set in a way consistent with condition (ii) on the wish list.

**Definition 4.1.3.** Let $E \in \mathcal{E}$ and suppose that it is represented as a disjoint union of rectangles $\{R_k\}_{k=1}^m$. We define the **content** of $E$ (also known as **Jordan content**) by

$$c(E) = \sum_{k=1}^{m} \mathrm{vol}\,(R_k).$$

For this definition to be meaningful, the content must be independent of the representation of the elementary set.

**Proposition 4.1.4.** *Suppose that a set $E \in \mathcal{E}$ is represented as a union of disjoint rectangles in two different ways:*

$$E = \bigcup_{k=1}^{m} R'_k \quad \textit{and} \quad E = \bigcup_{k=1}^{l} R''_k. \tag{4.1}$$

*Then $\sum_{k=1}^{m} vol(R'_k) = \sum_{k=1}^{l} vol(R''_k)$.*

*Proof.* Let $R_{ij} = R'_i \cap R''_j$, for $1 \leq i \leq m$ and $1 \leq j \leq l$. Then rectangles $\{R_{ij}\}$ are disjoint and

$$E = \bigcup_{i=1}^{m} \bigcup_{j=1}^{l} R_{ij}.$$

The result will follow from the equalities

$$\sum_{k=1}^{m} \mathrm{vol}(R'_k) = \sum_{i=1}^{m} \sum_{j=1}^{l} \mathrm{vol}\, R_{ij} \quad \text{and} \quad \sum_{k=1}^{l} \mathrm{vol}(R''_k) = \sum_{i=1}^{m} \sum_{j=1}^{l} \mathrm{vol}\, R_{ij}.$$

They can both be proved the same way, so we focus on the first one. It suffices to establish that for each fixed $i$, $1 \leq i \leq m$, $\mathrm{vol}(R'_i) = \sum_{j=1}^{l} \mathrm{vol}\, R_{ij}$. But, $R'_i \subset E$, so

$$R'_i = R'_i \cap E = R'_i \cap \bigcup_{j=1}^{l} R''_j = \bigcup_{j=1}^{l} (R'_i \cap R''_j) = \bigcup_{j=1}^{l} R_{ij},$$

and this is a disjoint union, so the result follows from Definition 4.1.3. □

It is worth noting that the function $c : \mathcal{E} \to \mathbb{R}$ has some useful properties.

**Theorem 4.1.5.** *Jordan content has the following properties:*

*(i)* ***Finite additivity:*** *For any disjoint collection $\{E_k\}_{k=1}^m \subset \mathcal{E}$,*

$$c\left(\bigcup_{k=1}^{m} E_k\right) = \sum_{k=1}^{m} c(E_k).$$

*(ii)* ***Subadditivity:*** *For any collection $\{E_k\}_{k=1}^m \subset \mathcal{E}$,*

$$c\left(\bigcup_{k=1}^{m} E_k\right) \leq \sum_{k=1}^{m} c(E_k).$$

(iii) ***Translation invariance:*** *For any* $y \in \mathbb{R}^2$ *and any* $E \in \mathcal{E}$, $E+y \in \mathcal{E}$ *and*

$$c(E+y) = c(E).$$

We leave the proof of Theorem 4.1.5 as Exercise 4.1.3.

*Remark* 4.1.6. Definition 4.1.3 aligns nicely with the proposed property (ii) of measure. However, it fails to take full advantage of this rule, restricting it to finite unions. We will consider countable unions and countable additivity in the next section.

Now that we have defined the content of an elementary set, our goal is to extend it to a larger class of sets. The idea that we will pursue is to approximate the given set by elementary sets, both using the inscribed and the circumscribed rectangles.

**Definition 4.1.7.** Given a bounded set $A \in \mathbb{R}^2$ we define the **inner content** of $A$ by

$$c_*(A) = \sup\{c(E) : E \in \mathcal{E},\ E \subset A\},$$

and the **outer content** of $A$ by

$$c^*(A) = \inf\{c(E) : E \in \mathcal{E},\ E \supset A\}.$$

The set $A$ is **Jordan measurable** if $c_*(A) = c^*(A)$ in which case we denote this number by $c(A)$ and call it the **content** of $A$. We will denote the set of all Jordan measurable sets by $\mathcal{J}$.

Several properties of the outer content, the inner content, and the content will be noted here without proof (see Exercise 4.1.8), while some can be found among the exercises.

**Theorem 4.1.8.** *Let* $A, B \in \mathcal{J}$.

(a) $c(\emptyset) = 0$.

(b) *If* $A \subset B$, *then* $c(A) \leq c(B)$ ***(monotonicity)***.

(c) *For any rectangle* $R$, $c(R) = vol(R)$.

The concept of the outer content for subsets of $[a, b]$ appeared in a paper by an Austrian mathematician Otto Stolz (1842–1905) in 1883 [116], who was inspired by several papers of Harnack. Later that year, a paper by Cantor [19] introduced the word *"inhalt"* (content) in $\mathbb{R}^n$, still meaning the outer content. His definition used multiple integrals (the theory of which was still very vague) and implicitly relied on the Heine–Borel theorem which had yet to be stated and proved. Stolz was working on defining the length of a curve, while Cantor was interested in defining the volume of an $n$-dimensional solid. In 1884, in [54] Harnack came up with the same definition as Stolz, but managed to prove some of Cantor's results, and also raised the important issue of whether one should consider infinite covers. Peano made many improvements, including the consideration of both the inner and outer content of a set in $\mathbb{R}^n$, for $n \leq 3$. Jordan found a more effective way to define these concepts, which works for any $n$, leading to the so-called Jordan content.

An alternative characterization of Jordan measurability is provided by the following result which gives a great geometric insight into the meaning of a Jordan set. It will also establish a link between measurability and integrability. (See Theorem 4.1.17.)

**Theorem 4.1.9.** *A bounded set $E$ is Jordan measurable if and only if its boundary has content 0.*

*Proof.* Suppose first that $E$ is Jordan measurable and let $\varepsilon > 0$. Then there exist elementary sets $E_1 \subset E \subset E_2$ such that $c(E_2) - c(E_1) < \varepsilon$. By splitting

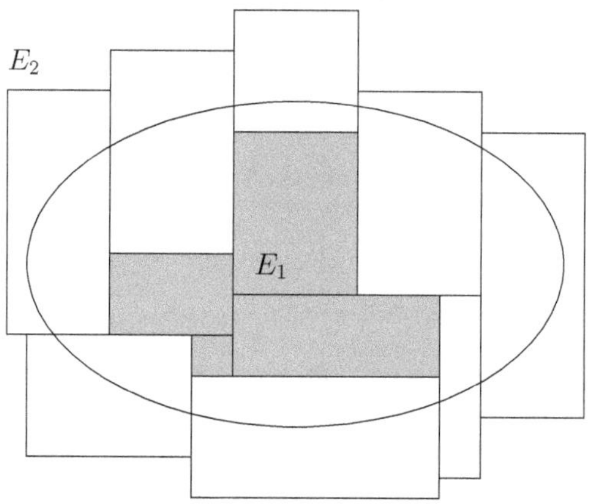


Figure 4.2: $E_1 = \bigcup_{k=1}^{3} R_k \subset E \subset \bigcup_{k=1}^{11} R_k = E_2$

each of the rectangles whose unions form $E_1$ and $E_2$, we can assume that we are dealing with a sequence of disjoint rectangles $\{R_k\}_{k=1}^{n}$ so that

$$E_1 = \bigcup_{k=1}^{m} R_k \subset E \subset \bigcup_{k=1}^{n} R_k = E_2,$$

for some $m \le n$. Since the content of any rectangle is independent of whether its boundary is included or not, we may assume that rectangles contained inside $E$ are all open and the others are closed. By definition, the boundary of $E$ satisfies

$$\partial E \subset \bigcup_{k=1}^{n} R_k \setminus \bigcup_{k=1}^{m} R_k = \bigcup_{k=m+1}^{n} R_k.$$

Thus, using Theorem 4.1.8,

$$c(\partial E) \le c\left(\bigcup_{k=m+1}^{n} R_k\right) = \sum_{k=m+1}^{n} \text{vol}(R_k) = c(E_2) - c(E_1) < \varepsilon.$$

In the opposite direction, suppose that $E$ is a bounded set and that the boundary of $E$ has content 0. Let $\varepsilon > 0$. Then there are closed rectangles $\{R_k\}_{k=1}^n$ such that

$$\partial E \subset \bigcup_{k=1}^n R_k \quad \text{and} \quad \sum_{k=1}^n \operatorname{vol}(R_k) < \varepsilon.$$

Since $E$ is bounded it is contained in a rectangle $R$ and we may assume that so are the rectangles $\{R_k\}_{k=1}^n$. Using the partition points of these rectangles we obtain a partition of $R$. Let $\{R'_k\}_{k=1}^m$ be the rectangles obtained this way that is contained in the interior of $E$ (Figure 4.3). Then

$$\left(\bigcup_{k=1}^m R'_k\right) \subset E \subset \left(\bigcup_{k=1}^m R'_k\right) \cup \left(\bigcup_{k=1}^n R_k\right),$$

so

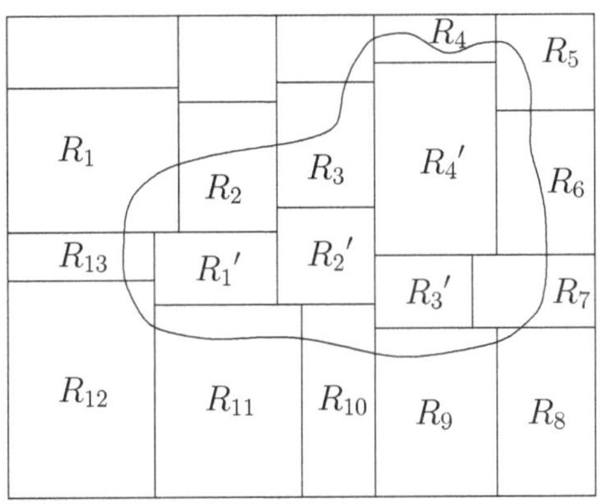


Figure 4.3: $\cup_{k=1}^4 R'_k \subset E \subset (\cup_{k=1}^4 R'_k) \cup (\cup_{k=1}^{12} R_k)$

$$\sum_{k=1}^m \operatorname{vol}(R'_k) \le c_*(E) \le c^*(E) \le \sum_{k=1}^m \operatorname{vol}(R'_k) + \sum_{k=1}^n \operatorname{vol}(R_k).$$

It follows that

$$c^*(E) - c_*(E) \le \sum_{k=1}^n \operatorname{vol}(R_k) < \varepsilon,$$

and the theorem is proved. □

We see that the content is a function $c : \mathcal{J} \to \mathbb{R}$. It turns out that Theorem 4.1.5 remains true when $\mathcal{E}$ is replaced by $\mathcal{J}$. Since $\mathcal{E} \subset \mathcal{J}$, a more precise statement is that the function $c$ can be extended to $\mathcal{J}$ while preserving the listed properties.

**Theorem 4.1.10.** *Jordan content has the following properties:*

(i) ***Finite additivity:*** *For any disjoint collection* $\{E_k\}_{k=1}^m \subset \mathcal{J}$,

$$c\left(\bigcup_{k=1}^m E_k\right) = \sum_{k=1}^m c(E_k).$$

(ii) ***Subadditivity:*** *For any collection* $\{E_k\}_{k=1}^m \subset \mathcal{J}$,

$$c\left(\bigcup_{k=1}^m E_k\right) \leq \sum_{k=1}^m c(E_k).$$

(iii) ***Translation invariance:*** *For any* $y \in \mathbb{R}^2$ *and any* $E \in \mathcal{J}$, $E+y \in \mathcal{J}$ *and*

$$c(E + y) = c(E).$$

*Proof.* In order to establish additivity, it suffices to consider the case $m = 2$. (The rest goes by induction.) Given $\varepsilon > 0$ there are elementary sets $E_1', E_1'', E_2', E_2''$ such that, for $k = 1, 2$,

$$E_k' \subset E_k \subset E_k'' \quad \text{and} \quad c(E_k'') - c(E_k') < \frac{\varepsilon}{2}.$$

Clearly,

$$E_1' \cup E_2' \subset E_1 \cup E_2 \subset E_1'' \cup E_2''.$$

Moreover, the sets $E_k'$ are disjoint so, using Theorem 4.1.5, $c(E_1' \cup E_2') = c(E_1') + c(E_2')$ and $c(E_1'' \cup E_2'') \leq c(E_1'') + c(E_2'')$, whence

$$c(E_1'' \cup E_2'') - c(E_1' \cup E_2') \leq c(E_1'') - c(E_1') + c(E_2'') - c(E_2') < \frac{\varepsilon}{2} + \frac{\varepsilon}{2} = \varepsilon.$$

We leave the proof of subadditivity as Exercise 4.1.10.

Finally, to establish translation invariance, let $E \in \mathcal{J}$ and let $\varepsilon > 0$. Then there are $E', E'' \in \mathcal{E}$ such that

$$E' \subset E \subset E'' \quad \text{and} \quad c(E'') - c(E') < \varepsilon.$$

By Theorem 4.1.5, $E'+y, E''+y \in \mathcal{E}$ and $c(E'+y) = c(E')$, $c(E''+y) = c(E'')$. Thus,

$$E' + y \subset E + y \subset E'' + y \quad \text{and} \quad c(E'' + y) - c(E' + y) < \varepsilon,$$

so (iii) holds. □

Theorems 4.1.5 and 4.1.10 show that the notion of Jordan's content is a step in the right direction because it has the features that we expect from a measure. Also, it passes a reliable test: the value of a mathematical concept is measured by how easy it is to obtain results about it.

The proofs of Theorem 4.1.9 and 4.1.10 deserve scrutiny, considering that they both contain another recurring theme: to establish a property of an object, we approximate it by other objects that have this property. For example, to prove the additivity over all Jordan sets, we approximated them by elementary sets, for which additivity is true. In Theorem 4.1.5 and 4.1.9 we went even further when we represented elementary sets as unions of rectangles and the success comes from manipulating the latter.

The collection of all Jordan sets, $\mathcal{J}$, has an interesting algebraic structure.

**Definition 4.1.11.** A nonempty collection $\mathcal{A}$ of subsets of some set $X$ is called a **ring of sets** if it is closed under the set-theoretic operations of finite union and taking differences, i.e., provided that

$$A, B \in \mathcal{A} \quad \Rightarrow \quad A \cup B \in \mathcal{A} \quad \text{and} \quad A \setminus B \in \mathcal{A}$$

*Remark* 4.1.12. It is not hard to see that every ring contains the empty set and that it is closed under the intersection (Exercise 4.1.30).

**Example 4.1.13. A ring of sets.**

On the real line $\mathbb{R}$ the family of sets consisting of the empty set and all finite unions of half-open intervals of the form $(a, b]$, with $a, b \in \mathbb{R}$, $a < b$, is a ring. ♦

*Remark* 4.1.14. Mathematicians are notorious for assigning the same name to a variety of objects. Rings are also used in algebra (with a different meaning), but even in set theory there is a different (non-equivalent definition) where the requirement is to be closed under finite unions and intersections. In this monograph we will have no use for these other concepts, and the ring will always carry the meaning as in Definition 4.1.11.

**Theorem 4.1.15.** *The collection of all Jordan measurable sets is a ring.*

*Proof.* Suppose that $A, B \in \mathcal{J}$. By Theorem 4.1.9, $c(\partial A) = c(\partial B) = 0$. It is not hard to see that $\partial(A \cup B) \subset \partial A \cup \partial B$ (Exercise 4.1.21) which implies, using the subadditivity of content, that $A \cup B \in \mathcal{J}$.

Similarly, in order to prove that $A \setminus B \in \mathcal{J}$ we rely on the inclusion $\partial(A \setminus B) \subset \partial A \cup \partial B$, established in Exercise 4.1.23, and the result follows in the same way. □

While the Jordan measure has some nice properties, there are many sets that are not Jordan measurable. For starters, every set in $\mathcal{J}$ must be bounded, so $\mathbb{R}^2$ is not measurable. Even if we stick to bounded sets there are those that are not measurable.

**Example 4.1.16. A set that is not a Jordan set.**

The set $D = \{(x, y) \in [0, 1] \times [0, 1] : x, y \in \mathbb{Q}\}$ is not a Jordan set.

**Solution.** It is not hard to see that the boundary of $D$ coincides with the square $[0, 1] \times [0, 1]$. Thus, it does not have content 0, hence $D$ is not a Jordan set. ♦

Example 4.1.16 shows that $\mathcal{J}$ is not closed under the formation of countable unions. (Every point of $D$ determines a one-point Jordan set.) This has an unfortunate effect on Riemann integrability. Let us explain the connection.

Recall that a function defined on a bounded set $D \subset \mathbb{R}^2$ is integrable if its extension to any rectangle containing $D$ is integrable. As a consequence, we have the following result.

**Theorem 4.1.17.** *A set $D \subset \mathbb{R}^2$ is a Jordan set if and only if the characteristic function $\chi_D$ is integrable. In that case*

$$c(D) = \iint_{\mathbb{R}^2} \chi_D(x)\, dA.$$

*Proof.* It follows straight from the definition that for any rectangle $R \subset D$,

$$\iint_R \chi_D(x)\, dA = \text{vol}\,(R).$$

Consequently, if $E \in \mathcal{E}$ and $E \subset D$, then

$$\iint_E \chi_D(x)\, dA = c(E).$$

Since the left-hand side represents a lower Darboux sum for $\chi_D$, taking supremum over all elementary sets $E \subset D$, we obtain that the lower Darboux integral of $\chi_D$ equals $c_*(D)$. A similar argument shows that the upper Darboux integral of $\chi_D$ equals $c^*(D)$, from which the assertion of the theorem follows easily. □

From here we deduce a result about Riemann integrability of a function of one variable. Roughly speaking, it says that a function is integrable if and only if the region under the graph is measurable.

**Theorem 4.1.18.** *Let $I$ be an interval and let $f$ be a bounded function on $I$. Then $f$ is Riemann integrable if and only if the set*

$$E = \{(x,t) : x \in I,\ 0 \le t \le f(x) \text{ or } f(x) \le t \le 0\}$$

*is Jordan measurable in $\mathbb{R}^2$, in which case*

$$\int_I f(x)\, dx = c(E). \tag{4.2}$$

*Proof.* Since $f$ is bounded, $E$ is a bounded set. If $A \in \mathcal{E}$ and $A \subset E$, then $c(A)$ is a lower Darboux sum for the integral in (4.2). It follows that $c_*(E)$ equals the lower Darboux integral of $f$ over $R$. A similar argument shows that $c^*(E)$ equals the upper Darboux integral of $f$ over $R$. □

Theorems 4.1.17 and 4.1.18 reveal a very important connection between the lower/upper Darboux sum for a function $f$ and the inner/outer content of the region below the graph of $f$. The ramifications of these two results are enormous. A better understanding of integration (and that is what prompted all this) is inevitably tied with the theory of measure. This was the main reason that the latter flourished, and why we have to become proficient with it before we can talk about integrals. Moreover, the shortcomings of Jordan content translate into flaws of the Riemann integral. Specifically, the class of Jordan sets is not big enough to include many interesting sets. For example, not every open set is Jordan measurable (Exercise 4.1.25), nor is the set of points with rational coordinates inside the square $[0, 1] \times [0, 1]$ (Example 4.1.16).

Therefore, the next natural step is to try to extend the function $c$ to a larger class of sets. A great feature of Jordan's theory is that it offers a path towards a generalization, and we will follow it in the next section.

Everything in this section was in Jordan's 1892 paper [66], although the concept of the content, as well as the inner and outer contents appear in Peano's 1887 book [83]. Jordan was likely familiar with Peano's work, or at least with Tannery's review of the book. Still, Jordan's publications turned out to be much more influential. To his credit, he was the first to define a measurable set and establish its importance. He was also the first to emphasize the relation between the additivity of content and the theory of multiple integrals. By setting his work in the space of $n$-dimensions (Peano dealt only with dimensions 1,2, and 3, and treated them separately), he created a solid foundation for further generalizations.

CAMILLE JORDAN (1838–1922) was a French mathematician, born and raised in Lyon, in a prominent family. He studied at Ecole Polytechnique where he graduated in 1861, with a doctoral dissertation which had algebra in the first part and analysis in the second part. He spent 15 years working as an engineer on jobs around France, doing mathematical research in his spare time. In 1876 he became a professor of analysis at his alma mater. Surprisingly, he did more research as an engineer. Jordan's contributions to mathematics are numerous. He introduced the concepts of homotopy of paths and of groups. On his own he developed the theory of finite groups. He discovered the *Jordan form* of a matrix. (However, he is not the Jordan of the Gauss–Jordan Elimination.) His study of transformation groups led to the theory of Lie groups. His masterpiece is a 3-volume book *Cours d'Analyse*, the first edition coming from 1882 to 1887. In this text he introduced functions of bounded variations and he proved the *Jordan curve theorem* (a simple closed curve divides the plane into exactly 2 regions). He was the editor of the *Journal de Mathématiques Pure et Appliquées* for 35 years until his death. He retired from the university positions in 1912. In WWI he lost 3 sons. He was in high esteem: a member of the "Académie des Sciences" and an officer of the "Légion d'Honneur", he was also the Honorary President of the International Congress of Mathematicians at Strasbourg in 1920.

## Exercises

4.1.1. Let $\{R_k\}_{k=1}^{\infty}$ be rectangles and let $\varepsilon > 0$. Prove that there exist open rectangles $\{R'_k\}_{k=1}^{\infty}$ such that $R_k \subset R'_k$, for all $k \in \mathbb{N}$, and such that $\sum_{k=1}^{\infty} \operatorname{vol}(R'_k) \le \sum_{k=1}^{\infty} \operatorname{vol}(R_k) + \varepsilon$.

4.1.2. Let $\{R_k\}_{k=1}^{\infty}$ be rectangles and let $\varepsilon > 0$. Prove that there exist closed rectangles $\{R'_k\}_{k=1}^{\infty}$ such that $R'_k \subset R_k$, for all $k \in \mathbb{N}$, and such that $\sum_{k=1}^{\infty} \operatorname{vol}(R_k) \le \sum_{k=1}^{\infty} \operatorname{vol}(R'_k) + \varepsilon$.

4.1.3. Prove Theorem 4.1.5.

4.1.4. Prove that if $A \subset B$ then $c^*(A) \le c^*(B)$ and $c_*(A) \le c_*(B)$.

4.1.5. Prove that $c^*(E) = c^*(\overline{E})$ for every set $E$.

4.1.6. Prove that $c_*(E) = c_*(\operatorname{int} E)$ for every set $E$.

4.1.7. Prove that $c^*(E) = \inf\{c^*(G) : G \text{ is open and } E \subset G\}$.

4.1.8. Prove Theorem 4.1.8.

4.1.9. Let $A \subset \mathbb{R}^2$ and $y \in \mathbb{R}^2$. Prove that $c^*(A+y) = c^*(A)$ and $c_*(A+y) = c_*(A)$.

4.1.10. Prove that if $\{E_k\}_{k=1}^m \subset \mathcal{J}$ then $c(\cup_{k=1}^m E_k) \leq \sum_{k=1}^m c(E_k)$.

4.1.11. Prove that the outer content $c^*$ is subadditive and that the inner content $c_*$ is superadditive.

4.1.12. Prove that if $E = E_1 \cup E_2$ and $d(E_1, E_2) > 0$, then $c^*(E) = c^*(E_1) + c^*(E_2)$.

4.1.13. Prove that there are bounded sets $A, B \subset \mathbb{R}^2$ such that $c_*(B) = 0$ and $c_*(A \cup B) \neq c_*(A)$.

4.1.14. Let $E$ be a set of Jordan content 0 and $A \subset E$. Prove that $A$ is Jordan measurable and $c(A) = 0$.

4.1.15. Let $\nu : \mathcal{J} \to \mathbb{R}$ be a map that is non-negative, finitely additive, and translation invariant. Show that there exists a constant $\alpha \geq 0$ such that $\nu(E) = \alpha c(E)$ for all Jordan measurable sets $E$.

4.1.16. Let $A$ be a line segment in the $xy$-plane. Prove that $A$ has content 0.

4.1.17. Let $f : [a, b] \to \mathbb{R}$ be a continuous function. Show that the graph $\{(x, f(x)) : x \in [a, b]\}$ is Jordan measurable in $\mathbb{R}^2$ with Jordan content zero.

4.1.18. Let $f : [a, b] \to \mathbb{R}$ be a continuous function. Show that the set $\{(x, t) : x \in [a, b], 0 \leq t \leq f(x)\}$ is Jordan measurable in $\mathbb{R}^2$.

4.1.19. Let $A, B, C$ be three points in $\mathbb{R}^2$. Show that the solid triangle with vertices $A, B, C$ is Jordan measurable.

4.1.20. Show that every compact polygon in $\mathbb{R}^2$ is Jordan measurable.

4.1.21. Prove that $\partial(A \cup B) \subset \partial A \cup \partial B$.

4.1.22. Prove that $\partial S = \overline{S} \cap \overline{S^C}$.

4.1.23. Use Exercise 4.1.22 to prove that $\partial(A \setminus B) \subset \partial A \cup \partial B$.

4.1.24.* Let $A$ and $B$ be sets such that their closures $\overline{A}, \overline{B}$ are Jordan sets. Prove that $\overline{A \cap B}$ is a Jordan set.

4.1.25.* Give an example of a bounded open set in $\mathbb{R}^2$ that is not a Jordan set.

4.1.26.** Prove or disprove: Every closed and bounded set in $\mathbb{R}^2$ is a Jordan set.

4.1.27. Prove or disprove: a Jordan set in $\mathbb{R}^2$ has content 0 if and only if it is nowhere dense.

4.1.28. Let $f$ be a bounded function on a Jordan set $D$, let $A \subset D$ be the set of all the points on which $f$ fails to be continuous, and suppose that $A$ has content 0. Prove that $f$ is integrable on $D$.

4.1.29. Prove that the set $\mathcal{E}$ is a ring.

4.1.30. Prove that every ring contains the empty set and that it is closed under intersection.

4.1.31. Give an example to show that a collection of sets that is closed under union and intersection need not be a ring.

4.1.32. Let $\mathfrak{R}$ be a ring of subsets of $X$ and let $\mathfrak{S}$ be a ring of subsets of $Y$. Determine whether the set

$$\{A \times B : A \in \mathfrak{R}, B \in \mathfrak{S}\}$$

is a ring.

4.1.33. Let $f : X \to Y$ and let $\mathfrak{R}$ be a ring of subsets of $X$. Determine whether the set $\{f(A) : A \in \mathfrak{R}\}$ is a ring.

4.1.34. Let $f : X \to Y$ and let $\mathfrak{S}$ be a ring of subsets of $Y$. Determine whether the set $\{f^{-1}(B) : B \in \mathfrak{S}\}$ is a ring.

## 4.2 Lebesgue Measure

We have seen that the definition of Jordan content applies to a very restrictive class of sets. For example, there are open sets that are not Jordan measurable (Exercise 4.1.25). The blame can be assigned to the absence of countable additivity. In this section we will demonstrate how its presence changes things and expands the class of measurable sets. In particular, it will include open and closed sets.

We start with open sets. By Theorem 3.1.3, every open set in $\mathbb{R}^2$ is a countable union of rectangles with disjoint interiors. In view of our commitment to include countable unions in the definition of content, we can enlarge the class of measurable sets using the following definition. Since this is a generalization of content we need a new name and we will call it measure.

**Definition 4.2.1.** If $G$ is an open set in $\mathbb{R}^2$ and $\{R_i\}_{i\in I}$ is a finite or countable collection of rectangles with disjoint interiors such that $G = \cup_{i\in I} R_i$, we define the **measure of an open set** $G$ by

$$m(G) = \sum_{i\in I} \operatorname{vol} R_i. \tag{4.3}$$

For a set $E \in \mathcal{J}$ that is not open we define $m(E) = c(E)$.

*Remark* 4.2.2. Since representation (4.3) is not unique, it is necessary to verify that two different representations yield the same value for $m(G)$. We leave it as Exercise 4.2.1.

Now, we have the situation as in Figure 4.4, with $m$ defined for open sets and $c$ for Jordan sets. What if a set in both classes? The answer is furnished by the following proposition, and we leave the proof as Exercise 4.2.2.

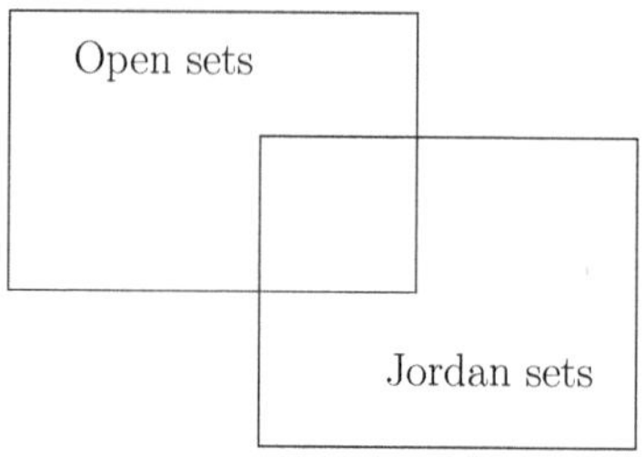


Figure 4.4: Relationship between Jordan sets and open sets

**Proposition 4.2.3.** *If $G$ is an open set in $\mathcal{J}$ then $m(G) = c(G)$.*

Thus, $m$ is an extension of $c$ and it is reasonable to ask which properties of $c$ carry over to $m$. The following propositions provide a partial answer. We leave their proofs to the reader. (See Exercises 4.2.3–4.2.4.)

**Proposition 4.2.4.** *If $G_1, G_2$ are open sets in $\mathbb{R}^2$ such that $G_1 \subset G_2$ then $m(G_1) \leq m(G_2)$.*

**Proposition 4.2.5.** *If $\{G_i\}_{i\in I}$ is a finite or a countable colection of open sets in $\mathbb{R}^2$ and $G = \cup_{i\in I} G_i$, then*

$$m(G) \leq \sum_{i\in I} m(G_i).$$

*If, in addition, the sets $\{G_i\}_{i\in I}$ are disjoint then $m(G) = \sum_{i\in I} m(G_i)$.*

> Open sets appear in the 1899 article *Sur les fonctions de variables réelles* (On functions of real variables) by René Baire.

Next, we turn our attention to closed bounded sets. If $F$ is such a set, then there is an open rectangle $R$ that contains $F$. It follows that $R \setminus F$ is an open set.

**Definition 4.2.6.** Let $F$ be a closed and bounded set in $\mathbb{R}^2$ and $R$ an open rectangle such that $F \subset R$. We define the **measure of a closed set** $F$ by

$$m(F) = m(R) - m(R \setminus F).$$

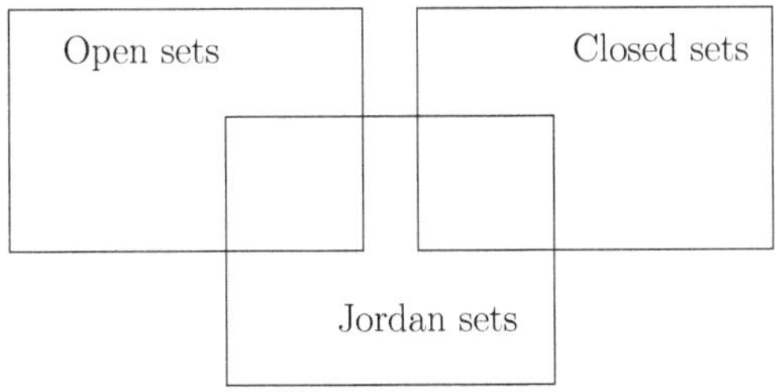


Figure 4.5: Relationship between Jordan sets, open sets, and closed sets

Once again, we leave the task of verifying that the definition is independent of the choice of rectangle $R$ (Exercise 4.2.5), and that for closed Jordan sets $m(F) = c(F)$ (Exercise 4.2.6). The relation between various types of sets is presented in Figure 4.5.

The reader is likely starting to get the flavor of measure theory by now. We started with rectangles, and we are slowly expanding the collection of measurable sets: elementary sets, Jordan sets, open sets, and closed sets. At every stage we need to verify that the new extension of $m$ preserves its important properties. Proofs of the following propositions are left to the reader (Exercises 4.2.7–4.2.8).

**Proposition 4.2.7.** *If $\{F_k\}_{k=1}^n$ is a finite colection of closed rectangles without common interior points and $F = \cup_{k=1}^n F_k$, then $m(F) = \sum_{k=1}^n m(F_k)$.*

**Proposition 4.2.8.** *If $F$ is a closed set and $G$ an open bounded set such that $F \subset G$, then $m(F) \leq m(G)$.*

> The closed sets ("abgeschlossen" in German) were introduced by Cantor in 1883 in [19] through his work on the Continuum Hypothesis. Namely, he proved that no closed set can have a cardinal number strictly between the cardinal numbers of $\mathbb{N}$ and $\mathbb{R}$.

It is not hard to see that the area of the open square $(0,1) \times (0,1)$ is the supremum of the areas of closed squares $[1/n, 1-1/n] \times [1/n, 1-1/n]$. It turns out that a similar approximation is available for any bounded open set in $\mathbb{R}^2$.

**Theorem 4.2.9.** *The measure of an open bounded set $G \subset \mathbb{R}^2$ is the supremum of the measures of all closed sets contained in $G$.*

*Proof.* Proposition 4.2.8 shows that if $F$ is a closed set contained in $G$, then $m(F) \leq m(G)$. Thus, $m(G)$ is an upper bound and the task is to establish that it is the *least* upper bound.

Let $\varepsilon > 0$. By Theorem 3.1.3, $G$ is a union of closed rectangles $\{R_n\}$ with disjoint interiors. By Definition 4.2.1, $m(G) = \sum \operatorname{vol} R_n$. There exists $N \in \mathbb{N}$ for which

$$\sum_{n=1}^{N} \operatorname{vol} R_n > m(G) - \varepsilon.$$

Since $F := \cup_{n=1}^{N} R_n$ is a closed set contained in $G$, we have that $m(G) - \varepsilon < m(F) \le m(G)$, and the theorem is proved. □

In a similar way one can prove the following result. We leave this task to the reader (Exercise 4.2.9).

**Theorem 4.2.10.** *The measure of a closed bounded set $F \subset \mathbb{R}^2$ is the infimum of the measures of all open sets that contain $F$.*

In order to further expand the collection of measurable sets, we will use a procedure similar to the development of Jordan content. Recall that Definition 4.1.7 allowed us to extend the notion of content from the class of elementary sets to Jordan sets. In the following definition, instead of elementary sets we use open and closed sets.

**Definition 4.2.11.** Let $A$ be a bounded subset of $\mathbb{R}^2$. The **(Lebesgue) outer measure** of $A$ is

$$m^*(A) := \inf\{m(G) : G \text{ is an open set and } A \subset G\}.$$

The **(Lebesgue) inner measure** of $A$ is

$$m_*(A) := \sup\{m(F) : F \text{ is a compact set and } F \subset A\}.$$

The set $A$ is **(Lebesgue) measurable** if $m^*(A) = m_*(A)$, in which case we denote this number (or $+\infty$) by $m(A)$ and call it the **(Lebesgue) measure** of $A$. If $A$ is unbounded it is measurable if $A \cap R$ is measurable for every rectangle $R$. The collection of Lebesgue measurable subsets of $\mathbb{R}^2$ will be denoted by $\mathcal{M}$, or $\mathcal{M}_2$ if we want to emphasize that these are subsets of $\mathbb{R}^2$.

Henri Lebesgue

HENRI LEBESGUE (1875–1941) was a French mathematician. He attended the Ecole Normale Supérieure in Paris and received a teaching diploma in mathematics in 1897. He spent the next 2 years reading papers by Baire on discontinuous functions, which served as an inspiration for his most important work. From 1899 to 1902 he taught in a high school in Nancy. His 1901 paper gave the definition of the integral that could be applied to a much larger class of functions than the Riemann integral. In particular, many discontinuous functions became integrable. His 1902 dissertation, with Borel as his doctoral advisor at the Sorbonne, is considered to be one of the finest ever written by a mathematician. It contains a full account of the theory of measure and the Lebesgue integral. In 1903, he published a paper where he applied his new integral to trigonometric series. It contains the Riemann–Lebesgue Lemma together with several results which are true for the Lebesgue integral but not for the Riemann integral. Upon graduating Lebesgue taught at the University of Rennes, the University of Poitiers, the Sorbonne and, from 1921, at the Collège de France, where he lectured and did research for the rest of his life. He made major contributions in other areas of mathematics, including topology. By 1922 he had written nearly 90 books and papers. After 1922 he remained active, but his contributions were directed towards pedagogical issues, historical work, and elementary geometry. He received a number of prizes and he was honored with election to many academies, including the Académie des Sciences, and the Royal Society.

It is a good idea to pause here and compare Definitions 4.1.7 and 4.2.11. The essential difference between them is the use of finite versus countable additivity. Namely, for Jordan content the supremum and infimum are taken over elementary sets which are *finite* unions of rectangles, whereas the Lebesgue measure is built through the use of open sets which are *countable unions* of rectangles. (Closed sets need not be, but their measure is defined via open sets.) The difference may not seem like much, but it will have a "butterfly effect" on the whole theory of integration.

It is not hard to prove the following fact (Exercise 4.2.10).

**Proposition 4.2.12.** *Let* $E \subset \mathbb{R}^2$*. Then*

$$c_*(E) \leq m_*(E) \leq m^*(E) \leq c^*(E).$$

From Proposition 4.2.12 we get two easy corollaries.

**Corollary 4.2.13.** *If a set* $A$ *has outer measure zero, then it is Lebesgue measurable and* $m(A) = 0$.

**Corollary 4.2.14.** *If a bounded set* $A$ *has Jordan content, then it is Lebesgue measurable and* $c(A) = m(A)$.

The converse to Corollary 4.2.14 is not true.

**Example 4.2.15. A Lebesgue measurable set that is not Jordan measurable.**

Let $D = \{(x, y) \in [0,1] \times [0,1] : x, y \in \mathbb{Q}\}$. Then $D$ is not a Jordan set but $m(D) = 0$.

**Solution:** We know from Example 4.1.16 that $D$ is not a Jordan set. Let $\varepsilon > 0$. Since the set $D$ is countable it is possible to arrange its elements in a sequence $\{(a_n, b_n)\}$. Let us define

$$R_n = \left(a_n - \frac{\sqrt{\varepsilon}}{2^{n+1}}, a_n + \frac{\sqrt{\varepsilon}}{2^{n+1}}\right) \times \left(b_n - \frac{\sqrt{\varepsilon}}{2^{n+1}}, b_n + \frac{\sqrt{\varepsilon}}{2^{n+1}}\right).$$

Then $D \subset \cup_{n=1}^{\infty} R_n$ and $\sum_{n=1}^{\infty} \operatorname{vol}(R_n) = \varepsilon/3$. Taking the infimum over all $\varepsilon > 0$ we obtain that $m^*(D) = 0$. By Corollary 4.2.13, $D$ is measurable and $m(D) = 0$. ♦

Example 4.2.15 gives a blueprint how to prove the following result. (See Exercise 4.2.14.)

**Proposition 4.2.16.** *Every countable set is measurable and it has measure 0.*

The converse is not true (see Exercise 4.1.17).

> The fact that every countable set has an outer measure zero was known to Harnack. Actually, when he considered infinite covers, he was discouraged by the conclusion that a topologically large set (like rationals—dense in $\mathbb{R}$) could be small in measure-theoretic way. It took Lebesgue, some 20 years later, to overcome this philosophical obstruction.

Now that we have a collection $\mathcal{M}$ and a function $m : \mathcal{M} \to [0, +\infty]$ our goal is to prove that they satisfy the conditions in our wish list. We start with some easy results that we leave to the reader to prove (Exercise 4.2.16).

**Proposition 4.2.17.** *Let $A, F, G, E \subset \mathbb{R}^2$.*

*(a) If $G$ is open then $m_*(G) = m^*(G) = m(G)$.*

*(b) If $F$ is closed then $m_*(F) = m^*(F) = m(F)$.*

*(c)* ***Monotonicity:*** *If $A \subset E$ then $m_*(A) \leq m_*(E)$ and $m^*(A) \leq m^*(E)$.*

It follows from Theorem 4.1.8 and Corollary 4.2.14 that the Lebesgue measure satisfies condition (i) of the wish list. Next, we turn our attention to property (ii), i.e., we will show that $m$ is additive. In that direction, our first step is to establish that the outer measure $m^*$ is subadditive.

**Theorem 4.2.18.** ***Subadditivity of outer measure.*** *If a set $E \subset \mathbb{R}^2$ is a finite or a countable union of the sets $\{E_i\}_{i\in I}$, $E = \cup_{i\in I} E_i$, then*

$$m^*(E) \leq \sum_{i\in I} m^*(E_i). \tag{4.4}$$

*Proof.* If at least one of the sets $E_i$ has infinite outer measure there is nothing to prove, so we assume that $m^*(E_i) < \infty$, for all $i \in I$. Let $\varepsilon > 0$. For every $i \in I$, let $G_i$ be an open set such that

$$E_i \subset G_i \quad \text{and} \quad m(G_i) < m^*(E_i) + 2^{-i}\varepsilon.$$

It follows that $E \subset \cup_{i\in I} G_i$, so the monotonicity of outer measure combined with Proposition 4.2.5 implies that

$$m^*(E) \leq \sum_{i\in I} m(G_i) < \sum_{i\in I} \left(m^*(E_i) + 2^{-i}\varepsilon\right) \leq \sum_{i\in I} m^*(E_i) + \varepsilon.$$

Since $\varepsilon$ is arbitrary, we obtain (4.4). □

Perhaps not surprisingly, the inner measure is not subadditive but superadditive, at least when it comes to *disjoint* unions. We leave the proof as Exercise 4.2.17.

**Theorem 4.2.19.** ***Superadditivity of inner measure.*** *If a set $E \subset \mathbb{R}^2$ is a finite or a countable disjoint union of the sets $\{E_i\}_{i\in I}$, $E = \cup_{i\in I} E_i$, then*

$$m_*(E) \geq \sum_{i\in I} m_*(E_i).$$

From Theorems 4.2.18 and 4.2.19 we get an easy consequence: Lebesgue measure $m$ satisfies condition (ii) on the wish list. More precisely, we have

**Corollary 4.2.20.** *If a set $E \subset \mathbb{R}^2$ is a finite or a countable disjoint union of measurable sets $\{E_i\}_{i\in I}$, $E = \cup_{i\in I} E_i$, then $E$ is measurable and*

$$m(E) = \sum_{i\in I} m(E_i).$$

We leave the proof to the reader (Exercise 4.2.26).

*Remark* 4.2.21. When the index set $I$ is countable, we refer to this property of measure as **$\sigma$-additivity**.

Corollary 4.2.20 is a huge accomplishment of Lebesgue theory. It validates his definition of measure by showing that it is $\sigma$-additive.

Another important result gives a connection between the inner and the outer measure.

**Theorem 4.2.22.** *Let $E$ be a bounded set in $\mathbb{R}^2$ and let $R$ be an open rectangle containing $E$. Then,*

$$m^*(E) + m_*(R \setminus E) = m(R).$$

*Proof.* First we will prove the inequality $m^*(E) + m_*(R \setminus E) \leq m(R)$. Let $\varepsilon > 0$. Let $F$ be a closed subset of $R \setminus E$ (Figure 4.6) such that

$$m(F) > m_*(R \setminus E) - \varepsilon.$$

Notice that $R \setminus F$ is an open set that contains $E$, so using Corollary 4.2.20,

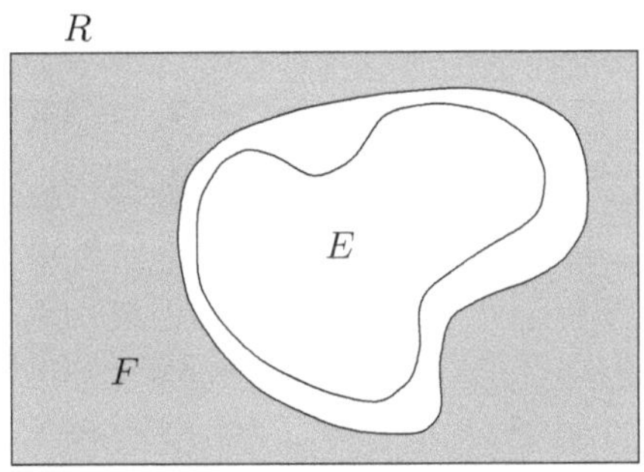


Figure 4.6: $F \subset R \setminus E$ is closed and $m(F) > m_*(R \setminus E) - \varepsilon$

$$m_*(E) \leq m(R \setminus F) = m(R) - m(F) < m(R) - m_*(R \setminus E) + \varepsilon.$$

Since $\varepsilon$ was arbitrary we obtain that

$$m^*(E) + m_*(R \setminus E) \leq m(R).$$

To prove the opposite inequality, once again let $\varepsilon > 0$ and let $G_0$ be a bounded open set such that

$$E \subset G_0 \quad \text{and} \quad m(G_0) < m^*(E) + \frac{\varepsilon}{2}.$$

(See Figure 4.7.) Let $R = (a, b) \times (c, d)$. We select $\eta = \varepsilon/[4(b - a + d - c)]$ and we define

$$A = a + \eta, \quad B = b - \eta, \quad C = c + \eta, \quad D = d - \eta.$$

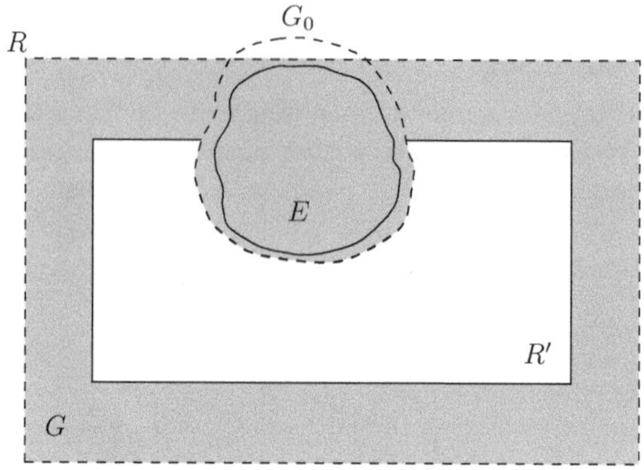


Figure 4.7: $G_0 \supset E$ is a bounded open set and $m(G_0) < m^*(E) + \varepsilon/2$

Let $R' = [A, B] \times [C, D]$ and let

$$G = (G_0 \cap R) \cup (R \setminus R').$$

It is not hard to see that $G$ is an open set that contains $E$ and that

$$m(G) \leq m^*(E) + \varepsilon.$$

Indeed,

$$\begin{aligned} m(G) &\leq m(G_0 \cap R) + m(R \setminus R') \leq m(G_0) + 2\eta(b - a + d - c) - 4\eta^2 \\ &< m^*(E) + \frac{\varepsilon}{2} + \frac{\varepsilon}{2}. \end{aligned}$$

Moreover, it is an exercise in set algebra to verify that $R \setminus G = R' \setminus G$, hence a closed set $F$. Further, $F \subset R \setminus E$, so

$$m_*(R \setminus E) \geq m(F) = m(R) - m(G) \geq m(R) - m^*(E) - \varepsilon.$$

Since $\varepsilon$ was arbitrary we obtain that

$$m^*(E) + m_*(R \setminus E) \geq m(R)$$

and the proof is complete. □

**Corollary 4.2.23.** *A set $E$ is measurable if and only if its complement is measurable.*

We leave the proof as Exercise 4.2.29.

The material in this section is taken from Lebesgue's 1902 dissertation [68]. As mentioned earlier, he followed the lead of his teacher Borel who created a theory of measure using an axiomatic approach. In his 1898 book [14], Borel compiled a list of properties that a measure should possess, one of them being the countable additivity. Nevertheless, Lebesgue's approach is much more precise and general, and, as we will see later, he found a way to apply the theory of measure to integration theory, something that was missing in Borel's monograph.

## Exercises

4.2.1.* Prove that the measure of an open set is independent of its representation as a disjoint union of rectangles.

4.2.2. Prove Proposition 4.2.3.

4.2.3. Prove Proposition 4.2.4.

4.2.4. Prove Proposition 4.2.5.

4.2.5.* Prove that the measure of a closed set is independent of the choice of an open rectangle that contains it.

4.2.6. Let $F$ be a closed Jordan set. Using only Definition 4.2.6, prove that $m(F) = c(F)$.

4.2.7. Prove Proposition 4.2.7.

4.2.8. Prove Proposition 4.2.8.

4.2.9. Prove Theorem 4.2.10.

4.2.10. Prove Proposition 4.2.12.

4.2.11. Prove that the empty set is measurable and that $m(\emptyset) = 0$.

4.2.12. Let $E$ be an elementary set in $\mathbb{R}^2$. Prove or disprove: $m_*(E) = c(E)$. What if $E$ is a Jordan measurable set?

4.2.13.* Let $E \subset \mathbb{R}^2$ be compact. Prove that $m(E) = c^*(E)$.

4.2.14. Prove Proposition 4.2.16.

4.2.15. Give an example of an unbounded set with Lebesgue outer measure 0.

4.2.16. Prove Proposition 4.2.17.

4.2.17. Prove Theorem 4.2.19.

4.2.18. An interval $I$ is unbounded if it is of the form $(-\infty, b)$, $(-\infty, b]$, $(a, +\infty)$, $[a, +\infty)$, or $(-\infty, \infty)$. A rectangle $R = I_1 \times I_2$ is unbounded if at least one of the intervals $I_1, I_2$ is unbounded. Prove that, if $R$ is an unbounded rectangle, then $m^*(R) = \mathrm{vol}(R) = \infty$.

4.2.19. Prove that if $m^*(A) = 0$, then $m^*(A \cup B) = m^*(B)$.

4.2.20. Prove that $|m^*(A) - m^*(B)| \leq m^*(A \triangle B)$ for any $A, B \subset \mathbb{R}^2$, provided that at least one of the numbers $m^*(A), m^*(B)$ is finite.

4.2.21. Let $A \sim B$ if and only if $m^*(A \triangle B) = 0$. Prove that:

(a) $\sim$ is an equivalence relation;

(b) Prove that $d([A],[B]) = m^*(A \triangle B)$ defines a metric on the set of equivalence classes in $\mathbb{P}(\mathbb{R}^2)$.

4.2.22.* **Nikodym (1929).** Prove that the metric space in Exercise 4.2.21 is complete.

4.2.23. Let $A$ be the set of points in $[0,1] \times [0,1]$ with both coordinates irrational. Prove that $m^*(A) = 1$.

4.2.24.* Let $A$ and $B$ be bounded subsets of $\mathbb{R}^2$ for which there is $\alpha > 0$ such that $\|a - b\| \geq \alpha$ for all $a \in A$, $b \in B$. Prove that $m^*(A \cup B) = m^*(A) + m^*(B)$.

4.2.25. Let $E \subset \mathbb{R}^2$ be a countable union of rectangles $\{R_k\}_{k=1}^{\infty}$ with disjoint interiors. Prove that $m(E) = \sum_{k=1}^{\infty} m(R_k)$.

4.2.26. Prove Corollary 4.2.20.

4.2.27.* Let $E \subset \mathbb{R}^2$ and let $\{R_k\}_{k=1}^{n}$ be disjoint open rectangles. Prove that:

(a) $\sum_{k=1}^{n} m_*(E \cap R_k) = m_*(E \cap (\cup_{k=1}^{n} R_k))$.

(b) $\sum_{k=1}^{n} m^*(E \cap R_k) = m^*(E \cap (\cup_{k=1}^{n} R_k))$.

4.2.28.* Give examples to show that each of the inequalities $c_*(A) \leq m_*(A)$ and $m^*(A) \leq c^*(A)$ can be strict.

4.2.29. Prove Corollary 4.2.23.

4.2.30.* Let $E \subset \mathbb{R}^2$ and let $A \in \mathcal{E}$. Prove that $c(A) = m_*(A \cap E) + m^*(A \cap E^C)$.

4.2.31.* Let $E \subset [0,1] \times [0,1]$ with the property that if $F$ is a closed measurable set of positive measure then $E \cap F \neq \emptyset$. Prove that $m^*(E) = 1$.

4.2.32. Let $\mathcal{F}$ be the collection of all closed subsets of $\mathbb{R}^2$, and let $\mathcal{F}' \subset \mathcal{F}$ consists of sets of positive measure. Prove that $\mathcal{F}'$ has the same cardinality as $\mathbb{R}$.

4.2.33. Let $E$ be a set in $\mathbb{R}^2$ for which $m^*(E) > 0$ and let $0 < \alpha < m^*(E)$. Prove that there is a compact set $K \subset E$ such that $m(K) = \alpha$.

4.2.34. Let $E$ be a subset of $\mathbb{R}^2$. Prove or disprove: $\operatorname{Int}(E) = \emptyset$ if and only if $m^*(E) = 0$.

4.2.35. Let $A \subset \mathbb{R}^2$ intersect every compact set in $\mathbb{R}^2$. Prove that $m^*(A) = \infty$.

4.2.36. Let $A \subset \mathbb{R}^2$ be a bounded set with the property that

$$\inf\{m(G) : A \subset G,\ G \text{ open in } \mathbb{R}^2\} = \sup\{m(F) : F \subset A,\ F \text{ closed in } \mathbb{R}^2\}.$$

Prove or disprove: $A$ is measurable.

4.2.37. Find the Lebesgue measure of a subset of $[0,1] \times [0,1]$ consisting of points $(x,y)$ such that $|\sin x| < 1/2$ and $\cos(x+y) \in \mathbb{Q}$.

## 4.3 The $\sigma$-Algebra of Lebesgue Measurable Sets

In this section, we will show that the collection $\mathcal{M}$ is closed under the standard set-theoretic operations. We have already established that it is closed under complements (Corollary 4.2.23), and under disjoint unions (Corollary 4.2.20). What remains is to prove that it is closed under unions even when the sets are not disjoint. We start with finite unions.

**Theorem 4.3.1.** *A finite union of measurable sets is a measurable set.*

*Proof.* By definition of measurability for unbounded sets, it suffices to prove the case when a set $E \subset \mathbb{R}^2$ is bounded. Let $\{E_k\}_{k=1}^n$ be measurable sets such that

$$E = \bigcup_{k=1}^{n} E_k.$$

Let $\varepsilon > 0$ and for every $1 \le k \le n$ choose an open bounded set $G_k$ and a closed set $F_k$ such that

$$F_k \subset E_k \subset G_k \quad \text{and} \quad m(G_k) - m(F_k) < \frac{\varepsilon}{n}.$$

Define $F = \cup_{k=1}^n F_k$ and $G = \cup_{k=1}^n G_k$. Then $F$ is a closed set, $G$ is a bounded open set, and $F \subset E \subset G$, whence

$$m(F) \le m_*(E) \le m^*(E) \le m(G). \tag{4.5}$$

The set $G \setminus F$ is open, hence measurable. By Corollary 4.2.20, $m(F) + m(G \setminus F) = m(G)$ and $m(F_k) + m(G_k \setminus F_k) = m(G_k)$, for $1 \le k \le n$. It follows that

$$m(G \setminus F) = m(G) - m(F) \quad \text{and} \quad m(G_k \setminus F_k) = m(G_k) - m(F_k). \tag{4.6}$$

It is not hard to verify that

$$G \setminus F \subset \bigcup_{k=1}^{n} (G_k \setminus F_k). \tag{4.7}$$

Finally, combining (4.6) and (4.7),

$$m(G) - m(F) \le \sum_{k=1}^{n} (m(G_k) - m(F_k)) < \varepsilon.$$

Now the result follows from (4.5). □

Next, we look at finite intersections. We get the result by combining De Morgan's Laws and Corollary 4.2.23 with Theorem 4.3.1.

**Theorem 4.3.2.** *A finite intersection of measurable sets is a measurable set.*

From Theorem 4.3.2 and Corollary 4.2.23 we obtain

**Theorem 4.3.3.** *A difference of measurable sets is a measurable set.*

Now we are ready to prove a generalization of Theorem 4.3.1.

**Theorem 4.3.4.** *A countable union of measurable sets is a measurable set.*

*Proof.* Once again, it suffices to consider the case when $E$ is a bounded set in $\mathbb{R}^2$. Let $\{E_k\}_{k=1}^{\infty}$ be measurable sets such that

$$E = \bigcup_{k=1}^{\infty} E_k.$$

We define a sequence $\{A_k\}_{k=1}^{\infty}$ by $A_1 = E_1$ and, for $k \geq 2$,

$$A_k = E_k \setminus (E_1 \cup E_2 \cup \cdots \cup E_{k-1}).$$

(See Figure 4.8.) It is easy to verify that $\{A_k\}_{k=1}^{\infty}$ is a sequence of pairwise dis-

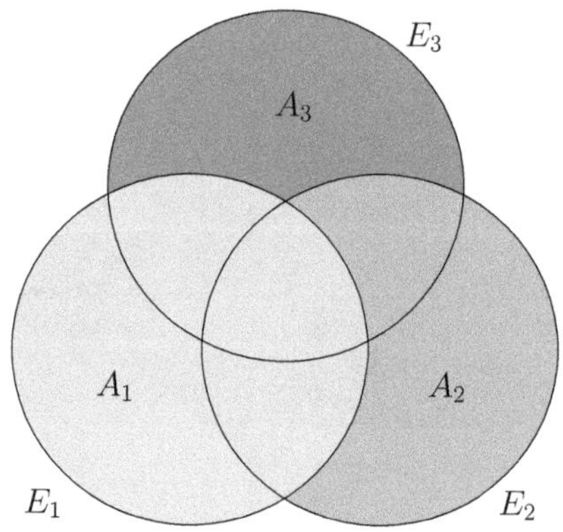


Figure 4.8: The sets $A_1$, $A_2$, and $A_3$

joint sets and that $E = \cup_{k=1}^{\infty} A_k$ (Exercise 1.1.24). Moreover, Theorems 4.3.1 and 4.3.3 show that each of them is measurable. Now the result follows from Corollary 4.2.20. □

In order to describe the structure of $\mathcal{M}$ we introduce the following definition.

**Definition 4.3.5.** A nonempty collection $\mathcal{A}$ of subsets of some set $X$ is called a **$\sigma$-algebra** of sets if it includes $X$ itself and if it is closed under the set-theoretic operations of complement and countable unions, i.e., such that

$$X \in \mathcal{A}, \qquad \{A_n\}_{n=1}^{\infty} \subset \mathcal{A} \Rightarrow \bigcup_{n=1}^{\infty} A_n \in \mathcal{A}, \qquad \text{and} \qquad A \in \mathcal{A} \Rightarrow A^C \in \mathcal{A}.$$

*Remark* 4.3.6. The definition implies that every $\sigma$-algebra also includes the empty set and that it is closed under countable intersections.

We infer from the definition of $\mathcal{M}$, Theorem 4.3.4, and Corollary 4.2.23 that

**Theorem 4.3.7.** *$\mathcal{M}$ is a $\sigma$-algebra.*

It is often the case in mathematics that we are looking for a set of specific natures that contains a given set. For example, the set of natural numbers is not a group with respect to addition, but it is contained in the set $\mathbb{Z}$ which *is* a group. The set of all intervals of the form $(x - a, x + a)$, where $x$ and $a$ are real numbers, is not a topology, but the usual topology of the real line contains all these sets. In the present situation, we started with elementary sets, and $\mathcal{E}$ is not a $\sigma$-algebra, but $\mathcal{M}$ is and it contains $\mathcal{E}$. It is almost always of interest to look for the *smallest* set of required nature that would contain given sets. We would say that such a structure is *generated* by given sets. It is tempting to try to visualize the process of creation. Don't!! At least, when it comes to a $\sigma$-algebra generated by a given collection of sets, there is no reasonable way to accomplish this.

> The expectation that measurable sets form a $\sigma$-algebra is in Borel's 1898 book [14], although the collection of measurables sets was not explicitly mentioned.

Since open and closed sets are measurable and $\mathcal{M}$ is closed under countable intersections and unions we obtain the following consequence of Theorem 4.3.7.

**Proposition 4.3.8.** *Every $G_\delta$ set is measurable. Every $F_\sigma$ set is measurable.*

The following result is a consequence of the definition of outer measure (see Exercise 4.3.3).

**Corollary 4.3.9.** *Let $E \subset \mathbb{R}^2$. There exists a $G_\delta$ set $G \in \mathbb{R}^2$ such that $E \subset G$ and $m(G) = m^*(E)$.*

So far, we have obtained a function $m : \mathcal{M} \to [0, +\infty]$ that satisfies conditions (i) and (ii) from the wish list on page 82. Indeed, (i) follows from the definition of $m$ and (ii) from Corollary 4.2.20. The only thing remaining on our list is translation invariance, so we turn our attention to this property.

**Theorem 4.3.10.** ***Translation invariance.*** *Let $E$ be a measurable set in $\mathbb{R}^2$ and let $a \in \mathbb{R}^2$. Then the set $E' := E+a$ is measurable and $m(E') = m(E)$.*

*Proof.* We will do the proof in stages, by increasing the generality of the set $E$. We start with rectangles. If $R$ is a rectangle in $\mathbb{R}^2$ then $R' := R + a$ is another rectangle, hence measurable, and it is obvious that $m(R') = m(R)$.

Next, let $G$ be an open set in $\mathbb{R}^2$. By Theorem 3.1.3, every open set in $\mathbb{R}^2$ is a countable union of rectangles with disjoint interiors,

$$G = \bigcup_{n=1}^{\infty} R_n.$$

Denoting $R'_n := R_n + a$, we obtain that $G' := G + a = \cup_{n=1}^{\infty} R_n + a = \cup_{n=1}^{\infty} R'_n$, so $G' \in \mathcal{M}$. Since the interiors of rectangles $\{R'_n\}$ are disjoint, Exercise 4.2.25 implies that

$$m(G') = \sum_{n=1}^{\infty} m(R'_n) = \sum_{n=1}^{\infty} m(R_n) = m(G). \tag{4.8}$$

Finally, let $E$ be an arbitrary bounded measurable set in $\mathbb{R}^2$ and let $\varepsilon > 0$. By definition, there is an open bounded set $G$ such that

$$E \subset G \quad \text{and} \quad m(G) < m(E) + \varepsilon.$$

It follows that $E' := E + a \subset G' := G + a$ and $m^*(E') \leq m(G') = m(G) < m(E) + \varepsilon$. Since $\varepsilon$ is arbitrary, we obtain that

$$m^*(E') \leq m(E). \tag{4.9}$$

Notice that translation maps open sets to open sets, hence closed sets to closed sets. Moreover, equality (4.8) shows that $m(F+a) = m(F)$ for any closed and bounded set $F$. Finally, let $F$ be a closed and bounded set such that

$$F \subset E \quad \text{and} \quad m(F) > m(E) - \varepsilon.$$

Then $F' := F + a \subset E' := E + a$ and $m_*(E') \geq m(F') = m(F) > m(E) - \varepsilon$, hence

$$m_*(E') \geq m(E).$$

Combining with (4.9), we get

$$m(E) \leq m_*(E') \leq m^*(E') \leq m(E),$$

so $E'$ is measurable and $m(E') = m(E)$.

The case when $E$ is unbounded can be easily reduced to the finite case by considering an arbitrary rectangle $R$ and the intersection $E \cap R$. □

This was the last piece of the puzzle and we now have a measure $m$ on a $\sigma$-algebra $\mathcal{M}$ with the following properties:

(i) For any rectangle $R$, $m(R) = \text{vol}(R)$.

(ii) **Countable additivity:** For any disjoint collection $\{E_n\}_{n\in\mathbb{N}}$,

$$m\left(\bigcup_{n=1}^{\infty} E_n\right) = \sum_{n=1}^{\infty} m(E_n).$$

(iii) **Translation invariance:** For any $y \in \mathbb{R}^2$ and any $E \subset \mathbb{R}^2$,

$$m(E + y) = m(E).$$

*Remark* 4.3.11. A careful reader has probably noticed that we did not examine whether the inclusion $\mathcal{M} \subset \mathbb{P}(\mathbb{R}^2)$ is proper. We will discuss this issue in Section 5.4.

We use this opportunity to remark that the content of this chapter can be almost verbatim applied to measure theory in other Euclidean spaces. We chose to present it in $\mathbb{R}^2$ because we felt that it gives the best insight into the Lebesgue theory.

In the next chapter, we will see that many of these ideas carry over to a much more general setting. The existence of such a generalization is another triumph of Lebesgue's theory and in particular his method of proof: *prove for a smaller class, extend it to a bigger class, and then to a yet bigger class.* As long as the smallest class has desirable properties and the intermediate classes allow their transference, we will be able to establish a measure theory in a variety of settings.

Everything in this section is in Lebesgue's dissertation [68] except that, just like Borel before him, he did not use words like "ring" or "algebra". Fréchet was the first one to talk about "additive families of sets" in 1915 in [38]. Initially the word "field" was borrowed from algebra (where it was coined by a German mathematician Richard Dedekind (1831–1916) in his lectures in 1858). However, a field of sets need not be closed under intersections. The word "algebra" seems to have been adopted from the related theory of Boolean algebras, and they are closed under intersections. The focus shifted to algebras, as opposed to fields, around 1950. The word "ring" was first used by Hilbert in number theory in 1892 (with the first appearance in a paper 5 years later) although the concept is also due to Dedekind from around 1870. The prefix $\sigma$ appears in the 1914 book of Hausdorff [56], when talking about $\sigma$-rings and $\sigma$-fields, because $\sigma$ is a traditional letter for sums (in this case countable). Dedekind was a student of Gauss, but during his academic career he learned a lot from Dirichlet and Riemann. The work on Dirichlet's notes inspired him to study algebraic number fields, and he introduced the concept of an ideal.

GIUSEPPE PEANO (1858–1932) was an Italian mathematician and linguist. His parents were farmers, and he was born in a farmhouse. His uncle was a priest in Turin and, realizing that Giuseppe was a very talented child, he took him to Turin where he attended high school and later the University of Turin. He received his doctorate in 1880 and joined the university the same year. The first part of his career was brilliant. Already in 1882 he discovered that the definition of the surface area in the standard textbook was incorrect, although Schwarz made the same discovery a year earlier. In 1886 he proved the existence of a solution of $dy/dx = f(x, y)$ when $f$ is continuous. Following the ideas of German mathematician and linguist Hermann Grassmann (1809–1877), he presented the first axiomatic definition of a real linear space in [84]. In that work he also introduced the modern symbols for the union and intersection of sets. Peano continued his work on the axiomatic systems and his axioms for positive integers appeared in 1889, in the book [85]. In 1890 he gave an example of a "space filling curve." Around that time he immersed himself in a new and extremely ambitious project known as *Formulario Mathematico*. The goal was to present the whole mathematics using the notation of mathematical logic. Although five thoroughly revised editions appeared between 1895 and 1908, the effort is nowadays considered misdirected. He also wrote in an international language *Latino sine flexione* (Latin without inflections), which is a simplified version of classical Latin. Most of his books and papers are in this language.

## Exercises

4.3.1. Let $A, B \subset \mathbb{R}^2$, let $A \in \mathcal{M}$, and suppose that $m^*(A \triangle B) = 0$. Prove that $B \in \mathcal{M}$ and $m(B) = m(A)$.

4.3.2. Prove that a set $B \subset \mathbb{R}^2$ is nonmeasurable if and only if there exists $\varepsilon > 0$ such that for every measurable set $A \subset B$, $m^*(B \setminus A) \geq \varepsilon$.

4.3.3. Prove Corollary 4.3.9.

4.3.4. Let $A, B \subset \mathbb{R}^2$ be sets of finite outer measure such that $A \cup B$ is measurable and $m(A \cup B) = m^*(A) + m^*(B)$. Prove that $A, B$ are both measurable sets.

4.3.5. Prove that the Lebesgue outer measure on $\mathbb{R}^2$ is translation invariant.

4.3.6.* Let $A \subset \mathbb{R}^2$ have a finite outer measure. Prove that $A^C - A^C = \mathbb{R}^2$.

4.3.7. The purpose of this exercise is to establish the following result: If $A \subset \mathbb{R}^2$ is a Lebesgue measurable set of positive measures, then the set $A - A := \{x - y : x \in A, y \in A\}$ contains a neighborhood of the origin.

(a) Prove that we may assume that the measure of $A$ is finite.

(b) Prove that $A$ can be covered by a countable union of rectangles $\{R_k\}$ for which

$$\frac{3}{4}\sum_{k=1}^{\infty} \operatorname{vol} R_k < \sum_{k=1}^{\infty} m(A \cap R_k) < \infty.$$

(c) Prove that there exists a rectangle $R_n$ for which

$$m(A \cap R_n) > (3/4)m(R_n).$$

(d) Prove that $[-(1/4)m(R_n), (1/4)m(R_n)] \times [-(1/4)m(R_n), (1/4)m(R_n)]$ contains a point $x$ such that the sets $A \cap R_n$ and $(A \cap R_n) + x$ are not disjoint.

(e) Prove that $[-(1/4)m(R_n), (1/4)m(R_n)] \times [-(1/4)m(R_n), (1/4)m(R_n)]$ is a subset of $A - A$.

4.3.8. Let $\mathfrak{M}$ be a $\sigma$-algebra of subsets of $X$ and let $f : X \to Y$. Prove or disprove: $f(\mathfrak{M})$ is a $\sigma$-algebra.

4.3.9. Let $\mathfrak{M}$ be a $\sigma$-algebra of subsets of $X$ and let $f : X \to Y$. Prove or disprove: $\mathfrak{N} := \{E : f^{-1}(E) \in \mathfrak{M}\}$ is a $\sigma$-algebra.

4.3.10. Let $f : X \to Y$ and let $\mathfrak{N}$ be a $\sigma$-algebra of subsets of $Y$. Prove or disprove: $\mathfrak{M} := \{f^{-1}(E) : E \in \mathfrak{N}\}$ is a $\sigma$-algebra.

4.3.11. Let $X = \cup_{n=1}^{\infty} X_n$ and let $\mathfrak{M}_n$ be a $\sigma$-algebra on $X_n$, for each $n \in \mathbb{N}$. Prove that the set

$$\mathfrak{M} = \{A \subset X : A \cap X_n \in \mathfrak{M}_n \text{ for all } n \in \mathbb{N}\}$$

is a $\sigma$-algebra on $X$.

4.3.12. Let $X$ be an uncountable set and let

$$\mathcal{A} = \{A \subset X : A \text{ is countable or } A^C \text{ is countable}\}.$$

Prove that $\mathcal{A}$ is a $\sigma$-algebra.

4.3.13. Let $\mathcal{A}_1, \mathcal{A}_2$ be $\sigma$-algebras on $X$. Prove or disprove: $\mathcal{A}_1 \cap \mathcal{A}_2$ is a $\sigma$-algebra.

4.3.14. Let $\mathcal{A}_1, \mathcal{A}_2$ be $\sigma$-algebras on $X$. Prove or disprove: $\mathcal{A}_1 \cup \mathcal{A}_2$ is a $\sigma$-algebra.

4.3.15.* Let $\{\mathcal{A}_n\}_{n=1}^{\infty}$ be an increasing sequence of $\sigma$-algebras. Prove or disprove: $\cup_{n=1}^{\infty} \mathcal{A}_n$ is a $\sigma$-algebra.

4.3.16.* Let $X \subset \mathbb{R}^2$ be a countable set and let

$$\mathcal{A} = \{A \subset \mathbb{R}^2 : A \subset X \text{ or } A^C \subset X\}.$$

Prove that $\mathcal{A}$ is the smallest $\sigma$-algebra that contains $\{x\}$ for each $x \in X$.

4.3.17. Let $\mathcal{S}$ be the collection of distinct subsets of $\mathbb{R}$ that are symmetric about the origin. (A set $A$ is symmetric if $x \in A$ if and only if $-x \in A$.) Prove that $\mathcal{S}$ is a $\sigma$-algebra.

4.3.18. If $\mathcal{S}$ is a collection of subsets of $\mathbb{R}^2$, we denote by $\sigma(\mathcal{S})$ the smallest $\sigma$-algebra that contains $\mathcal{S}$.

(a) Prove that if $\mathcal{S} \subset \mathcal{T}$ then $\sigma(\mathcal{S}) \subset \sigma(\mathcal{T})$.

(b) Prove that, if $\mathcal{S}$ and $\mathcal{T}$ are two collections of subsets of $\mathbb{R}$, then $\sigma(\mathcal{S} \cap \mathcal{T}) \subset \sigma(\mathcal{S}) \cap \sigma(\mathcal{T})$.

(c) Give an example for which $\sigma(\mathcal{S} \cap \mathcal{T}) \neq \sigma(\mathcal{S}) \cap \sigma(\mathcal{T})$.

4.3.19. Let $X$ be a set and $\mathcal{A}$ a subset of $\mathbb{P}(X)$. Prove that there is a unique smallest $\sigma$-algebra of subsets of $X$ that contains $\mathcal{A}$.

4.3.20. Let $X$ be a set and $\mathcal{A}$ a $\sigma$-algebra of subsets of $X$. Suppose that $T : X \to X$ is a bijection such that $T(A) \in \mathcal{A}$ for all $A \in \mathcal{A}$. Prove that $\{A \in \mathcal{A} : T(A) = A\}$ is a $\sigma$-algebra.

4.3.21. Let $X, Y$ be nonempty sets and let $\mathcal{A} \subset \mathbb{P}(X)$. Prove that $\sigma(\mathcal{A}) \times Y$ is a $\sigma$-algebra on $X \times Y$.

4.3.22. Let $X$ be a set and $\mathcal{A}$ a $\sigma$-algebra of subsets of $X$. Suppose that $S \subset X$ is a set that is not in $\mathcal{A}$. Prove that

$$\sigma(\mathcal{A} \cup \{S\}) = \{(A \cap S) \cup (B \cap S^C) : A, B \in \mathcal{A}\}.$$

4.3.23. Let $X$ be a set and $\mathcal{A}$ a nonempty subset of $\mathbb{P}(X)$. Prove that

$$\sigma(\mathcal{A}) = \cup\{\sigma(\mathcal{F}) : \mathcal{F} \text{ is a countable subset of } \mathcal{A}\}.$$

4.3.24. Let $X$ be a set and $\mathcal{A}$ a countable collection of disjoint subsets of $X$ such that their union covers $X$. Describe $\sigma(\mathcal{A})$.

4.3.25. Let $X \subset \mathbb{R}^2$, let $\mathcal{S}$ be a collection of subsets of $X$, and let $A \in \sigma(\mathcal{S})$. Prove that there is a countable subset $\mathcal{C}$ of $\mathcal{S}$ such that $A \in \sigma(\mathcal{C})$.

In Exercises 4.3.26–4.3.28, $Y$ is a nonempty subset of $X$ and, for each $\mathcal{A} \subset \mathbb{P}(X)$, $\mathcal{A}_Y = \{A \cap Y : A \in \mathcal{A}\}$.

4.3.26. Prove that if $\mathcal{A}$ is a $\sigma$-algebra then $\mathcal{A}_Y$ is a $\sigma$-algebra on $Y$.

4.3.27. Prove that $\{E \subset X : E \cap Y \in \sigma(\mathcal{A}_Y)\}$ is a $\sigma$-algebra.

4.3.28.* Prove that $\sigma(\mathcal{A}_Y) = \sigma(\mathcal{A})_Y$.

4.3.29.* Prove that a finitely generated $\sigma$-algebra is a finite set.

4.3.30.* Prove or disprove: if $\mathcal{A} = \sigma(\{A_n\}_{n \in \mathbb{N}})$ then $\mathcal{A}$ is uncountable.

# 5

# *Abstract Measure*

In the previous chapter we saw how Lebesgue developed measure theory in the plane. In his dissertation, he did similar work in one and three dimensions. Following his groundbreaking work, other mathematicians generalized his results. The idea to state results as general as possible was gaining momentum at the beginning of the 20th century, not only in measure theory but throughout all of mathematics. In particular, Carathéodory developed measure theory in Euclidean spaces, likely considering this to be the highest level of generality. Fréchet showed that no reliance on Euclidean properties is needed.

In this chapter, we will explore generalizations of the two-dimensional Lebesgue measure and we will see that a measure can be defined on any set. The process that led to the Lebesgue $\sigma$-algebra $\mathcal{M}$ and Lebesgue measure $m$ on $\mathbb{R}^2$ can work in a much more general situation and we will present the details of such a construction. We will also explore to what extent the more general $\sigma$-algebra and measure share the properties of $\mathcal{M}$ and $m$.

## 5.1 Measures and Measurable Sets

In this section, we will consider measures and measurable sets in the abstract setting.

**Definition 5.1.1.** Let $X$ be a set and let $\mathfrak{M}$ be a $\sigma$-algebra of subsets of $X$. The pair $(X, \mathfrak{M})$ is called a **measurable space**. A set $E \subset X$ is **measurable** if $E \in \mathfrak{M}$. A **measure** on $(X, \mathfrak{M})$ is an extended real-valued function $\mu : \mathfrak{M} \to [0, +\infty]$ for which $\mu(\emptyset) = 0$ and which is countably additive ($\sigma$-additive), i.e., if $\{E_k\}_{k=1}^{\infty}$ are disjoint sets in $\mathfrak{M}$ then

$$\mu\Big(\bigcup_{k=1}^{\infty} E_k\Big) = \sum_{k=1}^{\infty} \mu(E_k). \tag{5.1}$$

A **measure space** is the triple $(X, \mathfrak{M}, \mu)$.

*Remark* 5.1.2. The condition $\mu(\emptyset) = 0$ helps avoid the trivial case when $\mu(A) = +\infty$ for all $A \in \mathfrak{M}$. Indeed, if $A = B = \emptyset$ then $\mu(A \cup B) = \mu(A) + \mu(B)$, so $\mu(\emptyset) = 2\mu(\emptyset)$ and it follows that $\mu(\emptyset)$ is either 0 or $+\infty$. In the latter

DOI: 10.1201/9781032716497-5

case, for any $A \in \mathfrak{M}$,

$$\mu(A) = \mu(A \cup \emptyset) = \mu(A) + \mu(\emptyset) = \infty.$$

**Example 5.1.3. A measure space.**
Let $X = \mathbb{R}^2$, $\mathfrak{M} = \mathcal{M}$, $\mu = m$. As we saw in Chapter 4, $(\mathbb{R}^2, \mathcal{M}, m)$ is a measure space. ♦

**Example 5.1.4. A measure space.**
Let $X$ be a set and $\mathfrak{M} = \mathbb{P}(X)$. For $E \subset X$ we define $\mu(E) = n$ if $E$ has exactly $n$ elements and $\mu(E) = \infty$ if $E$ is an infinite set. Then $\mu$ is called the **counting measure** on $X$. The triple $(X, \mathbb{P}(X), \mu)$ is a measure space (Exercise 5.1.1). ♦

**Example 5.1.5. A measure space.** Let $x_0 \in X$ and let $\mu(E) = 1$ if $x_0 \in E$ and $\mu(E) = 0$ if $x_0 \notin E$. This is the **Dirac measure $\boldsymbol{\delta_{x_0}}$**. The triple $(X, \mathbb{P}(X), \delta_{x_0})$ is a measure space (Exercise 5.1.2). ♦

**Example 5.1.6. A measure space.**
Let $(X, \mathfrak{M}, \mu)$ be a measure space and let $E \in \mathfrak{M}$. The function $\mu_E$ on $\mathfrak{M}$ defined by $\mu_E(A) = \mu(E \cap A)$ is a measure. Also, the set $\mathfrak{M}_E = \{E \cap A : A \in \mathfrak{M}\}$ is a $\sigma$-algebra. Therefore, $(E, \mathfrak{M}_E, \mu_E)$ is a measure space. (See Exercise 5.1.3.) ♦

**Example 5.1.7. A finitely additive measure.**
Let $X$ be an infinite set, let $\mathfrak{M} = \mathbb{P}(X)$, and define $\mu(E) = 0$ if $E$ is finite, $\mu(E) = \infty$ if $E$ is infinite. Then $\mu$ is a finitely additive measure but not a measure. ♦

These examples show that an abstract measure may look nothing like Lebesgue measure. Nevertheless, we study them because they are important in applications. For example, the counting measure will make the theory of infinite series a part of the integration theory. In addition, as we will demonstrate, they share many properties of Lebesgue measure. In other words, these properties carry over from $\mathbb{R}^2$ to more general sets. (After all, $(\mathbb{R}^2, \mathcal{M}_2, m_2)$ is also a measure space.)

**Proposition 5.1.8.** *Let $(X, \mathfrak{M}, \mu)$ be a measure space.*

*(a)* ***Finite additivity.*** *For any finite disjoint collection $\{E_k\}_{k=1}^n$ of measurable sets,*

$$\mu\left(\bigcup_{k=1}^{n} E_k\right) = \sum_{k=1}^{n} \mu(E_k).$$

*(b)* ***Monotonicity.*** *If $A \subset B$, then $\mu(A) \leq \mu(B)$.*

*(c)* ***Excision Property.*** *If $A \in \mathfrak{M}$ has finite measure, $B \in \mathfrak{M}$, and $A \subset B$ then*

$$\mu(B \setminus A) = \mu(B) - \mu(A).$$

*Proof.* Finite additivity follows from countable additivity by setting $E_k = \emptyset$ for $k \geq n+1$. To prove monotonicity and excision property notice that the inclusion $A \subset B$ implies that $B \cap A = A$. By part (a),

$$\mu(B) = \mu(B \cap A) + \mu(B \cap A^C) = \mu(A) + \mu(B \setminus A). \tag{5.2}$$

From this monotonicity follows immediately. Finally, $\mu(A) < \infty$, so we can subtract it from both sides of (5.2). $\square$

The next result states a few additional properties of the measure. We say that a sequence of sets $\{E_k\}_{k=1}^{\infty}$ is increasing if $E_1 \subset E_2 \subset \dots$ and decreasing if $E_1 \supset E_2 \supset \dots$.

**Theorem 5.1.9.** ***Continuity of Measure.*** *Let $(X, \mathfrak{M}, \mu)$ be a measure space.*

*(i) If $\{A_n\}_{n=1}^{\infty}$ is an increasing sequence in $\mathfrak{M}$ then*

$$\mu\left(\bigcup_{n=1}^{\infty} A_n\right) = \lim_{n\to\infty} \mu(A_n). \tag{5.3}$$

*(ii) If $\{B_n\}_{n=1}^{\infty}$ is a decreasing sequence in $\mathfrak{M}$ and if $m(B_1) < \infty$ then*

$$\mu\left(\bigcap_{n=1}^{\infty} B_n\right) = \lim_{n\to\infty} m(B_n). \tag{5.4}$$

*Proof.* We will prove only (i) and leave (ii) as Exercise 5.1.7. Without loss of generality, we may assume that, for every $n \in \mathbb{N}$, $m(A_n) < \infty$. Let us define $A_0 = \emptyset$ and $C_n = A_n \setminus A_{n-1}$, for $n \in \mathbb{N}$. Then, the collection $\{C_n\} \subset \mathfrak{M}$ is disjoint and $\cup_{n=1}^{\infty} C_n = \cup_{n=1}^{\infty} A_n$. Now,

$$\begin{aligned}
\mu\left(\bigcup_{n=1}^{\infty} A_n\right) &= \mu\left(\bigcup_{n=1}^{\infty} C_n\right) \\
&= \sum_{n=1}^{\infty} \mu(C_n) \\
&= \sum_{n=1}^{\infty} \mu(A_n \setminus A_{n-1}) \\
&= \sum_{n=1}^{\infty} (\mu(A_n) - \mu(A_{n-1})) \\
&= \lim_n \sum_{k=1}^{n} (\mu(A_k) - \mu(A_{k-1})) \\
&= \lim_n (\mu(A_n) - \mu(A_0)) \\
&= \lim_n \mu(A_n).
\end{aligned}$$

$\square$

In the early days of measure theory, the word measure was used for geometric settings. In abstract framework, Fréchet used *fonction additive d'ensembles* (an additive function of sets). These words started slowly being replaced by "measure" in the 1930s, and were completely superseded by 1950.

Now that we have the definition of measure, we can define a very important quantifier.

**Definition 5.1.10.** Let $(X, \mathfrak{M}, \mu)$ be a measure space. If $E \in \mathfrak{M}$ we will say that a property P holds on $E$ **almost everywhere** if there exists a set $E_0 \subset E$ such that $\mu(E_0) = 0$ and $P$ holds on $E \setminus E_0$. We write **a.e.** $[\mu]$, or $\mu$**–a.e.**, or simply **a.e.** when there is no danger of ambiguity.

**Example 5.1.11. A function that is 0 almost everywhere.**

The Dirichlet function (the characteristic function of $\mathbb{Q} \cap [0,1]$) is 0 almost everywhere. ♦

The Lebesgue quantifier a.e. makes precise the way in which a set may be "negligible". In calculus, the Riemann integral over an interval $[a, b]$ is the same as over $(a, b)$. In fact, the value of the integral remains the same even if a finite number of points is removed from $[a, b]$. When we develop the Lebesgue integration theory we will see that an analogous property is shared by the new integral, and that sets of measure 0 can be removed without affecting the integral. Perhaps surprisingly, the use of the new quantifier is not limited to integrals. Some important facts are true only a.e. As an illustration, we preview a result from Chapter 9: every monotone increasing function is differentiable a.e. This assertion is false without the quantifier. Indeed, there are many monotone-increasing functions that are not even continuous everywhere, let alone differentiable.

Lebesgue introduced the expression "presque partout" (almost everywhere) in the early 1900s. One of the first appearances in print is in a work on singular integrals from 1909. Apparently, its use was met by a big opposition from his colleague Arnaud Denjoy (1884–1974), a student of Borel and a rising star, who claimed that it was too vague. Denjoy worked in harmonic analysis and differential equations. His integral was the first to be able to integrate all derivatives.

**Theorem 5.1.12.** ***Borel–Cantelli Lemma.*** *Let $(X, \mathfrak{M}, \mu)$ be a measure space. Let $\{E_n\}_{n=1}^{\infty} \subset \mathfrak{M}$ and suppose that $\sum_{n=1}^{\infty} m(E_n) < \infty$. Then almost all $x \in X$ belong to at most finitely many sets $E_n$.*

*Proof.* We will show that the set

$$E = \{x \in X : x \in E_n \text{ for infinitely many } n \in \mathbb{N}\}$$

has measure zero. So, let $x \in E$. By Exercise 1.1.28,

$$E = \limsup_{n \to \infty} E_n = \bigcap_{n=1}^{\infty} \bigcup_{k=n}^{\infty} E_k.$$

For every $n \in \mathbb{N}$,

$$\bigcap_{n=1}^{\infty} \bigcup_{k=n}^{\infty} E_k \subset \bigcup_{k=n}^{\infty} E_k,$$

so by the monotonicity and subadditivity of $\mu$ (Exercise 5.1.8),

$$\mu(E) \leq \mu\left(\bigcup_{k=n}^{\infty} E_k\right) \leq \sum_{k=n}^{\infty} \mu(E_k).$$

Taking the limit as $n \to \infty$, we get that $\mu(E) = 0$. □

> Borel was one of the proponents of the idea that probability is a branch of measure theory. Using measure-theoretic tools he proved the result in [15] in 1909. Francesco Paolo Cantelli (1875–1966) was an Italian mathematician who made significant contributions in probablity. However, his approach was very different and when he obtained the same result in [18] in 1917, it is likely that he was not familiar with Borel's work. He made contributions to celestial mechanics, probability theory, and actuarial science. He was the founder of the "Istituto Italiano degli Attuari" for the applications of mathematics and probability to economics.

**Definition 5.1.13.** Let $(X, \mathfrak{M}, \mu)$ be a measure space. The measure $\mu$ is **finite** if $\mu(X) < \infty$ and it is **$\sigma$-finite** if $X = \cup_{n=1}^{\infty} X_n$ and $\mu(X_n) < \infty$. A set $E \in \mathfrak{M}$ is of **finite measure** if $\mu(E) < \infty$ and it is **$\sigma$-finite** if $E = \cup_{n=1}^{\infty} E_n$ and $\mu(E_n) < \infty$.

**Example 5.1.14. A finite measure.**
Consider $(\mathbb{R}^2, \mathcal{M}, m)$ and define $X = [0,1] \times [0,1]$,

$$\mathfrak{M} = \mathcal{M}_{[0,1]\times[0,1]} = \{E \cap [0,1] \times [0,1] : E \in \mathcal{M}\},$$

and $\mu(E) = m_{[0,1]\times[0,1]}(E) = m(E \cap [0,1] \times [0,1])$. The measure $\mu$ on $(X, \mathfrak{M})$ is finite. ♦

**Example 5.1.15. A $\sigma$-finite measure.**
The Lebesgue measure $m$ on $(\mathbb{R}^2, \mathcal{M})$ is $\sigma$-finite because $\mathbb{R}^2$ can be represented as the union of rectangles $[-n, n] \times [-n, n]$. ♦

**Example 5.1.16. A measure that is not $\sigma$-finite.**
The counting measure on $(\mathbb{R}, \mathbb{P}(\mathbb{R}))$ is not $\sigma$-finite. ♦

We will see later that Lebesgue measure on any Euclidean space is $\sigma$-finite. The significance of this property is that many important results that hold in $\mathbb{R}^n$ carries over to more general measure spaces as long as they are $\sigma$-finite and may fail when they are not.

> The abstract exposition of measure and measurable sets appears for the first time in a 1914 paper of Carathéodory [21]. He presented his work in the framework of an abstract $q$-dimensional space. The following year Fréchet was the first to present measure theory in a completely abstract space in [38]. It should be said that although Carathéodory confines his theory to a finite dimensional space, much of it holds without this restriction.

EMILE BOREL (1871–1956) was a French mathematician. He attended high school in Montauben where he lived with his much older sister. Having shown a significant potential for mathematics he continued his education in Paris at the Collège Sainte-Barbe and Lycée Louis-le-Grand. In the latter, one of his schoolmates was a son of a renowned mathematician Jean-Gaston Darboux (1842–1917). Through him he met leading mathematicians among whom wasEmile Picard (1856–1941), a big influence to Borel. He took entrance exams for both the Ecole Normale Supérieure and the Ecole Polytechnique and ranked first on each one but chose the former. His doctoral dissertation, with Darboux as the advisor, contained many important results in function theory, including a special case of the famous Heine–Borel Theorem. He went to teach at the University of Lille and returned to the Ecole Normale in 1897 where he remained until 1941. He received numerous awards and by 1905 he was the president of the French Mathematical Society. During WWI he was the Secretary-General in the Cabinet Office, receiving the Croix de Guerre in 1918. He served in the French Chamber of Deputies (1924–36) and as Minister of the Navy (1925–40). During WWII, he was a member of the French Resistance, receiving the Grand Croix Légion d'Honneur in 1950. He became the president of the Science Committee of UNESCO in 1948. Together with Lebesgue, Borel is one of the creators of measure theory, and he pioneered its application in probability theory. He was the first to develop a systematic theory of divergent series in 1899. Between 1921 and 1927, he published a series of papers on game theory and became the first to define games of strategy. In [16] he introduced the amusing thought experiment that entered popular culture under the name *the infinite monkey theorem.*

## Exercises

5.1.1. Prove that the counting measure introduced in Example 5.1.4 is a measure on $\mathbb{P}(X)$.

5.1.2. Prove that the Dirac measure introduced in Example 5.1.5 is a measure on $\mathbb{P}(X)$.

5.1.3. Prove the assertions made in Example 5.1.6.

5.1.4. Let $\mu$ be a function defined on $\mathbb{P}(\mathbb{R})$ by

$$\mu(A) = \begin{cases} 1 & \text{if } (0,1) \subset A \\ 0 & \text{otherwise.} \end{cases}$$

Prove or disprove: $\mu$ is a measure.

5.1.5. Let $X = \{x_k\}_{k=1}^{\infty}$ and suppose that $\{p_k\}_{k=1}^{\infty}$ is a sequence of nonnegative numbers such that $\sum_{n=1}^{\infty} p_n = 1$. For any set $A \subset X$ we define $N_A = \{k \in \mathbb{N} : x_k \in A\}$ and

$$\mu(A) = \sum_{n \in N_A} p_n.$$

Prove that $\mu$ is a measure on $\mathbb{P}(X)$.

5.1.6. Let $\{\mu_n\}$ be a sequence of measures on a measurable space $(X, \mathfrak{M})$ for which there is a constant $M \geq 0$ such that $\mu_n(X) \leq M$, for all $n \in \mathbb{N}$. Define

$$\mu = \sum_{n=1}^{\infty} \frac{\mu_n}{2^n}.$$

Prove that $\mu$ is a measure on $\mathfrak{M}$.

5.1.7. Prove Theorem 5.1.9 (ii). Give an example to show that the assumption $\mu(B_1) < \infty$ cannot be omitted.

5.1.8. Let $(X, \mathfrak{M}, \mu)$ be a measure space. Prove that $\mu$ is subadditive on $\mathfrak{M}$.

5.1.9.* Let $\{\mu_n\}_{n=1}^{\infty}$ be a sequence of measures on a $\sigma$-algebra $\mathcal{A}$, and let $\{\mu_n\}$ converge to a function $\mu$. In other words, suppose that for every $A \in \mathcal{A}$, $\lim \mu_n(A) = \mu(A)$. Also, suppose that there is a finite measure $\nu$ on $\mathcal{A}$, such that for every $n \in \mathbb{N}$ and for every set $A \in \mathcal{A}$, $\mu_n(A) \leq \nu(A)$. Prove that $\mu$ is a measure of $\mathcal{A}$. Does this remain true if there is no such measure $\nu$?

5.1.10. Let $(X, \mathfrak{M})$ be a measurable space and let $\{E_n\} \subset \mathfrak{M}$. Prove that the set $A$ of all those elements of $X$ that belong to infinitely many sets from the sequence $\{E_n\}$ is measurable.

5.1.11. Let $\{\mu_n\}$ be a monotone sequence of measures on a measurable space $(X, \mathfrak{M})$ and let $\{\mu_n\}$ converge to a function $\mu$. Prove or disprove: $\mu$ is a measure on $\mathfrak{M}$.

5.1.12. Let $(X, \mathfrak{M}, \mu)$ be a measure space. Let $\mathfrak{M}'$ be the collection of all $E \subset X$ for which there exist $A, B \in \mathfrak{M}$ such that $A \subset E \subset B$ and $\mu(B \backslash A) = 0$. For such $E$ define $\mu(E) = \mu(A)$. Prove that $\mathfrak{M}'$ is a $\sigma$-algebra and that $\mu$ is a measure on $\mathfrak{M}'$.

5.1.13. Let $(X, \mathfrak{M}, \mu)$ be a $\sigma$-finite measure space and let $E \in \mathfrak{M}$ for which $\mu(E) \neq 0$. Prove that there exists a measurable set $A \subset E$ for which $0 < \mu(A) < \infty$.

5.1.14.* Define $E = \{x \in [0,1] : \left|x - \frac{p}{q}\right| < q^{-3}$ for infinitely many $p, q \in \mathbb{N}\}$. Prove that $m(E) = 0$.

5.1.15. Let $(X, \mathfrak{M}, \mu)$ be a measure space and suppose that there is $n \in \mathbb{N}$ and a set of real numbers $A = \{a_1, a_2, \ldots, a_n\}$ such that for all $E \in \mathfrak{M}$, $\mu(E) \in A$.

(a) Describe the measurable sets if $n = 4$.

(b) Is it possible that $n = 6$?

5.1.16. Let $(X, \mathfrak{M}, \mu)$ be a $\sigma$-finite measure space and let $\mathcal{A}$ be a disjoint family of subsets of $X$. Prove that there can be only countably many sets in $\mathcal{A}$ that have positive measure.

5.1.17. Let $\mathfrak{M}$ be a $\sigma$-algebra of subsets of $X$ and let $f : X \to Y$. Prove or disprove: If $\mu$ is a measure on $X$, then $\lambda(E) := \mu(f^{-1}(E))$ defines a measure on $Y$.

5.1.18. Let $(X, \mathfrak{M})$ be a measurable space and let $\mu$ be a finitely additive measure. Suppose that $\mu$ is $\sigma$-subadditive. Prove that $\mu$ is a measure.

5.1.19. Let $(X, \mathfrak{M})$ be a measurable space and let $\mu$ be a finitely additive measure. Suppose that $\mu$ has the following property: If $\{A_n\}$ is a decreasing sequence of sets in $\mathfrak{M}$ such that $\cap_n A_n = \emptyset$, then $\lim_n \mu(A_n) = 0$. Prove that $\mu$ is a measure.

5.1.20. Let $(X, \mathfrak{M}, \mu)$ be a finite measure space and let $\nu$ be a finitely additive measure. Suppose that $\nu$ has the following property: If $\{A_n\} \subset \mathfrak{M}$ such that $\lim_n \mu(A_n) = 0$, then $\lim_n \nu(A_n) = 0$. Prove that $\nu$ is a measure.

5.1.21. Let $(X, \mathfrak{M})$ be a measurable space and let $\mu$ be a finitely additive measure. Suppose that $\mu$ has the following property: If $\{A_n\}$ is an increasing sequence of sets in $\mathfrak{M}$, then $\mu\left(\bigcup_{n=1}^{\infty} A_n\right) = \lim_{n\to\infty} \mu(A_n)$. Prove that $\mu$ is a measure.

5.1.22. Let $(X, \mathfrak{M}, \mu)$ be a finite measure space and let $\{E_n\} \subset \mathfrak{M}$. Prove that

$$\limsup_n \mu(E_n) \leq \mu(\limsup E_n).$$

5.1.23. Let $(X, \mathfrak{M}, \mu)$ be a finite measure space and let $\{E_n\} \subset \mathfrak{M}$. Prove that

$$\liminf_n \mu(E_n) \geq \mu(\liminf E_n).$$

5.1.24. Let $(X, \mathfrak{M}, \mu)$ be a measure space and let $A, B$ be sets in $\mathfrak{M}$ such that $\mu(A \triangle B) = 0$. Prove that for any $E \in \mathfrak{M}$, $\mu(E \cap A) = \mu(E \cap B)$.

5.1.25. Let $\mu$ and $\nu$ be measured on a measurable space $(X, \mathfrak{M})$, and define $\lambda(E) = \min\{\mu(E), \nu(E)\}$ and $\omega(E) = \max\{\mu(E), \nu(E)\}$, for $E \in \mathfrak{M}$. Prove or disprove:

(a) $\lambda$ is a measure on $\mathfrak{M}$.

(b) $\omega$ is a measure on $\mathfrak{M}$.

5.1.26. Let $\mu$ and $\nu$ be measured on a measurable space $(X, \mathfrak{M})$, and let $\alpha \geq 0$. Prove that:

(a) $\mu + \nu$ is a measure on $\mathfrak{M}$.

(b) $\alpha\mu$ is a measure on $\mathfrak{M}$.

5.1.27.* Let $(X, \mathfrak{M}, \mu)$ be a probability space, i.e., a measure space such that $\mu(X) = 1$, and let $\{A_n\} \subset \mathfrak{M}$ be independent, meaning that for any $n \in \mathbb{N}$ and any positive integers $1 \leq k_1 < \cdots < k_n$,

$$\mu(A_{k_1} \cap \cdots \cap A_{k_n}) = \mu(A_{k_1}) \ldots \mu(A_{k_n}).$$

Prove that the sequence $\{A_n^C\}$ is also independent.

5.1.28. Let $(X, \mathfrak{M}, \mu)$ be a probability space, i.e., a measure space such that $\mu(X) = 1$, and let $\{A_n\} \subset \mathfrak{M}$. Prove that $\mu(\bigcap_{k=1}^{n} A_k) \geq \sum_{k=1}^{n} \mu(A_k) - (n-1)$.

5.1.29.* Let $\mu$ and $\nu$ be measures on a measurable space $(X, \mathfrak{M})$ and suppose that $\mu \geq \nu$. Prove that:

(a) There exists a measure $\lambda$ on $\mathfrak{M}$ for which $\mu = \nu + \lambda$.

(b) If $\nu$ is $\sigma$-finite the measure $\lambda$ is unique.

(c) If $\nu$ is not $\sigma$-finite the measure $\lambda$ need not be unique but there is always the smallest one.

5.1.30. Let $X$ be a nonempty set, let $x_0 \in X$, and consider the measure space $(X, \mathbb{P}(X), \delta_{x_0})$. Prove that two functions $f, g : X \to \mathbb{R}^*$ are equal a.e. if and only if $f(x_0) = g(x_0)$.

5.1.31. Let $X$ be a nonempty set, let $\mu$ be the counting measure on $X$, and consider the measure space $(X, \mathbb{P}(X), \mu)$. Prove that two functions $f, g : X \to \mathbb{R}^*$ are equal a.e. if and only if $f(x) = g(x)$ for every $x \in X$.

## 5.2 Carathéodory Extension of Measure

In Section 5.1 we developed the notion of an abstract measure on an abstract measurable space. Examples 5.1.4 and 5.1.5 show that, given a set $X$ and a $\sigma$-algebra $\mathfrak{M}$ of its subsets, there is always a measure $\mu$ on it, even when $\mathfrak{M}$ is as big as the power set $\mathbb{P}(X)$. A harder question is if there is a measure that has some desirable properties, such as (i) and (iii) on our wish list.

In this section, we will focus on the former. In the case $X = \mathbb{R}^n$, such a condition is natural: we would like $\mu$ to agree with volume on $n$-dimensional rectangles. In more general spaces, where there may not be rectangles, we will single out a ring of sets with preassigned values of $\mu$, and we will ask whether we can *extend* this "pre-measure" to a measure on a $\sigma$-algebra. So, let us define what we mean by a measure on a ring.

**Definition 5.2.1.** Let $\mathfrak{E}$ be a ring of subsets of $X$. A **measure** on $\mathfrak{E}$ is an extended real-valued function $\mu : \mathfrak{E} \to [0, +\infty]$ for which $\mu(\emptyset) = 0$ and if $\{E_k\}_{k=1}^{\infty}$ are disjoint sets in $\mathfrak{E}$ whose union is also in $\mathfrak{E}$, then

$$\mu\Big(\bigcup_{k=1}^{\infty} E_k\Big) = \sum_{k=1}^{\infty} \mu(E_k).$$

*Remark* 5.2.2. Notice that we had to postulate that the infinite union of sets $\{E_k\}$ is in $\mathfrak{E}$, because in a ring (as opposed to a $\sigma$-algebra) this is not automatic.

The main result of this section is that such a measure can be extended to a measure on a $\sigma$-algebra.

**Theorem 5.2.3.** *Let $\mathfrak{E}$ be a ring on $X$ and let $\mu$ be a measure on $\mathfrak{E}$. Then there exists a $\sigma$-algebra $\mathfrak{M}$ that contains $\mathfrak{E}$ and a measure $\mu'$ on $\mathfrak{M}$ such that the restriction of $\mu'$ to $\mathfrak{E}$ coincides with $\mu$.*

How does one prove an assertion like this? If we had no interest beyond Euclidean spaces, we could repeat the argument from Chapter 4. Unfortunately, it relies on open and closed sets so, when $X$ is not a topological space, we have to try something else. A modified approach that we will use is due to a Greek mathematician Carathéodory. We start by defining *outer measure.*

**Definition 5.2.4.** Let $X$ be a nonempty set. A function $\mu^* : \mathbb{P}(X) \to [0, +\infty]$ is an **outer measure** if:

(a) $\mu^*(\emptyset) = 0$.

(b) If $A \subset B$ then $\mu^*(A) \leq \mu^*(B)$.

(c) For any countable collection $\{A_n\}_{n=1}^{\infty}$ of sets in $X$,

$$\mu^* \left( \bigcup_{n=1}^{\infty} A_n \right) \leq \sum_{n=1}^{\infty} \mu^*(A_n).$$

Following Carathéodory we gave an abstract definition of outer measure. At first sight, this is very different from the approach in Chapter 4 (see Definition 4.2.11). The following example shows that the type we used there is relevant in the abstract framework as well. In fact, this will be the only one that we will use in this section.

**Example 5.2.5. An outer measure on an abstract set.**
Let $\mathfrak{E}$ be a ring on $X$ and for $A \subset X$ let

$$\mu^*(A) := \inf \left\{ \sum_{k=1}^{\infty} \mu(E_k) : \{E_k : k \in \mathbb{N}\} \subset \mathfrak{E} \text{ and } A \subset \bigcup_{k=1}^{\infty} E_k \right\}.$$

Then $\mu^*$ is an outer measure. We leave the proof as Exercise 5.2.1. It is important to notice that if $A \in \mathfrak{E}$ then $\mu^*(A) = \mu(A)$. ♦

Definition 5.2.4 was introduced by Carathéodory in his 1918 book [23].

As we saw in Chapter 4, at this point, Lebesgue introduced the inner measure. Instead, we will use a shortcut and formulate directly a definition of a measurable set.

**Definition 5.2.6.** A set $E \subset X$ is **measurable** if for any set $A \subset X$,

$$\mu^*(A) = \mu^*(A \cap E) + \mu^*(A \setminus E). \tag{5.5}$$

We denote by $\mathfrak{M}$ the collection of all measurable subsets of $X$.

*Remark* 5.2.7. In view of the subadditivity of $\mu^*$, we have that inequality always holds in (5.5). Thus, a real test for a set to be measurable is whether the opposite inequality holds, i.e., whether $\mu^*(A) \geq \mu^*(A \cap E) + \mu^*(A \setminus E)$.

*Remark* 5.2.8. Formula (5.5) can be viewed as a confirmation that $E$ "slices" every set $A$ "nicely" (Figure 5.1). This is not always the case but examples are not easy to come by. (See Theorem 5.4.3).

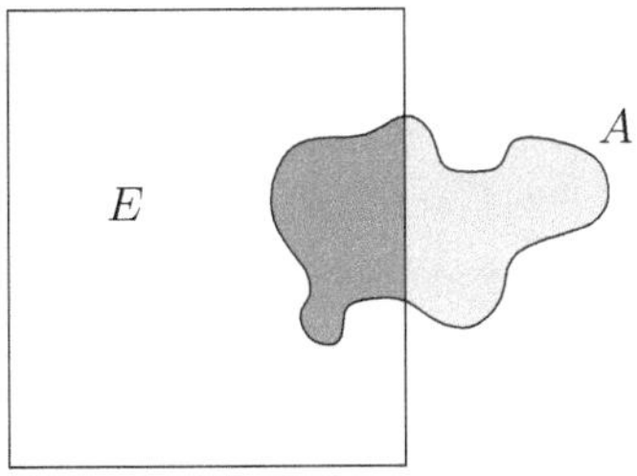


Figure 5.1: $E$ "slices" every set $A$ "nicely"

A big step towards proving Theorem 5.2.3 is to verify that a $\mathfrak{M}$ is a $\sigma$-algebra. First, we will show that it is an *algebra*.

**Definition 5.2.9.** A nonempty collection $\mathcal{A}$ of subsets of some set $X$ is called an **algebra of sets** if it includes $X$ itself and if it is closed under the operations of complement and finite unions, i.e., provided that

$$A, B \in \mathcal{A} \quad \Rightarrow \quad A \cup B \in \mathcal{A} \quad \text{and} \quad A^C \in \mathcal{A}$$

**Example 5.2.10. An algebra of sets.** Let $X$ be an arbitrary set. The collection of finite subsets of $X$ and their complements is an algebra of sets, (the so-called *finite-cofinite algebra*). ♦

**Example 5.2.11. An algebra of sets.** On the real line $\mathbb{R}$ the family of sets consisting of the empty set and all finite unions of half-open intervals of the form $(a, b]$, with $-\infty \leq a < b \leq +\infty$ is an algebra. ♦

*Remark* 5.2.12. In addition to an algebra of sets there is also the notion of an algebra over a field (see Definition 6.2.2).

**Proposition 5.2.13.** *The set $\mathfrak{M}$ is an algebra.*

*Proof.* It follows directly from the definition that $X \in \mathfrak{M}$ and that $E \in \mathfrak{M}$ implies that $E^C \in \mathfrak{M}$. Thus, we focus on finite unions.

Let $E_1, E_2 \in \mathfrak{M}$ and $A \subset X$. Then, using in order the measurability of $E_1$, De Morgan's laws, the measurability of $E_2$, and once again of $E_1$,

$$\begin{aligned}
&\mu^*(A \cap (E_1 \cup E_2)) + \mu^*(A \cap (E_1 \cup E_2)^C) \\
&= \mu^*(A \cap (E_1 \cup E_2) \cap E_1) + \mu^*(A \cap (E_1 \cup E_2) \cap E_1^C) + \mu^*(A \cap (E_1 \cup E_2)^C) \\
&= \mu^*(A \cap E_1) + \mu^*(A \cap E_2 \cap E_1^C) + \mu^*(A \cap E_1^C \cap E_2^C) \\
&= \mu^*(A \cap E_1) + \mu^*(A \cap E_1^C) = \mu^*(A).
\end{aligned}$$

□

Although the following result can be deduced from the definition, it is worth highlighting. In a nutshell, it states that $\mu^*$ is finitely additive on $\mathfrak{M}$. We leave its proof as Exercise 5.2.13.

**Proposition 5.2.14.** *If $E_1, E_2, \ldots, E_n$ are pairwise disjoint sets in $\mathfrak{M}$ and $A$ is any subset of $X$ then*

$$\mu^*\left(A \cap \left(\bigcup_{i=1}^{n} E_i\right)\right) = \sum_{i=1}^{n} \mu^*(A \cap E_i). \tag{5.6}$$

*In particular,* $\mu^*(\bigcup_{i=1}^{n} E_i) = \sum_{i=1}^{n} \mu^*(E_i)$.

Now we can get a big result.

**Theorem 5.2.15.** *The set $\mathfrak{M}$ is a $\sigma$-algebra.*

*Proof.* We have already established that $\mathfrak{M}$ is a ring that contains $X$, so it remains to demonstrate that it is closed under countable unions. To that end, let $\{E_i\}_{i=1}^{\infty}$ be a countable collection of sets in $\mathfrak{M}$, let $A$ be any subset of $X$, and let $E = \cup_{i=1}^{\infty} E_i$. By Remark 5.2.7, it suffices to show that

$$\mu^*(A) \geq \mu^*(A \cap E) + \mu^*(A \cap E^C).$$

To accomplish this goal, we define a sequence $\{E'_n\}$ by $E'_1 = E_1$ and, for every $n \geq 2$,

$$E'_n := E_n \setminus \bigcup_{i=1}^{n-1} E_i.$$

The sets $\{E'_n\}_{n=1}^{\infty}$ are all in $\mathfrak{M}$ (by Proposition 5.2.13), they are pairwise disjoint, and $\cup_{n=1}^{\infty} E'_n = E$ (see Exercise 1.1.24). Further, let

$$F_n = \cup_{i=1}^{n} E'_i, \quad \text{for } n = 1, 2, \ldots.$$

Then, for every $n \in \mathbb{N}$, $F_n \in \mathfrak{M}$ and $F_n \subset E$, hence $F_n^C \supset E^C$. Using Proposition 5.2.14 and the monotonicity of $\mu^*$, we have that, for every $n \in \mathbb{N}$,

$$\begin{aligned} \mu^*(A) &= \mu^*(A \cap F_n) + \mu^*(A \cap F_n^C) \\ &\geq \mu^*(A \cap F_n) + \mu^*(A \cap E^C) \\ &= \sum_{i=1}^{n} \mu^*(A \cap E'_i) + \mu^*(A \cap E^C). \end{aligned}$$

Remember, this is true for every $n \in \mathbb{N}$, so using the subadditivity of the outer measure we obtain

$$\begin{aligned} \mu^*(A) &\geq \sum_{i=1}^{\infty} \mu^*(A \cap E'_i) + \mu^*(A \cap E^C) \\ &\geq \mu^*\left(\bigcup_{i=1}^{\infty} (A \cap E'_i)\right) + \mu^*(A \cap E^C) \\ &= \mu^*(A \cap E) + \mu^*(A \cap E^C). \end{aligned}$$

□

We have already remarked that $\mu^*$ and $\mu$ coincide on $\mathfrak{E}$. In other words, $\mu^*$ is an extension of $\mu$ to $\mathbb{P}(X)$. We will now consider the restriction $\mu'$ of $\mu^*$ to $\mathfrak{M}$. In order to complete the proof of Theorem 5.2.3 we need to show that $\mathfrak{E} \subset \mathfrak{M}$ (so that $\mu'$ is an extension of $\mu$) and that $\mu'$ is a measure of $\mathfrak{M}$. We start with the latter.

**Proposition 5.2.16.** *$\mu'$ is a measure on $\mathfrak{M}$.*

*Proof.* Being a restriction of $\mu^*$, $\mu'$ is nonnegative and $\mu'(\emptyset) = 0$. Suppose that $\{E_n\}_{n=1}^{\infty}$ is a countable disjoint collection of sets in $\mathfrak{M}$, and let

$$E := \bigcup_{n=1}^{\infty} E_n.$$

Since $\mathfrak{M}$ is a $\sigma$-algebra, $E \in \mathfrak{M}$. Next,

$$\mu'(E) = \mu^*(E) \leq \sum_{n=1}^{\infty} \mu^*(E_n) = \sum_{n=1}^{\infty} \mu'(E_n),$$

so it remains to prove the opposite inequality. Let $N \in \mathbb{N}$. Then $\cup_{n=1}^{N} E_n \subset E$ so, using Proposition 5.2.14,

$$\mu'(E) = \mu^*(E) \geq \mu^*\left(\bigcup_{n=1}^{N} E_n\right) = \sum_{n=1}^{N} \mu^*(E_n) = \sum_{n=1}^{N} \mu'(E_n).$$

Since this holds for any $N$, we obtain the desired inequality. □

Thus, it remains to establish the following result.

**Proposition 5.2.17.** *The ring $\mathfrak{E}$ is contained in $\mathfrak{M}$ and $\mu' = \mu$ on $\mathfrak{E}$.*

*Proof.* First, we show that $\mathfrak{E} \subset \mathfrak{M}$. Let $E \in \mathfrak{E}$ and let $A \subset X$. Denote $A_1 = A \cap E$, $A_2 = A \cap E^C$. Once again, it suffices to show that

$$\mu^*(A_1) + \mu^*(A_2) \leq \mu^*(A).$$

Without loss of generality, we may assume that $\mu^*(A) < \infty$. Let $\varepsilon > 0$. By the way we defined outer measure (Example 5.2.5), there exists a collection $\{E_n\}_{n=1}^{\infty}$ of sets in $\mathfrak{E}$ covering $A$ such that

$$\sum_{n=1}^{\infty} \mu(E_n) \leq \mu^*(A) + \varepsilon.$$

We denote

$$E'_n = E_n \cap E \quad \text{and} \quad E''_n = E_n \cap E^C,$$

and we notice that, for every $n \in \mathbb{N}$, $E'_n \in \mathfrak{E}$ and $E''_n = E_n \setminus E \in \mathfrak{E}$. Since $A \subset \cup_{n=1}^{\infty} E_n$ we have that $A_1 \subset \cup_{n=1}^{\infty} E'_n$ and $A_2 \subset \cup_{n=1}^{\infty} E''_n$. Thus, $\mu^*(A_1) \leq \mu^*(\cup_{n=1}^{\infty} E'_n) \leq \sum_{n=1}^{\infty} \mu^*(E'_n)$ and, similarly, $\mu^*(A_2) \leq \sum_{n=1}^{\infty} \mu^*(E''_n)$. Next, we note that if $B \in \mathfrak{E}$ then $\mu^*(B) \leq \mu(B)$. This follows from our definition of outer measure (Example 5.2.5) by considering the covering $\{B, \emptyset, \emptyset, \dots\}$. In particular,

$$\mu^*(E_n') \leq \mu(E_n') \quad \text{and} \quad \mu^*(E_n'') \leq \mu(E_n'').$$

Thus,

$$\begin{aligned}
\mu^*(A_1) + \mu^*(A_2) &\leq \sum_{n=1}^{\infty} \mu^*(E_n') + \sum_{n=1}^{\infty} \mu^*(E_n'') \\
&= \sum_{n=1}^{\infty} (\mu^*(E_n') + \mu^*(E_n'')) \\
&= \sum_{n=1}^{\infty} (\mu(E_n') + \mu(E_n'')) \\
&= \sum_{n=1}^{\infty} \mu(E_n) \\
&\leq \mu^*(A) + \varepsilon.
\end{aligned}$$

Since $\varepsilon$ is arbitrary, we obtain the desired inequality. Thus, $\mathfrak{E} \subset \mathfrak{M}$.

It remains to prove that $\mu'(E) = \mu(E)$, for all $E \in \mathfrak{E}$. Given that $\mathfrak{E} \subset \mathfrak{M}$ and $\mu' = \mu^*$ on $\mathfrak{M}$, this is equivalent to proving that $\mu^*(E) = \mu(E)$, for all $E \in \mathfrak{E}$. As noted above, we know that

$$\mu^*(E) \leq \mu(E),$$

so it remains to prove the opposite inequality. Suppose that $\{E_n\}_{n=1}^{\infty}$ is a covering of $E$ by sets in $\mathfrak{E}$. If we define

$$A_1 = E_1 \cap E \quad \text{and} \quad A_{n+1} = \left(E_{n+1} \setminus \bigcup_{k=1}^{n} E_k\right) \cap E, \quad \text{for } n \in \mathbb{N},$$

then $\{A_n\}_{n=1}^{\infty}$ is a disjoint collection in $\mathfrak{E}$ with a union equal to $E$. (See Exercise 1.1.24.) Further, for every $n \in \mathbb{N}$, we have that $A_n \subset E_n$ hence $\mu(A_n) \leq \mu(E_n)$, because $\mu$ is a measure. (See Proposition 5.1.8.) Using the assumption that $E \in \mathfrak{E}$, it follows that

$$\mu(E) = \mu\left(\bigcup_{n=1}^{\infty} A_n\right) = \sum_{n=1}^{\infty} \mu(A_n) \leq \sum_{n=1}^{\infty} \mu(E_n).$$

Since the covering $\{E_n\}_{n=1}^{\infty}$ was arbitrary, we see that $\mu(E) \leq \mu^*(E)$ and the proposition is proved. □

*Remark* 5.2.18. Since $\mathfrak{M}$ is a $\sigma$-algebra we obtain that $\sigma(\mathfrak{E}) \subset \mathfrak{M}$.

Having proved Theorem 5.2.3 we can now apply it to an arbitrary set $X$ and any ring $\mathfrak{E}$ of subsets of $X$. Of special interest is the case when $X$ is a Euclidean space and we will do that in the next section.

Definition 5.2.6 appears in Carathéodory's 1918 book [23]. However, he did not consider extensions of measures. Fréchet did in his 1924 paper [39] but without the use of Carathéodory's definition of a measurable set. Hahn combined the two in 1933 in [50], thus paving a way for it to become a standard way of thinking, and giving rise to the expression *Carathéodory's extension.*

CONSTANTIN CARATHEODORY (1873–1950) was a Greek mathematician. He was born in Berlin where his father was a secretary in the Ottoman embassies, but grew up in Brussels. During his high school years, he won a prize twice as the best mathematics student in Belgium. He started a career as an engineer and worked on the construction of the Assiut dam in Egypt, all the while reading Jordan's *Cours d'Analyse.* He studied at the Universities of Berlin and Göttingen, where he received his doctorate in 1904 with Minkowski as advisor. He led a nomadic life teaching at several German and Greek universities, and he visited the United States twice. In 1928 he was the first visiting lecturer of the American Mathematical Society and had a visiting position at Harvard. In 1936-37 he gave a lecture at the meeting celebrating the tercentenary of Harvard University, then spent the winter semester at the University of Wisconsin. He was fluent in more than 8 languages and had a broad education. His most significant contributions were in calculus of variations, measure theory, and the theory of functions of a real and complex variable. He also created an axiomatic formulation of thermodynamics. He wrote many books including the first volume on Real functions in 1939 but the second volume was destroyed during the bombing of Leipzig in 1943. He was in high esteem and he was given the honor of presenting the first ever Fields Medals at the 1936 International Congress of Mathematicians in Oslo, Norway.

## Exercises

5.2.1. Prove that $\mu^*$ defined in Example 5.2.5 is an outer measure.

5.2.2. Let $(X, \mathfrak{M})$ be a measurable space and let $\{\mu_n^*\}$ be a sequence of outer measures on $(X, \mathfrak{M})$. Prove that $\sum_n \mu_n^*$ and $\sup_n \mu_n^*$ are also outer measures.

5.2.3. Let $\mathcal{A}$ be a collection of subsets of $X$ that includes the empty set, let $F : \mathcal{A} \to [0, +\infty]$ such that $F(\emptyset) = 0$, and let

$$\mu^*(E) := \inf \left\{ \sum_n F(E_n) : E \subset \bigcup_n E_n, \text{ and } \{E_n\} \subset \mathcal{A} \right\}.$$

Prove that $\mu^*$ is an outer measure.

In Exercises 5.2.4–5.2.7, $X$ is a set and $\mu^* : \mathbb{P}(X) \to [0, +\infty]$ is given. Prove that $\mu^*$ is an outer measure and determine the $\sigma$-algebra $\mathfrak{M}$ of all measurable sets.

5.2.4. $\mu^*(A)$ is the number of points in $A$ if $A$ is a finite set, and $\mu^*(A) = \infty$ when $A$ is an infinite set.

5.2.5. $\mu^*(\emptyset) = 0$ and $\mu^*(E) = 1$ for $E \neq \emptyset$.

5.2.6. $\mu^*(\emptyset) = 0$, $\mu^*(X) = 2$, and $\mu^*(E) = 1$ for all other subsets of $X$.

5.2.7. $\mu^*(A) = 0$ if $A$ is a countable set and $\mu^*(A) = 1$ if $A$ is uncountable.

5.2.8. Let $X$ be an infinite set, let $\nu$ be the counting measure on $X$, and let $F : \mathbb{P}(X) \to [0, +\infty]$ be defined by $F(A) = \sqrt{\nu(A)}$.

(a) Prove that $F$ is an outer measure.

(b) Find the corresponding $\sigma$-algebra $\mathfrak{M}$ of measurable sets and the measure $\mu$.

(c) Find the outer measure $\mu^*$ associated with $\mu$.

5.2.9. Let $\mu$ be a function defined on subsets of $\mathbb{R}$ by setting $\mu(A)$ to be the number of integers in $A$.

(a) Determine the outer measure $\mu^*$ induced by $\mu$ and the $\sigma$-algebra of measurable sets.

(b) Prove that the Carathéodory extension of $\mu$ is $\sigma$-additive.

5.2.10. Let $X$ be a nonempty set and let $\mu$ be a function defined on $\mathfrak{E} = \{\emptyset, X\}$ by $\mu(\emptyset) = 0$ and $\mu(X) = 1$. Determine the outer measure $\mu^*$ induced by $\mu$ and the $\sigma$-algebra of measurable sets.

5.2.11. Let $X$ be a set with at least 2 points, let $A$ be a proper, nonempty subset of $X$, and let $\mu$ be a function defined on $\mathfrak{E} = \{A, X\}$ by $\mu(A) = 1$ and $\mu(X) = 2$. Determine the outer measure $\mu^*$ induced by $\mu$ and the $\sigma$-algebra of measurable sets.

5.2.12.* Let $\mathfrak{E} = \{\emptyset, [0,1], [0,3], [2,3]\} \subset \mathbb{P}(\mathbb{R})$, and let $\mu$ be a function defined on $\mathfrak{E}$ by $\mu(\emptyset) = 0$, $\mu([0,1]) = \mu([0,3]) = \mu([2,3]) = 1$. Determine the outer measure $\mu^*$ induced by $\mu$ and the $\sigma$-algebra of measurable sets.

5.2.13. Prove Proposition 5.2.14. Does the result remain true if $n$ is replaced by $\infty$?

5.2.14. Let $\mu^*$ be an outer measure on a set $X$, let $\{A_n\}$ be a sequence of subsets of $X$, and let $\{E_n\}$ be a sequence of pairwise disjoint, measurable subsets of $X$ such that $A_n \subset E_n$, for all $n \in \mathbb{N}$. Prove that

$$\mu^*\left(\bigcup_{n=1}^{\infty} A_n\right) = \sum_{n=1}^{\infty} \mu^*(A_n).$$

5.2.15. Let $\mu^*$ be an outer measure and suppose that it is additive. Prove that it is countably additive, hence a measure.

5.2.16. Let $\mu^*$ be an outer measure on a set $X$ and let $A \subset X$ have outer measure 0. Prove that

$$\mu^*(B \cup A) = \mu^*(B \setminus A) = \mu^*(B),$$

for any set $B \subset X$.

5.2.17.* Let $\mu^*$ be an outer measure on a set $X$ and let $A \subset X$. Prove that $A$ is measurable if and only if for each $\varepsilon > 0$ there exists a measurable set $E$ such that $E \subset A$ and $\mu^*(A \setminus E) < \varepsilon$.

5.2.18.* Let $\mu^*$ be an outer measure on a set $X$ and let $A \subset X$. Suppose that for each $\varepsilon > 0$ there exists a measurable set $E$ such that $\mu^*(A \triangle E) < \varepsilon$. Prove that $A$ is measurable.

5.2.19. Let $\mu^*$ be an outer measure on a set $X$, let $A \subset X$, and let $E$ be a measurable subset of $X$. Prove that

$$\mu^*(E \cup A) + \mu^*(E \cap A) = \mu^*(E) + \mu^*(A).$$

5.2.20.* Let $(X, \mathfrak{M}_1, \mu_1)$ and $(X, \mathfrak{M}_2, \mu_2)$ be measure spaces. Prove that $\mu_1$ and $\mu_2$ generate the same outer measure on $X$ (in the sense of Example 5.2.5) if and only if both $\mu_1 = \mu_2^*$ on $\mathfrak{M}_1$ and $\mu_2 = \mu_1^*$ on $\mathfrak{M}_2$.

5.2.21. Let $(X, \mathfrak{M}, \mu)$ be a measure space and $\mu^*$ the outer measure on $X$ as defined in Example 5.2.5. Prove that if $E \subset X$ then there exists a measurable set $B$ that contains $E$ and $\mu(B) = \mu^*(E)$.

5.2.22.* Let $\mathfrak{C}$ be a ring on $X$, let $\mu$ be a measure on $\mathfrak{C}$, and let $\mu^*$ be the outer measure on $X$ as in Example 5.2.5. We will denote by $\mathfrak{C}_\sigma$ the collection of all countable unions of sets in $\mathfrak{C}$ and by $\mathfrak{C}_{\sigma\delta}$ the collection of all countable intersections of sets in $\mathfrak{C}_\sigma$. Prove that:

(a) For any $E \subset X$ and $\varepsilon > 0$ there exists $A \in \mathfrak{C}_\sigma$ such that $E \subset A$ and $\mu^*(A) \le \mu^*(E) + \varepsilon$.

(b) If $\mu^*(E) < \infty$ then $E$ is measurable if and only if there exists $B \in \mathfrak{C}_{\sigma\delta}$ such that $E \subset B$ and $\mu^*(B \setminus E) = 0$.

5.2.23.* Let $(X, \mathfrak{M}, \mu)$ be a measure space such that $\mu(X) = 1$ and $\mu^*$ an outer measure on $X$. Suppose that $E \subset X$ and $\mu^*(E) = 1$. Prove that if $A, B \in \mathfrak{M}$ and $A \cap E = B \cap E$ then $\mu(A) = \mu(B)$.

5.2.24. Prove that if $\mu^*(E) = 0$ then $E \in \mathfrak{M}$ and $\mu(E) = 0$.

5.2.25. Let $h : [a, b] \to \mathbb{R}$ be a bounded, left-continuous, increasing function. Prove that there is a $\sigma$-algebra $\mathfrak{M} \subset \mathbb{P}([a, b])$ and a measure $\mu$ on $\mathfrak{M}$ such that $\mu([x, y)) = h(y) - h(x)$, for all $x < y$ in $[a, b]$.

*Remark* 5.2.19. This measure is known as the **Lebesgue–Stieltjes measure** associated to $h$.

5.2.26.* Give an example of a collection $\mathcal{A}$ of subsets of $X$ and measures $\mu, \nu$ on $\sigma(\mathcal{A})$ such that $\mu(A) = \nu(A)$, for all $A \in \mathcal{A}$, but there is a set $E \in \sigma(\mathcal{A})$ for which $\mu(E) \neq \nu(E)$.

5.2.27.* Let $E \subset X$ be such that $E \cap A \in \mathfrak{M}$ whenever $A \in \mathfrak{M}$ and $\mu(A) < \infty$. Prove that $E \in \mathfrak{M}$.

5.2.28.* Prove that the Carathéodory extension of a measure $\mu'$ is unique if $\mu'$ is $\sigma$-finite on $\mathfrak{C}$.

5.2.29. Let $\mathfrak{C}$ be a ring that includes $X$ and suppose that $\mu$ is a finite measure on $\mathfrak{C}$.

(a) Prove that the Carathéodory extension of $\mu$ is a finite measure.

(b) Give an example that the statement is false if $X$ is not assumed to be in $\mathfrak{C}$.

5.2.30. Let $\{r_n\}$ be an arrangement of $\mathbb{Q}$ in a sequence, let $\alpha_n(E) = 1$ if $r_n \in E$ and $\alpha_n(E) = 0$ otherwise, and let $\mu$ be a function defined on subsets of $\mathbb{R}$ by setting $\mu(E) = \sum_{n=1}^{\infty} 2^{-n}\alpha_n(E)$. Is $\mu$ additive? Is it $\sigma$-additive?

5.2.31.** For $E \subset \mathbb{N}$ let $\alpha_k(E) = 1$ if $k \in E$ and $\alpha_k(E) = 0$ otherwise. Let $\mathfrak{M}$ be the collection of all sets for which $\mu(E)$ exists, where

$$\mu(E) := \lim_{n\to\infty} \frac{\alpha_1(E) + \alpha_2(E) + \cdots + \alpha_n(E)}{n}$$

(a) Is $\mathfrak{M}$ an algebra?

(b) Is $\mathfrak{M}$ a $\sigma$-algebra?

(c) Is $\mu$ additive?

(d) Is $\mu$ $\sigma$-additive?

5.2.32.** Let $\mathcal{A}$ be the collection of finite unions of sets of the form $(a, b] \cap \mathbb{Q}$, where $-\infty \le a \le b \le +\infty$, and define $\mu(\emptyset) = 0$ and $\mu(A) = \infty$ for $A \neq \emptyset$. Prove that:

(a) $\mathcal{A}$ is an algebra on $\mathbb{Q}$.

(b) $\sigma(\mathcal{A}) = \mathbb{P}(\mathbb{Q})$.

(c) $\mu$ is a measure of $\mathcal{A}$ that has more than one extension to $\mathbb{P}(\mathbb{Q})$.

5.2.33. Let $X$ be a nonempty set, let $\mathcal{A} \subset \mathbb{P}(X)$, and let $\mu : \mathcal{A} \to [0, +\infty]$. Prove or disprove: if $\mathfrak{M}$ is the $\sigma$-algebra obtained by Carathéodory's construction then $\mathcal{A} \subset \mathfrak{M}$.

5.2.34. Let $X$ be a nonempty set, let $E \subset X$, let $\mu^*$ be an outer measure on $\mathbb{P}(X)$, let $\mu$ be the restriction of $\mu^*$ to the $\sigma$-algebra $\mathfrak{M}$ of $\mu^*$-measurable sets, and let $\mu^+$ be the outer measure associated to $\mu$. Prove that:

(a) $\mu^+(E) \ge \mu^*(E)$.

(b) $\mu^+(E) = \mu^*(E)$ if and only if there is $A \in \mathfrak{M}$ such that $A \supset E$ and $\mu^*(A) = \mu^*(E)$.

5.2.35. Let $(X, \mathfrak{M})$ be a measurable space and $\mu^*$ an outer measure on $X$. Prove that $|\mu^*(A) - \mu^*(B)| \le \mu^*(A \triangle B)$ for any $A, B \subset X$, provided that at least one of the numbers $\mu^*(A), \mu^*(B)$ is finite.

5.2.36. Let $\mathfrak{C}$ be an algebra on $X$, let $\mu$ be a finite measure on $\mathfrak{C}$, let $\overline{\mu}$ be its Carathéodory extension on a $\sigma$-algebra $\mathfrak{M}$, let $E \in \mathfrak{M}$, and let $\varepsilon > 0$. Prove that there is a set $A \in \mathfrak{C}$ such that $\overline{\mu}(A \triangle E) < \varepsilon$.

5.2.37. Prove that every algebra is a ring. Give an example to show that the converse is not true.

5.2.38. Prove that every ring that contains $X$ is also an algebra.

5.2.39. Suppose that $\mathcal{R}$ is a ring. Prove that the set

$$\{A \subset X : A \in \mathcal{R} \text{ or } A^C \in \mathcal{R}\}$$

is an algebra.

5.2.40. Suppose that $\mathcal{R}$ is a ring. Prove that the set

$$\{A \subset X : A \cap E \in \mathcal{R} \text{ for all } E \in \mathcal{R}\}$$

is an algebra.

5.2.41.* Let $\{A_n\} \subset \mathbb{R}^2$ be an increasing sequence. Prove that $m^*(\cup_n A_n) = \lim m^*(A_n)$.

## 5.3 Lebesgue Measure on Euclidean Spaces

Theorem 5.2.3 is very general, but in this section, we will focus on its most interesting application, namely when $X = \mathbb{R}^n$. It allows us to establish the existence of the Lebesgue measure on $\mathbb{R}^n$ that agrees with the volume on rectangles. When $n = 1$, a rectangle is just a (finite) interval on the real line and its volume is its length. For $n \geq 2$, a rectangle in $\mathbb{R}^n$ is a Cartesian product of $n$ finite intervals, $R = I_1 \times \cdots \times I_n$, and the volume of $R$ is the product of the lengths of these intervals, $\operatorname{vol} R = \ell(I_1) \cdot \cdots \cdot \ell(I_n)$. To apply Theorem 5.2.3 we notice that elementary subsets of $\mathbb{R}^n$, i.e., sets that are finite unions of rectangles, form a ring and the function vol defined on rectangles extends naturally to this ring. Thus, we get the following result.

**Theorem 5.3.1.** *Let $\mathfrak{E}_n$ be the ring of all elementary subsets of $\mathbb{R}^n$ and let vol denote the volume defined on rectangles and extended to $\mathfrak{E}$. There exists a translation invariant measure $m_n$ on $\sigma$-algebra $\mathcal{M}_n$ that contains $\mathfrak{E}_n$ such that the restriction of $m_n$ to $\mathfrak{E}$ coincides with vol.*

The only part of the conclusion that does not follow directly from Theorem 5.2.3 concerns translation invariance which we leave as Exercise 5.3.5.

Although a huge accomplishment, Theorem 5.3.1 leaves open a number of issues. First and foremost, in the case $n = 2$ we now have two extensions of premeasure "vol", obtained in different ways. Much of this section is devoted to a demonstration that both constructions yield the same $\sigma$-algebra and the same measure. Second, some nice properties of the measure space $(\mathbb{R}^2, \mathcal{M}_2, m_2)$ were obtained in Chapter 4 from the construction and they are not obvious consequences of Theorem 5.3.1 or Carathéodory's approach. We will return to these questions toward the end of this section.

We will now embark on the task of establishing that the measured spaces $(\mathbb{R}^2, \mathcal{M}_2, m_2)$ and $(\mathbb{R}^2, \mathfrak{M}_2, \mu_2)$ obtained respectively in Chapter 4 and Section 5.2 are identical. Since this part of the section is devoted exclusively to $\mathbb{R}^2$, we will drop the subscripts and talk about $(\mathbb{R}^2, \mathcal{M}, m)$ and $(\mathbb{R}^2, \mathfrak{M}, \mu)$.

Let us start with the outer measure. On one hand, we have $m^*(A)$ introduced by Definition 4.2.11. On the other hand, there is $\mu^*(A)$ as in Example 5.2.5, with $\mathfrak{E}$ the ring of elementary subsets of $\mathbb{R}^2$.

**Proposition 5.3.2.** *If $A \subset \mathbb{R}^2$, then $m^*(A) = \mu^*(A)$.*

*Proof.* Since every open set $G$ is a countable union of rectangles, we have that

$$\{G : G \text{ is open and } A \subset G\} \subset \{\cup_{k=1}^{\infty} E_k : E_k \in \mathcal{E} \text{ and } A \subset \cup_{k=1}^{\infty} E_k\}.$$

Taking infimum over both sets, we conclude that

$$m^*(A) \geq \mu^*(A).$$

To prove the opposite inequality, let $\varepsilon > 0$ and let $\{E_k\}_{k=1}^{\infty}$ be a collection of elementary sets such that

$$A \subset \bigcup_{k=1}^{\infty} E_k \quad \text{and} \quad \sum_{k=1}^{\infty} \mu(E_k) < \mu^*(A) + \varepsilon.$$

By Proposition 4.1.2, each $E_k$ is a finite, disjoint union of rectangles. Thus, there exist rectangles $\{R_i\}_{i=1}^{\infty}$ such that

$$A \subset \bigcup_{i=1}^{\infty} R_i \quad \text{and} \quad \sum_{i=1}^{\infty} \mu(R_i) < \mu^*(A) + \varepsilon.$$

By Exercise 5.3.7, we can make the edges of each rectangle slightly longer to obtain *open* rectangles $\{R_i'\}_{i=1}^{\infty}$ such that

$$A \subset \bigcup_{i=1}^{\infty} R_i' \quad \text{and} \quad \sum_{i=1}^{\infty} \mu(R_i') < \mu^*(A) + 2\varepsilon.$$

The set $G = \cup_{i=1}^{\infty} R_i'$ is open, it contains $A$, and $m(G) < \mu^*(A) + 2\varepsilon$. It follows that

$$m^*(A) \leq \mu^*(A) + 2\varepsilon,$$

and since $\varepsilon$ is arbitrary, we get that $m^*(A) \leq \mu^*(A)$. □

Thus, both approaches lead to the same concept of outer measure. Since Definition 4.2.11 also employs the concept of inner measure, our next step will be to define the inner measure in a seemingly different way.

**Definition 5.3.3.** Let $A$ be a bounded subset of $\mathbb{R}^n$ and let $R$ be an open rectangle that contains $A$. The **inner measure** of $A$ is defined by

$$\mu_*(A) = \mu(R) - \mu^*(R \setminus A).$$

If $A$ is not bounded, and $\{R_k\}$ is a sequence of disjoint rectangles with union $\mathbb{R}^n$, then

$$\mu_*(A) = \sum_{k=1}^{\infty} \mu_*(A \cap R_k).$$

First, it is necessary to show that the definition is independent of the choice of rectangles $R$ and $\{R_k\}$ (Exercise 5.3.6). Second, we need to show that two definitions of inner measure are equivalent on $\mathbb{R}^2$.

**Proposition 5.3.4.** *If $A \subset \mathbb{R}^2$ then $\mu_*(A) = m_*(A)$.*

*Proof.* We will prove the bounded case only and leave the unbounded as Exercise 5.3.29. By Proposition 5.3.2, $m^*(A) = \mu^*(A)$. Further, $m$ and $\mu$ agree on elementary sets, hence on rectangles, so $m(R) = \mu(R)$. Thus,

$$\mu_*(A) = \mu(R) - \mu^*(R \setminus A) = m(R) - m^*(R \setminus A) = m_*(A),$$

where the last equality is Theorem 4.2.22. □

Recall that a set $A \subset \mathbb{R}^2$ belongs to $\mathcal{M}$ if and only if $m^*(A) = m_*(A) = m(A)$, and that for $A \in \mathfrak{M}$, $\mu(A) = \mu^*(A)$. Thus, our project will be complete once we show that the $\sigma$-algebras $\mathcal{M}$ and $\mathfrak{M}$ are the same when the underlying space is $\mathbb{R}^2$. To establish their equality we need a lemma first. By Exercise 5.2.21, given a set $A \subset \mathbb{R}^2$ there exists a $\mu$-measurable set $B$ such that

$$A \subset B \quad \text{and} \quad \mu^*(A) = \mu(B). \tag{5.7}$$

The following is a crucial property of the set $B$.

**Lemma 5.3.5.** *If $B$ is a $\mu$-measurable set satisfying (5.7), then*

$$\mu_*(A) + \mu^*(B \setminus A) \le \mu(B).$$

*Proof.* By definition of $m_*$ (which is also $\mu_*$),

$$\mu_*(A) = m_*(A) = \sup\{m(F) : F \text{ is a compact set and } F \subset A\}. \tag{5.8}$$

Since every open set in $\mathbb{R}^2$ is a countable union of rectangles (which are in $\mathfrak{M}$), it belongs to $\mathfrak{M}$, and so does every closed set. Also, closed sets are in $\mathcal{M}$. Thus, every closed set $F$ is measurable in both ways and $\mu^*(F) = m^*(F)$, hence $\mu(F) = m(F)$. Finally, if we consider $F$ to be a subset of $A$, then $F \subset B$, so

$$m(F) = \mu(F) = \mu(B) - \mu(B \setminus F). \tag{5.9}$$

Using (5.8) and (5.9), it follows that

$$\begin{aligned} \mu_*(A) &= \sup\{\mu(B) - \mu(B \setminus F) : F \text{ is a compact set and } F \subset A\} \\ &= \mu(B) - \inf\{\mu(B \setminus F) : F \text{ is a compact set and } F \subset A\}. \end{aligned} \tag{5.10}$$

Since $F \subset A$ we have that $B \setminus F \supset B \setminus A$ and the monotonicity of outer measure implies that $\mu^*(B \setminus F) \geq \mu^*(B \setminus A)$. Taking the infimum over all compact sets $F \subset A$, we obtain that

$$\inf\{\mu(B \setminus F) : F \text{ is a compact set and } F \subset A\} \geq \mu^*(B \setminus A). \tag{5.11}$$

Combining (5.10) and (5.11) yields

$$\mu_*(A) \leq \mu(B) - \mu^*(B \setminus A),$$

and the proposition is proved. □

Now we are ready for the main result.

**Theorem 5.3.6.** *In $\mathbb{R}^2$ we have that $\mathfrak{M} = \mathcal{M}$.*

*Proof.* First, we will show that $\mathfrak{M} \subset \mathcal{M}$. To that end, suppose that $E$ is a bounded set in $\mathfrak{M}$. By definition, for any open rectangle $R$ that contains $E$,

$$\mu^*(R) = \mu^*(R \cap E) + \mu^*(R \setminus E) = \mu^*(E) + \mu^*(R \setminus E).$$

Thus, $m^*(E) = \mu^*(E) = \mu^*(R) - \mu^*(R \setminus E) = \mu_*(E) = m_*(E)$, so $E \in \mathcal{M}$. We leave the unbounded case to the reader and focus on the opposite implication.

Let $E \in \mathcal{M}$. Let $B$ be a set in $\mathfrak{M}$ (provided by Exercise 5.2.21) such that

$$E \subset B \quad \text{and} \quad \mu^*(E) = \mu(B).$$

By Lemma 5.3.5,

$$\mu_*(E) + \mu^*(B \setminus E) \leq \mu(B).$$

Since $E \in \mathcal{M}$, Propositions 5.3.2 and 5.3.4 yield

$$\mu_*(E) = m_*(E) = m^*(E) = \mu^*(E) = \mu(B)$$

and we obtain that $\mu^*(B \setminus E) = 0$. Now, for any set $A \subset \mathbb{R}^2$, using Exercise 1.1.4,

$$\begin{aligned} \mu^*(A \cap E) + \mu^*(A \cap E^C) &\leq \mu^*(A \cap B) + \mu^*((A \setminus B) \cup (B \setminus E)) \\ &\leq \mu^*(A \cap B) + \mu^*(A \setminus B) + \mu^*(B \setminus E) \\ &= \mu^*(A \cap B) + \mu^*(A \setminus B) \\ &= \mu^*(A), \end{aligned}$$

where the last equality follows from the fact that $B \in \mathfrak{M}$. Thus, $E \in \mathfrak{M}$ and the theorem is proved. □

We close this section with a brief discussion of the Lebesgue measure $m_n$ on $\mathbb{R}^n$, for arbitrary $n$. Some of the questions have been answered in Chapter 4 for $\mathbb{R}^2$ but not for other Euclidean spaces. We list some of these properties here, but we leave the proofs to the reader. For several others, they can be either found among the exercises, or they are left to the reader to formulate and prove if needed.

**Theorem 5.3.7.** *All open and all closed subsets of $\mathbb{R}^n$ are Lebesgue measurable.*

**Theorem 5.3.8.** *Let $A \subset \mathbb{R}^n$ and $\varepsilon > 0$. The following are equivalent:*

*(a) $A \in \mathcal{M}_n$.*

*(b) There is an open set $G$ such that*

$$A \subset G \quad \text{and} \quad m_n^*(G \setminus A) < \varepsilon.$$

*(c) There is a $G_\delta$ set $G$ such that*

$$A \subset G \quad \text{and} \quad m_n^*(G \setminus A) = 0.$$

*(d) There is a closed set $F$ such that*

$$F \subset A \quad \text{and} \quad m_n^*(A \setminus F) < \varepsilon.$$

*(e) There is an $F_\sigma$ set $F$ such that*

$$F \subset A \quad \text{and} \quad m_n^*(A \setminus F) = 0.$$

**Theorem 5.3.9.** *Every countable set in $\mathbb{R}^n$ is measurable and has measure 0.*

Finally, there are some issues that we cannot address now postponing them until Chapter 8. One of them is the relation between measured spaces in various dimensions. For example, how can we formalize the connection between $m_1$ and $m_2$? Such questions will be one of the main topics of Chapter 8.

Also, Carathéodory's approach gave us one extension of $\mu$ from a ring to the $\sigma$-algebra generated by this ring. Is there another extension?

**Theorem 5.3.10.** *The measure $m_n$ is a unique extension of the function* vol, *defined on rectangles in $\mathbb{R}^n$, to the $\sigma$-algebra $\mathcal{M}_n$ generated by these rectangles.*

For bounded sets, Theorem 5.3.10 will be an easy consequence of Exercise 8.1.21. In the unbounded case, a set $E$ can be represented as $\cup(R_m \cap E)$, where $\{R_m\}$ is a disjoint collection of rectangles covering $\mathbb{R}^n$.

There are a few questions that we have carefully sidestepped. Is $\mathcal{M}_n = \mathbb{P}(\mathbb{R}^n)$? If not, is there a larger $\sigma$-algebra $\mathcal{M}'$ and a function $m' : \mathcal{M}' \to [0, +\infty]$ that extends $m_n$ and still satisfies our "wish list" (page 82)? These are the central topics of the next section.

AXEL HARNACK (1851–1888) was a German mathematician who was born in Dorpat (now Tartu, Estonia) where his father was a professor. He studied mathematics and physics at the University of Dorpat and earned his doctorate in 1875 at the University of Erlangen, advised by Felix Klein (1849–1925). He taught in Leipzig, Darmstadt, and Dresden, where he remained for the rest of his life. He emphasized clear teaching methods and authored significant texts, including a revised translation of Joseph Serret's calculus work. He also wrote a Calculus book of his own, and its English translation contains the first known use of the phrase *neighbourhood of a point.* However, Harnack's health declined from around 1883, leading to intermittent teaching breaks. Despite temporary recoveries, his health deteriorated further in 1887. He collapsed during a lecture in 1888 and passed away shortly after, leaving a profound impact on mathematical analysis. His notable contributions, including the Harnack Inequality and Harnack's Principle, revolutionized the theory of harmonic functions, earning widespread recognition. Although some of the results he published were false, nevertheless his work is important in the development of integration theory. His contributions remain fundamental in mathematical analysis, evidenced by the numerous publications referencing his work.

## Exercises

5.3.1. Prove that there is no Lebesgue measurable set $A \subset \mathbb{R}$ with the property that for any interval $I$,

$$m(A \cap I) = \frac{1}{2} m(I).$$

5.3.2. Let $E$ be a Lebesgue measurable set such that $0 < m(E) < \infty$ and let $\{x_n\}$ be a sequence in $[0,1]$. Prove that there exists $y \in \mathbb{R}$ and a subsequence $\{x_{n_k}\}$ such that

$$y + x_{n_k} \in E, \qquad \text{for all } k \in \mathbb{N}.$$

5.3.3. Show that if $A \subset [0,1]$ and $m(A) > 0$, then there are $x$ and $y$ in $A$ such that $|x - y|$ is an irrational number.

5.3.4. Let $f : [0,1] \to \mathbb{R}$ be a Lipschitz function with constant $K$, and let $E \subset [0,1]$. Prove that

$$m^*(f(E)) \leq K m^*(E).$$

5.3.5. Let $E \subset \mathbb{R}^n$, let $\alpha \in \mathbb{R} \setminus \{0\}$, let $\beta \in \mathbb{R}^n$, and let $\alpha E + \beta = \{\alpha x + \beta : x \in E\}$. Prove that:

(a) $m^*(\alpha E + \beta) = |\alpha|^n m^*(E)$.

(b) If $E$ is Lebesgue measurable then so is $\alpha E + \beta$, and $m(\alpha E + \beta) = |\alpha|^n m(E)$.

5.3.6.* Prove that Definition 5.3.3 supplies the same value of $\mu_*(A)$ regardless of the choice of a rectangle $R$ in the bounded case, and rectangles $\{R_k\}$ in the unbounded case.

5.3.7. Let $\{R_i\}_{i=1}^{\infty}$ be a collection of rectangles in $\mathbb{R}^n$, and let $\varepsilon > 0$. Prove that there exist open rectangles $\{R_i'\}$ such that $R_i \subset R_i'$, for all $i \in \mathbb{N}$, and

$$\sum_{i=1}^{\infty} \operatorname{vol} R_i' \leq \sum_{i=1}^{\infty} \operatorname{vol} R_i + \varepsilon.$$

In Exercises 5.3.8–5.3.10, the Lebesgue $\sigma$-algebra $\mathcal{M}_n$ and the $n$-dimensional Lebesgue measure $m_n$ were obtained using Caratheodory construction, starting with the ring $\mathcal{E}$ of elementary sets and defining the outer measure $m_n^*$ as in Example 5.2.5.

5.3.8. Prove Theorem 5.3.7.

5.3.9. Prove Theorem 5.3.8.

5.3.10. Prove Theorem 5.3.9.

5.3.11. Let $E \in \mathcal{M}$ and $0 < m(E) < \infty$. Prove that for every $\alpha \in (0,1)$ there exists an interval $I$ for which $m(E \cap I) > \alpha\, m(I)$.

5.3.12. Prove or disprove: If the boundary of a set $A \subset \mathbb{R}^2$ has outer measure zero, then $A$ is measurable.

5.3.13. Prove that the collection of all elementary subsets of $\mathbb{R}^n$ is a ring.

5.3.14. Let $A, B \subset \mathbb{R}$ such that $\inf\{|x-y| : x \in A,\, y \in B\} > 0$. Prove or disprove: $m^*(A \cup B) = m^*(A) + m^*(B)$.

5.3.15. Let $E \subset \mathbb{R}$. Prove that $m^*(E) = \sum_{n=-\infty}^{\infty} m^*(E \cap [n, n+1])$.

5.3.16.* Let $E \subset \mathbb{R}^n$ and let $\{G_n\}$ be open subsets of $\mathbb{R}^n$ such that $\cup_{n=1}^{\infty} G_n = \mathbb{R}^n$. Prove that $m^*(E) = \lim_{n\to\infty} m^*(E \cap (\cup_{k=1}^{n} G_k))$.

5.3.17.* Let $A, B \subset \mathbb{R}^n$, let $m^*(A) = m^*(B) = 1$ and $m^*(A \cup B) = 2$. Prove that $m^*(A \cap B) = 0$.

5.3.18. Prove that the outer measure of the set of irrational numbers in $[0,1]$ is 1.

5.3.19. Let $f : [0,1] \to \mathbb{R}$ be a Lipschitz function with constant $K$, and let $E$ be a Lebesgue measurable subset of $[0,1]$. Prove that $f(E)$ is a measurable set.

5.3.20. Construct a closed, uncountable subset of $[0,1]$ that contains no rational numbers.

5.3.21. Prove that a set $A \subset \mathbb{R}^n$ is measurable if and only if

$$\operatorname{vol}(R) = m^*(R \cap A) + m^*(R \setminus A),$$

for each rectangle $R$ in $\mathbb{R}^n$.

5.3.22. Let $\mathcal{A}$ be a $\sigma$-algebra contained in $\mathcal{M}_n$. Prove or disprove: the restriction $m_n|\mathcal{A}$ is $\sigma$-finite.

5.3.23. Let $E \subset \mathbb{R}$ have a finite outer measure and suppose that it is not measurable. Prove that there exists an open set $G \supset E$ that has finite outer measure and for which $m^*(G \setminus E) > m^*(G) - m^*(E)$.

5.3.24. Let $X$ be a proper subspace of $\mathbb{R}^n$. Prove that $m(X) = 0$, where $m$ is the $n$-dimensional Lebesgue measure.

5.3.25. Let $\mu$ be a translation invariant measure on $\mathcal{M}_n$. Prove that there exists a constant $\alpha \geq 0$ such that $\mu(E) = \alpha m_n(E)$ for all Lebesgue measurable sets $E \subset \mathbb{R}^n$.

5.3.26. Prove that Proposition 5.3.2 holds in every Euclidean space $\mathbb{R}^n$. That is, prove that $\mu^*(A) = m^*(A)$, where $m^*(A)$ is as in Definition 4.2.11 and $\mu^*(A)$ as in Example 5.2.5, with $\mathfrak{E}$ the ring of elementary subsets of $\mathbb{R}^n$.

5.3.27. Prove that Theorem 4.2.22 holds in every Euclidean space $\mathbb{R}^n$. That is, prove that if $E$ is a bounded set in $\mathbb{R}^n$ and $R$ an open rectangle containing $E$ then,

$$m^*(E) + m_*(R \setminus E) = m(R),$$

where $m^*$ and $m_*$ are defined analogously to Definition 4.2.11.

5.3.28.* Prove that Proposition 5.3.4 holds in every Euclidean space $\mathbb{R}^n$. That is, prove that $\mu_*(A) = m_*(A)$, where $\mu_*(A)$ is as in Definition 5.3.3 and $m^*(A)$ as in Definition 4.2.11.

5.3.29. Prove the unbounded case in Proposition 5.3.4.

5.3.30. Prove or disprove: If $A$ is a subset of $[0, 1]$ such that $m(\text{Int}\,(A)) = m(\overline{A})$, then $A$ is measurable.

5.3.31. Let $f : \mathbb{R} \to \mathbb{R}$ be continuous. Determine the measure of its graph.

5.3.32. Show that if a set $E \subset \mathbb{R}^n$ has a positive outer measure, then there is a bounded subset of $E$ that also has a positive outer measure.

5.3.33. Let $E$ be a subset of $\mathbb{R}$ such that $m^*(E) < \infty$ and let $\varepsilon > 0$. Prove that $E$ is a disjoint union of a finite number of sets, each of which has an outer measure at most $\varepsilon$. Does the assertion remain true in $\mathbb{R}^n$? In any measurable space?

5.3.34. Find the Lebesgue measure of a subset of $[0, 1]$ whose decimal expansion contains a digit 2 before a digit 3.

5.3.35. Prove that Cantor set $F$ is measurable and that $m(F) = 0$.

5.3.36. Let $F$ denote the Fat Cantor set obtained by deleting at step $n$ from each of the $2^{n-1}$ intervals the middle interval of length $\alpha^n I$, where $0 < \alpha \leq 1/3$. (In the classical Cantor set $\alpha = 1/3$.) Prove that $F$ is measurable and find its measure.

5.3.37. Let $0 < \lambda < 1$ and construct a set $K$ as follows: Remove from $[0, 1]$ its middle part of length $\lambda$; we are left with two intervals $(a_1, b_1)$ and $(a_2, b_2)$. Remove from each of them a middle part of length $\lambda(b_i - a_i)$, etc. Prove that $m(K) = 0$.

5.3.38. Give an example of a subset of $[0, 1]$ that is of the first category and of Lebesgue measure 1.

5.3.39. Prove that there exists an open subset of $\mathbb{R}$ such that its boundary has a positive measure.

5.3.40. Prove or disprove: There exists a dense uncountable subset of $\mathbb{R}$ with Lebesgue measure zero.

5.3.41. Prove or disprove: If $A$ is an open subset of $[0, 1]$, then $m(A) = m(\overline{A})$.

5.3.42.* Prove or disprove: There exists an open dense subset $A$ of $[0,1]\times[0,1]$ such that its complement $([0,1]\times[0,1])\setminus A$ has positive Lebesgue measure.

5.3.43. Given a nonempty set $A\subset\mathbb{R}$ construct a function $f:\mathbb{R}\to\mathbb{R}$ which is constant a.e. and $f(\mathbb{R})=A$.

5.3.44. Prove that the Lebesgue $\sigma$-algebra $\mathcal{M}$ has the same cardinality as $\mathbb{P}(\mathbb{R})$.

5.3.45. Let $\{r_n\}$ a be a sequence that is dense in $\mathbb{R}$, and let

$$A=\bigcap_{k=1}^{\infty}\bigcup_{n=1}^{\infty}(r_n-2^{-(n+k)},\,r_n+2^{-(n+k)}).$$

Prove that $A$ is an uncountable set of measure 0.

5.3.46. Let $A\subset[0,1]$ be a measurable set of positive measures. Show that there exist two distinct points $x,y\in A$ such that $x-y$ is a rational number.

5.3.47.* Prove that there exists a measurable set $A\subset[0,1]$ such that $0<m(A\cap I)<m(I)$ for every interval $I\subset[0,1]$.

## 5.4 Beyond Lebesgue $\sigma$-Algebra

The main result of the previous section is Theorem 5.3.1 which established the existence of the Lebesgue $\sigma$-algebra $\mathcal{M}_n$ on $\mathbb{R}^n$. Even though we know some of its properties, we do not have an answer to the question: Is $\mathcal{M}_n=\mathbb{P}(\mathbb{R}^n)$? In other words: Is every set in $\mathbb{R}^n$ measurable? The answer depends on whether one accepts the Axiom of Choice or not. We have made a commitment to accept it, so we will see that there are nonmeasurable sets. (Another argument for its detractors to boot it out.) This will bring up more questions: Can $m_n$ be extended to a larger $\sigma$-algebra? Can we obtain a "measure" defined on all of $\mathbb{P}(\mathbb{R}^n)$ if we sacrifice or relax some of the properties from the wish list?

Our first order of business is to establish the existence of a nonmeasurable set. We will set our proofs in $\mathbb{R}$ but a similar construction can be performed in every Euclidean space. We start with a lemma.

**Lemma 5.4.1.** *Let $E$ be a bounded measurable subset of $\mathbb{R}$ and suppose that $\Lambda$ is a bounded, countably infinite subset of $\mathbb{R}$ such that the family $\{E+\lambda:\lambda\in\Lambda\}$ is disjoint. Then $m(E)=0$.*

*Proof.* By the translation invariance of the measure, $E+\lambda$ is measurable for every $\lambda$. Since $\{E+\lambda:\lambda\in\Lambda\}$ is disjoint, using the $\sigma$-additivity of $m$ we have

$$m\left(\bigcup_{\lambda\in\Lambda}(E+\lambda)\right)=\sum_{\lambda\in\Lambda}m(E+\lambda)=\sum_{\lambda\in\Lambda}m(E).$$

This shows that either $m(E) = 0$ or $m(\cup_{\lambda\in\Lambda}(E+\lambda)) = \infty$. However, both sets $E$ and $\Lambda$ is bounded, so the same is true of $\cup_{\lambda\in\Lambda}(E+\lambda)$, whence its measure is finite. It follows that $m(E) = 0$. □

Now we are ready to prove that $m(E)$ cannot be defined for all $E \in \mathbb{P}(\mathbb{R})$. As we said before, we will rely on the Axiom of Choice.

**Theorem 5.4.2.** *If $E \subset \mathbb{R}$ and $m(E) > 0$ then there exists a set $A \subset E$ that is not measurable.*

*Proof.* By Exercise 5.1.13, we may assume that $E$ is bounded. Let $E \subset [-b, b]$. We define a relation $\sim$ on $E$ by

$$x \sim y \quad \text{if and only if} \quad x - y \in \mathbb{Q}.$$

It is easy to verify that $\sim$ is an equivalence relation (Exercise 5.4.1), so $E$ is a union of equivalence classes. By the Axiom of Choice, we can choose a representative from each equivalence class. Let $C_E$ be the collection of these representatives (exactly one from each class). We will show that $C_E$ is not measurable, meaning that $m(C_E)$ cannot be defined.

Let $\Lambda = [-2b, 2b] \cap \mathbb{Q}$. Notice that $\{C_E + \lambda : \lambda \in \Lambda\}$ is disjoint. Otherwise, we would have two distinct numbers $\lambda_1, \lambda_2 \in \Lambda$ such that

$$(C_E + \lambda_1) \cap (C_E + \lambda_2) \neq \emptyset,$$

so there would exist $x, y \in C_E$ such that $x + \lambda_1 = y + \lambda_2$. This would imply that $x - y \in \mathbb{Q}$, hence $x \sim y$, contradicting the assumption that each element of $C_E$ is a representative of a unique equivalence class. Further,

$$E \subset \bigcup_{\lambda\in\Lambda} (C_E + \lambda).$$

Indeed, if $x \in E$, then $x \sim z$ for some $z \in C_E \subset E$, so $x - z \in \mathbb{Q}$ and $|x-z| \le |x|+|z| \le 2b$. This shows that $\lambda := x-z \in \Lambda$, so $x = z+\lambda \in C_E+\lambda$.

Suppose that $m(C_E)$ exists, either as a finite real number or $\infty$. The sets $C_E$ and $\Lambda$ satisfy the hypotheses of Lemma 5.4.1 and it follows that $m(C_E) = 0$. By the translation invariance of $m$, $m(C_E + \lambda) = 0$, for every $\lambda \in \Lambda$. Therefore, using the monotonicity and the $\sigma$-additivity of $m$,

$$m(E) \le m\left(\bigcup_{\lambda\in\Lambda} (C_E + \lambda)\right) = \sum_{\lambda\in\Lambda} m(C_E + \lambda) = 0,$$

contradicting the assumption that $m(E) > 0$. □

> With minor modification, the example used in the proof of Theorem 5.4.2 was provided by Vitali in 1905 in [119]. He gave a sketch of the construction and of the proof, while a much more detailed version that we presented is due to Sierpiński. There are many examples of nonmeasurable sets and it is remarkable that they all rely on the Axiom of Choice. It is worth mentioning that nobody has ever proved that removing this axiom ensures measurability of every subset of $\mathbb{R}^n$.

The existence of nonmeasurable sets has an important consequence, namely that the outer measure $m^*$ is not additive. We state the result for $\mathbb{R}$ but it holds in any Euclidean space. The proof is left to the reader (Exercise 5.4.2).

**Theorem 5.4.3.** *There are disjoint sets $A, B \subset \mathbb{R}$ such that $m^*(A \cup B) < m^*(A) + m^*(B)$.*

Since we have established that $\mathcal{M}_n \neq \mathbb{P}(\mathbb{R}^n)$, it is a natural question whether there is a larger $\sigma$-algebra $\mathcal{M}'$ and a function $m' : \mathcal{M}' \to [0, +\infty]$ that is countably additive, translation invariant, and it coincides with Lebesgue measure on rectangles. The answer is yes, and there is more than one such measure. However, none of them is a "universal" measure, i.e., a measure defined on all of $\mathbb{P}(\mathbb{R}^n)$. Moreover, Theorem 5.3.8 shows that every Lebesgue measurable set can be approximated by open and closed sets, and this property characterizes $\mathcal{M}_n$, so it would be lost if we were to select one of these larger $\sigma$-algebras. Thus, we stick to $\mathcal{M}_n$ and $m_n$.

Another question is whether a universal measure could be obtained by relaxing one of the properties in the wish list. One such modification would be to require that the measure be merely additive (instead of countably additive). In 1923, Polish mathematician Banach proved that such a measure exists, both on the real line and in the plane. However, the result does not generalize to higher dimensions. Already in $\mathbb{R}^3$ there is no universal, finitely additive, isometrically invariant extension of the Lebesgue measure. (Isometricaly invariant means that if $T$ is an isometric transformation of $\mathbb{R}^3$, then the sets $E$ and $T(E)$ have the same measure.) Postulating the existence of such a universal measure would lead to the so-called Banach–Tarski paradox. Here, two sets are called congruent if and only if one can be transformed into the other by an isometry, i.e., a combination of rigid motions, namely a translation, a rotation, and a reflection.

**Theorem 5.4.4.** *Given any two bounded subsets $A$ and $B$ of an Euclidean space in at least three dimensions, both of which have a nonempty interior, there are partitions of $A$ and $B$ into a finite number of disjoint subsets, $A = A_1 \cup \cdots \cup A_k$, $B = B_1 \cup \cdots \cup B_k$ (for some integer $k$), such that for each $i$ between 1 and $k$, the sets $A_i$ and $B_i$ are congruent.*

As a consequence, we can take a unit ball in $\mathbb{R}^3$ and decompose it into a finite number of pieces (5 will suffice) and then assemble these pieces to obtain two copies of the unit ball. Of course, the reason for this phenomenon is that the pieces will not be measurable, even if we forego countable additivity and only require the measure to be finitely additive.

> The Banach–Tarski paradox was published in 1924 in [8] by Banach and his colleague Alfred Tarski (1901–1983). Karl Stromberg published in 1979 a much simpler proof (down to 10 from 30 pages) in [117]. Tarski was a Polish-American logician and mathematician, considered with Gödel the most important logicians of the 20th century. He was the youngest person ever to complete a doctorate at Warsaw University. From 1942 until his death he was on the faculty of UC Berkeley.

We were all the time concerned with extending the measure to a $\sigma$-algebra as large as possible. For many purposes, it is sufficient to consider the smallest one, i.e., the one generated by rectangles.

**Definition 5.4.5.** The **Borel $\sigma$-algebra** $\mathcal{B}$ is the smallest $\sigma$-algebra that contains all open sets. The sets in $\mathcal{B}$ are called **Borel sets**.

The $\sigma$-algebra $\mathcal{B}$ is generated by open sets. However, it can be also generated by several other classes of sets. (See Exercise 5.4.8.)

The following result can be deduced straight from the definition.

**Proposition 5.4.6.** *The $\sigma$-algebra $\mathcal{M}$ contains $\mathcal{B}$.*

It is easy to see that the restriction of $m$ to $\mathcal{B}$ the satisfies conditions of our wish list, so we have the option to choose $\mathcal{M}$ or $\mathcal{B}$ to work with. This immediately raises a question of whether the inclusion $\mathcal{B} \subset \mathcal{M}$ is proper. In order to demonstrate that it is, we will use the Cantor–Lebesgue function $\varphi$. By Theorem 3.1.7, $\varphi$ is an increasing, continuous function that maps $[0, 1]$ onto $[0, 1]$. Its derivative exists in $O$ (the complement of the Cantor set) and $\varphi' = 0$ on $O$. Further, Exercise 5.3.35 shows that the Cantor set has measure 0 so $m(O) = 1$. Actually, we will use the function $\psi$ on $[0, 1]$ defined by $\psi(x) = \varphi(x) + x$. The following theorem states the relevant properties of this function.

**Theorem 5.4.7.** *Let $\varphi$ be the Cantor–Lebesgue function and $\psi(x) = \varphi(x)+x$. Then $\psi$ has the following properties:*

*(a) It is a strictly increasing, continuous function;*

*(b) it maps $[0, 1]$ onto $[0, 2]$;*

*(c) it maps the Cantor set $C$ onto a set of measure 1;*

*(d) it maps some measurable subset $A$ of $C$ onto a nonmeasurable set.*

*Proof.* Assertion (a) follows from the analogous properties of $\varphi$ and $f(x) = x$. In view of (a), $\psi$ maps $[0, 1]$ onto the interval $[\psi(0), \psi(1)] = [0, 2]$, so (b) holds.

Next, the interval $[0, 2]$ is a disjoint union of $\psi(O)$ and $\psi(C)$. Thus, in order to establish (c) it suffices to show that $\psi(O)$ is a measurable set of measure 1. To that end, notice that $O$ is a countable union of intervals $\{O_n\}$ and, although these are not disjoint, $O$ can be represented as a disjoint union of open intervals that we will denote by $\{I_n\}_{n=1}^{\infty}$. Since $\psi$ is injective, Example 1.2.4 shows that $\psi(O)$ is a disjoint union of $\psi(I_n)$. Further, if $I_n = (a_n, b_n)$ then

$$\psi(I_n) = (\psi(a_n), \psi(b_n)) = (\varphi(a_n) + a_n, \varphi(b_n) + b_n).$$

The Cantor-Lebesgue function $\varphi$ is constant on $I_n$ (and continuous), and we get that $\psi(I_n)$ is an interval of the same length as $I_n$. Thus, $\psi(O)$ is a disjoint union of intervals $\{\psi(I_n)\}$, hence it is measurable and

$$m(\psi(O)) = m\left(\bigcup_{n=1}^{\infty} \psi(I_n)\right) = \sum_{n=1}^{\infty} m(\psi(I_n))$$
$$= \sum_{n=1}^{\infty} m(I_n) = m\left(\bigcup_{n=1}^{\infty} I_n\right) = m(O) = 1.$$

Finally, by Theorem 5.4.2, $\psi(C)$ contains a nonmeasurable subset $B$. It is easy to verify that the set $A = \psi^{-1}(B)$ satisfies (d) and the theorem is proved. □

Now we can at last prove that the inclusion $\mathcal{B} \subset \mathcal{M}$ is proper or, equivalently, that there is a Lebesgue measurable set that is not a Borel set.

**Theorem 5.4.8.** *There exists a measurable set that is not a Borel set.*

*Proof.* The function $\psi$ is increasing and continuous, so it maps a Borel set to a Borel set (Exercise 5.4.9). Let $A$ be the set guaranteed by part (d) of Theorem 5.4.7. Then $A \in \mathcal{M}$ but $\psi(A)$ is not a Borel set because it does not belong to $\mathcal{M}$. Therefore, $A \notin \mathcal{B}$. □

Theorem 5.4.8 shows that $\mathcal{B}$ is a proper subset of $\mathcal{M}$. In order to understand better the relation between these $\sigma$-algebras we need to introduce the notion of a *complete* measure space.

**Definition 5.4.9.** A measure space $(X, \mathfrak{M}, \mu)$ is **complete** if $A \subset E$ and $\mu(E) = 0$ implies that $\mu(A) = 0$.

**Example 5.4.10. A complete measure space.**
$(\mathbb{R}^n, \mathcal{M}_n, m_n)$ is complete. (See Exercise 5.2.24.) ♦

**Example 5.4.11. A measure space that is not complete.**
$(\mathbb{R}, \mathcal{B}, m|\mathcal{B})$ is not complete.

**Solution:** In the proof of Theorem 5.4.8 we exhibited a set $A$ that is not a Borel set. On the other hand, it is a subset of the Cantor set which belongs to $\mathcal{B}$ because it is a closed set, and it is a set of measure 0. Thus, $m|\mathcal{B}$ is not complete. ♦

**Example 5.4.12. A measure space that is not complete.**
Let $X = [0,1]$, let $\mathfrak{M}$ consist of the empty set and $X$, and let $\mu(\emptyset) = \mu(X) = 0$. The measure space $(X, \mathfrak{M}, \mu)$ is not complete. ♦

Suppose that a measure space $(X, \mathfrak{M}, \mu)$ is not complete. Can we make it complete? That is, does there exist a $\sigma$-algebra that contains $\mathfrak{M}$ so that $\mu$ extends to this larger $\sigma$-algebra and this larger measure space is complete? The following result shows that the answer is in the affirmative and moreover provides a blueprint for how to construct the extension. The proof is Exercise 5.4.24.

**Theorem 5.4.13.** *Let $(X, \mathfrak{M}, \mu)$ be a measure space. Define $\mathfrak{M}_0$ to be the collection of subsets $E$ of $X$ of the form $E = A \cup B$, where $B \in \mathfrak{M}$ and $A$ is contained in a set of $\mu$ measure 0. For such a set $E$ define $\mu_0(E) = \mu(B)$. Then $\mathfrak{M}_0$ is a $\sigma$-algebra that contains $\mathfrak{M}$, $\mu_0$ is a unique measure that extends $\mu$ to $\mathfrak{M}_0$, and $(X, \mathfrak{M}_0, \mu_0)$ is complete. Further, $\mathfrak{M}_0$ is the smallest such $\sigma$-algebra.*

**Definition 5.4.14.** The measure space $(X, \mathfrak{M}_0, \mu_0)$ constructed in Theorem 5.4.13 is called the **completion** of $(X, \mathfrak{M}, \mu)$.

With the established terminology we can finally formulate the relationship between $\mathcal{B}$ and $\mathcal{M}$. We leave the proof as Exercise 5.4.29.

**Theorem 5.4.15.** *The Lebesgue measure space $(\mathbb{R}^n, \mathcal{M}_n, m_n)$ is the completion of the Borel measure space $(\mathbb{R}^n, \mathcal{B}_n, m_n|\mathcal{B}_n)$.*

Fréchet discusses completeness of a measure in [38] from 1915. In his influential book [47] from 1921, Hahn stated and proved Theorem 5.4.13.

GIUSEPPE VITALI (1875–1932) was an Italian mathematician. He grew up in Ravenna and studied for two years at the University of Bologna. With Arzelà's support he received a scholarship to study at the Scuola Normale Superiore in Pisa with Luigi Bianchi (1856–1928) and Dini. He graduated in 1899 and assisted Dini for two years until he obtained a teaching diploma and became a secondary school teacher. He taught in Genoa until 1923. While there, he became involved in politics as a member of the Italian Socialist Party until it was disbanded by the fascists in 1922. During that time he established his most significant results: the theorem on discontinuity points of Riemann-integrable functions (Theorem 7.4.3), the first example of a nonmeasurable set (p.138), a characterisation of absolutely continuous functions (p.276), and the covering theorem (p.256). Starting in 1923 he held university positions in Modena, Padua, and Bologna. From 1926 he had a paralysed arm, meaning he could no longer write. Nevertheless, about half his research papers were written in the last four years of his life. He was an invited speaker at the International Congress of Mathematicians in 1928 and a member of the Academy of Sciences of Turin, the Lincean Academy, and the Academy of Bologna. He was only 56 when he collapsed and died in the street.

## Exercises

5.4.1. Prove that the relation $\sim$ introduced in the proof of Theorem 5.4.2 is an equivalence relation.

5.4.2. Prove Theorem 5.4.3.

5.4.3. Show that there is a continuous, strictly increasing function on the interval $[0, 1]$ that maps a set of positive measures onto a set of measure zero.

5.4.4. Let $f$ be a continuous function on a set $E \subset \mathbb{R}$, and let $A$ be a Lebesgue measurable subset of $\mathbb{R}$. Prove or disprove: $f^{-1}(A)$ is measurable.

5.4.5. Let $A$ be a nonmeasurable set constructed in the proof of Theorem 5.4.2 and let $B \subset A$ be measurable. Prove that $m(B) = 0$.

5.4.6. Let $C_E$ be as in Theorem 5.4.2.

(a) Prove that $m^*(C_E) > 0$.

(b) Give an example of a decreasing sequence $\{A_n\}$ such that $m^*(A_1)$ is finite and $m^*(\cap_{n=1}^{\infty} A_n) < \lim_n m^*(A_n)$.

(c) Prove that $m_*(C_E) = 0$.

5.4.7. Let $A \subset \mathbb{R}$ have the property that

$$\inf\{m(G) : A \subset G,\ G \text{ open in } \mathbb{R}\} = \sup\{m(F) : F \subset A,\ F \text{ closed in } \mathbb{R}\}.$$

Prove or disprove: $A$ is measurable.

5.4.8. Show that the Borel $\sigma$-algebra $\mathcal{B}$ is generated by any of the following collections of subsets of $\mathbb{R}^2$: (i) The closed sets. (ii) The compact sets. (iii) The open disks. (iv) The rectangles. (v) The elementary sets.

5.4.9. Let $f : \mathbb{R} \to \mathbb{R}$ be a continuous increasing function and let $E$ be a Borel set. Prove that $f(E)$ is also a Borel set.

5.4.10. Prove that if a $\sigma$-algebra of subsets of $\mathbb{R}$ contains all intervals of the form $(a, +\infty)$, then it contains all intervals.

5.4.11. Show that every interval is a Borel set.

5.4.12. Let $X = \mathbb{R}^2$, let $\mathfrak{M} = \mathcal{B}_2$, the $\sigma$-algebra of Borel subsets of $\mathbb{R}^2$, and let $\mu = m_2$, the two-dimensional Lebesgue measure on $\mathcal{B}_2$. Let $D = \{(x, x) : x \in \mathbb{R}\}$. Prove that $D \in \mathcal{B}_2$ and that $m_2(D) = 0$.

5.4.13. Prove that the $\sigma$-algebra $\mathcal{M}$ obtained in Exercise 5.2.25 contains all Borel subsets of $[a, b]$.

5.4.14.* Let $\mu$ be a finite Borel measure on $\mathbb{R}^n$, and for $x \in \mathbb{R}^n$ let $f(x) = \mu(B(x, 1))$, where $B(x, 1)$ denotes the open ball centered at $x$ of radius 1. Prove that $f$ attains its minimum on each compact set in $\mathbb{R}^n$.

5.4.15. Let $E \in \mathcal{B}_n, F \in \mathcal{B}_m$. Prove that $E \times F \in \mathcal{B}_{n+m}$.

5.4.16. Prove that every measurable set in $\mathbb{R}^n$ is a union of a Borel set and a set of measure 0.

5.4.17.* Let $\mathcal{A}$ be a countably generated $\sigma$-algebra and let $\mathcal{C} \subset \mathcal{A}$ be a $\sigma$-algebra. Proves or disprove: $\mathcal{C}$ is countably generated.

5.4.18.* Prove or disprove: there is a $\sigma$-finite measure on $\mathcal{B}(\mathbb{R}^n)$ such that every Borel set with a nonempty interior has infinite measure.

5.4.19. Let $\mathcal{F}$ be a family of open intervals such that their endpoints form a dense subset of $\mathbb{R}$. Prove or disprove: $\sigma(\mathcal{F}) = \mathcal{B}$.

5.4.20. Prove that the subset of $\mathbb{P}(\mathbb{R})$ consisting of the sets of the form $A \triangle B$, where $A \in \mathcal{B}$ and $B \in \mathcal{M}$ is a set of measure zero, is a $\sigma$-algebra.

5.4.21.* Let $\mu$ be a finite Borel measure on $\mathbb{R}$ and let $A$ be a closed subset of $\mathbb{R}$. Prove that $f : \mathbb{R} \to \mathbb{R}$ defined by $f(x) = \mu(A + x)$ is upper semicontinuous.

5.4.22. Suppose that $X$ is a metric space. Prove that every Borel measurable subset of $X$ can be approximated from the inside by a closed set, i.e., if $E \subset X$ is a Borel set of finite measure and $\varepsilon > 0$ then there exists a closed set $F \subset E$ such that $\mu(E \setminus F) < \varepsilon$.

5.4.23. Let $E \subset \mathbb{R}^n$ be a nonmeasurable set of finite outer measures. Show that there is a $G_\delta$ set $G$ such that

$$E \subset G, \quad m^*(E) = m^*(G), \quad \text{and} \quad m^*(G \setminus E) > 0.$$

5.4.24.* Prove Theorem 5.4.13.

5.4.25. Prove that any measure obtained from Carathéodory's extension theorem is complete. In particular, the Lebesgue measure is complete.

5.4.26. Let $(X, \mathfrak{M}, \mu)$ be a complete measure space. Prove that if $A \cup B \in \mathfrak{M}$ and $\mu(B) = 0$ then $A \in \mathfrak{M}$.

5.4.27. Let $(X, \mathfrak{M}, \mu)$ be a complete measure space. Prove that if $A, B \subset X$ are such that $A \in \mathfrak{M}$ and $\mu(A \triangle B) = 0$ then $B \in \mathfrak{M}$ and $\mu(B) = \mu(A)$.

5.4.28. Let $(X, \mathfrak{M}, \mu)$ be a measure space and let $(X, \mathfrak{M}_0, \mu_0)$ be its completion. Prove that $\mu^*(E) = \mu_0(E)$ for all $E \in \mathfrak{M}_0$.

5.4.29. Prove Theorem 5.4.15.

5.4.30. Let $(X, \mathfrak{M}, \mu)$ be a measure space. Define $\mathfrak{M}_1$ to be the collection of subsets $E$ of $X$ for which there exist sets $A, B \in \mathfrak{M}$ such that $A \subset E \subset B$ and $\mu(B \setminus A) = 0$. For such a set $E$ define $\mu_1(E) = \mu(A)$. Prove that $(X, \mathfrak{M}_1, \mu_1)$ is the (minimal) completion of $(X, \mathfrak{M}, \mu)$.

## 5.5 Signed Measures

As a further generalization, one may consider measures whose values are not necessarily nonnegative numbers. To remain within the realm of *real* analysis we will assume that the values of such a measure may be negative, but will not consider complex numbers. To distinguish them from the "usual" measures we will attach the adjective *signed* to them. (Those who use complex numbers call them complex measures.)

Why study signed measures? Let us consider the following situation. Suppose that $(X, \mathfrak{M})$ is a measurable space and that $\mu_1$, $\mu_2$ are two measures on $\mathfrak{M}$. Let $\alpha, \beta$ be real numbers. Is $\alpha\mu_1 + \beta\mu_2$ a measure? A cursory inspection seems to indicate that the answer is in the affirmative. Since $\mu_1(\emptyset) = 0$ and $\mu_2(\emptyset) = 0$ it follows that $(\alpha\mu_1 + \beta\mu_2)(\emptyset) = 0$. Also, if $\{E_n\}_{n=1}^{\infty}$ is a disjoint sequence in $\mathfrak{M}$ then

$$(\alpha\mu_1 + \beta\mu_2)\left(\bigcup_{n=1}^{\infty} E_n\right) = \alpha\mu_1\left(\bigcup_{n=1}^{\infty} E_n\right) + \beta\mu_2\left(\bigcup_{n=1}^{\infty} E_n\right)$$
$$= \alpha\sum_{n=1}^{\infty}\mu_1(E_n) + \beta\sum_{n=1}^{\infty}\mu_2(E_n)$$
$$= \sum_{n=1}^{\infty}(\alpha\mu_1 + \beta\mu_2)(E_n).$$

However, this calculation is justified when $\alpha, \beta \geq 0$. In general, for a set $E \in \mathfrak{M}$, the number $(\alpha\mu_1 + \beta\mu_2)(E)$ may be negative. Thus, $\alpha\mu_1 + \beta\mu_2$ may not be a measure but a signed measure.

**Definition 5.5.1.** **A signed measure** $\nu$ on a measurable space $(X, \mathfrak{M})$ is a function $\nu : \mathfrak{M} \to \mathbb{R}^*$ such that:

(i) $\nu(\emptyset) = 0$.

(ii) If $\{E_n\}_{n=1}^{\infty}$ is a disjoint sequence in $\mathfrak{M}$ then $\nu(\cup_{n=1}^{\infty} E_n) = \sum_{n=1}^{\infty} \nu(E_n)$.

**Example 5.5.2. A signed measure on $\mathbb{R}$.**
Let $m$ be Lebesgue measure on the real line and define, for a set $E \in \mathcal{M}$,

$$\mu_1(E) = m(E \cap (-1, 0)), \qquad \mu_2(E) = m(E \cap (0, 1)).$$

Then $\mu_1 - \mu_2$ is a signed measure. ♦

The next example shows another way in which a signed measure may appear.

**Example 5.5.3. A signed measure on $\mathcal{M}|[0, 2\pi]$.**
For every interval $I = [a, b] \subset [0, 2\pi]$ let $\mu(I) = \int_a^b \sin x\, dx$. Clearly, $\mu$ takes both positive and negative values as is well known to anyone who has gone past elementary calculus. It is not hard to see that $\mu = \mu_1 - \mu_2$, where $\mu_1(I) = \mu(I \cap [0, \pi])$ and $\mu_2(I) = -\mu(I \cap [\pi, 2\pi])$ and while $\mu_1, \mu_2$ are not measures (they are defined only on intervals), they can be extended to measures on $\mathcal{M}|[0, 2\pi]$. Thus, $\mu$ extends to a signed measure on $\mathcal{M}|[0, 2\pi]$. ♦

We make several observations here. First, it is essential that a signed measure assumes at most one of the values $+\infty, -\infty$. Indeed, if $\nu(A) = +\infty$ and $\nu(B) = -\infty$, then what is $\nu(A \cap B)$? It cannot be a finite number because we would have $\nu(A \setminus B) = +\infty$ and $\nu(B \setminus A) = -\infty$, so $\nu(A \triangle B)$ would be undefined. If it were $+\infty$, then we would run into a snag with the formula $\nu(B) = \nu(B \setminus A) + \nu(A \cap B)$. A similar argument shows that $\nu(A \cap B)$ cannot be $-\infty$. We conclude that

**Proposition 5.5.4.** *A signed measure assumes at most one of the values $+\infty, -\infty$.*

In view of Proposition 5.5.4 we will make a standing assumption that $\nu(E) > -\infty$ for all measurable sets $E$.

*Remark* 5.5.5. A definition where the infinite value is $-\infty$ would be perfectly acceptable and some authors do just that.

Our second observation concerns the $\sigma$-additivity of $\nu$. Since the measure of $\cup_{n=1}^{\infty} E_n$ does not depend on the order of sets $E_n$, neither should the sum $\sum_{n=1}^{\infty} \nu(E_n)$. In other words, we would hope that the series converges absolutely.

**Proposition 5.5.6.** *Let $\nu$ be a signed measure on a measurable space $(X, \mathfrak{M})$ and let $\{E_n\}_{n=1}^{\infty}$ be a disjoint sequence in $\mathfrak{M}$ such that $\nu(\cup_{n=1}^{\infty} E_n)$ is a finite number. Then the series $\sum_{n=1}^{\infty} \nu(E_n)$ converges absolutely.*

*Proof.* Let $I = \{n \in \mathbb{N} : \nu(E_n) \geq 0\}$ and $J = \{n \in \mathbb{N} : \nu(E_n) < 0\}$. If we denote $A := \cup_{n\in I} E_n$ and $B := \cup_{n\in J} E_n$, then $\nu(A \cup B) = \nu(A) + \nu(B)$, so the finiteness of $\nu(A \cup B)$ implies the same for $\nu(A)$ and $\nu(B)$. Let $m \in \mathbb{N}$ and let $C = \{1, 2, \dots, m\}$. Then

$$\begin{aligned}\sum_{n=1}^{m} |\nu(E_n)| &= \sum_{n\in I\cap C} |\nu(E_n)| + \sum_{n\in J\cap C} |\nu(E_n)| = \sum_{n\in I\cap C} \nu(E_n) - \sum_{n\in J\cap C} \nu(E_n) \\ &\leq \sum_{n\in I} \nu(E_n) - \sum_{n\in J} \nu(E_n) = \nu(A) - \nu(B).\end{aligned}$$

Since this holds for any $m \in \mathbb{N}$ we get that the series $\sum_{n=1}^{\infty} |\nu(E_n)|$ converges. $\square$

Another property of signed measures has already been brought forward at the beginning of the section.

**Proposition 5.5.7.** *Let $\nu_1, \nu_2$ be two signed measure on a measurable space $(X, \mathfrak{M})$ such that at least one of them is finite, and let $\alpha, \beta \in \mathbb{R}$. Then $\alpha\nu_1 + \beta\nu_2$ is a signed measure.*

Our goal in this section will be to show that every signed measure resembles the one in Examples 5.5.2 and 5.5.3, i.e., that there is a difference of the two *measures*. To that end, we start with a definition.

**Definition 5.5.8.** A set $A$ is **positive** (resp., **negative, null**) with respect to $\nu$, if it is measurable and $\nu(E) \geq 0$ (resp., $\nu(E) \leq 0$, $\nu(E) = 0$) for every measurable set $E \subset A$.

In Example 5.5.2 positive sets are, for example, $[0, 1/2]$, $(3, 5]$, $(1/3, +\infty)$. Some negative sets are $(-3, -2)$ and $(-2, 0)$. In fact, the set of real numbers is a disjoint unuon of sets $(-\infty, 0)$ and $[0, +\infty)$, the former being a negative set, the latter a positive set for $\mu_1 - \mu_2$. This motivates the following definition.

**Definition 5.5.9.** Let $(X, \mathfrak{M})$ be a measurable space and let $\nu : \mathfrak{M} \to (-\infty, +\infty]$ be a signed measure. Suppose that $N, P \in \mathfrak{M}$. The ordered pair $(P, N)$ is a **Hahn decomposition** for $\nu$ if $P$ is a positive set for $\nu$, $N$ is a negative set for $\nu$, and $X$ is a disjoint union of $P$ and $N$.

Our goal is to prove that for every signed measure there are such two sets. First, we need to establish a necessary and sufficient condition for a pair of sets to be a Hahn decomposition.

**Lemma 5.5.10.** *Let $(X, \mathfrak{M})$ be a measurable space and $\nu : \mathfrak{M} \to (-\infty, +\infty]$ a signed measure. Suppose that $N$ and $P := N^C$ are in $\mathfrak{M}$. Then the following are equivalent.*

*(a) The ordered pair $(P, N)$ is a Hahn decomposition for $\nu$.*

*(b) For each $E \in \mathfrak{M}$, we have $\nu(E) \geq \nu(N)$.*

*Proof.* (a) $\Rightarrow$ (b) Let $E \in \mathfrak{M}$. Then $\nu(P \cap E) \geq 0 \geq \nu(N \cap E^C)$ and it follows that

$$\nu(E) = \nu(N \cap E) + \nu(P \cap E) \geq \nu(N \cap E) + \nu(N \cap E^C) = \nu(N).$$

(b) $\Rightarrow$ (a) Let $E \in \mathfrak{M}$. By assumption in (b), $\nu(N \cap E^C) \geq \nu(N)$, so

$$\nu(N) + \nu(N \cap E) \leq \nu(N \cap E^C) + \nu(N \cap E) = \nu(N).$$

Further, $-\infty < \nu(N) \leq \nu(\emptyset) = 0 < \infty$, so $\nu(N \cap E) \leq 0$, whence $N$ is a negative set. Next,

$$\nu(N) \leq \nu(N \cup E) = \nu(N) + \nu(E \cap N^C),$$

which shows that, defining $P = N^C$, $\nu(P \cap E) = \nu(E \cap N^C) \geq 0$, whence $P$ is a positive set. □

**Theorem 5.5.11.** ***The Hahn Decomposition Theorem.*** *Let $\nu : \mathfrak{M} \to (-\infty, +\infty]$ be a signed measure on $(X, \mathfrak{M})$. Then there is a positive set $P$ for $\nu$ and a negative set $N$ for $\nu$ such that*

$$X = P \cup N \quad \text{and} \quad P \cap N = \emptyset.$$

In the language of Definition 5.5.9, the theorem asserts that every signed measure has a Hahn decomposition.

*Proof.* Let $\lambda = \inf\{\nu(E) : E \in \mathfrak{M}\}$. If $\lambda \geq 0$ we take $P = X$. So, without loss of generality, we will assume that $\lambda < 0$.

By Lemma 5.5.10, it suffices to exhibit a set $N \in \mathfrak{M}$ such that $\nu(N) = \lambda$. To that end, let $\{D_n\}_{n=1}^{\infty}$ be a sequence in $\mathfrak{M}$ such that $0 > \lim \nu(D_n) = \lambda$. Let $\mathcal{A}_n$ be a $\sigma$-algebra generated by $\{D_k\}_{k=1}^{n}$ and denote by $\nu_n$ the restriction of $\nu$ to $\mathcal{A}_n$. By Exercise 4.3.29, $\mathcal{A}_n$ is finite, so $\nu_n$ attains its minimum at a set $C_n \in \mathcal{A}_n$. Let

$$A_m = \bigcap_{k=m}^{\infty} C_k \quad \text{and } A_{mn} = \bigcap_{k=m}^{n} C_k, \quad \text{for } n \geq m.$$

For a fixed $m$, the sequence $\{A_{mn}\}_{n=m}^{\infty}$ is decreasing and

$$-\infty < \nu(A_{mm}) = \nu(C_m) \leq \nu(D_m) < 0,$$

so the continuity of a signed measure (Exercise 5.5.1) implies that

$$\lim_n \nu(A_{mn}) = \nu(A_m). \tag{5.12}$$

Since $\nu_n$ attains the minimum at $C_n$, Lemma 5.5.10 implies that $(C_n, X \setminus C_n)$ is a Hahn decomposition for $\nu_n$, i.e., $C_n$ is a negative set and $X \setminus C_n$ is a positive set for $\nu_n$. For $n > m$,

$$\begin{aligned}
\nu(A_{m,n-1}) = \nu_n(A_{m,n-1}) &= \nu_n\left(\bigcap_{k=m}^{n-1} C_k\right) \\
&= \nu_n\left(\bigcap_{k=m}^{n-1} C_k \cap C_n\right) + \nu_n\left(\bigcap_{k=m}^{n-1} C_k \cap (X \setminus C_n)\right) \\
&\geq \nu_n\left(\bigcap_{k=m}^{n-1} C_k \cap C_n\right) = \nu_n(A_{m,n}) = \nu(A_{m,n}).
\end{aligned}$$

It follows that

$$\nu(D_m) \geq \nu(C_m) = \nu(A_{mm}) \geq \nu(A_{m,m+1}) \geq \nu(A_{m,m+2}) \geq \dots$$

Therefore, using (5.12),

$$\nu(D_m) \geq \nu(A_m) \geq \lambda,$$

so $\lim_m \nu(A_m) = \lambda$. Now, we define

$$N = \bigcup_{m=1}^{\infty} A_m.$$

The sequence $\{A_m\}_{m=1}^{\infty}$ is increasing, so $\nu(N) = \lim \nu(A_m) = \lambda$. □

*Remark* 5.5.12. Hahn decomposition is not unique. If $\nu(E) = 0$ then $(N \setminus E, P \cup E)$ is another Hahn decomposition of the same space. In fact, that is the only way this can happen (see Exercise 5.5.5).

In Example 5.5.3, Hahn's Decomposition of $[0, 2\pi]$ is $([0, \pi], (\pi, 2\pi])$. We again emphasize that not only the integral over $[0, \pi]$ is positive, but so is the integral over any of its subintervals. And a similar statement holds for the negative set $(\pi, 2\pi]$.

The Hahn Decomposition Theorem appears in Hahn's 1921 book [47]. Sierpiński published in 1924 in [112] a much simpler proof.

A Hahn decomposition partitions the underlying set $X$. This leads to a decomposition of the signed measure $\nu$. Before we can state this result precisely we need to expand our vocabulary.

**Definition 5.5.13.** Two signed measures $\nu_1, \nu_2$ on $(X, \mathfrak{M})$ are **mutually singular** if $X$ is a disjoint union of measurable sets $A, B$ and $\nu_1$ is zero on all measurable subsets of $B$ while $\nu_2$ is zero on all measurable subsets of $A$. In that case, we write $\nu_1 \perp \nu_2$.

**Example 5.5.14. Two mutually singular measures.**

Let $\nu_1 = m$, Lebesgue measure on $\mathbb{R}$ and $\nu_2 = \delta_{x_0}$, Dirac measure. These measures are mutually singular.

Reason: take $A = \mathbb{R}\backslash\{x_0\}$ and $B = \{x_0\}$. Then $\nu_1$ is zero on all measurable subsets of $B$ while $\nu_2$ is zero on all measurable subsets of $A$. ♦

We now introduce some important measures that can be associated with a signed measure $\nu$.

**Definition 5.5.15.** Let $\nu$ be a signed measure on $(X, \mathfrak{M})$ and let $(P, N)$ be a Hahn decomposition for $\nu$. The functions

$$\nu^+(E) = \nu(E \cap P), \quad \nu^-(E) = -\nu(E \cap N), \quad |\nu|(E) = \nu^+(E) + \nu^-(E), \tag{5.13}$$

are called the **positive variation**, the **negative variation**, and the **total variation** of $\nu$, respectively.

*Remark* 5.5.16. It is not hard to see that the values of these functions do not depend on the chosen Hahn decomposition, and that each of them is a measure (not a signed measure).

**Example 5.5.17. Positive and negative variation.** Let $\mu, \mu_1, \mu_2$ be as in Example 5.5.3. Then $\nu^+(E) = \nu(E \cap [0, \pi])$ and $\nu^-(E) = -\nu(E \cap [\pi, 2\pi])$. ♦

Now we can state the result about a decomposition of a signed measure.

**Theorem 5.5.18.** ***The Jordan Decomposition Theorem.*** *Let $\nu$ be a signed measure on $(X, \mathfrak{M})$. Then there are two mutually singular measures $\nu^+, \nu^-$ on $(X, \mathfrak{M})$ for which $\nu = \nu^+ - \nu^-$. Moreover, there is only one such pair of measures.*

*Proof.* It is easy to verify that the positive and the negative variation of $\nu$ satisfy the conclusions of the theorem (Exercise 5.5.12).

For uniqueness, suppose that $\nu$ is the difference of mutually singular measures $\nu = \lambda - \mu$. By definition, $X$ is a disjoint union of measurable sets $A, B$

and $\lambda$ is zero on all measurable subsets of $B$ while $\mu$ is zero on all measurable subsets of $A$. If $E \in \mathfrak{M}$, then

$$\nu(E\cap A) = \lambda(E\cap A)-\mu(E\cap A) = \lambda(E\cap A) = \lambda(E\cap A)+\lambda(E\cap B) = \lambda(E) \geq 0, \tag{5.14}$$

which shows that $A$ is a positive set for $\nu$. A similar argument yields that

$$-\nu(E \cap B) = \mu(E), \tag{5.15}$$

so $B$ is a negative set for $\nu$ whence $\{A, B\}$ is a Hahn decomposition of $X$. If $\{P, N\}$ is the decomposition used in (5.13), Exercise 5.5.5 shows that

$$\nu(P\triangle A) = \nu(N\triangle B) = 0.$$

That way, if $E \in \mathfrak{M}$, then

$$\nu(E \cap P) = \nu(E \cap A) \quad \text{and} \quad \nu(E \cap N) = \nu(E \cap B). \tag{5.16}$$

(See Exercise 5.1.24.) Combining (5.13)–(5.16) we obtain that

$$\nu^+(E) = \nu(E \cap P) = \nu(E \cap A) = \lambda(E), \text{ and}$$
$$\nu^-(E) = -\nu(E \cap N) = -\nu(E \cap B) = \mu(E).$$

In other words, $\lambda = \nu^+$ and $\mu = \nu^-$, so Jordan's decomposition is unique. □

In his paper [73] from 1910, Lebesgue presented this decomposition in the special case when $f$ is a function on $\mathbb{R}^n$ and $\nu$ is defined by means of an integral. (More about what that means in Chapter 7. See Example 7.3.7.) He also introduced the adjective *singular* when applied to a measure. The name *Jordan's Theorem* is probably due to the analogy with Theorem 9.3.11 (which *is* due to Jordan).

HANS HAHN (1879–1934) was an Austrian mathematician. He studied at Strasbourg, Munich, Göttingen, and Vienna where he was awarded a Ph.D. in 1902. He then held positions at Innsbruck, Czernowitz (a city in Austria-Hungary until 1914, in Romania after World War I, in USSR since 1940, and in Ukraine since 1991), Bonn, and finally Vienna where one of his most famous students was Kurt Gödel (1906-1978). He served in the Austro-Hungarian army in World War I and was severely wounded. Hahn was a pioneer in set theory and functional analysis, and he made important contributions to the calculus of variations, but he is best remembered for the Hahn-Banach theorem (Theorem 12.2.2). He also wrote an influential book on real functions [47] and worked on the theory of abstract measures, in particular product measures. During the 1920s, together with Philipp Frank, Otto Neurath, and Moritz Schlick, he founded the *Vienna Circle of Logical Positivists*, a discussion group of gifted scientists and philosophers who met regularly. Hahn received many honours for his achievements including the Lieben Prize in 1921. He was elected to the Austrian Academy of Sciences, he was the president of the German Mathematical Society, and in 1928 he was an Invited Speaker at the International Congress of Mathematicians in Bologna. The book *Set Functions* (co-authored by a German mathematician Arthur Rosenthal (1887–1959)) was published in 1948, fourteen years after his death. It continued to develop material which first appeared in Saks's [103].

## Exercises

5.5.1. Prove that a signed measure has the continuity property (as in Theorem 5.1.9).

5.5.2.* Let $\nu : \mathfrak{M} \to (-\infty, +\infty]$ be a signed measure on $(X, \mathfrak{M})$, let $A \in \mathfrak{M}$ so that $\nu(A) < \infty$, and let $\varepsilon > 0$. Prove that there exists a measurable set $B \subset A$ for which $\nu(B) \leq \nu(A)$ and $\nu(E) \leq \varepsilon$ for all measurable $E \subset B$.

5.5.3. Let $\nu$ be a signed measure on $(X, \mathfrak{M})$. Prove that:

(a) Every measurable subset of a positive set is positive.

(b) The union of a countable collection of positive sets is positive.

(c) Give an example to show that assertion (b) need not be true for an uncountable collection.

5.5.4. Let $\nu : \mathfrak{M} \to (-\infty, +\infty]$ be a signed measure on $(X, \mathfrak{M})$. Prove that $\nu$ is bounded below.

5.5.5. Let $\nu : \mathfrak{M} \to (-\infty, +\infty]$ be a signed measure on $(X, \mathfrak{M})$ and let $(P, N)$ and $(P', N')$ be two Hahn decompositions of $X$. Prove that $P \triangle P'$ and $N \triangle N'$ are null sets with respect to $\nu$.

5.5.6. Let $\nu : \mathfrak{M} \to (-\infty, +\infty]$ be a signed measure on $(X, \mathfrak{M})$ and let $E \in \mathfrak{M}$ so that $\nu(E) > 0$. Prove that there exists a positive set $A \subset E$ for which $\nu(A) > 0$.

5.5.7. Let $(X, \mathfrak{M}, \mu)$ be a measure space and let $\mu_1$ and $\mu_2$ be mutually singular measures on $(X, \mathfrak{M})$ such that $\mu = \mu_1 - \mu_2$. Prove that $\mu_2 = 0$.

5.5.8. Let $\mu$ and $\nu$ be two measures on a measurable space $(X, \mathfrak{M})$. Suppose that for every $\varepsilon > 0$, there exists $E \in \mathfrak{M}$ such that $\mu(E) < \varepsilon$ and $\nu(E^C) < \varepsilon$. Prove that $\mu \perp \nu$.

5.5.9. Consider the measure space $([0,1], \mathcal{M}|[0,1], m|[0,1])$. Define a set function $\lambda$ on all subintervals $[a,b] \subset [0,1]$ by $\lambda([a,b]) = \varphi(b) - \varphi(a)$, where $\varphi$ is the Cantor–Lebesgue function on $[0,1]$. If $\mu$ is the Caratheodory extension of $\lambda$ to $\mathcal{M}|[0,1]$, prove that $\mu$ is singular with respect to $m|[0,1]$.

5.5.10. Let $(X, \mathfrak{M})$ be a measurable space and let $\mu, \nu$ be finite measures on $\mathfrak{M}$. Prove that either $\nu \perp \mu$ or there exist $\varepsilon > 0$ and $E \in \mathfrak{M}$ such that $\mu(E) > 0$ and $\nu \geq \varepsilon\mu$ on $E$.

5.5.11. Let $\nu$ be a signed measure on $(X, \mathfrak{M})$ and let $E \in \mathfrak{M}$. Prove that:

(a) $\nu^+(E) = \sup\{\nu(F) : F \in \mathfrak{M}, F \subset E\}$.

(b) $\nu^-(E) = \sup\{-\nu(F) : F \in \mathfrak{M}, F \subset E\}$.

5.5.12. Prove that the positive variation $\nu^+$ and the negative variation $\nu^-$ of a signed measure $\nu$ are mutually singular and that $\nu = \nu^+ - \nu^-$.

5.5.13. Let $\nu$ be a signed measure on $(X, \mathfrak{M})$ and let $E \in \mathfrak{M}$. Prove that

$$|\nu|(E) = \sup\{\sum |\nu(F_i)|\},$$

where the supremum is taken over all finite disjoint collections $\{F_i\} \subset \mathfrak{M}$ such that $\cup F_i \subset E$.

5.5.14. Let $(X, \mathfrak{M})$ be a measurable space, let $\nu$ be a signed measure on $\mathfrak{M}$, and let $\lambda$ and $\mu$ be measures on $\mathfrak{M}$ such that $\nu = \lambda - \mu$. Prove that

$$\lambda(E) \geq \nu^+(E), \quad \mu(E) \geq \nu^-(E), \qquad \text{for all } E \in \mathfrak{M}.$$

In Exercises 5.5.15–5.5.19, $\mu$ and $\nu$ are finite signed measures. We define

$$\mu \wedge \nu := \frac{\mu + \nu - |\mu - \nu|}{2} \quad \text{and} \quad \mu \vee \nu := \mu + \nu - \mu \wedge \nu. \tag{5.17}$$

5.5.15. Prove that $\mu \wedge \nu$ is a signed measure that is smaller than $\mu$ and $\nu$ but larger than any other signed measure that is smaller than $\mu$ and $\nu$.

5.5.16. Prove that $\mu \vee \nu$ is a signed measure that is larger than $\mu$ and $\nu$ but smaller than any other signed measure that is larger than $\mu$ and $\nu$.

5.5.17. Let $\mu$ and $\nu$ be measures. Prove that they are mutually singular if and only if $\mu \wedge \nu = 0$.

5.5.18. Let $m$ be Lebesgue measure on $\mathbb{R}$ and let $\mu = \delta_1$, the Dirac measure concentrated at 1. Describe $\mu \wedge \nu$ and $\mu \vee \nu$.

5.5.19. Let $m$ be Lebesgue measure on $\mathbb{R}$ and let $\mu$ be the measure counting the number of integers in the set. Describe $\mu \wedge \nu$ and $\mu \vee \nu$.

5.5.20. Let $(X, \mathfrak{M})$ be a measurable space and let $\nu$ be a finite signed measure on $\mathfrak{M}$, i.e. $-\infty < \nu(E) < \infty$ for all $E \in \mathfrak{M}$. Prove that the total variation $|\nu|$ is a finite measure.

5.5.21. Let $(X, \mathfrak{M})$ be a measurable space and let $\nu$ be a $\sigma$-finite signed measure on $\mathfrak{M}$. Prove that the total variation $|\nu|$ is a $\sigma$-finite measure on $X$, and so are $\nu^+$ and $\nu^-$.

5.5.22. Let $\nu$ be a signed measure on $(X, \mathfrak{M})$, let $E \in \mathfrak{M}$, and let $|\nu(E)| < \infty$. Prove that if $A$ is a measurable subset of $E$ then $|\nu(A)| < \infty$.

5.5.23. Let $m$ be Lebesgue measure on $\mathbb{R}$ and let $\mu = \delta_1$, the Dirac measure concentrated at 1. Find the total variation $|\nu|$ of $\nu := m - \mu$ and calculate $|\nu|([0, 1])$.

5.5.24. Let $m$ be Lebesgue measure on $\mathbb{R}$ and let $\mu$ be the measure counting the number of integers in the set. Find the total variation $|\nu|$ of $\nu := m - \mu$ and calculate $|\nu|([0, 1])$.

5.5.25. Let $\mu, \nu$ be measured on $(X, \mathfrak{M})$ such that $\mu(X) = \nu(X) = 1$ and let $\lambda = \mu - \nu$. Prove that $|\lambda|(X) = 2 \sup\{|\lambda(E)| : E \in \mathfrak{M}\}$.

5.5.26. Let $\mu, \nu$ be finite signed measures on $(X, \mathfrak{M})$. Prove that $|\mu + \nu|(A) \leq |\mu|(A) + |\nu|(A)$, for any $A \in \mathfrak{M}$.

5.5.27. Let $\mu, \nu$ be signed measures on $(X, \mathfrak{M})$. Prove that the following are equivalent:

(a) $\nu \perp \mu$.

(b) $|\nu| \perp \mu$.

(c) $\nu^+ \perp \mu$ and $\nu^- \perp \mu$.

5.5.28. Let $\nu$ be a finite signed measure on $(X, \mathfrak{M})$ and let $\alpha \in \mathbb{R}$. Express $(\alpha\nu)^+$, $(\alpha\nu)^-$, and $|\alpha\nu|$ in terms of $\nu^+$, $\nu^-$, and $|\nu|$.

5.5.29. Let $(X, \mathfrak{M}, \mu)$ be a measure space and let $\nu$ be a signed measure on $(X, \mathfrak{M})$. Prove that $|\nu| \leq \mu$ if and only if $\nu \leq \mu$ and $-\nu \leq \mu$.

5.5.30.* Let $\nu$ be a signed measure on $(X, \mathfrak{M})$. Prove that a set $E$ is a null set for $\nu$ if and only if $|\nu|(E) = 0$.

# 6

# *Measurable Functions*

The Riemann integral was designed at a time when the main role in calculus was played by functions that are continuous or at the very least piecewise continuous. As the progress of science brought forward functions that were far from continuous, a new theory of integration was needed with a widely expanded class of integrable functions. Now that we have laid the groundwork with the new measure theory we can conduct a meaningful study of a more general type of functions, the so-called *measurable* functions. Throughout this chapter we will assume that functions are *extended real-valued*, meaning that their range is contained in $[-\infty, +\infty]$. This is not an unreasonable assumption because many applications require the use of sequences of real-valued functions, hence not taking infinite values, yet these may converge at some points to $+\infty$ or $-\infty$.

## 6.1 Definition and Basic Facts

Our goal is to describe a class of functions that would be a suitable candidate to replace Riemann-integrable functions. Such a class should certainly contain all continuous functions. It should also include characteristic functions of measurable sets, because (as we will see) the measure of such a set $A$ is the integral of its characteristic function $\chi_A$. Naturally, it should be an algebra (Definition 6.2.2), i.e., it should be closed under addition, multiplication. and under multiplication by a scalar. Finally, we would want it to be closed under pointwise limits, meaning that if it contains a sequence $\{f_n\}$ and $\{f_n(x)\}$ converges to $f(x)$ for a.e. $x$, then $f$ is also in this class. It is remarkable that the following definition will accomplish all that. We will see later though (Remark 6.2.14) the motivation for it.

**Definition 6.1.1.** Let $(X, \mathfrak{M})$ be a measurable space and $f : X \to [-\infty, +\infty]$. Then, $f$ is **measurable** if for each open set $G \subset [-\infty, +\infty]$, the inverse image $f^{-1}(G)$ is measurable.

*Remark* 6.1.2. It is possible to consider a more general situation when $(X, \mathfrak{M})$ and $(Y, \mathfrak{N})$ are measure spaces and $f : X \to Y$. Such a function is considered measurable if for each $E \in \mathfrak{N}$, $f^{-1}(E) \in \mathfrak{M}$. When $Y$ is a topological space

DOI: 10.1201/9781032716497-6

the standard choice for $\mathfrak{N}$ is the Borel $\sigma$-algebra $\mathcal{B}$, i.e., the smallest one that contains all open sets. For the most part, we will work with $Y = \mathbb{R}$ and $\mathfrak{N} = \mathcal{M}$. Nevertheless, it is worth noting that the vast majority of results for real-valued functions carry over to the more general case.

*Remark* 6.1.3. Just like with measurable sets, the name *measurable* function may be slightly confusing, because neither one has anything to do with a measure. Rather, a set is measurable if it belongs to the relevant $\sigma$-algebra, and a function is measurable if the inverse image of every open set belongs to this $\sigma$-algebra.

In the special case when $X = \mathbb{R}^n$ and $\mathfrak{M} = \mathcal{M}_n$, the $\sigma$-algebra of Lebesgue measurable sets on $\mathbb{R}^n$, we talk about **Lebesgue measurable** functions. Similarly, if $\mathfrak{M} = \mathcal{B}_n$, the $\sigma$-algebra of Borel measurable sets on $\mathbb{R}^n$, we talk about **Borel measurable** functions.

Notice the similarity with the definition of a continuous function. The next result follows directly from the definition.

**Proposition 6.1.4.** *Every continuous function is measurable.*

The converse is not true (see Exercise 6.1.8).

Next, we present a more practical definition of a measurable function.

**Proposition 6.1.5.** *Let $(X, \mathfrak{M})$ be a measurable space and suppose that $f : X \to [-\infty, +\infty]$. The following are equivalent:*

*(a) $f$ is measurable,*

*(b) $\forall \alpha \in \mathbb{R}$, $\{x \in X : f(x) > \alpha\} \in \mathfrak{M}$.*

*(c) $\forall \alpha \in \mathbb{R}$, $\{x \in X : f(x) \geq \alpha\} \in \mathfrak{M}$.*

*(d) $\forall \alpha \in \mathbb{R}$, $\{x \in X : f(x) < \alpha\} \in \mathfrak{M}$.*

*(e) $\forall \alpha \in \mathbb{R}$, $\{x \in X : f(x) \leq \alpha\} \in \mathfrak{M}$.*

*Proof.*

(b) $\Leftrightarrow$ (e) $\quad \{x \in X : f(x) \leq \alpha\} = X \setminus \{x \in X : f(x) > \alpha\}$

(c) $\Leftrightarrow$ (d) $\quad$ Similarly.

(b) $\Rightarrow$ (c) $\quad \{x \in E : f(x) \geq \alpha\} = \bigcap_{n=1}^{\infty} \{x \in E : f(x) > \alpha - 1/n\}$

(c) $\Rightarrow$ (b) $\quad \{x \in E : f(x) > \alpha\} = \bigcup_{n=1}^{\infty} \{x \in E : f(x) \geq \alpha + 1/n\}$

(a) $\Rightarrow$ (b) $\quad \{x \in X : f(x) > \alpha\}$ is the inverse image of the open set $(\alpha, +\infty]$.

(b) and (d) $\Rightarrow$ (a) Let $G$ be an open subset of $[-\infty, +\infty]$. By Theorem 3.1.2, $G$ is a countable disjoint union of open intervals $\{I_k\}_{k=1}^{\infty}$. It follows that

$$f^{-1}(G) = f^{-1}\left(\bigcup_{k=1}^{\infty} I_k\right) = \bigcup_{k=1}^{\infty} f^{-1}(I_k),$$

(see Exercise 1.2.7) so we may assume that $G$ is an open interval. If $G = (a, b)$, with $a, b \in \mathbb{R}$ and $a < b$, then $G = [-\infty, b) \cap (a, +\infty]$, and by Exercise 1.2.4,

$$f^{-1}(G) = f^{-1}([-\infty, b)) \cap f^{-1}((a, +\infty]) \in \mathfrak{M}.$$

A similar argument applies to other forms of $G$. □

**Example 6.1.6. A Lebesgue measurable function on $\mathbb{R}$.**

The function

$$f(x) = \begin{cases} 0 & \text{if } x < 0 \\ +\infty & \text{if } x \geq 0 \end{cases}$$

is Lebesgue measurable on $\mathbb{R}$. Indeed, let $E_\alpha = \{x \in \mathbb{R} : f(x) > \alpha\}$. Then $E_\alpha = \mathbb{R}$ if $\alpha < 0$ and $E_\alpha = [0, +\infty)$ if $\alpha \geq 0$. Thus, $f$ is measurable since both $\mathbb{R}$ and $[0, +\infty)$ are Lebesgue measurable sets. ♦

**Example 6.1.7. A function on $\mathbb{R}$ that is not Lebesgue measurable.**

Let $E$ be a non-measurable subset of $\mathbb{R}$ and let $f$ be the characteristic function of $E$, $f = \chi_E$. Then,

$$\{x \in \mathbb{R} : f(x) > 0\} = E$$

which is not measurable, so $f$ is not measurable. ♦

**Example 6.1.8. A function that is not measurable.**

Let $X = [0, 1]$, and let $\mathfrak{M} = \{\emptyset, [0, 1]\}$. Let $f(x) = x$. Then $f$ is not measurable because

$$\{x \in [0, 1] : f(x) > 1/2\} = (1/2, 1] \notin \mathfrak{M}.$$

We notice that in this situation only constant functions are measurable. ♦

**Example 6.1.9. A measurable space such that every function is measurable.**

If $\mathfrak{M} = \mathbb{P}(X)$, then every function is measurable. ♦

**Example 6.1.10. Measurable functions on Borel $\sigma$-algebra.**

Let $X$ be a topological space and $\mathfrak{M} = \mathcal{B}(X)$, the smallest $\sigma$-algebra that contains all open sets. Then continuous functions are measurable functions. (See Exercise 6.1.1.) ♦

*Remark* 6.1.11. Another equivalent definition of measurable functions is contained in Exercise 6.2.25. However, instead of focusing on the properties of an individual function, it describes the set of all measurable functions. Namely, it is an algebra, just like $C(X)$, the set of continuous functions on $X$. Unlike $C(X)$, this algebra is closed under pointwise limits. This is a huge improvement, and it will be instrumental when we study the interplay of integrals and limits. (An Achilles heel of the Riemann integral.) This duality of definitions is somewhat reminiscent of the measurability of sets in $\mathbb{R}^n$ and the equivalence formulated in Theorem 5.3.8.

> The concept of a measurable function was introduced by Lebesgue in his dissertation [68] in 1902. In 1913, Radon generalized Lebesgue's definition to functions with domain in any Euclidean space [87]. The first text that considers real valued functions on an abstract measure space is Hahn's 1921 book [47].

When the measure is complete, the measurability of a set is not affected when it is changed by a set of measure zero. This leads to a result that concerns the measurability of a function, namely that it is not affected if a set of measure 0 is added to the domain or removed from it.

**Proposition 6.1.12.** *Let $(X, \mathfrak{M}, \mu)$ be a complete measure space and $X_0 \subset X$ a measurable set for which $\mu(X \setminus X_0) = 0$. Then an extended real-valued function $f$ on $X$ is measurable if and only if its restriction to $X_0$ is measurable. In particular, if $g$ and $h$ are extended real-valued functions on $X$ for which $g = h$ a.e. on $X$, then $g$ is measurable if and only if $h$ is measurable.*

*Proof.* Let $f_0 = f|X_0$, let $c \in \mathbb{R}$, and let $E = (c, +\infty)$.

Suppose first that $f$ is measurable. Then, using Exercise 1.2.6, $f_0^{-1}(E) = f^{-1}(E) \cap X_0 \in \mathfrak{M}$ as an intersection of measurable sets. Thus, $f_0$ is measurable.

If $f_0$ is measurable, then $f^{-1}(E) = f_0^{-1}(E) \cup A$ for some $A \subset X \setminus X_0$. (See Exercise 1.2.6.) Since $(X, \mathfrak{M}, \mu)$ is complete, $A \in \mathfrak{M}$ so $f^{-1}(E) \in \mathfrak{M}$. Thus, $f$ is measurable.

The second assertion follows from the first one. Let

$$X_0 = \{X \in X : g(x) = h(x)\},$$

and suppose that $g$ is measurable. By the first part, $g|X_0$ is measurable, which implies that $h|X_0$ is measurable, whence the result follows by another application of the first part. $\square$

*Remark* 6.1.13. Proposition 6.1.12 shows that when discussing the measurability of a function, we may ignore sets of measure zero.

Although the class of measurable functions contains continuous functions as a proper subset, many results about the former class are established based on the strong ties between the two. To paraphrase English mathematician John Littlewood, "every measurable function is nearly continuous". This is made precise by the following theorem. Once again, it holds in every Euclidean space, but we will state and prove it in $\mathbb{R}^2$.

**Theorem 6.1.14.** ***Luzin's Theorem.*** *Let $E$ be a measurable subset of $\mathbb{R}^2$ and $f$ a real-valued measurable function on $E$. For each $\varepsilon > 0$ there is a measurable set $F \subset E$ such that $m(E \setminus F) < \varepsilon$ and the restriction of $f$ to $F$ is continuous.*

*Proof.* Let $\{I_n\}_{n=1}^{\infty}$ be the collection of all open intervals with rational endpoints in $\mathbb{R}$. For every $n \in \mathbb{N}$, let $A_n$ and $B_n$ be closed sets in $\mathbb{R}^2$ (Figure 6.1) such that

$$A_n \subset E \cap f^{-1}(I_n), \quad B_n \subset E \setminus f^{-1}(I_n) \quad \text{and} \quad m(E \setminus (A_n \cup B_n)) < \frac{\varepsilon}{2^n}.$$

(Use Exercise 1.1.12.) Let

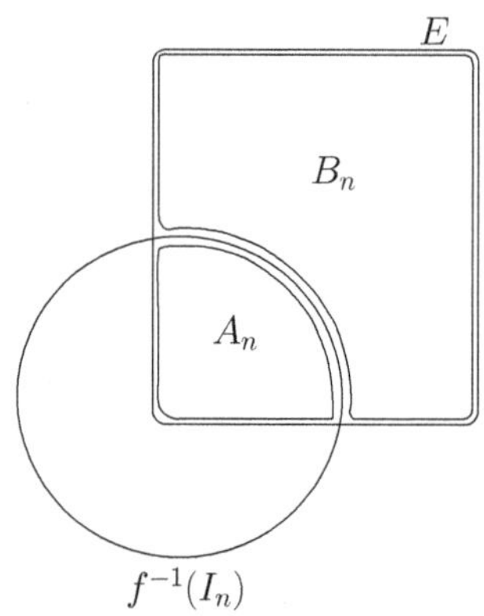


Figure 6.1: $A_n \subset E \cap f^{-1}(I_n)$, $B_n \subset E \setminus f^{-1}(I_n)$ and $m(E \setminus (A_n \cup B_n)) < \varepsilon/2^n$

$$F = \bigcap_{n=1}^{\infty} (A_n \cup B_n).$$

Clearly, $F$ is a measurable subset of $E$. It is not hard to establish (see Exercise 1.1.2) that

$$E \setminus F = \bigcup_{n=1}^{\infty} (E \setminus (A_n \cup B_n)),$$

whence

$$m(E \setminus F) \leq \sum_{n=1}^{\infty} m(E \setminus (A_n \cup B_n)) < \sum_{n=1}^{\infty} \frac{\varepsilon}{2^n} = \varepsilon.$$

Finally, let us prove that the restriction of $f$ to $F$ is continuous. Let $y \in f(F)$ and let $x \in F$ for which $y = f(x)$. Let $G$ be an open set containing $y$ (Figure 6.2). We will show that

$$f^{-1}(G \cap f(F))$$

is a relatively open subset of $F$.

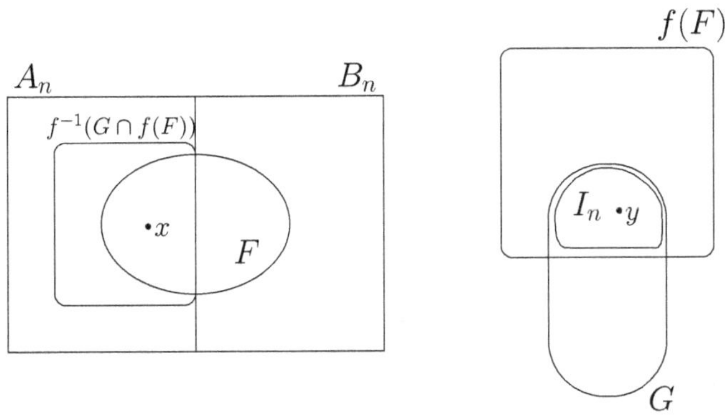


Figure 6.2: $f^{-1}(G \cap f(F))$ is a relatively open subset of $F$

Since $G$ is open and contains $y = f(x)$, there is $n \in \mathbb{N}$ for which $y \in I_n \subset G$. Further, $x \in F \subset E$ so $x \in A_n$ and $x \notin B_n$. Thus, $F \cap B_n^C$ is a relatively open subset of $F$ that contains $x$. It remains to show that it is contained in $f^{-1}(G \cap f(F))$ or, equivalently, that

$$f\left(F \cap B_n^C\right) \subset G \cap f(F). \tag{6.1}$$

Indeed, in that case $x$ is an interior point of $f^{-1}(G \cap f(F))$ in the relative topology on $F$. Focusing on (6.1), the inclusion $f\left(F \cap B_n^C\right) \subset f(F)$ is obvious. Further,

$$F \cap B_n^C = \bigcap_{k=1}^{\infty} (A_k \cup B_k) \cap B_n^C \subset (A_n \cup B_n) \cap B_n^C = A_n \cap B_n^C \subset A_n \subset f^{-1}(I_n),$$

and it follows that

$$f(F \cap B_n^C) \subset I_n \subset G.$$

Thus, $f$ is continuous on $F$ and the proof is complete. □

We have stated the theorem for $\mathbb{R}^2$ but the same argument applies to $\mathbb{R}^n$. Beyond that, the proof will work as long as every measurable set can be approximated from the inside by closed sets. Such is the case, for example, when $X$ is a metric space, $\mathfrak{M} = \mathcal{B}(X)$, and $\mu$ a $\sigma$-finite measure. (See Exercise 5.4.22.)

> Lebesgue announced Theorem 6.1.14 in 1904 [70], but without proof. Vitali proved it next year in [120], and Luzin published his proof in 1912. Yet, even Vitali refers to it in his 1935 book as "Luzin's theorem".

*Remark* 6.1.15. Although Luzin's theorem supplies the set $F$ such that $f|F$ is continuous at every $x \in F$, the function $f$ need not be continuous at $x$. For example, the Dirichlet function satisfies the hypotheses of Luzin's theorem but it is not continuous at any point.

When Luzin published his work in French, he signed it as *Lusin*, perhaps because French pronounce the letter "s" as "z" when it is between two vowels. However, in his native Russian language his name is spelled *Luzin*. American authors seem to prefer the French spelling.

NIKOLAI LUZIN (1883–1950) was a Russian mathematician. He was born in Irkutsk, and attended high school in Tomsk. It appears that mathematics there required a lot of mechanical memorization so he blossomed only after a tutor recognized his inventive problem-solving abilities. After enrolling at Moscow University he had a personal crisis and contemplated abandoning mathematics. With support from friends, he returned to mathematics, completing his master's thesis under Egorov's guidance in 1910. Their joint papers on function theory mark the beginning of the Moscow school of function theory. From 1910 to 1914 he studied at Göttingen, where he was influenced by a German mathematician Edmund Landau (1877–1938). Luzin's doctoral thesis held monumental significance in function theory. A set of problems formulated in this thesis attracted attention from mathematicians for a long time. He became a professor at Moscow University, nurturing a group of brilliant mathematicians known as *Luzitania*. The list of students included Pavel Alexandrov (1896–1982), Andrey Kolmogorov (1903–1987), Aleksandr Khinchin (1894–1959), and Pavel Urysohn (1898–1924). Luzin made important contributions to the foundations of mathematics, measure theory, and topology. He was bestowed many honors but in 1936, he was accused of being an "enemy under the mask of a Soviet citizen." He was found guilty and punished with the closure of his department in the Steklov Institute and the loss of all his official positions. The decision was finally reversed only in 2012. Despite this, his influence in mathematics and teaching persisted until his death.

## Exercises

6.1.1. Let $X$ be a topological space and $\mathfrak{M} = \mathcal{B}(X)$, the smallest $\sigma$-algebra that contains all open sets. Prove that a function $f : X \to [-\infty, +\infty]$ is measurable if it is continuous. Give an example that the converse is not true.

6.1.2. Suppose that $f : X \to [-\infty, +\infty]$, let $\alpha \in [-\infty, +\infty]$, and let $E_\alpha = \{x \in X : f(x) = \alpha\}$. Prove that if $f$ is measurable then the set $E_\alpha$ is measurable.

6.1.3. Suppose that $f : \mathbb{R} \to [-\infty, +\infty]$, let $\alpha \in [-\infty, +\infty]$, and let $E_\alpha = \{x \in \mathbb{R} : f(x) = \alpha\}$. Give an example to show that even if $E_\alpha$ is measurable for every $\alpha \in [-\infty, +\infty]$, $f$ need not be measurable.

6.1.4. Let $f : \mathbb{R} \to \mathbb{R}$ be Lebesgue measurable and let $c \in \mathbb{R}$. Prove that the function $g$, defined by $g(x) = f(x + c)$ is Lebesgue measurable.

6.1.5. Give an example to show that Proposition 6.1.12 is not true when the measure space is not complete.

6.1.6. Let $(X, \mathfrak{M}, \mu)$ be a measure space, let $f : X \to \mathbb{R}$, and let $A, B, C \in \mathfrak{M}$ such that $C$ is a disjoint union of $A$ and $B$. Prove that the restriction of $f$ to $C$ is measurable if and only if both the restriction to $A$ and the restriction to $B$ are measurable.

6.1.7. Let $(X, \mathfrak{M})$ be a measurable space and $f : X \to [-\infty, +\infty]$. Prove that $f$ is measurable if and only if for each Borel set $B \subset [-\infty, +\infty]$, the inverse image $f^{-1}(B)$ is measurable.

6.1.8. Give an example of a Lebesgue measurable function on $\mathbb{R}$ that is not continuous at any point.

6.1.9.* Give an example of a measurable function $f : [0,1] \to [-\infty, +\infty]$ for which there is no continuous function $g$ such that $f = g$ a.e.

6.1.10.* Let $x = 0.x_1x_2x_3\ldots$ and $y = 0.y_1y_2y_3\ldots$ be decimal expansions of real numbers $x, y \in [0,1]$. Let $f(x,y) = k$ if $x_k = y_k$ and $x_i \neq y_i$ for all $i < k$. If $x_k \neq y_k$ for all $k \in \mathbb{N}$, we define $f(x,y) = \infty$. Prove that $f$ is Lebesgue measurable and finite a.e.

6.1.11. Let $A$ be a measurable subset of $\mathbb{R}$ and $\mathbf{f} : A \to \mathbb{R}^n$. Denote by $f_k(x)$ the component of $(x)$ in the direction of standard basis vector $e_k$. Prove that $\mathbf{f}$ is measurable if and only if each $f_k$ is measurable. (See Exercise 6.2.8.)

6.1.12.* Let $f : \mathbb{R} \to \mathbb{R}$ be a Lebesgue measurable function. Prove that there exists a Borel measurable function $g : \mathbb{R} \to \mathbb{R}$ such $f = g$ a.e. (with respect to the Lebesgue measure).

6.1.13. Suppose that $X$ is a metric space and $(X, \mathfrak{M})$ a measurable space. Let $f$ be a function on $X$ that is continuous except at a finite number of points. Prove or disprove: $f$ is measurable.

6.1.14. Suppose that $(X, \mathfrak{M}, \mu)$ is a complete measure space and let $f$ be a function on $X$ that is continuous a.e. Prove or disprove: $f$ is measurable.

6.1.15. Suppose that $X$ is a metric space and $(X, \mathfrak{M})$ a measurable space. Let $f$ be a function on $X$ such that $\{x \in X : f(x) > r\}$ is measurable for each rational number $r$. Prove or disprove: $f$ is measurable.

6.1.16. Let $f, g : \mathbb{R} \to \mathbb{R}$, and suppose that $f$ is measurable and that $g$ is a bijection, with an inverse that is a Lipschitz function. Prove that the composition $f \circ g$ is measurable.

6.1.17. Let $(X, \mathfrak{M}, \mu)$ be a finite measure space and let $f$ be a measurable function on $X$ that is finite a.e. Prove that for every $\varepsilon > 0$ there exists a measurable set $A$ such that $f$ is bounded on $A$ and $\mu(A^C) < \varepsilon$.

6.1.18. Let $\{E_n\}_{n \in \mathbb{N}}$ be a family of measurable subsets of a measurable space $(X, \mathfrak{M})$, let $E = \cup_{n=1}^{\infty} E_n$, and let $f : E \to [-\infty, +\infty]$. Prove that $f$ is a measurable function on $E$ if and only if it is measurable on each $E_n$.

6.1.19. Prove that every monotone function is Borel measurable.

6.1.20. Let $f : \mathbb{R} \to \mathbb{R}$ be upper semicontinuous. Prove that it is measurable.

6.1.21. Let $(X, \mathfrak{M}, \mu)$ be a measure space and let $f$ be a measurable function on $X$. Suppose that $\mu^*(B) > 0$ and that $f(x) = 0$ for all $x \in B$. If $A$ is a measurable set that contains $B$ and $\mu(A) = \mu^*(B)$, prove that $f(x) = 0$ for a.e. $x \in A$.

6.1.22. Let $f : \mathbb{R} \to \mathbb{R}$ be a function that maps every measurable set to a measurable set. Prove or disprove: $f$ is measurable.

6.1.23.* Let $f : [0,1] \to \mathbb{R}$. Prove that there exists a measurable function $g$ and a Borel function $h$ for which $f = g \circ h$.

6.1.24. Let $(X, \mathfrak{M})$ be a measurable space and $f : X \to [-\infty, +\infty]$. Suppose that $f^{-1}(I) \in \mathfrak{M}$ for any finite interval $I$. Prove or disprove: $f$ is measurable.

6.1.25. Let $(X, \mathfrak{M})$ be a measurable space and $f : X \to [-\infty, +\infty]$. Suppose that $x, y$ are two points in $X$ such that every set in $\mathfrak{M}$ either contains both of them or neither one. Prove that $f(x) = f(y)$.

6.1.26. Let $X$ be a disjoint union of a countable collection $\{A_k\}$ and let $f : X \to [-\infty, +\infty]$ be measurable with respect to $\sigma(\{A_k\})$. Prove that there exist real numbers $\{a_k\}$ such that $f = \sum_k a_k \chi_{A_k}$.

6.1.27. Let $X$ be a set and $\mathcal{A}$ a $\sigma$-algebra of subsets of $X$. Suppose that $T : X \to X$ is a bijection such that $T(A) \in \mathcal{A}$ for all $A \in \mathcal{A}$. Prove that:

(a) $\mathcal{F} := \{A \in \mathcal{A} : T(A) = A\}$ is a $\sigma$-algebra.

(b) A function $f : X \to [-\infty, +\infty]$ is measurable with respect to $\mathcal{F}$ if and only if it is measurable with respect to $\mathcal{A}$ and satisfies $f = f \circ T$.

6.1.28.* Let $\mathfrak{M}$ be the $\sigma$-algebra on $\mathbb{R}$ consisting of sets that are either countable or their complement is (see Exercise 4.3.12), and let $\mu$ be the measure for which the measure of a countable set is 0, and the sets whose complements are countable and have infinite measure. Prove that if a function is measurable then it is constant a.e.

6.1.29. Prove that the characteristic function of a set $E$ is measurable if and only if $E$ is measurable.

6.1.30. Let $(X, \mathfrak{M})$ and $(Y, \mathfrak{N})$ be measurable spaces. Suppose that $X$ is a topological space, with topology $\tau$, and that $\mathfrak{M}$ is the $\sigma$-algebra generated by $\tau$. Further, suppose that there is a bijection $f : X \to Y$ such that both $f$ and $f^{-1}$ are measurable. Let $\tau' = \{f(V) : V \in \tau\}$. Prove that:

(a) $\tau'$ is a topology on $Y$;

(b) $\mathfrak{N}$ is generated by $\tau'$.

6.1.31. Let $g : \mathbb{R} \to \mathbb{R}$ be a measurable function, let $\{f_i\}_{i \in I}$ be continuous functions, $f_i : \mathbb{R} \to \mathbb{R}$, and let $f = \sup_i f_i$. Prove that $f \circ g$ is a measurable function.

## 6.2 Fundamental Properties of Measurable Functions

We start with operations on measurable functions that preserve measurability. Naturally, we are interested in deciding whether basic arithmetic operations do that. The following result will establish even more.

**Theorem 6.2.1.** *Let $(X, \mathfrak{M}, \mu)$ be a measure space, let $f, g : X \to \mathbb{R}$, and suppose that each of them is measurable. Let $F : \mathbb{R}^2 \to \mathbb{R}$ be a continuous function. Then the function*

$$h(x) := F(f(x), g(x)), \quad \textit{for } x \in X,$$

*is measurable.*

*Proof.* Let $G$ be an open subset of $\mathbb{R}$. Then $F^{-1}(G)$ is an open set in $\mathbb{R}^2$ and by Theorem 3.1.3, it is a countable union of closed rectangles $\{R_n\}_{n=1}^{\infty}$. If we represent

$$R_n = [a_n, b_n] \times [c_n, d_n],$$

then $(x, y) \in F^{-1}(G)$ if and only if there is $n \in \mathbb{N}$ such that $x \in [a_n, b_n]$ and $y \in [c_n, d_n]$. It follows that

$$\begin{aligned} x \in h^{-1}(G) \quad &\Leftrightarrow \quad \exists n \in \mathbb{N} \quad f(x) \in [a_n, b_n], \quad g(x) \in [c_n, d_n] \\ &\Leftrightarrow \quad \exists n \in \mathbb{N} \quad x \in f^{-1}([a_n, b_n]), \quad x \in g^{-1}([c_n, d_n]) \\ &\Leftrightarrow \quad \exists n \in \mathbb{N} \quad x \in f^{-1}([a_n, b_n]) \cap g^{-1}([c_n, d_n]). \end{aligned}$$

Since $f, g$ are measurable functions we see that

$$h^{-1}(G) = \bigcup_{n=1}^{\infty} \left( f^{-1}([a_n, b_n]) \cap g^{-1}([c_n, d_n]) \right) \in \mathfrak{M},$$

and the theorem is proved. □

Theorem 6.2.1 will allow us to establish several important results. In particular, we will show that the set of measurable functions is an *algebra*.

**Definition 6.2.2.** Let $X, Y$ be arbitrary sets and let $\mathcal{F} = \{f : X \to Y\}$. The collection $\mathcal{A} \subset \mathcal{F}$ is an **algebra of functions** if it is closed under addition, under multiplication, and under multiplication by a scalar, i.e., provided that

$$f, g \in \mathcal{A}, \quad c \in \mathbb{R} \quad \Rightarrow \quad f + g \in \mathcal{A}, \quad fg \in \mathcal{A}, \quad \text{and} \quad cf \in \mathcal{A}.$$

*Remark* 6.2.3. When it is clear from the context that we are not talking about an algebra of sets, we will simply call such a collection **algebra**.

From Theorem 6.2.1 it follows easily that the set of measurable functions has some important properties.

**Corollary 6.2.4.** *Let $(X, \mathfrak{M}, \mu)$ be a complete measure space and $\{f_k\}_{k=1}^{n}$ measurable functions on $X$ with values in $[-\infty, +\infty]$ that are finite a.e. If $\alpha \in \mathbb{R}$, $\alpha f_1$ is measurable on $X$ and so are $f_1 + f_2 + \cdots + f_n$, $f_1 f_2 \cdots f_n$, $\max\{f_1, \ldots, f_n\}$, and $\min\{f_1, \ldots, f_n\}$. In particular, the collection of measurable functions on $X$ is an algebra of functions.*

*Proof.* In view of Remark 6.1.13, without loss of generality, we may assume that functions $\{f_k\}_{k=1}^n$ are finite on $X$. For two functions, the measurability of the sum and the product is a direct consequence of Theorem 6.2.1. For more than two, the proof goes by induction. For the maximum, notice that

$$\{x \in X : \max\{f_1(x), \ldots, f_n(x)\} > c\} = \bigcup_{k=1}^{n} \{x \in X : f_k(x) > c\}, \tag{6.2}$$

and $\min\{f_1, \ldots, f_n\} = -\max\{-f_1, \ldots, -f_n\}$. □

**Example 6.2.5. Some important measurable functions.**
Let $(X, \mathfrak{M}, \mu)$ be a complete measure space and $f : X \to [-\infty, +\infty]$. Define

$$f^+(x) = \max\{f(x), 0\}, \quad f^-(x) = \max\{-f(x), 0\}.$$

If $f$ is measurable, then $f^+$ and $f^-$ are *positive* measurable functions and $f = f^+ - f^-$. Also, it is not hard to see that $|f| = f^+ + f^-$. ♦

Corollary 6.2.4 is an important result if one is interested in finite sequences of measurable functions. Since we will often deal with *infinite* sequences, we will need the following generalization.

**Proposition 6.2.6.** *Let $(X, \mathfrak{M}, \mu)$ be a complete measure space and suppose that $\{f_k\}_{k=1}^\infty$ are measurable functions on $X$ with values in $[-\infty, +\infty]$ that are finite a.e. The following functions are measurable: $g_1(x) = \sup_n\{f_n(x)\}$, $g_2(x) = \inf_n\{f_n(x)\}$, $g_3(x) = \limsup f_n(x)$, and $g_4 = \liminf f_n(x)$.*

*Proof.* The proof that $g_1$ and $g_2$ are measurable is the infinite analogue of the proof in Corollary 6.2.4. For $g_3$ and $g_4$ we use the identities

$$\limsup f_n(x) = \inf_{k\geq 1} \sup_{n\geq k} f_n(x) \quad \text{and} \quad \liminf f_n(x) = \sup_{k\geq 1} \inf_{n\geq k} f_n(x)$$

established in Theorem 2.4.5. □

From this proposition we get an immediate consequence.

**Corollary 6.2.7.** *Let $(X, \mathfrak{M}, \mu)$ be a complete measure space and $\{f_k\}_{k=1}^\infty$ measurable functions on $X$ with values in $[-\infty, +\infty]$ that are finite a.e. If $\{f_k\}_{k=1}^\infty$ converges to a function $f$ a.e. on $X$, then $f$ is measurable.*

*Remark* 6.2.8. Corollary 6.2.7 is not true if $(X, \mathfrak{M}, \mu)$ is not complete. (See Exercise 6.2.3.) However, there is a measurable function $g$ such that $\{f_k\}_{k=1}^\infty$ converges to $g$ a.e. on $X$. (See Exercise 6.2.4.) Because of this, we will assume that, even when $(X, \mathfrak{M}, \mu)$ is not complete, $\{f_n\}$ converges to some measurable function a.e. All this is caused by convergence *almost everywhere.* If a sequence of measurable functions converges *everywhere*, the limit is also a measurable function.

Proposition 6.2.6 and Corollary 6.2.7 do not hold if the word *measurable* is replaced by *continuous*. This justifies the selection of the class of measurable functions to deal with. To be honest, we lose something in this trade-off. Namely, the composition of two continuous functions is continuous in its domain, whereas the composition of two measurable functions need not be measurable.

**Example 6.2.9. A composition of measurable functions need not be measurable.**

It was established in the proof of Theorem 5.4.8 that there exists a measurable set $A \subset [0,1]$ and a strictly increasing continuous function $\psi$ defined on $[0,1]$, such that $\psi(A)$ is not measurable. We will extend $\psi$ to an increasing continuous bijection on $\mathbb{R}$. Then $\psi^{-1}$ exists and it is also a continuous, hence a measurable function.

Let $\chi_A$ be the characteristic function of the set $A$. Then

$$\{x \in \mathbb{R} : \chi_A(x) < c\} = \begin{cases} \mathbb{R} & \text{if } c > 1 \\ A^C & \text{if } 0 < c \leq 1 \\ \emptyset & \text{if } c \leq 0, \end{cases}$$

so $\chi_A$ is measurable.

However, by Exercise 1.2.9, $\chi_A \circ \psi^{-1}(x) = \chi_{\psi(A)}$. It follows that

$$\left\{x \in \mathbb{R} : \chi_A \circ \psi^{-1}(x) > \frac{1}{2}\right\} = \left\{x \in \mathbb{R} : \chi_{\psi(A)}(x) > \frac{1}{2}\right\} = \psi(A),$$

so $\chi_A \circ \psi^{-1}$ is not measurable. ♦

First examples of a nonmeasurable composition of measurable functions were given by Sierpiński in 1916 [111] and Carathéodory in 1918 [23].

It is useful to recall that when we defined the concept of a measurable set we developed it gradually, starting with rectangles, and then moving to open sets, and eventually arriving at the definition of a measurable set (Definition 4.2.11). This principle—going from simpler objects towards more general ones—will be on display when we present integration theory in Chapter 7. Namely, in order to define the integral of a measurable function we will need to start with a more manageable class of functions, which we now introduce.

**Definition 6.2.10.** A real-valued function $s$ defined on a measurable space $(X, \mathfrak{M})$ is called **simple** if it is measurable and takes only a finite number of values. If $s$ takes distinct values $c_1, c_2, \dots, c_n$ on the sets $E_1, E_2, \dots, E_n$, then $s = \sum_{k=1}^{n} c_k \chi_{E_k}$. We will denote the class of simple functions on $X$ by $\mathcal{S}(X)$.

It is not hard to verify that $\mathcal{S}(X)$ is an algebra of functions. (See Exercise 6.2.22.) Here are some examples of simple functions.

**Example 6.2.11. A simple function.**

The Dirichlet function is a simple function because it is the characteristic function of rationals. ♦

**Example 6.2.12. A class of simple functions.**

Every step function is a simple function, with sets $E_k = [x_{k-1}, x_k)$, $1 \leq k \leq n$. ♦

Example 6.2.12 shows that simple functions are a generalization of step functions where the intervals (or rectangles in higher dimensions) are replaced by any measurable sets. Just like step functions are essential in the definition of the Riemann integral, simple functions will play an important role regarding the so-called Lebesgue integral. In particular, using inscribed rectangles we obtained a sequence of step functions, smaller than the continuous function $f$ that we are trying to approximate, that converges to $f$. The following does the same for measurable functions and simple functions.

**Theorem 6.2.13.** *Let $(X, \mathfrak{M})$ be a measurable space and $f : X \to [0, +\infty]$ a measurable function. There exists a sequence $\{s_n\}_{n=1}^{\infty}$ of nonnegative simple functions on $X$ such that:*

*(a) $s_n(x) \leq s_{n+1}(x)$, for all $x \in X$ and all $n \in \mathbb{N}$.*

*(b) $\lim_n s_n(x) = f(x)$, for all $x \in X$.*

*(c) If $f$ is a bounded function, the convergence is uniform.*

*Proof.* For $n \in \mathbb{N}$ and $k = 1, 2, \dots, n2^n$ we define

$$E_{nk} = \left\{ x \in X : \frac{k-1}{2^n} \leq f(x) < \frac{k}{2^n} \right\}, \qquad F_n = \{x \in X : f(x) \geq n\}, \quad \text{and}$$

$$s_n(x) = \sum_{k=1}^{n2^n} \frac{k-1}{2^n} \chi_{E_{nk}}(x) + n\chi_{F_n}(x).$$

(See Figure 6.3.) The measurability of $f$ implies that the sets $E_{nk}$ and $F_n$ are measurable, so $\{s_n\}$ is indeed a sequence of simple functions, and it is obvious that each one is nonnegative. Further, it is not hard to see that, when passing from $n$ to $n + 1$, the set $E_{nk}$ splits into 2 parts, i.e.,

$$E_{nk} = E_{n+1,2k-1} \cup E_{n+1,2k}.$$

Meanwhile, the set $F_n$ splits into $F_{n+1}$ (where $f(x) \geq n + 1$) and $2^{n+1}$ pieces (where $n \leq f(x) < n + 1$), i.e.,

$$F_n = F_{n+1} \cup \left( \bigcup_{k=n2^{n+1}+1}^{(n+1)2^{n+1}} E_{n+1,k} \right).$$

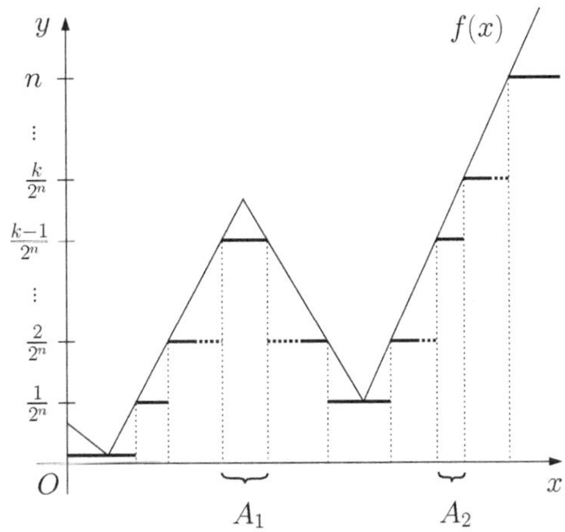


Figure 6.3: $E_{nk} = \{x : (k-1)/2^n \leq f(x) < k/2^n\} = A_1 \cup A_2$

Now, if $x \in E_{nk}$ then $s_n(x) = (k-1)/2^n$ and $s_{n+1}(x)$ equals $(2k-2)/2^{n+1}$ or $(2k-1)/2^{n+1}$, so

$$s_{n+1}(x) \geq \frac{2k-2}{2^{n+1}} = \frac{k-1}{2^n} = s_n(x).$$

On the other hand, if $x \in F_n$ then $s_n(x) = n$ and $s_{n+1}(x)$ is one of the numbers $n+1$ or $(k-1)/2^{n+1}$, for $n2^{n+1}+1 \leq k \leq (n+1)2^{n+1}$. It follows that

$$s_{n+1}(x) \geq \frac{[n2^{n+1}+1]-1}{2^{n+1}} = n = s_n(x).$$

Either way, we have that (a) holds.

Let $x \in X$. If $f(x) = +\infty$, then $x \in F_n$, for all $n \in \mathbb{N}$, so $s_n(x) = n$ and $\lim_n s_n(x) = f(x)$. If $f(x)$ is a finite number, we consider $n > f(x)$. For such $n$ there exists $k$ such that $x \in E_{nk}$ so

$$s_n(x) = \frac{k-1}{2^n} \leq f(x) < \frac{k}{2^n}.$$

It follows that, for all $n \geq f(x)$,

$$|f(x) - s_n(x)| < \frac{1}{2^n} \to 0, \quad n \to \infty, \tag{6.3}$$

so (b) is proved.

Finally, if $f$ is bounded, say $|f(x)| \leq M$, then (6.3) holds for all $n \geq M$ and all $x \in X$, so the sequence $\{s_n\}$ converges uniformly to $f$. □

We point out an essential difference between step functions and simple functions constructed in the proof of Theorem 6.2.13. For step functions the sets $\{E_k\}$ are obtained by partitioning an interval on the $x$-axis, whereas sets $\{S_{nk}\}$ and $\{F_n\}$ corresponded to the values of the function $f$. In other words, they came as a result of partitioning the $y$-axis. Step functions approximate continuous functions very well, but they are inadequate to deal with measurable functions in general which can fail to be continuous at any point.

*Remark* 6.2.14. The sets $E_{nk}$ and $F_n$ in the proof of Theorem 6.2.13 offer a clue about the motivation for the definition of measurable functions, in particular the equivalent formulations in Proposition 6.1.5. The sequence $\{s_n\}$ will be useful only when these sets are measurable, and as such they will be a crucial ingredient in defining the Lebesgue integral.

While the monotonicity of the sequence of step functions was almost irrelevant, the fact that the constructed sequence of simple functions is increasing will be priceless. It will allow us to interchange the limit and the integral thereby reducing questions about the integral of an arbitrary measurable function to questions about the integral of a simple function. From that point of view, Theorem 6.2.13 is one of the cornerstones of Lebesgue's integration theory. More about it in Chapter 7.

## Exercises

6.2.1. Let $(X, \mathfrak{M}, \mu)$ be a measure space, let $f : X \to \mathbb{R}$ be measurable, and let $g : \mathbb{R} \to \mathbb{R}$ be a continuous function. Prove that the composition $h(x) = g(f(x))$ is measurable. Conclude that if $f$ is measurable then so is $|f|^p$, for any $p > 0$.

6.2.2. Let $f, g$ be Borel measurable functions on $\mathbb{R}$. Prove that their composition is Borel measurable.

6.2.3. Give an example to show that Corollary 6.2.7 is not true if $(X, \mathfrak{M}, \mu)$ is not complete.

6.2.4.* Let $(X, \mathfrak{M}, \mu)$ be a measure space, and let $\{f_n\}$ be a sequence of measurable functions that converge a.e. Prove that its limit is equal to a measurable function a.e.

6.2.5. Let $(X, \mathfrak{M})$ be a measurable space and $f : X \to [-\infty, +\infty] \setminus \{0\}$. Prove that $1/f$ is measurable.

6.2.6. Let $(X, \mathfrak{M})$ be a measurable space and $f : X \to [-\infty, +\infty]$. Determine all mutually prime positive integers $m, n$ such that the measurability of $f^{m/n}$ implies the measurability of $f$.

6.2.7.** Let $f$ be a continuous function on $[a, b]$ and let $N(c)$ denote the number of solutions of the equation $f(x) = c$. Prove that the function $N$ is Lebesgue measurable.

6.2.8. Let $A$ be a measurable subset of $\mathbb{R}$ and $\mathbf{f} : A \to \mathbb{R}^n$. Prove that the measurability of each component of $\mathbf{f}$ does not depend on the choice of a basis.

6.2.9.** Let $x = 0.n_1n_2\ldots$ be a decimal expansion of a real number $x \in [0, 1]$ and let $f(x) = \max_k n_k$. Prove that $f$ is measurable and constant a.e. [Note: some numbers have two different decimal representations. For example, $0.5 = 0.49999\ldots$. In a situation like this we will take the one that does **not** end with the digit 9 repeated infinitely many times. Thus $f(0.5) = f(0.4999999\ldots) = 5$.]

6.2.10.** Let $x = 0.n_1n_2\ldots$ be a decimal expansion of a real number $x \in [0, 1]$ and let $f(x) = \limsup_k n_k$. Prove that $f$ is measurable and constant a.e. [Note: some numbers have two different decimal representations. For example, $0.5 = 0.49999\ldots$. In a situation like this we will take the one that does **not** end with the digit 9 repeated infinitely many times. Thus $f(0.5) = f(0.4999999\ldots) = 0$.]

6.2.11. Does there exist a non-measurable function $f \geq 0$ such that $\sqrt{f}$ is measurable?

6.2.12. Let $\{E_n\}_{n\in\mathbb{N}}$ be a family of measurable subsets of a measurable space $(X, \mathfrak{M})$, let $k$ be a positive integer or $\infty$, and let

$$E = \{x \in X : x \text{ belongs to at least } k \text{ of the sets } E_n\}.$$

Prove that $E$ is measurable.

6.2.13. Let $f : \mathbb{R}^2 \to [-\infty, +\infty]$. Suppose that for every fixed $y$, $f$ is a measurable function of $x$ and that for every fixed $x$, $f$ is a continuous function of $y$. Prove that the functions

$$\limsup_{y\to\infty} f(x, y) \quad \text{and} \quad \liminf_{y\to\infty} f(x, y)$$

are measurable.

6.2.14. Let $E$ be a measurable subset of $\mathbb{R}^2$ and $f$ a real-valued function on $E$. Suppose that for each $\varepsilon > 0$ there is a measurable set $F \subset E$ such that $m(E \setminus F) < \varepsilon$ and the restriction of $f$ to $F$ is continuous. Prove that $f$ is measurable.

6.2.15. Let $f : \mathbb{R}^n \to \mathbb{R}$ be a continuous function and let $g_k : \mathbb{R} \to \mathbb{R}$, $1 \leq k \leq n$, be measurable functions. Prove that $h(x) := f(g_1(x), g_2(x), \ldots, g_n(x))$ is a measurable function.

6.2.16. Let $(X, \mathfrak{M}, \mu)$ be a measure space, let $A \in \mathfrak{M}$, and let $f : A \to \mathbb{R}^*$ and $g : A^C \to \mathbb{R}^*$ be measurable. Prove that the function $h$, defined by $h(x) = f(x)$ if $x \in A$ and $h(x) = g(x)$ if $x \notin A$, is measurable.

6.2.17. Let $f : \mathbb{R} \to \mathbb{R}$ be measurable, and let $g : \mathbb{R} \to \mathbb{R}$ be a continuous function. Prove or disprove: the composition $h(x) = f(g(x))$ is measurable.

6.2.18. Give an example of an uncountable family of measurable functions $\mathcal{F}$ for which $\sup\{f : f \in \mathcal{F}\}$ is not a measurable function.

6.2.19. Let $f, g : \mathbb{R} \to \mathbb{R}$ and suppose that $f$ is Borel measurable and that $g$ is Lebesgue measurable. Prove or disprove: $f \circ g$ is Lebesgue measurable.

6.2.20. Prove that the set of all continuous functions $f : [0,1] \to \mathbb{R}$ is an algebra of functions.

6.2.21. Let $\mathcal{A}$ consist of all functions $f : [0,1] \to \mathbb{R}$ such that $f(0) = 0$. Prove that $\mathcal{A}$ is an algebra of functions.

6.2.22. Prove that the class of simple functions on $X$ is an algebra of functions.

6.2.23. Let $(X, \mathfrak{M}, \mu)$ be a $\sigma$-finite measure space and $f : X \to \mathbb{R}$ a nonnegative measurable function. Prove that there exists an increasing sequence $\{s_n\}_{n=1}^{\infty}$ of simple functions on $X$ such that:

(a) for each $n \in \mathbb{N}$, the set of points where $s_n \neq 0$ has a finite measure.

(b) $\lim_n s_n(x) = f(x)$, for all $x \in X$.

6.2.24.* Let $(X, \mathfrak{M}, \mu)$ be a measure space and let $(X, \mathfrak{M}_0, \mu_0)$ be its completion. If $f$ is a function that is measurable with respect to $\mathfrak{M}_0$, prove that there exists a function $g$ that is measurable with respect to $\mathfrak{M}$ such that $f = g$ a.e. $[\mu]$.

6.2.25. Prove that the collection of Lebesgue measurable functions on $\mathbb{R}^n$ is the smallest algebra that contains characteristic functions of all Lebesgue measurable sets and is closed under pointwise limits a.e.

6.2.26. Let $(X, \mathfrak{M})$ be a measurable space and $f : X \to [0, +\infty]$ a bounded measurable function. There exists a sequence $\{s_n\}_{n=1}^{\infty}$ of nonnegative simple functions on $X$ such that:

(a) $s_n(x) \geq s_{n+1}(x)$, for all $x \in X$ and all $n \in \mathbb{N}$.

(b) $\lim_n s_n(x) = f(x)$, and the convergence is uniform on $X$.

6.2.27. Let $\{f_n\}$ be a sequence of measurable functions. Prove that the set of points $x$ such that $\lim f_n(x)$ exists, is measurable.

6.2.28. Let $f$ be a function that is defined and differentiable on $\mathbb{R}$. Prove that its derivative $f'$ is a measurable function.

6.2.29. Let $f : [a,b] \to \mathbb{R}$ be a bounded Borel measurable function. Prove that $\{x \in [a,b] : f \text{ is continuous at } x\}$ is Borel measurable.

6.2.30. Let $g : \mathbb{R} \to \mathbb{R}$ be a measurable function, let $G \subset \mathbb{R}$ be a nonempty open set, and let $h(t) = t \sin t$. Define

$$f(x) = \inf\{t \geq 0 : h(t)g(x) \in G\}.$$

Prove that $f$ is a real-valued measurable function.

## 6.3 Sequences of Measurable Functions

In this section, we will consider various modes in which a sequence of functions can converge.

The sequence of simple functions in Theorem 6.2.13 converges uniformly to $f$ whenever $f$ is bounded. This is not true for an arbitrary sequence of functions.

**Example 6.3.1. An increasing sequence of measurable functions that converges to a bounded function pointwise but not uniformly.**

Let $X = [0,1]$ and let $\mathfrak{M}$ be the $\sigma$-algebra of Lebesgue measurable subsets of $[0,1]$. Let $f_n(x) = 1 - x^n$. Then $\{f_n\}$ is an increasing sequence of nonnegative, continuous (hence measurable) functions that converges pointwise to a bounded function

$$f(x) = \begin{cases} 1 & \text{if } 0 \le x < 1 \\ 0 & \text{if } x = 1. \end{cases}$$

If the convergence was uniform, $f$ would have to be continuous which is not the case here. ♦

Notice that $\{f_n\}$ is "almost" uniformly convergent. It can be shown that, for any $c \in (0,1)$, the sequence $\{f_n\}$ converges uniformly to 1 on $[0,c]$. Indeed,

$$\sup_{x\in[0,c]} |f_n(x) - 1| = \sup_{x\in[0,c]} |x^n| = c^n \to 0, \quad n \to \infty.$$

The following theorem shows that such a behaviour of $\{f_n\}$ is to be expected. To quote Littlewood once again, "every convergent sequence of functions is nearly uniformly convergent".

**Theorem 6.3.2.** ***Egorov's Theorem.*** *Let $(X, \mathfrak{M}, \mu)$ be a finite measure space and $\{f_k\}_{k=1}^{\infty}$ a sequence of measurable functions on $X$ with values in $[-\infty, +\infty]$ that are finite a.e. If $\{f_n\}$ converges to a function $f$ a.e., then for any $\varepsilon > 0$ there exists a measurable set $F \subset X$ such that $\mu(X \setminus F) < \varepsilon$ and $\{f_n\}$ converges uniformly on $F$.*

*Proof.* In view of Remark 6.2.8, we may assume without loss of generality that $f$ is measurable. Further, the set where at least one of the functions $f_n$ is infinite is a measurable set of measure 0, and so is the set where $\{f_n\}$ does not converge to $f$. Thus, we will assume that each function $f_n$ is a real-valued function and that $\{f_n\}$ converges to $f$ everywhere.

Let $\delta > 0$ be fixed. For any $n \in \mathbb{N}$ we define

$$F_{\delta,n} = \{x \in X : |f_k(x) - f(x)| < \delta, \quad \text{for all } k \ge n\}, \quad \text{and} \quad E_{\delta,n} = X \setminus F_{\delta,n}.$$

Since each function in sight is measurable, each set $F_{\delta,n}$ and $E_{\delta,n}$ is measurable. The sets $\{F_{\delta,n}\}_{n=1}^{\infty}$ form an increasing sequence of measurable sets whose union is $X$. Therefore, $\{E_{\delta,n}\}_{n=1}^{\infty}$ is a decreasing sequence of measurable sets (of finite measure) with empty intersection. By the continuity of measure, $\lim \mu(E_{\delta,n}) = 0$.

Let $k \in \mathbb{N}$ and $\varepsilon > 0$. Then there exists $n_k \in \mathbb{N}$ such that $\mu(E_{1/k,n_k}) < 2^{-k}\varepsilon$. We define

$$E = \bigcup_{k=1}^{\infty} E_{1/k,n_k} \quad \text{and} \quad F = X \setminus E.$$

Let us verify that $F$ satisfies all conclusions of the theorem. First,

$$\mu(X \setminus F) = \mu(E) = \mu\left(\bigcup_{k=1}^{\infty} E_{1/k,n_k}\right) \leq \sum_{k=1}^{\infty} \mu(E_{1/k,n_k}) < \sum_{k=1}^{\infty} 2^{-k}\varepsilon = \varepsilon.$$

Second, using De Morgan's Laws, for any $m \in \mathbb{N}$,

$$F = \bigcap_{k=1}^{\infty} (X \setminus E_{1/k,n_k}) = \bigcap_{k=1}^{\infty} F_{1/k,n_k} \subset F_{1/m,n_m},$$

which implies that, for every $x \in F$,

$$|f_k(x) - f(x)| < \frac{1}{m}, \quad \text{for all } k \geq n_m,$$

whence $\{f_n\}$ converges uniformly on $F$. □

> The first proof of the theorem was given by an Italian mathematician Carlo Severini (1872–1951) in 1910 [110]. He used the result as a tool in his research on series of orthogonal functions. His work remained apparently unnoticed outside Italy, probably due to the fact that it is written in Italian, appeared in a scientific journal with small circulation, and was considered only as a means to obtain other theorems. A year later Egorov published his independently proved results [30], and the theorem became widely known under his name. However, it is not uncommon to find references to this theorem as the Severini–Egoroff theorem or Severini–Egorov Theorem. Before all of them, Lebesgue had stated the result in 1904 [70] but without proof. The first mathematicians to prove independently the theorem in the nowadays common abstract measure space setting were Riesz (1922, 1928) [97, 98], and Sierpiński (1928) [113].

In advanced calculus, you learned that a series of functions can be integrated term-by-term when it converges uniformly. Unfortunately, in real life applications very few series converge uniformly, so one needs to consider other modes of convergence. Even pointwise a.e. convergence is sometimes too much to ask. This motivated the invention of yet another one, which may be even weaker. (See Proposition 6.3.8 and Example 6.3.9 below.)

**Definition 6.3.3.** Let $(X, \mathfrak{M}, \mu)$ be a measure space, let $\{f_n\}_{n=1}^{\infty}$, $f$ be measurable functions on $X$ with values in $[-\infty, +\infty]$ that are finite a.e. The sequence $\{f_n\}$ **converges to $f$ in measure** on $X$ if, for all $\eta > 0$,

$$\lim_n \mu\left(\{x \in X : |f_n(x) - f(x)| > \eta\}\right) = 0.$$

We write $f_n \xrightarrow{\mu} f$.

*Remark* 6.3.4. Strictly speaking, the difference $f_n(x) - f(x)$ is undefined if both $f_n(x)$ and $f(x)$ happen to be infinite. However, the set of such elements is a measurable subset of $X$ and has measure 0, so we conveniently ignore it when considering $\{x \in X : |f_n(x) - f(x)| > \eta\}$.

*Remark* 6.3.5. The idea of this type of convergence is that we require $f_k(x)$ and $f(x)$ to be close except on a set that has smaller and smaller measures as $k$ increases.

Now that we have a new mode of convergence, it is natural to ask how it compares to pointwise and uniform convergence. As we hinted, uniform convergence is stronger. The proof is straightforward and we leave it to the reader (Exercise 6.3.1).

**Proposition 6.3.6.** *Let $(X, \mathfrak{M}, \mu)$ be a measure space, let $\{f_n\}_{n=1}^{\infty}$ be a sequence of measurable functions on $X$ with values in $[-\infty, +\infty]$ that are finite a.e. If $\{f_n\}$ converges uniformly on $X$ to a real-valued function $f$, then $\{f_n\}$ converges to $f$ in measure on $X$.*

The following example shows that the converse fails.

**Example 6.3.7. A sequence that converges in measure but not uniformly.**

Let $X = [0,1]$, let $\mathfrak{M}$ be the $\sigma$-algebra of Lebesgue measurable subsets of $[0,1]$, and let $\mu$ be the restriction of Lebesgue measure $m$ to $[0,1]$. Let $f_n : [0,1] \to \mathbb{R}$ be defined by $f_n(x) = \chi_{[0,1/n]}(x)$ and $f(x) = 0$. Then, for all $\eta > 0$,

$$\begin{aligned}\lim_n \mu\left(\{x \in X : |f_n(x) - f(x)| > \eta\}\right) &= \lim_n m\left(\{x \in [0,1] : |\chi_{[0,1/n]}(x)| > \eta\}\right)\\ &\leq \lim_n \frac{1}{n} = 0.\end{aligned}$$

On the other hand,

$$\sup\{|f_n(x) - 0| : x \in [0,1]\} = \sup\{|\chi_{[0,1/n]}(x)| : x \in [0,1]\} = 1,$$

so $\{f_n\}$ does not converge to 0 uniformly. ♦

When $\mu(X) < \infty$, even the pointwise a.e. convergence is stronger than convergence in measure.

**Proposition 6.3.8.** *Let $(X, \mathfrak{M}, \mu)$ be a finite measure space and let $\{f_n\}_{n=1}^{\infty}$, $f$ be measurable functions on $X$ with values in $[-\infty, +\infty]$ that are finite a.e. If $\{f_n\}$ converges to a real-valued function $f$ a.e. on $X$, then $\{f_n\}$ converges to $f$ in measure on $X$.*

*Proof.* Using the argument outlined in Remarks 6.3.4 and 6.2.8 we may assume that functions $\{f_n\}$ are finite and converge to a measurable function $f$.

Let $\varepsilon > 0$. By Egorov's Theorem, there exists a measurable set $F \subset X$ such that $\mu(X \setminus F) < \varepsilon$ and $\{f_n\}$ converge to $f$ uniformly on $F$. By definition of uniform convergence, there exists $N \in \mathbb{N}$ such that

$$n \geq N \quad \Rightarrow \quad \{x \in X : |f_n(x) - f(x)| > \eta\} \subset X \setminus F.$$

It follows that for such $n$,

$$\mu\left(\{x \in X : |f_n(x) - f(x)| > \eta\}\right) < \varepsilon.$$

□

The assumption that $\mu(X) < \infty$ in Proposition 6.3.8 cannot be omitted (see Exercise 6.3.2).

Although Proposition 6.3.8 ensures that convergence a.e. implies convergence in measure, it is instructive to try to prove for several sequences (which converge a.e.) that they converge in measure (Exercises 6.3.3–6.3.5).

Proposition 6.3.8 with a proof is in Lebesgue's book on trigonometric series from 1906 [71], although he has formulated it as early as 1903 and used without proof in 1904. However, the first systematic treatment of convergence in measure is in Riesz paper [90] from 1909. This concept has become indispensable in probability where it is known under the name "convergence in probability". In fact, when Riesz introduced it, one of the two main applications was the proof of the so-called *Law of large numbers.*

While the converse of Proposition 6.3.8 is not true, examples are harder to construct.

**Example 6.3.9. A sequence that converges in measure but not a.e.** Let $X = [0, 1]$, let $\mathfrak{M}$ be the $\sigma$-algebra of Lebesgue measurable subsets of $[0, 1]$, and let $\mu$ be the restriction of Lebesgue measure $m$ to $[0, 1]$. We define a sequence of functions

$$\begin{array}{llll} f_1 = \chi_{[0,1]}, & & & \\ f_2 = \chi_{[0,1/2]}, & f_3 = \chi_{[1/2,1]}, & & \\ f_4 = \chi_{[0,1/3]}, & f_5 = \chi_{[1/3,2/3]}, & f_6 = \chi_{[2/3,1]}, & \\ f_7 = \chi_{[0,1/4]}, & f_8 = \chi_{[1/4,1/2]}, & f_9 = \chi_{[1/2,3/4]}, & f_{10} = \chi_{[3/4,1]}, \end{array}$$

and so on. The sequence $\{f_n\}$ converges to 0 in measure on $[0, 1]$. Indeed, let $0 < \eta < 1$. Then, for $n = 1, 2, 3, \dots$, the measure of the set $\{x \in [0, 1] : |f_n(x)| > \eta\}$ is

$$1, \frac{1}{2}, \frac{1}{2}, \frac{1}{3}, \frac{1}{3}, \frac{1}{3}, \frac{1}{4}, \frac{1}{4}, \frac{1}{4}, \frac{1}{4}, \dots.$$

Thus, $\lim_n m(\{x \in [0, 1] : |f_n(x)| > \eta\}) = 0$.

On the other hand, $\{f_n\}$ does not converge to 0 a.e. on $[0, 1]$. In fact, for any $x \in [0, 1]$ there are infinitely many $n$ such that $f_n(x) = 1$. ♦

Example 6.3.9 shows that the converse of Proposition 6.3.8 is not true, i.e., pointwise convergence almost everywhere *is* stronger than convergence in measure (at least on finite measure spaces). Regarding the said converse, the following theorem is the best one can say.

**Theorem 6.3.10.** ***(Riesz)*** *Let $(X, \mathfrak{M}, \mu)$ be a measure space and let $\{f_n\}_{n=1}^{\infty}$, $f$ be measurable functions on $X$ with values in $[-\infty, +\infty]$ that are finite a.e. If $\{f_n\}$ converges to $f$ in measure on $X$, then there exists a subsequence $\{f_{n_k}\}$ that converges to $f$ a.e. on $X$.*

*Proof.* For any $k \in \mathbb{N}$ we select a positive integer $n_k$ (bigger than $n_{k-1}$) so that, for any $j \geq n_k$,

$$m\left(\left\{x \in X : |f_j(x) - f(x)| > \frac{1}{k}\right\}\right) < \frac{1}{2^k}.$$

In particular, for $j = n_k$ we have the set

$$E_k = \left\{x \in X : |f_{n_k}(x) - f(x)| > \frac{1}{k}\right\} \quad \text{and} \quad m(E_k) < \frac{1}{2^k},$$

so $\sum_{k=1}^{\infty} m(E_k) < \infty$. Applying Borel–Cantelli Lemma we obtain that almost all $x \in X$ belong to at most finitely many sets $E_k$. In other words, there exists a set $X_0 \subset X$ such that $\mu(X \setminus X_0) = 0$ and, if $x \in X_0$ then $x$ belongs to at most finitely many sets $E_k$. For such $x$ there is $K = K(x)$ with the property that if $k \geq K(x)$, then $x \notin E_k$. In other words,

$$\forall x \in X_0 \quad \exists K = K(x) \quad k \geq K \quad \Rightarrow \quad |f_{n_k}(x) - f(x)| \leq \frac{1}{k},$$

and it follows that $\{f_{n_k}\}$ converges to $f$ a.e. on $X$. □

DMITRI EGOROV (1869–1931) was a Russian mathematician. His father taught at the Moscow Teachers' Institute and Egorov was educated at home until secondary school. At Moscow University, where he studied mathematics and physics, his exceptional mathematical talent manifested by his undergraduate paper and thesis. After getting a doctoral degree in 1901 he traveled abroad, attending lectures by leading mathematicians in Berlin, Paris, and Göttingen. In 1903, he obtained a position at Moscow University. His marriage to a daughter of a renowned violinist, connected him with Moscow's cultural elite, where music frequently filled their home. Egorov's significant contributions lay in differential geometry and mathematical analysis. He was a dedicated educator, teaching at various institutions and lecturing extensively on mathematical subjects. He was praised for the rigor and depth of his lectures. In addition to research papers he published books, textbooks, and lecture notes. He mentored Luzin, jointly founding the influential Moscow School of Pure Mathematics. Egorov's administrative roles in mathematical societies and institutes showcased his leadership. However, his deeply religious beliefs clashed with Soviet ideologies, leading to his dismissal, arrest, and persecution. In 1930, after being expelled from the Moscow Mathematical Society, Egorov was arrested as a "religious sectarian" and imprisoned. His health deteriorated, and he passed away in 1931 while under arrest.

## Exercises

6.3.1. Prove Proposition 6.3.6.

6.3.2. Prove that Proposition 6.3.8 is not true if $\mu(X) = \infty$.

In Exercises 6.3.3–6.3.5 prove that the sequence $\{f_n\}$ converges in (Lebesgue) measure on $X$, without using Propositions 6.3.6 or 6.3.8.

6.3.3. $f_n(x) = x^n$, $X = [0, 1]$.

6.3.4. $f_n(x) = \frac{1}{n} \tan x$, $X = (-\pi/2, \pi/2)$.

6.3.5. $f_n(x) = n\chi_{[0,1/n]}(x)$, $\mathbb{R}$.

6.3.6. Find a subsequence of the sequence $\{f_n\}$ in Example 6.3.9 that converges to 0 a.e. on $[0, 1]$.

6.3.7. Let $\{f_n\}$ be a sequence of measurable functions and suppose that $f_n \to f$ a.e. and $f_n \to g$ a.e. Prove that $f = g$ a.e.

6.3.8. Let $\{r_n\}$ be an enumeration of rational numbers in $[0,1]$ and let $r_n = p_n/q_n$, where $p_n, q_n$ are mutually prime positive integers. Let

$$f_n(x) = e^{-(p_n - xq_n)^2}.$$

Prove that $\{f_n\}$ converges to 0 in Lebesgue measure.

6.3.9. Let $\{f_n\}$ be a sequence of measurable functions and suppose that $f_n \to f$ in measure and $f_n \to g$ in measure. Prove that $f = g$ a.e.

6.3.10. Suppose that $(X, \mathfrak{M}, \mu)$ is not complete. Show that there is a sequence of measurable functions on $X$ that converges pointwise a.e. on $X$ to a function that is not measurable.

6.3.11. Let $\{f_n\}$, $\{g_n\}$ be two sequences of measurable functions that converge in measure on a set of finite measure E, and let $f$, $g$ denote their limits. If $\alpha, \beta \in \mathbb{R}$, prove that:

(a) $\{\alpha f_n + \beta g_n\}$ converges in measure to $\alpha f + \beta g$;

(b) $\{f_n g_n\}$ converges in measure to $fg$.

(c) Determine whether (a) and (b) remain true when $E$ is a set of infinite measure.

6.3.12. Assume $E$ has a finite measure. Let $\{f_n\}$ be a sequence of measurable functions on $E$ and $f$ measurable on $E$ for which $f$ and each $f_n$ is finite a.e. on $E$. Prove that $\{f_n\} \to f$ in measure on $E$ if and only if every subsequence of $\{f_n\}$ has in turn a further subsequence that converges to $f$ pointwise a.e. on $E$.

6.3.13. Prove that Egorov's Theorem need not hold if $\mu(X) = \infty$.

6.3.14.* A sequence of measurable functions is **Cauchy in measure** if, for all $\eta > 0$, there is $N \in \mathbb{N}$ such that

$$m, n \geq N \quad \Rightarrow \quad \mu\left(\{x \in X : |f_n(x) - f_m(x)| > \eta\}\right) < \eta.$$

Prove that:

(a) if a sequence $\{f_n\}$ is Cauchy in measure then there is a measurable function $f$ on $X$ to which $\{f_n\}$ converges in measure;

(b) if a sequence of measurable functions $\{f_n\}$ converges in measure then it is Cauchy in measure.

6.3.15. Let $(X, \mathfrak{M}, \mu)$ be a finite measure space. Prove or disprove: There exist a sequence $\{f_k\}_{k=1}^{\infty}$ of measurable functions on $X$ with values in $[-\infty, +\infty]$ that are finite a.e. and that converge to 0 in measure, but no subsequence converges uniformly on any subset of positive measure.

6.3.16. Prove that $f_n(x) = e^{-n|1-\sin x|}$ converges in measure to $f(x) = 0$ on $[a, b]$. Is this true if the interval is infinite?

6.3.17. Let $f : \mathbb{R}^n \to [-\infty, +\infty]$ be a measurable function. Prove that there exists a sequence of continuous functions on $\mathbb{R}^n$ that converges to $f$ a.e.

6.3.18. Let $(X, \mathfrak{M}, \mu)$ be a measure space and let $\{f_n\}_{n=1}^\infty$ be measurable functions on $X$ with values in $[-\infty, +\infty]$ that are finite a.e. If $\{f_n\}$ converges to $f$ in measure on $X$, prove that $f$ is measurable.

6.3.19. Let $(X, \mathfrak{M}, \mu)$ be a finite measure space, let $\{f_n\}_{n=1}^\infty$, $f$ be measurable functions on $X$ such that $\{f_n\}$ converges to $f$ in measure on $X$, and let $F : \mathbb{R} \to \mathbb{R}$ be a continuous function. Prove that $\{F \circ f_n\}$ converges in measure to $F \circ f$.

6.3.20. Prove the converse of Luzin's theorem: If $f : \mathbb{R}^n \to \mathbb{R}$ has the property that for every $\varepsilon > 0$ there is an open set $G$ with $m_n(G) < \varepsilon$ and $f$ is continuous on $\mathbb{R}^n \setminus G$ then $f$ is measurable.

6.3.21. **(Lebesgue–Hausdorff)** Prove that the collection of Borel measurable functions on $\mathbb{R}^n$ is the smallest algebra that contains continuous functions and is closed under pointwise limits.

6.3.22. Let $(X, \mathfrak{M}, \mu)$ be a measure space and let $\{E_n\} \subset \mathfrak{M}$ for which $\lim_n \mu(E_n) = 0$. Prove that $\{\chi_{E_n}\}$ has a subsequence that converges to the zero function a.e.

6.3.23.* Prove that the Borel $\sigma$-algebra on a metric space is the smallest $\sigma$-algebra with respect to which all bounded, continuous functions are measurable.

6.3.24.* Give an example of a sequence of measurable functions for which the set of points where the convergence is uniform is nowhere dense.

6.3.25.* Let $(X, \mathfrak{M}, \mu)$ be a measure space, let $f_n : X \to [0, +\infty]$ be a measurable function for each $n \in \mathbb{N}$, and for each $\epsilon > 0$ and each $n \in \mathbb{N}$ define $E_{n,\epsilon} = \{x \in X : f_n(x) \geq \epsilon\}$. Prove that the sequence $\{f_n\}$ converges to 0 a.e. if and only if $\mu(\limsup_n E_{n,\epsilon}) = 0$ for all $\epsilon > 0$.

6.3.26.* Let $(X, \mathfrak{M}, \mu)$ be a measure space, let $\{f_n\}, f$ be measurable functions on $X$, let $\{c_n\}$ be real numbers, and let $E_n = \{x \in X : |f_n(x)| \leq c_n\}$, for $n \in \mathbb{N}$. Suppose that $\sum_{n=1}^\infty \mu(E_n^C) < \infty$ and $\lim_n f_n \chi_{E_n} = f$ a.e. Prove that $\lim_n f_n = f$ a.e.

6.3.27. Let $(X, \mathfrak{M}, \mu)$ be a finite measure space, let $\{f_n\}, f$ be nonnegative measurable functions on $X$ such that $\{f_n\}$ converges to $f$ in measure, and let $a > 0$. Prove that $\{f_n^a\}$ converges to $f^a$ in measure. Give an example to show that the result is not true if $\mu(X) = \infty$.

6.3.28. Let $\mu$ be counting measure on $\mathbb{N}$ and let $\{f_n\}$ be functions on $\mathbb{N}$. Prove that $\{f_n\}$ converges to a function $f$ in measure if and only if it converges to $f$ uniformly.

6.3.29. Let $(X, \mathfrak{M}, \mu)$ be a measure space and let $f : X \to \mathbb{R}$ be a measurable function. Prove that there exists a sequence of measurable functions $\{f_n\}$ that converge uniformly to $f$ on $X$ and each $f_n$ takes at most a countable number of values.

6.3.30.* Let $(X, \mathfrak{M}, \mu)$ be a $\sigma$-finite measure space and let $\{f_n\}$ be a sequence of measurable functions that converges pointwise to a function $f : X \to \mathbb{R}$. Prove that $X$ is a countable union of measurable sets $\{X_n\}$ where $\mu(X_1) = 0$ and $\{f_n\}$ converge uniformly to $f$ on $X_k$ for $k \geq 2$.

# Part III

# Integration Theory

# 7

# *The Integral*

From the very beginning our goal was to develop a solid theory of integration that would overcome deficiencies of the Riemann integral. In the previous chapters, we laid the foundation. First, we developed a measure theory. Reason: in order for $\chi_A$ to be integrable (and knowing that its integral should equal the "measure" of $A$), we needed as many sets as possible to be measurable. Next, we studied measurable functions. Reason: many functions that appear in applications are measurable but not Riemann integrable. In this chapter we will define the Lebesgue integral of a measurable function and learn how to calculate it. An additional benefit that validates Lebesgue theory is that we will get some formulas that have no analogue in the Riemann theory of integration. In particular, the Lebesgue integral is much better suited to the interchange of the integral and the limit process and we will obtain results that are either very hard to prove in the case of Riemann integral or even incorrect.

One of the underlying themes of Lebesgue's approach is the build-up from easy cases to harder ones. We have already encountered it in Chapter 4, when developing the concept of a measure of a set, from rectangles via elementary sets, etc. We employ a similar strategy in this chapter. Namely, we start with the integral of a simple function, then use it to define the integral of a non-negative function, which is then used to define the integral of any measurable function.

## 7.1 About Riemann Integral

In this section, as a prelude to the Lebesgue integration theory, we will remind ourselves of the reasons why a new type of integral is a welcome contribution. The following list presents some of the shortcomings of the Riemann integral:

- It does not commute with the limit process (including derivatives and integrals)
- The derivative of a differentiable function need not be integrable
- The limit of a sequence of integrable functions need not be integrable

DOI: 10.1201/9781032716497-7

We support our criticism of Riemann integral with several examples.

**Example 7.1.1. A sequence of Riemann integrable functions $\{f_n\}$ such that $\int_0^1 \lim f_n(x)\,dx \neq \lim \int_0^1 f_n(x)\,dx$.**

Let $f_n(x) = n^2 x e^{-nx}$, for $n \in \mathbb{N}$ and $x \in (0,1)$. Then $\lim f_n = 0$ so $\int_0^1 \lim f_n(x)\,dx = 0$.
But, $\int f_n(x)\,dx = -e^{-nx}(1+nx)$ so $\int_0^1 f_n(x)\,dx = 1 - e^{-n}(n+1) \to 1$. ♦

For the next example, we need a function that is often associated with German mathematician Carl Thomae. The Thomae function is is also known under more picturesque names: the popcorn function, the raindrop function, the countable cloud function, or even the Stars over Babylon. It is defined by

$$f(x) = \begin{cases} \frac{1}{q}, & \text{if } x = \frac{p}{q},\ p \in \mathbb{Z},\ q \in \mathbb{N}, \text{ and } p, q \text{ are mutually prime} \\ 0, & \text{if } x \text{ is irrational or } x = 0. \end{cases}$$

**Example 7.1.2. A function $f$ such that $\left(\int_0^x f(t)\,dt\right)' \neq f(x)$**

Let $f$ be the Thomae function. Then $\int_0^x f(t)\,dt = 0$. The proof that $f$ is integrable can be found in [86, Example 6.3.6]. Moreover, this proof also shows that $\int_0^1 f(x)\,dx = 0$. Since $f(x) \geq 0$ for all $x$, it follows that $\int_0^x f(t)\,dt = 0$ for $x \in [0,1]$. Finally, equality $\int_n^{n+1} f(x)\,dx = 0$ can be established in almost verbatim the same way as for $n = 0$. ♦

**Example 7.1.3. A function $f : (0,1) \times (0,1)$ such that the iterated integrals $\int_0^1 \int_0^1 f(x,y)\,dy\,dx$ and $\int_0^1 \int_0^1 f(x,y)\,dx\,dy$** exist but are not equal.

Let $f(x,y) = \dfrac{x^2 - y^2}{(x^2+y^2)^2}$. Then

$$\frac{\pi}{4} = \int_0^1 \int_0^1 f(x,y)\,dy\,dx \neq \int_0^1 \int_0^1 f(x,y)\,dx\,dy = -\frac{\pi}{4}.$$ ♦

**Example 7.1.4. A sequence of integrable functions that converges pointwise to a nonintegrable function.**

Let $\{r_n\}$ be an enumeration of $\mathbb{Q} \cap [0,1]$. For every $n \in \mathbb{N}$, define

$$f_n(x) = \begin{cases} 1 & \text{if } x \in \{r_k\}_{k=1}^n \\ 0 & \text{otherwise.} \end{cases}$$

Then each $f_n$ is integrable but the limit is the Dirichlet function which is not Riemann integrable. ♦

**Example 7.1.5. A function $f$ that is differentiable on $[-1,1]$, but $f'$ is not integrable on $[-1,1]$.**

Let $f(x) = \begin{cases} x^2 \sin \frac{1}{x^2}, & \text{if } x \neq 0 \\ 0, & \text{if } x = 0. \end{cases}$ Then $f'(x)$ is unbounded, hence not integrable. ♦

The reason that the derivative in Example 7.1.5 is not integrable is that it is unbounded. This raises the question if one can manufacture a function whose derivative is bounded but not integrable. The following function is a good candidate. It is known as Volterra's function, named for Vito Volterra, who presented the construction in 1881 in[124].

**Example 7.1.6. Volterra's function.**
We start with the Fat Cantor set $F$ (p. 74) and the function

$$f(x) = \begin{cases} x^2 \sin \frac{1}{x}, & \text{if } x \neq 0 \\ 0, & \text{if } x = 0. \end{cases}$$

First, let $x_1 = \sup\{x \in [0, 1/8] : f'(x) = 0\}$ and consider the function $f$ restricted to the interval $[0, x_1]$. Next, extend this restriction to the right with a constant value of $f(x_1)$ up to and including the point $1/8$. Once this is done, a mirror image of the function can be created on the interval $[1/8, 1/4]$. We then translate this graph to the interval $[3/8, 5/8]$ so that the resulting function, which we call $f_1$, is nonzero only on the middle interval of the complement of the set $F_1$. To construct $f_2$ , let $x_2 = \sup\{x \in [0, 1/32] : f'(x) = 0\}$, consider the function $f$ restricted to the interval $[0, x_2]$, and extend this restriction to the right with a constant value of $f(x_2)$ up to and including the point $1/32$. A mirror image of the function can be created on the interval $[1/32, 1/16]$ and two translated copies (to $[5/32, 7/32]$ and $[25/32, 27/32]$) are added to $f_1$ to produce the function $f_2$, which is nonzero only on $F_2^C$. Volterra's function $V$ then results by repeating this procedure for every interval removed in the construction of the Fat Cantor set; namely, $V(x) = \lim f_n(x)$. ♦

This function has the desired properties.

**Theorem 7.1.7.** *The Volterra's function $V$ is differentiable on $[0, 1]$ and its derivative $V'$ is bounded but not integrable.*

*Proof.* Let $x \in [0, 1]$. If $x \in F^C$ then there is $\eta > 0$ such that $(x-\eta, x+\eta) \subset F^C$ and on that interval $V$ is obtained from $f$ by translation or symmetry, so it is differentiable at $x$. Moreover, since $|f'(x)| = |2x \sin(1/x) - \cos(1/x)| \leq 3$ we see that $|V'(x)| \leq 3$. On the other hand, if $x \in F$, $V(x) = 0$, by construction. The difference quotient $[V(y) - V(x)]/(y - x) = V(y)/(y - x)$ depends on $y$. Namely, if $y \in F$ then $V(y)/h = 0$. If $y \notin F$, let $z$ be the point in $F$ closest to $y$. By construction, $V(y)$ is either $f(y-z)$ or $f(z-y)$, depending on whether $y$ is a point on the part of the graph of $V$ obtained using merely translation or both translation and symmetry. Since $|f(x)| \leq x^2$, we have that $V(y) = |y - z|^2$. Consequently, $|V(y)/(y - x)| \leq |V(y)/(y - z)| \leq |y - z|^2/|y - z| = |y - z|$. Either way, $V'(x) = 0$ on $F$.

Finally, we show that $V'$ is not integrable. It is not hard to see that $\limsup_{x\to 0} f'(x) = 1$ and $\liminf_{x\to 0} f'(x) = -1$. Since each point of $F$ is a cluster point of $F^C$, we get that if $a \in F$ then $\limsup_{x\to a} V'(x) = 1$ and $\liminf_{x\to a} V'(x) = -1$. This shows that $V'$ is discontinuous on the Fat Cantor set $F$, a set of positive measures so, by Theorem 7.4.3, it is not integrable. (An alternative but arduous way would be to write down a sequence of partitions and show that the upper and lower Darboux integrals are not equal.) □

Many more examples can be found in the classical book [43].

VITO VOLTERRA (1860–1940) was an Italian mathematician and physicist. After graduating from the University of Pisa, he held several academic positions, settling as professor of mathematical physics at the University of Rome La Sapienza. He is considered one of the founders of functional analysis because his early work centered around theory of functionals and their applications in integral and integro-differential equations. His opus is summarized in his book "Theory of functionals and of Integral and Integro-Differential Equations" (1930). Like his mentor Enrico Betti, Volterra was an enthusiastic supporter of the unification of Italy. He served as a senator of the Kingdom of Italy in 1905. During World War I, he originated the idea of using inert helium rather than flammable hydrogen in hot air balloons. After World War I, Volterra turned his attention to the application of his mathematical ideas to biology. In 1922, he joined the opposition to the Fascist regime of Benito Mussolini and in 1931 he was one of only 12 out of 1,250 professors who refused to take a mandatory oath of loyalty. As a result, he was compelled to resign his university post and his membership in scientific academies, and during the following years, he lived largely abroad, returning to Rome just before his death. Volterra was an International Member of the United States National Academy of Sciences and the American Philosophical Society. He is the only person who was a plenary speaker in the International Congress of Mathematicians four times (1900, 1908, 1920, 1928).

## 7.2 Integration of Nonnegative Measurable Functions

Having developed the theory of measure of sets and functions, we will now introduce the integral. We start with our building blocks, characteristic functions of measurable sets. It is natural to define the integral of $\chi_A$ to be $\mu(A)$. Naturally, we next look at linear combinations of building blocks—the simple functions. In the following definition, we will restrict our attention to *nonnegative* simple functions.

**Definition 7.2.1.** Let $(X, \mathfrak{M}, \mu)$ be a measure space and $s$ a nonnegative simple function on $X$. If $s = 0$ we define $\int_X s\, d\mu = 0$. If $s = \sum_{k=1}^{n} c_k \chi_{E_k}$, where $\{E_k\}_{k=1}^{n}$ are disjoint measurable sets and $\{c_k\}_{k=1}^{n}$ are nonnegative real numbers, then its Lebesgue integral is

$$\int_X s\, d\mu = \sum_{k=1}^{n} c_k \mu(E_k).$$

When there is no danger of ambiguity we write simply $\int_X s$ or even $\int s$. If $E \in \mathfrak{M}$, then $\int_E s\, d\mu = \int_X s\, \chi_E\, d\mu$. Just like in the general case both $E$ and $d\mu$ can be omitted when it is clear from the context what they are.

*Remark* 7.2.2. In the special case when $(X, \mathfrak{M}, \mu)$ is $(\mathbb{R}, \mathcal{M}, m)$ and the sets $\{E_k\}$ are intervals, a simple function is a step function and its Riemann and Lebesgue integrals are equal.

*Remark* 7.2.3. Definition 6.2.10 stipulates that a simple function takes only a *finite number* of values. Therefore, in Definition 7.2.1 we do not need to worry about infinite sums. Because of that, proofs of results that involve the integration of simple functions (such as Proposition 7.2.6 below) typically turn out to be exercises in measure theory. Also, it makes every measurable simple function on a set of finite measures automatically integrable (meaning that its integral is finite).

*Remark* 7.2.4. By definition, a simple function $s$ takes finitely many different values $c_1, c_2, \ldots, c_p$. If we denote $E_k = \{x \in X : s(x) = c_k\}$ then $s = \sum_{k=1}^{p} c_k \chi_{E_k}$. Such a form is often called the standard form of a simple function. Clearly, the standard form is unique.

It is important to ensure that if a simple function is represented in two different ways they yield the same value of the integral. The next result will confirm that. Even though in this section we are interested only in *nonnegative* simple functions, such a hypothesis is not needed here.

**Lemma 7.2.5.** *Suppose that a simple function $s$ has two different representations:*

$$s = \sum_{k=1}^{n} a_k \chi_{A_k} \quad \text{and} \quad s = \sum_{k=1}^{m} b_k \chi_{B_k},$$

*where $\{A_k\}_{k=1}^{n}$ is a collection of pairwise disjoint sets and so is $\{B_k\}_{k=1}^{m}$. Then*

$$\sum_{k=1}^{n} a_k \mu(A_k) = \sum_{k=1}^{m} b_k \mu(B_k). \tag{7.1}$$

*Proof.* It suffices to prove (7.1) under the assumption that $\sum_{k=1}^{m} b_k \chi_{B_k}$ is the standard form, i.e., that all coefficients $\{b_k\}$ are distinct and the sets $\{B_k\}$ are disjoint.

Let $I_j$ be a subset of $\{1, 2, \ldots, n\}$ so that, for each $i \in I_j$, $a_i = b_j$. Then $\{1, 2, \ldots, n\} = \cup_{j=1}^{m} I_j$. Also, $B_k$ is a disjoint union

$$B_k = \bigcup_{i \in I_k} A_i,$$

so $m(B_k) = \sum_{i \in I_k} m(A_i)$. It follows that

$$\sum_{k=1}^{n} a_k \mu(A_k) = \sum_{j=1}^{m} \left( \sum_{i \in I_j} a_i \mu(A_i) \right) = \sum_{j=1}^{m} b_j \sum_{i \in I_j} \mu(A_i) = \sum_{j=1}^{m} b_j \mu(B_j).$$

□

The integral of a simple function has several fundamental properties that one expects from an integral.

**Proposition 7.2.6.** *Let $(X, \mathfrak{M}, \mu)$ be a measure space and $s, t$ nonnegative simple functions on $X$.*

*(i) If $\alpha, \beta$ are positive real numbers,*

$$\int_X (\alpha s + \beta t)\, d\mu = \alpha \int_X s\, d\mu + \beta \int_X t\, d\mu.$$

*(ii) If $A, B$ are disjoint measurable subsets of $X$, then*

$$\int_{A \cup B} s\, d\mu = \int_A s\, d\mu + \int_B s\, d\mu.$$

*(iii) If $X_0$ is a measurable subset of $X$ and $\mu(X \setminus X_0) = 0$, then $\int_X s\, d\mu = \int_{X_0} s\, d\mu$.*

*(iv) If $s \leq t$ a.e. on $X$, then $\int_X s\, d\mu \leq \int_X t\, d\mu$.*

*Proof.*
(i) If either $s$ or $t$ is nonzero on a set of infinite measures then both sides of the asserted equality are $+\infty$. Otherwise, we partition $X = \cup_{k=1}^n E_k$ so that on $E_k$, $s = a_k$ and $t = b_k$. By Exercise 6.2.22, $\alpha s + \beta t$ is a simple function that takes the value $\alpha a_k + \beta b_k$ on $E_k$. Then, using Lemma 7.2.5,

$$\begin{aligned}
\int_X (\alpha s + \beta t) &= \sum_{k=1}^n (\alpha a_k + \beta b_k) m(E_k) \\
&= \alpha \sum_{k=1}^n a_k m(E_k) + \beta \sum_{k=1}^n b_k m(E_k) \\
&= \alpha \int_X s + \beta \int_X t.
\end{aligned}$$

(ii) Follows from (i) because $s\chi_{A\cup B} = s\chi_A + s\chi_B$ whenever $A \cap B = \emptyset$.
(iii) Follows from (ii) with $A = X_0$, $B = X \setminus X_0$. (Exercise 7.2.1 is also needed.)
(iv) By (iii), we may assume that $s \leq t$ on $X$. If $s = \sum_{k=1}^n a_k \chi_{E_k}$ and $t = \sum_{k=1}^n b_k \chi_{E_k}$, then $a_k \leq b_k$, for all $1 \leq k \leq n$, so $\sum_{k=1}^n a_k \mu(E_k) \leq \sum_{k=1}^n b_k \mu(E_k)$. □

*Remark* 7.2.7. Lemma 7.2.5 shows that, when calculating the Lebesgue integral of a simple function, it is not necessary that its values be distinct. Proposition 7.2.6 shows that sets $\{E_k\}$ in Definition 7.2.1 need not be disjoint. Indeed, the linearity of the integral easily extends to a finite sum of simple functions, hence to any finite sum of characteristic functions.

Having defined the integral of a simple function, we now have yet another example of a measure.

**Example 7.2.8. A measure on $(X, \mathfrak{M}, \mu)$.**

Let $s$ be a nonnegative simple function on $X$, and let $\lambda$ be a function on the $\sigma$-algebra $\mathfrak{M}$ defined by

$$\lambda(E) = \int_E s\,d\mu.$$

Then $\lambda$ is a measure.

**Proof:** We write $s = \sum_{k=1}^n a_k \chi_{A_k}$ where $a_k \geq 0$ and $A_k \in \mathfrak{M}$, for all $1 \leq k \leq n$. Then,

$$\begin{aligned}\lambda(E) &= \int_E s\,d\mu = \int_X \chi_E s\,d\mu = \int_X \chi_E \left(\sum_{k=1}^n a_k \chi_{A_k}\right) d\mu \\ &= \int_X \sum_{k=1}^n a_k \chi_{A_k \cap E}\,d\mu = \sum_{k=1}^n a_k \mu(A_k \cap E),\end{aligned}$$

so $\lambda$ is a linear combination (with nonnegative coefficients) of measures (see Example 5.1.6). Therefore, $\lambda$ is a measure based on the argument at the beginning of Section 5.5. ♦

After simple functions, our next level of generality is nonnegative measurable functions. Throughout this section the word nonnegative will mean a function $f : X \to [0, +\infty]$.

**Definition 7.2.9.** Let $(X, \mathfrak{M}, \mu)$ be a measure space and $f$ a nonnegative measurable function. The **integral** of $f$ over $X$ with respect to $\mu$ is

$$\int_X f\,d\mu = \sup\left\{\int_X s : s \in \mathcal{S}(X),\, 0 \leq s \leq f \text{ on } X\right\}.$$

We say that $f$ is **integrable** over $X$ if $\int_X f\,d\mu < \infty$. We use notation $f \in L(X, \mu)$ or simply $f \in L$ if there is no danger of ambiguity. Just like with the integral of a simple function, both $X$ and $d\mu$ can be omitted when it is clear from the context what they are.

*Remark* 7.2.10. Since the integral of $f$ is defined as a supremum of a set of positive numbers, it always exists either as a real number or $+\infty$. We prefer when it is finite and for such a function $f$ we say that it is *integrable.*

When Lebesgue introduced this integral it was for functions defined on the real line. Nevertheless, we call it **Lebesgue integral** when $X = \mathbb{R}^n$, for all $n$.

> In a 1926 address in Copenhagen, Lebesgue presented the following example as an analogy for the relation between the Riemann and the Lebesgue integrals. Suppose you have the task of calculating the total value of a sequence of coins. The "Riemann" way is to add them as you go along the sequence. The "Lebesgue" way is to sort them based on their value (pennies, nickels, dimes, etc.) and find the sum as $1 \times$ (the number of pennies) $+\ 5 \times$ (the number of nickels) $+ \ldots$.

We summarize some fundamental properties of the integral in the following proposition and leave the proof to the reader (Exercise 7.2.3).

**Proposition 7.2.11.** *Let $(X, \mathfrak{M}, \mu)$ be a measure space and $f, g$ nonnegative measurable functions on $X$.*

*(i) If $\alpha \geq 0$, then $\int_X \alpha f\, d\mu = \alpha \int_X f\, d\mu$.*

*(ii) If $f \leq g$ a.e. on $X$, then $\int_X f\, d\mu \leq \int_X g\, d\mu$.*

*(iii) If $X_0$ is a measurable subset of $X$ and $\mu(X \setminus X_0) = 0$, then $\int_X f\, d\mu = \int_{X_0} f\, d\mu$.*

Comparing Proposition 7.2.11 with Proposition 7.2.6, we see that the next goal might be to establish the linearity of the integral of a nonnegative function. However, it will be easier to prove this fact once we have a convergence theorem in place.

**Theorem 7.2.12.** ***Monotone Convergence Theorem.*** *Let $(X, \mathfrak{M}, \mu)$ be a measure space and $\{f_n\}$ an increasing sequence of nonnegative measurable functions on $X$ such that $\lim_n f_n = f$ a.e. Then*

$$\lim \int_X f_n = \int_X f.$$

*Proof.* Since $\{f_n\}$ is an increasing sequence, we make two conclusions. First, it converges to a measurable function $f = \sup f_n$, for all $x \in X$. Second, Proposition 7.2.11 (ii) implies that $\{\int_X f_n\, d\mu\}$ is increasing. As such, it converges to $L = \sup\{\int_X f_n\}$. Finally, $f_n \leq f$ implies that $\int_X f_n\, d\mu \leq \int_X f\, d\mu$, so

$$L \leq \int_X f\, d\mu. \tag{7.2}$$

To prove the opposite inequality, let $0 < \alpha < 1$, let $s$ be a simple function such that $0 \leq s \leq f$ on $X$, and for every $n \in \mathbb{N}$, define

$$E_n = \{x \in X : f_n(x) \geq \alpha f(x)\}.$$

It is not hard to see that $\{E_n\}$ is an increasing sequence of measurable sets whose union is $X$. Indeed, if $f(x) = 0$ then $x \in E_1$. If $f(x) > 0$ then $f(x) > \alpha f(x)$ so there is $n$ for which $f_n(x) > \alpha f(x)$. Now

$$\int_X f_n\, d\mu \geq \int_X \chi_{E_n} f_n\, d\mu = \int_{E_n} f_n\, d\mu \geq \int_{E_n} \alpha f\, d\mu \geq \alpha \int_{E_n} s\, d\mu. \tag{7.3}$$

By Example 7.2.8, $\int_{E_n} s\, d\mu$ defines a measure, and by the continuity of a measure

$$\lim_n \int_{E_n} s\, d\mu = \int_X s\, d\mu.$$

Therefore, taking the limit as $n \to \infty$ in (7.3) we obtain that

$$L \geq \alpha \int_X s\, d\mu.$$

Since this holds for any $0 < \alpha < 1$, letting $\alpha \to 1$, we get

$$L \geq \int_X s\, d\mu.$$

Finally, taking supremum over all simple functions satisfying $0 \leq s \leq f$, we get that $\int_X f\, d\mu \leq L$ and it follows from (7.2) that $\lim_n \int_X f_n\, d\mu = \int_X f\, d\mu$. □

*Remark* 7.2.13. An analogous theorem for Riemann integrals on $\mathbb{R}$ is false because the limit function $f$ need not be Riemann integrable (Exercise 7.4.6). If we make an additional assumption that $f$ is also integrable then the result is true. However, the known proofs that do not rely on Theorem 7.2.12 are very hard. On the other hand, a proof using Lebesgue theory will be quite easy once we establish a connection between Riemann and Lebesgue integrals of a function (Theorem 7.4.1).

Definition 7.2.9 introduced the Lebesgue integral of a nonnegative function, based on the integrals of simple functions. Lurking in the background is Theorem 6.2.13 asserts that every measurable function $f$ is the limit of a sequence $\{s_n\}$ of simple functions. At least intuitively, it justifies the close proximity of the integrals of $f$ and $s_n$. The crucial ingredient is the fact that the sequence $\{s_n\}$ is *monotone* as evidenced by the Monotone Convergence Theorem (MCT).

Now we can prove that the integral of a nonnegative function is linear.

**Theorem 7.2.14.** ***Linearity of Integration.*** *Let $(X, \mathfrak{M}, \mu)$ be a measure space and $f, g$ nonnegative measurable functions on $X$. Then, for any positive numbers $\alpha, \beta$,*

$$\int_X (\alpha f + \beta g) = \alpha \int_X f + \beta \int_X g.$$

*Proof.* In view of Proposition 7.2.11, we may assume that $\alpha = \beta = 1$. By Theorem 6.2.13, there are two increasing sequences of nonnegative simple functions $\{s_n\}$ and $\{t_n\}$, such that $\lim s_n = f$ and $\lim t_n = g$ on $X$. Then $\{s_n + t_n\}$ is an increasing sequence of nonnegative simple functions that converges to $f + g$. Applying the Monotone Convergence Theorem and Proposition 7.2.6, we have

$$\begin{aligned}\int_X (f+g)\, d\mu &= \int_X \lim_n (s_n + t_n)\, d\mu = \lim_n \int_X (s_n + t_n)\, d\mu \\ &= \lim_n \left( \int_X s_n\, d\mu + \int_X t_n\, d\mu \right) = \int_X f\, d\mu + \int_X g\, d\mu.\end{aligned}$$

□

Using induction, one can show that Theorem 7.2.14 applies to finite sums. In fact, it remains true for infinite sums as well. We leave the proof as Exercise 7.2.11.

**Corollary 7.2.15.** *Let $(X, \mathfrak{M}, \mu)$ be a measure space and $\{f_n\}$ a sequence of nonnegative measurable functions on $X$. Then,*

$$\int_X \sum_{n=1}^{\infty} f_n \, d\mu = \sum_{n=1}^{\infty} \int_X f_n \, d\mu.$$

It seems impossible to exaggerate the significance of Monotone Convergence Theorem. We have already seen it in action in Theorem 7.2.14, and its impact will be felt throughout the course. Corollary 7.2.15 is deceptively simple, but it allows term-by-term integration of any series with nonnegative terms. This is false in the case of the Riemann integral (see Exercise 7.4.6), because in general $\sum f_n$ need not be Riemann integrable. If we were to assume its integrability then Corollary 7.2.15 would hold for the Riemann integral, but the known proofs tend to be more complicated than the proof of Monotone Convergence Theorem.

We have seen in Example 7.2.8 that each simple function leads to a new measure. The same is true of nonnegative functions.

**Example 7.2.16. A measure on $(X, \mathfrak{M}, \mu)$.**

Let $f$ be a nonnegative measurable function on $X$, and let $\lambda$ be a function on the $\sigma$-algebra $\mathfrak{M}$ defined by

$$\lambda(E) = \int_E f \, d\mu.$$

Then $\lambda$ is a measure.

**Proof.** It is easy to see that $\lambda$ is nonnegative and $\lambda(\emptyset) = 0$. Suppose that $E$ is a disjoint union of mesaurable sets $\{E_n\}_{n=1}^{\infty}$. Using Corollary 7.2.15,

$$\begin{aligned} \lambda(E) &= \int_X \chi_E f \, d\mu = \int_X \sum_{n=1}^{\infty} \chi_{E_n} f \, d\mu = \sum_{n=1}^{\infty} \int_X \chi_{E_n} f \, d\mu \\ &= \sum_{n=1}^{\infty} \int_{E_n} f \, d\mu = \sum_{n=1}^{\infty} \lambda(E_n). \end{aligned}$$

♦

The monotonicity assumption is quite powerful and allows us to prove various results. In the situation when the sequence is *not* monotone, we can apply the following result.

**Theorem 7.2.17. *Fatou's Lemma.*** *Let $(X, \mathfrak{M}, \mu)$ be a measure space and $\{f_n\}$ a sequence of nonnegative measurable functions on $X$. Then*

$$\int_X \liminf f_n \, d\mu \le \liminf \int_X f_n \, d\mu.$$

*Proof.* For $x \in X$ and $n \in \mathbb{N}$ we define

$$g_n(x) = \inf_{k \ge n} f_k(x).$$

Then $g_n \leq f_n$ on $X$ and $\{g_n\}$ is an increasing sequence of measurable functions that converges to

$$\sup_n g_n = \sup_n \inf_{k \geq n} f_k = \liminf_n f_n.$$

It follows, using Monotone Convergence Theorem, that

$$\begin{aligned}\int_X \liminf_n f_n \, d\mu &= \int_X \lim_n g_n \, d\mu = \lim_n \int_X g_n \, d\mu = \liminf_n \int_X g_n \, d\mu \\ &\leq \liminf \int_X f_n \, d\mu.\end{aligned}$$

□

**Example 7.2.18. The strict inequality in Fatou's Lemma.**
Let $f_n(x) = \chi_{[n,n+1)}$. Clearly, $f := \lim f_n = 0$ on $\mathbb{R}$. Yet, $\int_{\mathbb{R}} f_n = 1$ and $\int_{\mathbb{R}} f = 0$. ♦

Fatou's Lemma appears for the first time in Fatou's memoir on trigonometirc series [32], published in 1906. However, he proved the result for bounded functions. Riesz disposed of this additional assumption in 1910 in a paper [91] written in Hungarian.

It is well known that the Riemann integral of a nonnegative, continuous function $f$ is 0 if and only if $f$ is the zero function. We will prove that the same is true for the Lebesgue integral, and we will use the following result which has a much wider significance.

**Theorem 7.2.19.** ***Chebyshev's Inequality.*** *Let $(X, \mathfrak{M}, \mu)$ be a measure space, $f$ a nonnegative measurable function on $X$, and $\lambda > 0$. Then,*

$$\mu\{x \in X : f(x) \geq \lambda\} \leq \frac{1}{\lambda} \int_X f \, d\mu.$$

*Proof.* Denote $E_\lambda = \{x \in X : f(x) \geq \lambda\}$. We will prove that

$$\lambda \mu(E_\lambda) \leq \int_X f.$$

**Case I**: $\mu(E_\lambda) < \infty$.
The simple function $h = \lambda \chi_{E_\lambda}$ satisfies $0 \leq h \leq f$ on $X$. By definition,

$$\lambda \mu(E_\lambda) = \int_X h \leq \int_X f.$$

**Case II**: $\mu(E_\lambda) = \infty$.
We will show that $\int_X f = \infty$. Let $n \in \mathbb{N}$ and define $E_{\lambda,n} = E_\lambda \cap [-n, n]$ and $s_n = \lambda \chi_{E_{\lambda,n}}$. Then $s_n$ is a simple function such that

$$\int_X s_n = \lambda \mu(E_{\lambda,n}), \qquad \text{and } 0 \leq s_n \leq f \text{ on } X.$$

It follows that, $\int_X s_n \leq \int_X f$. By continuity of measure, when $n \to \infty$, $\mu(E_{\lambda,n}) \to \mu(E_\lambda) = \infty$, so $\int_X f = \infty$. □

The theorem is named after Russian mathematician Chebyshev, although it was first formulated (without proof) by his friend and colleague Bienaymé in 1853 [10]. Chebyshev proved it in 1867 [24]. His student Markov provided another proof in his 1884 Ph.D. thesis. In the language of probability theory, Chebyshev's inequality (also called the Bienaymé–Chebyshev inequality) guarantees that, for a wide class of probability distributions, no more than a certain fraction of values can be more than a certain distance from the mean. A French statistician Irénée-Jules Bienaymé (1796–1878) was among the last in a long line of French probabilists. After him, progress in statistics took place in the United Kingdom and Russia. As a supporter of Napoleon Bonaparte, his career had a setback whenever the political situation in France changed. He and Cauchy created different regression methods at about the same time and there was a dispute over the superiority of one method over the other. A Russian mathematician Andrey Markov (1856–1922) is best known for his work on stochastic processes. A primary subject of his research later became known as the Markov chain. He was also a strong, close to master-level chess player.

Chebyshev's Inequality allows us to prove several useful results.

**Proposition 7.2.20.** *Let $(X, \mathfrak{M}, \mu)$ be a measure space and $f$ a nonnegative measurable function on $X$. Then $\int_X f = 0$ iff $f = 0$ a.e. on $X$.*

*Proof.* ($\Rightarrow$) By Chebyshev's Inequality, $\mu\{x \in X : f(x) \geq 1/n\} = 0$, for all $n \in \mathbb{N}$. Since

$$\{x \in X : f(x) > 0\} = \bigcup_{n=1}^{\infty} \left\{x \in X : f(x) \geq \frac{1}{n}\right\},$$

we obtain that $\mu\{x \in X : f(x) > 0\} = 0$, i.e., $f = 0$ a.e.

($\Leftarrow$) Let $s \in \mathcal{S}(X)$ such that $0 \leq s \leq f$. Then $s = 0$ a.e. on $X$ so $\int_X s = 0$. Taking the supremum over all such simple functions, we obtain that $\int_X f = 0$. □

**Proposition 7.2.21.** *Let $(X, \mathfrak{M}, \mu)$ be a measure space and $f$ a nonnegative measurable function on $X$ for which $\int_X f\,d\mu < \infty$ . Then $f$ is finite on $X$ a.e. and $\{x \in X : f(x) > 0\}$ is $\sigma$-finite.*

*Proof.* Let $n \in \mathbb{N}$. Using Chebyshev's Inequality,

$$\mu\{x \in X : f(x) = \infty\} \leq \mu\{x \in X : f(x) \geq n\} \leq \frac{1}{n}\int_X f\,d\mu \to 0, \quad \text{as } n \to \infty,$$

so $f$ is finite a.e.

Let $E = \{x \in X : f(x) > 0\}$ and $E_n = \{x \in X : f(x) \geq 1/n\}$. Clearly, $E = \cup_{n=1}^{\infty} E_n$. If we denote $s_n = \chi_{E_n}$ then $s_n \leq nf$, so

$$\mu(E_n) = \int_X s_n\,d\mu \leq \int_X nf\,d\mu = n\int_X f\,d\mu < \infty,$$

and it follows that $E$ is $\sigma$-finite. □

Chebyshev's Inequality is a useful result, even more so in the theory of probability. Nevertheless, Propositions 7.2.20 and 7.2.21 show that this inequality is a very effective tool in integration theory as well.

Lebesgue developed his integration theory for functions of one real variable. In 1915, Fréchet published the first paper devoted to integrals over abstract measure sets [38]. Nevertheless, the shift toward abstract did not happen immediately. In the first edition of his monograph *Theory of the integral* that came out in 1933, Saks presents the integration within Euclidean space and abstract integration is relegated to a fifteen-page appendix. In the second edition [103] dated as 1937, the abstract Lebesgue integral comes at the beginning, as the topic of Chapter I.

PAFNUTY CHEBYSHEV (1821–1894) was a Russian mathematician. He had one leg shorter than the other and he was educated at home where he was tutored by one of the best teachers in Moscow. During his student years at Moscow University a big influence on him was Nikolai Brashman (1796–1866), a professor of applied mathematics. Chebyshev's masters thesis was on probability, and his doctoral dissertation at the University of Saint Petersburg was on number theory. He remained at Saint Petersburg for the rest of his life, retiring in 1882, always praised as a teacher. He was never married and lived alone in a large house with ten rooms. He had a daughter whom he refused to officially acknowledge, but supported her nevertheless. He is considered a founding father of Russian mathematics, although he wanted to be considered as a "world-wide mathematician". His papers were written in French and published in the most prestigious journals in France and Germany. He traveled to Western Europe almost every summer in order to maintain contact with leading mathematicians. His many honors include the memberships in the Saint Petersburg Academy of Sciences, the Berlin Academy of Sciences, and the Royal Society of London. He was also awarded the French Légion d'Honneur. In Number Theory he proved that between $n$ and $2n$ there is always a prime number. In Probability he proved the Chebyshev's Inequality (Theorem 7.2.19), generalized the Central Limit Theorem, and introduced the concept of a random variable and the expectation. He constructed a calculating machine, and seven of his mechanical inventions were exhibited at the World's Exposition in Chicago in 1893.

## Exercises

7.2.1. Let $(X, \mathfrak{M}, \mu)$ be a measure space and $s$ a simple function on $X$. If $E$ is a measurable set of measure 0, prove that $\int_E s\, d\mu = 0$.

7.2.2. Prove that if $E \subset X$ is measurable then $\int_E f\, d\mu = \int_X f\chi_E\, d\mu$.

7.2.3. Prove Proposition 7.2.11.

7.2.4. Let $(X, \mathfrak{M}, \mu)$ be a measure space and $f$ a nonnegative measurable function on $X$. If $A, B$ are disjoint measurable subsets of $X$, prove that $\int_{A\cup B} f\, d\mu = \int_A f\, d\mu + \int_B f\, d\mu$.

7.2.5. Let $(X, \mathfrak{M}, \mu)$ be a measure space and $f$ a nonnegative, bounded, measurable function on $X$. Prove that $f \in L(X, \mu)$ if and only if the series

$$\sum_{n=0}^{\infty} \frac{1}{2^n}\, \mu\left(\left\{x \in X : f(x) > \frac{1}{2^n}\right\}\right)$$

converges.

7.2.6. Let $(X, \mathfrak{M}, \mu)$ be a finite measure space and $f$ a nonnegative, bounded and measurable function on $X$. Prove that

$$\int_X f\, d\mu = \inf\left\{\int_X s : s \in \mathcal{S}(X),\ f \leq s \text{ on } X\right\}.$$

Does the assertion remain true if $\mu(X) = \infty$?

7.2.7. Let $\lambda$ be a set function on all subintervals $[a, b] \subset [0, 1]$ defined by $\lambda([a, b]) = \varphi(b) - \varphi(a)$, where $\varphi$ is the Cantor–Lebesgue function on $[0, 1]$. Let $\mu$ be the Caratheodory extension of $\lambda$ to $\mathcal{M}|[0, 1]$. Calculate $\int_0^1 x\, d\mu$.

7.2.8. Let $(X, \mathfrak{M}, \mu)$ be a measure space and $\{f_n\}$ a sequence of nonnegative measurable functions on $X$ that converges a.e. to a function $f$. Prove or disprove: The limit

$$\lim_{n\to\infty} \int_X f_n\, d\mu$$

exists (finite or infinite).

7.2.9. Find a sequence of real valued nonnegative function $f_n$ on $[0, 1]$ so that

$$\limsup_{n\to+\infty} f_n(x) = +\infty, \quad \forall x \in [0, 1]$$

and

$$\lim_{n\to+\infty} \int_{[0,1]} f_n(x) dx = 0.$$

7.2.10. Let $E$ be a subset of $\mathbb{R}$ that is of finite Lebesgue measure and let $f$ be a measurable function on $E$ that is a.e. strictly positive. Prove that for every $0 < \alpha < m(E)$

$$\inf\left\{\int_A f : A \subset E,\, m(A) \geq \alpha\right\} > 0.$$

7.2.11. Prove Corollary 7.2.15.

7.2.12. Let $(X, \mathfrak{M}, \mu)$ be a measure space and $f : X \to [0, +\infty)$ a measurable function. Prove that $f \in L(X, \mu)$ if and only if $\sup \int_A f\, d\mu < \infty$, where the supremum is taken over all sets $A$ of finite measure on which $f$ is bounded above.

7.2.13. Let $(X, \mathfrak{M}, \mu)$ be a measure space, let $\{A_n\} \subset \mathfrak{M}$ such that $\mu(\cup_{n=1}^{\infty} A_n)$ is finite, and let $f$ be a measurable function that is strictly positive a.e. and such that $\lim_n \int_{A_n} f\, d\mu = 0$. Prove that $\lim_n \mu(A_n) = 0$.

7.2.14. Let $(X, \mathfrak{M}, \mu)$ be a finite measure space. Prove that a nonnegative measurable function $f$ is integrable over $X$ if and only if

$$\sum_{n=1}^{\infty} 2^n \mu\left(\{x \in X : f(x) \geq 2^n\}\right) < \infty.$$

7.2.15. Let $(X, \mathfrak{M}, \mu)$ be a measure space, and let $f$ be a nonnegative measurable function that is finite a.e., and let $\{f_n\}$ be defined by

$$f_n(x) = \begin{cases} f(x) & \text{if } f(x) \leq n \\ 0 & \text{if } f(x) > n. \end{cases}$$

Prove that the sequence $\{\int_X f_n\, d\mu\}$ is monotone increasing and that it converges to $\int_X f\, d\mu$. Will the assertion still hold if $f$ is not assumed to be finite a.e.?

7.2.16. Let $\{f_n\}$ be a sequence of nonnegative Lebesgue measurable functions on $(0, 10)$ that converges to a function $f$ for almost all $x \in (0, 10)$. Let $F(x) = \int_0^x f\, dm$ and $F_n(x) = \int_0^x f_n\, dm$. Prove that

$$\int_0^{10} (f + F)\, dm \leq \liminf \int_0^{10} (f_n + F_n)\, dm.$$

7.2.17. Show that the Monotone Convergence Theorem may not hold for decreasing sequences of functions.

7.2.18. Let $\{f_n\}$ be a decreasing sequence of nonnegative integrable functions on $(X, \mathfrak{M}, \mu)$ that converges to $f$ a.e. Prove that $\lim f_n = 0$ a.e. if and only if $\lim \int f_n\, d\mu = 0$.

7.2.19. Show that the Monotone Convergence Theorem holds for a decreasing sequence of functions $\{f_n\}$ if $f_1 \in L(X, \mu)$.

7.2.20. Let $(X, \mathfrak{M}, \mu)$ be a measure space and $f, g$ be nonnegative functions in $L(X, \mu)$ and define

$$\nu(E) = \int_E g\, d\mu \quad \text{for all } E \in \mathfrak{M}.$$

Prove that

$$\int_X f\, d\nu = \int_X fg\, d\mu.$$

7.2.21. Let $\{f_n\}$ be a sequence of nonnegative measurable functions that converge pointwise to a function $f$ on $\mathbb{R}$. Suppose that

$$\lim \int_{\mathbb{R}} f_n = \int_{\mathbb{R}} f < \infty.$$

Prove that

$$\lim \int_A f_n = \int_A f$$

for every measurable set $A$.

7.2.22.* Let $f$ be a function defined on $[0, 1]$ in the following way. If $x$ belongs to the Cantor set, then $f(x) = 0$. If $x$ belongs to a complementary interval of length $3^{-k}$, then $f(x) = k$. Find $\int_{[0,1]} f$.

7.2.23. Let $\mu$ and $\nu$ be measures on the measurable space $(X, \mathfrak{M})$, let $\lambda = \mu + \nu$, and let $f$ be a nonnegative measurable function on $X$. Prove that, for any $E \in \mathfrak{M}$,

$$\int_E f\, d\lambda = \int_E f\, d\mu + \int_E f\, d\nu.$$

7.2.24. Let $\{r_n\}$ be an enumeration of rational numbers in $(0, 1)$, let $Q_n = \{r_1, r_2, \ldots, r_n\}$, and let

$$f_n(x) = \begin{cases} [d(x, Q_n)]^{-1/2} & \text{if } x \notin Q_n \\ 1/n & \text{if } x \in Q_n. \end{cases}$$

Prove that each $f_n$ is measurable and find

$$\lim_n \int_0^1 f_n.$$

7.2.25. Let $(X, \mathfrak{M}, \mu)$ be a finite measure space and let $f$ be a measurable, nonnegative function on $X$ such that $f^n \in L(X, \mu)$, for all $n \in \mathbb{N}$.

(a) Prove that, if $\lim_n \int_X f^n \, d\mu$ exists then $f \leq 1$ a.e.

(b) Prove that, if $\int_X f^n \, d\mu = C$ for all $n \in \mathbb{N}$ then there exists a measurable set $E$ for which $f = \chi_E$ a.e.

7.2.26. (**Beppo Levi's Theorem.**) Let $(X, \mathfrak{M}, \mu)$ be a measure space and let $\{f_n\}$ be an increasing sequence of nonnegative measurable functions such that $\sup_n \int_X f_n \, d\mu < \infty$. Prove that

(a) $\{f_n\}$ converges pointwise to a function $f \in L(X, \mu)$.

(b) $\lim \int_X f_n \, d\mu = \int_X f \, d\mu$.

7.2.27. Let $(X, \mathfrak{M}, \mu)$ be a measure space and let $E$ be a set of measure 0. Suppose that $f = \infty$ on $E$. Prove that $\int_E f = 0$.

7.2.28. Let $(X, \mathfrak{M}, \mu)$ be a measure space, let $\mathfrak{M}_0$ be a $\sigma$-algebra contained in $\mathfrak{M}$, let $\mu_0$ be the restriction of $\mu$ to $\mathfrak{M}_0$, and let $f$ be a nonnegative function that is measurable with respect to $\mathfrak{M}_0$. Prove that $f$ is measurable with respect to $\mathfrak{M}$ and

$$\int_X f \, d\mu_0 \leq \int_X f \, d\mu.$$

Can the inequality be strict?

7.2.29.* Prove or disprove: there exists a nonnegative $f \in L(\mathbb{R})$ such that for any $a < b$ and any $M > 0$ the set $(a, b) \cap \{x \in \mathbb{R} : f(x) \geq M\}$ has positive measure.

7.2.30. Let $(X, \mathfrak{M}, \mu)$ be a measure space and let $f$ be a nonnegative function that is integrable over $X$. Prove that for every $\varepsilon > 0$ there exists a set $A \in \mathfrak{M}$ such that $\mu(A) < \infty$ and $\int_A f > \int_X f - \varepsilon$.

7.2.31.* Let $(X, \mathfrak{M}, \mu)$ be a measure space, let $f$ be a nonnegative function that is integrable over $X$, and let $\varepsilon > 0$ . Prove that there exists a set $A \in \mathfrak{M}$ such that $\mu(A) < \infty$ and a simple function $s$ that vanishes outside of $A$ and satisfies $0 \leq s \leq f$ and $\int_X |f - s| \, d\mu < \varepsilon$.

7.2.32. Prove the Monotone Convergence Theorem under a weaker assumption that $\lim f_n = f$ in measure (instead of a.e.).

7.2.33. Let $(X, \mathfrak{M}, \mu)$ be a measure space, let $f$ be a nonnegative function on $X$, and let $\{X_n\}$ be an increasing sequence of measurable sets such that $X = \cup_{n=1}^{\infty} X_n$. Prove that $f$ is integrable over $X$ if and only if the sequence $\{\int_{X_n} f \, d\mu\}$ is bounded.

7.2.34. Let $(X, \mathfrak{M}, \mu)$ be a finite measure space. For any measurable functions $f, g$ on $X$ we define

$$d(f, g) = \int_X \frac{|f - g|}{1 + |f - g|} \, d\mu.$$

Prove that $d$ has all properties of distance except that $d(f, g) = 0$ does not imply that $f = g$.

Exercise 7.2.34 motivates the construction of a metric space $M(X, \mu)$ whose elements are equivalence classes of measurable functions. ($f \sim g$ if and only if $d(f, g) = 0$.) Assuming that $(X, \mathfrak{M}, \mu)$ is a finite measure space, this is indeed a metric space. In Exercises 7.2.35–7.2.38, we will explore some of its properties.

7.2.35.* Prove that convergence in the space $M(X, \mu)$ is equivalent to the convergence in measure.

7.2.36.* Prove that the space $M(X, \mu)$ is complete.

7.2.37.* Prove that the function

$$\rho(f, g) = \int_X \arctan|f - g|\, d\mu$$

defines another metric in the space $M(X, \mu)$ and that convergence in the space $M(X, \mu)$ with respect to $\rho$ is equivalent to the convergence in measure.

7.2.38.* Prove that the only continuous linear map $F : M(X, \mu) \to \mathbb{R}$ is the zero map.

7.2.39. Let $(X, \mathfrak{M}, \mu)$ be a measure space. Prove or disprove: a sequence of real-valued measurable functions $\{f_n\}$ converges in measure to $f$ if and only if

$$\lim_n \int_X \frac{|f_n - f|}{1 + |f_n - f|}\, d\mu = 0.$$

7.2.40. Suppose that every point in $[0, 1]$ belongs to at least $k$ sets of the family of Lebesgue measurable sets $\{E_k\}_{k=1}^N$. Prove that for some $n \in \{1, 2, \dots, N\}$, $m(E_n) \geq k/N$.

7.2.41. Let $(X, \mathfrak{M}, \mu)$ be a measure space. Prove that $X$ is $\sigma$-finite if and only if there is a strictly positive, real-valued function $f \in L(X, \mu)$.

7.2.42. Let $(X, \mathfrak{M}, \mu)$ be a measure space and let $\{A_n\} \subset \mathfrak{M}$ such that $\sum_{n=1}^{\infty} \mu(A_n) < \infty$. For each $m \in \mathbb{N}$ let $B_m$ consist of those points of $X$ that belong to $A_n$ for at least $m$ distinct values of $n$. Prove that each $B_m$ is measurable and that $\mu(B_m) \leq (1/m) \sum_{n=1}^{\infty} \mu(A_n)$.

## 7.3 The Integral of a Real-Valued Function

In this section we will extend the concept of integrability to the most general case: functions that are not necessarily nonnegative. The key formula is $f = f^+ - f^-$, where $f^+ = \max\{f, 0\}$ and $f^- = \max\{-f, 0\}$. Since these functions are nonnegative, the material from the previous section applies to them.

**Proposition 7.3.1.** *Let $(X, \mathfrak{M}, \mu)$ be a measure space and let $f$ be a measurable function on $X$. Then $f^+, f^-$ are integrable on $X$ if and only if $|f|$ is integrable on $X$.*

*Proof.* ($\Rightarrow$) A direct consequence of linearity of the integral and the formula $|f| = f^+ + f^-$.

($\Leftarrow$) $0 \le f^+, f^- \le |f|$, so the result follows from the monotonicity of the integral. □

In view of the equality $f = f^+ - f^-$, it would be natural to define $\int f = \int f^+ - \int f^-$. In order to avoid indeterminate form $(\infty - \infty)$ we will require that both $f^+$ and $f^-$ be integrable.

**Definition 7.3.2.** Let $(X, \mathfrak{M}, \mu)$ be a measure space. A measurable function $f$ on $X$ is **integrable** over $X$ with respect to $\mu$ if $|f| \in L(X, \mu)$. For such a function we define

$$\int_X f\, d\mu = \int_X f^+\, d\mu - \int_X f^-\, d\mu. \tag{7.4}$$

If $E \subset X$ is measurable then $f$ is integrable over $E$ if $f\chi_E \in L(X, \mu)$. In that case,

$$\int_E f\, d\mu = \int_X f\chi_E\, d\mu.$$

As usual, when there is no danger of ambiguity, both the domain of integration and $d\mu$ can be omitted.

*Remark* 7.3.3. Definition 7.3.2 specifies $f$ to be integrable if and only if it is absolutely integrable. This is not the case for the Riemann integral. For example, the function that takes 1 at rational points and -1 at irrational points is not Riemann integrable, but its absolute value is.

Definition 7.3.2 establishes a connection between integrals of nonnegative and real-valued functions. Our first result extends Proposition 7.2.21 and Proposition 7.2.11 (iii).

**Proposition 7.3.4.** *Let $(X, \mathfrak{M}, \mu)$ be a measure space and let $f \in L(X, \mu)$. Then $f$ is finite a.e. on $X$. Further, if $X_0 \subset X$ and $\mu(X_0) = 0$ then*

$$\int_X f\, d\mu = \int_{X \setminus X_0} f\, d\mu. \tag{7.5}$$

*Proof.* By definition, $|f| \in L(X, \mu)$ so Proposition 7.2.21 implies that $|f|$ is finite a.e., whence $f$ is finite a.e. By Proposition 7.2.11 (iii),

$$\int_X f^+\, d\mu = \int_{X \setminus X_0} f^+\, d\mu \quad \text{and} \quad \int_X f^-\, d\mu = \int_{X \setminus X_0} f^-\, d\mu.$$

Using (7.4), we obtain (7.5). □

Before we can prove some basic properties of the integral, it will be useful to establish the following test for integrability.

**Proposition 7.3.5. *The Integral Comparison Test.*** *Let $(X, \mathfrak{M}, \mu)$ be a measure space and $f$ a measurable function on $X$. If $g \in L(X, \mu)$ and $|f| \leq g$ a.e. on $X$, then $f \in L(X, \mu)$ and*

$$\left|\int_X f\, d\mu\right| \leq \int_X |f|\, d\mu \leq \int_X g\, d\mu.$$

*Proof.* The monotonicity of the integral (Proposition 7.2.11) implies that $|f| \in L(X, \mu)$, so the same is true of $f$. Also,

$$\left|\int_X f\, d\mu\right| = \left|\int_X f^+\, d\mu - \int_X f^-\, d\mu\right| \leq \int_X f^+\, d\mu + \int_X f^-\, d\mu = \int_X |f|\, d\mu. \quad \square$$

Our next goal is to establish the linearity of the integral. Since this is the general case, we must consider linear combinations with coefficients of either sign. In particular, we will want to prove the equality

$$\int_X f\, d\mu - \int_X g\, d\mu = \int_X (f - g)\, d\mu,$$

and there may be a problem if $f(x) = g(x) = \infty$ for some $x \in X$. The good news is that it does not matter. If $f, g$ are integrable, then Proposition 7.3.4 guarantees that this can happen only on a set $X_0$ of measure zero, so (7.5) comes to rescue.

**Theorem 7.3.6. *Linearity and Monotonicity of Integration.*** *Suppose that $(X, \mathfrak{M}, \mu)$ is a measure space, let $f, g \in L(X, \mu)$, and let $\alpha, \beta \in \mathbb{R}$. Then:*

*(a) $\alpha f + \beta g$ is integrable over $X$ and*

$$\int_X (\alpha f + \beta g)\, d\mu = \alpha \int_X f\, d\mu + \beta \int_X g\, d\mu.$$

*(b) If $f \leq g$ on $X$, then*

$$\int_X f\, d\mu \leq \int_X g\, d\mu.$$

*(c) If $A, B$ are disjoint measurable subsets of $X$, then*

$$\int_{A \cup B} f\, d\mu = \int_A f\, d\mu + \int_B f\, d\mu.$$

*Proof.* (a) Let $\alpha > 0$. Then $(\alpha f)^+ = \alpha f^+$ and $(\alpha f)^- = \alpha f^-$, so

$$\begin{aligned}\int_X \alpha f\, d\mu &= \int_X (\alpha f)^+\, d\mu - \int_X (\alpha f)^-\, d\mu \\ &= \alpha \left(\int_X f^+\, d\mu - \int_X f^-\, d\mu\right) = \alpha \int_X f\, d\mu.\end{aligned}$$

The case $\alpha < 0$ is similar, except that $(\alpha f)^+ = -\alpha f^+$ and $(\alpha f)^- = -\alpha f^-$. As usual, the case $\alpha = 0$ is trivial.

Since $f, g$ are integrable, the same is true of $|f|, |g|$, hence of $|f| + |g|$. By the Integral Comparison Test, so is $|f+g|$, whence $f+g$ is integrable. Further, by Proposition 7.3.4, we may assume that $f, g$ are finite on $X$. Notice that $(f+g)^+ - (f+g)^- = f + g = f^+ - f^- + g^+ - g^-$, so

$$(f+g)^+ + f^- + g^- = (f+g)^- + f^+ + g^+.$$

Since all terms in the last equality are nonnegative functions, it follows that

$$\int_X (f+g)^+ \, d\mu + \int_X f^- \, d\mu + \int_X g^- \, d\mu = \int_X (f+g)^- \, d\mu + \int_X f^+ \, d\mu + \int_X g^+ \, d\mu,$$

whence

$$\begin{aligned}\int_X (f+g) \, d\mu &= \int_X (f+g)^+ \, d\mu - \int_X (f+g)^- \, d\mu \\ &= \int_X f^+ \, d\mu - \int_X f^- \, d\mu + \int_X g^+ \, d\mu - \int_X g^- \, d\mu \\ &= \int_X f \, d\mu + \int_X g \, d\mu.\end{aligned}$$

(b) By excising a set of measure 0, we may assume that the function $g - f$ is well defined, nonnegative measurable function, so linearity and Proposition 7.2.11 imply that

$$\int_X g \, d\mu - \int_X f \, d\mu = \int_X (g - f) \, d\mu \geq 0.$$

(c) Inequalities $|f\chi_A| \leq |f|$ and $|f\chi_B| \leq |f|$ imply that $f\chi_A$ and $f\chi_B$ are integrable functions, so the same is true for their sum. Now the result follows from the equality $\chi_{A \cup B} = \chi_A + \chi_B$ and linearity of integration. □

The established properties of the integral allow us to generalize Example 7.2.16.

**Example 7.3.7. A signed measure on $(X, \mathfrak{M}, \mu)$.**

Let $f$ be a measurable function on $X$ such that at least one of the functions $f^+$ and $f^-$ is integrable, and let $\nu$ be a function on the $\sigma$-algebra $\mathfrak{M}$ defined by

$$\nu(E) = \int_E f \, d\mu.$$

Then $\nu$ is a signed measure.

**Proof.** Let

$$\nu^+(E) := \int_E f^+ d\mu \quad \text{and} \quad \nu^-(E) := \int_E f^- d\mu.$$

By Example 7.2.16, both $\nu^+$ and $\nu^-$ are measures and at least one of them is finite. Since $\nu = \nu^+ - \nu^-$, it is a signed measure by Proposition 5.5.7. ♦

*Remark* 7.3.8. If we define

$$P = \{x \in \mathbb{R} : f(x) \geq 0\}, \quad N = \{x \in \mathbb{R} : f(x) < 0\}$$

then $\{P, N\}$ is a Hahn decomposition of $X$ with respect to $\nu$ and $\nu = \nu^+ - \nu^-$ is the Jordan decomposition of $\nu$.

As stated before, a major flaw of the Riemann integral is that it does not commute with the limit process. The following theorem is another illustration of the superiority of the Lebesgue integral. It is considered the central achievement of the Lebesgue integration theory.

**Theorem 7.3.9.** ***The Lebesgue Dominated Convergence Theorem.*** *Let $(X, \mathfrak{M}, \mu)$ be a measure space and $\{f_n\}$ a sequence of measurable functions on $X$ that converges a.e. on $X$ to a measurable function $f$. Let $g$ be a nonnegative function that is integrable over $X$. If $|f_n| \leq g$ a.e on $X$, for all $n \in \mathbb{N}$, then $f \in L(X, \mu)$ and*

$$\lim \int_X f_n \, d\mu = \int_X f \, d\mu.$$

*Proof.* The inequality $|f_n| \leq g$ a.e. implies that $|f| \leq g$ a.e. on $X$. By the Integral Comparison Test, $\{f_n\}$ and $f$ are integrable over $X$. Proposition 7.3.4 shows that we may assume that $f, g$ are finite on $X$. Notice that $\{g - f_n\}$ and $g - f$ are nonnegative functions and $\lim(g - f_n) = g - f$ a.e. on $X$, so Fatou's Lemma implies that

$$\begin{aligned} \int_X g \, d\mu - \int_X f \, d\mu &= \int_X (g - f) \, d\mu \leq \liminf \int_X (g - f_n) \, d\mu \\ &= \int_X g \, d\mu - \limsup \int_X f_n \, d\mu. \end{aligned}$$

Consequently,

$$\limsup \int_X f_n \, d\mu \leq \int_X f \, d\mu. \tag{7.6}$$

Similarly, $\{g + f_n\}$ is a sequence of nonnegative functions converging to $g + f$ a.e. on $X$, so another application of Fatou's Lemma yields

$$\int_X (g + f) \, d\mu \leq \liminf \int_X (g + f_n) \, d\mu,$$

and we obtain that

$$\int_X f \, d\mu \leq \liminf \int_X f_n \, d\mu. \tag{7.7}$$

The result now follows from (7.6) and (7.7). □

We echo Remark 7.2.13 here. The analogue of the Lebesgue Dominated Convergence Theorem for Riemann integrals is false. Reason: neither the functions $\{f_n\}$ nor their limit are necessarily Riemann integrable. As an example, take all of them to be equal to Dirichlet's function. If we impose a much stronger condition that they are all Riemann integrable as well, then we would get a result that is correct but not easy to prove without any reference to Theorem 7.3.9. Such a result and proof can be found in a paper by Italian mathematician Arzelà (page 359) from 1895.

**Example 7.3.10. Applying Dominated Convergence Theorem.** Calculate

$$\lim_n \int_0^1 \frac{n\sin x}{1+n^2x^2}\,dx.$$

Notice that even if we considered given integrals as Riemann integrals, we would not have the Fundamental Theorem of Calculus at our disposal. However, we can use the Dominated Convergence Theorem. Indeed, for every $x \in [0,1]$ and $n \in \mathbb{N}$, we have that

$$f_n(x) := \frac{n\sin x}{1+n^2x^2} \le \frac{nx}{1+n^2x^2} \le \frac{1}{2},$$

because $(1-nx)^2 \ge 0$. Since each $f_n$ is measurable and $\lim f_n = 0$ we obtain that

$$\lim_n \int_0^1 \frac{n\sin x}{1+n^2x^2}\,dx = \int_0^1 0\,dx = 0.$$ ♦

Many results that hold for bounded functions can be extended to unbounded functions using the following construction which is another application of Dominated Convergence Theorem.

**Example 7.3.11. Approximation by truncated functions.**

Let $(X, \mathfrak{M}, \mu)$ be a measure space, let $f : X \to \mathbb{R}$ be a nonnegative integrable function, and let $\{f_n\}$ be a sequence of truncated functions defined by

$$f_n = \begin{cases} f, & \text{if } f \le n, \\ n, & \text{if } f > n. \end{cases} \tag{7.8}$$

(See Figure 7.1.) Since $\lim f_n = f$ and $f_n \le f$, each function $f_n$ is integrable

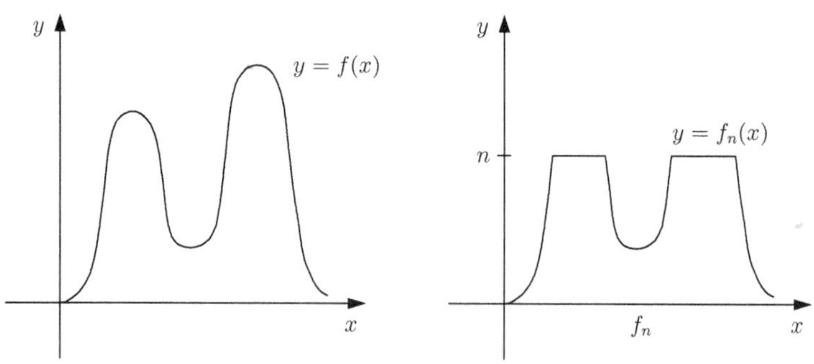


Figure 7.1: Approximating with a truncated function

and Dominated Convergence Theorem implies that

$$\lim \int_X f_n\,d\mu = \int_X f\,d\mu.$$ ♦

The Dominated Convergence Theorem appears in Lebesgue's paper [73] in 1910, albeit in the context of the integration on Euclidean spaces.

PIERRE FATOU (1878–1929) was a French mathematician and astronomer who worked in several branches of analysis. His father was a navy captain, who wanted his son to follow in his footsteps, but his health was not sufficiently good to pursue a military career. He excelled at Lorient Lycée, and later pursued advanced studies in Paris, at Ecole Normale Supérieure. After graduating in 1901, he joined the Paris Observatory where he remained for the rest of his life. At this job he was able to follow his true passion and work on mathematical research. His dissertation in 1906 was the first application of the Lebesgue integral to concrete problems of analysis, mainly to the study of analytic and harmonic functions in the unit disc. His theorem about bounded analytic functions in the unit disk and their radial limits is the foundation of a whole research area that blossomed in the second part of the 20th century. Another great accomplishment of his was the groundbreaking work on the so-called holomorphic dynamics, presented in 1917. However, a similar result was obtained by Gaston Julia (1893–1978), who was able to establish priority and claim the Academy of Sciences 1918 Grand Prix. Fatou was awarded the Becquerel prize in 1918 and the Legion of Honour in 1923. His involvement in the French Mathematical Society was significant, and he became its President in 1926-27. In 1929, during a holiday in Pornichet, he passed away suddenly, leaving behind a legacy of extraordinary mathematical contributions coupled with a quiet and modest life.

The following property of the integral is known as *absolute continuity*. It will be in the spotlight in Chapter 9.

**Theorem 7.3.12.** *Let $(X, \mathfrak{M}, \mu)$ be a measure space and let $f : X \to \mathbb{R}$ be an integrable function. Then, for every $\varepsilon > 0$ there exists $\delta > 0$ such that for every $E \in \mathfrak{M}$,*

$$\mu(E) < \delta \quad \Rightarrow \quad \int_E |f| \, d\mu < \varepsilon.$$

*Proof.* We may assume that $f \geq 0$, because $f$ is integrable if and only if $|f|$ is. Let $\varepsilon > 0$ and let $\{f_n\}$ be a sequence as in Example 7.3.11. Choose $N \in \mathbb{N}$ so that

$$\int_X (f - f_N) < \frac{\varepsilon}{2}.$$

Let $\delta = \varepsilon/(2N)$ and let $E$ be any measurable set such that $\mu(E) < \delta$. Since $f_N \leq N$,

$$\int_E f \, d\mu = \int_E f_N \, d\mu + \int_E (f - f_N) \, d\mu \leq N \, \mu(E) + \int_X (f - f_N) \, d\mu < \frac{\varepsilon}{2} + \frac{\varepsilon}{2} = \varepsilon.$$

□

Another result that will be useful in Chapter 9 is the following lemma.

**Lemma 7.3.13.** *Let $(X, \mathfrak{M}, \mu)$ be a measure space and let $f : X \to \mathbb{R}$ be an integrable function. Then $f = 0$ a.e. on $X$ if and only if $\int_E f \, d\mu = 0$, for every $E \in \mathfrak{M}$.*

*Proof.* Clearly, if $f = 0$ a.e. on $X$ and $E \in \mathfrak{M}$ then $\int_E f \, d\mu = 0$, by Proposition 7.2.20.

In the opposite direction, denoted by $A$ and $B$, respectively, the sets where $f \geq 0$ and $f < 0$. In other words, $f = f^+$ on $A$ and $f = -f^-$ on $B$. Since both $A, B$ are measurable, it follows that both $\int_A f\,d\mu = 0$ and $\int_B f\,d\mu = 0$. It remains to apply Proposition 7.2.20 to nonnegative functions $f^+$ and $f^-$, and conclude that $f = 0$ a.e. on $X$. □

## Exercises

7.3.1. Let $(X, \mathfrak{M}, \mu)$ be a measure space and $\{f_n\}$ a sequence of measurable functions on $X$. Let $g$ be a nonnegative function that is integrable over $X$ and $|f_n| \leq g$ a.e on $X$, for all $n \in \mathbb{N}$. Prove that each $f_n$ is integrable over $X$ and so are the limit inferior and the limit superior of $\{f_n\}$.

7.3.2. Let $(X, \mathfrak{M}, \mu)$ be a finite measure space and let $f \in L(X, \mu)$. Let $P = \{\dots, x_{-2}, x_{-1}, x_0, x_1, x_2, \dots\}$ be a partition of $\mathbb{R}$, and let $\xi_k \in [x_k, x_{k+1}]$. Prove that

$$\int_X f\,d\mu = \lim_{\|P\| \to 0} \sum_{k=-\infty}^{\infty} \xi_k\, \mu(\{x \in X : x_k \leq f(x) \leq x_{k+1}\}).$$

7.3.3. Prove that the assertion in Exercise 7.3.2 remains true if $\mu(X) = \infty$, provided that $\xi_k = 0$ for all $k$ for which $0 \in [x_k, x_{k+1}]$.

7.3.4. Let $(X, \mathfrak{M}, \mu)$ be a measure space and let $f$ be a bounded measurable function on $X$. Suppose that there are constants $A > 0$ and $\alpha < 1$ such that, for every $\varepsilon > 0$,

$$\mu\left(\{x \in X : |f(x)| > \varepsilon\}\right) < \frac{A}{\varepsilon^\alpha}.$$

Prove that $f$ is integrable over $X$.

7.3.5. Let $(X, \mathfrak{M}, \mu)$ be a finite measure space and let $f$ be a measurable function on $X$. Suppose that there are constants $A > 0$ and $\alpha > 1$ such that, for every $M > 0$,

$$\mu\left(\{x \in X : |f(x)| > M\}\right) < \frac{A}{M^\alpha}.$$

Prove that $f$ is integrable over $X$.

7.3.6. Suppose that a sequence of measures $\{\mu_i\}$ on a measurable space $(X, \mathfrak{M})$ converges to a measure $\mu$, meaning that for every $A \in \mathfrak{M}$, $\lim \mu_n(A) = \mu(A)$. Prove or disprove: if $f$ is integrable with respect to each $\mu_n$ then it is integrable with respect to $\mu$.

7.3.7. Let $(X, \mathfrak{M}, \mu)$ be a $\sigma$-finite measure space, let $f : X \to \mathbb{R}$ be measurable and suppose that there is $C > 0$ such that for every set $E$ of finite measure, $|\int_E f\,d\mu| \leq C$. Prove that $f$ is integrable over $X$.

7.3.8. Let $(\mathbb{R}, \mathfrak{M}, \mu)$ be a finite measure space, and $f : \mathbb{R} \to \mathbb{R}^*$ a measurable function. Prove that $f \in L(\mathbb{R}, \mu)$ if and only if

$$\sum_{n=1}^{\infty} \mu\{x \in X : |f(x)| \geq n\} < \infty.$$

7.3.9. Prove that in Egorov's Theorem the hypothesis that $(X, \mathfrak{M}, \mu)$ is a finite measure space can be replaced by $|f_n| \le g$ for all $n \in \mathbb{N}$ where $g$ is integrable over $X$.

7.3.10. Let $(X, \mathfrak{M}, \mu)$ be a measure space, let $f : X \to \mathbb{R}$ be measurable, and let $\{X_n\}$ be disjoint measurable sets such that $X = \cup_{n=1}^{\infty} X_n$. Characterize the integrability of $f$ over $X$ in terms of the integrability of $f$ over $X_n$ and $\int_{X_n} f\,d\mu$.

7.3.11. Let $(X, \mathfrak{M}, \mu)$ be a measure space and $\{f_n\}$ a sequence of measurable functions on $X$. Let $g$ be a nonnegative function that is integrable over $X$ and $|f_n| \le g$ a.e on $X$, for all $n \in \mathbb{N}$. Prove that

$$\int_X \liminf_{n\to\infty} f_n\,d\mu \le \liminf_{n\to\infty} \int_X f_n\,d\mu \le \limsup_{n\to\infty} \int_X f_n\,d\mu \le \int_X \limsup_{n\to\infty} f_n\,d\mu\,.$$

7.3.12. Let $(X, \mathfrak{M}, \mu)$ be a measure space and $\{f_n\}$ a sequence of measurable functions on $X$ that converges in measure to a measurable function $f$. Let $g$ be a function that is integrable over $X$ such that $f_n \ge -g$ for all $n$. Prove that

$$\int_X f\,d\mu \le \liminf_{n\to\infty} \int_X f_n\,d\mu\,.$$

7.3.13. Let $(X, \mathfrak{M}, \mu)$ be a finite measure space and let $\{f_n\}$ be a sequence of integrable function over $X$ that converges uniformly to a function $f$. Prove that

$$\int_X f\,d\mu = \lim \int_X f_n\,d\mu.$$

What if $\mu(X) = \infty$?

7.3.14.** Let $f : [a, b] \to \mathbb{R}$ be a Lebesgue integrable function. Prove that there exists a constant $C_0$ such that

$$\inf_C \int_{[a,b]} |f(x) - C|\,dx = \int_{[a,b]} |f(x) - C_0|\,dx.$$

Prove that such $C_0$ is unique if the function $f$ is continuous.

7.3.15. Let $f : [0, +\infty) \to \mathbb{R}$ be a continuous function ans suppose that $\lim_{x\to+\infty} f(x) = \gamma$. Prove that for any $a > 0$,

$$\lim_{n\to\infty} \int_0^a f(nx)\,dx = a\gamma.$$

7.3.16. Let $(X, \mathfrak{M}, \mu)$ be a measure space, let $\{f_n\}, \{g_n\}, f, g$ be integrable functions over $X$ such that $\lim f_n = f$ and $\lim g_n = g$ a.e. on $X$. Suppose that $|f_n| \le g_n$ a.e on $X$, for all $n \in \mathbb{N}$, and that $\lim \int_X g_n\,d\mu = \int_X g\,d\mu$. Prove that

$$\lim \int_X f_n\,d\mu = \int_X f\,d\mu.$$

7.3.17.* Let $\{f_n\}$ be a sequence of integrable functions on $[0, 1]$, let $f$ be an integrable function on $[0, 1]$, and suppose that $f_n \to f$ a.e. on $[0, 1]$. Prove that $\lim_{n\to\infty} \int_{[0,1]} |f - f_n| = 0$ if and only if $\lim_{n\to\infty} \int_{[0,1]} |f_n| = \int_{[0,1]} |f|$.

7.3.18. Let $\{f_n\}$ be a sequence of Lebesgue measurable functions on $[0,1]$ that converges to a function $f$ a.e. and suppose that there is $M > 0$ such that for all $n \in \mathbb{N}$, $|f_n| \leq M$ a.e. Prove that the sequence

$$g_n(x) = \int_0^x (f_n(t) - f(t))\, dt$$

converges uniformly to 0 on $[0,1]$.

7.3.19. Let $f$ be an integrable function on a Lebesgue measurable set $E \subset \mathbb{R}^n$.

(a) Find

$$\lim_n n \int_E \left( \left| 1 + \frac{f}{n} \right| - 1 \right).$$

(b) If $m(E) < \infty$ prove that

$$\int_E f = 0 \quad \text{if and only if} \quad \int_E |1 + \lambda f| \geq m(E), \text{ for all } \lambda \in \mathbb{R}.$$

7.3.20. Suppose that $f$ is integrable over $[0,1]$. Calculate

$$\lim_n \int_0^1 n \ln \left( 1 + \frac{|f(x)|^2}{n^2} \right) dx.$$

7.3.21. Suppose that $f$ is integrable over $[0,1]$ and that $\lim_{x \to 1-} f(x) = A$. Prove that

$$\lim_n \int_0^1 nx^n f(x)\, dx = A.$$

7.3.22.* Let $X = [0,1]$, $\mathfrak{M} = \mathcal{M}|[0,1]$, and $\mu = m|[0,1]$. Let $f$ be integrable over $X$. Find

$$\lim_{n \to \infty} \frac{1}{n} \int_0^1 \ln(1 + e^{nf(x)})\, dx.$$

7.3.23. Let $(X, \mathfrak{M}, \mu)$ be a measure space, let $\{f_n\}$ be an increasing sequence of integrable functions, and suppose that $\sup_n \int_X f_n\, d\mu < \infty$. Prove that $f := \lim_n f_n$ is integrable and $\lim_n \int_X f_n\, d\mu = \int_X f\, d\mu$.

7.3.24. Let $\phi : \mathbb{R} \to \mathbb{R}$ be a bounded function in $C^1(\mathbb{R})$ with a bounded derivative, let $\phi(0) = 0$, and let $g$ be a function that is integrable over $\mathbb{R}$. Prove that

$$f(t) = \int_{\mathbb{R}} \phi(tg(x))\, dx$$

is well-defined and differentiable for all $t \in \mathbb{R}$.

7.3.25. Let $f_n : [0,1] \to [0,\infty)$ be Lebesgue integrable for $n = 1, 2, \ldots$. Suppose that for every $n \in \mathbb{N}$,

$$\int_0^1 f_n = 1 \quad \text{and} \quad \int_{1/n}^1 f_n < \frac{1}{n}.$$

Finally, let $g(x) = \sup\{f_n(x) : n \in \mathbb{N}\}$. Prove that $\int_0^1 g = \infty$.

7.3.26. Let $[a_{ij}]$ be an infinite matrix of real numbers such that $|a_{ij}| \leq 1$ for all $i, j \in \mathbb{N}$. Suppose that the limit along each row is 0, i.e., $\lim_{j\to\infty} a_{ij} = 0$ for all $i \in \mathbb{N}$. If $p > 1$ prove that $\lim_{j\to\infty} \sum_{i=1}^{\infty} a_{ij}/i^p = 0$.

7.3.27. Let $(X, \mathfrak{M}, \mu)$ be a measure space, let $\{f_n\}, f$ be nonnegative integrable functions over $X$ such that $\lim f_n = f$ a.e. on $X$ and $\lim \int_X f_n \, d\mu = \int_X f \, d\mu$. Prove that

$$\lim \int_X |f_n - f| \, d\mu = 0.$$

7.3.28. Let $(X, \mathfrak{M}, \mu)$ be a measure space, let $f \in L(X, \mu)$, and let $A_n = \{x \in X : 1/n \leq |f(x)| \leq n\}$, for $n \in \mathbb{N}$. Prove that

$$\lim_{n\to\infty} \int_{A_n^C} |f| \, d\mu = 0.$$

7.3.29. Let $(X, \mathfrak{M}, \mu)$ be a measure space, let $\{f_n\}$ be integrable functions over $X$ such that $\sum \int_X |f_n| < \infty$. Prove that $\sum_1^\infty f_n$ converges a.e. to an integrable function and $\int_X \sum_1^\infty f_n \, d\mu = \sum_1^\infty \int_X f_n \, d\mu$.

7.3.30.* Let $f : [0,1] \times [0,1] \to \mathbb{R}$ be a measurable function of $x$, a continuous the function of $y$, and let $g : [0,1] \to \mathbb{R}$ be integrable over $[0,1]$ satisfying $|f(x,y)| \leq g(x)$ for $(x,y) \in [0,1] \times [0,1]$. Prove that $h(y) := \int_0^1 f(x,y) \, dx$ is continuous.

7.3.31. Prove the Dominated Convergence Theorem under a weaker assumption that $\lim f_n = f$ in measure (instead of a.e.).

7.3.32. Let $f : \mathbb{R} \to \mathbb{R}^*$ be a continuous function, let $a \in \mathbb{R}$, and let

$$\mu(E) := \int_E f \, dm.$$

Prove that

$$\lim_{\varepsilon\to 0^+} \frac{\mu((a-\varepsilon, a+\varepsilon))}{m((a-\varepsilon, a+\varepsilon))} = f(a).$$

7.3.33. Let $\delta_0$ be the Dirac measure on $\mathbb{R}$ concentrated at $x = 0$ and let $f$ be a function on $\mathbb{R}$. Calculate

$$\int_{\mathbb{R}} f \, d\delta_0.$$

7.3.34. Let $f$ be a Lebesgue integrable function on $\mathbb{R}$ and suppose that $\int_E f = 0$ for every measurable set $E \subset \mathbb{R}$ with $m(E) = \pi$. Prove or disprove: $f = 0$ a.e.

7.3.35.* Let $(X, \mathfrak{M}, \mu)$ be a measure space, let $f : X \to \mathbb{R}$ be measurable, and let $F$ be a closed subset of $\mathbb{R}$ such that for every measurable set $E$ of positive measure

$$\frac{1}{\mu(E)} \int_E f \, d\mu \in F.$$

Prove that $\mu\{x \in X : f(x) \notin F\} = 0$.

7.3.36. Let $f : \mathbb{R} \to \mathbb{R}$ be a measurable function and let $a, b \in \mathbb{R}$, $a \neq 0$. Prove that

$$\int_{\mathbb{R}} f(ax+b)\,dx = \frac{1}{|a|}\int_{\mathbb{R}} f(x)\,dx.$$

7.3.37. Let $a, b, c \in \mathbb{R}$ satisfy $a < b$ and $c > 0$, and let $f$ be integrable over $[a+c, b+c]$. Prove that $\int_a^b f(x+c)\,dx = \int_{a+c}^{b+c} f(x)\,dx$.

7.3.38. Let $(\mathbb{R}, \mathfrak{M}, \mu)$ be a measure space, and for each function $f : \mathbb{R} \to \mathbb{R}^*$ and $a \in \mathbb{R}$ let $f_a(x) = f(x-a)$. Prove that $\mu$ is translation invariant if and only if for each $f \in L(\mathbb{R}, \mu)$ and each $a \in \mathbb{R}$, the function $f_a$ belongs to $L(\mathbb{R}, \mu)$ and $\int_{\mathbb{R}} f\,d\mu = \int_{\mathbb{R}} f_a\,d\mu$.

7.3.39.* Let $(X, \mathfrak{M}, \mu)$ be a measure space, let $f, g \in L(X, \mu)$, and for $E \in \mathfrak{M}$ define $\lambda(E) = \int_E f\,d\mu$, $\nu(E) = \int_E g\,d\mu$, $\omega(E) = \int_E \max\{f, g\}\,d\mu$, $\rho(E) = \int_E \min\{f, g\}\,d\mu$. Prove that $\omega = \lambda \vee \nu$ and $\rho = \lambda \wedge \nu$. (See (5.17).)

7.3.40. Let $(X, \mathfrak{M}, \mu)$ be a measure space, let $f : X \to \mathbb{R}^*$ be a measurable function, and suppose that $f = g + h$, where $g \in L(X, \mu)$ is finite and $h \geq 0$ on $X$. If we define $\int_X f\,d\mu = \int_X g\,d\mu + \int_X h\,d\mu$, prove that the definition does not depend on the particular choice of a finite, integrable function $g$ and a nonnegative function $h$.

7.3.41.* Let $(X, \mathfrak{M}, \mu)$ be a measure space, let $f : X \to \mathbb{R}$ be a bounded function that vanishes outside of a set of finite measure, and suppose that

$$\sup\left\{\int_X s : s \in \mathcal{S}(X),\ s \leq f \text{ on } X\right\} = \inf\left\{\int_X t : t \in \mathcal{S}(X),\ t \geq f \text{ on } X\right\}.$$

Prove that $f$ is measurable with respect to the completion of $(X, \mathfrak{M}, \mu)$.

7.3.42. Let $(X, \mathfrak{M}, \mu)$ be a measure space, and let $f$ be a function integrable over $X$, and let $\nu$ be a signed measure defined by

$$\nu(A) = \int_A f\,d\mu.$$

Prove that

$$|\nu|(A) = \int_A |f|\,d\mu, \quad \text{for each } A \in \mathfrak{M}.$$

7.3.43. Let $X = [0, 2\pi]$, $\mathfrak{M} = \mathcal{M}_{[0,2\pi]}$, and $\mu = m|[0, 2\pi]$. Define a signed measure $\nu$ on the measurable space $(X, \mathfrak{M})$ by

$$\nu(E) = \int_E \sin x\,dm, \quad \text{for } E \subset [0, 2\pi].$$

Let $C = [4\pi/3, 5\pi/3]$ and $\varepsilon > 0$. Find a measurable set $C' \subset C$ such that $\nu(C') \geq \nu(C)$ and $\nu(E) > -\varepsilon$ for every measurable subset $E$ of $C'$.

7.3.44. Let $(X, \mathfrak{M}, \mu)$ be a measure space and let $(X, \mathfrak{M}_0, \mu_0)$ be its completion. If $f$ is a function in $L(X, \mu)$ prove that $f \in L(X, \mu_0)$ and that $\int_X f\,d\mu = \int_X f\,d\mu_0$. Moreover, if $f \in L(X, \mu_0)$ then there is a $\mathfrak{M}$-measurable function $g$ such that $g = f$ a.e. $[\mu]$ and $\int_X g\,d\mu = \int_X g\,d\mu_0$.

## 7.4 Computing Lebesgue Integral

Just like in the case of the Riemann integral, computing the Lebesgue integral using the definition is not a very practical way. In this section we will look at some options when calculating the integral. Although our techniques can be easily generalized to $\mathbb{R}^n$ for any $n \in \mathbb{N}$, in the interest of simplicity we will consider only the case $n = 1$. That is, we will consider the functions of one real variable and their integrals.

It would be foolish not to try to use well-developed techniques for the computation of Riemann integrals such as substitution, integration by parts, etc. Many of them carry over, and the proofs are in this section. Since we will be dealing with both Riemann and Lebesgue integrals in this section, in order to tell them apart we will use notation $\int_a^b$ for the former and $\int_{[a,b]}$ for the latter.

Our first result shows that the Lebesgue integral is a generalization of the Riemann integral.

**Theorem 7.4.1.** *Let $f$ be a Riemann integrable function over $[a, b]$. Then $f$ is Lebesgue integrable over $[a, b]$ and the two integrals are equal.*

*Proof.* Recall that every Riemann integrable function is bounded, i.e., there is $M > 0$ such that $|f| \leq M$. Since constant functions (such as $M$) are Lebesgue integrable over $[a, b]$, the Dominated Convergence Theorem will imply that $f$ is integrable, as soon as we establish that $f$ is measurable. To that end, we consider a sequence $\{P_k\}_{k=1}^{\infty}$ of partitions of $[a, b]$ such that $P_k \subset P_{k+1}$ for all $k \in \mathbb{N}$. Let

$$P_k = \{x_0^{(k)}, x_1^{(k)}, \dots, x_{n_k}^{(k)}\}, \quad m_i^{(k)} = \inf\{f(x) : x \in [x_{i-1}^{(k)}, x_i^{(k)}]\},$$
$$M_i^{(k)} = \sup\{f(x) : x \in [x_{i-1}^{(k)}, x_i^{(k)}]\}.$$

For every $k \in \mathbb{N}$, we define simple functions $s_k$ and $S_k$ by

$$s_k(x) = m_i^{(k)} \quad \text{and} \quad S_k(x) = M_i^{(k)}, \quad \text{for} \quad x \in [x_{i-1}^{(k)}, x_i^{(k)}].$$

It is not hard to see that each $s_k$ as well as each $S_k$ is a step function. Since each partition $P_{k+1}$ is obtained from $P_k$ by partitioning some of the existing subintervals, we obtain that $\{s_k\}$ and $\{S_k\}$ are increasing and decreasing sequences, respectively. Therefore, both are convergent and we denote their limits by $s$ and $S$, respectively. Notice that, for any $k$, $s_k(x) \leq f(x) \leq S_k(x)$, whence

$$s(x) \leq f(x) \leq S(x). \tag{7.9}$$

The monotonicity of Riemann integral implies that

$$\lim_k \int_a^b s_k(x)\,dx \leq \int_a^b f(x)\,dx \leq \lim_k \int_a^b S_k(x)\,dx. \tag{7.10}$$

By Monotone Convergence Theorem and the monotonicity of Lebesgue integral ,

$$\lim_k \int_{[a,b]} s_k(x)\,dx = \int_{[a,b]} s(x)\,dx \le \int_{[a,b]} S(x)\,dx = \lim_k \int_{[a,b]} S_k(x)\,dx. \quad (7.11)$$

Moreover, Riemann integrals of $s_k$ and $S_k$ are respectively the lower and the upper Darboux sums:

$$\int_a^b s_k(x)\,dx = L(f, P_k), \quad \int_a^b S_k(x)\,dx = U(f, P_k).$$

Since $f$ is Riemann integrable these converge to the same limit, so both inequalities in (7.10) are equalities:

$$\lim_k \int_a^b s_k(x)\,dx = \int_a^b f(x)\,dx = \lim_k \int_a^b S_k(x)\,dx. \quad (7.12)$$

Further, Riemann and Lebesgue integrals of a step function are equal (Exercise 7.4.7), so the inequality in (7.11) is also an equality:

$$\int_{[a,b]} s(x)\,dx = \int_{[a,b]} S(x)\,dx.$$

It follows that the nonnegative function $S - s$ has the (Lebesgue) integral equal to 0. By Proposition 7.2.20, $S = s$ a.e. on $[a, b]$ and (7.9) yields $f = s = S$ a.e. The measurability of $f$ now follows from the fact that each $s_k$ is measurable and Corollary 6.2.7.

Finally, the monotonicity of Lebesgue integral yields

$$\lim_k \int_{[a,b]} s_k(x)\,dx \le \int_{[a,b]} f(x)\,dx \le \lim_k \int_{[a,b]} S_k(x)\,dx,$$

so, comparing with (7.12), we obtain the equality of Riemann and Lebesgue integrals of $f$. □

By Theorem 7.4.1, every Riemann integrable function is also Lebesgue integrable. However, the converse is not true, so the Lebesgue integral is a proper generalization of the Riemann integral.

**Example 7.4.2. A function that is Lebesgue integrable but not Riemann integrable.**

The Dirichlet function is not Riemann integrable. Since it equals $\chi_{[0,1]\cap\mathbb{Q}}$ it is Lebesgue integrable and

$$\int_{[0,1]} \chi_{[0,1]\cap\mathbb{Q}} = m([0,1] \cap \mathbb{Q}) = 0. \qquad \blacklozenge$$

The existence of Lebesgue integrable functions that are not Riemann integrable is tied with the fact that Riemann integral is defined with continuous functions in mind. In fact, Cauchy and his contemporaries did not even bother to integrate discontinuous functions. Riemann's definition of integral allowed the existence of integrable functions that are not continuous, but do not provide an answer to the question: how discontinuous can a Riemann integrable function be? Building on the ideas presented in the proof of Theorem 7.4.1, we will now apply Lebesgue's ideas to characterize Riemann's integrability.

**Theorem 7.4.3.** *Let $f$ be a bounded function on $[a,b]$. Then $f$ is Riemann integrable over $[a,b]$ if and only if it is continuous a.e. on $[a,b]$.*

*Proof.* We will use the same notation as in the proof of Theorem 7.4.1. Let $A$ is the subset of $[a,b]$ that consists of points that are *not* partition points for any $P_k$. We will show that

$$f \text{ is continuous at } x_0 \in A \quad \text{if and only if} \quad s(x_0) = S(x_0). \tag{7.13}$$

Suppose that we have proved (7.13). If $f$ is Riemann integrable over $[a,b]$, we have seen in the proof of Theorem 7.4.1 that $s = S$ a.e. on $[a,b]$, and all the more a.e. on $A$. By (7.13), $f$ is continuous a.e. on $A$. Since the set $[a,b] \setminus A$ is countable, we obtain that $f$ is continuous a.e. on $[a,b]$.

In the other direction, suppose that $f$ is continuous a.e. on $[a,b]$. Then $f$ is continuous a.e. on $A$ so (7.13) implies that $s = S$ a.e. on $A$, hence a.e. on $[a,b]$. Now the result follows from (7.10) and (7.11).

Thus, it remains to prove (7.13). Since $x_0$ is not a partition point of any $P_k$, for every $k \in \mathbb{N}$ it lies in one of the open intervals created by $P_k$, and we will denote this interval by $I_k$.

$\Leftarrow$ Let $\varepsilon > 0$. By assumption,

$$\lim_n (S_n(x_0) - s_n(x_0)) = S(x_0) - s(x_0) = 0,$$

so there exists a positive integer $N$ for which $S_N(x_0) - s_N(x_0) < \varepsilon$. Both $S_N$ and $s_N$ are constant on $I_N$ so

$$S_N(x) - s_N(x) < \varepsilon, \qquad \text{for all } x \in I_N.$$

Further, for any $x \in I_N$, $f(x)$ and $f(x_0)$ lie both between $s_N(x) = s_N(x_0)$ and $S_N(x) = S_N(x_0)$, whence $|f(x) - f(x_0)| < \varepsilon$, which means that $f$ is continuous at $x_0$.

$\Rightarrow$ Let $\varepsilon > 0$. By assumption, there is $\delta > 0$ such that

$$x \in (x_0 - \delta, x_0 + \delta) \quad \Rightarrow \quad f(x) \in (f(x_0) - \varepsilon, f(x_0) + \varepsilon).$$

The sequence $\{I_k\}$ is decreasing and each set contains $x_0$. Since $\lim_n \ell(I_n) = 0$, there exists $K \in \mathbb{N}$ for which $I_K \subset (x_0 - \delta, x_0 + \delta)$. It follows that, for any $x \in I_K$,

$$\begin{aligned} f(x_0) - \varepsilon &\le \inf\{f(x) : x \in I_K\} = s_K(x) \\ &\le S_K(x) = \sup\{f(x) : x \in I_K\} \le f(x_0) + \varepsilon, \end{aligned}$$

whence $S_K(x) - s_K(x) \le 2\varepsilon$. In particular, this holds for $x = x_0$, and because of the monotonicity of sequences $\{s_k\}$ and $\{S_k\}$ it holds for $k \ge K$. In other words,

$$S_k(x_0) - s_k(x_0) \le 2\varepsilon, \qquad \text{for all} \quad k \ge K,$$

implying that $S(x_0) - s(x_0) \le 2\varepsilon$, whence $S(x_0) = s(x_0)$. □

Theorem 7.4.3 shows that a Riemann integrable function must be continuous a.e. On the other hand, for a bounded function to be Lebesgue integrable over a finite interval it is sufficient that it is measurable.

> Lebesgue proved Theorem 7.4.3 in his dissertation. Ever since Riemann's integral was defined, mathematicians were intrigued by the question: How discontinuous can a Riemann integrable function be? In a paper from 1870 (small circulation—reprinted in 1882 [53]), a German mathematician Hermann Hankel (1839–1873) made some progress by considering the jumps of a function, and claiming that in every interval the set of points where the jump exceeds a fixed number is finite. This reasoning was consolidated by Lebesgue in his proof. Independently, Vitali proved the result in 1903. Hankel restated Riemann's integration theory in terms of measure theoretic concepts. However, he erroniusly believed that if a set is nowhere dense it must have measure 0. He is remembered for the Hankel transform and the Hankel matrix (any square matrix with constant antidiagonals).

While Theorem 7.4.1 is a powerful tool to compute the value of a Lebesgue integral it applies only to functions that are Riemann integrable. In particular, it cannot be used for unbounded functions, even if their improper Riemann integrals converge. In such a situation we need a more delicate approach.

One learns in calculus that the restriction to bounded functions and finite intervals can be easily dispensed with, at least for functions of one variable. Improper and infinite integrals are a useful way to deal with functions that do not conform to the rigid requirements of Riemann integration theory. This works well when the domain of integration is one-dimensional but not in higher dimensions. Let us explain why. Suppose that we are looking for the Riemann integral of a function $f$ over $(0, +\infty)$. By definition

$$\int_0^{\infty} f(x)\, dx = \lim_{b\to\infty} \int_0^b f(x)\, dx.$$

In other words, we are using the fact that $(0, +\infty)$ is the union of an increasing sequence of intervals $\{(0, b_n)\}$, with $\lim b_n = \infty$. Certainly, intervals are most manageable, but what if we use a different sequence of measurable sets whose union is $(0, +\infty)$? It is possible that we will get a different result. However, it is natural to stick with intervals and we do.

There are no such outstanding sets in $\mathbb{R}^2$. For example, if the domain is the whole plane, do we use rectangles or disks? And if we choose rectangles, do we insist on the ratio between its sides? Trouble is that switching from one to another may give us a different value of the integral. (See Exercise 7.4.12.) This highlights the advantage of working with the Lebesgue integral because we can ignore all these issues. Remember, a function is Lebesgue integrable if and only if it is absolutely integrable, and if $\{A_k\}$ is an increasing sequence of sets with union equal to $\mathbb{R}^n$, the Monotone Convergence Theorem ensures that

$$\lim_k \int_{A_k} f^+ = \int_{\mathbb{R}^n} f^+ \quad \text{and} \quad \lim_k \int_{A_k} f^- = \int_{\mathbb{R}^n} f^-.$$

Admittedly, the absolute convergence renders some functions not Lebesgue integrable, even though their improper/infinite Riemann integral might converge. This is a small price to pay because this is possible only when $f$ is a function of 1 variable, and only when its improper/infinite Riemann integral converges conditionally (and not absolutely). When the integrand is a nonnegative function this cannot happen, and we have the following theorem.

**Theorem 7.4.4.** *Let $f$ be a nonnegative function on $[a, b]$ such that $f$ is Riemann integrable over every interval $[a', b] \subset (a, b]$ and the improper Riemann integral of $f$ over $[a, b]$ converges. Then $f$ is Lebesgue integrable over $[a, b]$ and its Lebesgue integral is equal to the improper Riemann integral of $f$.*

*Proof.* Let $\{a_n\}_{n=1}^\infty$ be a decreasing sequence of real numbers in $[a, b]$ that converges to $a$, and let $E_n = [a_n, b]$. By assumption, $f$ is Riemann integrable over $E_n$ so by Theorem 7.4.1, it is Lebesgue integrable and

$$\int_{[a_n,b]} f(x)\, dx = \int_{a_n}^b f(x)\, dx. \tag{7.14}$$

Moreover,

$$\lim_n \int_{a_n}^b f(x)\, dx = \int_a^b f(x)\, dx < \infty.$$

Thus, it suffices to demonstrate that the left-hand side of (7.14) converges to the Lebesgue integral of $f$ over $[a, b]$. To that end, notice that

$$\int_{[a_n,b]} f(x)\, dx = \int_{[a,b]} \chi_{E_n}(x) f(x)\, dx.$$

The sequence $\{\chi_{E_n} f\}$ is an increasing sequence of Riemann integrable, hence measurable functions that converge to $f$. Therefore, $f$ is measurable by Corollary 6.2.7, and by Monotone Convergence Theorem,

$$\lim_n \int_{[a_n,b]} f(x)\, dx = \int_{[a,b]} f(x)\, dx.$$

□

**Example 7.4.5. Computing a Lebesgue integral.**
Compute the Lebesgue integral

$$\int_0^1 \frac{1}{\sqrt{x}}\, dx.$$

**Solution:** Since the function $f(x) = 1/\sqrt{x}$ is not bounded on $(0, 1)$ it is not Riemann integrable. Nevertheless, it satisfies the hypotheses of the Theorem 7.4.4 so it is Lebesgue integrable. The (improper) Riemann integral of $f$ equals 2, and therefore the Lebesgue integral has a value 2 as well. ♦

**Example 7.4.6. Applying Dominated Convergence Theorem.**
Calculate

$$\lim_n \int_0^1 \frac{n^{3/2}x}{1+n^2x^2}\,dx.$$

The strategy employed in Example 7.3.10 does not work here. Namely, if

$$f_n(x) = \frac{n^{3/2}x}{1+n^2x^2},$$

then the sequence $\{f_n\}$ is not uniformly bounded. Indeed, on the interval $(0,1)$, the function $f_n$ attains the maximum at $x = 1/n$ and

$$f_n\left(\frac{1}{n}\right) = \frac{\sqrt{n}}{2} \to \infty, \quad n \to \infty.$$

However, $|f_n(x)| \leq 1/\sqrt{x}$, for all $x \in (0,1]$. To justify the last claim, we fix $x \in (0,1]$ and we consider the function

$$g(y) = \frac{y^{3/2}x}{1+y^2x^2}.$$

For a fixed $x$ this function attains its maximum at $y = \sqrt{3}/x$ and

$$g\left(\frac{\sqrt{3}}{x}\right) = \frac{3^{3/4}}{4\sqrt{x}} < \frac{1}{\sqrt{x}}.$$

By Example 7.4.5, the function $f(x) = 1/\sqrt{x}$ is Lebesgue integrable on $(0,1)$. Also, the sequence $\{f_n\}$ converges to 0 on $(0,1)$ so

$$\lim_n \int_0^1 \frac{n^{3/2}x}{1+n^2x^2}\,dx = 0. \qquad \blacklozenge$$

Theorem 7.4.4 provides a sufficient condition for an unbounded function to be Lebesgue integrable. However, it requires the improper Riemann integral to exist. What if this is not the case?

**Example 7.4.7. A continuous function that is not Lebesgue integrable.**
The function $f(x) = 1/x$ is not Lebesgue integrable over $(0,1)$.

**Proof.** It suffices to exhibit a sequence $\{s_n\}$ of simple functions such that $0 \leq s_n \leq f$ on $(0,1)$ and

$$\lim_n \int_0^1 s_n(x)\,dx = +\infty. \tag{7.15}$$

Such a sequence is, for example,

$$s_n(x) = \sum_{k=1}^{n} k\chi_{E_k}(x), \quad \text{where} \quad E_k = \left(\frac{1}{k+1}, \frac{1}{k}\right).$$

Let us verify that $s_n \leq f$ on $(0,1)$. If $x \in (0,1)$ it can be only in one interval $E_k$, and if $x \in E_k$ then $s_n(x) = k$ and $f(x) = 1/x \geq k$.

In order to prove (7.15), notice that

$$\int_0^1 s_n = \int_0^1 \sum_{k=1}^n k\chi_{E_k} = \sum_{k=1}^n k\, m(E_k) = \sum_{k=1}^n k\left(\frac{1}{k} - \frac{1}{k+1}\right) = \sum_{k=1}^n \frac{1}{k+1},$$

so (7.15) holds because the last sum is a partial sum of the Harmonic series.♦

Example 7.4.7 is indicative of a more general result (see Exercise 7.4.32).

An analogue of Theorem 7.4.4 for infinite intervals can be formulated as follows. We leave the proof as Exercise 7.4.8.

**Theorem 7.4.8.** *Let $f$ be a nonnegative function on $(a, +\infty)$ such that $f$ is Riemann integrable over every interval $(a, b)$ and the improper Riemann integral of $f$ over $(a, +\infty)$ converges. Then $f$ is Lebesgue integrable over $(a, +\infty)$ and its Lebesgue integral is equal to the improper Riemann integral of $f$.*

In Theorems 7.4.4 and 7.4.8 we have made the assumption that $f$ is nonnegative. What if $f$ takes the values of both signs in its domain? We can still apply the same ideas but we need to exercise caution because an improper Riemann integral may converge conditionally but not absolutely. That is to say, we may have $\int_a^\infty f(x)\,dx$ converge but $\int_a^\infty |f(x)|\,dx$ diverge. A well-known example is $f(x) = \sin x/x$. On the other hand, this cannot happen with Lebesgue integral because $f$ is integrable if and only if $|f|$ is integrable. With this in mind, we leave the modifications of Theorems 7.4.4 and 7.4.8 as Exercises 7.4.9 and 7.4.10.

WACLAW SIERPINSKI (1882–1969) was a Polish mathematician who worked in the area of set theory, point set topology, and number theory. His mathematical talent was evident from an early age. This was the period of the Russian occupation of Poland which meant that all lectures at the University of Warsaw were in Russian. In protest, he failed the mandatory Russian language course but a compassionate lecturer changed the grade, enabling him to graduate in 1904. After a brief stint as a school teacher he went to Kraków to study for his doctorate. From 1907 he immersed himself in the study of set theory, and discovered the so-called Sierpiński curve and Sierpiński triangle. He graduated in 1908 and was appointed to the University of Lwów where he launched a prolific career. After World War I he moved to the University of Warsaw and was one of the founders of Fundamenta Mathematicae in 1920. He was elected to the Polish Academy in 1921 and became chairman of the Polish Mathematical Society in 1928. During World War II he contributed to the "Underground Warsaw University"—when education was banned for non-German population. Sierpiński's output is astonishing, with more than 700 papers and 50 books. After he retired in 1960, he continued to give a seminar on the theory of numbers at the Polish Academy of Sciences up to 1967. He earned numerous honors, including membership in many foreign Academies of Science and honorary degrees from many universities.

## Exercises

7.4.1. Prove that the characteristic function of the Cantor set is Riemann integrable over $[0, 1]$.

7.4.2. Let $F$ be a closed subset of $[0,1]$ whose Lebesgue measure is 0. Prove that the characteristic function of $F$ is Riemann integrable over $[0,1]$.

7.4.3. Give an example of functions $f, g : [0,1] \to [0,1]$ such that $f$ is continuous, $g$ is Riemann integrable, but $g \circ f$ is not Riemann integrable.

7.4.4. Give an example of a bounded, Lebesgue integrable function $f : [0,1] \to \mathbb{R}$ for which there is no Riemann integrable function $g : [0,1] \to \mathbb{R}$ that is equal to $f$ a.e.

7.4.5. Let $f, g$ be functions on $[a,b]$ that are both Riemann integrable over $[a,b]$. Prove that their product is Riemann integrable over $[a,b]$.

7.4.6. Give an example of an increasing sequence of Riemann integrable functions on $[0,1]$ that converges to a function that is not Riemann integrable.

7.4.7. Prove that the Riemann and the Lebesgue integral of a step function are equal.

7.4.8. Prove Theorem 7.4.8.

7.4.9. Let $f : [a,b] \to \mathbb{R}$ be Riemann integrable over every interval $[a', b] \subset [a,b]$ and the improper Riemann integral of $f$ over $[a,b]$ converges absolutely. Prove that $f$ is Lebesgue integrable over $[a,b]$.

7.4.10. Let $f : (a, +\infty) \to \mathbb{R}$ be Riemann integrable over every interval $(a,b)$ and the improper Riemann integral of $f$ over $(a, +\infty)$ converges absolutely. Prove that $f$ is Lebesgue integrable over $(a, +\infty)$.

7.4.11. Let $f : [0,1] \to \mathbb{R}$ be continuous on $(0,1]$ and Lebesgue integrable over $[0,1]$. Prove that the improper Riemann integral of $f$ over $[0,1]$ converges absolutely.

7.4.12. Give an example of a function $f : \mathbb{R}^2 \to \mathbb{R}$ and two sequences of rectangles $\{R_k\}_{k=1}^{\infty}$ and $\{S_k\}_{k=1}^{\infty}$ such that

$$\lim_n \int_{R_n} f \neq \lim_n \int_{S_n} f.$$

7.4.13. Let $f : (a, +\infty) \to \mathbb{R}$ be continuous on every finite interval $(a,b)$ and Lebesgue integrable over $(a, +\infty)$. Prove that the improper Riemann integral of $f$ over $(a, +\infty)$ converges absolutely.

7.4.14. Let

$$f(x) = \frac{(-1)^{\lfloor x \rfloor}}{1 + \lfloor \frac{x}{2} \rfloor} = \begin{cases} \dfrac{1}{n+1}, & \text{if} \quad 2n \leq x < 2n+1 \\ \dfrac{-1}{n+1}, & \text{if} \quad 2n+1 \leq x < 2n+2. \end{cases}$$

Prove that the improper Riemann integral over $[0, +\infty)$ exists but the Lebesgue integral does not.

7.4.15. Let $f : [0,1] \to \mathbb{R}$ be nonnegative and continuous on $(0,1]$, and suppose that

$$\lim \int_{[1/n,1]} f$$

exists. Prove or disprove: $f$ is integrable over $[0,1]$. What if the nonnegativity assumption is dropped?

7.4.16. Let $f$ be a function on $[a,b]$ such that $f$ is Riemann integrable over every interval $[a',b] \subset (a,b]$ and the improper Riemann integral of $f$ over $[a,b]$ converges. Suppose that $f$ is also Lebesgue integrable over $[a,b]$. Prove that the improper Riemann integral of $f$ and the Lebesgue integral of $f$ over $[a,b]$ are equal.

7.4.17. Give an example of a bounded function $f$ on $[0,+\infty)$ whose improper Riemann integral converges but $|f|^p$ is not Lebesgue integrable for any $p \geq 1$.

7.4.18.* Let $f, g : [a,b] \to \mathbb{R}$ and let $h = \max\{f, g\}$. For what type of integral (Riemann or Lebesgue, proper or improper) does the integrability of $f$ and $g$ imply the integrability of $h$?

7.4.19. Calculate

$$\lim_{n\to\infty} \int_0^{n^2} e^{-x^2} n \sin\frac{x}{n}\, dx.$$

7.4.20. Find

$$\lim_{n\to\infty} \int_{\mathbb{R}} \frac{n\sin(x/n)}{x(x^2+1)}\, dx.$$

7.4.21.* Find

$$\lim_{n\to\infty} \int_0^{\infty} (1+(x/n))^{-n} \sin(x/n)\, dx.$$

7.4.22.* Find

$$\lim_{n\to\infty} \int_0^{\infty} (1+nx^2)(1+x^2)^{-n}\, dx.$$

7.4.23. Calculate

$$\lim_{n\to\infty} \int_0^{\infty} \frac{1}{1+x^{\frac{\sqrt{n}}{\ln(n+4)}}}\, dx.$$

7.4.24. Calculate

$$\lim_{n\to\infty} \int_0^{\infty} \frac{\sin x^n}{x^n}\, dx.$$

7.4.25. Give an example of a function $f : \mathbb{R} \to \mathbb{R}$ that is differentiable on $\mathbb{R}$ but its derivative is not Lebesgue integrable over $\mathbb{R}$.

7.4.26. Find the sum of the series

$$\sum_{n=0}^{\infty} \int_0^{\pi/3} (1-\sqrt{\sin x})^n \cos x\, dx.$$

7.4.27.* Find

$$\lim_{n\to\infty}\int_0^1 \frac{n^3x^{3/4}}{1+n^4x^2}\,dx,$$

and prove carefully that your result is correct.

7.4.28. Let $\{r_k\}$ denote all rational numbers in $[0,1]$ and let $\{a_k\}$ be a sequence of real numbers such that $\sum |a_k|$ converges. Prove that $\sum a_k/\sqrt{|x-r_k|}$ converges absolutely a.e. on $[0,1]$.

7.4.29.* Let $f(x) = 1/\sqrt{x}$ if $0 < x < 1$ and $f(x) = 0$ otherwise, let $\{r_k\}$ denote all rational numbers, and let $g(x) = \sum_{n=1}^{\infty} f(x-r_n)/2^n$. Prove that:

(a) $g$ is integrable over $\mathbb{R}$.

(b) $g$ is unbounded on every nondegenerate interval.

(c) $g$ is discontinuous at every point.

(d) If $h$ is a function such that $h = g$ a.e. then $h$ has properties in (b) and (c) as well.

(e) $g^2$ is not integrable over any interval.

7.4.30. Find

$$\lim_{n\to\infty}\int_a^{\infty} n(1+n^2x^2)^{-1}\,dx.$$

7.4.31.* Show that $f(x) = \sin x/x$ is not Lebesgue integrable over $(0,+\infty)$, although the infinite Riemann integral exists.

7.4.32.* Let $f$ be a function on $[a,b]$ such that $f$ is Riemann integrable over every interval $[a',b] \subset (a,b]$ and the improper Riemann integral of $f$ over $[a,b]$ does not converge. Prove that $f$ is not Lebesgue integrable over $[a,b]$.

7.4.33.* Let $f$ be a function on $(a,+\infty)$ such that $f$ is Riemann integrable over every interval $(a,b)$ and the improper Riemann integral of $f$ over $(a,+\infty)$ does nor converge. Prove that $f$ is not Lebesgue integrable over $(a,+\infty)$.

7.4.34. Prove that

$$\int_0^1 \frac{\ln x}{1+x^3}\,dx = \sum_{n=0}^{\infty}\frac{(-1)^{n+1}}{(3n+1)^2}.$$

7.4.35. Find

$$\lim_{n\to\infty}\int_0^n \left(1+\frac{x}{n}\right)^n e^{-2x}\,dx,$$

and prove carefully that your result is correct.

7.4.36.* Let $f(x) = x^{\alpha}\sin(x^{\beta})$.

(a) For what values of $\alpha, \beta$ is $f$ Lebesgue integrable over $(0,1]$?

(b) For what values of $\alpha, \beta$ the improper Riemann integral of $f$ over $(0,1]$ converges?

7.4.37. What is the Lebesgue integral of $f(x) = x^2$ over $[0,1] \setminus C$, where $C$ is the Cantor set?

In Exercises 7.4.38–7.4.39, find the Lebesgue integral $\int_0^{\pi/2} f$:

7.4.38.

$$f(x) = \begin{cases} \sin x & \text{if } x \in \mathbb{Q} \\ \cos x & \text{if } x \notin \mathbb{Q}. \end{cases}$$

7.4.39.

$$f(x) = \begin{cases} \sin x & \text{if } \cos x \in \mathbb{Q} \\ \sin^2 x & \text{if } \cos x \notin \mathbb{Q}. \end{cases}$$

7.4.40. Find the Lebesgue integral of

$$f(x,y) = \begin{cases} 1 & \text{if } xy \notin \mathbb{Q} \\ 0 & \text{if } xy \in \mathbb{Q}, \end{cases}$$

over the unit square $[0,1] \times [0,1]$.

7.4.41. Find the Lebesgue integral

$$\int_{\mathbb{R}} \frac{1}{1+x^2}.$$

7.4.42. Let $f$ be a function defined on $[0,1]$ that is integrable over $[0,1]$, differentiable at $x = 0$, and $f(0) = 0$. Let

$$g(x) = \begin{cases} x^{-3/2} f(x), & \text{if } 0 < x \leq 1, \\ 0, & \text{if } x = 0. \end{cases}$$

Prove that $g$ is integrable over $[0,1]$.

7.4.43. Give an example that the Monotone Convergence Theorem does not hold without the assumption that each $f_n$ is nonnegative.

# 8

# *Integration on Product Spaces*

In Chapter 5 we developed measure theory on $\mathbb{R}^n$. However, if $i + j = n$ then $\mathbb{R}^n = \mathbb{R}^i \times \mathbb{R}^j$ and our approach did not include questions about the relation between measures $m_i, m_j, m_n$ nor between $\sigma$-algebras $\mathcal{M}_i, \mathcal{M}_j, \mathcal{M}_n$. Integration brings forward even more interesting questions. In the case of a double Riemann integral, a recommended method is to reduce it to two iterated integrals, and this carries over to higher dimensions. (A triple integral is reduced to two integrals, one of them double, etc.) In this chapter, we will present analogous results for Lebesgue integrals. As usual, we will consider a very abstract set-up, a Cartesian product of abstract measure spaces. We will present measures and integrals over the product space, and their relation with the component measure spaces.

## 8.1 Measurability on Cartesian Products

As we said, the goal is to reduce the calculation of an integral over $X \times Y$ to integrals over $X$ and over $Y$. For that to happen, we need to have a meaningful concept of the integral on both $X$ and $Y$. Therefore, we will assume that there are $\sigma$-algebras $\mathfrak{M}$ and $\mathfrak{N}$ on them, respectively. In other words, we have measurable spaces $(X, \mathfrak{M})$ and $(Y, \mathfrak{N})$. Clearly, we should have measures on them too, but we will postpone the conversation about them until the next section.

We would like to have measurable sets on $X \times Y$ as well but in a way that aligns with their definitions on $X$ and $Y$. For intuition, we go back to Chapter 4, where basic measurable sets in $\mathbb{R}^2$ are rectangles $R = I_1 \times I_2$, their measures are given by $m_2(R) = m(I_1)m(I_2)$, and the smallest $\sigma$-algebra that contains all rectangles is the Borel $\sigma$-algebra $\mathcal{B}_2$. In a more abstract situation, we have the following definition.

**Definition 8.1.1.** Let $(X, \mathfrak{M})$ and $(Y, \mathfrak{N})$ be measurable spaces. If $A \in \mathfrak{M}$ and $B \in \mathfrak{N}$ are nonempty sets then $A \times B$ is called a **measurable rectangle**. The smallest $\sigma$-algebra in $\mathbb{P}(X \times Y)$ that contains every measurable rectangle is denoted by $\boldsymbol{\mathfrak{M} \otimes \mathfrak{N}}$.

DOI: 10.1201/9781032716497-8

Our first task is to explore the connection between $\mathfrak{M} \otimes \mathfrak{N}$ and $\sigma$-algebras $\mathfrak{M}$ and $\mathfrak{N}$. It will be helpful to introduce the following terminology.

**Definition 8.1.2.** Let $E \subset X \times Y$, $x \in X$, and $y \in Y$. The ***x*-section** (resp., the ***y*-section**) of $E$ is $E_x = \{y \in Y : (x, y) \in E\}$ (resp., $E^y = \{x \in X : (x, y) \in E\}$).

**Example 8.1.3. The sections of a set.**

If $E = A \times B$ with $A \subset X$ and $B \subset Y$ then $E_x = B$ if $x \in A$, otherwise $E_x = \emptyset$. Similarly, $E^y = A$ if $y \in B$, otherwise $E^y = \emptyset$. ♦

Example 8.1.3 shows that sections of a measurable rectangle are measurable. Our next result shows that this property holds for all sets in $\mathfrak{M} \otimes \mathfrak{N}$, thus establishing a strong tie between $\sigma$-algebras $\mathfrak{M}$, $\mathfrak{N}$, and $\mathfrak{M} \otimes \mathfrak{N}$.

**Theorem 8.1.4.** *If $E \in \mathfrak{M} \otimes \mathfrak{N}$ then $E_x \in \mathfrak{N}$ and $E^y \in \mathfrak{M}$, for all $x \in X, y \in Y$.*

*Proof.* Let $\mathcal{A}$ denote the collection of sets that have the desired property, i.e.,

$$\mathcal{A} = \{E \in \mathfrak{M} \otimes \mathfrak{N} : E_x \in \mathfrak{N} \text{ and } E^y \in \mathfrak{M}, \text{ for all } x \in X, y \in Y\}.$$

By Example 8.1.3, $\mathcal{A}$ contains every measurable rectangle. We will show that $\mathcal{A}$ is a $\sigma$-algebra. The theorem will follow from there because $\mathcal{A} \subset \mathfrak{M} \otimes \mathfrak{N}$ and $\mathfrak{M} \otimes \mathfrak{N}$ is the smallest $\sigma$-algebra that contains every measurable rectangle.

Clearly, $\mathcal{A}$ contains $X \times Y$. We will show that it is closed under the operations of taking complements and countable unions.

Suppose first that $E \in \mathcal{A}$. Then $E \in \mathfrak{M} \otimes \mathfrak{N}$ so $E^C \in \mathfrak{M} \otimes \mathfrak{N}$ and, using Exercise 8.1.1,

$$(E^C)_x = \{y \in Y : (x, y) \in E^C\} = \{y \in Y : (x, y) \in E\}^C = (E_x)^C. \tag{8.1}$$

By definition of $\mathcal{A}$, $E_x \in \mathfrak{N}$. Since $\mathfrak{N}$ is a $\sigma$-algebra, $(E_x)^C \in \mathfrak{N}$, so $(E^C)_x \in \mathfrak{N}$. A similar argument shows that $(E^C)^y \in \mathfrak{M}$, and we conclude that $E^C \in \mathcal{A}$.

Finally, let $\{E_k\}_{k=1}^\infty \subset \mathcal{A}$ and let $E = \cup_{k=1}^\infty E_k$. Then, using Exercise 8.1.2,

$$\begin{aligned} E_x &= \left\{y \in Y : (x, y) \in \bigcup_{k=1}^\infty E_k\right\} \\ &= \bigcup_{k=1}^\infty \{y \in Y : (x, y) \in E_k\} \\ &= \bigcup_{k=1}^\infty (E_k)_x \in \mathfrak{N}. \end{aligned} \tag{8.2}$$

Similarly, $E^y \in \mathfrak{M}$ and the theorem is proved. □

*Remark* 8.1.5. The converse of Theorem 8.1.4 does not hold. (See Exercise 8.1.4.)

In order to explore further the relationship between $\mathfrak{M}$, $\mathfrak{N}$, and $\mathfrak{M} \otimes \mathfrak{N}$ we will establish an alternate characterization of the last one (Theorem 8.1.16 below). On the path toward this goal, we start by introducing a generalization of the concept of an elementary set.

**Definition 8.1.6.** A set $Q \subset X \times Y$ is an **elementary set** if it is a finite disjoint union of measurable rectangles. We denote by $\mathcal{E}$ the collection of elementary sets.

*Remark* 8.1.7. In the special case when $X = Y = \mathbb{R}$ and $\mathfrak{M} = \mathfrak{N} = \mathcal{M}$, Definitions 8.1.6 and 4.1.1 do not introduce the same collection of sets. For example, the set $\mathbb{Q} \times \mathbb{Q}$ meets the requirements of the former but not of the latter. Although Definition 4.1.1 is more natural in the context of $\mathbb{R}^2$ it cannot be easily generalized to an abstract measure space where there need not be such outstanding sets as intervals. Nevertheless, both classes serve the same purpose—they are the simplest $\mathfrak{M} \otimes \mathfrak{N}$-measurable sets.

In addition, each of them is a ring.

**Proposition 8.1.8.** *The collection $\mathcal{E}$ is a ring.*

*Proof.* By Exercise 1.1.1,

$$\begin{aligned} (A_1 \times B_1) \cap (A_2 \times B_2) &= (A_1 \cap A_2) \times (B_1 \cap B_2) \in \mathcal{E} \\ (A_1 \times B_1) \setminus (A_2 \times B_2) &= [(A_1 \setminus A_2) \times B_1] \cup [(A_1 \cap A_2) \times (B_1 \setminus B_2)] \in \mathcal{E}. \end{aligned} \tag{8.3}$$

Equalities in (8.3) imply that

$$P, Q \in \mathcal{E} \quad \Rightarrow \quad P \cap Q \in \mathcal{E} \quad \text{and} \quad P \setminus Q \in \mathcal{E}. \tag{8.4}$$

(See Exercise 8.1.5.) It follows that

$$P, Q \in \mathcal{E} \quad \Rightarrow \quad P \cup Q = (P \setminus Q) \cup Q \in \mathcal{E}. \tag{8.5}$$

□

We remark here that $\mathcal{E}$ is in fact an algebra (Definition 5.2.9), which was not true for the collection of elementary sets as defined in Chapter 4, but we emphasize that it is a ring to alert the reader to the analogy with the material in Section 4.1. Next, we introduce a concept that was not used earlier.

**Definition 8.1.9.** We say that $\mathcal{G}$ is a **monotone class** if whenever $\{A_k\}_{k=1}^\infty$ is an increasing and $\{B_k\}_{k=1}^\infty$ a decreasing sequence of sets in $\mathcal{G}$, then $\cup_{k=1}^\infty A_k$ and $\cap_{k=1}^\infty B_k$ belong to $\mathcal{G}$ as well.

**Example 8.1.10. A monotone class.**
Let $X$ be a set. Then $\mathbb{P}(X)$ is a monotone class. ♦

**Example 8.1.11. A monotone class.**
Let $\mathcal{G} = \{[-n, n] : n \in \mathbb{N}\}$. Then $\mathcal{G}$ is a monotone class. ♦

**Example 8.1.12. A collection that is not a monotone class.**
Let $\mathcal{A} = \{(1/n, 1) : n \in \mathbb{N}\}$. Then $\mathcal{A}$ is not a monotone class. Indeed, the sequence $\{(1/n, 1)\}_{n=1}^{\infty}$ is increasing but $\cup_{n=1}^{\infty}(1/n, 1) = (0, 1) \notin \mathcal{A}$. ♦

When we add the requirements that $\mathcal{G}$ be closed under the operations of union and difference we get a $\sigma$-algebra.

**Lemma 8.1.13.** *Let $\mathcal{G}$ be a monotone class on $X \times Y$ that contains $\mathcal{E}$ and is also a ring. Then $\mathcal{G}$ is a $\sigma$-algebra.*

*Proof.* First, $X \times Y \in \mathcal{E} \subset \mathcal{G}$. Next, if $E \in \mathcal{G}$, then $E^C = (X \times Y) \setminus E \in \mathcal{G}$ as a difference of two sets in $\mathcal{G}$. Finally, let

$$\{E_k\}_{k=1}^{\infty} \subset \mathcal{G} \quad \text{and} \quad E = \bigcup_{k=1}^{\infty} E_k.$$

Since $\mathcal{G}$ is closed under union, the sets $F_k := E_1 \cup \cdots \cup E_k \in \mathcal{G}$, for all $k \in \mathbb{N}$. Further, the sequence $\{F_k\}_{k=1}^{\infty}$ is increasing and $\cup_{k=1}^{\infty} F_k = E$ so, by the definition of a monotone class, $E \in \mathcal{G}$. □

Examples 8.1.10 and 8.1.11 show that a monotone class may be a $\sigma$-algebra but it need not be.

Lemma 8.1.13 has as one of its hypotheses that $\mathcal{G}$ is a ring. When $\mathcal{G}$ is the *smallest* monotone class that contains elementary sets we get this condition for free.

**Proposition 8.1.14.** *Let $\mathcal{G}$ be the smallest monotone class on $X \times Y$ that contains $\mathcal{E}$. Then $\mathcal{G}$ is a ring.*

*Proof.* If $P \subset X \times Y$ we define

$$\mathcal{A}_P = \{Q \in X \times Y : P \setminus Q \in \mathcal{G},\ Q \setminus P \in \mathcal{G},\ P \cup Q \in \mathcal{G}\}. \tag{8.6}$$

By Proposition 8.1.8, $\mathcal{E}$ is a ring. It follows that if $T \in \mathcal{E}$, then $\mathcal{E}$ is contained in $\mathcal{A}_T$. We leave the following two properties to the reader (see Exercise 8.1.12):
(a) $Q \in \mathcal{A}_P \quad \Leftrightarrow \quad P \in \mathcal{A}_Q$;
(b) $\mathcal{A}_P$ is a monotone class.
From (b) we deduce that if $T \in \mathcal{E}$ then $\mathcal{E}$ is contained in the monotone class $\mathcal{A}_T$, while it is contained in $\mathcal{G}$ by hypothesis. Since $\mathcal{G}$ is the smallest, we obtain that

$$\mathcal{G} \subset \mathcal{A}_T, \quad \text{for all } T \in \mathcal{E}. \tag{8.7}$$

This implies that, if $Q \in \mathcal{G}$ and $T \in \mathcal{E}$ then $Q \in \mathcal{A}_T$ and, by the equivalence (a), $T \in \mathcal{A}_Q$. Consequently, $\mathcal{E}$ is contained in a monotone class $\mathcal{A}_Q$ as well as in a monotone class $\mathcal{G}$. Since $\mathcal{G}$ is the smallest, we obtain that

$$\mathcal{G} \subset \mathcal{A}_Q, \quad \text{for all } Q \in \mathcal{G}.$$

In other words, for any $P, Q \in \mathcal{G}$, $P \in \mathcal{A}_Q$ and (8.6) implies that $P \setminus Q \in \mathcal{G}$ and $P \cup Q \in \mathcal{G}$, whence $\mathcal{G}$ is a ring, and the proposition is proved. □

From Lemma 8.1.13 and Proposition 8.1.14 we obtain an immediate corollary.

**Corollary 8.1.15.** *Let $\mathcal{G}$ be the smallest monotone class on $X \times Y$ that contains $\mathcal{E}$. Then $\mathcal{G}$ is a $\sigma$-algebra.*

Finally, we get the desired characterization of $\mathfrak{M} \otimes \mathfrak{N}$.

**Theorem 8.1.16.** *$\mathfrak{M} \otimes \mathfrak{N}$ is the smallest monotone class in $\mathbb{P}(X \times Y)$ that contains all elementary sets.*

*Proof.* Let $\mathcal{G}$ be the smallest monotone class that contains $\mathcal{E}$. By Corollary 8.1.15, $\mathcal{G}$ is a $\sigma$-algebra and it contains $\mathcal{E}$, hence all measurable rectangles. By definition, $\mathfrak{M} \otimes \mathfrak{N}$ is the smallest such $\sigma$-algebra, so $\mathfrak{M} \otimes \mathfrak{N} \subset \mathcal{G}$. Since , $\mathfrak{M} \otimes \mathfrak{N}$ is a monotone class that contains $\mathcal{E}$, and we assumed $\mathcal{G}$ to be the smallest, we get the equality $\mathfrak{M} \otimes \mathfrak{N} = \mathcal{G}$ and the theorem is proved. □

We finish this section with a few words about the measurability of functions. If $f$ is a function defined on $X \times Y$, $f_x$ is a function on $Y$ defined by $f_x(y) = f(x, y)$. Similarly, $f^y$ is a function on $X$ defined by $f^y(x) = f(x, y)$.

**Theorem 8.1.17.** *Let $f$ be a $(\mathfrak{M} \otimes \mathfrak{N})$-measurable function on $X \times Y$. Then:*

*(a) For every $x \in X$, $f_x$ is a $\mathfrak{N}$-measurable function on $Y$.*

*(b) For every $y \in Y$, $f^y$ is a $\mathfrak{M}$-measurable function on $X$.*

*Proof.* We will prove only (a) and leave (b) to the reader. Let $x \in X$ and $c \in \mathbb{R}$. Then,

$$\begin{aligned}\{y \in Y : f_x(y) > c\} &= \{y \in Y : f(x, y) > c\} \\ &= \{y \in Y : (x, y) \in f^{-1}(c, +\infty)\} = E_x,\end{aligned}$$

where $E = f^{-1}(c, +\infty) \in \mathfrak{M} \otimes \mathfrak{N}$. By Theorem 8.1.4, $E_x \in \mathfrak{N}$. □

An astute reader may have recognized the importance of Theorems 8.1.4 and 8.1.17. When reducing the integral over $E \subset X \times Y$ to iterated integrals, for a fixed $x \in X$ we will be integrating $f_x$ over $E_x$, and these two theorems ensure that both are measurable.

> The abstract approach to the measure and integration on product spaces can be found in the work of Hahn [50] from 1933.

## Exercises

8.1.1. Prove the second equality in (8.1).

8.1.2. Prove the second equality in (8.2).

8.1.3. Let $(X, \mathfrak{M})$ and $(Y, \mathfrak{N})$ be measurable spaces. Prove or disprove: $A \times B \in \mathfrak{M} \otimes \mathfrak{N}$ if and only if $A \in \mathfrak{M}$ and $B \in \mathfrak{N}$.

8.1.4.* Let $\mathcal{B}$ denote the $\sigma$-algebra of Borel subsets of $\mathbb{R}$. Prove that there exists a set $E \subset \mathbb{R} \times \mathbb{R}$ such that $E_x \in \mathcal{B}$ and $E^y \in \mathcal{B}$ for every $x, y \in \mathbb{R}$, but $E \notin \mathcal{B} \otimes \mathcal{B}$.

8.1.5.* Prove the implication in (8.4).

8.1.6. Prove or disprove: If $A, B$ are Lebesgue measurable subsets of $\mathbb{R}^2$ then so is $A + B$. Here $A + B = \{x + y : x \in A, y \in B\}$.

8.1.7. Suppose $(X, \mathfrak{M})$ is a measurable space. Prove that if $E \in \mathfrak{M} \otimes \mathfrak{M}$, then $\{x \in X : (x, x) \in E\} \in \mathfrak{M}$.

8.1.8. Let $(X, \mathfrak{M})$ and $(Y, \mathfrak{N})$ be measurable spaces and let $\mathcal{A} = \{E \subset Y : X \times E \in \mathfrak{M} \otimes \mathfrak{N}\}$. Prove that $\mathcal{A}$ is a $\sigma$-algebra that contains $\mathfrak{N}$.

8.1.9.* Let $(X_k, \mathfrak{M}_k)$, for $k = 1, 2, 3$ be measurable spaces. Prove that $\mathfrak{M}_1 \otimes (\mathfrak{M}_2 \otimes \mathfrak{M}_3) = (\mathfrak{M}_1 \otimes \mathfrak{M}_2) \otimes \mathfrak{M}_3$.

8.1.10.* Let $X, Y$ be nonempty sets and let $\mathcal{A} \subset \mathbb{P}(X)$. Prove that $\sigma(\mathcal{A}) \times Y = \sigma(\mathcal{A} \times Y)$.

8.1.11. Let $A$ be a Borel set in $\mathbb{R}^p$. Prove that $A \times \mathbb{R}^q$ is a Borel set in $\mathbb{R}^{p+q}$.

8.1.12. Let $\mathcal{G}$ be a monotone class in $X \times Y$, let $P \subset X \times Y$, and

$$\mathcal{A}_P = \{Q \in X \times Y : P \setminus Q \in \mathcal{G}, Q \setminus P \in \mathcal{G}, P \cup Q \in \mathcal{G}\}.$$

Prove that:

(a) $Q \in \mathcal{A}_P \quad \Leftrightarrow \quad P \in \mathcal{A}_Q$;

(b) $\mathcal{A}_P$ is a monotone class.

8.1.13. Let $\mathcal{G}$ be a monotone class that is closed under finite disjoint unions. Prove that it is closed under countable disjoint unions.

8.1.14. Give an example of a monotone class $\mathcal{G}$ on $\mathbb{R}$ that satisfies

(a) $\mathbb{R}$ belongs to $\mathcal{G}$, and

(b) if $A \in \mathcal{G}$ then its complement $A^c$ is in $\mathcal{G}$,

but it is not a $\sigma$-algebra.

8.1.15. Let $\mathcal{G}$ be a monotone class and let $\mathcal{F} = \{A \in \mathcal{G} : A^C \in \mathcal{G}\}$. Prove that $\mathcal{F}$ is a monotone class.

8.1.16. Let $\mathcal{G}$ be a monotone class on $X$, let $\mathcal{C}$ be a collection of subsets of $X$, and let $\mathcal{A} = \{A \in \mathcal{G} : A \cap C \in \mathcal{G},\ \text{for all } C \in \mathcal{C}\}$. Prove that $\mathcal{A}$ is a monotone class.

8.1.17. Let $\mathcal{G}$ be a monotone class, let $\mathcal{S} \subset \mathcal{G}$ be an algebra, and let $\mathcal{F} = \{A \in \mathcal{G} : A \cup B \in \mathcal{G},\ \text{for all } B \in \mathcal{S}\}$. Prove that $\mathcal{F}$ is a monotone class.

8.1.18. Let $\mathcal{G}$ be a monotone class on $X$ that is also an algebra. Prove that $\mathcal{G}$ is a $\sigma$-algebra.

8.1.19.* Let $\mathcal{S}$ be an algebra, let $\mathcal{G}$ be the smallest monotone class that contains $\mathcal{S}$, and let $\mathcal{T} = \{A \in \mathcal{G} : A \cup B \in \mathcal{G},\ \text{for all } B \in \mathcal{G}\}$. Prove that $\mathcal{T}$ is a monotone class that contains $\mathcal{S}$.

8.1.20.* Let $\mathcal{A}$ be an algebra on $X$. Prove that the smallest monotone class that contains $\mathcal{A}$ is the smallest $\sigma$-algebra that contains $\mathcal{A}$.

8.1.21.* Let $\mathcal{A}$ be an algebra on $X$, let $\sigma(\mathcal{A})$ be the $\sigma$-algebra generated by $\mathcal{A}$, and let $\mu, \nu$ be finite measures on $X$ that agree on $\mathcal{A}$.

(a) Prove that they agree on $\sigma(\mathcal{A})$.

(b) Prove that (a) remains true if $\mu, \nu$ are $\sigma$-finite measures.

(c) Does the result hold when $\mu, \nu$ are not required to be $\sigma$-finite measures?

8.1.22. Let $X$ be a disjoint union $X = \cup_{i=1}^{\infty} X_i$, and for each $i \in \mathbb{N}$ let $\mathcal{G}_i$ be a monotone class on $X_i$. Prove or disprove: $\cup_{i=1}^{\infty} \mathcal{G}_i$ is a monotone class.

8.1.23. Let $\{\mathcal{G}_i\}_{i=1}^{\infty}$ be nested monotone classes on $X$, i.e., $\mathcal{G}_i \subset \mathcal{G}_{i+1}$ for all $i \in \mathbb{N}$. Prove or disprove: $\cup_{i=1}^{\infty} \mathcal{G}_i$ is a monotone class.

8.1.24. Prove assertion (b) of Theorem 8.1.17.

8.1.25. Let $(X, \mathfrak{M})$ and $(Y, \mathfrak{N})$ be measurable spaces, let $f : X \to \mathbb{R}$ be $\mathfrak{M}$-measurable, let $g : Y \to \mathbb{R}$ be $\mathfrak{N}$-measurable, and let $h : X \times Y \to \mathbb{R}$ be defined by $h(x, y) = f(x)g(y)$. Prove that $h$ is $\mathfrak{M} \otimes \mathfrak{N}$-measurable.

8.1.26. Let $(X, \mathfrak{M})$ be a measurable space, and let let $f : X \to \mathbb{R}$ be a nonnegative function. Prove that $f$ is measurable if and only if the set $\{(x, y) \in X \times \mathbb{R} : 0 \leq y \leq f(x)\}$ is measurable, meaning that it belongs to $\mathfrak{M} \otimes \mathcal{M}$.

8.1.27.* Let $f : [a, b] \times \mathbb{R} \to \mathbb{R}$ such that $f_x$ is continuous for every $x \in [a, b]$ and $f^y$ is Lebesgue measurable for every $y \in \mathbb{R}$. Prove that $f$ is $\mathcal{B}([a, b]) \otimes \mathcal{M}$-measurable.

8.1.28. Let $(X, \mathfrak{M})$ and $(Y, \mathfrak{N})$ be measurable spaces, let $f : X \to Y$, and suppose that the graph $f$ is $\mathfrak{M}\otimes\mathfrak{N}$-measurable. Prove or disprove: $f$ is measurable, meaning that if $A \in \mathfrak{N}$ then $f^{-1}(A) \in \mathfrak{M}$.

8.1.29. Let $(X, \mathfrak{M})$ and $(Y, \mathfrak{N})$ be measurable spaces and let $f : X \to Y$ be measurable, meaning that if $A \in \mathfrak{N}$ then $f^{-1}(A) \in \mathfrak{M}$. Prove or disprove: the graph $f$ is $\mathfrak{M} \otimes \mathfrak{N}$-measurable.

8.1.30. Let $(X, \mathfrak{M})$ and $(Y, \mathfrak{N})$ be measurable spaces. Prove that $\mathfrak{M} \otimes \mathfrak{N}$ is the smallest $\sigma$-algebra that makes projections measurable. (Projections $P_1$ and $P_2$ are defined by $P_1(x, y) = x$ and $P_2(x, y) = y$.)

8.1.31. Let $(X, \mathfrak{M})$ be a measurable space and let $f : X \times [0, 1] \to \mathbb{R}$ be a function such that $f^y$ is measurable for each $y \in [0, 1]$ and $f_x$ is continuous for each $x \in X$. Show that the set $E_{\varepsilon,\delta} := \{x : |f(x, y) - f(x, 0)| \geq \varepsilon, \text{ for all } y < \delta\}$ is measurable.

8.1.32. Let $f : \mathbb{R}^2 \to \mathbb{R}$ be a function for which both $f_x$ and $f^y$ are Borel measurable. Prove or disprove: $f$ is Borel measurable.

8.1.33. Let $X$ and $Y$ be sets, let $E \subset X \times Y$, and let $f = \chi_E$. Prove that $f_x = \chi_{E_x}$ and $f^y = \chi_{E^y}$.

## 8.2 Product Measures

In the previous section we established that $(X \times Y, \mathfrak{M} \otimes \mathfrak{N})$ is a measurable space, and our next goal is to define a measure on it. For our goals, it will be useful to assume that we already have measures on $X$ and $Y$, i.e., that we have measure spaces $(X, \mathfrak{M}, \mu)$ and $(Y, \mathfrak{N}, \lambda)$ and that they are both $\sigma$-finite. Continuing the analogy with rectangles in $\mathbb{R}^2$, a measure on $(X \times Y, \mathfrak{M} \otimes \mathfrak{N})$ should assign to a measurable rectangle $A \times B$ the value $\mu(A)\lambda(B)$. Exercise 5.2.28 ensures that there can be no more than one such $\sigma$-finite measure, so the main topic of this section is to establish that there exists one. After that, we will look at the specific situation when $\mu$ and $\nu$ are Lebesgue measures on Euclidean spaces.

For $Q \in \mathfrak{M} \otimes \mathfrak{N}$ and $x \in X$, $y \in Y$ we will use the notation

$$\varphi(x) = \lambda(Q_x) \quad \text{and} \quad \psi(y) = \mu(Q^y). \tag{8.8}$$

By Theorem 8.1.4, $Q_x \in \mathfrak{N}$ and $Q^y \in \mathfrak{M}$, so $\varphi$ and $\psi$ are well-defined. The main result of this section is the following theorem.

**Theorem 8.2.1.** *Let $(X, \mathfrak{M}, \mu)$ and $(Y, \mathfrak{N}, \lambda)$ be $\sigma$-finite measure spaces and let $Q \in \mathfrak{M} \otimes \mathfrak{N}$. If $\varphi$ and $\psi$ are defined by (8.8), then $\varphi$ is $\mathfrak{M}$-measurable, $\psi$ is $\mathfrak{N}$-measurable and*

$$\int_X \varphi \, d\mu = \int_Y \psi \, d\lambda. \tag{8.9}$$

Before we embark on the proof of Theorem 8.2.1, notice that

$$\lambda(Q_x) = \int_Y \chi_{Q_x}(y) \, d\lambda(y) = \int_Y \chi_Q(x, y) \, d\lambda(y), \quad \text{and}$$
$$\mu(Q^y) = \int_X \chi_{Q^y}(x) \, d\mu(x) = \int_X \chi_Q(x, y) \, d\mu(x).$$

Therefore, (8.9) is equivalent to

$$\int_X d\mu(x) \int_Y \chi_Q(x, y) \, d\lambda(y) = \int_Y d\lambda(y) \int_X \chi_Q(x, y) \, d\mu(x).$$

Let us denote

$$\mathcal{A} = \{Q \in \mathfrak{M} \otimes \mathfrak{N} : \varphi \text{ is } \mathfrak{M}\text{-measurable, } \psi \text{ is } \mathfrak{N}\text{-measurable and (8.9) holds}\}.$$

In other words, $\mathcal{A}$ consists of all sets in $\mathfrak{M} \otimes \mathfrak{N}$ which satisfy the conclusions of Theorem 8.2.1. Naturally, we want to show that $\mathfrak{M} \otimes \mathfrak{N} \subset \mathcal{A}$. The idea is to establish that $\mathcal{A}$ is a $\sigma$-algebra that contains all elementary sets and take advantage of the fact that $\mathfrak{M} \otimes \mathfrak{N}$ is the *smallest* such $\sigma$-algebra. Here is where Theorem 8.1.16 will show its power: all we need to do is demonstrate that $\mathcal{A}$ is a monotone class and that it contains all elementary sets. First, we consider the case when $\lambda$ and $\mu$ are finite measures.

**Theorem 8.2.2.** *Let $(X, \mathfrak{M}, \mu)$ and $(Y, \mathfrak{N}, \lambda)$ be finite measure spaces. Then $\mathcal{A}$ is a monotone class that contains all elementary sets.*

*Proof.* Let $\{Q_k\}_{k=1}^\infty$ be an increasing sequence in $\mathcal{A}$ and $Q = \cup_{k=1}^\infty Q_k$. This implies that functions $\varphi_k(x) = \lambda((Q_k)_x)$ and $\psi_k(y) = \mu((Q_k)^y)$ are measurable and satisfy (8.9) for all $k \in \mathbb{N}$. Moreover, they converge monotonically to $\varphi(x) = \lambda(Q_x)$ and $\psi(y) = \mu(Q^y)$, respectively. Thus, both $\varphi$ and $\psi$ are measurable and equality (8.9) is a consequence of the Monotone Convergence Theorem. The argument for a decreasing sequence of sets is almost the same, except that now the Monotone Convergence Theorem requires that $\varphi_1$ and $\psi_1$ be integrable (Exercise 7.2.17), which is true since $\mu$ and $\lambda$ are finite measures.

Next, we show that every measurable rectangle belongs to $\mathcal{A}$. Let $A \in \mathfrak{M}$, $B \in \mathfrak{N}$, and $Q = A \times B$. Then,

$$\varphi(x) = \lambda(Q_x) = \begin{cases} \lambda(B) & \text{if } x \in A, \\ \lambda(\emptyset) & \text{if } x \notin A \end{cases} = \lambda(B)\chi_A(x),$$

hence $\varphi$ is $\mathfrak{M}$-measurable. Similarly, $\psi(y) = \mu(A)\chi_B(y)$ implying that $\psi$ is $\mathfrak{N}$-measurable. Moreover,

$$\begin{aligned} \int_X \varphi(x)\, d\mu(x) &= \int_X \lambda(B)\chi_A(x)\, d\mu(x) = \lambda(B)\mu(A) \\ &= \int_Y \mu(A)\chi_B(y)\, d\lambda(y) = \int_Y \psi(y)\, d\lambda(y). \end{aligned}$$

Finally, if $Q \in \mathcal{E}$ then $Q$ is a finite, disjoint union of measurable rectangles. It suffices to consider $n = 2$, i.e., $Q = Q_1 \cup Q_2$ with $Q_1, Q_2$ disjoint. This implies that $\chi_Q = \chi_{Q_1} + \chi_{Q_2}$, so both $\varphi$ and $\psi$ are the sums of two measurable functions, hence measurable, and (8.9) follows from the additivity of the integral. ☐

*Remark* 8.2.3. The finiteness of $\mu$ and $\lambda$ was used only when proving that $\mathcal{A}$ is closed under descending intersections. Everything else holds without this assumption. In particular, it is worth stating the conclusion of the last paragraph of the proof: $\mathcal{A}$ is closed under finite disjoint unions.

It remains to address the $\sigma$-finite case. In this situation we can write $X = \cup X_n$ and $Y = \cup Y_n$, where $\{X_n\}$ are disjoint, $\{Y_n\}$ are disjoint, $\mu(X_n) < \infty$, and $\lambda(Y_n) < \infty$. For $Q \in \mathfrak{M} \otimes \mathfrak{N}$ we consider sets $Q_{mn} = Q \cap (X_n \times Y_m)$. Let

$$\mathcal{G} = \{Q \in \mathfrak{M} \otimes \mathfrak{N} : Q_{mn} \in \mathcal{A}, \text{ for all } m, n \in \mathbb{N}\}.$$

**Proposition 8.2.4.** *$\mathcal{G}$ is a monotone class that contains elementary sets.*

*Proof.* Let $\{Q_k\}$ be an increasing sequence in $\mathcal{G}$, and $Q = \cup Q_k$. For any $m, n \in \mathbb{N}$, $Q_{mn} = \cup_k (Q_k)_{mn}$ and $\{(Q_k)_{mn}\}$ is an increasing sequence in $\mathcal{A}$. By Theorem 8.2.2, $Q_{mn} \in \mathcal{A}$ so $Q \in \mathcal{G}$, by the definition of $\mathcal{G}$. The same argument can be used when $\{Q_k\}$ is a decreasing sequence. We conclude that $\mathcal{G}$ is a monotone class.

Suppose next that $A \times B$ is a measurable rectangle. Using (8.3),

$$(A \times B)_{mn} = (A \cap X_n) \times (B \cap Y_m),$$

hence a measurable rectangle. For fixed $m, n \in \mathbb{N}$, $\lambda$ and $\mu$ are finite measures on $X_n \times Y_m$ so Theorem 8.2.2 implies that $(A \times B)_{mn} \in \mathcal{A}$. Since this is true for all $m, n \in \mathbb{N}$, we conclude that $A \times B \in \mathcal{G}$.

Finally, it remains to show that if $P^{(1)}, P^{(2)}, \dots, P^{(N)}$ are pairwise disjoint measurable rectangles in $\mathcal{G}$ then so is their union. By definition of $\mathcal{G}$, for any $m, n \in \mathbb{N}$ and $1 \leq j \leq N$, $P^{(j)}_{mn} \in \mathcal{A}$. Once again relying on the finiteness of $\lambda$ and $\mu$ on $X_n \times Y_m$, Remark 8.2.3 shows that

$$\left(\bigcup_{j=1}^{N} P^{(j)}\right)_{mn} = \bigcup_{j=1}^{N} P^{(j)}_{mn} \in \mathcal{A},$$

so $\cup_{j=1}^{N} P^{(j)} \in \mathcal{G}$, and the proof is complete. □

At long last, we can prove Theorem 8.2.1.

*Proof.* Clearly, it suffices to prove that $\mathfrak{M} \otimes \mathfrak{N} \subset \mathcal{A}$. To that end, let $Q \in \mathfrak{M} \otimes \mathfrak{N}$. By Proposition 8.2.4, $\mathcal{G}$ is a monotone class that contains elementary sets so Theorem 8.1.16 implies that $\mathfrak{M} \otimes \mathfrak{N} \subset \mathcal{G}$. Thus, $Q \in \mathcal{G}$ so $Q_{mn} \in \mathcal{A}$, for all $m, n \in \mathbb{N}$. Clearly, the sets $\{Q_{mn}\}$ are pairwise disjoint. By Remark 8.2.3, any finite union of these sets is in $\mathcal{A}$. Moreover, the set $Q = \cup_{m,n} Q_{mn}$ can be represented as a union of an increasing sequence of sets in $\mathcal{A}$. It follows from Theorem 8.2.2, with the aid of Remark 8.2.3, that $Q \in \mathcal{A}$. Therefore, $\mathfrak{M} \otimes \mathfrak{N} \subset \mathcal{A}$ and the proof is complete. □

Based on equality (8.9) we can now define a measure of the product $\sigma$-algebra $\mathfrak{M} \otimes \mathfrak{N}$.

**Definition 8.2.5.** Let $(X, \mathfrak{M}, \mu)$ and $(Y, \mathfrak{N}, \lambda)$ be $\sigma$-finite measure spaces and let $Q \in \mathfrak{M} \otimes \mathfrak{N}$. We define the product of $\mu$ and $\lambda$ by

$$(\mu \times \lambda)(Q) = \int_X \lambda(Q_x)\, d\mu(x) = \int_Y \mu(Q^y)\, d\lambda(y).$$

*Remark* 8.2.6. The definition is correct because the second equality is (8.9).

We leave the verification of the fact that $\mu \times \lambda$ is a measure to the reader. (See Exercise 8.2.1).

*Remark* 8.2.7. If we were only interested in the existence of a measure on $\mathfrak{M} \otimes \mathfrak{N}$, we could have easily established this fact using the Carathéodory construction (Theorem 5.2.3). The strength of Theorem 8.2.1 is in the relation between the product measure $\mu \times \lambda$ and the measures $\mu$ and $\lambda$. This will become obvious in the next section.

Suppose now that $(X, \mathfrak{M}, \mu)$ and $(Y, \mathfrak{N}, \lambda)$ are complete measure spaces. Definition 8.2.5 gives the product measure $\mu \times \lambda$ on the $\sigma$-algebra $\mathfrak{M} \otimes \mathfrak{N}$. That way, we have a measure space $(X \times Y, \mathfrak{M} \otimes \mathfrak{N}, \mu \times \lambda)$ and one may ask whether it is complete. It turns out that the product measure is almost never complete. Here is an illuminating example.

**Example 8.2.8. The product measure need not be complete.**
Let $X = Y = [0,1]$, $\mathfrak{M} = \mathfrak{N} = \mathcal{M}$, and $\mu = \lambda = m$. Let $A = \{x\} \subset [0,1]$ and let $B$ be a nonmeasurable subset of $[0,1]$. The set $A \times B$ is contained in the set $A \times [0,1]$ of $\mu \times \lambda$-measure 0, but $A \times B \notin \mathfrak{M} \otimes \mathfrak{N}$ (Theorem 8.1.4).♦

Example 8.2.8 shows that the measure $m \times m$ is not complete. On the other hand, the Lebesgue measure $m_2$ on $\mathbb{R}^2$ is complete. It turns out that $m_2$ is the completion of $m \times m$. In fact, the same is true in a product of any Euclidean spaces. In order to establish this result we start with the following theorem. In what follows we will denote by $\mathcal{B}_n$ and $\mathcal{M}_n$ the $\sigma$-algebras of Borel measurable and Lebesgue measurable sets, respectively, on $\mathbb{R}^n$. As usual, $m_n$ is the Lebesgue measure on $\mathbb{R}^n$.

**Theorem 8.2.9.** *Let $p, q, r$ be positive integers satisfying $p + q = r$. Then*
*(a) $\mathcal{B}_r \subset \mathcal{M}_p \otimes \mathcal{M}_q \subset \mathcal{M}_r$.*
*(b) $\mathcal{M}_p \otimes \mathcal{M}_q \neq \mathcal{M}_r$.*

*Proof.* We will prove only part (a) and leave (b) to the reader (Exercise 8.2.36). First, every rectangle in $\mathbb{R}^r$ belongs to the $\sigma$-algebra $\mathcal{M}_p \otimes \mathcal{M}_q$. Since $\mathcal{B}_r$ is the smallest $\sigma$-algebra that contains all rectangles, we obtain that $\mathcal{B}_r \subset \mathcal{M}_p \otimes \mathcal{M}_q$.

Thus, we focus on the second inclusion. Since $\mathcal{M}_p \otimes \mathcal{M}_q$ is the smallest $\sigma$-algebra that contains $\mathcal{M}_p \times \mathcal{M}_q$, it suffices to prove the inclusion

$$\mathcal{M}_p \times \mathcal{M}_q \subset \mathcal{M}_r.$$

To that end, let $E \in \mathcal{M}_p$ and $F \in \mathcal{M}_q$ so that $E \times F \in \mathcal{M}_p \times \mathcal{M}_q$. Since $E \times F = (E \times \mathbb{R}^q) \cap (\mathbb{R}^p \times F)$ it suffices to establish that $E \times \mathbb{R}^q$ and $\mathbb{R}^p \times F$ belong to $\mathcal{M}_r$. We will prove the first one and leave the second one to the reader.

By Exercise 5.3.9, there exist Borel sets $A_p, B_p$ such that

$$A_p \subset E \subset B_p, \quad m_p(E \setminus A_p) = 0, \quad \text{and} \quad m_p(B_p \setminus E) = 0.$$

By Exercise 8.1.11, the sets $A_p \times \mathbb{R}^q$ and $B_p \times \mathbb{R}^q$ are in $\mathcal{B}_r$, and so is

$$(B_p \setminus A_p) \times \mathbb{R}^q = (B_p \times \mathbb{R}^q) \setminus (A_p \times \mathbb{R}^q).$$

Also,

$$A_p \times \mathbb{R}^q \subset E \times \mathbb{R}^q \subset B_p \times \mathbb{R}^q.$$

Measures $m_p \times m_q$ and $m_r$ agree on rectangles, hence on Borel sets. Thus,

$$\begin{aligned} m_r((B_p \times \mathbb{R}^q) \setminus (A_p \times \mathbb{R}^q)) &= m_r((B_p \setminus A_p) \times \mathbb{R}^q) \\ &= (m_p \times m_q)((B_p \setminus A_p) \times \mathbb{R}^q) = 0, \end{aligned}$$

because $m_p(B_p \setminus A_p) = 0$. (See Exercise 8.2.20.) Using Exercise 5.3.9 once again, we obtain that $E \times \mathbb{R}^q \in \mathfrak{M}_r$, and the theorem is proved. □

Now we can establish the completion result.

**Theorem 8.2.10.** *Let $p, q, r$ be positive integers satisfying $p + q = r$. Then $(\mathbb{R}^r, \mathcal{M}_r, m_r)$ is the completion of $(\mathbb{R}^r, \mathcal{M}_p \otimes \mathcal{M}_q, m_p \times m_q)$.*

*Proof.* First we will demonstrate that $(\mathbb{R}^r, \mathcal{M}_r, m_r)$ is a completion (not necessarily the minimal) of $(\mathbb{R}^r, \mathcal{M}_p \otimes \mathcal{M}_q, m_p \times m_q)$. Since $(\mathbb{R}^r, \mathcal{M}_r, m_r)$ is a complete measure space and $\mathcal{M}_p \otimes \mathcal{M}_q \subset \mathcal{M}_r$ (by Theorem 8.2.9), we need to verify that $m_r$ and $m_p \times m_q$ agree on $\mathcal{M}_p \otimes \mathcal{M}_q$. To that end, let $E \in \mathcal{M}_p \otimes \mathcal{M}_q$. Then $E \in \mathcal{M}_r$, so Exercise 5.3.9 ensures that there are Borel sets $A, B \in \mathcal{B}_r$ such that

$$A \subset E \subset B \quad \text{and} \quad m_r(B \setminus A) = 0.$$

Since $m_p \times m_q$ and $m_r$ agree on Borel sets we have that $m_p \times m_q(B \setminus A) = 0$ and it follows that $m_p \times m_q(E \setminus A) = 0$. Now, using completeness of $m_r$,

$$\begin{aligned} m_p \times m_q(E) &= m_p \times m_q(A) + m_p \times m_q(E \setminus A) = m_p \times m_q(A) \\ &= m_r(A) = m_r(A) + m_r(E \setminus A) = m_r(E). \end{aligned}$$

In order to establish that $(\mathbb{R}^r, \mathcal{M}_r, m_r)$ is the minimal completion of the measure space $(\mathbb{R}^r, \mathcal{M}_p \otimes \mathcal{M}_q, m_p \times m_q)$, we will show that $\mathcal{M}_r$ is contained in the minimal completion of $\mathcal{M}_p \otimes \mathcal{M}_q$ as defined in Exercise 5.4.30. Namely, we will prove that for any $E \in \mathcal{M}_r$ there exist $A, B \in \mathcal{M}_p \otimes \mathcal{M}_q$ such that

$$A \subset E \subset B \quad \text{and} \quad m_p \times m_q(B \setminus A) = 0.$$

So, let $E \in \mathcal{M}_r$. By Exercise 5.3.9, there exist Borel sets $A, B \in \mathcal{B}_r$ such that

$$A \subset E \subset B \quad \text{and} \quad m_r(B \setminus A) = 0.$$

Theorem 8.2.9 shows that $\mathcal{B}_r \subset \mathcal{M}_p \otimes \mathcal{M}_q$ and it follows that $A, B \in \mathcal{M}_p \otimes \mathcal{M}_q$. Finally, $m_p \times m_q$ and $m_r$ agree on Borel sets so $m_p \times m_q(B \setminus A) = 0$, and the proof is complete. □

We summarize the relations between various $\sigma$-algebras featured above in the following:

$$\mathcal{B}_p \otimes \mathcal{B}_q = \mathcal{B}_r \subsetneq \mathcal{M}_p \otimes \mathcal{M}_q \subsetneq \mathcal{M}_r.$$

The inclusions are proved in Theorem 8.2.9. The equality is relegated to Exercise 8.2.37, and allows us to deduce easily that the first inclusion must be proper. So is the second inclusion (Theorem 8.2.9). The most important issue here is that $\mathcal{M}_r$ is the completion of $\mathcal{M}_p \otimes \mathcal{M}_q$, which is Theorem 8.2.10.

The presentation in this section is based on the 1950 book [51] by a Hungarian-born American mathematician Paul Halmos (1916–2006). A very similar one can be found in the monograph [103] by Saks from 1937. Saks did not use monotone classes in his proofs, and it appears that Halmos has distilled them from the work of Hahn and Saks, and coined the name. Halmos was an assistant to John von Neumann at the Institute for Advanced Study. He was considered a great mathematical expositor. He is credited with the invention of "iff" notation for the words "if and only if" and the "tombstone" symbol ■ (sometimes called a halmos) to signify the end of a proof.

## Exercises

8.2.1. Prove that $\mu \times \lambda$, as presented in Definition 8.2.5, is a measure.

8.2.2. Let $(X, \mathfrak{M}, \mu)$ and $(Y, \mathfrak{N}, \lambda)$ be $\sigma$-finite measure spaces. Prove that $\mu\times\lambda$, as presented in Definition 8.2.5, is $\sigma$-finite.

8.2.3. Let $\mu$ be a counting measure on $\mathbb{P}(\mathbb{N})$. Describe the product measure $\mu \times \mu$.

8.2.4. Let $(X, \mathfrak{M}, \mu)$ and $(Y, \mathfrak{N}, \lambda)$ be finite measure spaces such that $\mu(X) = \lambda(Y) = 1$ and let $E \in \mathfrak{M} \otimes \mathfrak{N}$. Prove that if $(\mu \times \lambda)(E) = 1$ then $\lambda(E_x) = 1$ for a.e. $x \in X$.

8.2.5. Let $(X_k, \mathfrak{M}_k, \mu_k)$, for $k = 1, 2, 3$ be $\sigma$-finite measure spaces. Prove that the measures $(\mu_1 \times \mu_2) \times \mu_3$ and $\mu_1 \times (\mu_2 \times \mu_3)$ are identical measures on $X_1 \times X_2 \times X_3$.

8.2.6. Let $(X, \mathfrak{M}, \mu)$ and $(Y, \mathfrak{N}, \lambda)$ be $\sigma$-finite measure spaces, let $E, F \in \mathfrak{M} \otimes \mathfrak{N}$, and suppose that $\lambda(E_x) = \lambda(F_x)$ for a.e. $[\mu]$ $x \in X$. Prove that $(\mu\times\lambda)(E) = (\mu \times \lambda)(F)$.

8.2.7. Let $(X, \mathfrak{M}, \mu)$ and $(Y, \mathfrak{N}, \lambda)$ be $\sigma$-finite measure spaces, let $E \in \mathfrak{M} \otimes \mathfrak{N}$, and suppose that $\lambda(E_x) = 0$ for a.e. $[\mu]$ $x \in X$. Prove that $\mu(E^y) = 0$ for a.e. $[\lambda]$ $y \in Y$.

8.2.8. Let $(X, \mathfrak{M}, \mu)$ and $(Y, \mathfrak{N}, \lambda)$ be $\sigma$-finite measure spaces, and let $\omega$ be a measure on $\mathfrak{M} \otimes \mathfrak{N}$ such that $\omega(A \times B) = \mu(A)\lambda(B)$ for all $A \in \mathfrak{M}$ and all $B \in \mathfrak{N}$. Prove that $\omega = \mu \times \nu$.

8.2.9. Let $(X, \mathfrak{M}, \mu)$ and $(Y, \mathfrak{N}, \lambda)$ be finite measure spaces such that $\mu(X) = \lambda(Y) = 1$, let $E \subset X \times Y$ such that $(\mu \times \lambda)(E) = \eta^2 < 1$, and let $A = \{x \in X : \lambda(E_x) \geq \eta\}$. Find the sharp upper bound for $\mu(A)$.

8.2.10. Let $(X, \mathfrak{M}, \mu)$ and $(Y, \mathfrak{N}, \lambda)$ be finite measure spaces such that $\mu(X) = \lambda(Y) = 1$, let $E \subset X \times Y$ such that $\mu(E^y) \geq \alpha$, for all $y \in Y$, let $\beta < \alpha$, and let $A = \{x \in X : \lambda(E_x) \geq \beta\}$. Prove that $\mu(A) \geq (\alpha - \beta)/(1 - \beta)$.

8.2.11. Let $(X, \mathfrak{M}, \mu)$ and $(Y, \mathfrak{N}, \lambda)$ be finite measure spaces such that $\mu(X) = \lambda(Y) = 1$, and let $E \in \mathfrak{M}\otimes\mathfrak{N}$ such that $\lambda(E_x) \leq 1/2$, for a.e. $[\mu]$ $x \in X$. Prove that $\lambda(\{y \in Y : \mu(E^y) = 1\}) \leq 1/2$.

8.2.12. Let $(X, \mathfrak{M}, \mu)$ be a $\sigma$-finite measure space and let $f : X \to \mathbb{R}$ be a nonnegative, measurable function with graph $G = \{(x, y) \in X\times\mathbb{R} : y = f(x)\}$. Prove that $(\mu \times m)(G) = 0$.

8.2.13. Let $(X, \mathfrak{M}, \mu)$ and $(Y, \mathfrak{N}, \nu)$ be finite measure spaces such that $\mu(X) = \nu(Y) = 1$, and let $A \in \mathfrak{M} \otimes \mathfrak{N}$. Prove that

$$\mu(\{x \in X : \nu(\{y \in Y : (x, y) \in A\} = 1\}) \leq (\mu \times \nu)(A).$$

8.2.14. Let $(X, \mathfrak{M}, \mu)$ be a $\sigma$-finite measure space, let $f : X \to \mathbb{R}$ be a nonnegative, measurable function, and let

$$B = \{(x, y) \in X \times \mathbb{R} : 0 \leq y \leq f(x), \text{ if } f(x) < \infty \text{ or } 0 \leq y < \infty \text{ if } f(x) = \infty\}.$$

Prove that $B \in \mathfrak{M} \otimes \mathcal{B}$ and relate $(\mu \times m)(B)$ to the integral of $f$.

8.2.15. Let $\delta$ be the Dirac measure concentrated at $x = 0$, and let $E = \{(x, y) \in \mathbb{R}^2 : |x - y| < 1\}$. Find $(\delta \times m)(E)$.

8.2.16.* Let $(X, \mathfrak{M}, \mu)$ and $(X, \mathfrak{M}, \lambda)$ be nontrivial $\sigma$-finite measure spaces such that $(\mu \times \lambda)((X \times X) \setminus \Delta) = 0$, where $\Delta = \{(x, y) \in X \times X : x = y\}$. Prove that there exists $a \in X$ and $\alpha, \beta \in \mathbb{R}$ such that $\mu = \alpha\delta_a$ and $\lambda = \beta\delta_a$.

8.2.17.* The purpose of this exercise is to show that in the absence of $\sigma$-finiteness the product measure need not be unique. Let $\Delta = \{(x, y) \in [0, 1]^2 : x = y\}$, let $m$ be the Lebesgue measure on $\mathcal{B}([0, 1])$, let $\nu$ be the counting measure on $\mathbb{P}([0, 1])$, and let $m \times \nu$ be the Carathéodory extension of the measure $\omega$ defined on measurable rectangles by $\omega(E \times F) = m(E)\nu(F)$. Define $\lambda$ on $\mathcal{B}([0, 1]) \otimes \mathbb{P}([0, 1])$ by $\lambda(E) = \sup\{\sum_{y \in A} m_y(([0, 1] \times \{y\}) \cap E) : A \text{ is a finite subset of } [0, 1]\}$, where $m_y$ is the Lebesgue measure on $[0, 1] \times \{y\}$. Prove that:

(a) $\lambda$ is a measure.

(b) $m \times \nu$ and $\lambda$ agree on measurable rectangles.

(c) $(m \times \nu)(\Delta) \neq \lambda(\Delta)$.

8.2.18. Let $(X, \mathfrak{M}, \mu)$ and $(Y, \mathfrak{N}, \lambda)$ be measure spaces and let $E \in \mathfrak{M} \otimes \mathfrak{N}$. By $\mathcal{E}_\sigma$ we will denote the collection of countable unions of measurable rectangles and by $\mathcal{E}_{\sigma\delta}$ the collection of countable intersections of sets in $\mathcal{E}_\sigma$. Prove that there exists a set $F \in \mathcal{E}_{\sigma\delta}$ such that $E \subset F$ and $(\mu \times \lambda)(E) = (\mu \times \lambda)(F)$.

8.2.19. Let $(X, \mathfrak{M}, \mu)$ and $(X, \mathfrak{M}, \lambda)$ be $\sigma$-finite measure spaces, and let $E \in \mathfrak{M} \otimes \mathfrak{N}$ be such that $(\mu \times \lambda)(E) = 0$. Prove that for $\mu$-a.e. $x \in X$, we have $\lambda(E_x) = 0$.

8.2.20. Suppose that $(X, \mathfrak{M}, \mu)$ and $(Y, \mathfrak{N}, \lambda)$ are $\sigma$-finite measure spaces. Let $A \in \mathfrak{M}$ and $B \in \mathfrak{N}$ satisfy $\mu(A) = 0$ and $\lambda(B) = \infty$. Prove that $(\mu \times \lambda)(A \times B) = 0$.

8.2.21. Let $E$ be a measurable subset of $[0, 1] \times [0, 1]$ with the property that it its intersection with every vertical line is at most countable. Prove or disprove: the two-dimensional Lebesgue measure of $E$ is 0.

8.2.22.** Prove that there exists a non-measurable set $E \subset [0, 1] \times [0, 1]$ with the property that it intersects every vertical line at exactly one point.

8.2.23. Let $f : \mathbb{R}^2 \to \mathbb{R}$ be $\mathcal{M} \otimes \mathcal{M}$-measurable, let $A = \{(x, y) \in \mathbb{R}^2 : f(x, y) \neq 0\}$, and suppose that $m \times m(A) = 0$. Prove that $m(A_x) = 0$ for a.e. $x \in \mathbb{R}$, and that $m(A^y) = 0$ for a.e. $y \in \mathbb{R}$.

8.2.24.* Let $f$ be a Lebesgue measurable function on $\mathbb{R}$ and let $F(x, y) = f(x - y)$. Prove that $F$ is Lebesgue measurable on $\mathbb{R}^2$.

8.2.25. Let $A \subset \mathbb{R}^{p+q}$ and $m_{p+q}(A) = 0$. Prove that for almost every $x \in \mathbb{R}^p$ the set $A_x$ belongs to $\mathcal{M}_q$ and $m_q(A_x) = 0$.

8.2.26. Let $(X, \mathfrak{M}, \mu)$ and $(Y, \mathfrak{N}, \lambda)$ be finite measure spaces such that $\mu(X) = \lambda(Y) = 1$, and let $E \subset X \times Y$ such that $\lambda(E_x) = 0$ for a.e. $[\mu]$ $x \in X$. Suppose that $\lambda(\{y \in Y : \mu(X \setminus E^y) < 1\}) > 0$. Prove that $E \notin (\mathfrak{M} \otimes \mathfrak{N})$.

8.2.27. Let $(X, \mathfrak{M}, \mu)$ be a $\sigma$-finite measure space, let $A \in \mathfrak{M}$, let $0 \leq \alpha \leq \infty$, and define $A_\alpha = \{(x, y) : x \in A,\ 0 \leq y \leq \alpha\}$ and $A_\infty = \{(x, y) : x \in A,\ 0 \leq y < \infty\}$. Prove that $A_\alpha \in \mathfrak{M} \otimes \mathcal{B}$ and find $\mu \times m(A_\alpha)$.

8.2.28.* Let $f$ be a Borel measurable function on $\mathbb{R}$ and $E = \{(x + f(x), x - f(x)) : x \in \mathbb{R}\}$. Prove or disprove: $m_2(E) = 0$.

8.2.29. Let $f$ and $g$ be real-valued integrable functions on a measure space $(X, \mathfrak{M}, \mu)$ and define

$$F_t = \{x \in X : f(x) > t\}, \quad G_t = \{x \in X : g(x) > t\}.$$

Prove that

$$\int_X |f - g|\, d\mu = \int_{\mathbb{R}} \mu(F_t \triangle G_t)\, dt.$$

8.2.30. By Example 1.4.13, there is a well-ordered set $W$ and a bijection $j : [0, 1] \to W$ such that for any $x \in [0, 1]$, $j(x)$ has at most a countable number of predecessors. Let $Q$ be the set of points $(x, y)$ in $[0, 1] \times [0, 1]$ such that $j(x)$ precedes $j(y)$, and let $f = \chi_Q$. Prove that (8.9) is not true for $(X, \mathfrak{M}, \mu) = (Y, \mathfrak{N}, \lambda) = [0, 1]$ with Lebesgue measure. Explain why it does not contradict Theorem 8.2.1.

8.2.31.* Give an example of a set $Q \subset [0, 1] \times [0, 1]$ for which both $\varphi(x)$ and $\psi(y)$ (as in (8.8)) exist, but only one of them is measurable.

8.2.32.* Prove or disprove: there exists a set $Q \subset [0, 1] \times [0, 1]$ for which $\varphi(x) = 1$ and $\psi(y) = 0$, where $\varphi$ and $\psi$ are as in (8.8).

8.2.33. Let $(X, \mathfrak{M}, \mu)$ and $(Y, \mathfrak{N}, \lambda)$ be $\sigma$-finite measure spaces such that $(X \times Y, \mathfrak{M} \otimes \mathfrak{N}, \mu \times \lambda)$ is complete. Suppose that $B \subset \mathfrak{N}$ and $\lambda(B) = 0$. Prove that $\mathfrak{M} = \mathbb{P}(X)$.

8.2.34.* Suppose that $(X, \mathfrak{M}, \mu)$ and $(Y, \mathfrak{N}, \lambda)$ are measure spaces and let $(X, \overline{\mathfrak{M}}, \overline{\mu})$ and $(Y, \overline{\mathfrak{N}}, \overline{\lambda})$ denote their completions. Also, we denote by $(X \times Y, \overline{\mathfrak{M} \otimes \mathfrak{N}}, \overline{\mu \times \lambda})$ the completion of $(X \times Y, \mathfrak{M} \otimes \mathfrak{N}, \mu \times \lambda)$. Finally, let $(X \times Y, \overline{\overline{\mathfrak{M}} \otimes \overline{\mathfrak{N}}}, \overline{\overline{\mu} \otimes \overline{\lambda}})$ denote the completion of $(X \times Y, \overline{\mathfrak{M}} \otimes \overline{\mathfrak{N}}, \overline{\mu} \otimes \overline{\lambda})$. Prove that

$$(X \times Y, \overline{\mathfrak{M} \otimes \mathfrak{N}}, \overline{\mu \times \lambda}) = (X \times Y, \overline{\overline{\mathfrak{M}} \otimes \overline{\mathfrak{N}}}, \overline{\overline{\mu} \otimes \overline{\lambda}}).$$

8.2.35. Let $N \subset \mathbb{R}^p$ satisfy $m_p(N) = 0$ and let $B \subset \mathbb{R}^q$. Prove that $m_{p+q}(N \times B) = 0$.

8.2.36. Prove part (b) of Theorem 8.2.9.

8.2.37. Let $p, q, r$ be positive integers satisfying $p+q = r$. Prove that $\mathcal{B}_p \otimes \mathcal{B}_q = \mathcal{B}_r$. Conclude that $\mathcal{B}_r \neq \mathcal{M}_p \otimes \mathcal{M}_q$.

## 8.3 The Fubini Theorem

In the previous section we have established the formula

$$\int_X d\mu(x) \int_Y \chi_Q(x,y)\, d\lambda(y) = \int_Y d\lambda(y) \int_X \chi_Q(x,y)\, d\mu(x) \tag{8.10}$$

and we defined this quantity to be

$$(\mu \times \lambda)(Q) = \int_{X \times Y} \chi_Q(x,y)\, d(\mu \times \lambda).$$

In this section, we will show that (8.10) holds when $\chi_Q$ is replaced by a more general function.

First, we will demonstrate that it holds for nonnegative functions.

**Theorem 8.3.1.** ***Tonelli's Theorem.*** *Let $(X, \mathfrak{M}, \mu)$ and $(Y, \mathfrak{N}, \lambda)$ be $\sigma$-finite measure spaces, let $f$ be a nonnegative, $\mathfrak{M} \otimes \mathfrak{N}$-measurable function on $X \times Y$, and let*

$$\varphi(x) = \int_Y f_x\, d\lambda, \quad \psi(y) = \int_X f^y\, d\mu.$$

*Then $\varphi$ is $\mathfrak{M}$-measurable, $\psi$ is $\mathfrak{N}$-measurable and*

$$\int_X \varphi\, d\mu = \int_{X \times Y} f\, d(\mu \times \lambda) = \int_Y \psi\, d\lambda. \tag{8.11}$$

*Proof.* By Theorem 8.1.17, $f_x$ and $f^y$ are nonnegative measurable functions, so $\varphi$ and $\psi$ are extended real-valued functions. The case when $Q \in \mathfrak{M} \otimes \mathfrak{N}$ and $f = \chi_Q$ follows from Theorem 8.2.1 and Definition 8.2.5. It follows that the theorem is true when $f$ is a simple $\mathfrak{M} \otimes \mathfrak{N}$-measurable function. In the general case, Theorem 6.2.13 supplies an increasing sequence $\{s_n\}$ of nonnegative simple functions that converge to $f$. Let

$$\varphi_n(x) = \int_Y (s_n)_x\, d\lambda.$$

Then $\{\varphi_n\}$ is an increasing sequence that converges to $\int_Y f_x\, d\lambda = \varphi(x)$. It follows that $\varphi$ is $\mathfrak{M}$-measurable. Moreover, (8.11) holds for simple functions $s_n$, so

$$\int_X \varphi_n\, d\mu = \int_{X \times Y} s_n\, d(\mu \times \lambda),$$

whence Monotone Convergence Theorem yields the first equality in (8.11). The proof of the second one goes along the same lines. □

*Remark* 8.3.2. Equality (8.11) means that the *iterated* integrals

$$\int_X d\mu(x) \int_Y f(x,y)\, d\lambda(y) = \int_Y d\lambda(y) \int_X f(x,y)\, d\mu(x)$$

are equal. We refer to $\int_{X\times Y} f\, d(\mu\times\lambda)$ as "double integral".

When $f$ takes values of both signs the situation is more complicated. We will see later (Example 8.3.5) that without the nonnegativity assumption Theorem 8.3.1 need not be true. For such functions, we need an additional assumption. Since several measure spaces and measures are present, to avoid ambiguity we will augment the expressions such as "$f$ is integrable" "with respect to $\mu$", or we will say "$f$ is $\mu$-integrable".

**Theorem 8.3.3.** ***Fubini's Theorem.*** *Let $(X,\mathfrak{M},\mu)$ and $(Y,\mathfrak{N},\lambda)$ be $\sigma$-finite measure spaces, let $f$ be a real-valued, $\mathfrak{M}\otimes\mathfrak{N}$-measurable function on $X\times Y$, let*

$$\varphi(x) = \int_Y f_x\, d\lambda, \quad \psi(y) = \int_X f^y\, d\mu,$$

*and suppose that $f$ is $\mu\times\lambda$-integrable. Then $\varphi$ is $\mu$-integrable, $\psi$ is $\lambda$-integrable, and*

$$\int_X \varphi\, d\mu = \int_{X\times Y} f\, d(\mu\times\lambda) = \int_Y \psi\, d\lambda. \tag{8.12}$$

*Proof.* We will prove that $\varphi\in L(X,\mu)$ and the first equality in (8.12). The rest can be proved in exactly the same way. Let

$$\varphi_1(x) = \int_Y (f^+)_x\, d\lambda, \quad \varphi_2(x) = \int_Y (f^-)_x\, d\lambda. \tag{8.13}$$

Applying Tonelli's Theorem to nonnegative functions $f^+$ and $f^-$ we obtain that $\varphi_1$ and $\varphi_2$ are $\mathfrak{M}$-measurable, and

$$\int_X \varphi_1\, d\mu = \int_{X\times Y} f^+\, d(\mu\times\lambda), \quad \int_X \varphi_2\, d\mu = \int_{X\times Y} f^-\, d(\mu\times\lambda). \tag{8.14}$$

Further, $f^+, f^- \le |f|$ and $f$ is $\mu\times\lambda$-integrable, so $\varphi_1$ and $\varphi_2$ are $\mu$-integrable, hence finite a.e. $[\mu]$. Notice that $\varphi = \varphi_1 - \varphi_2$ a.e. $[\mu]$ (excluding the points where they are not finite), so $\varphi\in L(X,\mu)$. Subtracting equations in (8.14) we obtain the first equality in (8.12). □

It is remarkable how simple the proofs of Tonelli's and Fubini's theorems are, following the thematic approach: prove first for characteristic functions (Theorem 8.2.1 and Definition 8.2.5), then routinely extend to simple functions, and finally, prove Tonelli's theorem by approximating the nonnegative integrand by a sequence of simple functions and applying Monotone Convergence Theorem. From here, it is just one small step to the proof of Fubini's theorem, namely represent the integrand as a difference of 2 nonnegative functions. In summary, the bulk of the work was done on the basic case $f = \chi_Q$ in Sections 8.1 and 8.2.

In applications, these two theorems are often used in succession. First, Tonelli's theorem is applied to ensure that $f$ is $\mu \times \lambda$-integrable, and then Fubini's theorem is used to reverse the order of integration. Here is an example.

**Example 8.3.4. Evaluating a double integral**

Evaluate

$$\int_0^1 \int_y^1 x^{-3/2} \cos\left(\frac{\pi y}{2x}\right) dxdy.$$

**Solution:** First, applying Tonelli's theorem,

$$\begin{aligned}\int_0^1 \int_y^1 \left|x^{-3/2} \cos\left(\frac{\pi y}{2x}\right)\right| dxdy &\leq \int_0^1 \int_y^1 x^{-3/2}\, dx\, dy \\ &= \int_0^1 \int_0^x x^{-3/2}\, dy\, dx = \int_0^1 x^{-1/2} = 2 < \infty,\end{aligned}$$

so $f(x,y) := \chi_{[0,1]\times[y,1]}(x,y)x^{-3/2}\cos(\pi y/(2x))$ is integrable over $[0,1]\times[0,1]$. (Strictly speaking, $f$ is not defined for $x = 0$ but, as always, we ignore sets of measure 0.) Next, we apply Fubini's theorem:

$$\begin{aligned}\int_0^1 \int_y^1 x^{-3/2} \cos\left(\frac{\pi y}{2x}\right) dxdy &= \int_0^1 \int_0^x x^{-3/2} \cos\left(\frac{\pi y}{2x}\right) dydx \\ &= \int_0^1 x^{-3/2} \frac{2x}{\pi} \sin\left(\frac{\pi y}{2x}\right)\Big|_0^x dx = \frac{2}{\pi}\int_0^1 x^{-1/2}\, dx = \frac{4}{\pi}.\end{aligned}$$ ♦

How important are the conditions of Theorem 8.3.1?

**Example 8.3.5. Theorem 8.3.1 fails when $f$ is neither nonnegative nor integrable.**

Let $X = Y = [0,1]$ and $\mu = \lambda = m$. Let $\{\delta_n\}$ be an increasing sequence in $(0,1)$ that converges to 1 and $\{g_n\}$ a sequence of nonnegative continuous functions on $[0,1]$ where, for each $n$, $g_n = 0$ outside of $[\delta_n, \delta_{n+1}]$ and $\int_0^1 g_n(t)\, dt = 1$. Finally, we define

$$f(x,y) = \sum_{n=1}^{\infty} [g_n(x) - g_{n+1}(x)]g_n(y).$$

Notice that at each point of $[0,1] \times [0,1]$ at most one of the terms is nonzero, so the series converges and $f$ is well-defined. We have that

$$\begin{aligned}\int_0^1 dx \int_0^1 f(x,y)\, dy &= \int_0^1 dx \int_0^1 \sum_{n=1}^{\infty} [g_n(x) - g_{n+1}(x)]g_n(y)\, dy \\ &= \int_0^1 dx \sum_{k=1}^{\infty} \int_{\delta_k}^{\delta_{k+1}} \sum_{n=1}^{\infty} [g_n(x) - g_{n+1}(x)]g_n(y)\, dy \\ &= \int_0^1 dx \sum_{k=1}^{\infty} \int_{\delta_k}^{\delta_{k+1}} [g_k(x) - g_{k+1}(x)]g_k(y)\, dy\end{aligned}$$

$$
\begin{aligned}
&= \int_0^1 dx \sum_{k=1}^{\infty} [g_k(x) - g_{k+1}(x)] \int_{\delta_k}^{\delta_{k+1}} g_k(y)\, dy \\
&= \int_0^1 dx \sum_{n=1}^{\infty} [g_n(x) - g_{n+1}(x)] \int_0^1 g_n(y)\, dy \\
&= \int_0^1 \sum_{n=1}^{\infty} [g_n(x) - g_{n+1}(x)]\, dx \\
&= \sum_{k=1}^{\infty} \int_{\delta_k}^{\delta_{k+1}} \sum_{n=1}^{\infty} [g_n(x) - g_{n+1}(x)]\, dx \\
&= \int_{\delta_1}^{\delta_2} g_1(x)\, dx + \int_{\delta_2}^{\delta_3} [g_2(x) - g_2(x)]\, dx + \int_{\delta_3}^{\delta_4} [g_3(x) - g_3(x)]\, dx + \dots \\
&= 1.
\end{aligned}
$$

On the other hand,

$$
\begin{aligned}
\int_0^1 dy \int_0^1 f(x,y)\, dx &= \sum_{n=1}^{\infty} \int_{\delta_n}^{\delta_{n+1}} dy \int_0^1 f(x,y)\, dx \\
&= \sum_{n=1}^{\infty} \int_{\delta_n}^{\delta_{n+1}} dy \int_0^1 g_n(y)[g_n(x) - g_{n+1}(x)]\, dx \\
&= \sum_{n=1}^{\infty} \int_{\delta_n}^{\delta_{n+1}} g_n(y)\, dy \int_0^1 [g_n(x) - g_{n+1}(x)]\, dx \\
&= 0,
\end{aligned}
$$

because, for any $n \in \mathbb{N}$, $\int_0^1 g_n(x)\, dx = \int_0^1 g_{n+1}(x)\, dx = 1$.

It is fairly easy to see that Tonelli's Theorem does not apply because $f$ takes the values of both signs. If $y_0 \in (\delta_1, \delta_2)$ and $g(y_0) > 0$, then $f(x, y_0) = [g_1(x) - g_2(x)]g(y_0)$. This number will be positive if $x \in (\delta_1, \delta_2)$ and $g_1(x) > 0$, and negative if $x \in (\delta_2, \delta_3)$ and $g_2(x) > 0$.

Finally, Theorem 8.3.3 does not apply because $f$ is not $m \times m$-integrable. Indeed,

$$
\begin{aligned}
\int_0^1 dx \int_0^1 |f(x,y)|\, dy &= \int_0^1 dx \sum_{k=1}^{\infty} \int_{\delta_k}^{\delta_{k+1}} |f(x,y)|\, dy \\
&= \int_0^1 dx \sum_{k=1}^{\infty} \int_{\delta_k}^{\delta_{k+1}} \left| \sum_{n=1}^{\infty} [g_n(x) - g_{n+1}(x)] g_n(y) \right| dy \\
&= \int_0^1 dx \sum_{k=1}^{\infty} \int_{\delta_k}^{\delta_{k+1}} |[g_k(x) - g_{k+1}(x)] g_k(y)|\, dy \\
&= \int_0^1 \sum_{k=1}^{\infty} |g_k(x) - g_{k+1}(x)|\, dx
\end{aligned}
$$

$$= \sum_{j=1}^{\infty} \int_{\delta_j}^{\delta_{j+1}} \sum_{k=1}^{\infty} |g_k(x) - g_{k+1}(x)|\, dx$$
$$= \int_{\delta_1}^{\delta_2} |g_1(x)|\, dx + \int_{\delta_2}^{\delta_3} 2|g_2(x)|\, dx + \int_{\delta_3}^{\delta_4} 2|g_3(x)|\, dx + \dots,$$

and the last series diverges. ♦

Throughout this section, we made the assumption that $(X, \mathfrak{M}, \mu)$ and $(Y, \mathfrak{N}, \lambda)$ are $\sigma$-finite measure spaces. This was by design because, without this hypothesis, the result need not hold anymore.

**Example 8.3.6. Theorem 8.3.1 fails when one of the measures is not $\sigma$-finite.**

Let $X = Y = [0, 1]$, $\mu = m$, and $\lambda$ a counting measure. We define

$$f(x, y) = \begin{cases} 1 & \text{if } x = y \\ 0 & \text{if } x \neq y. \end{cases}$$

Then

$$\int_0^1 f(x, y)\, dm(x) = m(\{y\}) = 0 \quad \text{and} \quad \int_0^1 f(x, y)\, d\lambda(y) = \lambda(\{x\}) = 1.$$

Therefore,

$$\int_0^1 d\lambda(y) \int_0^1 f(x, y)\, dm(x) = 0 \neq 1 = \int_0^1 dm(x) \int_0^1 f(x, y)\, d\lambda(y).$$

Notice that all assumptions (other than $\lambda$ being $\sigma$-finite) are satisfied. Namely, $f$ is nonnegative and $\mathfrak{M} \otimes \mathfrak{N}$-measurable, if $\mathfrak{M} = \mathcal{M}$ and $\mathfrak{N} = \mathbb{P}([0, 1])$. Indeed, $f = \chi_D$, where $D = \{(x, x) : x \in [0, 1]\}$. For $n \in \mathbb{N}$, let

$$Q_n = [0, 1/n] \times [0, 1/n] \cup [1/n, 2/n] \times [1/n, 2/n] \cup \dots \cup [(n-1)/n, 1] \times [(n-1)/n, 1].$$

The each $Q_n \in \mathcal{E}$ and $D = \cap Q_n$. ♦

The Fubini Theorem is the main result of the whole chapter and one of the most important results of the Lebesgue theory of integration. The analogue for the Riemann integral is much weaker. Namely, known results typically rely on the continuity of the integrand, and in the case of improper/infinite integrals on some form of uniform convergence of the integral. In contrast, for Lebesgue integrals, we require either that the integrand is a nonnegative measurable function (Tonelli's Theorem) or that it is integrable with respect to the product measure (Fubini's Theorem). In addition, in both results, we assume that the underlying measure spaces are $\sigma$-finite. This last assumption is not a big restriction because almost all measure spaces of practical interest are $\sigma$-finite.

It was known to Euler that for a continuous function of 2 variables on a bounded rectangle, the double integral can be reduced to two iterated integrals. In his dissertation Lebesgue proved the same result under the weaker assumption that the integrand is measurable and bounded. Fubini in 1907 [42] and another Italian mathematician, Leonida Tonelli (1885–1946) in 1909 [118] made further generalizations, but still within the realm of a Euclidean space. An abstract version can be found in the work of Hahn from 1933 [50]. In the same year, a Polish-American mathematician, Salomon Bochner (1899–1982) [12] made it even further by considering functions with values in an abstract vector spaces instead of $\mathbb{R}$. Tonelli was a student of Arzelà at Bologna. He made significant contributions in calculus of variation. The department of mathematics at the University of Pisa, where he taught, is named after him. Bochner was a student of Erhard Schmidt. He held a position at the University of Munich but in 1933 fled to Princeton, and after retiring there became a chair at Rice University. He worked on Fourier series, differential geometry, and several complex variables.

GUIDO FUBINI (1879–1943) was an Italian mathematician, a son of a mathematics teacher. He grew up in Venice and demonstrated his talent while in high school as well as at the Scuola Normale Superiore di Pisa where his teachers were Dini and Bianchi. With the latter Fubini conducted research in geometry and graduated with a doctoral degree in 1900. Starting with 1901 he held several university positions, first in Sicily, then in Genoa, and finally at the University of Turin in 1908, where he taught for the next 30 years. He conducted research in several fields including analysis, with most important contibutions in tensor analysis, but also in differential equations, complex analysis, and the calculus of variations. Upon Bianchi's death, Fubini succeeded him as coeditor of the *Annali di matematica pura ed applicata*. He was a member of several Italian scientific academies including the Lincean Academy, one of the oldest and most prestigious European scientific institutions. Fleeing from the growing anti-Semitism in the Fascist Italy, he immigrated in 1939 to the United States and worked successively at the Institute of Advanced Studies, Princeton, and New York University. He became interested in engineering problems during the First World War, and his continued success in this direction led him to write a book based on such research. Already in poor health when he left Italy, he died from heart problems at the age of sixty four, and the book, jointly authored with Giuseppe Albenga, an Italian civil engineer, only appeared in 1954, eleven years after his death.

## Exercises

8.3.1. Let $m$ be the Lebesgue measure on $\mathbb{R}$ and $E = \{(x, y) : 0 \le y \le x\}$. Calculate

$$\int_E \frac{y}{x} e^{-x}\, d(m \times m).$$

8.3.2. Let $(X, \mathfrak{M}, \mu)$ be a $\sigma$-finite complete measure space and let $f : X \to [0, \infty)$ be measurable. Prove that

$$\int_X f\, d\mu = \int_0^\infty \mu\left(\{x \in X : f(x) \ge t\}\right) dt.$$

8.3.3. Let $(X, \mathfrak{M}, \mu)$ be a measure space, let $f$ be integrable over $X$, let $\lambda > 0$, and define

$$\varphi(\lambda) = \mu(\{x \in X : f(x) > \lambda\}), \quad \psi(\lambda) = \mu(\{x \in X : f(x) < -\lambda\}).$$

Prove that $\varphi$ and $\psi$ are Borel measurable and that $\int_X |f|\, d\mu = \int_0^\infty (\varphi(t) + \psi(t))\, dt$.

8.3.4. Calculate

$$\int_0^\infty \int_0^\infty \frac{1}{(1+y)(1+x^2y)}\, dm(x)\, dm(y)$$

and use it to find the values of

$$\int_0^\infty \frac{\ln x}{x^2-1}\, dm(x) \quad \text{and} \quad \int_0^1 \frac{\ln x}{x^2-1}\, dm(x).$$

8.3.5. Let $g : [0,1] \to \mathbb{R}$ be a Lebesgue measurable function such that the function $f(x,y) = 2g(x) - 3g(y)$ is Lebesgue integrable over $[0,1] \times [0,1]$. Prove that $g$ is Lebesgue integrable over $[0,1]$.

8.3.6. Let $f, g$ be nonnegative integrable functions over $\mathbb{R}$, let $A, B$ be sets of positive measures in $\mathbb{R}$, and suppose that $f$ is strictly positive on $A$ and $g$ is strictly positive on $B$. Let

$$f * g(x) = \int_{\mathbb{R}} f(t)g(x-t)\, dt.$$

Prove that $f * g > 0$ on a set of positive measure.

8.3.7. Let $X = [0,1]$, $Y = [0,\infty)$, $f(u,v) = 1/(1+u^2v^2)$, $g(x,y,t) = f(x,t)f(y,t)$, for $(x,y,t) \in X \times X \times Y := Z$.

(a) Prove that $g$ is integrable on $Z$. Using Tonelli's theorem on $Y \times X \times X$ show that

$$A =: \int_Z g\, dt\, dx\, dy = \int_Y \left(\frac{\arctan t}{t}\right)^2 dt.$$

(b) Using Tonelli's theorem on $X \times X \times Y$ show that

$$A = \frac{\pi}{2} \int_{X \times X} \frac{1}{x+y}\, dx dy.$$

(c) Using Tonelli's theorem again show that $A = \pi \ln 2$.

8.3.8. Let $X = Y = [-\pi/2, \pi/2]$, with Lebesgue measure and

$$f(x,y) = \frac{\sin x \sin y}{x^2+y^2}.$$

Prove that both iterated integrals exist and that they are equal. Does the double integral of $f$ exist?

8.3.9. Let $m$ be the Lebesgue measure on $\mathbb{R}$ and $E = \{(x,y) : 0 \leq y \leq \sqrt{x}\}$. Calculate

$$\int_E \frac{y}{x} e^{-x} \sin x\, d(m \times m).$$

8.3.10. Calculate

$$\int_0^2 \int_{y^2}^4 y \cos x^2\, dx\, dy.$$

8.3.11. Calculate

$$\int_0^\infty \frac{e^{-x} - e^{-2x}}{x}\, dx.$$

8.3.12. Let $(X, \mathfrak{M}, \mu)$ and $(Y, \mathfrak{N}, \lambda)$ be measure spaces. Suppose that $g : X \to \mathbb{R}$ is a $\mu$-integrable function and that $h : Y \to \mathbb{R}$ is a $\lambda$-integrable function. Define $f : X \times Y \to \mathbb{R}$ by $f(x, y) = g(x)h(y)$. Prove that $f$ is $\mu \times \lambda$ integrable and that

$$\int_{X\times Y} f\, d(\mu \times \lambda) = \left(\int_X g\, d\mu\right)\left(\int_Y h\, d\lambda\right).$$

8.3.13. Let $(X, \mathfrak{M}, \mu)$ be a finite measure space with $\mu(X) = 1$ and let $f, g : X \to \mathbb{R}$ be integrable functions that satisfy

$$(f(x) - f(y))\,(g(x) - g(y)) \geq 0,$$

for all $x, y \in X$. Prove that $\int_X f\, d\mu \int_X g\, d\mu \leq \int_X fg\, d\mu$.

8.3.14. Prove that, for $s > 0$,

$$\int_0^\infty e^{-sxy} \frac{\sin^2 x}{x}\, dx = \frac{1}{4} \ln\left(1 + \frac{4}{s^2}\right).$$

8.3.15. Let $f$ be Lebesgue integrable over $(0, a)$ and let $g(x) = \int_x^a t^{-1} f(t)\, dt$. Prove that $g$ is integrable over $(0, a)$ and $\int_0^a g(x)\, dx = \int_0^a f(x)\, dx$.

8.3.16.* Let $f : [0, 1] \times [0, 1] \to \mathbb{R}$ be a Lebesgue measurable function such that the integrals $\int_0^1 f(x, y)\, dx$, $\int_0^1 f(x, y)\, dy$, and $\int_0^1 dy \int_0^1 f(x, y)\, dx$ exist and are finite. Prove or disprove: $\int_0^1 dx \int_0^1 f(x, y)\, dy$ exists.

8.3.17.** Give an example of a bounded nonnegative function $f$ on $[0, 1] \times [0, 1]$ that is not Lebesgue integrable but both iterated integrals exist and are equal to 0.

8.3.18. Let $X = Y = [0, 1]$, with Lebesgue measure and

$$f(x, y) = \frac{x^2 - y^2}{(x^2 + y^2)^2}.$$

Prove that both iterated integrals exist but that they are not equal.

8.3.19. Let $X = Y = [-1, 1]$, with Lebesgue measure and

$$f(x, y) = \frac{xy}{(x^2 + y^2)^2}.$$

Prove that both iterated integrals exist and that they are equal but the double integral of $f$ does not exist.

8.3.20. Let $X = Y = (0, 1)$ and let $m$ be the Lebesgue measure on $(0, 1)$. Find an example of a positive continuous function $f : (0, 1) \times (0, 1) \to \mathbb{R}$, such that

$$\int_{X\times Y} f\, d(m \times m)$$

is finite but such that $\int_Y f_x\, dy$ is infinite for some $x \in (0, 1)$.

8.3.21. Consider the measure space $(\mathbb{N}, \mathbb{P}(\mathbb{N}), \mu)$, where $\mu$ is the counting measure, and let

$$f(n,k) = \begin{cases} 1 & \text{if } n = k \\ -1 & \text{if } n = k+1 \\ 0 & \text{otherwise.} \end{cases}$$

Prove that $f$ is not integrable with respect to $\mu \times \mu$, and that both iterated integrals exist but are not equal.

8.3.22. Let $f$ be a bounded, Borel measurable function on $\mathbb{R}^2$ with the property that both $f_x$ and $f^y$ are integrable and $\int_{\mathbb{R}} f_x = \int_{\mathbb{R}} f^y = 0$. Prove or disprove: $f$ is integrable over $\mathbb{R}^2$ with respect to Lebesgue measure on $\mathcal{B}(\mathbb{R}^2)$.

8.3.23.* Prove that $(\mu \times \lambda)(D) = \infty$ in Example 8.3.6.

8.3.24. Let $f_n(x) = ae^{-nax} - be^{-nbx}$, where $0 < a < b$. Prove that

(a) $\sum_{n=1}^{\infty} \int_0^{\infty} |f_n(x)|\, dx = +\infty$

(b) $\sum_{n=1}^{\infty} \int_0^{\infty} f_n(x)\, dx \neq \int_0^{\infty} \sum_{n=1}^{\infty} f_n(x)\, dx$.

8.3.25. Let $X = Y = [0,1]$, with Lebesgue measure and $f(x,y) = (x - 1/2)^{-3}$ if $0 < y < |x - 1/2|$, $f(x,y) = 0$ otherwise. Find the iterated integrals and the double integral of $f$ over $X \times Y$.

8.3.26.* Give an example of a nonmeasurable function $f$ such that $|f|$ is integrable and both repeated integrals are well-defined but have different values.

8.3.27. Let $f \in L(\mathbb{R})$, and $\beta \in (0,1)$. Prove that

$$\int_{\mathbb{R}} \frac{|f(x)|}{|x-a|^{\beta}}\, dx < \infty$$

for (Lebesgue) a.e. $a \in \mathbb{R}$.

8.3.28. Let $(X, \mathfrak{M}, \mu)$ be a $\sigma$-finite measure space and, for each $t \in \mathbb{R}$, let $e_t = \chi_{(t,\infty)}$. Prove that if $f, g : X \to \mathbb{R}$ are measurable, then $\int_X |f - g|\, d\mu = \int_{\mathbb{R}} dt \int_X |e_t \circ f - e_t \circ g|\, d\mu$.

8.3.29. Let $f \in L((0,\infty) \times \mathbb{R})$. Define the sequence $g_n : \mathbb{R} \to \mathbb{R}$ by

$$g_n(x) = \int_0^{\infty} e^{-y} f(ny, x)\, dy, \quad n \in \mathbb{N}.$$

Prove that:

(a) $\lim_{n\to\infty} \int_0^{\infty} |g_n| = 0$.

(b) $\lim_n g_n = 0$ a.e.

8.3.30. Let $(X, \mathfrak{M}, \mu)$ and $(Y, \mathfrak{N}, \lambda)$ be complete measure spaces, and let $(X \times Y, \overline{\mathfrak{M} \otimes \mathfrak{N}}, \overline{\mu \times \lambda})$ denote the completion of $(X \times Y, \mathfrak{M} \otimes \mathfrak{N}, \mu \times \lambda)$. Suppose that $f$ is a real-valued, $\overline{\mathfrak{M} \otimes \mathfrak{N}}$-measurable function on $X \times Y$ that is $\overline{\mu \times \lambda}$-integrable. Prove that conclusions of Fubini's Theorem hold for $f$.

# 9

# *Differentiation and Integration*

As we stated earlier, one of the main criticisms of the Riemann integral was that it does not commute with the differentiation process. In this chapter, we will address the question of whether the Lebesgue integral is an improvement in this regard. When studying derivatives it is most convenient to stay within $\mathbb{R}^n$, preferably for $n = 1$, and we will do just that. That is, we will consider the measure space $(\mathbb{R}, \mathcal{M}, m)$ and functions $f : E \to \mathbb{R}$, where $E \in \mathcal{M}$. Our goal will be to characterize the class of functions for which

$$\int_a^b f'(x)\,dx = f(b) - f(a). \tag{9.1}$$

## 9.1 Dini Derivatives

We start our journey towards formula (9.1) by considering the following generalization of the derivative that is due to Dini.

**Definition 9.1.1.** Let $f$ be a real-valued function defined on an interval $[a, b]$ and let $x_0 \in (a, b)$. The four one-sided limits

$$D^+f(x_0) = \limsup_{x\to x_0+} \frac{f(x) - f(x_0)}{x - x_0}, \quad \text{(the upper right derivative of } f \text{ at } x_0\text{)},$$

$$D_+f(x_0) = \liminf_{x\to x_0+} \frac{f(x) - f(x_0)}{x - x_0}, \quad \text{(the lower right derivative of } f \text{ at } x_0\text{)},$$

$$D^-f(x_0) = \limsup_{x\to x_0-} \frac{f(x) - f(x_0)}{x - x_0}, \quad \text{(the upper left derivative of } f \text{ at } x_0\text{)},$$

$$D_-f(x_0) = \liminf_{x\to x_0-} \frac{f(x) - f(x_0)}{x - x_0}, \quad \text{(the lower left derivative of } f \text{ at } x_0\text{)}.$$

are called **Dini derivatives** of $f$ at $x_0$. When all four are finite and equal we say that $f$ is **differentiable** at $x$ and we denote the common value by $f'(x)$.

> Dini used symbols $\Lambda$, $\lambda$, $\Lambda'$, and $\lambda'$. The modern notation was invented by Scheeffer and introduced in 1884 [107]. However, it took several decades for it to become commonly accepted.

DOI: 10.1201/9781032716497-9

Regardless of whether $f$ is differentiable or not (or even whether $f$ is continuous or not), the four Dini derivatives always exist and take values in $[-\infty, +\infty]$. Therefore, by focusing on them we get information that may be lacking in the situation when $f'$ does not exist.

**Example 9.1.2. A function that is neither left nor right differentiable but has all Dini derivatives finite.**

Let

$$f(x) = \begin{cases} x \sin \frac{1}{x} & \text{if } x \neq 0 \\ 0 & \text{if } x = 0. \end{cases}$$

At $x = 0$ we have that

$$\frac{f(0+x) - f(0)}{x} = \frac{x \sin 1/x - 0}{x} = \sin \frac{1}{x}.$$

When $x \to 0$ neither the left nor the right limit exists. On the other hand,

$$D^+ f(0) = \limsup_{x \to 0+} \sin \frac{1}{x} = \lim_{h \to 0} \left[ \sup_{0 < x \leq h} \sin \frac{1}{x} \right] = 1,$$

and similarly $D_+ f(0) = -1$, $D^- f(0) = 1$, and $D_- f(0) = -1$ (Figure 9.1). ♦

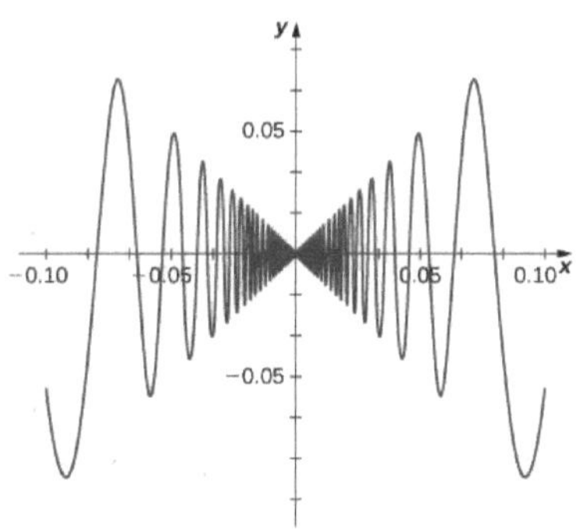


Figure 9.1: Seeing Dini derivatives in the graph of a function

*Remark* 9.1.3. Definition 9.1.1 characterizes the derivative $f'$ in a way that is different from the standard one. Nevertheless, the two concepts lead to the same derivative (see Exercise 9.1.7).

We know that if a measurable function is differentiable, then its derivative is measurable as well (Exercise 6.2.28). This remains true if we replace the derivative by any of the Dini derivatives. We will prove the result for monotone functions. (See also Exercises 9.1.28 and 9.1.29.)

**Proposition 9.1.4.** *If $f$ is monotone the four Dini derivatives of $f$ are measurable.*

*Proof.* We will prove that $D^+f$ is measurable. For the other three see Exercise 9.1.17. Let

$$f_h(x) = \sup_{0<\delta<h} \frac{f(x+\delta)-f(x)}{\delta}.$$

Notice that

$$D^+f = \lim_n f_{1/n}.$$

By Corollary 6.2.7, it suffices to prove that $f_h$ is measurable for any $h > 0$. Let $c \in \mathbb{R}$ and $A = \{x : f_h(x) > c\}$. We will show that $A$ is a measurable set. To that end, let $x_0 \in A$. This means that there is a $\delta \in (0, h)$ such that

$$\frac{f(x_0+\delta)-f(x_0)}{\delta} > c.$$

Suppose first that $f$ is increasing. Choose a $\delta' \in (\delta, h)$ such that

$$\frac{f(x_0+\delta)-f(x_0)}{\delta'} > c, \tag{9.2}$$

and let $x \in [x_0 + \delta - \delta', x_0]$. (See Figure 9.2.) Then

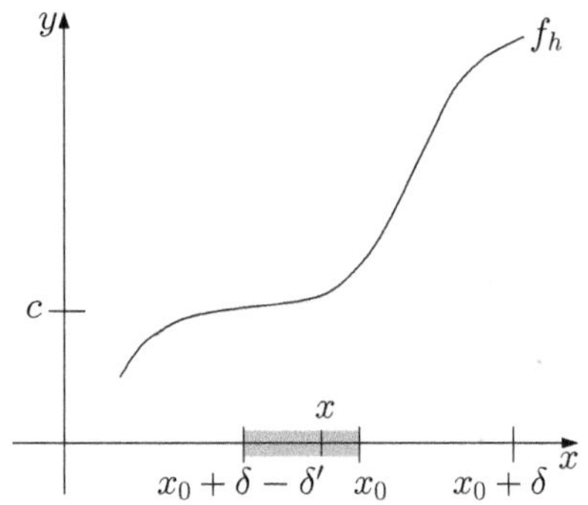


Figure 9.2: $x$ is chosen in the interval $[x_0 + \delta - \delta', x_0]$

$$c < \frac{f(x_0+\delta)-f(x_0)}{\delta'} \le \frac{f(x+\delta')-f(x)}{\delta'},$$

because $f(x_0 + \delta) \le f(x + \delta')$ and $f(x_0) \ge f(x)$. Thus, $x \in A$ which implies that $[x_0+\delta-\delta', x_0] \subset A$. Similarly, if $f$ is decreasing we can choose $\delta' \in (0, \delta)$ satisfying (9.2) and show that $[x_0, x_0 + \delta - \delta'] \subset A$. In both cases, we get that $A$ is a union of nondegenerate closed intervals. By Exercise 3.1.3, $A$ is a countable (disjoint) union of intervals, and therefore measurable. □

Dini derivatives share another property of the derivative. For more on this direction see Exercise 9.1.19.

**Theorem 9.1.5.** *Let $f : (a, b) \to \mathbb{R}$ be a continuous function and $D^+f \ge 0$ on $(a, b)$. Then $f$ is nondecreasing on $(a, b)$.*

*Proof.* **Case I.** $D^+f > 0$ on $(a,b)$.
Suppose to the contrary that $f$ is not nondecreasing. Then there are $\alpha, \beta \in (a,b)$ such that

$$\alpha < \beta \quad \text{and} \quad f(\alpha) > f(\beta).$$

Let $y_0$ be a number such that $f(\beta) < y_0 < f(\alpha)$ and $E = \{x \in [a,b] : f(x) > y_0\}$. (See Figure 9.3.) The set $E$ is nonempty since it contains $\alpha$ and it is

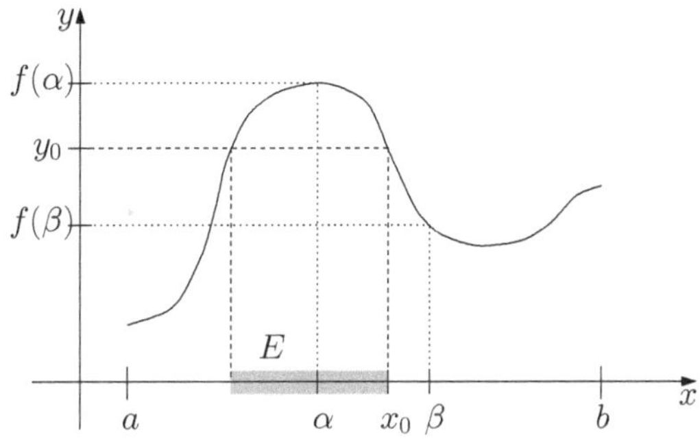


Figure 9.3: $E = \{x \in [a,b] : f(x) > y_0\}$

bounded as a subset of $[a,b]$. Therefore $x_0 := \sup E$ is a real number in $[a,b]$. We will show that $y_0 = f(x_0)$. Once we accomplish this, we will use the fact that $x > x_0$ implies that $x \notin E$ and it will follow that

$$D^+f(x_0) = \lim_{h \to 0+} \sup_{x_0 < x \leq x_0 + h} \frac{f(x) - f(x_0)}{x - x_0} = \lim_{h \to 0+} \sup_{x_0 < x \leq x_0 + h} \frac{f(x) - y_0}{x - x_0} \leq 0,$$

contradicting the assumption of case I.

Thus, we need to demonstrate that $y_0 = f(x_0)$. Suppose that this is not true. If $f(x_0) > y_0$ then $x_0 \neq \beta$, so $(x_0, \beta)$ is a nondegenerate interval. For any $x \in (x_0, \beta)$ we have that $x > x_0$, so $x \notin E$. This implies that $f(x) \leq y_0$ so

$$\lim_{x \to x_0^+} f(x) \leq y_0 < f(x_0),$$

contradicting the assumption that $f$ is continuous. If $f(x_0) < y_0$ a symmetric argument shows that $f$ cannot be continuous from the left.
**Case II.** $D^+f \geq 0$ on $(a,b)$.
Let $\varepsilon > 0$ and define $f_\varepsilon(x) = f(x) + \varepsilon x$. It is not hard to verify (see also Exercise 9.1.15) that $D^+f_\varepsilon = D^+f + \varepsilon > 0$, so Case I implies that $f_\varepsilon$ is nondecreasing. Let $x < y$. Then

$$0 \leq f_\varepsilon(y) - f_\varepsilon(x) = f(y) - f(x) + \varepsilon(y - x).$$

Letting $\varepsilon \to 0$ we obtain that $f$ is nondecreasing. □

It is worth emphasizing once again that we can get information about a function, such as that it is monotone on an interval, even when $f'$ does not exist, by relying on Theorem 9.1.5.

ULISSE DINI (1845–1918), was an Italian mathematician and politician from Pisa. He attended a teacher's college *Scuola Normale Superiore*, attached to the University of Pisa. A scholarship allowed him to continue his studies in Paris. During 1865-66 his research blossomed and he published 7 papers. In 1866 he returned to the University of Pisa to teach advanced algebra and the theory of geodesy. In 1871 he was promoted to a professor of analysis and higher geometry. These were turbulent years in Italy which was undergoing the process of unification. Dini was elected to the Pisa City Council and in 1880 to the National Parliament, as a representative from Pisa. In 1890 he became a senator. At the professional level, he served as the rector of the University of Pisa 1888-90 and as the director of the *Scuola Normale Superiore*, from 1908 until his death. Dini did some significant research in the field of analysis during a time when a major goal was to determine precisely when the theorems (which had earlier been stated and proved in an imprecise way) were valid. To achieve this aim mathematicians tried to see how far results could be generalized and they needed to find pathological counterexamples to show the limits to which generalization was possible. Dini was one of the greatest masters of generalization and constructing counterexamples. In addition to one surface, several results were named after him: Dini's criterion for the pointwise convergence of Fourier series, Dini's Theorem on uniform convergence, and the Implicit Function Theorem is known as Dini's Theorem in Italy.

During the 18th century, mathematicians worked mostly with elementary functions that are differentiable except at some isolated points. This resulted in a naive belief that continuity ensures some degree of differentiability. The example of Weierstrass (a continuous, nowhere differentiable function) put an end to it, although there remained a feeling that such a function was an aberration. We will now see that this is far from true and that differentiable functions are an oddity. In fact, this extends to functions whose only finite derivative occurs at one of the endpoints.

**Theorem 9.1.6.** *Functions that have a finite derivative in at least one point of the interval $[a, b]$ form a set of first categories in $C[a, b]$.*

*Proof.* Let us denote by $A$ the set of functions that have a finite derivative in at least one point of the interval $[a, b]$. If $f \in A$ and it has a finite derivative at $x \in [a, b]$, then there exists $m, n \in \mathbb{N}$ such that

$$\left|\frac{f(x+h)-f(x)}{h}\right| \leq n, \qquad \text{whenever } |h| < \frac{1}{m}. \tag{9.3}$$

We will denote by $A_{m,n}$ the subset of $C[a, b]$ consisting of all such functions. Namely, $f \in A_{m,n}$ if there is $x \in [a, b]$ such that (9.3) holds. It is not hard to see that

$$A \subset \bigcup_{n=1}^{\infty} \bigcup_{m=1}^{\infty} A_{m,n},$$

so it suffices to prove that each set $A_{m,n}$ is nowhere dense.

First we show that $A_{m,n}$ is a closed set in $C[a, b]$, i.e., in the topology of uniform convergence. Let $\{f_k\}$ be a sequence in $A_{m,n}$ that converges to a function $f \in C[a, b]$. By definition of $A_{m,n}$, for every $k \in \mathbb{N}$ there exists $x_k \in [a, b]$ such that

$$\left|\frac{f_k(x_k+h)-f_k(x_k)}{h}\right| \leq n, \qquad \text{whenever } |h| < \frac{1}{m}. \tag{9.4}$$

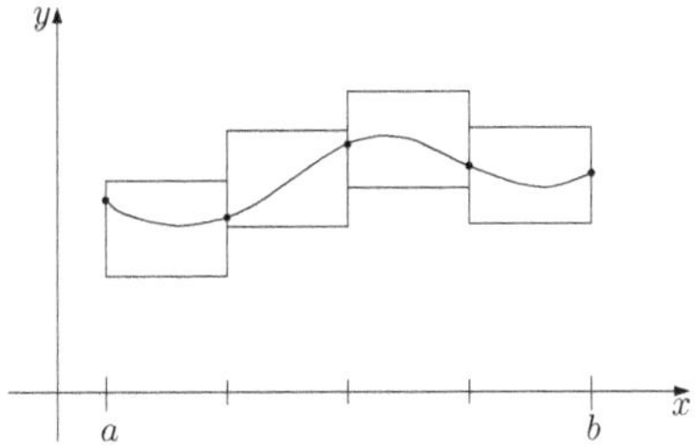


Figure 9.4: The graph of $f$ is covered by rectangles $R_k$, ...

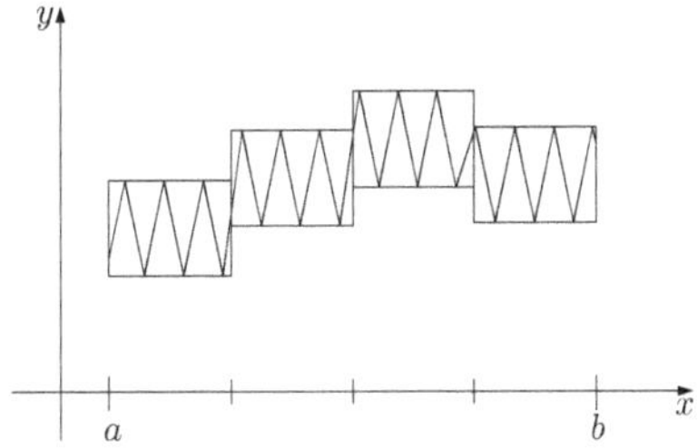


Figure 9.5: ...and so is the graph of a polygonal line

The sequence $\{x_k\}$ is bounded so it has a convergent subsequence. To simplify the notation we will discard all other members of the sequence and do the same for $\{f_k\}$. That way, we may assume that in (9.4) we have $\lim f_k = f$ and $\lim x_k = x \in [a,b]$. The uniform convergence then implies that $\lim f_k(x_k) = f(x)$ (Exercise 2.3.39) and $\lim f_k(x_k + h) = f(x+h)$, so (9.4) implies that $f \in A_{m,n}$, whence it is a closed set.

It remains to establish that the interior of each $A_{m,n}$ is empty. To that end, we fix $m$ and $n$, and we choose an arbitrary $f \in A_{m,n}$. Let $\varepsilon > 0$. We will find a continuous function $g$ in $B(f,\varepsilon)$ that is not in $A_{m,n}$. Since $f$ is uniformly continuous there is $N \in \mathbb{N}$, and a partition $P = \{a = x_0, x_1, \dots, x_{N-1}, x_N\}$ of $[a,b]$, where $x_k = a + k\,\frac{b-a}{N}$, $k = 0,1,2,\dots,N$, so that $|f(u) - f(v)| < \varepsilon/8$ whenever $u, v$ are in the same subinterval. It follows that the graph of $f$ is covered by rectangles

$$R_k = \left\{(x,y) : x_{k-1} \le x \le x_k,\ f(x_{k-1}) - \frac{\varepsilon}{4} \le y \le f(x_{k-1}) + \frac{\varepsilon}{4}\right\},$$

for $k = 1,2,\dots,N$. (See Figure 9.4.) Moreover, if the graph of another function $g$ is covered by the same rectangles, then at any point $x \in [a,b]$ we have that $|g(x) - f(x)| \le \varepsilon/2$, so $g \in B(f,\varepsilon)$. Such is, for example, a polygonal line whose each piece has a slope $2n$ or $-2n$ (Figure 9.5), and it clearly does not belong to $A_{m,n}$. □

We finish this section by stating without proof one of the most powerful results about Dini derivatives. Denjoy (1915) proved the theorem for continuous functions, Young (1917) extended it to measurable functions, and Saks (1924) extended it to arbitrary functions. For the proof, we refer the reader to [103, Chapter IX].

**Theorem 9.1.7.** ***The Denjoy–Young–Saks Theorem.*** *Let $f$ be a real valued function defined on $[a,b]$. Then almost every point $x \in [a,b]$ is in one of four sets:*

*(a) $A_1$ on which $f$ has a finite derivative;*

*(b) $A_2$ on which $D^+f = D_-f$ is finite, $D^-f = \infty$, and $D_+f = -\infty$;*

*(c) $A_3$ on which $D^- f = D_+ f$ is finite, $D^+ f = \infty$, and $D_- f = -\infty$;*

*(d) $A_4$ on which $D^- f = D^+ f = \infty$ and $D_- f = D_+ f = -\infty$.*

## Exercises

9.1.1. Calculate all Dini derivatives of $\chi_{\mathbb{Q}}$, the characteristic function of rationals.

9.1.2. Let

$$f(x) = \begin{cases} |x| & \text{if } x \in \mathbb{Q}, \\ |2x| & \text{if } x \notin \mathbb{Q}. \end{cases}$$

Calculate all four Dini derivatives of $f$ at $x = 0$ and at $x = 1$.

9.1.3. Calculate all Dini derivatives of $x\chi_{\mathbb{Q}}(x) - x\chi_{\mathbb{Q}^c}(x)$ at $x = 0$ and at $x = 1$.

9.1.4. Let $f(x) = 2^{-n}$ if $x \in (2^{-n}, 2^{-n+1}]$ and $f(0) = 0$. Calculate all four Dini derivatives of $f$ at $x = 0$.

9.1.5. Calculate all Dini derivatives of $\chi_{[0,1]}(x)$ at $x = 0$ and at $x = 1$.

9.1.6. Construct an increasing function $f : [-1, 1] \to \mathbb{R}$ such that none of the four Dini derivatives at 0 are equal.

9.1.7. Prove that a function $f$ is differentiable at $x_0$ (in the standard meaning) if and only if all 4 Dini derivatives are equal and finite, in which case $f'(x_0) = D^+ f(x_0) = D_+ f(x_0) = D^- f(x_0) = D_- f(x_0)$.

9.1.8. If $f$ is defined and finite on $[-c, c]$, prove the following:

(a) $D^+(-f(x)) = -D_+ f(x)$, $D^-(-f(x)) = -D_- f(x)$.

(b) $D^+(f(-x)) = -D_- f(x)$, $D_+(f(-x)) = -D^- f(x)$.

(c) $D^-(f(-x)) = -D_+ f(x)$, $D_-(f(-x)) = -D^+ f(x)$.

9.1.9. If $f$ and $g$ are defined and finite on $[a, b]$, prove that for every $x$, $D^+(f + g)(x) \le D^+ f(x) + D^+ g(x)$. Give an example to show that the inequality may be strict.

9.1.10. Let $f, g : \mathbb{R} \to \mathbb{R}$ be continuous and suppose that for every $x \in \mathbb{R}$, $D^+ f(x) = D^+ g(x) < \infty$. Prove that $f$ and $g$ differ by a constant.

9.1.11. Give an example to show that the chain rule for Dini derivatives $D^+(f \circ g)(x) = D^+ f(g(x)) D^+ g(x)$ does not hold.

9.1.12.* Let $A$ be a subset of $\mathbb{R}$ of measure 0. Prove that there exists a continuous increasing function $f : \mathbb{R} \to \mathbb{R}$ for which each Dini derivative is infinite on each point of $A$.

9.1.13. Let $f : (a, b) \to \mathbb{R}$ be a continuous function. Prove that $f$ is nonincreasing on $(a, b)$ if and only if $D^+ f \le 0$ on $(a, b)$.

9.1.14. Let $f, g : [a,b] \to \mathbb{R}$ be continuous functions such that $D^+f(t) \leq 0$ and $D^+g(t) \leq 0$ on $(a,b)$. If $h(t) = \max\{f(t), g(t)\}$, prove that $D^+h(t) \leq 0$ on $(a,b)$.

9.1.15. If $f$ and $g$ are defined and finite on $[a,b]$, prove the following:

(a) If $g$ has a right derivative $g'(c+)$ at $c \in [a,b]$, then

$$D^+(f+g)(c) = D^+f(c) + g'(c+) \quad \text{and} \quad D_+(f+g)(c) = D_+f(c) + g'(c+).$$

(b) If $g$ has a left derivative $g'(c-)$ at $c \in [a,b]$, then

$$D^-(f+g)(c) = D^-f(c) + g'(c-) \quad \text{and} \quad D_-(f+g(c)) = D_-f(c) + g'(c-).$$

9.1.16. Let $f$ and $g$ be nonnegative and continuous at $c$.

(a) Prove that $D^+(fg)(c) \leq g(c)D^+f(c) + f(c)D^+g(c)$.

(b) Give an example to show that the inequality in (a) may be strict.

(c) Give an example to show that the inequality in (a) is not true without the continuity and nonnegativity assumptions.

9.1.17. Complete the proof of Proposition 9.1.4 by showing that if $f$ is monotone then the remaining three Dini derivatives of $f$ are measurable ($D_+$, $D^-$, and $D_-$).

9.1.18. Suppose that $D_+f(x_0)$ is finite. Prove that $f$ is continuous from the right at $x_0$.

9.1.19. Suppose $f$ is continuous on $(a,b)$. Prove that $f$ is increasing on $(a,b)$ if and only if at least one of the four Dini derivatives of $f$ is nonnegative on $(a,b)$.

9.1.20. Let $f$ be an increasing function on $[a,b]$ and let $x_0 \in (a,b)$. Prove that none of the Dini derivatives can be negative (or equal to $-\infty$) at $x_0$.

9.1.21.* (**Dini, 1878**) Let $f : [a,b] \to \mathbb{R}$ be a continuous function. Prove that the bounds of each of the Dini derivatives are equal to the bounds of the difference quotient

$$\frac{f(y) - f(x)}{y - x},$$

as $x, y \in [a,b]$. Use this to conclude that if a single Dini derivative of a continuous function $f$ is continuous at a point $x_0$, then so are the other three. Show that in this case all four Dini derivatives are equal, so $f$ is differentiable at $x_0$.

9.1.22. Give an example to show that the conclusions of Exercise 9.1.21 need not hold if the continuity assumption on $f$ is dropped.

9.1.23.** Let $f$ be continuous on $[a,b]$, and let a constants $M$ and a set $E \subset [a,b]$ of positive measure satisfy $D^+f(x) \geq M$ for all $x \in E$ (or $D^-f(x) \geq M$ for all $x \in E$). Prove that for every $\varepsilon > 0$ there exists a finite collection of disjoint intervals $(a_1,b_1)$, ..., $(a_n,b_n)$ in $[a,b]$ such that $\sum_{i=1}^{n}(b_i - a_i) > m(E) - \varepsilon$ and

$$f(b_i) - f(a_i) \geq (M - \varepsilon)(b_i - a_i), \quad i = 1, 2, \ldots, n.$$

9.1.24.* Let $f$ be a real-valued function. Prove that the set of points where $D^+f(x) < D_-f(x)$ is countable.

9.1.25. Give an example showing that it is possible that the set $A_1$ in Theorem 9.1.7 have Lebesgue measure $b - a$. Do the same for sets $A_2$ and $A_3$.

9.1.26. Let $f$ be continuous on $[a,b]$ and suppose that it has a local maximum at $c \in (a,b)$. Prove that

$$D_+f(c) \leq D^+f(c) \leq 0 \leq D_-f(c) \leq D^-f(c).$$

9.1.27. Let $f : \mathbb{R} \to \mathbb{R}$ be continuous and suppose that $D^+f$ is Riemann integrable over $[a,b]$. Prove that

$$\int_a^b D^+f(x)\,dx = f(b) - f(a).$$

9.1.28.** If $f$ is continuous on $[a,b]$ prove that the four Dini derivatives of $f$ are measurable.

9.1.29.** If $f$ is measurable prove that the four Dini derivatives of $f$ are measurable.

9.1.30. A function $f \in C[a,b]$ is called **Hölder continuous** of order $\delta$, notation $f \in \Lambda_\delta$, if there is a constant $C$ such that $|f(x) - f(y)| \leq C|x - y|^\delta$ for all $x, y \in [a,b]$. Prove that the Hölder continuous functions form a set of first categories in $C[a,b]$.

## 9.2 Monotone Functions

It is not sufficient that $f$ is continuous for (9.1) to hold, because there are continuous functions that are not differentiable. Such an example was given by Weierstrass. A scrutiny of his function shows that it oscillates without control in the vicinity of every point. In order to avoid such a predicament, we start from the opposite extreme, i.e., we will consider monotone functions. By monotone, we mean either increasing ($x < y \Rightarrow f(x) \leq f(y)$) or decreasing ($x < y \Rightarrow f(x) \geq f(y)$). The main result of this section (Theorem 9.2.8 below) will show that such functions are differentiable a.e. Yet, they need not be differentiable at *every* point.

**Example 9.2.1. An increasing continuous function that is not differentiable at a point.**

The function

$$f(x) = \begin{cases} x & \text{if } x < 0, \\ 2x & \text{if } x \geq 0, \end{cases}$$

is increasing and continuous but not differentiable at $x = 0$ (Figure 9.6). ♦

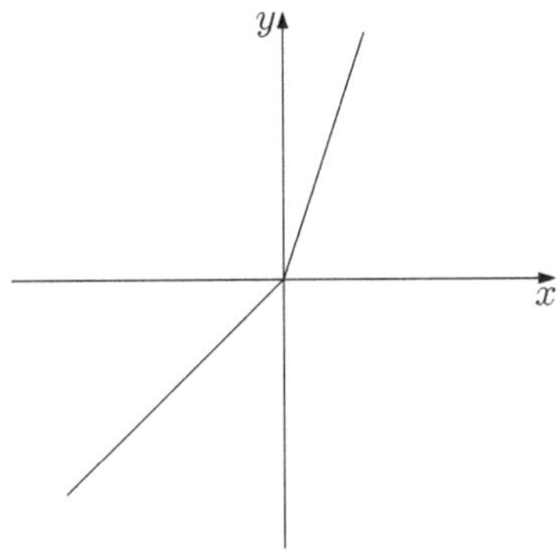


Figure 9.6: $f$ is increasing and continuous but not differentiable at $x = 0$

Notice that in this example $f$ fails to be differentiable only at $x = 0$. It is not hard to make examples of functions of this type with any number of singularities, even infinitely many. Our goal is to show that, in general, the set of such points cannot be too big. We start with a definition.

**Definition 9.2.2.** A collection $\mathcal{F}$ of closed, bounded, nondegenerate intervals covers a set $E$ in the **sense of Vitali** if

$$\forall x \in E, \quad \forall \varepsilon > 0, \quad \exists I \in \mathcal{F} \quad \text{such that } x \in I, \text{ and } \ell(I) < \varepsilon.$$

For such a collection we often say that it is a **Vitali covering** of $E$.

**Example 9.2.3. A Vitali covering of** $[0, 1]$**.**

The collection $\mathcal{F} = \{[p, q] : 0 \leq p < q \leq 1,\, p, q \in \mathbb{Q}\}$ is a Vitali covering of $[0, 1]$. We leave the verification as Exercise 9.2.1. ♦

It appears that a Vitali covering of a set $E$ has to be very large. Certainly, it is an infinite collection. The following result shows that a set can be "almost" covered with a finite subcollection. Although it holds in any Euclidean space, we will state and prove it in $\mathbb{R}$.

**Theorem 9.2.4.** ***The Vitali Covering Lemma.*** *Let $E$ be a set of finite outer measure and $\mathcal{F}$ a collection of closed, bounded, nondegenerate intervals that covers $E$ in the sense of Vitali. Then for any $\varepsilon > 0$, there is a finite disjoint subcollection $\{I_k\}_{k=1}^{n}$ of $\mathcal{F}$ for which*

$$m^*\left(E \setminus \bigcup_{k=1}^{n} I_k\right) < \varepsilon. \tag{9.5}$$

*Proof.* **Step I.** Reducing $\mathcal{F}$ to uniformly bounded sets.

Since $m^*(E) < \infty$, there exists an open set $G \supset E$ such that $m(G) < \infty$. We will discard all sets in $\mathcal{F}$ that are not contained in $G$. Let us show that the remaining members of $\mathcal{F}$ are still a Vitali covering of $E$. Indeed, let $x \in E$ and let $\varepsilon > 0$. Then, there exists an interval $(x - \varepsilon, x + \varepsilon) \subset G$. By definition of $\mathcal{F}$, there exists $I \in \mathcal{F}$ (possibly one of the discarded intervals) such that $x \in I$ and $\ell(I) < \varepsilon$. It follows that $I \subset (x - \varepsilon, x + \varepsilon) \subset G$ (so $I$ was not one of the discarded intervals). Thus, we may assume that each set in $\mathcal{F}$ is contained in $G$.

**Step II.** Selecting a countable collection of disjoint sets in $\mathcal{F}$.

We will define the desired subcollection $\{I_k\}_{k=1}^{\infty}$ inductively. The idea of the proof is the following: start with $I_1 \in \mathcal{F}$, select $I_2 \in \mathcal{F}$ as large as possible but not intersecting $I_1$, then select $I_3 \in \mathcal{F}$ as large as possible but not intersecting $I_1 \cup I_2$, etc.

So, let $I_1 \in \mathcal{F}$ be arbitrary and suppose that disjoint intervals $I_1, \dots, I_n$ have been chosen. We will denote $A_n = \cup_{k=1}^{n} I_k$. If $E \subset A_n$ then (9.5) holds and the proof is complete, so we will assume that this is not true. We continue by defining

$$s_n = \sup\{\ell(I) : I \in \mathcal{F} \text{ and } I \cap A_n = \emptyset\}.$$

Since $E$ is not contained in the compact set $A_n$, by definition the covering

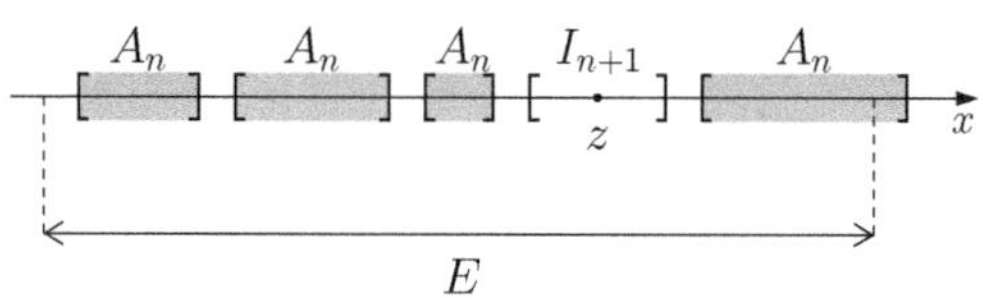


Figure 9.7: Constructing the interval $I_{n+1}$

in the sense of Vitali, there exists a point $z \in E$ and an interval $I_z$ in $\mathcal{F}$ containing $z$, so that $I_z \cap A_n = \emptyset$ (Figure 9.7). Consequently, $s_n \neq 0$ and we can choose $I_{n+1} \in \mathcal{F}$ so that $I_{n+1} \cap A_n = \emptyset$ and

$$\ell(I_{n+1}) > s_n/2. \tag{9.6}$$

Thus, the inductive process yields a countable collection $\{I_n\}$. It follows from the construction that $\cup_{n=1}^{\infty} I_n \subset G$, so

$$\sum_{n=1}^{\infty} \ell(I_n) < \infty. \tag{9.7}$$

Notice that $\{I_n\}$ need not be a Vitali covering anymore.

**Step III.** Selecting a finite subcollection of $\{I_n\}$ so that (9.5) holds.

Let $\varepsilon > 0$. Then there exists $N \in \mathbb{N}$ such that

$$\sum_{n=N+1}^{\infty} \ell(I_n) < \frac{\varepsilon}{5}.$$

Let

$$R = E \setminus \bigcup_{n=1}^{N} I_n. \tag{9.8}$$

It remains to prove that $m^*(R) < \varepsilon$. We will show that

$$R \subset \bigcup_{n=N+1}^{\infty} 5I_n, \tag{9.9}$$

where $5I_n$ denotes the interval with the same midpoint as $I_n$ but 5 times longer. Assuming that (9.9) holds, we have

$$m^*(R) \leq \sum_{n=N+1}^{\infty} m(5I_n) = 5 \sum_{n=N+1}^{\infty} m(I_n) < \varepsilon.$$

Thus, we focus on (9.9).

Let $x \in R$. In view of (9.8), $x$ does not belong to the closed set $A_N$, so $d := \operatorname{dist}(x, A_N) > 0$. Thus, there exists $I \in \mathcal{F}$ such that $x \in I$ and $\ell(I) < d$. It follows that $I \cap A_N = \emptyset$, whence $\ell(I) \leq s_N$. If we had $I \cap A_m = \emptyset$ for every $m \geq N$, then we would have, using (9.6) and (9.7), that

$$\ell(I) \leq s_m < 2\ell(I_{m+1}) \to 0$$

which is impossible. Therefore, there exists $m \geq N$ such that $I \cap A_m \neq \emptyset$. Let $M$ denotes the smallest number such that $I \cap A_M \neq \emptyset$. Of course, $M > N$ and

$$I \cap A_{M-1} = \emptyset, \quad I \cap A_M \neq \emptyset,$$

so $I \cap I_M \neq \emptyset$. Also,

$$\ell(I) \leq s_{M-1} \leq 2\ell(I_M). \tag{9.10}$$

Finally, $x \in I$ and $I \cap I_M \neq \emptyset$. If we denote by $\operatorname{mid} I_M$ the midpoint of the

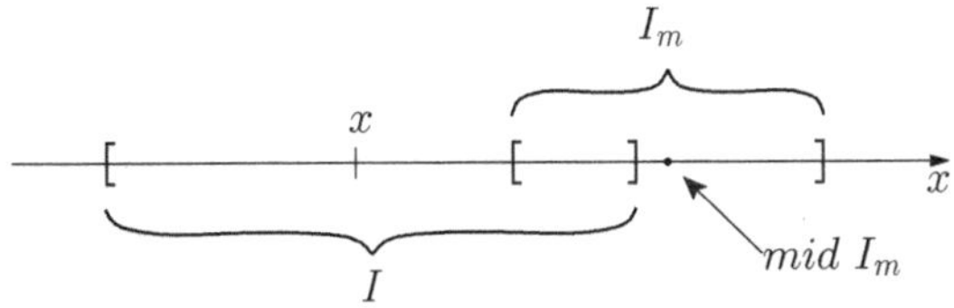


Figure 9.8: $\operatorname{dist}(x, \operatorname{mid}(I_M)) \leq \ell(I) + \frac{1}{2}\ell(I_M) \leq \frac{5}{2}\ell(I_M)$

interval $I_M$ (Figure 9.8), using (9.10) we obtain that

$$\operatorname{dist}(x, \operatorname{mid}(I_M)) \leq \ell(I) + \frac{1}{2}\ell(I_M) \leq \frac{5}{2}\ell(I_M),$$

and it follows that $x \in 5I_M$. This establishes (9.9) and the proof is complete. □

Vitali proved this theorem in 1908 [123] for sets in $\mathbb{R}^n$ with the ssumption that the covering sets are cubes (all edges of equal length). In 1910, Lebesgue [73] showed that the same argument can be used for more general covering sets. The proof that we presented is from 1924 and is due to Banach [3].

Our goal is to demonstrate that a monotone function is differentiable a.e. By definition, we need to show that all four Dini derivatives are equal and finite almost everywhere. We start with the latter task. Without loss of generality, we may assume that $f$ is an increasing function in which case none of the Dini derivatives can be equal to $-\infty$. (See Exercise 9.1.20). The following lemma ensures that the set of points where $D^+f(x) = +\infty$ has measure zero.

**Lemma 9.2.5.** *Let $f$ be an increasing function on $[a, b]$ and*

$$S = \{x \in (a, b) : D^+f(x) = +\infty\}.$$

*Then, $m^*(S) = 0$.*

*Proof.* Suppose that $\alpha := m^*(S) > 0$. Let

$$A = \frac{2}{\alpha}(f(b) - f(a)). \tag{9.11}$$

By definition of $D^+$, for any $x \in S$ there exists a decreasing sequence $\{\beta_n(x)\}$ converging to $x$ such that

$$\frac{f(\beta_n(x)) - f(x)}{\beta_n(x) - x} > A. \tag{9.12}$$

Clearly, the collection $\mathcal{F} := \{[x, \beta_n(x)] : x \in S,\ n \in \mathbb{N}\}$ is a Vitali covering of $S$ (Exercise 9.2.2). By the Vitali Covering Lemma, there exists a finite collection $\{I_k\}_{k=1}^n$ of disjoint intervals in $\mathcal{F}$ whose total length is bigger than $\alpha/2$. Let $I_k = [a_k, b_k]$. Then, using (9.11) and (9.12),

$$\sum_{k=1}^{n} [f(b_k) - f(a_k)] > A \sum_{k=1}^{n} (b_k - a_k) > \frac{\alpha A}{2} = f(b) - f(a).$$

However, this is impossible because $f$ is increasing, so that

$$\sum_{k=1}^{n} [f(b_k) - f(a_k)] \le f(b) - f(a).$$

□

*Remark* 9.2.6. Since $D_+f(x) \le D^+f(x)$, Lemma 9.2.5 implies that, when $f$ is an increasing function, $D_+f(x) < +\infty$ a.e. Under the same assumption, $D^-f(x) < +\infty$ a.e. (Exercise 9.2.12) whence the inequality $D_-f(x) \le D^-f(x)$ implies that none of the Dini derivatives can be equal to $+\infty$ except on a set of measure zero.

Next, we will show that all Dini derivatives are equal a.e. We already know that $D_+f(x) \leq D^+f(x)$ and $D_-f(x) \leq D^-f(x)$. Thus, it suffices to establish that $D^+f(x) \leq D_-f(x)$ a.e. and $D^-f(x) \leq D_+f(x)$ a.e. We will now prove the first of these inequalities and leave the second one to the reader (Exercise 9.2.13).

**Lemma 9.2.7.** *Let $f$ be an increasing function on $[a,b]$ and*

$$T = \left\{x \in (a,b) : D^+f(x) > D_-f(x)\right\}.$$

*Then, $m^*(T) = 0$.*

*Proof.* Suppose that $m^*(T) > 0$. Let

$$T(u,v) = \{x \in (a,b) : D^+f(x) > u > v > D_-f(x)\}. \tag{9.13}$$

Then each set $T(u,v)$ is measurable by Proposition 9.1.4, and

$$T = \bigcup_{u,v\in\mathbb{Q}} T(u,v).$$

This is a countable union so not each set $T(u,v)$ can be of measure zero. It follows that there exists $p, q \in \mathbb{Q}$ such that the set $A := T(p,q)$ has positive outer measure $\gamma$. By Exercise 9.1.20, $p, q > 0$.

Let $\varepsilon > 0$. There exists an open set $G$ containing $A$ such that

$$m(G) < \gamma + \varepsilon. \tag{9.14}$$

Let $x \in A$. Since $D_-f(x) < q$, there exists an increasing sequence $\{\alpha_n(x)\}_{n=1}^\infty$ converging to $x$ such that

$$\frac{f(x) - f(\alpha_n(x))}{x - \alpha_n} < q. \tag{9.15}$$

Clearly, the collection

$$\mathcal{F} := \{[\alpha_n(x), x] : x \in A,\, n \in \mathbb{N}\}$$

is a Vitali covering of $A$. By the Vitali Covering Lemma and the conclusion reached in Step I of its proof, there exists a finite collection $\{I_k\}_{k=1}^n$ of disjoint intervals in $\mathcal{F}$ that are all contained in $G$, whose total length is bigger than $\gamma - \varepsilon$. Let $I_k = [a_k, b_k]$. Then, using (9.14) and (9.15),

$$\sum_{k=1}^n [f(b_k) - f(a_k)] < q \sum_{k=1}^n (b_k - a_k) < q\, m(G) < q(\gamma + \varepsilon). \tag{9.16}$$

Let $B$ be the part of $A$ that is covered by $\{I_k\}_{k=1}^n$. If $x \in B$ then $D^+f(x) > p$, so there exists a decreasing sequence $\{\beta_n(x)\}_{n=1}^\infty$ converging to $x$ such that

$$\frac{f(\beta_n(x)) - f(x)}{\beta_n(x) - x} > p. \tag{9.17}$$

The collection

$$\mathcal{F}' := \{[x, \beta_n(x)] : x \in B, \, n \in \mathbb{N}\} \tag{9.18}$$

is a Vitali covering of $B$. By the Vitali Covering Lemma (and the first part of its proof), there exists a finite collection $\{J_k\}_{k=1}^m$ of disjoint intervals in $\mathcal{F}'$ that are all contained in $\cup\{I_k\}_{k=1}^n$, and whose total length is bigger than $\gamma - 2\varepsilon$. Let $J_k = [c_k, d_k]$. Then, using (9.17),

$$\sum_{k=1}^{m} [f(d_k) - f(c_k)] > p \sum_{k=1}^{n} (d_k - c_k) > p(\gamma - 2\varepsilon). \tag{9.19}$$

The total length of intervals $\{J_k\}$ is less than the total length of $\{I_k\}$ which implies that

$$\sum_{k=1}^{m} [f(d_k) - f(c_k)] < \sum_{k=1}^{n} [f(b_k) - f(a_k)] \, .$$

Combining with (9.16) and (9.19) we obtain $p(\gamma - 2\varepsilon) < q(\gamma + \varepsilon)$. When $\varepsilon \to 0$ this yields $p\gamma \leq q\gamma$. Dividing by $\gamma > 0$ we obtain $p \leq q$ which is a contradiction. □

From Lemmas 9.2.5 and 9.2.7 we obtain a very important result.

**Theorem 9.2.8.** ***Lebesgue's Theorem.*** *If $f$ is monotone on $(a, b)$ then it is differentiable a.e. on $(a, b)$.*

*Proof.* Without loss of generality, we may assume that $f$ is increasing. If $f$ fails to be differentiable at a point then either at least one of the Dini derivatives is infinite, or they are all finite but at least two of them are different. Regarding the former, Lemmas 9.2.5 and Remark 9.2.6 ensure that the set of such points is of measure 0. On the set where all four Dini derivatives are finite, Lemma 9.2.7 shows that

$$D^+ f(x) \leq D_- f(x) \quad \text{almost everywhere.}$$

It follows from the same lemma (Exercise 9.2.13) that

$$D^- f(x) \leq D_+ f(x) \quad \text{almost everywhere.} \tag{9.20}$$

Together with the obvious inequalities $D_+ f(x) \leq D^+ f(x)$ and $D_- f(x) \leq D^- f(x)$ this leads to

$$D^+ f(x) \leq D_- f(x) \leq D^- f(x) \leq D_+ f(x) \leq D^+ f(x) \quad \text{almost everywhere,}$$

and it follows that $f$ is differentiable a.e. on $(a, b)$. □

Lebesgue's Theorem states that there may be a set of measure 0 on which $f$ is not differentiable. Can this result be strengthtened? The answer is no! (See Exercise 9.2.14.)

Lebesgue proved Theorem 9.2.8 towards the end of his 1904 book [69], as a consequence of his whole theory of integration. He also used the hypothesis that the function is continuous. Several authors gave different proofs, without the continuity assumption and independent of integration theory but using the axiom of choice. Among the first were a German mathematician Georg Faber (1877–1966) in 1910 [31] and Young and Young (1911) [126]. The proof we presented is due to Riesz and it is in his 1932 paper [100]. Faber's most important work was on the polynomial expansion of functions. Only in the 1980s was his idea recognized as an important ingredient for the efficient solution of partial differential equations. He verified the conjecture that "given a fixed area of ox-hide to make a drum, the ground tone is lowest if you make your drum circular".

From Lebesgue's Theorem we derive one half of (9.1).

**Corollary 9.2.9.** *Let $f$ be an increasing function on $[a,b]$. Then $f'$ is integrable over $[a,b]$ and*

$$\int_a^b f' \leq f(b) - f(a). \tag{9.21}$$

*Proof.* We extend $f$ to the interval $[b, b+1]$ by $f(x) = f(b)$, for $x \in [b, b+1]$. Then, $f$ is increasing on $[a, b+1]$. By Exercise 9.2.20, it is measurable. By Lebesgue's Theorem,

$$g_n(x) := \frac{f(x+1/n) - f(x)}{1/n} \to f'(x), \quad n \to \infty,$$

for a.e. $x \in (a,b)$. Exercise 6.1.4 implies that $\{g_n\}_{n=1}^{\infty}$ is a sequence of measurable functions, so it follows from Fatou's Lemma that

$$\int_a^b f' \leq \liminf_n \int_a^b g_n. \tag{9.22}$$

On the other hand, each $g_n$ as well as $f$ is Riemann integrable (Exercise 9.2.7) so we can calculate the last integral using standard techniques. We obtain (Exercise 9.2.8):

$$\int_a^b g_n(x)\, dx \leq f(b) - f(a), \tag{9.23}$$

whence the result follows from (9.22). □

**Example 9.2.10. A monotone function for which a strict inequality holds in (9.21).**

Let $\varphi$ denote the Cantor–Lebesgue function. Then $\varphi' = 0$ a.e. so $\int_0^1 \varphi' = 0$. On the other hand, $\varphi(1) - \varphi(0) = 1$. ♦

*Remark* 9.2.11. Example 9.2.10 shows that monotone functions are not the class characterized by (9.1).

Lebesgue's Theorem implies that every monotone function is continuous a.e. In fact, we have a stronger result.

**Theorem 9.2.12.** *Let $f$ be a monotone function on $(a,b)$. Then $f$ has at most a countable number of jump discontinuities on $(a,b)$.*

We leave to the reader the proofs of Theorem 9.2.12 (Exercise 9.2.9) and Theorem 9.2.13 (Exercise 9.2.10) which can be viewed as a converse to the former.

**Theorem 9.2.13.** *Let $\{x_n\}_{n=1}^{\infty}$ be a countable subset of $[a,b]$ and let $\{c_n\}_{n=1}^{\infty}$ be a sequence of positive numbers for which the series $\sum_{n=1}^{\infty} c_n$ converges. Given $x \in [a,b]$, let $E(x)$ be a subset of $\mathbb{N}$ consisting of those integers $n$ for which $x_n < x$ and let*

$$f(x) := \sum_{n \in E(x)} c_n,$$

*and $f(x) = 0$ if $E(x) = \emptyset$. Then:*

*(a) $f$ is a monotone-increasing function on $[a,b]$.*

*(b) $f$ has a jump discontinuity at each $x_n$.*

*(c) The jump at $x = x_n$ is $c_n$.*

*(d) $f$ is continuous at every other point of $[a,b]$.*

WILLAIM HENRY YOUNG (1863–1942) was an English mathematician. Although a son of a grocer, he showed exceptional mathematical talent which was recognized both at the City of London School, by the principal Edwin Abbott (1838–1926), the author of *Flatland*, and in Cambridge. However, he was not focused on mathematics and even won a theology prize. After graduation, he remained at Cambridge, coaching students for mathematical exams. His interest in mathematical research was sparked by his subsequent marriage to Grace Chisholm in 1896, who earned a doctorate herself in Göttingen with Klein as advisor. He independently discovered Lebesgue integration, albeit two years after Lebesgue. In 1915, after living in Göttingen, Italy, and Geneva, the Youngs settled in Lausanne, Switzerland. He established a distinguished teaching career and held part-time positions at Calcutta University and the University of Liverpool but was unable to get a more prestigious position. The recognition came in the form of the membership in the Royal Society, the De Morgan Medal, the Sylvester Medal, and he served as the president of the London Mathematical Society from 1922 to 1924 and the International Mathematical Union from 1929 to 1936. In addition to research papers co-authored with his wife, his most valued contribution is the treatise "The fundamental theorems of the differential calculus" from 1910.

## Exercises

9.2.1. Let $\mathcal{F} = \{[p,q] : 0 \le p < q \le 1,\ p,q \in \mathbb{Q}\}$. Prove that $\mathcal{F}$ is a Vitali cover for $[0,1]$.

9.2.2. Let $E \subset \mathbb{R}$, and for every $x \in E$ let $\{\beta_n(x)\}$ be a strictly decreasing sequence of real numbers that converges to $x$. Prove that $\mathcal{F} = \{[x, \beta_n(x)] : x \in E\}$ is a Vitali cover for $E$.

9.2.3. Let $E \subset \mathbb{R}$ be a union (not necessarily countable and not necessarily disjoint) of nondegenerate intervals (open, closed, half open and half closed, infinite, etc). Show that $E$ is measurable.

9.2.4. Show that the Vitali Covering Lemma extends to the case in which the covering collection consists of nondegenerate intervals that are not necessarily closed.

9.2.5. Show that the Vitali Covering Lemma does not extend to the case in which the covering collection contains some intervals that are not necessarily nondegenerate.

9.2.6. Prove that a set $E \subset \mathbb{R}$ has measure 0 if and only if there is a countable collection of open intervals $\{I_k\}_{k=1}^{\infty}$ for which each point in $E$ belongs to infinitely many of these intervals and $\sum_{k=1}^{\infty} \ell(I_k) < \infty$.

9.2.7. Prove that each function $g_n$ defined in the proof of Corollary 9.2.9 is Riemann integrable.

9.2.8. Prove inequality (9.23).

9.2.9. Prove Theorem 9.2.12.

9.2.10.* Prove Theorem 9.2.13.

9.2.11. Let $E$ be a subset of $\mathbb{R}$ and $A$ a countable subset of $E$. Prove or disprove: There is a monotone function on $E$ that is continuous only at points in $E \setminus A$.

9.2.12. Let $f$ be an increasing function on $[a, b]$. Prove that

$$m^* \left(\{x \in (a, b) : D^- f(x) = +\infty\}\right) = 0.$$

9.2.13. Prove inequality (9.20).

9.2.14.* Let $E \subset [a, b]$ be a set of measure 0. Prove that there exists a continuous increasing function $f$ such that $f'(x) = +\infty$ for all $x \in E$.

9.2.15. Prove or disprove: the derivative of a continuous, monotone function is differentiable a.e.

9.2.16. Let $f$ be continuously on $(a, b)$ and let $E = \{x \in (a, b) : D_- f(x) < D^+ f(x)\}$. Prove that $E$ cannot contain a nondegenerate interval.

9.2.17. Let $f : \mathbb{R} \to \mathbb{R}$ be monotone increasing and let $g(x) = \lim_{h \to 0+} f(x + h)$. Prove that the function $g$ has the following properties:

(a) it is defined everywhere;

(b) it is monotone increasing;

(c) it is right continuous;

(d) it is differentiable a.e.;

(e) $f' = g'$ a.e.

9.2.18.* Let $f : [a, b] \to \mathbb{R}$ be strictly increasing and $E \subset [a, b]$ such that at every point $x \in E$, $D^+ f(x) \leq M$. Prove that $m^* f(E) \leq M m^*(E)$.

9.2.19.* Let $f : [a, b] \to \mathbb{R}$ be strictly increasing and $E \subset [a, b]$ such that at every point $x \in E$, $D_- f(x) \geq C$. Prove that $m^* f(E) \geq C m^*(E)$.

9.2.20. Prove that every monotone function defined on an interval is Borel measurable.

9.2.21. Let $f$ be continuous on $\mathbb{R}$. Prove or disprove: there exists an open interval on which $f$ is monotone.

9.2.22. Let $f$ be differentiable on $\mathbb{R}$, let $x_0 \in \mathbb{R}$, and suppose that $f'(x_0) > 0$. Prove or disprove: there exists an open interval $I$ containing $x_0$ such that $f$ is monotone on $I$.

9.2.23.* **(Fubini)** Let $f_k : [a, b] \to \mathbb{R}$ be monotone increasing functions for $k = 1, 2, \ldots$ and assume that the series $f(x) = \sum_{k=1}^{\infty} f_k(x)$ converges pointwise on $[a, b]$. Prove that $f'(x) = \sum_{k=1}^{\infty} f_k'(x)$ for almost all $x \in [a, b]$.

9.2.24. Let $f_k : (0, 1) \to \mathbb{R}$ be monotone increasing for $k = 1, 2, \ldots$ and suppose that $\lim_n f_n(x) = 1$ for a.e. $x \in (0, 1)$. Prove that $\liminf_n f_n'(x) = 0$ for a.e. $x \in (0, 1)$.

9.2.25. Give an example of functions $f, g : [0, 1] \to \mathbb{R}$ such that $f$ is increasing and continuous, $g$ is a polynomial, and

$$\int_{[0,1]} f g' \neq f(1)g(1) - f(0)g(0) - \int_{[0,1]} f' g.$$

9.2.26.* Prove or disprove: if $f : \mathbb{R}^2 \to \mathbb{R}$ is increasing in each variable then it is Borel measurable.

9.2.27.* Let $\phi$ denote the Cantor-Lebesgue function and set $\phi(x) = 0$ for $x < 0$ and $\phi(x) = 1$ for $x > 1$. Let $\{[a_n, b_n]\}$ be the collection of all closed subintervals of $[0, 1]$ with rational endpoints, let

$$F_n(x) = \phi\left(\frac{x - a_n}{b_n - a_n}\right),$$

and let $G = \sum_1^{\infty} 2^{-n} F_n$. Prove that $G$ is continuous and strictly increasing on $[0, 1]$ and that $G' = 0$ a.e.

9.2.28.** The purpose of this exercise is to establish the so-called Helly Selection Principle: If $\{f_n\}$ is a uniformly bounded sequence of increasing functions on $[0, 1]$ then there exists a subsequence $\{f_{n_k}\}$ that converges pointwise at every point of $[0, 1]$ to an increasing function $f$.

(a) Prove that there exists a subsequence $\{g_n\}$ of $\{f_n\}$ that converges pointwise at every rational point of $[0, 1]$.

(b) Extend $f$ to irrational points in $[0, 1]$ so that it is increasing on $[0, 1]$.

(c) Prove that the subsequence $\{g_n\}$ obtained in (a) converges at every point of continuity of $f$.

(d) Use the approach as in (a) and Theorem 9.2.12 to obtain a subsequence of $\{g_n\}$ that converges at every point of $[0,1]$.

9.2.29.* Let $\{I_n\}$ be a sequence of subintervals of $[0,1]$ with rational endpoints. For each $I_n$, let $E_n$ be the set of all continuous functions on $[0,1]$ that are monotone on $I_n$. Using these sets proves that there exists a continuous function on $[0,1]$ that is not monotone on any subinterval of $[0,1]$.

9.2.30. Let $A \subset [0,1]$ be a set of the first category, and $f : A \to \mathbb{R}$ a strictly monotone function. Prove that $f(A)$ is a set of first categories.

## 9.3 Functions of Bounded Variation

Lebesgue's Theorem shows that every monotone function is differentiable a.e. The same is true if $f$ is a difference of two monotone functions. In this section, our goal is to characterize such functions, namely those that can be represented as a difference of two monotone functions.

**Definition 9.3.1.** Let $f$ be a real-valued function on $[a,b]$ and let $P = \{x_0, x_1, \ldots, x_n\}$ be a partition of $[a,b]$. The **variation** of $f$ with respect to $P$ is

$$V(f,P) = \sum_{k=1}^{n} |f(x_k) - f(x_{k-1})|,$$

and the **total variation** of $f$ on $[a,b]$ is

$$V_a^b(f) = \sup\{V(f,P) : P \text{ is a partition of } [a,b]\}.$$

If $V_a^b(f)$ is finite we say that $f$ is of **bounded variation** on $[a,b]$. We write $f \in BV[a,b]$.

The easiest way to determine whether a specific function has bounded variation is to determine whether it satisfies some of the sufficient conditions for membership in $BV[a,b]$. The following proposition contains one such condition.

**Proposition 9.3.2.** *If $f$ is an increasing function on $[a,b]$ then $f \in BV[a,b]$ and $V_a^b(f) = f(b) - f(a)$.*

*Proof.* By definition, using the fact that $f$ is increasing,

$$\begin{aligned} V(f,P) &= \sum_{k=1}^{n} |f(x_k) - f(x_{k-1})| = \sum_{k=1}^{n} [f(x_k) - f(x_{k-1})] \\ &= f(x_n) - f(x_0) = f(b) - f(a), \end{aligned}$$

so $f$ is of bounded variation and $V_a^b(f) = f(b) - f(a)$. □

**Example 9.3.3. A function of bounded variation.**
Let $f(x) = x^2$ on $[0,1]$. Then $f \in BV[0,1]$ and $V_0^1(f) = 1^2 - 0^2 = 1$. ♦

Here is another sufficient condition for a function to be of bounded variation.

**Proposition 9.3.4.** *If $f$ is Lipschitz on $[a,b]$ then it is of bounded variation on $[a,b]$.*

*Proof.* Let $P = \{x_0, x_1, \ldots, x_n\}$ be a partition of $[a,b]$ and suppose that $f$ is a Lipschitz function with constant $M$. Then

$$V(f,P) = \sum_{k=1}^{n} |f(x_k) - f(x_{k-1})| \leq \sum_{k=1}^{n} M|x_k - x_{k-1}| = M(b-a).$$

Taking supremum over all possible partitions of $[a,b]$ we get $V_a^b(f) \leq M(b-a)$, so $f$ is of bounded variation on $[a,b]$. □

**Example 9.3.5. A function of bounded variation.**
Let $f(x) = \begin{cases} x^2 \sin \frac{\pi}{2x} & \text{if } 0 < x \leq 1, \\ 0 & \text{if } x = 0. \end{cases}$ Then $f \in BV[0,1]$.

**Proof:** For $x \neq 0$, $f'(x) = 2x\sin(\pi/(2x)) - \pi\cos(\pi/(2x))/2$, and $f'(0) = 0$. Therefore, $|f'(x)| \leq 2 + \pi/2$ on $[0,1]$. By Theorem 2.4.20, $f$ is Lipschitz on $[0,1]$ so Proposition 9.3.4 implies that $f \in BV[0,1]$. ♦

When the goal is to establish that a function is *not* of bounded variation it is frequently necessary to exhibit a sequence of partitions $\{P_n\}$ and then demonstrate that $\{V(f,P_n)\}$ is not bounded.

**Example 9.3.6. A continuous function that is not of bounded variation.**
Let

$$f(x) = \begin{cases} x\cos\frac{\pi}{2x} & \text{if } 0 < x \leq 1, \\ 0 & \text{if } x = 0. \end{cases}$$

It is continuous (easy). However, let $n \in \mathbb{N}$ and

$$P_n = \left\{0, \frac{1}{2n}, \frac{1}{2n-1}, \ldots, \frac{1}{3}, \frac{1}{2}, 1\right\}.$$

Now,

$$\left|f\left(\frac{1}{2n}\right) - f(0)\right| = \left|\frac{1}{2n}\cos n\pi\right| = \frac{1}{2n}$$

$$\left|f\left(\frac{1}{2n-1}\right) - f\left(\frac{1}{2n}\right)\right| = \left|\frac{1}{2n-1}\cos\left(\frac{2n-1}{2}\pi\right) - \frac{1}{2n}\cos n\pi\right| = \frac{1}{2n}$$

$$\left|f\left(\frac{1}{2n-2}\right) - f\left(\frac{1}{2n-1}\right)\right|$$

$$= \left|\frac{1}{2n-2}\cos(n-1)\pi - \frac{1}{2n-1}\cos\left(\frac{2n-1}{2}\pi\right)\right| = \frac{1}{2n-2},$$

and so forth. It follows that

$$V(f, P_n) = 1 + 1/2 + \cdots + 1/n \to \infty, \quad \text{as } n \to \infty. \qquad \blacklozenge$$

One of the useful properties of the class of functions of bounded variation is that it is stable under the usual operations. More precisely, we have the following theorem.

**Theorem 9.3.7.** *Let $f, g$ be functions of bounded variation on $[a, b]$. Then the same is true of: (a) $f + g$; (b) $\alpha f$ for any $\alpha \in \mathbb{R}$; (c) $fg$; (d) $f/g$, assuming that $1/g$ is bounded on $[a, b]$.*

*Proof.* We will prove the assertion in (a) and leave the rest to the reader. Let $P = \{x_0, \ldots, x_n\}$ be a partition of $[a, b]$. Then

$$|(f+g)(x_k) - (f+g)(x_{k-1})| \le |f(x_k) - f(x_{k-1})| + |g(x_k) - g(x_{k-1})|,$$

so $V(f+g, P) \le V(f, P) + V(g, P) \le V_a^b(f) + V_a^b(g)$. Taking supremum over all possible partitions of $[a, b]$ we get $V_a^b(f+g) \le V_a^b(f) + V_a^b(g)$, so $f + g$ is of bounded variation on $[a, b]$. □

An important question that is relevant for every interesting class of functions concerns its behaviour under limits. In particular, if $\{f_n\}$ is a sequence in $BV[a, b]$ that converges pointwise to a function $f$, we ask whether $f \in BV[a, b]$. In general, the answer is no.

**Example 9.3.8. A pointwise convergent sequence of functions of bounded variation whose limit is not of bounded variation.**

Let $f_n(x) = \chi_{[1/n,1]}(x)/x$. By Theorem 9.3.7, each $f_n$ is in $BV[0, 1]$. On the other hand, $\{f_n\}$ converges to $f(x) = \chi_{(0,1]}(x)/x$ which is not even bounded so it is not of bounded variation on $[0, 1]$. ♦

Example 9.3.8 shows that the class $BV[0, 1]$ is not closed under pointwise limits. This can be remedied if we include an additional hypothesis.

**Proposition 9.3.9.** *Let $\{f_k\}_{k=1}^{\infty}$ be a sequence of functions of bounded variation on $[a, b]$ and suppose that $V_a^b(f_k) \le M < \infty$ for all $k$. Also, suppose that $f_k \to f$ pointwise on $[a, b]$. Then $f$ is of bounded variation and $V_a^b(f) \le M$.*

*Proof.* Let $P = \{a = x_0, x_1, \ldots, x_N = b\}$ be a partition of $[a, b]$. Then

$$\begin{aligned} V(f, P) &= \sum_{k=1}^{N} |f(x_k) - f(x_{k-1})| = \lim_n \sum_{k=1}^{N} |f_n(x_k) - f_n(x_{k-1})| \\ &= \lim_n V(f_n, P) \le \limsup_n V_a^b(f_n) \le M. \end{aligned}$$

Taking supremum over all possible partitions of $[a, b]$ we get $V_a^b(f) \le M$, whence $f$ is a function of bounded variation on $[a, b]$. □

In addition to the total variation $V_a^b(f)$ it is useful to consider the so-called **total variation function** $V_a^x(f)$, defined for $x \in [a, b]$. We present here one of its important properties—additivity.

**Proposition 9.3.10.** *Let $f : [a, b] \to \mathbb{R}$ and suppose that $a < c < b$. Then $V_a^b(f) = V_a^c(f) + V_c^b(f)$.*

*Proof.* ($\leq$) Let $P$ be a partition of $[a, b]$ and $Q = P \cup \{c\}$. Then $Q = Q' \cup Q''$, where $Q'$ is a partition of $[a, c]$ and $Q''$ is a partition of $[c, b]$. We have that $V(f, P) \leq V(f, Q) = V(f, Q') + V(f, Q'') \leq V_a^c(f) + V_c^b(f)$.

($\geq$) Let $P'$ be a partition of $[a, c]$ and $P''$ a partition of $[c, b]$. Let $P = P' \cup P''$. Then $V(f, P') + V(f, P'') = V(f, P) \leq V_a^b(f)$. □

At the beginning of this section, we set the goal to characterize those functions that can be represented as a difference of two monotone functions. Proposition 9.3.2 shows that every monotone function has a bounded variation. It turns out that the same is true for a difference of two monotone functions and that the converse holds as well.

**Theorem 9.3.11.** ***Jordan's Theorem.*** *A function $f$ is of bounded variation on $[a, b]$ if and only if it is a difference of two increasing functions on $[a, b]$.*

*Proof.*
($\Rightarrow$) First, the total variation function $x \mapsto V_a^x(f)$ is increasing because, if $a \leq u < v \leq b$ then, using Propositon 9.3.10,

$$V_a^v(f) - V_a^u(f) = V_u^v(f) \geq 0.$$

Since $f(x) = [f(x) + V_a^x(f)] - V_a^x(f)$, it suffices to establish that $f(x) + V_a^x(f)$ is an increasing function. To that end, let $a \leq u < v \leq b$. Then,

$$f(u) - f(v) \leq |f(v) - f(u)| = V(f_{[u,v]}, \{u, v\}) \leq V_u^v(f) = V_a^v(f) - V_a^u(f),$$

and it follows that $f(u) + V_a^u(f) \leq f(v) + V_a^v(f)$, so $f(x) + V_a^x(f)$ is increasing.
($\Leftarrow$) Suppose that $f = g - h$ and that $g, h$ are increasing functions. By Exercises 9.3.15 and 9.3.16, $V_a^b(f) \leq V_a^b(g) + V_a^b(h) < \infty$, because each monotone function is of bounded variation (Proposition 9.3.2). □

**Corollary 9.3.12.** *If $f \in BV([a, b])$ then $f$ is differentiable a.e. on $(a, b)$ and $f' \in L([a, b], m)$.*

Functions of bounded variation of a single variable were first introduced by Jordan in 1881 [65]. He generalized the classical result of a German mathematician Lejeune Dirichlet (1805–1859) that *if $f$ is piecewise monotone and has at most finitely many discontinuity points on on $[-\pi, \pi]$ then its Fourier series converges to $[f(x + 0) + f(x - 0)]/2$ for all $x \in [-\pi, \pi]$*. Namely, he showed that the same conclusion holds if $f$ is merely of bounded variation. The study of this class was progressively expanded through several editions of his book *Cours d'analyse*. Dirichlet's family last name is corrupted *le jeune de Richellet* (a youth from Richellet) and it reflects the fact that his grandfather came from Richellet in Belgium. At the age of 20 he produced a proof of Fermat's last theorem for the case $n = 5$. He is credited with being one of the first to give the modern formal definition of a function.

## Exercises

9.3.1. Let $f$ be differentiable on $[a, b]$ with bounded derivative. Prove that $f$ is of bounded variation on $[a, b]$.

9.3.2. Prove that the Dirichlet function is not of bounded variation.

9.3.3. Let $f$ be a function of bounded variation on $[a, b]$ and $[c, d] \subset [a, b]$. Prove that $f$ is of bounded variation on $[c, d]$.

9.3.4. Prove that every function of bounded variation is bounded.

9.3.5. Let $f, g : [a, b] \to \mathbb{R}$ be of bounded variation. Prove that so is $fg$.

9.3.6. Let $f : [a, b] \to \mathbb{R}$ be of bounded variation. Prove that so is $|f|$.

9.3.7. Let $f, g : [a, b] \to \mathbb{R}$ be of bounded variation and suppose that $1/g$ is bounded on $[a, b]$. Prove that $f/g$ is of bounded variation.

9.3.8. Let $f : [a, b] \to \mathbb{R}$ and suppose that $f$ is of bounded variation on every closed subinterval of $(a, b)$. Prove or disprove: $f$ is of bounded variation on $[a, b]$.

9.3.9. Let

$$f(x) = \begin{cases} \sqrt[3]{x} \sin(\pi/x), & \text{for } x \neq 0 \\ 0 & \text{for } x = 0. \end{cases}$$

Determine whether $f$ is of bounded variation on $[0, 1]$.

9.3.10.* Let

$$f(x) = \begin{cases} x^\alpha \sin(x^\beta), & \text{if } 0 < x \leq 1 \\ 0 & \text{if } x = 0. \end{cases}$$

Prove that $f$ is of bounded variation on $[0, 1]$ if and only if $\alpha + \beta > 0$ or both $\alpha + \beta = 0$ and $\beta \geq 0$.

9.3.11. Let $g$ be a function of bounded variation on $[0, 1]$ and let $f$ be a continuous function on $\mathbb{R}$. Prove or disprove: $f \circ g$ is of bounded variation on $[0, 1]$.

9.3.12.* Let $g$ be a function of bounded variation on $[0, 1]$ and let $f$ be of bounded variation on $\mathbb{R}$. Prove or disprove: $f \circ g$ is of bounded variation on $[0, 1]$.

9.3.13. Let $f$ be a function on $[0, 1]$ that is not of bounded variation. Prove that there is a point $x_0 \in [0, 1]$ such that $f$ fails to be of bounded variation on each subinterval of $[0, 1]$ that contains $x_0$ in its interior.

9.3.14.* Give an example of a function that is differentiable on $[0, 1]$ but not of bounded variation on $[0, 1]$.

9.3.15. Let $f, g : [a, b] \to \mathbb{R}$. Prove that $V_a^b(f + g) \leq V_a^b(f) + V_a^b(g)$.

9.3.16. Let $f : [a, b] \to \mathbb{R}$ and let $\alpha \in \mathbb{R}$. Prove that $V_a^b(\alpha f) = |\alpha| V_a^b(f)$.

9.3.17. Let $f : [a, b] \to \mathbb{R}$. Prove that $V_a^b(f) = 0$ if and only if $f$ is constant.

9.3.18. Let $f : [a,b] \to \mathbb{R}$ be a function of bounded variation and let $x_0 \in (a,b)$. Prove that if $f$ is continuous from the left (right) at $x_0$, then so is $V_a^x(f)$.

9.3.19. For a function $f : [a,b] \to \mathbb{R}$ the length of the curve $y = f(x)$, $a \le x \le b$ is defined by

$$L(f) = \sup\left\{\sum_{k=1}^{n} \sqrt{(x_k - x_{k-1})^2 + (f(x_k) - f(x_{k-1}))^2}\right\},$$

where the supremum is taken over all partitions $P = \{x_0, x_1, \dots, x_n\}$ of $[a,b]$. Prove that $L(f)$ is finite if and only $f$ is of bounded variation on $[a,b]$.

9.3.20. Let $f$ be differentiable on $[a,b]$ with derivative $f'$ Riemann integrable over $[a,b]$. Prove that

$$V_a^b(f) = \int_a^b |f'|.$$

What if the integral of $|f'|$ is a convergent improper integral?

9.3.21. Let $f$ be continuous on $[a,b]$ and suppose that $D^+f$ is Riemann integrable over $[a,b]$. Prove that

$$V_a^b(f) = \int_a^b |D^+f|.$$

What if the integral of $|D^+f|$ is a convergent improper integral?

9.3.22. Let $f, g$ be differentiable on $[a,b]$ and let $C > 0$ such that $|f'| \le C|g'|$ on $[a,b]$. Suppose that either $g'$ is Riemann integrable over $[a,b]$ or the integral of $|g'|$ is a convergent improper integral. Prove that $f$ is of bounded variation on $[a,b]$.

In Exercises 9.3.23–9.3.26 find the total variation:

9.3.23. $V_0^{50}(e^x)$.

9.3.24. $V_1^2(\ln x)$.

9.3.25. $V_0^{4\pi}(\cos x)$.

9.3.26. $V_{-1}^1(x - x^3)$.

9.3.27. Let $f, g$ be continuous functions of bounded variation on $[0,1]$. Prove that the set $\{(f(x), g(x)) : x \in [0,1]\}$ cannot be the square $[0,1] \times [0,1]$. What if $f, g$ are merely continuous?

9.3.28.* Let $\{f_n\}$ be a sequence of real-valued functions on $[a,b]$ that converges pointwise on $[a,b]$ to the real-valued function $f$. Prove that

$$V_a^b f \le \liminf V_a^b f_n.$$

9.3.29.* Find all functions $f$ of bounded variation on $[0,1]$ such that

$$f(x) + (V_0^x f)^{1/2} = 1, \quad \text{for all } x \in [0,1], \quad \text{and} \quad \int_0^1 f(x)\,dx = \frac{1}{3}.$$

9.3.30. Let $f$ be a function of bounded variation on a closed and bounded interval $[a, b]$. Prove that there exists a sequence of partitions $\{P_n\}$ of $[a, b]$ for which the sequence $\{V(f, P_n)\}$ is increasing and converges to $V_a^b(f)$.

9.3.31.* Let $f : [a, b] \to \mathbb{R}$ be a continuous function of bounded variation and let $P = \{x_0, x_1, \ldots, x_n\}$ be a partition of $[a, b]$. For $1 \leq k \leq n$, let $M_k$ and $m_k$ denote the maximum and the minimum, respectively, of $f$ on $[x_{k-1}, x_k]$, and $\Omega(f, P) = \sum_{k=1}^{n}(M_k - m_k)$. Prove that

$$\lim_{\|P\|\to 0} V(f, P) = V_a^b f \quad \text{and} \quad \lim_{\|P\|\to 0} \Omega(f, P) = V_a^b f.$$

9.3.32. Prove or disprove: There exists a continuous function that is not of bounded variation on any interval.

9.3.33. Let $f(x) = \sin x$ on $[0, 2\pi]$. Find two increasing functions $h$ and $g$ for which $f = h - g$ on $[0, 2\pi]$.

9.3.34. Let $f(x) = 3x^2 - 2x^3$ on $[-2, 2]$. Find two increasing functions $h$ and $g$ for which $f = h - g$ on $[-2, 2]$.

9.3.35. Let $f(x) = \lfloor x \rfloor - x$ on $[0, 2]$. Find two increasing functions $h$ and $g$ for which $f = h - g$ on $[0, 2]$.

9.3.36. Let $f$ be a function of bounded variation on a closed and bounded interval $[a, b]$. Prove or disprove: $f$ is a measurable function.

9.3.37. Let $f$ be a function of bounded variation on a closed and bounded interval $[a, b]$. Prove that there exist two bounded increasing functions $f_1$ and $f_2$ such that $f = f_1 - f_2$ and $f_1 + f_2 = V_a^b f$.

9.3.38. Let $f$ be a function of bounded variation on a closed and bounded interval $[a, b]$. Prove that $|f'(x)| \leq (V_a^x(f))'$ for a.e. $x \in [a, b]$.

9.3.39. Let $f$ be a function of bounded variation on a closed and bounded interval $[a, b]$ and suppose that $f$ has the intermediate value property, i.e., if $x, y \in [a, b]$ and $C$ is a number between $f(x)$ and $f(y)$ then there is $z \in [x, y]$ such that $f(z) = C$. Prove that $f$ is continuous.

9.3.40. Let $f$ be a continuous function on $[a, b]$ and suppose that $|f|$ has a bounded variation on $[a, b]$. Prove that $f$ has bounded variation on $[a, b]$.

## 9.4 Absolutely Continuous Functions

In this section, we will answer the question raised at the beginning of the chapter, namely what class of functions is characterized by (9.1). We commence by introducing a property that is a special type of continuity.

**Definition 9.4.1.** A real-valued function $f$ on $[a,b]$ is **absolutely continuous** on $[a,b]$ if for every $\varepsilon > 0$ there exists $\delta > 0$ such that, for every collection $\{[a_k, b_k]\}_{k=1}^n$ of closed nonoverlapping intervals in $[a,b]$,

$$\sum_{k=1}^{n}(b_k - a_k) < \delta \quad \Rightarrow \quad \sum_{k=1}^{n}|f(b_k) - f(a_k)| < \varepsilon.$$

We write $f \in AC[a,b]$.

*Remark* 9.4.2. For $n = 1$ this is the definition of uniform continuity. Thus $AC[a,b] \subset UC[a,b] \subset C[a,b]$.

*Remark* 9.4.3. The intervals $\{[a_k, b_k]\}$ are assumed to be nonoverlapping (meaning with disjoint interiors). Some texts require them to be disjoint which is not really necessary. To see why nonoverlapping is important, see Exercise 9.4.14.

In order to determine whether a given function is absolutely continuous it is helpful, just like with functions of bounded variation, to establish a sufficient condition for membership in $AC[a,b]$.

The next result follows from the definition.

**Proposition 9.4.4.** *If $f$ is Lipschitz on $[a,b]$ then it is absolutely continuous on $[a,b]$.*

**Example 9.4.5. An absolutely continuous function.**

Let $f(x) = x^2$ on $[0,1]$. It is easy to see that $f'(x) = 2x$ is a bounded function on $[0,1]$, so $f$ must be Lipschitz, hence absolutely continuous. ♦

The condition that $f$ be Lipschitz is sufficient but not necessary for a function to be absolutely continuous. (See Exercise 9.4.1.) For a more precise relation between these concepts see Exercise 9.4.18.

> The name "absolutely continuous" was coined by Vitali in 1905 [121]. The concept was well known to mathematicians since 1880s. It was introduced by Harnack in [55] as a property of some definite integrals.

When the goal is to establish that a specific function is *not* absolutely continuous, the best strategy is to look at some necessary conditions for a function to belong to $AC[a,b]$. We have seen that (uniform) continuity is such. Here we present another one.

**Theorem 9.4.6.** *If $f$ is an absolutely continuous function on $[a,b]$ then $f$ is of bounded variation on $[a,b]$.*

*Proof.* Let $\varepsilon = 1$ and let $\delta$ be the corresponding positive number given by the definition, i.e.,

$$\sum_{k=1}^{n}(b_k - a_k) < \delta \quad \Rightarrow \quad \sum_{k=1}^{n}|f(b_k) - f(a_k)| < 1. \tag{9.24}$$

Let $P = \{x_0, x_1, \ldots, x_n\}$ be a partition of $[a, b]$ such that $\|P\| < \delta$. Let $P_k$ be any partition of $[x_{k-1}, x_k]$ (see Figure 9.9). It follows from (9.24) that $V(f_{[x_{k-1},x_k]}, P_k) < 1$ so $V_{x_{k-1}}^{x_k}(f) \le 1$, whence $V_a^b(f) \le n$ and $f \in BV[a, b]$. □

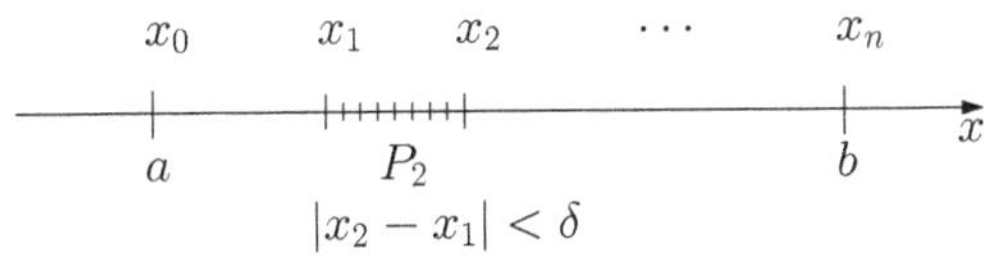


Figure 9.9: $P_2$ is a partition of the interval $[x_1, x_2]$

**Example 9.4.7. A continuous function that is not absolutely continuous.**

Let

$$f(x) = \begin{cases} x \cos \frac{\pi}{2x} & \text{if } 0 < x \le 1, \\ 0 & \text{if } x = 0. \end{cases}$$

It was shown in Example 9.3.6 that $f$ is a continuous function that is not of bounded variation. By Theorem 9.4.6, it cannot be absolutely continuous. ♦

Combining Theorem 9.4.6 and Corollary 9.3.12 we obtain

**Corollary 9.4.8.** *If $f$ is an absolutely continuous function on $[a, b]$ then $f$ is differentiable a.e. on $(a, b)$ and $f' \in L([a, b], m)$.*

Is the converse of Theorem 9.4.6 true? In order to facilitate the answer to this question, we will establish a useful property of absolutely continuous functions.

**Proposition 9.4.9.** *Let $f : [a, b] \to \mathbb{R}$ be absolutely continuous and suppose that $f' = 0$ a.e. on $[a, b]$. Then $f$ is constant.*

*Proof.* It suffices to prove that $f(a) = f(b)$, because the assumptions of the proposition are true on any interval $[a, x] \subset [a, b]$. Let $\varepsilon > 0$. By definition, there exists $\delta > 0$ such that, for every collection $\{[c_k, d_k]\}$ of closed nonoverlapping intervals in $[a, b]$,

$$\sum_{k=1}^{n} (d_k - c_k) < \delta \quad \Rightarrow \quad \sum_{k=1}^{n} |f(d_k) - f(c_k)| < \varepsilon. \tag{9.25}$$

Let $\eta > 0$ and let $E$ be the set of measures $b - a$ on which $f' = 0$. For every $x \in E$ there exists a decreasing sequence $\{\beta_n(x)\}$ that converges to $x$ and satisfies

$$\left| \frac{f(\beta_n) - f(x)}{\beta_n - x} \right| < \eta. \tag{9.26}$$

The collection

$$\mathcal{F} = \{[x, \beta_n(x)] : x \in E,\ n \in \mathbb{N}\}$$

is a Vitali covering of $E$. By the Vitali Covering Lemma, there exists a finite disjoint collection $\{I_k\}_{k=1}^n$ such that the collection $\{J_k\}_{k=1}^m$ of complementing intervals have a total length of less than $\delta$. If we denote $I_k = [a_k, b_k]$ and $J_k = [c_k, d_k]$, using (9.25) and (9.26), as well as the fact that intervals $\{I_k\}_{k=1}^n$ and $\{J_k\}_{k=1}^m$ cover $[a, b]$ and that they have disjoint interiors, we obtain

$$\begin{aligned}
|f(b) - f(a)| &\leq \sum_{k=1}^{n} |f(b_k) - f(a_k)| + \sum_{k=1}^{m} |f(d_k) - f(c_k)| \\
&< \sum_{k=1}^{n} \eta |b_k - a_k| + \varepsilon \\
&\leq \eta |b - a| + \varepsilon.
\end{aligned}$$

Letting $\eta \to 0$ and $\varepsilon \to 0$ yields $f(a) = f(b)$. □

Now it is easy to show that the converse of Theorem 9.4.6 does not hold. Namely, there are functions of bounded variation that do not satisfy the conclusions of Proposition 9.4.9, so they are not absolutely continuous (Exercise 9.4.19).

Next, we introduce the concept of *indefinite* integral.

**Definition 9.4.10.** Let $f$ be an integrable function over $[a, b]$. We say that $F$ is the **indefinite (Lebesgue) integral** of $f$ over $[a, b]$ if

$$F(x) = F(a) + \int_a^x f, \qquad \text{for all } x \in [a, b]. \tag{9.27}$$

It is well known that if a function is Riemann integrable, its indefinite integral must be continuous. Actually, more is true even under a weaker hypothesis that the function are Lebesgue integrable.

**Proposition 9.4.11.** *Let $f : [a, b] \to \mathbb{R}$ be integrable. Then the indefinite integral of $f$ over $[a, b]$ is absolutely continuous.*

*Proof.* Let $\varepsilon > 0$. Since $f \in L([a, b])$ we can apply Theorem 7.3.12 and we obtain that there exists $\delta > 0$ such that if $E \subset [a, b]$ is a measurable set,

$$m(E) < \delta \quad \Rightarrow \quad \int_E |f| < \varepsilon.$$

Let $\{[a_k, b_k]\}$ be a disjoint collection of closed intervals in $[a, b]$ of total length less than $\delta$ and $E = \cup_{k=1}^n [a_k, b_k]$. Then, $m(E) < \delta$ so $\int_E |f| < \varepsilon$. Denoting the indefinite integral of $f$ by $F$, we have

$$\sum_{k=1}^{n} |F(b_k) - F(a_k)| = \sum_{k=1}^{n} \left| \int_{a_k}^{b_k} f \right| \leq \sum_{k=1}^{n} \int_{a_k}^{b_k} |f| = \int_E |f| < \varepsilon,$$

so $F$ is absolutely continuous. □

The following lemma is another important step on the way to establishing the Fundamental Theorem of Calculus for Lebesgue integral.

**Lemma 9.4.12.** *Let $f \in L([a,b])$. Then $f = 0$ a.e. on $[a,b]$ if and only if $\int_{[c,d]} f = 0$, for every $[c,d] \subset [a,b]$.*

*Proof.* It is easy to see that, if $f = 0$ a.e. on $[a,b]$ then $\int_{[c,d]} f = 0$, for every $[c,d] \subset [a,b]$. In the opposite direction, Lemma 7.3.13 shows that it is sufficient to establish that

$$\int_E f = 0, \tag{9.28}$$

for every measurable set $E \subset [a,b]$. Notice that (9.28) holds for every open set $E$, because every open set is a countable disjoint union of intervals. By the continuity of integration, it holds whenever $E$ is a $G_\delta$ set. Finally, Exercise 5.3.9 shows that every measurable subset $E$ of $[a,b]$ is of the form $G \setminus E_0$, where $G$ is a $G_\delta$ set and $m(E_0) = 0$. Thus,

$$\int_E f = \int_G f - \int_{E_0} f = 0 - 0 = 0.$$

□

Now we are ready for the main theorem in this chapter.

**Theorem 9.4.13.** *Let $f : [a,b] \to \mathbb{R}$ be integrable and let $F$ denote the indefinite integral of $f$. Then $F$ is differentiable a.e. on $(a,b)$ and $F' = f$ a.e. on $(a,b)$.*

*Proof.* Proposition 9.4.11 shows that $F$ is absolutely continuous, hence of bounded variation (Theorem 9.4.6) and, by Corollary 9.3.12, differentiable a.e. on $(a,b)$. It remains to demonstrate that $F' = f$ a.e. on $(a,b)$ and for that it suffices to consider the situation when $f$ is nonnegative on $[a,b]$.

**Case I.** $f$ is bounded on $[a,b]$.

Let $|f| \leq M$ on $[a,b]$. Let $\{h_n\}_{n=1}^{\infty}$ be a sequence of positive numbers converging to 0 and

$$g_n(x) := \frac{F(x+h_n) - F(x)}{h_n}, \quad n = 1, 2, \dots.$$

Clearly, $\lim g_n = F'$ a.e. on $[a,b]$. Moreover,

$$|g_n(x)| = \left| \frac{1}{h_n} \int_x^{x+h_n} f \right| \leq \frac{1}{h_n} \int_x^{x+h_n} |f| \leq \frac{1}{h_n} \int_x^{x+h_n} M = M.$$

By Dominated Convergence Theorem, for any $c \in [a,b]$,

$$\int_a^c F' = \int_a^c \lim_n g_n = \lim_n \int_a^c g_n = \lim_n \frac{1}{h_n} \int_a^c (F(x+h_n) - F(x))\, dx.$$

Using Exercise 9.4.22 and the substitution $t = x + h_n$,

$$\int_a^c F(x + h_n)\,dx = \int_{a+h_n}^{c+h_n} F(t)\,dt,$$

whence

$$\int_a^c F' = \lim_n \frac{1}{h_n}\left(\int_{a+h_n}^{c+h_n} F - \int_a^c F\right) = \lim_n \frac{1}{h_n}\left(\int_c^{c+h_n} F - \int_a^{a+h_n} F\right).$$

Notice that

$$\min_{c\le x\le c+h_n} F(x) \le \frac{1}{h_n}\int_c^{c+h_n} F \le \max_{c\le x\le c+h_n} F(x),$$

so the continuity of $F$ implies that

$$\lim_n \frac{1}{h_n}\int_c^{c+h_n} F = F(c),$$

and similarly

$$\lim_n \frac{1}{h_n}\int_a^{a+h_n} F = F(a).$$

Thus,

$$\int_a^c F' = F(c) - F(a) = \int_a^c f.$$

Since this is true for all $c \in [a,b]$, it follows that, for any $[c,d] \subset [a,b]$,

$$\int_c^d (F' - f) = \int_a^d (F' - f) - \int_a^c (F' - f) = 0 - 0 = 0.$$

By Lemma 9.4.12, we have that $F' = f$ a.e. on $[a,b]$.

**Case II.** $f$ need not be bounded.

Let $\{f_n\}_{n=1}^\infty$ be a sequence defined in Example 7.3.11. The function $\int_a^x (f - f_n)$ is increasing because $f - f_n \ge 0$. By Lebesgue's Theorem, it is differentiable a.e. on $[a,b]$ and by Exercise 9.1.20 its derivative cannot be negative. Further, each $f_n$ is bounded so Case I implies that

$$f_n = \frac{d}{dx}\int_a^x f_n.$$

Since

$$F(x) = F(a) + \int_a^x f = F(a) + \int_a^x (f - f_n) + \int_a^x f_n$$

we conclude that $F$ is differentiable a.e. and

$$F'(x) = \frac{d}{dx}\int_a^x f \ge \frac{d}{dx}\int_a^x f_n = f_n \quad \text{a.e. on } [a,b].$$

Letting $n \to \infty$ we obtain that

$$F'(x) \ge f(x) \text{ a.e. on } [a,b]. \tag{9.29}$$

Now,

$$\int_a^b F' \geq \int_a^b f = F(b) - F(a).$$

Combining with Corollary 9.2.9 ($F$ is increasing because $f \geq 0$), we get that

$$\int_a^b (F' - f) = 0,$$

whence the result follows from Proposition 7.2.20 and (9.29). □

Now we get the desired implication in (9.1). The proof is left as Exercise 9.4.23.

**Corollary 9.4.14.** *Let $f : [a, b] \to \mathbb{R}$ be a function that is differentiable a.e. with an integrable derivative. If*

$$\int_a^x f' = f(x) - f(a), \quad \forall x \in [a, b],$$

*then $f$ is absolutely continuous on $[a, b]$.*

As promised earlier, the converse is also true.

**Theorem 9.4.15.** *Let $f \in AC[a, b]$. Then $f$ is differentiable a.e. on $(a, b)$, $f' \in L([a, b])$ and*

$$\int_a^x f' = f(x) - f(a), \quad a \leq x \leq b. \tag{9.30}$$

*Proof.* By Theorem 9.4.6, $f$ is of bounded variation on $[a, b]$, so by Corollary 9.3.12, $f'$ exists a.e. and it is integrable over $[a, b]$.

Since $f'$ is integrable, Proposition 9.4.11 implies that the function

$$g(x) := f(x) - \int_a^x f'$$

is a difference of two absolutely continuous functions, hence absolutely continuous itself (Exercise 9.4.2). By Theorem 9.4.13, $g' = 0$ a.e. on $[a, b]$. Now Proposition 9.4.9 implies that $g$ is constant on $[a, b]$, hence

$$C = f(x) - \int_a^x f'.$$

Taking $x = a$ we obtain (9.30). □

The following result is the crown of our efforts throughout this chapter. The proof is left as Exercise 9.4.24.

**Corollary 9.4.16.** *A function $f$ is absolutely continuous on $[a, b]$ if and only if it is an indefinite integral over $[a, b]$.*

Lebesgue announced Corollary 9.4.16 in a footnote on the last page of his 1904 book [69]. The significance of this result and the class of absolutely continuous functions was recognized by Vitali who gave the proof in 1905 [121]. Lebesgue gave a much simpler proof in 1907 [72].

This result reveals the existence of another counterintuitive class of functions. Suppose that $f$ is a continuous function of bounded variation. By Example 9.2.10, (9.30) need not be true, so let us assume that $f$ does not satisfy (9.30). By Corollary 9.3.12, $f'$ is integrable so the function

$$h(x) = f(x) - f(a) - \int_a^x f' \tag{9.31}$$

is well-defined. It is not hard to see that $h$ is a continuous function of bounded variation (Proposition 9.4.11 and Exercise 9.3.15). Also, Theorem 9.4.13 implies that $h' = 0$ a.e. Yet, $h$ is not a constant function. Indeed, $h(a) = 0$, but $h$ is not the zero function because $f$ does not satisfy (9.30). We give a name to all such functions.

**Definition 9.4.17.** A function $f : [a, b] \to \mathbb{R}$ is called **singular** if it has the following properties:

- $f$ is continuous on $[a, b]$.
- $f$ is of bounded variation on $[a, b]$.
- The derivative of $f$ vanishes almost everywhere.
- $f$ is non-constant on $[a, b]$.

*Remark* 9.4.18. Some authors do not require that a singular function be continuous.

Thus, the function $h$ defined by (9.31) is a singular function, which leads us to the next result.

**Theorem 9.4.19.** *A continuous function $f$ of bounded variation on $[a, b]$ can be represented in a unique way as $f = g + h$ where $g$ is absolutely continuous on $[a, b]$ satisfying $g(a) = f(a)$, and $h$ is either a singular function or the zero function.*

*Proof.* Let

$$g(x) = f(a) + \int_a^x f' \quad \text{and} \quad h = f - g.$$

Then $f = g + h$, $g$ is absolutely continuous (Corollary 9.3.12 and Proposition 9.4.11), $g(a) = f(a)$, and $h$ is singular or the zero function. Indeed, both $f$ and $g$ are continuous and of bounded variation ($f$ by assumption, and $g$ because it is absolutely continuous), so the same is true of their difference $h$. Further, Theorem 9.4.13 implies that $g' = f'$ a.e., so $h' = 0$. Finally, $h(a) = g(a) - f(a) = 0$ so the only way for $h$ to be a constant function is if it is the zero function.

Thus, it remains to establish the uniqueness of decomposition. Suppose that there are $g_1, g_2 \in AC[a,b]$ such that $g_1(a) = g_2(a) = f(a)$, and $h_1, h_2$ such that each of them is either a singular function or a zero function. If

$$f = g_1 - h_1 = g_2 - h_2,$$

then $g_1 - g_2 = h_1 - h_2$ so $(g_1 - g_2)' = (h_1 - h_2)' = 0$ a.e. on $[a,b]$. Since $g_1 - g_2$ is absolutely continuous, Proposition 9.4.9 shows that $g_1 - g_2 = C$. By assumption $g_1(a) = f(a)$ and $g_2(a) = f(a)$ whence $C = 0$. Thus $g_1 = g_2$ and $h_1 = h_2$. □

Lebesgue introduced the concept of a singular function in his 1910 article [73]. (See Exercise 9.4.36.)

## Exercises

9.4.1. Prove that $f(x) = \sqrt{x}$ is absolutely continuous on $[0,1]$ but not Lipschitz.

9.4.2. Let $f, g : [a,b] \to \mathbb{R}$ be absolutely continuous and let $\alpha, \beta \in \mathbb{R}$. Prove that $\alpha f + \beta g$ is absolutely continuous on $[a,b]$.

9.4.3. Let $f, g : [a,b] \to \mathbb{R}$ be absolutely continuous. Prove that $fg$ is absolutely continuous on $[a,b]$.

9.4.4. Let $f : [a,b] \to \mathbb{R}$ be absolutely continuous and suppose that $f(x) \neq 0$ for all $x \in [a,b]$. Prove that $1/f$ is absolutely continuous on $[a,b]$.

9.4.5. Let $g$ be a strictly increasing absolutely continuous function on $[a,b]$ and let $f$ be an absolutely continuous function on $[g(a), g(b)]$. Prove that $f \circ g$ is absolutely continuous on $[a,b]$.

9.4.6. Prove that $f$ is absolutely continuous on $[a,b]$ if and only if for every $\varepsilon > 0$ there exists $\delta > 0$ such that, for every collection $\{[a_k, b_k]\}_{k=1}^{\infty}$ of nonoverlapping closed intervals in $[a,b]$,

$$\sum_{k=1}^{\infty}(b_k - a_k) < \delta \quad \Rightarrow \quad \sum_{k=1}^{\infty}|f(b_k) - f(a_k)| < \varepsilon.$$

9.4.7. Let $\{G_n\}$ be a decreasing sequence of open subsets of $[0,1]$ such that $m(G_n) < 1/2^n$, and let $f(x) = \sum_{n=1}^{\infty} m(G_n \cap [0,x])$. Prove that:

(a) $f \in AC[0,1]$;

(b) $f$ is Lipschitz on $[0,1]$ if and only if there is $N \in \mathbb{N}$ such that $G_n = \emptyset$ for all $n \geq N$.

9.4.8. Let $\{f_n\}$ be a sequence of absolutely continuous functions on $[0,1]$ that converge to a function $f$ uniformly on $[0,1]$. Prove or disprove: $f$ is absolutely continuous.

**Definition 9.4.20.** A real-valued function defined on $[a, b]$ has the **Luzin N property**, if it maps every set of measure zero to a set of measure 0.

9.4.9. Let $f$ be a Lipschitz function on $[a, b]$ with constant $M$ and let $E \subset [a, b]$. Prove that $m(f(E)) \leq Mm(E)$. Conclude that if $f$ is a Lipschitz function on $[a, b]$ then it has the Luzin N property. Is the converse true?

9.4.10. Prove that every Lipschitz function maps measurable sets to measurable sets.

9.4.11. Give an example of a function without the Luzin N property.

9.4.12.* **(Banach–Zarecki)** Prove that a function $f$ on the interval $[a, b]$ is absolutely continuous if and only if it is continuous, is of bounded variation, and has the Luzin N property.

9.4.13. Prove that an absolutely continuous function maps a measurable set to a measurable set.

9.4.14.* Prove that if the word *nonoverlapping* is deleted in Definition 9.4.1 the function must be Lipschitz. More precisely, let $f : [a, b] \to \mathbb{R}$ be a function with the property that for every $\varepsilon > 0$ there exists $\delta > 0$ such that, for every collection $\{[a_k, b_k]\}_{k=1}^n$ of closed intervals in $[a, b]$,

$$\sum_{k=1}^{n}(b_k - a_k) < \delta \quad \Rightarrow \quad \sum_{k=1}^{n}|f(b_k) - f(a_k)| < \varepsilon.$$

Then $f$ is Lipschitz on $[a, b]$.

9.4.15.* Let $f : [0, 1] \to \mathbb{R}$ be an increasing continuous function that is absolutely continuous on $[\varepsilon, 1]$ for each $\varepsilon \in (0, 1)$. Prove that $f$ is absolutely continuous on $[0, 1]$.

9.4.16.* Let $f(x) = \sqrt{x}$ and

$$g(x) = \begin{cases} x^2 \sin\left(\frac{1}{x}\right), & \text{if } x \neq 0, \\ 0, & \text{if } x = 0. \end{cases}$$

Prove that $f$ and $g$ are absolutely continuous on $[0, 1]$ but $h := f \circ g$ is not even of bounded variation on $[0, 1]$.

9.4.17.* Suppose that $f$ is a function on $[a, b]$, and let $V_a^x(f)$ be the total variation function for $f$. Prove that $V_a^x$ is absolutely continuous on $[a, b]$ if and only if $f$ is. Conclude that every absolutely continuous function is the difference of increasing absolutely continuous functions.

9.4.18. Prove that a function $f$ is Lipschitz on $[a, b]$ if and only if it is absolutely continuous function on $[a, b]$ with $|f'|$ essentially bounded. (A function is *essentially bounded* if it is equal a.e. to a bounded function.)

9.4.19. Give an example of a function of bounded variation that is not absolutely continuous.

9.4.20.* Let $f$ be a function of bounded variation on $[a,b]$. Prove that $V_a^b(f) = \int_a^b |f'|$ if and only if $f$ is absolutely continuous on $[a,b]$.

9.4.21. Let $g : [a,b] \to \mathbb{R}$ be a strictly increasing, absolutely continuous function. Prove that for any measurable set $E \subset [a,b]$

$$m(g(E)) = \int_E g'(x)\,dx.$$

9.4.22. Let $f : [a,b] \to \mathbb{R}$, be integrable and let $g : [\alpha,\beta] \to \mathbb{R}$ be a strictly increasing, absolutely continuous function for which $g(\alpha) = a$, $g(\beta) = b$. Prove that

$$\int_a^b f(x)\,dx = \int_\alpha^\beta f(g(t))g'(t)\,dt.$$

9.4.23. Prove Corollary 9.4.14.

9.4.24. Prove Corollary 9.4.16.

9.4.25.* If $f : [a,b] \to \mathbb{R}$ is continuous and real-valued, $f'$ integrable on $[a,b]$, and $\int_a^b f' = f(b) - f(a)$, must $f$ be absolutely continuous? What if $f$ is monotone increasing?

9.4.26. Let $f : \mathbb{R} \to \mathbb{R}$ be a function that is absolutely continuous on each finite interval, and suppose that both $f$ and $f'$ are integrable over $\mathbb{R}$. Prove that:

(a) $\int_{\mathbb{R}} f' = 0$;

(b) $f(x) \to 0$ when $|x| \to \infty$.

9.4.27. Let $\{f_n\} \subset AC[0,1]$ and suppose that

(a) $f = \sum_{n=1}^\infty f_n$ converges for every $x \in [0,1]$;

(b) $\int_0^1 \sum_{n=1}^\infty |f_n'|$ is finite.

Prove that $f \in AC[0,1]$.

9.4.28.* The purpose of this exercise is to establish the following result: If $f$ is differentiable *for all* $x \in [a,b]$ and $f'$ is integrable over $[a,b]$ then (9.30) holds.

(a) Let

$$g_n(x) = \begin{cases} f'(x), & \text{if } f'(x) \le n, \\ n, & \text{if } f'(x) > n, \end{cases} \qquad \text{and} \qquad G_n(x) = f(x) - \int_a^x g_n.$$

Prove that $G_n$ is an increasing function, for each $n \in \mathbb{N}$.

(b) Use the result in (a) to prove that

$$\int_a^b f' \le f(b) - f(a).$$

(c) Replace $g_n$ and $G_n$ in (a) by

$$h_n(x) = \begin{cases} f'(x), & \text{if } f'(x) \geq -n, \\ -n, & \text{if } f'(x) < -n, \end{cases} \quad \text{and} \quad H_n(x) = f(x) - \int_a^x h_n.$$

and show that $\{H_n\}$ is a decreasing sequence.

(d) Use (c) to obtain

$$\int_a^b f' \geq f(b) - f(a).$$

9.4.29. Let $f$ be differentiable on $[a, b]$ with derivative $f'$ integrable over $[a, b]$. Prove that

$$V_a^b(f) = \int_a^b |f'|.$$

9.4.30. Let $f$ be a function on $[a, b]$ for which $D^+f$ is integrable over $[a, b]$. Prove or disprove: $V_a^b(f) = \int_a^b |D^+f|$.

9.4.31. Let $f$ and $g$ be absolutely continuous functions on $[a, b]$. Show that

$$\int_a^b fg' = f(b)g(b) - f(a)g(a) - \int_a^b f'g.$$

9.4.32. Let $f$ be integrable over $\mathbb{R}$. Prove that for a.e. $x \in \mathbb{R}$

$$\lim_{h\to 0+} \frac{1}{h} \int_{(x-h,x+h)} |f - f(x)|\, dm = 0.$$

9.4.33. Let $f$ be integrable over $\mathbb{R}$ and suppose that $f$ is continuous at $x = a$. Prove that the indefinite integral of $f$ is differentiable at $x = a$.

9.4.34. Let $A$ be a measurable subset of $\mathbb{R}$ and suppose that there is $\alpha > 0$ such that $m(A \cap [a, b]) \geq \alpha(b - a)$, whenever $a < b$. Prove that $m(A^C) = 0$

9.4.35. Let $f$ be absolutely continuous and singular on $[a, b]$. Show that $f$ is constant.

9.4.36.* When Lebesgue defined a singular function it was the following way: $F_s$ is the singular function associated with a function of bounded variation $F$, if the difference $F - F_s$ is absolutely continuous and among all the functions $G_s$ such that $F - G_s$ is absolutely continuous, $F_s$ has the smallest total variation, and $F_s(a) = 0$. Prove that the decomposition $F = (F - F_s) + F_s$ is the same as the one provided by Theorem 9.4.19.

9.4.37.* Let $A$ be a measurable subset of $[0, 1]$ such that $0 < m(A \cap I) < m(I)$ for every interval $I \subset [0, 1]$ (see Exercise 5.3.47), and let $F(x) := m([0, x] \cap A)$, $G(x) := m([0, x] \cap A) - m([0, x] \cap A^C)$. Prove that:

(a) $F$ is absolutely continuous and strictly increasing on $[0, 1]$, but $F' = 0$ on a set of positive measure.

(b) $G$ is absolutely continuous on $[0, 1]$, but it is not monotone on any subinterval of $[0, 1]$.

9.4.38. Show that the condition imposed on the set $A$ in Exercise 9.4.37 (a) is not necessary, i.e., that there is a set that does not satisfy it, yet $F(x) := m([0, x] \cap A)$ is an absolutely continuous, strictly increasing function on $[0, 1]$ such that $F' = 0$ on a set of positive measures.

9.4.39.* **(Vitali)** Let $f$ be an increasing function on a closed and bounded interval $[a, b]$. Prove that $f$ is singular if and only if for any $\varepsilon, \delta > 0$ there is a finite disjoint collection $\{(a_k, b_k)\}_{k=1}^{n}$ of open intervals in $(a, b)$ for which

$$\sum_{k=1}^{n}(b_k - a_k) < \delta \quad \Rightarrow \quad \sum_{k=1}^{n}[f(b_k) - f(a_k)] > f(b) - f(a) - \varepsilon.$$

9.4.40. Give an example of a function $f : [a, b] \to \mathbb{R}$ that is differentiable on $[a, b]$ but its derivative is not integrable over $[a, b]$.

9.4.41. Let $f : \mathbb{R} \to \mathbb{R}$ be an increasing function such that $f(-\infty) = 0$ and $f(\infty) = 1$. Prove that $f$ is absolutely continuous on every closed finite interval if and only if $\int_{\mathbb{R}} f' = 1$.

## 9.5 The Radon–Nikodym Theorem

Let $(X, \mathfrak{M}, \mu)$ be a measure space and $f$ a nonnegative measurable function on $X$. Let

$$\nu(E) = \int_E f \, d\mu, \tag{9.32}$$

for $E \in \mathfrak{M}$. We have seen in Example 7.2.16 that $\nu$ is a measure. An important consequence of (9.32) is that if $\mu(E) = 0$ then $\nu(E) = 0$. Such a relationship between the two measures is the main focus of this section. We start with a formal definition.

**Definition 9.5.1.** Let $\mu, \nu$ be measured on a measurable space $(X, \mathfrak{M})$. We say that $\nu$ is **absolutely continuous with respect to** $\mu$, and we write $\nu \ll \mu$, if $\mu(E) = 0$ implies that $\nu(E) = 0$. If $\mu$ and $\nu$ satisfy both $\nu \ll \mu$ and $\mu \ll \nu$ we say that they are **equivalent** measures.

*Remark* 9.5.2. The concept of absolute continuity extends verbatim to the case when $\nu$ is a *signed measure.* (See also Exercise 9.5.1.) The definition extends also to the case when $\mu$ is a signed measure, although we will not be interested in such a scenario. (See Exercise 9.5.5.)

The following proposition gives an equivalent definition of this property.

**Proposition 9.5.3.** *Let $(X, \mathfrak{M}, \mu)$ be a measure space and $\nu$ a finite measure on $(X, \mathfrak{M})$. Then $\nu$ is absolutely continuous with respect to $\mu$ if and only if*

$$\forall \varepsilon > 0, \quad \exists \delta > 0, \quad \forall E \in \mathfrak{M}, \quad \mu(E) < \delta \quad \Rightarrow \quad \nu(E) < \varepsilon. \tag{9.33}$$

*Proof.* ($\Leftarrow$) For $n \in \mathbb{N}$, let $\varepsilon_n = 1/n$ and $\delta_n = \delta(\varepsilon_n)$. If $\mu(E) = 0$ then $\mu(E) < \delta_n$, so $\nu(E) < 1/n$. Since the last inequality holds for every $n \in \mathbb{N}$, we conclude that $\nu(E) = 0$.

($\Rightarrow$) Suppose that (9.33) does not hold. Then

$$\exists \varepsilon_0, \quad \forall \delta > 0, \quad \exists E \in \mathfrak{M}, \quad \mu(E) < \delta \quad \text{and} \quad \nu(E) \geq \varepsilon_0. \tag{9.34}$$

Let $\delta_n = 1/2^n$. By (9.34), there exists a set $E_n \in \mathfrak{M}$ such that $\mu(E_n) < 1/2^n$ and $\nu(E_n) \geq \varepsilon_0$.

Let $A_n = \cup_{k=n}^{\infty} E_k$. Then

$$\mu(A_n) \leq \sum_{k=n}^{\infty} \mu(E_k) < \sum_{k=n}^{\infty} \frac{1}{2^k} = \frac{1}{2^{n-1}}.$$

Let $A = \cap_{n=1}^{\infty} A_n$. The sequence $\{A_n\}$ is decreasing, $\mu(A_1) < 1$, so the continuity of measure implies that $\mu(A) = \lim \mu(A_n) = 0$. On the other hand, $\nu(A_n) \geq \nu(E_n) \geq \varepsilon_0$ and $\nu(A_1)$ is finite by assumption, and it follows from the continuity of measure that $\nu(A) = \lim \nu(A_n) \geq \varepsilon_0$. This shows that $\nu$ is not absolutely continuous with respect to $\mu$. □

*Remark* 9.5.4. It is not coincidental that we are using the same name for a possible relationship between two measures and a property that a function may have. We will postpone the explanation for a moment and return to it later (Theorem 9.5.10).

*Remark* 9.5.5. In many situations there are two measures on the same $\sigma$-algebra and one of them is absolutely continuous with respect to the other. One such example will occur in Chapter 12, in the study of the so-called duality between $L^p$ spaces.

We have seen that if $\nu$ is defined by (9.32) then $\nu \ll \mu$. It turns out that this goes in the opposite direction as well.

**Theorem 9.5.6.** ***The Radon–Nikodym Theorem.*** *Let $(X, \mathfrak{M}, \mu)$ be a $\sigma$-finite measure space and $\nu$ a $\sigma$-finite measure on $(X, \mathfrak{M})$ that is absolutely continuous with respect to $\mu$. Then there exists a nonnegative measurable function $f$ on $X$ such that (9.32) holds for every $E \in \mathfrak{M}$. The function $f$ is unique, i.e., if $g$ has the same property then $f = g$ a.e. $[\mu]$.*

*Proof.* We will assume that measures $\mu, \nu$ are finite and we will treat the general case at the end of the proof. Also, we will assume that $\nu \neq 0$. (Otherwise, take $f = 0$.)

**Step I**: There exists a nonnegative measurable function $f$ on $X$ such that

$$\int_X f\,d\mu > 0 \quad \text{and} \quad \int_E f\,d\mu \le \nu(E), \quad \forall E \in \mathfrak{M}. \tag{9.35}$$

**Proof:** Let $\lambda > 0$. Then $\nu - \lambda\mu$ is a signed measure so $X$ is a disjoint union $X = P_\lambda \cup N_\lambda$, where $P_\lambda$ and $N_\lambda$ are positive and negative sets for this signed measure. We will show that there exists $\lambda > 0$ such that $\mu(P_\lambda) > 0$. Indeed, suppose that $\mu(P_\lambda) = 0$, for every positive $\lambda$. If $E \in \mathfrak{M}$ and $E \subset P_\lambda$ then $\mu(E) = 0$ and $\nu(E) = 0$ because $\nu \ll \mu$. It follows that

$$\nu(E) \le \lambda\mu(E), \tag{9.36}$$

for every measurable set $E \subset P_\lambda$. By definition of $N_\lambda$, (9.36) also holds when $E \subset N_\lambda$, and we conclude that (9.36) is true for every $E \in \mathfrak{M}$. Letting $\lambda \to 0$ we obtain that $\nu(E) = 0$, contradicting our assumption that $\nu \ne 0$. Thus, there exists $\lambda_0 > 0$ such that $\mu(P_{\lambda_0}) > 0$. We define

$$f = \lambda_0 \chi_{P_{\lambda_0}}.$$

Clearly, $\int_X f\,d\mu = \lambda_0\mu(P_{\lambda_0}) > 0$. Also, if $E \in \mathfrak{M}$, then

$$\int_E f\,d\mu = \lambda_0\mu(P_{\lambda_0} \cap E) \le \nu(P_{\lambda_0} \cap E) \le \nu(E),$$

because $\nu - \lambda_0\mu \ge 0$ on any subset of $P_{\lambda_0}$ and $\nu$ is a measure so it is monotone.

**Step II**: There exists a function $f$ that satisfies the conclusions of the theorem.

**Proof:** Let

$$\mathcal{F} = \left\{f : \text{ nonnegative, measurable, and} \int_E f\,d\mu \le \nu(E),\ \forall E \in \mathfrak{M}\right\}, \quad \text{and}$$

$$M = \sup\left\{\int_X f\,d\mu : f \in \mathcal{F}\right\}.$$

By Step I, $\mathcal{F}$ is nonempty, and for any $f \in \mathcal{F}$, $\int_X f\,d\mu \le \nu(X) < \infty$ so $M$ is a (finite) positive number. We will show that there exists $f \in \mathcal{F}$ such that $\int_X f\,d\mu = M$.

Clearly, there exists a sequence $\{f_n\}$ in $\mathcal{F}$ such that $\lim \int_X f_n = M$. Unfortunately, there is no guarantee that the sequence $\{f_n\}$ converges, and even if it does $\int_X \lim f_n$ may be smaller than $M$. Instead, we will define a new sequence by $g_n = \max\{f_1, \dots, f_n\}$. It is increasing, so it converges to a function $f$. Since $g_n \ge f_n$, the Monotone Convergence Theorem implies that

$$\int_X f = \lim \int_X g_n \ge \lim \int_X f_n = M. \tag{9.37}$$

Further, $\{g_n\} \subset \mathcal{F}$. To establish this claim, we must show that if $g, h \in \mathcal{F}$ then $\max\{g, h\} \in \mathcal{F}$. To that end, let $E \in \mathfrak{M}$ and let $E_1 = \{x \in E : g < h\}$, $E_2 = \{x \in E : g \ge h\}$. Then,

$$\int_E \max\{g, h\}\,d\mu = \int_{E_1} h\,d\mu + \int_{E_2} g\,d\mu \le \nu(E_1) + \nu(E_2) = \nu(E),$$

so $\max\{g,h\} \in \mathcal{F}$. It follows that

$$\int_X f = \lim \int_X g_n \leq M.$$

Combining with (9.37), we get $\int_X f = M$. Moreover, $\{g_n\} \subset \mathcal{F}$ so $\int_E g_n \, d\mu \leq \nu(E)$ for all $n \in \mathbb{N}$ and all $E \in \mathfrak{M}$. Passing to the limit and using Monotone Convergence Theorem, we obtain that

$$\int_E f \, d\mu \leq \nu(E), \text{ for all } E \in \mathfrak{M}, \tag{9.38}$$

so $f \in \mathcal{F}$.

It remains to show that equality holds in (9.38). Let $\eta$ be a measure defined by $\eta(E) = \nu(E) - \int_E f \, d\mu$. It is easy to see that $\eta \ll \mu$. Suppose that $\eta \neq 0$. In that case, $\mu$ and $\eta$ (instead of $\nu$) satisfy the hypotheses of the theorem, and Step I would yield a nonnegative measurable function $\hat{f}$ on $X$ satisfying (9.35), with $\eta$ instead of $\nu$. Then we would have that $f + \hat{f} \in \mathcal{F}$ because

$$\int_E (f + \hat{f}) \, d\mu = \int_E f \, d\mu + \int_E \hat{f} \, d\mu \leq \int_E f \, d\mu + \eta(E) = \nu(E).$$

On the other hand, we would have that

$$\int_X (f + \hat{f}) \, d\mu > \int_X f \, d\mu = M,$$

which is impossible. Thus, we must have $\eta = 0$, i.e., $\nu(E) = \int_E f \, d\mu$, for every $E \in \mathfrak{M}$.

**Step III**: $f$ is unique.

Suppose that $\int_E f_1 \, d\mu = \int_E f_2 \, d\mu$, for every $E \in \mathfrak{M}$. Then, $\int_E (f_1 - f_2) \, d\mu = 0$, so $f_1 - f_2 = 0$ a.e. (Lemma 7.3.13).

**Step IV**: The theorem remains true if $\mu$ is $\sigma$-finite.

By definition, $X$ is a countable union of sets $\{X_n\}$ of finite measure $\mu$. We may assume that the sets are disjoint, otherwise, we can replace the sets $\{X_n\}$ by $X_1$, $X_2 \setminus X_1$, $X_3 \setminus (X_1 \cup X_2)$, etc. Since $\mu$ is finite on $X_n$, from what was proved in Step II we obtain a nonnegative measurable function $g_n$ on $X_n$ such that

$$\nu(E) = \int_E g_n \, d\mu,$$

for every measurable set $E \subset X_n$. Moreover, the sets $\{X_n\}$ are disjoint so we can define a function $g$ by $g = g_n$ on $X_n$, i.e., $g = \sum_{n=1}^{\infty} g_n \chi_{X_n}$. Clearly, $g$ is nonnegative and measurable and, if $E \in \mathfrak{M}$,

$$\nu(E) = \nu(E \cap X) = \nu\left(E \cap \bigcup_{n=1}^{\infty} X_n\right) = \nu\left(\bigcup_{n=1}^{\infty} (E \cap X_n)\right) = \sum_{n=1}^{\infty} \nu(E \cap X_n)$$

$$= \sum_{n=1}^{\infty} \int_{E \cap X_n} g \, d\mu = \int_E g \, d\mu.$$

**Step V**: The theorem remains true if both $\mu$ and $\nu$ are $\sigma$-finite.

Let us write $X$ as a countable disjoint union of sets $\{X_n\}$ for which $\nu(X_n) < \infty$ and let us apply Step IV to each $X_n$. That way we obtain, for each $n$, a nonnegative measurable function $h_n$ on $X_n$ such that, if $E$ is a measurable subset of $X_n$,

$$\nu(E) = \int_E h_n \, d\mu.$$

We define a nonnegative measurable function $h$ on $X$ by $h = h_n$ on $X_n$. Then, for any $E \in \mathfrak{M}$,

$$\nu(E) = \nu(E \cap X) = \nu\left(E \cap \bigcup_{n=1}^{\infty} X_n\right) = \nu\left(\bigcup_{n=1}^{\infty} (E \cap X_n)\right) = \sum_{n=1}^{\infty} \nu(E \cap X_n)$$
$$= \sum_{n=1}^{\infty} \int_{E \cap X_n} h \, d\mu = \int_E h \, d\mu. \qquad \square$$

*Remark* 9.5.7. The function $f$ is called the **Radon–Nikodym derivative of $\nu$ with respect to $\mu$**, and it is often denoted by $d\nu/d\mu$.

> In 1913, Radon proved Theorem 9.5.6 [87] in the special case when $\mu$ is a Borel measure on a Euclidean space. The general case was established by Nikodym in 1930 [79].

We proved the Radon–Nikodym theorem for the case when the measures $\mu, \nu$ are $\sigma$-finite. Is it possible to generalize to measures that do not have this property?

**Example 9.5.8. The Radon–Nikodym theorem fails when $(X, \mathfrak{M}, \mu)$ is not a $\sigma$-finite measure space.**

Let $X = [0, 1]$, $\mathfrak{M} = \mathcal{M}$, and $\mu$ the counting measure. Then $m \ll \mu$ but there is no $f \geq 0$ such that $m(E) = \int_E f \, d\mu$.

**Proof.** Indeed, if $\mu(E) = 0$, then $E = \emptyset$, so $m(E) = 0$, and it follows that $m \ll \mu$. To prove that there is no $f$ with desired properties, we argue by contradiction. Let $x_0 \in X$ and $E = \{x_0\}$. Then $m(E) = 0$. On the other hand,

$$\begin{aligned}\int_E f \, d\mu &= \sup\left\{\int_E \varphi \, d\mu : \varphi \leq f\right\} \\ &= \sup\{\varphi(x_0)\mu\{x_0\} : \varphi \leq f\} \\ &= \sup\{\varphi(x_0) : \varphi \leq f\} \\ &= f(x_0).\end{aligned}$$

Thus, $f(x_0) = 0$ and it follows that $f$ is the zero function. From here we obtain a contradiction:

$$1 = m(X) = \int_X 0 \, d\mu = 0. \qquad \blacklozenge$$

*Remark* 9.5.9. The Radon–Nikodym theorem carries over to the case when $\nu$ is a signed measure. In that case, there exists a function $f$ (not necessarily nonnegative) satisfying (9.32) for every $E \in \mathfrak{M}$. We leave the details as Exercise 9.5.28.

We finish this section with a discussion of the connection between the absolute continuity of a function and of a measure.

**Theorem 9.5.10.** *A finite measure $\mu$ on Borel subsets of the interval $[a, b]$ is absolutely continuous with respect to Lebesgue measure $m$ if and only if the function*

$$h(x) = \mu[a, x], \qquad \text{for } x \in [a, b],$$

*is absolutely continuous on $[a, b]$.*

*Proof.* If $\mu \ll m$ the Radon–Nikodym theorem implies that there is a measurable function $g$ such that

$$\mu(E) = \int_E g\,dm, \quad \text{for any measurable set } E. \tag{9.39}$$

In particular, taking $E = [a, x]$ we get that

$$h(x) = \int_a^x g\,dm, \quad \text{for all } x \in [a, b].$$

In other words, $h$ is an indefinite integral of $g$. Since $\mu$ is finite, $g$ must be integrable (take $E = X$ in (9.39)), so $h$ is absolutely continuous by Proposition 9.4.11.

In the other direction, suppose that $h \in AC[a, b]$. Let $\varepsilon > 0$. By Exercise 9.4.6, there exists $\delta > 0$ such that, for any nonoverlapping intervals $\{[x_n, y_n]\}$,

$$\sum_{k=1}^{\infty} |y_k - x_k| < \delta \quad \Rightarrow \quad \sum_{k=1}^{\infty} |h(y_k) - h(x_k)| < \varepsilon.$$

Let $E$ be a measurable subset of $[a, b]$ with $m(E) < \delta/2$. We will show that $\mu(E) < \varepsilon$ and the result will follow from Proposition 9.5.3.

By Exercise 5.3.9, there exists an open set $G \supset E$ such that $m(G) < \delta$. Theorem 3.1.2 implies that $G$ is a disjoint union of open intervals $\{I_n\}$. If we write $I_n = (a_n, b_n)$ then

$$\sum_{n=1}^{\infty} (b_n - a_n) < \delta.$$

Notice that $h(b_n) - h(a_n) = \mu(\,(a_n, b_n]\,)$, and using Exercise 9.5.6 we get that $h(b_n) - h(a_n) = \mu(\,(a_n, b_n)\,)$. Thus,

$$\mu(E) \leq \mu(G) = \sum_{n=1}^{\infty} \mu(\,(a_n, b_n)\,) = \sum_{n=1}^{\infty} (h(b_n) - h(a_n)) < \varepsilon,$$

and the theorem is proved. □

A careful reader has certainly noticed that the function $g$ in the proof of Theorem 9.5.10 is the Radon–Nikodym derivative of $\mu$. Thus, Theorem 9.4.15 is a special case of the Radon–Nikodym theorem and (9.30) is a special case of (9.32). There is a similar relationship between Theorem 9.4.19 and the following result.

**Theorem 9.5.11.** ***The Lebesgue Decomposition Theorem.*** *Suppose that* $(X, \mathfrak{M}, \mu)$ *is a* $\sigma$*-finite measure space and* $\nu$ *a* $\sigma$*-finite measure on* $(X, \mathfrak{M})$*. Then there is a measure* $\nu_0$ *on* $\mathfrak{M}$ *that is singular with respect to* $\mu$*, and a measure* $\nu_1$ *that is absolutely continuous with respect to* $\mu$*, for which* $\nu = \nu_0 + \nu_1$*. The measures* $\nu_0, \nu_1$ *are unique.*

*Proof.* Let $\lambda = \mu + \nu$. Since both $\mu, \nu$ are $\sigma$-finite measures, then so is $\lambda$. Also, $\mu \ll \lambda$ and $\nu \ll \lambda$. By the Radon–Nikodym theorem, there exist nonnegative measurable functions $f$ and $g$ such that

$$\mu(E) = \int_E f \, d\lambda \quad \text{and} \quad \nu(E) = \int_E g \, d\lambda, \tag{9.40}$$

for any $E \in \mathfrak{M}$. Let

$$X_+ = \{x \in X : f(x) > 0\}, \quad X_0 = \{x \in X : f(x) = 0\}.$$

The sets $X_+, X_0$ are measurable disjoint sets and $X = X_0 \cup X_+$. Let

$$\nu_0(E) = \nu(E \cap X_0), \quad \nu_1(E) = \nu(E \cap X_+), \tag{9.41}$$

for $E \in \mathfrak{M}$. Then $\nu_0, \nu_1$ are measures on $\mathfrak{M}$ (Example 5.1.6) and, for any $E \in \mathfrak{M}$,

$$\nu(E) = \nu(E \cap X) = \nu(E \cap (X_0 \cup X_+)) = \nu(E \cap X_0) + \nu(E \cap X_1) = \nu_0(E) + \nu_1(E).$$

Therefore, $\nu = \nu_0 + \nu_1$.

Next, we show that $\nu_0$ is singular with respect to $\mu$. By (9.40) and (9.41),

$$\mu(X_0) = \int_{X_0} f \, d\lambda = 0 \quad \text{and} \quad \nu_0(X_+) = \nu(X_+ \cap X_0) = 0,$$

so $\nu_0 \perp \mu$.

Finally, $\nu_1$ is absolutely continuous with respect to $\mu$. Indeed, suppose that $\mu(E) = 0$. By (9.40), $\int_E f \, d\lambda = 0$ which means that $f = 0$ a.e. $[\lambda]$ on $E$ (Proposition 7.2.20). Consequently, $\lambda(E \cap X_+) = 0$. Further, $\nu \ll \lambda$ so $\nu_1(E) = \nu(E \cap X_+) = 0$, whence $\nu_1$ is absolutely continuous with respect to $\mu$.

Thus, it remains to prove the uniqueness of decomposition. Suppose to the contrary that there is an alternate decomposition $\nu = \nu_1' - \nu_2'$. Then $\nu_1 - \nu_2 = \nu_1' - \nu_2'$, so

$$\nu_1 - \nu_1' = \nu_2 - \nu_2'.$$

By Exercises 9.5.2 and 9.5.3, the signed measure $\nu_1 - \nu_1' = \nu_2 - \nu_2'$ is both absolutely continuous and singular with respect to $\mu$. The result now follows from Exercise 9.5.4. □

JOHANN RADON (1887–1956), was born in Tetschen in the Bohemia region of Austria-Hungary (now part of the Czech Republic). In high school he excelled in all subjects, but chose to study mathematics at the University of Vienna where his teachers were Hahn, Wilhelm Wirtinger (1865–1945), and a Polish mathematician Franz Mertens (1840–1927). Upon receiving a doctorate in 1910 he attended lectures by Hilbert in Göttingen and held several teaching positions. His highly praised habilitation thesis in 1913 was on additive set functions (the old name for measures, see p.114). Due to his nearsightedness, he was exempt from the military service. After the war, he held several university positions eventually settling in Breslau where he remained until 1945 when it became a part of Poland. He worked on the calculus of variations, differential geometry, and measure theory. The *Radon transform* was not initially in the spotlight until Cormack and Hounsfield received the *Nobel prize in Physiology or Medicine* in 1979, reinventing the inverse Radon transform. After WWII, when Breslau became a part of Poland, he spent a year in Innsbruck then returned to his alma mater where he became dean and later president of the University of Vienna. Radon was elected a corresponding member of the Austrian Academy of Sciences in 1939 and a full member in 1947. He was chairman of the Mathematical and Physical Section of the Academy (1952–1956) and the president of the Austrian Mathematical Society (1948–1950). He loved music, and he played music with friends at home, being an excellent violinist himself, and a good singer.

## Exercises

9.5.1. Let $(X, \mathfrak{M}, \mu)$ be a measure space and $\nu$ be a signed measure on $(X, \mathfrak{M})$. Prove that the following are equivalent:

(a) $\nu$ is absolutely continuous with respect to $\mu$.

(b) Both $\nu^+$ and $\nu^-$ are absolutely continuous with respect to $\mu$.

(c) $|\nu|$ is absolutely continuous with respect to $\mu$.

9.5.2. Let $(X, \mathfrak{M}, \mu)$ be a measure space, let $\{\nu_k\}$ be measures on $(X, \mathfrak{M})$, and let $\{a_k\}$ be nonnegative real numbers. Prove that

$$\nu_k \ll \mu \text{ for all } n \in \mathbb{N} \quad \Rightarrow \quad \sum_{k=1}^{\infty} a_k \nu_k \ll \mu.$$

Show that the assertion remains true when $\{a_k\}$ are not necessarily nonnegative if we add the assumption that the series $\sum_{k=1}^{\infty} a_k \nu_k$ converges.

9.5.3. Let $(X, \mathfrak{M}, \mu)$ be a measure space, let $\{\nu_k\}$ be measured on $(X, \mathfrak{M})$, and let $\{a_k\}$ be nonnegative real numbers. Prove that

$$\nu_k \perp \mu \text{ for all } n \in \mathbb{N} \quad \Rightarrow \quad \sum_{k=1}^{\infty} a_k \nu_k \perp \mu.$$

Show that the assertion remains true when $\{a_k\}$ are not necessarily nonnegative if we add the assumption that the series $\sum_{k=1}^{\infty} a_k \nu_k$ converges.

9.5.4. Let $(X, \mathfrak{M}, \mu)$ be a measure space and $\nu$ be a signed measure on $(X, \mathfrak{M})$. Suppose that $\nu$ is absolutely continuous with respect to $\mu$ and that $\mu$ and $\nu$ are mutually singular. Prove that $\nu$ is the zero measure, i.e., $\nu(E) = 0$ for all $E \in \mathfrak{M}$.

9.5.5. Let $(X, \mathfrak{M}, \mu)$ be a measure space and $f_1, f_2$ two integrable functions on $X$. For $k = 1, 2$, let $A_k = \{x \in X : f_k(x) \neq 0\}$, and define signed measures $\nu_1, \nu_2$ by

$$\nu_k(E) = \int_E f_k \, d\mu.$$

(a) Prove that if $\nu_1$ and $\nu_2$ are equivalent then $\mu(A_1 \triangle A_2) = 0$.

(b) Assuming that $f_1, f_2$ are nonnegative prove that $\mu(A_1 \triangle A_2) = 0$ implies the equivalence of $\nu_1$ and $\nu_2$.

(c) Give an example to show that (b) is false without the nonnegativity assumption.

9.5.6. Let $h$ be an absolutely continuous function on $[a, b]$ and suppose that $\mu$ is a measure on Borel subsets of the interval $[a, b]$ such that $\mu[a, x] = h(x)$, for $x \in [a, b]$. Prove that for any $c \in [a, b]$, $\mu(\{c\}) = 0$.

9.5.7. Let $(X, \mathfrak{M}, \mu)$ be a $\sigma$-finite measure space. Prove that there exists a finite measure $\nu$ on $\mathfrak{M}$ that is equivalent to $\mu$.

9.5.8. Let $(X, \mathfrak{M})$ be a measurable space and $\mu$, $\nu$ finite measures on $\mathfrak{M}$. Prove that the following are equivalent:

(a) $\mu \perp \nu$;

(b) $\mu - (\mu - \nu)^+ = 0$;

(c) $\mu + (\nu - \mu)^+ = \mu + \nu$.

9.5.9.* Let $\{r_n\}$ be an ordering of rational numbers, let $f_n : \mathbb{R} \to \mathbb{R}$ be nonnegative, Borel-measurable functions such that $f_n = 0$ outside of the interval $(r_n - 2^{-n}, r_n + 2^{-n})$ and $\int_{\mathbb{R}} f_n = 1$, and let $\mu(E) = \int_E \sum_n f_n$ for all $E \in \mathcal{B}$. Prove that:

(a) $\sum_n f_n(x)$ is finite a.e.

(b) $\mu$ is a $\sigma$-finite measure on $\mathcal{B}$.

(c) $\mu(G) = \infty$ for every nonempty, open set $G$.

(d) $\mu \ll m$.

9.5.10. Give an example to show that Proposition 9.5.3 fails if $\nu$ is a $\sigma$-finite measure.

9.5.11. Give an example of 2 $\sigma$-algebras $\mathfrak{M}$ and $\mathfrak{N}$ on a set $X$ and 2 measures $\mu$ and $\lambda$ such that $\lambda \ll \mu$ on $(X, \mathfrak{M})$ but not on $(X, \mathfrak{N})$.

9.5.12.* Let $\mu$ be a nonzero finite Borel measure on $\mathbb{R}$ and for every $t \in \mathbb{R}$, let $\mu_t$ be a measure defined by $\mu_t(A) = \mu(A + t)$. Suppose that each measure $\mu_t$ is equivalent to $\mu$. Prove that $\mu$ is equivalent to the Lebesgue measure on $\mathbb{R}$.

9.5.13. Prove or disprove: there exists a measure $\nu$ on Lebesgue $\sigma$-algebra $\mathcal{M}$ of subsets of $\mathbb{R}$ that is absolutely continuous with respect to Lebesgue measure and $\nu(A) = \infty$ for any $A \in \mathcal{M}$ with a nonempty interior.

9.5.14. Let $(X, \mathfrak{M}, \mu)$ be a finite measure space, let $\{E_k\}_{k=1}^n$ a collection of measurable sets, and $\{c_k\}_{k=1}^n$ a collection of real numbers. For $E \in \mathfrak{M}$ define

$$\nu(E) = \sum_{k=1}^{n} c_k \, \mu(E \cap E_k).$$

(a) Prove that $\nu$ is a measure on $(X, \mathfrak{M})$;

(b) prove that $\nu$ is absolutely continuous with respect to $\mu$;

(c) find the Radon–Nikodym derivative $d\nu/d\mu$.

9.5.15. Let $\mu$, $\nu$, and $\lambda$ be $\sigma$-finite measures on the measurable space $(X, \mathfrak{M})$.

(a) If $\nu \ll \mu$ and $f$ is a measurable function on $X$, prove that

$$\int_X f \, d\nu = \int_X f \frac{d\nu}{d\mu} \, d\mu.$$

(b) If $\nu \ll \mu$ and $\lambda \ll \mu$ prove that

$$\frac{d(\nu + \lambda)}{d\mu} = \frac{d\nu}{d\mu} + \frac{d\lambda}{d\mu}, \qquad \text{a.e. } [\mu].$$

(c) If $\nu \ll \mu \ll \lambda$, prove that

$$\frac{d\nu}{d\lambda} = \frac{d\nu}{d\mu} \frac{d\mu}{d\lambda}, \qquad \text{a.e. } [\lambda].$$

(d) If $\nu \ll \mu$ and $\mu \ll \nu$, prove that

$$\frac{d\nu}{d\mu} \frac{d\mu}{d\nu} = 1, \qquad \text{a.e. } [\mu].$$

9.5.16. Let $\mu_1, \nu_1$ be measured on a measurable space $(X, \mathcal{S})$ and let $\mu_2, \nu_2$ be measured on a measurable space $(Y, \mathcal{T})$. Suppose that all 4 measures are $\sigma$-finite and that $\nu_k \ll \mu_k$, $k = 1, 2$.

(a) Prove that $\nu_1 \times \nu_2 \ll \mu_1 \times \mu_2$.

(b) Find $\frac{d(\nu_1 \times \nu_2)}{d(\mu_1 \times \mu_2)}$.

9.5.17.* Let $(X, \mathfrak{M})$ be a measurable space and $\mu$, $\nu$ $\sigma$-finite measures on $\mathfrak{M}$ such that $\nu \ll \mu$, let $\lambda = \mu + \nu$, and let $f = d\nu/d\lambda$. Prove that $0 \le f < 1$ a.e. $[\mu]$ and $d\nu/d\mu = f/(1 - f)$.

9.5.18. Let $(X, \mathfrak{M}, \mu)$ be a measure space, $\mathfrak{N}$ a subalgebra of $\mathfrak{M}$, $\nu$ a restriction of $\mu$ to $\mathfrak{N}$, and $f$ a $\mu$-integrable function on $X$. Prove that:

(a) there exists a $\nu$-integrable function function $g$ on $X$ such that $\int_E f\,d\mu = \int_E g\,d\nu$ for all $E \in \mathfrak{N}$;

(b) $g$ is unique, meaning that if $h$ satisfies (a) then $g = h$ a.e. $[\nu]$.

9.5.19. Let $m$ be Lebesgue measure on the real line and for each Lebesgue measurable subset $E$ of $\mathbb{R}$ define

$$\mu(E) = \int_E \frac{1}{1+x^2}\,dm.$$

Show that $m$ is absolutely continuous with respect to $\mu$, and compute the Radon–Nikodym derivative $dm/d\mu$.

9.5.20. Let $m$ denote the Lebesgue measure on $[0, +\infty]$ and

$$\mu_1(E) = \sum_{n=1}^{\infty} \frac{1}{n^3} \int_{E\cap[n,n+1]} x\,dm, \quad \mu_2(E) = \int_{E\cap[1,+\infty]} \frac{1}{x^2}\,dm,$$

for any Lebesgue measurable set $E \subset [0, +\infty]$. Prove or disprove:

(a) $m \ll \mu_2$;

(b) $\mu_2 \ll \mu_1$;

(c) $\mu_1 \ll \mu_2$.

Find the Radon–Nikodym derivative in the cases when it exists.

9.5.21. Let $(X, \mathfrak{M})$ be a measurable space and $\mu$, $\nu$ measures on $\mathfrak{M}$ such that $\mu$ is $\sigma$-finite and $\nu \ll \mu$, let $f = d\nu/d\mu$, and suppose that there is a constant $C > 0$ such that $\nu(E) > C\mu(E)$ for every set $E$ of finite measure. Prove that $f > 0$ for $\mu$–a.e. $x \in X$.

9.5.22. Let $\nu$ be a finite signed measure on a measurable space $(X, \mathfrak{M})$. Prove that there exists a measurable function $f$ on $X$ such that $|f(x)| = 1$ for a.e. $x \in X$, and

$$\nu(E) = \int_E f\,d|\nu|, \quad \text{for any } E \in \mathfrak{M}.$$

9.5.23.* Let $(X, \mathfrak{M})$ be a measurable space and let $\mu, \nu$ be two $\sigma$-finite measures on $\mathfrak{M}$ such that $\mu \ll \nu$ and there is a set $A \in \mathfrak{M}$ which is a $\nu$-atom (meaning that for any measurable subset $B \subset A$ we have either $\nu(B) = 0$ or $\nu(A \setminus B) = 0$). Determine the values of the Radon-Nikodym derivative $d\mu/d\nu$ on $A$.

9.5.24. Let $X$ be an uncountable set, and let $\mathfrak{M}$ consist of those sets $A$ such that either $A$ or $A^C$ is countable. Define two measures $\mu, \nu$ on $\mathfrak{M}$ by $\mu(A) = \nu(A) = 0$ if $A$ is countable, and $\mu(A) = 1$, $\nu(A) = \infty$ if $A^C$ is countable.

(a) Find a measurable function $f : X \to [0, +\infty]$ such that $\nu(E) = \int_E f\,d\mu$, for any $E \in \mathfrak{M}$.

(b) Prove that there does not exist a measurable function $g : X \to [0, +\infty]$ such that $\mu(E) = \int_E g\,d\nu$, for any $E \in \mathfrak{M}$.

9.5.25. Let $(X, \mathfrak{M}, \mu)$ be a measure space and let $\lambda, \nu$ be finite signed measures on $\mathfrak{M}$. Prove that $\lambda \perp \nu$ if and only if $|\lambda| \perp |\nu|$.

9.5.26. Let $(X, \mathfrak{M})$ be a measurable space, let $E \in \mathfrak{M}$, and let $\nu$ be a signed measure on $\mathfrak{M}$. Prove that

$$|\nu|(E) = \sup\left\{\left|\int_E f\,d\nu\right| : |f| \le 1\right\}.$$

9.5.27. Let $\mu$ and $\nu$ be two $\sigma$-finite positive measures on a measurable space $(X, \mathfrak{M})$. Prove that there exists $A, B \in \mathfrak{M}$ such that $A \cap B = \emptyset$, $A \cup B = X$, $\mu$ and $\nu$ are equivalent on $(A, A \cap \mathfrak{M})$ and $\mu \perp \nu$ on $(B, B \cap \mathfrak{M})$.

9.5.28. Prove Radon–Nikodym theorem in the case when $\nu$ is a signed measure.

9.5.29. Let $(X, \mathfrak{M}, \mu)$ be a measure space and $f, g : X \to [0, +\infty]$ measurable functions such that the associated measures $\nu(E) = \int_X f\,d\mu$ and $\lambda(E) = \int_X g\,d\mu$ are equal on $\mathfrak{M}$. Prove or disprove: $f = g$ a.e. $[\mu]$.

9.5.30. Give an example that the Radon–Nikodym theorem fails when $\nu$ is not $\sigma$-finite.

9.5.31. Prove or disprove: The Lebesgue Decomposition Theorem remains true if measures $\mu$ and $\nu$ are not assumed to be $\sigma$-finite.

9.5.32. Show that the Lebesgue decomposition of a function of bounded variation is unique if the singular function is required to vanish at $x = a$.

9.5.33. Let $(X, \mathfrak{M}, \mu)$ be a measure space and $f$ a nonnegative function that is integrable over $X$ with respect to $\mu$. Find the Lebesgue decomposition with respect to $\mu$ of the measure $\nu$ defined by $\nu(E) = \int_E f\,d\mu$, for $E \in \mathfrak{M}$.

9.5.34. Let $(X, \mathfrak{M})$ be a measurable space and $\mu$, $\nu$ two $\sigma$-finite measures on $\mathfrak{M}$. Prove that $\mu \perp \nu$ if and only if there is no nonzero $\sigma$-finite measure $\lambda$ on $(X, \mathfrak{M})$ such that $\lambda \ll \mu$, $\lambda \ll \nu$.

9.5.35. Let $\{a_n\}$ be a fixed sequence of real numbers and let $\{p_n\}$ be a sequence of positive real numbers. Define a set function $\nu : \mathcal{M} \to \mathbb{R}$ by

$$\nu(E) = \sum_{a_n \in E} p_n, \qquad \text{for } E \in \mathcal{M}.$$

(a) Show that $\nu$ is a $\sigma$-finite measure on $\mathcal{M}$.

(b) Find the Lebesgue decomposition of $\nu$ with respect to the Lebesgue measure $m$.

9.5.36. Let $f(x) = x^2 - 5x + 6$, and define a set function $\nu : \mathcal{B} \to \mathbb{R}$ by

$$\nu(E) = \int_E f\,dm, \qquad \text{for } E \in \mathcal{M}.$$

(a) Show that $\nu$ is a $\sigma$-finite signed measure on $\mathcal{B}$;

(b) Find the Hahn decomposition of $\mathbb{R}$ with respect to $\nu$;

(c) Find the Jordan decomposition of $\nu$;

(d) Find the Lebesgue decomposition of $\nu$ with respect to the Lebesgue measure on $\mathcal{M}$.

9.5.37. Let $(X, \mathfrak{M})$ and $(Y, \mathfrak{N})$ be measurable spaces, let $\mu, \nu$ be measures on $\mathfrak{M}$, let $\mu', \nu'$ be measures on $\mathfrak{N}$, and suppose that $\mu$ and $\mu'$ are $\sigma$-finite. Determine the Lebesgue decomposition of $\mu \times \mu'$ with respect to $\nu \times \nu'$, in terms of the Lebesgue decompositions of $\mu$ with respect to $\nu$ and $\mu'$ with respect to $\nu'$.

9.5.38. Prove or disprove: the Lebesgue Decomposition Theorem holds when $\nu$ is a signed measure.

# Part IV

# An Introduction to Functional Analysis

# 10

# *Banach Spaces*

We have seen that if $(X, \mu)$ is a measure space and $f, g$ are integrable functions over $X$ then so is $\alpha f + \beta g$, for $\alpha, \beta \in \mathbb{R}$. In other words, $L(X, \mu)$ is a vector space. The same is true, but harder to prove, for the set of functions $f$ for which $|f|^p$ is integrable when $p > 1$. Functions that are continuous on a given set also form a vector space. A very interesting special case is when $X = \mathbb{N}$ and the functions are sequences. In this chapter, we will present a unified approach to the study of the mentioned vector spaces. Namely, we will introduce a norm and the metric properties of the norm will allow us to define a topology. Then we will explore the topological properties of such vector spaces, such as completeness and separability. In connection, with the latter, we will describe some countable dense sets (when they exist).

## 10.1 Normed Linear Spaces

In Chapter 2 we briefly studied the $n$-dimensional Euclidean space $\mathbb{R}^n$ which is a vector space with a norm. However, in this course we are more interested in vector spaces whose elements are functions. It is well known that if $f, g$ are functions on $X$ and $\alpha \in \mathbb{R}$ then $f + g$ and $\alpha f$ are also functions on $X$, defined by

$$(f+g)(x) = f(x) + g(x), \quad \text{and} \quad (\alpha f)(x) = \alpha f(x).$$

Inspired by $\mathbb{R}^n$, we would like to define a norm on a space of functions. Of course, the first step is to articulate what properties such a norm must have. We model our definition of the norm on the Euclidean norm and Theorem 2.1.6.

**Definition 10.1.1.** Let $X$ be a vector space. A function $\| \; \| : X \to \mathbb{R}$ is a norm on $X$ if, for any $x, y \in X$ and $\alpha \in \mathbb{R}$,

(i) $\|x\| \geq 0$ and $\|x\| = 0$ if and only if $x = 0$

(ii) $\|\alpha x\| = |\alpha| \|x\|$

(iii) $\|x + y\| \leq \|x\| + \|y\|$

DOI: 10.1201/9781032716497-10

The pair $(X, \| \|)$ is called a **normed linear space** or a **normed vector space**. Most of the time we will call it just a **normed space**.

> The symbol $\| \|$ for a norm was used by a German mathematician Schmidt in 1908 [108]. He did not call it a norm but refered to vectors of norm 1 as "normirt" (which one may translate as "normalized"). Gauss in 1832 referred to $a^2 + b^2$ as "norma" of $a + ib$. The word "norm" and Definition 10.1.1 appear in Banach's article [2] from 1922.

We will present several normed spaces and leave it to the reader to verify that the proposed norms satisfy conditions (i)-(iii) of Definition 10.1.1.

**Example 10.1.2. The space $\ell^1$.**
Let $X$ be the set of sequences $\{a_n\}_{n=1}^{\infty}$, where $a_n \in \mathbb{R}$, for all $n \in \mathbb{N}$. Define a norm by $\|a\|_1 = \sum_{n=1}^{\infty} |a_n|$. The space $\ell^1$ consists of all sequences for which $\|a\|_1$ is finite. (See Exercise 10.1.1.) ♦

**Example 10.1.3. The space $\ell^\infty$.**
Let $X$ be the set of sequence $\{a_n\}_{n=1}^{\infty}$, where $a_n \in \mathbb{R}$, for all $n \in \mathbb{N}$. Define a norm by $\|a\|_\infty := \sup_{n\in\mathbb{N}} |a_n|$. The space $\ell^\infty$ consists of all sequences for which $\|a\|_\infty$ is finite. (See Exercise 10.1.2.) ♦

**Example 10.1.4. The space $C[a, b]$.**
This is the vector space of continuous functions on the interval $[a, b]$. The norm is given by $\|f\|_{max} = \max_{x\in[a,b]} |f(x)|$. (See Exercise 10.1.3.) ♦

> The space $C[a, b]$ with the metric $d(f, g) = \max\{|f(x) - g(x)| : x \in [a, b]\}$ was one of the examples in the seminal work of Fréchet [35] in 1906. Riesz studied the same space in a work from 1918 [96] and used the word norm and its symbol for $\|f\| = \sup |f(x)|$. The paper was dated as 1916, but the completed volume appeared only in 1918, delayed by the circumstances of World War I. A Hungarian version of the paper appeared in 1917. Although Riesz exclusively used the space $C[a, b]$, he stated that almost all of his results remain true in abstract Banach spaces.

We make an important observation. If $X$ is a normed space, then it is automatically a metric space, with the metric defined by $d(x, y) = \|x - y\|$. However, the converse is not true.

**Example 10.1.5. A metric on a vector space that does not induce a norm.**
Let $X = \mathbb{R}$ and $d(x, y) = |x - y|^{1/2}$. Then $d$ is a metric, but $\|x\| := d(x, 0)$ does not define a norm.

**Proof:** It is straightforward to verify that $d$ is a metric. (See Exercise 2.1.11.) On the other hand, equality $\|x\| = d(x, 0)$ does not define a norm, because we would have

$$\|4x\| = d(4x, 0) = |4x|^{1/2} = 2\sqrt{x} = 2d(x, 0) = 2\|x\|.$$ ♦

Since every normed space is a metric space (Figure 10.1), everything said about metric spaces in Chapter 2 carries over to normed spaces. In particular, the norm can be used to define a topology, which we will refer to as **norm topology**. Many interesting normed spaces are *complete* in this topology (see

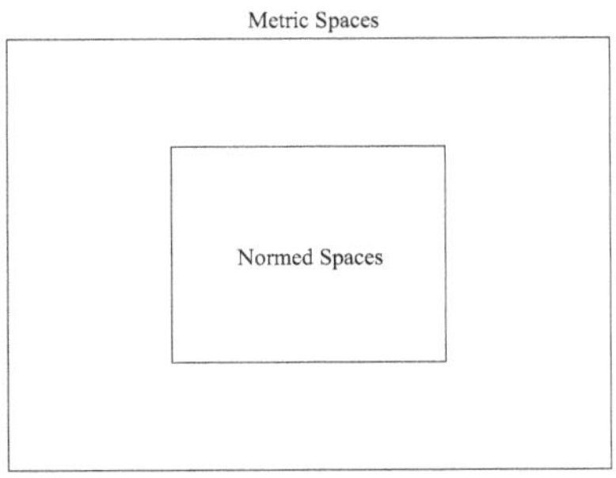


Figure 10.1: Every normed space is a metric space

Definition 2.2.10). This is a desirable property because it guarantees, for example, that if a sequence of functions converges then the limit function belongs to the same class. Throughout the rest of this monograph the emphasis is going to be on complete normed spaces.

**Definition 10.1.6.** A complete normed vector space is called a **Banach space.**

*Remark* 10.1.7. By definition, when establishing the completeness of a normed space we need to show that every Cauchy sequence *converges in norm topology,* or simply *converges in the norm.* Equivalently, we need to show that, given a Cauchy sequence $\{x_n\}_{n=1}^{\infty}$ in a normed linear space $X$, there exists $x \in X$ such that $\lim_n \|x - x_n\| = 0$.

We already know that $\mathbb{R}^n$ is a Banach space (Example 2.2.12). Here is a few more.

**Example 10.1.8.** $C[a, b]$ **is a Banach space.**

The fact that $C[a, b]$ is a normed vector space is the content of Example 10.1.4. We will show that with respect to the norm

$$\|f\|_{max} := \max\{|f(x)| : x \in [a, b]\}$$

$C[a, b]$ is complete. If $\{f_n\}$ is a Cauchy sequence in $C[a, b]$ then

$$\forall \varepsilon > 0 \quad \exists N \in \mathbb{N} \quad m \geq n \geq N \quad \Rightarrow \quad \|f_m - f_n\|_{max} < \varepsilon.$$

Let $x_0 \in [a, b]$. Then

$$|f_m(x_0) - f_n(x_0)| \leq \|f_m - f_n\|_{max} < \varepsilon, \tag{10.1}$$

so $\{f_n(x_0)\}$ is a Cauchy sequence of real numbers, hence convergent. Let $y_0 = \lim_n f_n(x_0)$. Now we have a function $f : [a, b] \to \mathbb{R}$ that to each $x_0 \in [a, b]$ assigns $y_0 \in \mathbb{R}$. Letting $m \to \infty$ in (10.1), we obtain $|f(x_0) - f_n(x_0)| \leq \varepsilon$, for all $x_0 \in [a, b]$ and all $n \geq N$. Therefore, $\{f_n\}$ converges uniformly to a function $f$ that is necessarily continuous. ♦

**Example 10.1.9. $c$ is a Banach space.**

Let $c$ denote the normed space of convergent sequences of real numbers with norm

$$\|(x_1, x_2, x_3, \dots)\| = \sup\{|x_n| : n \in \mathbb{N}\}. \tag{10.2}$$

This is a Banach space.

**Solution:** We leave the verification that $c$ is a vector space and that (10.2) defines a norm to the reader (Exercise 10.1.8). To establish the completeness let $\{\mathbf{x}_n\}$ be a Cauchy sequence in $c$. By definition, for every $\varepsilon > 0$ there is $N \in \mathbb{N}$ such that if $m \geq n \geq N$ then $\|\mathbf{x}_m - \mathbf{x}_n\| < \varepsilon/2$. In view of (10.2), the last inequality implies that for each $k \in \mathbb{N}$, $|\mathbf{x}_m(k) - \mathbf{x}_n(k)| < \varepsilon/2$. Thus, for each $k$, $\{\mathbf{x}_n(k)\}$ is a Cauchy sequence of real numbers, hence convergent. Let $\mathbf{x}(k) = \lim_n \mathbf{x}_n(k)$.

First, we need to establish that $\{\mathbf{x}_n\}$ converges to $\mathbf{x}$ in the norm of $c$. Since $|\mathbf{x}_m(k) - \mathbf{x}_n(k)| < \varepsilon/2$ for all $k \in \mathbb{N}$ and all $m \geq n \geq N$, letting $m \to \infty$ yields $|\mathbf{x}(k) - \mathbf{x}_n(k)| \leq \varepsilon/2 < \varepsilon$ for all $k \in \mathbb{N}$ and all $n \geq N$, whence $\lim_n \|\mathbf{x} - \mathbf{x}_n\| = 0$

We also need to prove that $\mathbf{x}$ is a convergent sequence, i.e., that $\mathbf{x} \in c$. Let $L_n = \lim_{k\to\infty} \mathbf{x}_n(k)$. Then $\{L_n\}$ is a Cauchy sequence, assuming that $\{\mathbf{x}_n\}$ is. Indeed, let $\varepsilon > 0$. As we noticed, there is $N \in \mathbb{N}$ such that if $m \geq n \geq N$ then $\|\mathbf{x}_m(k) - \mathbf{x}_n(k)\| < \varepsilon/3$, for all $k$. Let $m \geq n \geq N$ be fixed and choose $k$ so that $|\mathbf{x}_m(k) - L_m| < \varepsilon/3$ and $|\mathbf{x}_n(k) - L_n| < \varepsilon/3$. Then,

$$|L_m - L_n| \leq |\mathbf{x}_m(k) - L_m| + |\mathbf{x}_m(k) - \mathbf{x}_n(k)| + |\mathbf{x}_n(k) - L_n| < \frac{\varepsilon}{3} + \frac{\varepsilon}{3} + \frac{\varepsilon}{3} = \varepsilon.$$

Thus, $\{L_n\}$ is a Cauchy sequence hence convergent. Let $L = \lim_n L_n$. We will show that $\lim_{k\to\infty} \mathbf{x}(k) = L$. Let $\varepsilon > 0$. There is $n \in \mathbb{N}$ for which $\|\mathbf{x} - \mathbf{x}_n\| < \varepsilon/3$ and $|L - L_n| < \varepsilon/3$. For this $n$ we have that $\lim_{k\to\infty} \mathbf{x}_n(k) = L_n$, so there is $K \in \mathbb{N}$ such that if $k \geq K$ then $|\mathbf{x}_n(k) - L_n| < \frac{\varepsilon}{3}$. For such $k$, we have that

$$|\mathbf{x}(k) - L| \leq |\mathbf{x}(k) - \mathbf{x}_n(k)| + |\mathbf{x}_n(k) - L_n| + |L_n - L| < \frac{\varepsilon}{3} + \frac{\varepsilon}{3} + \frac{\varepsilon}{3} = \varepsilon.$$

Thus, $\lim_{k\to\infty} \mathbf{x}(k) = L$ whence $\mathbf{x} \in c$. ♦

**Example 10.1.10. A normed space that is not a Banach space.** Let $(X, \| \|_{max})$ be the normed space of all polynomials on $[a, b]$. Then $X$ is not a Banach space.

**Solution:** Let $f$ be a continuous function that is not a polynomial. (E.g., $f(x) = e^x$.) By Weierstrass Theorem, polynomials are dense in $C[a, b]$, so there is a sequence of polynomials $\{p_n\}$ that converges in the norm to $f$. It follows that $\{p_n\}$ is a Cauchy sequence in $X$, but it converges to a function that is not an element of $X$. ♦

*Remark* 10.1.11. Examples of normed spaces, either complete or not, can be found in the exercises. However, understanding some important Banach spaces requires the material that has yet to be covered, so they are postponed until Section 10.3.

Frequently, properties of a vector space are inherited by its proper subspace. The same is true for Banach spaces.

**Definition 10.1.12.** Let $X$ be a Banach space and $Y \subset X$. If $Y$ is itself a Banach space with respect to the algebraic and metric structure inherited from $X$, we say that it is a **Banach subspace** of $X$.

**Example 10.1.13. A Banach subspace of $C[0,1]$.**
The set $Y \subset C[0,1]$ consisting of functions that vanish at $x = 0$ is a Banach subspace of $C[0,1]$.

**Proof:** It is easy to see that $Y$ is a vector subspace of $C[0,1]$. If $\{f_n\} \subset Y$ is a Cauchy sequence, the completeness of $C[0,1]$ implies that it converges to a function $f \in C[0,1]$. Since $f_n(0) = 0$ it follows that $f(0) = 0$, so $f \in Y$. ♦

**Example 10.1.14. A vector subspace of $\ell^1$ that is not a Banach subspace.**
The set $Y \subset \ell^1$ consists of finite sequences, i.e., sequences that have a finite number of nonzero terms (but that number can be arbitrarily large) is not a Banach subspace of $\ell^1$.

**Proof:** It is easy to see that $Y$ is a vector subspace of $\ell^1$ but we will demonstrate that it is not a Banach subspace. Indeed, for each $n \in \mathbb{N}$ let

$$y_n = \left(\frac{1}{2}, \frac{1}{2^2}, \frac{1}{2^3}, \ldots, \frac{1}{2^{n-1}}, 0, 0, 0, \ldots\right).$$

Then, each $y_n \in Y$, and $\{y_n\}_{n=1}^{\infty}$ is a Cauchy sequence that converges to $y = \{1/2^n\}_{n=1}^{\infty}$ which is a sequence with infinitely many nonzero members, therefore not in $Y$. ♦

Some of the simplest examples of subspaces are those that are generated by a subset $S$ of a vector space $X$. Namely, the set of all linear combinations of vectors in $S$ is a subspace of $X$, and it is the smallest subspace that contains $S$. We refer to it as the **span** of $S$ and we write $\vee(S)$. Such a subspace is not closed and therefore not a Banach subspace of $X$ except in rare circumstances. If we take the closure of the span of $S$ we obtain the **closed span** of $S$. The following proposition shows that in this situation, and more generally whenever we take the closure of a subspace, we get a Banach subspace.

**Proposition 10.1.15.** *Let $Y$ be a vector subspace of a Banach space $X$. Then the closure $\overline{Y}$ of $Y$ is a Banach subspace of $X$.*

*Proof.* In view of Theorem 2.2.14, the only thing that needs to be proved is that $\overline{Y}$ is a vector subspace, i.e., that it is closed under linear operations. To that end, suppose that $x, y \in \overline{Y}$ and $\alpha, \beta \in \mathbb{R}$. Then there are sequences $\{x_n\}, \{y_n\} \in Y$ such that $\lim x_n = x$ and $\lim y_n = y$. It follows that $\alpha x_n + \beta y_n$ belongs to $Y$, for all $n \in \mathbb{N}$, and the sequence $\{\alpha x_n + \beta y_n\}$ converges to $\alpha x + \beta y$. Thus, $\alpha x + \beta y$ is in $\overline{Y}$ and $\overline{Y}$ is a vector subspace of $X$. □

Complete normed vector spaces were introduced by Banach in his 1922 dissertation [2]. In his 1928 book [40] Fréchet suggested the name *Banach space*. A very detailed presentation of Banach space theory appeared in 1932 in the book of Banach [5]. However, already in 1921, Helly used an axiomatically defined norm (which he called *Abstandsfunktion* meaning "distance function") in general sequence spaces in [60]. Around the same time, American mathematician Norbert Wiener (1894–1964) gave a similar set of axioms in [125]. The concept of an abstract vector space seems to have been a novelty when Banach presented it, even though they show up in a work of Grassmann [44] in 1844 and Peano [84] in 1888.

STEFAN BANACH (1892–1945) was a Polish mathematician. He was abandoned by his mother days after birth and received little support from his father. He made a living working as a tutor while studying engineering at Lemberg Technical University in Ukraine, and on road construction and as a high school teacher during World War I. He was exempt from army service because of poor vision. In 1916, a fortuitous encounter with Steinhaus led to Banach's first mathematical paper, sparking a prolific research career. Together with Steinhaus, Nikodym and Witold Wilkosz (1891–1941), a Polish mathematician, physicist, philosopher, and popularizer of science, he founded the Mathematical Society of Kraków, later renamed the Polish Mathematical Society. Banach's dissertation at Lwów Technical University with Steinhaus as advisor marked the birth of functional analysis, defining what we now know as a Banach space. He also introduced the so-called Banach algebras and produced fundamental results on normed linear spaces. His 1932 book [5] is one of the classics. He also contributed to measure theory, integration, set theory, and orthogonal series. In 1929, he and Steinhaus started a new journal Studia Mathematica and became the first editors. The way that Banach worked was unconventional. He liked to do mathematics with his colleagues in the cafés of Lwów. Sessions in the Scottish Café became legendary. During World War II, Banach was forced to work feeding lice in German institute dealing with infectious diseases. He planned to return to Kraków post-war but tragically succumbed to lung cancer in 1945.

## Exercises

10.1.1. Prove that $\ell^1$ is a vector space and that $\|a\|_1 = \sum_{k=1}^{\infty} |a_k|$ defines a norm on $\ell^1$.

10.1.2. Prove that $\ell^\infty$ is a vector space and that $\|a\|_\infty := \sup_{n\in\mathbb{N}} |a_n|$ defines a norm on $\ell^\infty$.

10.1.3. Prove that $C[a,b]$ is a vector space and that $\|f\|_{max} = \max\{|f(x)| : x \in [a,b]\}$ defines a norm on $C[a,b]$.

10.1.4. Let $AC[0,1]$ denote the vector space of absolutely continuous functions on $[0,1]$. (See Exercise 9.4.2.) Prove that $\|f\| := |f(0)| + \int_0^1 |f'(t)|\,dt$ defines a norm on $AC[0,1]$.

10.1.5. Let $BV[0,1]$ denote the vector space of functions of bounded variation on $[0,1]$. (See Exercise 9.3.15.) Prove that $\|f\| := |f(0)| + V_0^1(f)$ defines a norm on $BV[0,1]$.

10.1.6. Let $\mathrm{Lip}[0,1]$ denote the vector space of Lipschitz functions on $[0,1]$. For $f \in \mathrm{Lip}[0,1]$ define

$$\|f\| := |f(0)| + \sup\{\frac{|f(x) - f(y)|}{|x-y|} : x, y \in [0,1],\ x \neq y\}.$$

Prove that $\|\cdot\|$ is a norm.

10.1.7. Let $X$ be a finite-dimensional vector space with a basis $\{x_k\}_{k=1}^n$. Given $x = \sum_{k=1}^n c_k x_k$ define

$$\|x\|_\infty = \max\{|c_k| : 1 \le k \le n\}.$$

Prove that $\|\ \|_\infty$ is a norm on $X$.

10.1.8. Prove that $c$ is a vector space and that $\|(x_1, x_2, x_3, \dots)\| = \sup\{|x_n| : n \in \mathbb{N}\}$ defines a norm on $c$.

10.1.9. Let $X$ be a vector space and suppose that $\|\cdot\|_\alpha$ and $\|\cdot\|_\beta$ are both norms on $X$. Prove that so is $\|\cdot\| := \|\cdot\|_\alpha + \|\cdot\|_\beta$.

10.1.10.* Let $X$ be a normed linear space. Prove that $X$ is a Banach space if and only if for any $\{x_n\} \subset X$ the convergence of $\sum_{n=1}^\infty \|x_n\|$ implies the convergence of $\sum_{n=1}^\infty x_n$.

10.1.11.* Let $X$ be a normed linear space. Prove that $X$ is a Banach space if and only if the set $\{x \in X : \|x\| = 1\}$ is a complete metric space with respect to distance $d(x, y) := \|x - y\|$.

10.1.12. Prove that a normed linear space is complete if and only if every decreasing sequence of closed balls has a nonempty intersection.

10.1.13.** Let $X$ be a normed space and $V$ a closed subspace of $X$. Prove that:

(a) The relation $\sim$ on $X$, defined by $x \sim y$ if $x - y \in V$, is an equivalence relation.

(b) The quotient space $X/\sim$ is a vector space with respect to operations $[x] + [y] = [x + y]$ and $\alpha[x] = [\alpha x]$.

(c) $\|[x]\| := \inf\{\|x + y\| : y \in V\}$ is a norm on $X/\sim$.

(d) If $X$ is a Banach space then so is $X/\sim$ with this norm.

(e) For any $\varepsilon > 0$ there exists $x \in X$ such that $\|x\| = 1$ and $\|[x]\| \ge 1 - \varepsilon$.

10.1.14. Prove that $C^1[a, b]$ is a vector space and that the formula $\|f\|_{max} = \max_{x \in [a,b]} |f(x)|$ defines a norm on $C^1[a, b]$, but that it is not a Banach space.

10.1.15. Let $c_0$ denote the subspace of $c$ consisting of sequences converging to 0. Prove that $c_0$ is a Banach space with respect to the norm (10.2).

10.1.16. Let $B(X)$ be the set of all bounded, real-valued functions on a set $X$. Prove that $B(X)$ is a Banach space with respect to the norm $\|f\| = \sup\{|f(x)| : x \in X\}$.

10.1.17. For $f \in C[a, b]$ define

$$\|f\|_1 = \int_a^b |f|. \tag{10.3}$$

(a) Show that this is a norm on $C[a, b]$.

(b) Show that there is no number $c$ such that

$$\|f\|_{\max} \le c\|f\|_1, \quad \text{for all } f \in C[a, b].$$

(c) Show that there is $C$ such that

$$\|f\|_1 \le C\|f\|_{\max}, \qquad \text{for all } f \in C[a,b].$$

(d) Determine whether $C[a,b]$ is a Banach space with respect to this norm.

In Exercises 10.1.18–10.1.21 prove that the set of $x = \{x_n\}$ for which $\|x\| < \infty$ is a Banach subspace of $c$:

10.1.18. $\|x\| := |x_1| + \sum_{n=1}^{\infty} |x_n - x_{n+1}|$.

10.1.19. $\|x\| := \sum_{n=1}^{\infty} |x_n - x_{n+1}|$.

10.1.20.* $\|x\| := \sup_n |\sum_{k=1}^{n} x_k|$.

10.1.21.* $\|x\| := \sup_n |\sum_{k=1}^{n} x_k|$ and the space consists of $x \in c$ for which $\sum_{n=1}^{\infty} x_n$ converges.

10.1.22.* Let $(X, \mathfrak{M})$ be a measurable space and let $M(\mathfrak{M})$ be the vector space of all finite signed measures on $\mathfrak{M}$. Prove that:

(a) $\|\nu\| := |\nu|(X)$ is a norm on $M(\mathfrak{M})$.

(b) With this norm $M(\mathfrak{M})$ is a Banach space.

10.1.23. Prove that $\|f\| := \max_{x\in[0,1]} |f(x)| + \max_{x\in[0,1]} |f'(x)|$ defines a norm on $C^1[0,1]$ and that it is a Banach space with respect to the norm.

10.1.24. Let $X$ be a vector space of polynomials on $\mathbb{R}$ and, for a polynomial $p(x) = \sum_{k=0}^{\deg p} c_k x^k$, define

$$\|p\| = \sum_{k=0}^{\deg p} |c_k|.$$

Prove that $\| \ \|$ is a norm and that $X$ is not complete with respect to this norm.

10.1.25. Let $X, Y$ be Banach spaces and for $(x,y) \in X \times Y$ define $\|(x,y)\| = \|x\|_X + \|y\|_Y$. Prove that $\| \ \|$ is a norm on $X \times Y$ which makes $X \times Y$ a Banach space.

10.1.26. Let $C_b(\mathbb{R})$ be the set of all bounded functions in $C(\mathbb{R})$ Prove that $C_b(\mathbb{R})$ is a Banach space.

10.1.27.* Let $C_0(\mathbb{R}) \subset C(\mathbb{R})$ be defined by $f \in C_0(\mathbb{R})$ if and only if for every $\varepsilon > 0$ the set $\{x \in \mathbb{R} : |f(x)| \ge \varepsilon\}$ is compact. Prove that $C_0(\mathbb{R})$ is a Banach space.

10.1.28. Let $X$ be a subset of $\ell^1$ consisting of sequences $\{a_n\}$ for which $\sum n|a_n| < \infty$. Prove that $X$ is a subspace of $\ell^1$ that is not closed.

10.1.29. Let $X$ be a normed space and $V, W$ two subspaces of $X$. Prove that $V + W$ is a subspace of $X$ and that $V + W = \vee(V \cup W)$.

10.1.30. Let $X$ be a normed space and $\{V_i\}_{i\in I}$ a collection of subspaces of $X$. Prove that $\cap_{i\in I} V_i$ is a subspace of $X$.

10.1.31. Prove that every convergent sequence in a normed space $X$ is a Cauchy sequence.

10.1.32. Let $X$ be a normed space and $f : X \to \mathbb{R}$ a uniformly continuous function. Prove that there are $A, B > 0$ such that for any $x \in X$,

$$|f(x)| \leq A + B\|x\|.$$

10.1.33. Prove that the set consisting of the sequence $e = (1, 1, 1, \dots)$ and the sequences $e_n = (0, 0, \dots, 0, 1, 0, 0, \dots)$ (with 1 in the $n$th position) has a dense span in $c$. (See Example 10.1.9.)

10.1.34. Let $e_n = (0, 0, \dots, 0, 1, 0, 0, \dots)$ (with 1 in the $n$th position). Prove that the set $\{e_n\}_{n=1}^{\infty}$ has a dense span in $c_0$. (See Exercise 10.1.15.)

10.1.35. Let $X$ be a normed linear space. Prove that the function $F : X \to \mathbb{R}$ defined by $F(x) = \|x\|$ is continuous.

10.1.36. Let $X$ be a normed space. Prove that the closed unit ball is a closed subset of $X$.

10.1.37. Let $X$ be a normed space. Prove that the closed unit ball is the closure of the open unit ball.

10.1.38. Let $X$ be a normed space. Prove that the unit ball of $X$ has a unique center and radius.

10.1.39. Let $X$ be a Banach space and define $c_0(X)$ to be the space of all sequences $\{x_n\} \subset X$ such that $\lim_n \|x_n\| = 0$. Prove that $\|\{x_n\}\| := \sup_n \|x_n\|$ defines a norm on $c_0(X)$ and that it makes $c_0(X)$ a Banach space.

## 10.2 The Space $L^p(X, \mu)$

In applications, one encounters various exponents $p$ for which a function $f^p$ is integrable, so it makes sense to study all of them together. Here, we mean integrability as defined by Lebesgue. However, in order to talk about a normed space of $p$-integrable functions we need to be careful about sets of measure 0.

Let $(X, \mathfrak{M}, \mu)$ be a measure space. We consider the family of functions

$$\mathcal{F} = \{f : X \to [-\infty, +\infty] \text{ measurable, finite a.e. on } X\}.$$

By now we are used to ignoring sets of measure 0 which often have no effect on important properties of functions. It makes sense to define a relation on $\mathcal{F}$ by

$$f \sim g \quad \text{if} \quad f(x) = g(x) \text{ a.e. on } X.$$

It is straightforward to verify that this is an equivalence relation. Let $\mathcal{F}/\sim$ be the set of equivalence classes. Since $\mathcal{F}$ is a vector space, so is $\mathcal{F}/\sim$. Further, if $f \sim g$ and $f$ is integrable, then $g$ is integrable and $\int_X |f|\,d\mu = \int_X |g|\,d\mu$. In fact,

$$\int_X |f|^p\,d\mu = \int_X |g|^p\,d\mu,$$

for any $p \in [1, +\infty)$, either as equality of real numbers or if both sides are infinite. Here and throughout this chapter we will omit $d\mu$ in the integrals whenever there is no ambiguity with respect to which measure the integrals are defined.

**Definition 10.2.1.** The set $L^p(X, \mu)$ is a subset of $\mathcal{F}/\sim$ that consists of equivalence classes $[f]$ such that $\int_X |[f]|^p < \infty$. We often omit [ ] and write $\int_X |f|^p$. When it is clear what underlying measurable set is considered, we will write just $L^p$.

First we will show that $L^p$ is a vector space. For the operations with equivalence classes see Exercise 10.2.1.

**Proposition 10.2.2.** *For $p \in [1, +\infty)$, $L^p$ is a vector space.*

*Proof.* If $\alpha \in \mathbb{R}$ and $f \in L^p$, then $[\alpha f] = \alpha[f]$, so

$$\int_X |\alpha f|^p = \int_X |[\alpha f]|^p = \int_X |\alpha[f]|^p = |\alpha|^p \int_X |[f]|^p = |\alpha|^p \int_X |f|^p < \infty.$$

This shows that $\alpha f \in L^p$.

Notice that for any real numbers $a, b$,

$$|a + b| \le |a| + |b| \le 2 \max\{|a|, |b|\}, \quad \text{and} \quad (\max\{|a|, |b|\})^p \le |a|^p + |b|^p.$$

Therefore,

$$|a + b|^p \le 2^p (\max\{|a|, |b|\})^p \le 2^p (|a|^p + |b|^p). \tag{10.4}$$

If $f, g \in L^p(X)$, then $[f + g] = [f] + [g]$, so

$$\begin{aligned}\int_X |f + g|^p &= \int_X |[f + g]|^p = \int_X |[f] + [g]|^p \\ &\le \int_X 2^p \left(|[f]|^p + |[g]|^p\right) = 2^p \left(\int_X |f|^p + \int_X |g|^p\right) < \infty,\end{aligned}$$

whence $f + g \in L^p$. ∎

We proved that $L^p$ is a vector space when $p \in [1, +\infty)$. This leaves open the question about the case $p = \infty$. First we need to make a reasonable definition of the space $L^\infty$.

**Definition 10.2.3.** A function $f$ is **essentially bounded** on $X$ if there is $M > 0$ such that $|f| \le M$ a.e. on $X$. Any such $M$ is called an **essential bound** for $f$.

It is not hard to see that if $f = g$ a.e. and $f$ is essentially bounded on $X$, then so is $g$. This makes the following concept well-defined.

**Definition 10.2.4.** The set $L^\infty(X,\mu)$ is a subset of $\mathcal{F}/\sim$ that consists of equivalence classes $[f]$ such that $f$ is essentially bounded. When there can be no confusion, we simply write $L^\infty$ for $L^\infty(X,\mu)$.

With this definition, $L^\infty$ is indeed a vector space.

**Proposition 10.2.5.** *$L^\infty(X,\mu)$ is a vector space.*

The proof is straightforward and we leave it to the reader (Exercise 10.2.2).

Now that we know that $L^p$ is a vector space for every $p \in [1,+\infty]$, we would like to introduce a norm in each of them. We start with the case $p = 1$.

**Proposition 10.2.6.** *$L^1(X,\mu)$ is a normed vector space with the norm defined by $\|f\|_1 := \int_X |f|\,d\mu$.*

We leave the verification to the reader (Exercise 10.2.3), with the comment that property (i) of the norm here means that $\|f\|_1 = 0$ if and only if $[f] = 0$, i.e., if and only if $f = 0$ a.e. on $X$.

Next we turn our attention to the case $p = +\infty$. We define

$$\|f\|_\infty = \inf\{M : |f| \le M \text{ a.e. } [\mu]\}.$$

We often denote this quantity by suppress $f$ and call it the **essential supremum** of $f$.

Since $\|\ \|_\infty$ is defined as the infimum of essential bounds, i.e., numbers which dominate the function only a.e., it is natural to ask whether it is an essential bound itself.

**Proposition 10.2.7.** *Let $f$ be a measurable function on $X$. Then $|f| \le \|f\|_\infty$ a.e.*

*Proof.* Let $E_n = \{x \in X : |f(x)| > \|f\|_\infty + 1/n\}$. Then $\mu(E_n) = 0$, otherwise $\|f\|_\infty + 1/n$ would not be an essential bound for $f$. Therefore, $\mu(\cup_{n=1}^\infty E_n) = 0$. Finally, $|f(x)| > \|f\|_\infty$ if and only if $x \in E_n$ for some $n \in \mathbb{N}$. Therefore, $|f| \le \|f\|_\infty$ a.e. $\square$

Not only that the essential supremum an essential bound, it also defines a norm on $L^\infty$.

**Proposition 10.2.8.** *$L^\infty(X,\mu)$ is a normed vector space with norm $\|\ \|_\infty$.*

*Proof.* By Proposition 10.2.5, $L^\infty(X,\mu)$ is a vector space so we only need to verify that $\|\ \|_\infty$ satisfies properties (i)–(iii) of Definition 10.1.1. (i) $\|f\|_\infty \ge 0$ is obvious. If $f = 0$ a.e., then $|f| = 0$ a.e., so $\|f\|_\infty = 0$. In the other direction, if $\|f\|_\infty = 0$, Proposition 10.2.7 implies that $f = 0$ a.e.

(ii) If $\alpha = 0$, the assertion follows from (i). If $\alpha \neq 0$, then

$$\begin{aligned}\|\alpha f\|_\infty &= \inf\{M : |\alpha f| \leq M \text{ a.e.}\}\\ &= \inf\{M : |f| \leq M/|\alpha| \text{ a.e.}\}\\ &= |\alpha| \inf\{M/|\alpha| : |f| \leq M/|\alpha| \text{ a.e.}\}\\ &= |\alpha| \|f\|_\infty.\end{aligned}$$

(iii) By Proposition 10.2.7, the inequalities

$$|f+g| \leq |f| + |g| \leq \|f\|_\infty + \|g\|_\infty,$$

hold a.e. on $X$, so $\|f\|_\infty + \|g\|_\infty$ is an essential bound for $f+g$, hence at least as big as $\|f+g\|_\infty$. □

We saw that $L^1$ and $L^\infty$ are normed spaces. Our next goal is to prove that the same is true for any $p$ between these two exponents.

**Theorem 10.2.9.** *Let $(X, \mathfrak{M}, \mu)$ be a measure space and $1 \leq p < \infty$. Then $L^p(X, \mu)$ is a normed space with the norm*

$$\|f\|_p := \left(\int_X |f|^p \, d\mu\right)^{1/p}. \tag{10.5}$$

While showing that $\| \, \|_p$ as defined in (10.5) possesses the first two properties of the norm is almost trivial, proving the triangle inequality is more complicated. In fact, the triangle inequality is known as Minkowski's Inequality and for $\mathbb{R}^n$ it is proved in advanced calculus. We will establish here a version of this inequality for functions in $L^p$. The proof follows a path similar to the one for Euclidean norm.

For numbers $p, q \in (1, \infty)$ we say that they are **conjugate numbers** if $\frac{1}{p} + \frac{1}{q} = 1$. We extend this to the pair $\{1, \infty\}$ as well.

**Theorem 10.2.10.** ***Young's Inequality.*** *If $a, b > 0$ and $p, q \in (1, \infty)$ are conjugate numbers, then*

$$ab \leq \frac{a^p}{p} + \frac{b^q}{q}.$$

*Proof.* Let $x, y \in \mathbb{R}$ and $0 \leq \lambda \leq 1$. It is easy to see that the point $\lambda x + (1-\lambda)y$ lies in the interval with endpoints $x$ and $y$. The function $f(x) = \ln x$ is concave down. This implies that,

$$\ln(\lambda x + (1-\lambda)y) \geq \lambda \ln x + (1-\lambda) \ln y. \tag{10.6}$$

Using simple algebra this becomes

$$\lambda x + (1-\lambda)y \geq x^\lambda y^{1-\lambda}.$$

Now the result follows by taking $x = a^p$, $y = b^q$, and $\lambda = 1/p$. □

**Theorem 10.2.11.** ***Hölder's Inequality.*** *Let $(X, \mathfrak{M}, \mu)$ be a measure space, $1 \leq p < \infty$ and $q$ the conjugate of $p$. If $f \in L^p(X, \mu)$ and $g \in L^q(X, \mu)$ then $fg$ is integrable over $X$ and*

$$\int_X fg\, d\mu \leq \|f\|_p \|g\|_q.$$

*Proof.* If $p = 1$ then $g \in L^\infty$ so $|g| \leq \|g\|_\infty$ a.e. It follows that

$$\left| \int_X fg \right| \leq \int_X |fg| \leq \int_X |f| \|g\|_\infty = \|f\|_1 \|g\|_\infty.$$

Suppose now that $p > 1$. We may assume that $\|f\|_p = \|g\|_q = 1$. (Otherwise replace $f$ and $g$ by $f/\|f\|_p$ and $g/\|g\|_q$.) The functions $|f|^p$ and $|g|^q$ are integrable over $X$, hence finite a.e. on $X$. (This avoids a potential issue with a product $0 \cdot \infty$.) Thus, using Young's Inequality,

$$|fg| = |f||g| \leq \frac{|f|^p}{p} + \frac{|g|^q}{q}, \qquad \text{a.e. on } X,$$

so $fg$ is integrable over $X$ and

$$\int_X |fg| \leq \frac{1}{p} \int_X |f|^p + \frac{1}{q} \int_X |g|^q = \frac{1}{p} + \frac{1}{q} = 1 = \|f\|_p \|g\|_q. \quad \square$$

The Cauchy–Schwarz Inequality was proved by Cauchy, in his *Course d'Analyse*. In 1859 a Russian mathematician Viktor Bunyakovsky (1804–1889) obtained the integral form of this inequality [17]. In 1885, Schwarz discovered the modern proof (in the context of double integrals) which can be used to establish the inequality in any *inner product space* [109]. All three of them were only dealing with the case $p = 2$. The Cauchy–Schwarz Inequality is sometimes called the Cauchy–Bunyakovsky Inequality, and they are all frequently referred to as the Hölder's Inequality, by a German mathematician Otto Hölder (1859 –1937). While working on the convergence of Fourier series in 1889 [64], he gave a simpler proof of an inequality established a year earlier by a British mathematician Leonard James Rogers (1862–1933). This inequality is stated for finite sums and it looks very different from the one above (when reduced to sums), but the latter can be derived from it. Riesz was the first to appreciate the significance of the modern form which he presented in 1910 [92], and he proved Theorem 10.2.11, albeit his proof is more of a hint. Bunyakovsky received his bachelor's degree from the Sorbonne. Continuing his research, he wrote three doctoral dissertations under Cauchy's supervision by the spring of 1825, in theoretical physics, theoretical mechanics, and mathematical physics. He was the Vice President of the St. Petersburg Academy of Sciences for 25 years, with around 150 research works and a number of books. Hölder's contribution were mostly in algebra. At the age of 74, he signed the "Vow of allegiance of the Professors of the German Universities and High-Schools to Adolf Hitler and the National Socialistic State". Rogers was a man gifted in many fields: mathematics, music, linguistics, skater, and a maker of rock gardens. He is also known for Rogers-Ramanujan identities, later rediscovered by Ramanujan.

**Theorem 10.2.12.** ***Minkowski's Inequality.*** *Let $(X, \mathfrak{M}, \mu)$ be a measure space and $1 \leq p \leq \infty$. If $f, g \in L^p(X, \mu)$ then so is $f + g$ and*

$$\|f + g\|_p \leq \|f\|_p + \|g\|_p. \tag{10.7}$$

*Proof.* The cases $p = 1$ and $p = \infty$ are done. (Propositions 10.2.6–10.2.8.) So, let $p \in (1, \infty)$. If $\|f + g\|_p = 0$ the inequality is obvious, so we will assume that $\|f + g\|_p \neq 0$.

By Proposition 10.2.2, $f+g \in L^p(X)$. Then,

$$\begin{aligned} \|f+g\|_p^p &= \int |f+g|^p \\ &= \int |f+g| \cdot |f+g|^{p-1} \\ &\leq \int (|f|+|g|)|f+g|^{p-1} \\ &= \int |f||f+g|^{p-1} + \int |g||f+g|^{p-1} \\ &\leq \|f\|_p \|(f+g)^{p-1}\|_q + \|g\|_p \|(f+g)^{p-1}\|_q, \end{aligned} \tag{10.8}$$

where we have used Hölder's Inequality in the last step. Now, it is not hard to check that $(p-1)q = p$ so

$$\|(f+g)^{p-1}\|_q^q = \int_X |f+g|^{(p-1)q} = \int_X |f+g|^p. \tag{10.9}$$

Further, $p/q = p-1$ and $(p-1)q = p$. It follows that

$$\|(f+g)^{p-1}\|_q = \left(\int_X |f+g|^p\right)^{1/q} = \|f+g\|_p^{p/q} = \|f+g\|_p^{p-1}.$$

Combining with (10.8–10.9), we get

$$\|f+g\|_p^p \leq (\|f\|_p + \|g\|_p) \|f+g\|_p^{p-1},$$

from which the result follows easily. □

*Remark* 10.2.13. This establishes Theorem 10.2.9, so $L^p$ is a normed space, for all $p \in [1, +\infty]$. On the other hand, without identifying functions that are equal a.e., the space of $p$-integrable functions is not a normed space (see Exercise 10.2.4).

> Minkowski proved the inequality above for sums in his book [77] from 1896. Riesz generalized it to integrals in 1910 [92], leaving the proof to the reader.

**Example 10.2.14. An $L^p$ space.**

Let $X = \mathbb{N}$, $\mathfrak{M} = \mathbb{P}(\mathbb{N})$, and $\mu$ the counting measure. For each $p \geq 1$, every function in $L^p(X, \mu)$ is a sequence $\{a_n\}_{n=1}^\infty$ such that

$$\|\{a_n\}\|_p := \left(\sum |a_n|^p\right)^{1/p} < \infty. \tag{10.10}$$

For $p = 1$, we recognize the space $\ell^1$ introduced in Example 10.1.2. Similarly, for $p > 1$ the $L^p$ space as above is denoted by $\ell^p$. ♦

The following result tells us that when $X$ is a set of *finite* measures, the $L^p$ spaces are nested inside each other. (See Exercise 10.2.17.)

**Proposition 10.2.15.** *Let $(X, \mathfrak{M}, \mu)$ be a finite measure space and $1 \leq p_1 < p_2 \leq \infty$. Then $L^{p_2}(X, \mu) \subset L^{p_1}(X, \mu)$. Furthermore,*

$$\|f\|_{p_1} \leq C\|f\|_{p_2}, \quad \textit{for all } f \in L^{p_2}(X, \mu),$$

*where*

$$C = \begin{cases} (\mu(X))^{(p_2-p_1)/(p_1p_2)}, & \textit{if } p_2 < \infty \\ (\mu(X))^{1/p_1} & \textit{if } p_2 = \infty. \end{cases}$$

One may ask whether the inclusions in the previous result are all proper. In general, this is not true (see Exercise 10.2.18). In the case of Lebesgue measure on the real line, it is true when restricted to a *finite* interval.

**Example 10.2.16. The inclusions $L^{p_2}(E) \subset L^{p_1}(E)$ are all proper.**
Let $E = (0, 1]$, $-1/p_1 < \alpha < -1/p_2$ and $f(x) = x^\alpha$. Then $f \in L^{p_1}(E) \setminus L^{p_2}(E)$ (see Exercise 10.2.5). ♦

*Remark* 10.2.17. If $m(E) = \infty$, no inclusion holds between $L^{p_1}(E)$ and $L^{p_2}(E)$ (see Exercise 10.2.19).

The spaces $L^p(a, b)$ for $1 < p < \infty$ were introduced by Riesz in 1910 in his very important article [92]. The space $\ell^2$ featured in the work of Hilbert and the formal definition can be found in the paper of his student Schmidt from 1908 [108]. In that article, Schmidt also defined the $\ell^2$-norm and verified that it satisfies the triangle inequality.

Now that we have established that $L^p$ spaces are normed vector spaces, we will explore the properties that they may share with another well-known normed space, namely $\mathbb{R}^n$.

HERMANN SCHWARZ (1843–1921) was a German mathematician. He studied chemistry in Berlin, before Ernst Kummer (1810–1893) and Karl Weierstrass (1815–1897) persuaded him to change to mathematics. He received his Ph.D. from the University of Berlin, with these two as advisors. To a great degree, Weierstrass influenced the research interests of Schwarz. After graduating, he held positions at the University of Halle, ETF in Zürich, and the University of Göttingen. During these decades his research blossomed. His ideas were usually inspired by geometry and he excelled in Complex Analysis, especially in the theory of *conformal mappings*. Among other things, he improved the proof of the Riemann mapping theorem, developed a special case of the Cauchy–Schwarz inequality, and gave a proof that the ball has less surface area than any other body of equal volume. His work on the latter inspired Picard's proof of the existence of solutions of differential equations. By the time he returned to the University of Berlin in 1892 his best days were behind him. In the words of German mathematician Ludwig Bieberbach (1886–1982), "Schwarz retired to Berlin". This may have been a contributing factor of Göttingen overtaking Berlin as the leading German university. It is possible that he did not have time for research because he was a dedicated teacher (with 20 Ph.D. students in 25 years), the captain of the local Voluntary Fire Brigade and, more surprisingly, he assisted the stationmaster at the local railway station by closing the doors of the trains.

## Exercises

10.2.1. Let $(X, \mathfrak{M}, \mu)$ be a measure space, let $1 \leq p < \infty$, and for a function $f : X \to \mathbb{R}^*$ let $[f]$ denote the equivalence class consisting of all functions that are equal to $f$ a.e. Prove that operations defined by formulas $[f]+[g]=[f+g]$ and $[\alpha f]=\alpha[f]$, where $\alpha\in\mathbb{R}$, are well defined and the same is true of $\|[f]\|_p=\|f\|_p$.

10.2.2. Prove Proposition 10.2.5.

10.2.3. Prove Proposition 10.2.6.

10.2.4. Let $(X,\mathfrak{M},\mu)$ be a measure space, let $1\leq p<\infty$, and let $\mathcal{L}^p$ be the set of all functions such that $\int_X |f|^p\,d\mu<\infty$. Prove that $d(f,g):=\left(\int_X |f-g|^p\right)^{1/p}$ does not define a distance on $\mathcal{L}^p$. Formulate and prove a similar result for the space of all bounded functions.

10.2.5. Verify the assertion made in Example 10.2.16.

10.2.6. For $f\in L^1[a,b]$ define $\|f\|=\int_a^b x^2|f(x)|\,dx$. Prove that $\|\ \|$ is a norm on $L^1[a,b]$.

10.2.7. Let $(X,\mathfrak{M},\mu)$ be a finite measure space and let $f\in L^\infty(X,\mu)$. Prove that

$$\|f\|_1=\lim_{p\to 1}\|f\|_p.$$

10.2.8. Show that if $f(x)=\ln(1/x)$ for $x\in(0,1]$, then $f$ belongs to $L^p(0,1]$ for all $1\leq p<\infty$ but not to $L^\infty(0,1]$.

10.2.9.* Let $f$ be an absolutely continuous function on $[0,1]$, such that $f(0)=0$ and

$$\int_0^1 |f'(x)|^2\,dx<\infty.$$

Prove that

$$\lim_{x\to 0+}\frac{f(x)}{\sqrt{x}}$$

exists and calculate the value of this limit.

10.2.10. Let $(X,\mathfrak{M},\mu)$ be a measure space for which $\mu(X)=1$ and let $f$ be a measurable function on $X$. Prove that the function $F(p):=\|f\|_p$ is an increasing function on $[1,\infty)$. Under what circumstances is $F$ strictly increasing?

10.2.11. Prove or disprove: there exists a function $f$ that is bounded and continuous on $(0,+\infty)$ and $\lim_{x\to\infty}f(x)=0$ but $f\notin L^p(0,\infty)$ for any $p>0$.

10.2.12. Prove that the equality in Young's Inequality holds if and only if $b=a^{p/q}$.

10.2.13. Prove that the equality holds in Hölder's Inequality, i.e., $\int_X |fg|=\|f\|_p\|g\|_q$, if and only if $\|g\|_q^q|f|^p=\|f\|_p^p|g|^q$ a.e.

10.2.14. Prove that, for $1<p<\infty$, the equality in Minkowski's Inequality (10.7) holds if and only if there are nonnegative numbers $a,b$, not both zero, such that $af=bg$ a.e.

10.2.15. Prove that, for $p=1$, the equality in Minkowski's Inequality (10.7) holds if and only if $fg\geq 0$ a.e.

10.2.16. Let $(X, \mathfrak{M}, \mu)$ be a measure space, let $1 \leq p < q < r < \infty$, and let $f \in L^q(X, \mu)$. Prove that for every $t > 0$ there are functions $g \in L^p(X, \mu)$ and $h \in L^r(X, \mu)$ such that $f = g + h$ and

$$\|g\|_p^p \leq t^{p-q}\|f\|_q^q, \quad \|h\|_r^r \leq t^{r-q}\|f\|_q^q.$$

10.2.17. Prove Proposition 10.2.15.

10.2.18. Give an example of a measure space and numbers $p_1 > p_2 \geq 1$ such that $L^{p_1}(X)$ and $L^{p_2}(X)$ are equal.

10.2.19. Verify the assertion made in Remark 10.2.17.

10.2.20. Let $0 < \alpha \leq \beta$ and $f(x) = 1/(x^\alpha + x^\beta)$. Determine for what values of $p$ the function $f$ is in $L^p[0, +\infty)$.

10.2.21. Let $p, q, r$ be positive numbers satisfying $1/p + 1/q + 1/r = 1$ and suppose that $f \in L^p(X, \mu)$, $g \in L^q(X, \mu)$, $h \in L^r(X, \mu)$. Prove that the function $fgh$ is integrable and that $\|fgh\|_1 \leq \|f\|_p\|g\|_q\|h\|_r$.

10.2.22. Let $1 \leq p \leq r \leq q \leq \infty$. Prove that

$$L^p(X, \mu) \cap L^q(X, \mu) \subset L^r(X, \mu),$$

and that, if $f \in L^p(X, \mu) \cap L^q(X, \mu)$ then

$$\|f\|_r \leq \|f\|_p^\alpha \|f\|_q^\beta, \quad \text{where} \quad \alpha = \frac{r^{-1} - q^{-1}}{p^{-1} - q^{-1}}, \ \beta = \frac{p^{-1} - r^{-1}}{p^{-1} - q^{-1}},$$

and $q^{-1}$ is replaced by 0 if $q = \infty$.

10.2.23.* Let $(X, \mathfrak{M}, \mu)$ be a finite measure space and let $f \in L^\infty(X, \mu)$. Prove that

$$\|f\|_\infty = \lim_{p\to\infty} \|f\|_p = \lim_{p\to\infty} \frac{\|f\|_{p+1}^{p+1}}{\|f\|_p^p}.$$

10.2.24. Let $\{f_n\}$ be a sequence of functions on $[0, 1]$ such that:

(a) $f_n$ is absolutely continuous, for each $n \in \mathbb{N}$;

(b) $f_n(0) = 0$, for each $n \in \mathbb{N}$;

(c) the sequence $\{f_n'\}$ converges in $L^1([0, 1])$.

Prove that:

(A) the sequence $\{f_n(x)\}$ is convergent for all $x \in [0, 1]$;

(B) the limit of $\{f_n\}$ is absolutely continuous.

10.2.25.* For $p \in [1, +\infty]$, find the values of the parameter $\lambda$ for which

$$\lim_{\varepsilon\to 0^+} \frac{1}{\varepsilon^\lambda} \int_0^\varepsilon f = 0, \quad \text{for all } f \in L^p[0, 1].$$

10.2.26. Let $(X, \mathfrak{M}, \mu)$ be a finite measure space and for a measurable function $f$ let $E_n = \{x \in X : n-1 \leq |f(x)| < n\}$. Prove that $f \in L^p(X, \mu)$, $1 \leq p < \infty$, if and only if

$$\sum_{n=1}^{\infty} n^p \mu(E_n) < +\infty.$$

Is the finitness of $\mu$ necessary?

10.2.27. Let $(X, \mathfrak{M}, \mu)$ be a measure space and let $0 < p < 1$. Prove that $\|\cdot\|_p$ is not a norm on $L^p(X, \mu)$.

10.2.28. Let $(X, \mathfrak{M}, \mu)$ be a measure space such that $\mu(X) = 1$, let $f, g \in L^2(X, \mu)$, and suppose that $\int_X f\, d\mu = 0$. Prove that

$$\left(\int_X fg\, d\mu\right)^2 \leq \left(\int_X g^2\, d\mu - \left(\int_X g\, d\mu\right)^2\right) \int_X f^2\, d\mu.$$

10.2.29. Let $(X, \mathfrak{M}, \mu)$ be a measure space such that $\mu(X) = 1$. For $1 < p < \infty$ let

$$\mathcal{A}_p = \left\{f \in L^1(X, \mu) : \|f\|_1 = 1,\ \|f\|_p = 2^{1/p}\right\}.$$

Prove that for each $0 < \varepsilon < 1$ there exists $\delta_p > 0$ such that

$$\mu\left(\{x \in X : |f(x)| > \varepsilon\}\right) \geq \delta_p$$

for each $f \in \mathcal{A}_p$.

10.2.30. Let $(X, \mathfrak{M}, \mu)$ be a measure space and let $f$ be a nonnegative measurable function on $X$. Suppose that

$$\int_X f^2\, d\mu = \int_X f^3\, d\mu = \int_X f^4\, d\mu.$$

Prove that $f = f^2$ a.e.

10.2.31. Let $\{f_n\} \subset L^p(0,1)$, $\{g_n\} \subset L^q(0,1)$, $1/p + 1/q = 1$, and suppose that $\{f_n\}$ converges to $f$ in $L^p(0,1)$ and that $\{g_n\}$ converges to $g$ in $L^q(0,1)$. Prove or disprove:

(a) The sequence $\{f_n g_n\}$ converges to $fg$ in $L^1(0,1)$.

(b) The sequence $\{f_n g_n\}$ converges in measure to $fg$.

10.2.32. Let $\{f_n\}$ be a sequence of functions in $L^\infty(X)$ that converges a.e. to a function $f$, and suppose that $\|f_n\|_\infty \leq M$ for all $n \in \mathbb{N}$. Prove that $\|f\|_\infty \leq M$.

10.2.33. Let $E$ be a measurable subset of $[0,1]$ for which $m(E) > 0$, and let $p < q$. Prove or disprove: There exists a function that belongs to $L^p(0,1)$ but not to $L^q(E)$.

10.2.34.* Let $E$ be a measurable set and let $g$ be a function defined on $E$. Prove that the product $fg$ belongs to $L^1(E)$ for every function $f \in L^1(E)$ if and only if $g \in L^\infty(E)$.

10.2.35. Let $f$ be a function that belongs to both $L^4(\mathbb{R})$ and $L^8(\mathbb{R})$. Must $f$ belong to $L^5(\mathbb{R})$? To $L^2(\mathbb{R})$?

10.2.36.* Prove or disprove: If $f \in L^1(0, +\infty)$ then there is an interval $(a, b)$ such that $f \in L^2(a, b)$.

10.2.37. Let $(X, \mathfrak{M}, \mu)$ be a finite measure space for which $\mu(X) = 1$ and let $p, q$ be two distinct numbers in $[1, +\infty]$. Characterize those measurable, real-valued functions $f$ for which $\|f\|_p = \|f\|_q$.

10.2.38. Let $1 \le p < q \le \infty$. Prove that $\ell^p \subset \ell^q$. Is the inclusion strict?

10.2.39.* Let $(X, \mathfrak{M}, \mu)$ be a measure space, let $1 \le p < q \le \infty$, and suppose that $f \in L^r(X, \mu)$ for all $r \in (p, q)$. Prove that

$$\lim_{r \to q^-} \|f\|_r = \|f\|_q, \quad \text{and} \quad \lim_{r \to p^+} \|f\|_r = \|f\|_p.$$

10.2.40.* Let $(X, \mathfrak{M}, \nu)$ be a finite measure space, and let $\mu$ be a signed measure on $\mathfrak{M}$ such that $|\mu|$ is $\sigma$-finite and $\mu \ll \nu$. Prove that $d\mu/d\nu \in L^p(X, \nu)$ if and only if there is a constant $C > 0$ such that

$$\sum_{k=1}^{n} \frac{|\mu|(A_k)^p}{\nu(A_k)^{p-1}} \le C$$

for any finite disjoint collection $\{A_k\} \subset \mathfrak{M}$ satisfying $X = \cup_{k=1}^n A_k$ and $\nu(A_k) > 0$, for $1 \le k \le n$. Determine the relationship between the smallest such constant $C$ and $\|d\mu/d\nu\|_p$.

10.2.41.* Let $1 \le p < \infty$ and let $f \in L^p(\mathbb{R})$. Find $\lim_{|y| \to \infty} \int_{\mathbb{R}} |f(x+y) + f(x)|^p \, dx$.

10.2.42. Let $1 \le p < \infty$. Prove or disprove: $\cap_{1 \le r < p} L^r[0,1] \subset L^p[0,1]$.

10.2.43. Let $1 \le p < \infty$. Prove or disprove: $\cup_{r > p} L^r[0,1] \supset L^p[0,1]$.

10.2.44. Let $1 \le p < \infty$. Prove or disprove: $\cap_{r > p} L^r(\mathbb{R}) \subset L^p(\mathbb{R})$.

10.2.45. Let $1 \le p < \infty$. Prove or disprove: $\cup_{1 \le r < p} L^r(\mathbb{R}) \supset L^p(\mathbb{R})$.

10.2.46.* Let $(X, \mathfrak{M}, \mu)$ be a measure space and $f \in L^\infty(X, \mu)$. The **essential range** of $f$ is the set $R_f$ of all $y \in \mathbb{R}$ for which the set $\{x : |f(x) - y| < \varepsilon\}$ has positive measure for every $\varepsilon > 0$. Prove that $R_f$ is compact and that $\|f\|_\infty = \sup\{|y| : y \in R_f\}$.

10.2.47.* Let $(X, \mathfrak{M}, \mu)$ be a finite measure space, let $1 \le p < \infty$, let $\theta > 0$, and let $\{f_n\}$ be a bounded sequence in $L^{p+\theta}(X, \mu)$ that converges a.e. to a function $f$. Prove that $\{f_n\}$ converges to $f$ in the norm of $L^p(X, \mu)$.

10.2.48. Prove that $\sqrt{\cos t \sin^3 t} + \sqrt{\sin t \cos^3 t} \le 1$, for all $t \in [0, 2\pi]$.

10.2.49. Prove that for any $1 < p < \infty$,

$$\bigcup_{q<p} \ell^q \ne \ell^p.$$

10.2.50. For $p \in [1, +\infty]$, prove that the closed unit ball of $\ell^p$ is not compact.

## 10.3 Completeness of $L^p(X, \mu)$

In the previous sections we established that $L^p$ is a normed vector space for all $p \in [1, +\infty]$. This allows us to consider the topology induced by this norm and in this section, we will focus on the completeness of $L^p$. We stated earlier that this property is important because it is natural to hope that if a sequence of functions $\{f_n\}$ converges to $f$ and $\{f_n\} \subset L^p$ then $f \in L^p$ as well. Although pointwise and uniform convergence of a sequence of functions will play important roles, we will be primarily interested in the *convergence in norm* which we have already defined in Section 10.1.

The main result of this section is that all $L^p$ spaces are complete. Some of the groundwork has been done in Chapter 2 and we will rely on it. We will also need the following characterization of norm convergence in $L^\infty$.

**Theorem 10.3.1.** *Let $(X, \mathfrak{M}, \mu)$ be a measure space and let $\{f_n\}$, $f$ be in $L^\infty(X, \mu)$. Then $\{f_n\}$ converges to $f$ in the norm of $L^\infty(X)$ if and only if it converges uniformly to $f$ except on a set of measure zero.*

*Proof.* Suppose first that $\{f_n\}$ converges uniformly except on a set $E$ for which $\mu(E) = 0$. Let $\varepsilon > 0$. Let $\alpha_n = \sup\{|f(x) - f_n(x)| : x \in E^C\}$. Since $\{f_n\}$ converges uniformly to $f$ there is $N \in \mathbb{N}$ such that

$$n \geq N \Rightarrow \alpha_n \leq \frac{\varepsilon}{2}.$$

Now, for $n \geq N$, $|f - f_n| \leq \varepsilon/2$ a.e. so

$$\|f - f_n\|_\infty = \inf\{M : |f_n(x) - f(x)| \leq M \text{ a.e.}\} \leq \frac{\varepsilon}{2} < \varepsilon.$$

It follows that $\lim_n \|f_n - f\|_\infty = 0$, i.e., $\{f_n\}$ converges to $f$ in the norm of $L^\infty(X)$.

In the other direction, suppose that $\lim_n \|f_n - f\|_\infty = 0$ and let $\varepsilon > 0$. Then

$$\exists N \in \mathbb{N} \quad n \geq N \Rightarrow \|f_n - f\|_\infty < \varepsilon.$$

By definition, for $n \geq N$ there is a set $E_n$ with $\mu(E_n) = 0$ such that $|f_n(x) - f(x)| < \varepsilon$ on $E^C$. Let $E = \cup_{n \geq N} E_n$. Then $\mu(E) = 0$ and for $x \in E^C = \cap_{n \geq N} E_n^C$, $|f_n(x) - f(x)| < \varepsilon$. Thus, $\{f_n\}$ converges uniformly to $f$ except on $E$. □

Now we are ready for the crucial step in the proof of completeness of $L^p$ spaces.

**Theorem 10.3.2.** *Let $(X, \mathfrak{M}, \mu)$ be a measure space and $1 \leq p \leq \infty$. Then every rapidly Cauchy sequence in $L^p(X, \mu)$ converges both in the norm of $L^p(X, \mu)$ and pointwise a.e. on $X$ to a function in $L^p(X, \mu)$.*

*Proof.* Let $\{f_n\}$ be a rapidly Cauchy sequence in $L^p$. By definition, there is a sequence $\{\varepsilon_n\}$ such that

$$\sum_{k=1}^{\infty} \varepsilon_k < \infty \quad \text{and} \quad \|f_{k+1} - f_k\|_p \leq \varepsilon_k^2.$$

Notice that both the hypotheses and the conclusions of the theorem allow us to ignore sets of measure 0. Since every integrable function is finite a.e., and this also holds for functions in $L^\infty$, we may assume that each $f_n$ is finite on $X$, so expressions like $f_{k+1} - f_k$ are always defined.

**Case I.** $p = \infty$.

By Proposition 10.2.7, for any $m, n \in \mathbb{N}$ and a.e. $x \in X$,

$$|f_m(x) - f_n(x)| \leq \|f_m - f_n\|_\infty \leq \sum_{k=n}^{m-1} \|f_{k+1} - f_k\|_\infty \leq \sum_{k=n}^{m-1} \varepsilon_k^2 \leq \sum_{k=n}^{\infty} \varepsilon_k^2. \tag{10.11}$$

The convergence of the series $\sum_{k=1}^{\infty} \varepsilon_k$ implies the convergence of $\sum_{k=1}^{\infty} \varepsilon_k^2$, so its tail goes to 0. Consequently, $\{f_n(x)\}$ is a Cauchy sequence of real numbers for a.e. $x \in X$, so $\{f_n\}$ converges pointwise a.e. to a function $f$. Moreover, letting $m \to \infty$ in (10.11), we obtain

$$|f(x) - f_n(x)| \leq \sum_{k=n}^{\infty} \varepsilon_k^2.$$

on a set $E$ for which $\mu(X \setminus E) = 0$. In other words, the sequence $\{f_n\}$ converges uniformly to $f$ except on a set of measure zero, whence the result follows from Theorem 10.3.1. (Of course, uniform convergence implies pointwise convergence.)

**Case II.** $1 \leq p < \infty$.

We have that

$$\int_X |f_{k+1} - f_k|^p = \|f_{k+1} - f_k\|_p^p \leq \varepsilon_k^{2p}, \qquad \text{for all } k \in \mathbb{N}.$$

Let $k \in \mathbb{N}$ be fixed. Then, using Chebyshev's Inequality,

$$\begin{aligned}
m\left(\{x \in X : |f_{k+1}(x) - f_k(x)| \geq \varepsilon_k\}\right)& \\
&= m\left(\{x \in X : |f_{k+1}(x) - f_k(x)|^p \geq \varepsilon_k^p\}\right) \\
&\leq \frac{1}{\varepsilon_k^p} \int_X |f_{k+1} - f_k|^p \\
&\leq \frac{1}{\varepsilon_k^p} \varepsilon_k^{2p} \\
&= \varepsilon_k^p.
\end{aligned}$$

Since $\sum_{k=1}^{\infty} \varepsilon_k < \infty$, we have that $\sum_{k=1}^{\infty} \varepsilon_k^p < \infty$, so Borel–Cantelli's Lemma implies that there is a set $E_0 \subset X$ such that $m(E_0) = 0$ and for any $x \in X \setminus E_0$ there exists $K(x)$ such that

$$|f_{k+1}(x) - f_k(x)| < \varepsilon_k, \qquad \text{for all } k \geq K(x).$$

Let $x \in X \setminus E_0$ and $m \geq n \geq K(x)$. Then

$$|f_m(x) - f_n(x)| \leq \sum_{j=n}^{m-1} |f_{j+1}(x) - f_j(x)| \leq \sum_{j=n}^{\infty} \varepsilon_j,$$

so $\{f_n(x)\}$ is a Cauchy sequence of real numbers, hence a convergent sequence. Let $f(x) = \lim f_n(x)$, for $x \in X \setminus E_0$, i.e., for a.e. $x \in X$.

Next, we will establish that the sequence $\{f_n\}$ converges in the norm of $L^p$. Using Minkowski's inequality,

$$\int_X |f_{n+k} - f_n|^p = \|f_{n+k} - f_n\|_p^p \leq \left( \sum_{j=n}^{n+k-1} \|f_{j+1} - f_j\|_p \right)^p \leq \left( \sum_{j=n}^{\infty} \varepsilon_j^2 \right)^p.$$

Since $\lim_k f_{n+k} = f$ a.e. on $X$, Fatou's Lemma implies that, for any fixed $n \in \mathbb{N}$,

$$\int_X |f - f_n|^p \leq \liminf_k \int_X |f_{n+k} - f_n|^p \leq \left( \sum_{j=n}^{\infty} \varepsilon_j^2 \right)^p,$$

and it follows that $\|f_n - f\|_p \to 0$. Finally, there is $n \in \mathbb{N}$ such that $f - f_n \in L^p$ and $f_n \in L^p$, so $f \in L^p$. ☐

Now we can state the so-called Riesz–Fischer Theorem which ensures that every $L^p$ space is complete. The proof follows from the previous result and Propositions 2.2.17 and 2.2.21, so we leave it to the reader as Exercise 10.3.3.

**Theorem 10.3.3.** ***The Riesz–Fischer Theorem.*** *Let $(X, \mathfrak{M}, \mu)$ be a measure space and $1 \leq p \leq \infty$. Then $L^p(X, \mu)$ is a Banach space. Moreover, if the sequence $\{f_n\}$ converges to $f$ in $L^p(X, \mu)$, then there is a subsequence $\{f_{n_k}\}$ that converges pointwise a.e. on $X$ to $f$.*

The Riesz–Fischer Theorem is a very important result because it removes one of the major flaws of the Riemann integral. Namely, it is possible for a sequence of Riemann integrable functions to converge to a function that is not Riemann integrable. (See Example 7.1.4.) As we have just witnessed, this is impossible in the case of Lebesgue integrable functions.

Riesz [88] and Fischer [34] proved the same result independent of each other in 1907, both for $L^2(a,b)$. Actually, Riesz proved a result from which he showed 3 years later that completeness of $L^2(a,b)$ follows. It is still not known whether his result implies completeness in an abstract inner product space. The completeness of $L^p(a,b)$ was obtained by Riesz in 1910 [92]. The paper [63] from 1926, by an English mathematician Ernest William Hobson (1856–1933), contains the proof that works "in any number of dimensions" and "over a set of finite or infinite measure", and in fact it can be used verbatim in a general measure space. In 1941, Carathéodory included in a paper about ergodic theory a proof of the completeness of the general $L^2$ space [22]. Hobson was Sadleirian Professor of Pure Mathematics at the University of Cambridge from 1910 to 1931. In a paper in 1907 he realized that many problems with the change of order of integration disappear when Lebesgue integral is introduced. He failed to realize that the conclusions he drew for specific examples are true in general.

FRIGYES RIESZ (1880–1956) was a Hungarian mathematician of Jewish origin. After receiving his Ph.D. in 1902, he wrote several influential papers in function theory, containing among other results the Riesz–Fischer Theorem. From 1911 he was a professor in Kolozsvár which in 1920 became a part of Romania so the university moved to Szeged, Hungary. He established the János Bolyai Mathematical Institute there and became an editor of Acta Scientiarum Mathematicarum. He did some of the fundamental work in developing functional analysis and his work has a number of important applications in physics. In 1918 he came close to an axiomatic theory for Banach spaces, two years before Banach. He used the novel concept of a metric space to provide a link between Lebesgue's work and the area of integral equations developed by Hilbert and his student Schmidt. Riesz also worked in ergodic theory, topology, and orthonormal series. He had an uncommon method of giving lectures: an assistant would read from Riesz's handbook while another wrote the appropriate equations on the blackboard. He would stand aside, nodding occasionally. In 1945, he joined the University of Budapest. He was a member of the Hungarian Academy of Sciences, and the Paris Academy of Sciences, and, in 1949, he was awarded the Kossuth Prize. His book "Functional Analysis" remains one of the necessary readings for a young analyst.

While the convergence in the norm of $L^p$ does not imply the convergence almost everywhere, at least there is a subsequence that does. What about the converse? Does the convergence almost everywhere imply convergence in $L^p$?

**Example 10.3.4. A sequence that converges almost everywhere but not in the norm of $L^p(X,\mu)$.**

Let $f_n = \chi_{[n,n+1]}$ in $L^p(\mathbb{R})$. Then $\{f_n\}$ converges almost everywhere but not in the norm of $L^p(\mathbb{R})$.

**Proof:** For a fixed $x$, the sequence $\{f_n(x)\}$ has all zeros starting from $n = \lfloor x \rfloor + 1$ so $\lim_n f_n(x) = 0$, for all $x \in \mathbb{R}$. On the other hand, $\|f_n\|_p = 1$, so $\{f_n\}$ does not converge to 0 in the norm of $L^p(\mathbb{R})$. ♦

Part of the reason that Example 10.3.4 worked was that the sequence of norms did not converge to the norm of the limit function. The following theorem shows that when this additional condition is met we get the norm convergence.

**Theorem 10.3.5.** *Let $(X, \mathfrak{M}, \mu)$ be a measure space and $1 \le p < \infty$. Suppose that $\{f_n\}$ is a sequence in $L^p(X,\mu)$ that converges a.e. on $X$ to a function $f$ that belongs to $L^p(X,\mu)$. Then*

$$\lim_n \|f_n - f\|_p = 0 \quad \text{if and only if} \quad \lim_n \|f_n\|_p = \|f\|_p.$$

*Proof.* The "only if" part follows from the inequality

$$\big|\|f_n\|_p - \|f\|_p\big| \leq \|f_n - f\|_p.$$

In the other direction, the result follows from the General Dominated Convergence (Exercise 7.3.16), with $|f_n - f|^p$, 0, $2^p(|f_n|^p + |f|^p)$, and $2^{p+1}|f|^p$ instead of $f_n$, $f$, $g_n$, and $g$, respectively. The only item that needs verification is the hypothesis $|f_n| \leq g_n$ a.e., which now reads $|f_n - f|^p \leq 2^p(|f_n|^p + |f|^p)$ and this follows from (10.4). □

## Exercises

10.3.1. Let $(X, \mathfrak{M}, \mu)$ be a finite measure space and let $\{f_n\}$ be a sequence in $L^1(X)$ that converges uniformly to a function $f$. Prove that $\{f_n\}$ converges to $f$ in the norm of $L^1(X)$. Does this still hold if $\mu(X) = \infty$?

10.3.2. Let $(X, \mathfrak{M}, \mu)$ be a finite measure space and let $\{f_n\}$ be a sequence in $L^p(X, \mu)$ that it is uniformly Cauchy. Prove that it converges to a function $f$ that belongs to $L^p$.

10.3.3. Prove the Riesz–Fischer theorem.

10.3.4. Let $(X, \mathfrak{M}, \mu)$ be a measure space and $\{f_n\}_{n=1}^{\infty} \subset L^p(X, \mu)$, $1 \leq p \leq \infty$. Suppose that $\{f_n\}$ converges to 0 in the norm of $L^p$. Prove that it converges to 0 in measure.

10.3.5.* Let $(X, \mathfrak{M}, \mu)$ be a measure space, let $\{f_n\}_{n=1}^{\infty}$ a sequence of measurable functions that converges in measure to a function $f$, and let $1 \leq p < \infty$.

(a) If there exists $g \in L^p(X, \mu)$ such that $|f_n| \leq g$ for all $n \in \mathbb{N}$, prove that $\{f_n\}$ converges to $f$ in the norm of $L^p(X, \mu)$.

(b) Give an example to show that the implication in (a) is false when $p = \infty$.

10.3.6. Give an example of a sequence that converges in measure but not in the norm of $L^p(X, \mu)$, $1 \leq p \leq \infty$.

10.3.7. Let $(X, \mathfrak{M}, \mu)$ be a finite measure space and let $\{f_n\}$ be a sequence of nonnegative integrable functions such that $\lim \int_X f_n \, d\mu = 0$. Prove or disprove: $\lim f_n = 0$ a.e.

10.3.8. Give an example of a sequence $\{f_n\}$ in $L^1[0, 1]$ that has the following properties:

(i) $\lim f_n(x) = 0$ for all $x \in [0, 1]$;

(ii) there is $C > 0$ such that $\int_0^1 |f_n| \leq C$, for all $n \in \mathbb{N}$;

(iii) $\{f_n\}$ does not converge in the norm of $L^1[0, 1]$.

10.3.9. Give an example to show that Theorem 10.3.5 is false when $p = \infty$.

10.3.10. Let $(X, \mathfrak{M}, \mu)$ be a measure space and let $\{f_n\} \in L^p(X,\mu) \cap L^r(X,\mu)$, for some $p, r \in [1, +\infty)$. Suppose that $\{f_n\}$ converges to a function $g$ in the norm of $L^p(X,\mu)$ and to a function $h$ in the norm of $L^r(X,\mu)$. Prove or disprove: $g = h$ a.e.

10.3.11.** Let

$$X = L^1(\mathbb{R}) + L^2(\mathbb{R}) = \{g + h : g \in L^1(\mathbb{R}) \text{ and } h \in L^2(\mathbb{R})\}.$$

Given $p \in [1, \infty)$, define

$$\|f\|_p^{inf} := \inf\{(\|g\|_1^p + \|h\|_2^p)^{1/p} : f = g + h, g \in L^1(\mathbb{R}) \text{ and } h \in L^2(\mathbb{R})\}.$$

Show that $\|f\|_p^{inf}$ is a norm for each $p$, and that $X$ is a Banach space with each of these norms.

10.3.12.** Let

$$X = L^{p_1}(\mathbb{R}) + L^{p_2}(\mathbb{R}) = \{g + h : g \in L^{p_1}(\mathbb{R}) \text{ and } h \in L^{p_2}(\mathbb{R})\}.$$

Given $p_1, p_2 \in [1, \infty)$, define

$$\|f\|_{p_1,p_2} := \inf\{\|g\|_{p_1} + \|h\|_{p_2} : f = g + h, g \in L^{p_1}(\mathbb{R}) \text{ and } h \in L^{p_2}(\mathbb{R})\}.$$

Show that $\|f\|_{p_1,p_2}$ is a norm for each pair $p_1, p_2$, and that $X$ is a Banach space with each of these norms.

10.3.13.* Let $X = \mathbb{R}^{n+1}$ with the product measure $m_n \times m$ on the $\sigma$-algebra $\mathcal{M}_n \otimes \mathcal{M}$, and let $L^{p,r}$ be the space of equivalence classes of jointly measurable functions $f(x,t)$, $x \in \mathbb{R}^n$, $t \in \mathbb{R}$, for which the norm is defined by

$$\|f\|_{p,r} = \left(\int_{\mathbb{R}} \left(\int_{\mathbb{R}^n} |f|^p \, dm_n\right)^{r/p} dm\right)^{1/r}$$

Prove that $\|f\|_{p,r}$ is a norm for each pair $1 \le p, r < \infty$ and that with this norm $L^{p,r}$ is a Banach space.

10.3.14. Let $1 \le p \le r \le q < \infty$. Prove that $L^p(0,1) \cap L^q(0,1)$ is a dense subspace of $L^r(0,1)$. Does the result hold if $q = \infty$?

10.3.15. Let $(X, \mathfrak{M}, \mu)$ be a measure space, let $1 \le p < \infty$, and let $\{f_n\} \in L^p(X,\mu)$. Suppose

$$0 < c = \inf_k \|f_k\|_p \le \sup_k \|f_k\|_p = C < \infty \quad \text{and} \quad f_i f_j = 0 \quad \text{for } i \ne j.$$

Let $\{a_k\}$ be a sequence of real numbers such that $\sum_k |a_k|^p < \infty$ and let $f = \sum_k a_k f_k$. Prove that:

(a) $f \in L^p(X,\mu)$;

(b) $\|f\|_p \le C(\sum_k |a_k|^p)^{1/p}$;

(c) $\|f\|_p \ge c(\sum_k |a_k|^p)^{1/p}$.

10.3.16.* Let $(X, \mathfrak{M}, \mu)$ be a measure space, let $1 \le p \le \infty$, and let $\{f_n\} \in L^p(X,\mu)$ such that $\lim_n \|f_n\|_p = 0$. Prove that there exists a subsequence $\{f_{n_k}\}$ and a nonnegative function $h \in L^p(X,\mu)$ such that $|f_{n_k}| \le h$ a.e. $[\mu]$ and $\lim_k f_{n_k} = 0$ a.e. $[\mu]$.

10.3.17. Give an example of a sequence $\{f_n\}$ in $L^p(0,1)$ that converges to a function $f$ in the norm of $L^p(0,1)$ but not pointwise at any point of $[0,1]$.

10.3.18. Let $1 \le p < \infty$, and let $E_p = \{f \in L^p(\mathbb{R}) : \int_{\mathbb{R}} f\, d\mu = 0\}$. Prove that $E_p$ is closed if and only if $p = 1$.

10.3.19. Let $A = \{f \in L^2(0,1) : \exists \text{ interval } I_f \subset [0,1],\ 1/2 \in I_f,\ f = 0 \text{ a.e. on } I_f\}$. Prove or disprove: $A$ is a closed subset of $L^2$.

10.3.20. Let $\mathcal{F} = \{f \in L^2(0,1) : f([0,1]) \subset [0,1] \text{ a.e.}\}$. Prove or disprove: $\mathcal{F}$ is a closed subset of $L^2$.

10.3.21. Let $\{f_n\} \subset AC[0,1]$, let $f_n(0) = 0$ for all $n \in \mathbb{N}$, and suppose that $\{f_n'\}$ is a Cauchy sequence in $L^1(0,1)$. Prove that there exists $f \in AC[0,1]$ such that $\{f_n\}$ converges to $f$ uniformly on $[0,1]$.

10.3.22. Let $AC[0,1]$ denote the normed space of absolutely continuous functions on $[0,1]$ with the norm $\|f\| := |f(0)| + \int_0^1 |f'(t)|\, dt$. Prove that $AC[0,1]$ a Banach space with this norm.

10.3.23.* Let $BV[0,1]$ denote the normed space of functions of bounded variation on $[0,1]$ with the norm $\|f\| := |f(0)| + V_0^1(f)$. Prove that $BV[0,1]$ is a Banach space with this norm.

10.3.24. Let Lip$[0,1]$ denote the vector space of Lipschitz functions on $[0,1]$ with the norm

$$\|f\| := |f(0)| + \sup\{\frac{|f(x) - f(y)|}{|x - y|} : x, y \in [0,1],\ x \ne y\}.$$

Prove that Lip$[0,1]$ is a Banach space with this norm.

10.3.25. Prove that $\ell^p$ is a Banach space.

10.3.26. Prove that $\ell^\infty$ is a Banach space.

10.3.27. Let $1 \le p < q \le \infty$.

(a) Is the closed unit ball of $L^p[0,1]$ a closed subset of $L^q[0,1]$?

(b) Is the closed unit ball of $L^q[0,1]$ a closed subset of $L^p[0,1]$?

10.3.28. Let $1 \le p \le \infty$ and let $\{f_n\}$ be a sequence in $L^p[0,1]$ such that

$$f(x) = \sum_{n=1}^{\infty} f_n(x) \quad \text{a.e. } x \in [0,1] \qquad \text{and} \qquad \sum_{n=1}^{\infty} \|f_n\|_p < \infty.$$

(a) Prove that for a.e. $x \in [0,1]$

$$\sum_{n=1}^{\infty} |f_n(x)| < \infty.$$

(b) Prove that $f \in L^p[0,1]$ and that

$$\|f\|_p \leq \sum_{n=1}^{\infty} \|f_n\|_p < \infty.$$

10.3.29. Let $1 \leq p \leq \infty$, let $g \in L^p[0,1]$, and let $\mathcal{F}$ be the subset of $L^p[0,1]$ consisting of functions $f$ that are not greater than $g$ a.e. Prove that $\mathcal{F}$ is a closed, nowhere dense subset of $L^p[0,1]$. Conclude that the set of all nonnegative functions in $L^p[0,1]$ is a closed, nowhere dense subset of $L^p[0,1]$.

10.3.30. Give an example of a sequence that converges uniformly but not in the norm of $L^p(X,\mu)$, $1 \leq p < \infty$.

10.3.31. Let $(X, \mathfrak{M}, \mu)$ be a measure space, let $1 \leq p < \infty$, and let $f$ and $\{f_n\}$ be functions in $L^p(X,\mu)$ such that $\lim_n \|f_n\|_p = \|f\|_p$ and $\lim_n f_n = f$ in measure. Prove that $\{f_n\}$ converges to $f$ in the norm of $L^p(X,\mu)$.

10.3.32.* Let $(X, \mathfrak{M}, \mu)$ be a finite measure space, let $1 \leq r < p < \infty$, and let $f$ and $\{f_n\}$ be measurable functions such that $\sup_n \|f_n\|_p < \infty$ and $\lim_n f_n = f$ a.e. Prove that $\{f_n\}$ converges to $f$ in the norm of $L^r(X,\mu)$.

10.3.33. Let $(X, \mathfrak{M}, \mu)$ be a measure space, let $\{E_n\} \subset \mathfrak{M}$ such that $\mu(E_n) < \infty$ for all $n \in \mathbb{N}$, and suppose that $\{\chi_{E_n}\}$ converges to a function $f$ in the norm of $L^1(X,\mu)$. Prove that $f$ is equal a.e. to the characteristic function of a measurable set.

10.3.34.* Let $(X, \mathfrak{M}, \mu)$ be a measure space, let $0 < p < 1$, and for $f, g \in L^p(X,\mu)$ let

$$d(f,g) := \int_X |f-g|^p \, d\mu.$$

Prove that $d$ is a metric with respect to which $L^p(X,\mu)$ is complete.

10.3.35. By Hölder's Inequality, we have $L^2[a,b] \subset L^1[a,b]$. Show that the set $L^2[a,b]$, considered as a subset of a complete metric space $L^1[a,b]$, is of the first category.

10.3.36.* Prove that $\|f\| = \|f\|_1 + \|f'\|_1$ defines a norm on $AC[0,1]$ and that it is a Banach space with respect to the norm.

10.3.37. Let $V$ be a closed linear subspace of $L^1[0,1]$. Suppose that for each $f \in V$ there exists a $p > 1$ such that $f \in L^p[0,1]$. Show that there exists a $q > 1$ such that $V \subset L^q[0,1]$.

10.3.38. Let $(X, \mathfrak{M}, \mu)$ be a measure space, let $E$ be a nonempty measurable subset of $X$, and let $V$ consist of those functions in $L^p(X,\mu)$ that vanish on $E$. Prove that $V$ is a closed subspace of $L^p(X,\mu)$.

10.3.39. Prove or disprove: There is a nonempty closed set in $L^p([0,1])$ that does not contain an element of smallest norm.

10.3.40. For $p \in [1, +\infty]$, prove that the closed unit ball of $L^p([0,1])$ is not compact.

10.3.41. Let $X$ be a Banach space, let $1 \le p < \infty$ and define $\ell^p(X)$ to be the space of all sequences $\{x_n\} \subset X$ such that $\sum_n \|x_n\|^p < \infty$. Prove that $\|\{x_n\}\| := (\sum_n \|x_n\|^p)^{1/p}$ defines a norm on $\ell^p(X)$ and that it makes $\ell^p(X)$ a Banach space.

## 10.4 Dense Sets in $L^p(X, \mu)$

It is always beneficial if we can identify a set of elements that is dense in a given Banach space, especially when these elements belong to a class that is more manageable. For example, rational numbers are dense in $\mathbb{R}$ and so are irrational numbers. More of interest to us, the Weierstrass Theorem claims that polynomials are dense in $C[a, b]$. This means that if $f \in C[a, b]$ then there exists a sequence of polynomials $\{p_n\}$ that converges uniformly to $f$ on $[a, b]$, or equivalently, $\|p_n - f\|_{\max} \to 0$. It is even better when the dense set is countable, i.e., when the normed space is a separable metric space. This principle is in action whenever we approximate real numbers with rationals, and we have encountered other situations in which it is preferable to deal with a countable set. In this section we will explore these approximating sets in $L^p$ spaces. In the exercise section we will take a look at some other Banach spaces.

We start with a fundamental result in the theory of $L^p$ spaces.

**Proposition 10.4.1.** *Let $(X, \mathfrak{M}, \mu)$ be a measure space and $1 \le p \le \infty$. The collection $\mathcal{S}$ of simple functions in $L^p(X, \mu)$ is dense in $L^p(X, \mu)$.*

*Proof.* We consider separately the cases of a finite and infinite $p$.

**Case I:** $p = \infty$.

Let $g \in L^\infty(X, \mu)$. By definition, there exists a set $E_0 \subset X$ of measure 0 such that $g$ is bounded on $X \setminus E_0$. By Theorem 6.2.13, there exists a sequence $\{s_n\}$ of simple functions that converge uniformly to $g$ on $X \setminus E_0$. Theorem 10.3.1 now implies that $\{s_n\}$ converges to $g$ in the norm of $L^\infty$.

**Case II:** $1 \le p < \infty$.

Let $g \in L^p(X, \mu)$. By Theorem 6.2.13, there exist nonnegative sequences $\{s_n\}, \{t_n\} \in \mathcal{S}$ that converge on $X$ to $g^+$ and $g^-$, respectively. The sequence $\{s_n - t_n\}$ of simple functions converges to $g$ on $X$. Further, $0 \le s_n \le g^+$ and $0 \le t_n \le g^-$, hence

$$|s_n - t_n| \le |s_n| + |t_n| = s_n + t_n \le g^+ + g^- = |g|.$$

It follows that $s_n - t_n \in L^p$, and

$$|s_n - t_n - g|^p \le 2^p \left(|s_n - t_n|^p + |g|^p\right) \le 2^{p+1}|g|^p.$$

By the Lebesgue Dominated Convergence Theorem, $s_n - t_n \to g$ in $L^p$. □

Proposition 10.4.1 does not guarantee that an $L^p$ space is separable because the collection $\mathcal{S}$ is not countable. If we look for a countable dense subset of $\mathcal{S}$, we can start by considering those simple functions whose values are rational numbers, i.e., functions of the form $\sum c_k \chi_{E_k}$, with $c_k \in \mathbb{Q}$. The question then comes down to whether the sets $\{E_k\}$ can be selected from some countable collection. The following definition specifies the structure of the underlying measure space that would allow us to do so.

**Definition 10.4.2.** The triple $(X, \mathfrak{M}, \mu)$ is called a **separable measure space** if there exists a countable collection $\mathcal{A} \subset \mathfrak{M}$ such that for every $E \in \mathfrak{M}$ and every $\varepsilon > 0$ there exists $A \in \mathcal{A}$ such that $\mu(A \triangle E) < \varepsilon$.

Warning! Separability of a measure space $(X, \mathfrak{M}, \mu)$ has nothing to do with topological properties of the space $X$ (which may not have any topology on it).

*Remark* 10.4.3. It is possible to define an equivalence relation on a $\sigma$-algebra $\mathfrak{M}$ by

$$A \sim B \Leftrightarrow \mu(A \triangle B) = 0.$$

It is an equivalence relation (Exercise 10.4.1). If we define $d : \mathfrak{M}/\sim \times \mathfrak{M}/\sim \to \mathbb{R}$ by $d([A],[B]) = \mu(A \triangle B)$ then $(\mathfrak{M}/\sim, d)$ is a metric space (Exercise 10.4.2) and it is known as the **Nikodym metric space**. In this context, $(X, \mathfrak{M}, \mu)$ is a separable measure space exactly when $(\mathfrak{M}/\sim, d)$ is a separable metric space. Even though $d$ is not a metric on $\mathfrak{M}$, the distinction between $\mathfrak{M}/\sim$ and $\mathfrak{M}$ is often blurred, much like in the case of functions in $L^p$ spaces. Thus, we will follow this less pedantic practice and write $d(A, B)$ instead of $d([A],[B])$.

**Example 10.4.4. A separable measure space.**

Let $X = [0,1]$, let $\mathfrak{M} = \mathcal{M}|[0,1]$, and let $\mu = m|[0,1]$. Then $(X, \mathfrak{M}, \mu)$ is a separable measure space.

**Proof:** We will show that the collection $\mathcal{A}$ of all finite unions of open intervals with rational endpoints is dense in $\mathcal{M}$. Let $E \in \mathcal{M}|[0,1]$ and let $\varepsilon > 0$. By Exercise 5.3.9, there is an open set $G$ containing $E$ such that $m(G \setminus E) < \varepsilon/4$, and therefore

$$d(E, G) = m(E \triangle G) < \frac{\varepsilon}{4}.$$

By Theorem 3.1.2, $G$ is a disjoint countable union of open intervals $I_k$. Furthermore, $m(G) \leq 1$ implies that the series $\sum_k m(I_k)$ converges, so there is $n \in \mathbb{N}$ for which $\sum_{k=n+1}^{\infty} m(I_k) < \varepsilon/4$. If we denote $I = \cup_{k=1}^{n} I_k$ then

$$d(G, I) < \frac{\varepsilon}{4}.$$

Finally, let $I_k = (a_k, b_k)$. Since rational numbers are dense in $\mathbb{R}$, for every $k \in \mathbb{N}$ there is an interval $J_k = (\alpha_k, \beta_k) \subset I_k$ with rational endpoints $\alpha_k, \beta_k$

such that $|\alpha_k - a_k| < \varepsilon/4^{k+2}$ and $|\beta_k - b_k| < \varepsilon/4^{k+2}$. Then $d(I_k, J_k) < \varepsilon/4^{k+1}$ and

$$d(I, \cup_{k=1}^n J_k) = d(\cup_{k=1}^n I_k, \cup_{k=1}^n J_k) \le \sum_{k=1}^n d(I_k, J_k) < \sum_{k=1}^n \frac{\varepsilon}{4^{k+1}} < \frac{\varepsilon}{4}.$$

Thus, $A := \cup_{k=1}^n J_k \in \mathcal{A}$ and $d(E, A) \le d(E, G) + d(G, I) + d(I, A) < \varepsilon$. ♦

A similar proof can be used to establish that $(\mathbb{R}^n, \mathcal{M}_n, m_n)$ is separable for every $n \in \mathbb{N}$ (see Exercise 10.4.3).

**Example 10.4.5. A nonseparable measure space.**

Let $\mu$ be the counting measure on $\mathbb{N}$. Then $(\mathbb{N}, \mathbb{P}(\mathbb{N}), \mu)$ is a nonseparable measure space.

**Proof:** Let $\varepsilon = 1/2$. Since $\mu(A \triangle B) < 1/2$ if and only if $A = B$, we see that the only set that is dense in $\mathbb{P}(\mathbb{N})$ is $\mathbb{P}(\mathbb{N})$ itself, and it is not countable. ♦

As we hinted, the separability of the underlying measure space is a key ingredient for an $L^p$ space to be separable.

**Theorem 10.4.6.** *Let $(X, \mathfrak{M}, \mu)$ be a separable measure space and let $1 \le p < \infty$. Then $L^p(X, \mu)$ is separable.*

*Proof.* Suppose that there exists a countable collection $\mathcal{A} \subset \mathfrak{M}$ that is dense in the metric space $(\mathfrak{M}, d)$. Let

$$S = \left\{ s \in L^p(X, \mu) : s = \sum_{k=1}^n a_k \chi_{A_k}\, a_k \in \mathbb{Q},\, A_k \in \mathcal{A},\, n \in \mathbb{N} \right\}.$$

It is clear that $S$ is a countable subset of $L^p(X, \mu)$, which we will from now on denote by $L^p$. We will show that $S$ is dense in $L^p$.

Let $f \in L^p$ and $\varepsilon > 0$. Our goal is to show that there exists $s \in S$ such that $\|f - s\|_p < \varepsilon$. By Proposition 10.4.1, it suffices to prove this assertion in the case when $f$ is a simple function in $L^p$,

$$f = \sum_{k=1}^n c_k \chi_{E_k}. \tag{10.12}$$

Without loss of generality we may assume that $\{E_k\}$ is a disjoint family.

First, we will show that there exist a function $g \in L^p$ of the form

$$g = \sum_{k=1}^n a_k \chi_{E_k},$$

with rational coefficients $\{a_k\}$ and satisfying $\|f - g\|_p < \varepsilon$. Since $|f|^p$ is integrable over $X$, we must have $\mu(E_k) < \infty$, for $1 \le k \le n$, and therefore

$$M := \left( \sum_{k=1}^n \mu(E_k) \right)^{1/p} < \infty.$$

Let $\{a_k\}_{k=1}^n$ be rational numbers such that

$$|c_k - a_k| < \frac{\varepsilon}{M}, \qquad \text{for } 1 \le k \le n.$$

If we define $g = \sum_{k=1}^n a_k \chi_{E_k}$, then it is not hard to verify that $g \in L^p$. Further,

$$|f - g| = \left| \sum_{k=1}^n c_k \chi_{E_k} - \sum_{k=1}^n a_k \chi_{E_k} \right| \le \sum_{k=1}^n |c_k - a_k| \, \chi_{E_k} < \frac{\varepsilon}{M} \sum_{k=1}^n \chi_{E_k}.$$

Therefore,

$$\|f - g\|_p < \frac{\varepsilon}{M} \left\| \sum_{k=1}^n \chi_{E_k} \right\|_p = \frac{\varepsilon}{M} \left( \int_X |\sum_{k=1}^n \chi_{E_k}|^p \, d\mu \right)^{1/p}.$$

Since the integrand takes only values 1 and 0, we have that

$$\|f - g\|_p < \frac{\varepsilon}{M} \left( \int_X \sum_{k=1}^n \chi_{E_k} \, d\mu \right)^{1/p} = \frac{\varepsilon}{M} \left( \sum_{k=1}^n \mu(E_k) \right)^{1/p} = \varepsilon.$$

Thus, it remains to approximate $g$ by a function in $S$. To that end, let $L = \max\{|a_k| : 1 \le k \le n\}$ and let $\{A_k\}_{k=1}^n$ be sets in $\mathcal{A}$ such that

$$\mu(A_k \triangle E_k) < \left(\frac{\varepsilon}{nL}\right)^p.$$

By Exercise 1.2.9, $|\chi_A - \chi_B| = \chi_{A \triangle B}$, so

$$\begin{aligned} \|\chi_A - \chi_B\|_p &= \left( \int_X |\chi_A - \chi_B|^p \right)^{1/p} = \left( \int_X |\chi_{A \triangle B}|^p \right)^{1/p} \\ &= \left( \int_X \chi_{A \triangle B} \right)^{1/p} \mu(A \triangle B)^{1/p}. \end{aligned}$$

If we define $s = \sum_{k=1}^n a_k \chi_{A_k}$, then it is easy to see that $s \in S$ and

$$|g - s| = \left| \sum_{k=1}^n a_k \chi_{E_k} - \sum_{k=1}^n a_k \chi_{A_k} \right| \le \sum_{k=1}^n |a_k| \, |\chi_{E_k} - \chi_{A_k}|.$$

Therefore,

$$\|g - s\|_p \le L \sum_{k=1}^n \|\chi_{E_k} - \chi_{A_k}\|_p = L \sum_{k=1}^n \mu(E_k \triangle A_k)^{1/p} < L \sum_{k=1}^n \frac{\varepsilon}{nL} = \varepsilon,$$

and it follows that $L^p(X, \mu)$ is separable. □

*Remark* 10.4.7. The implication opposite to the one in Theorem 10.4.6 is not always true. See Exercises 10.4.10 and 10.4.11.

In the special case when the measure space is $(\mathbb{R}, \mathcal{M}, m)$ we obtain the following corollary (see Exercise 10.4.12).

**Corollary 10.4.8.** *Let $E$ be a Lebesgue measurable subset of $\mathbb{R}$ and $1 \le p < \infty$. Then $L^p(E)$ is separable.*

The case $p = \infty$ is dramatically different from $p < \infty$.

**Theorem 10.4.9.** *Let $(X, \mathfrak{M}, \mu)$ be a measure space and suppose that there is a countable disjoint collection $\{E_n\}_{n=1}^{\infty} \subset \mathfrak{M}$, such that $\mu(E_n) \neq 0$ for all $n \in \mathbb{N}$. Then $L^\infty(X, \mu)$ is not separable.*

*Proof.* For any real number $t \in (0,1)$, let $t = 0.t_1t_2t_3\ldots$ be its representation in the binary system. That way, we can associate to each $t \in (0,1)$ a function

$$f_t = \sum_{k=1}^{\infty} t_k \chi_{E_k}.$$

It is not hard to verify that the collection $\{f_t : t \in (0,1)\}$ is an uncountable family in $L^\infty(X)$. Further, for any two distinct real numbers $t', t'' \in (0,1)$,

$$\|f_{t'} - f_{t''}\|_\infty \ge 1.$$

Indeed, $t' \neq t''$ implies that there exists $k \in \mathbb{N}$ for which $t'_k \neq t''_k$, so $|f_{t'} - f_{t''}| = 1$ on $E_k$. Therefore, $\{f_t : t \in (0,1)\}$ is an uncountable discrete family so, by Theorem 2.5.18, $L^\infty(X, \mu)$ is not separable. □

Once again the result for $(\mathbb{R}, \mathcal{M}, m)$ follows easily from the general case (see Exercise 10.4.14).

**Corollary 10.4.10.** *Let $E$ be a Lebesgue measurable subset of $\mathbb{R}$ of positive measure. Then $L^\infty(E, m)$ is not separable.*

Theorem 10.4.9 seems to be pretty general. The following example shows that it is not universal.

**Example 10.4.11. A measure space on which $L^\infty$ is separable.**
Let $X$ be a finite set, let $\mathfrak{M}$ be the power set of $X$, and let $\mu$ be the counting measure on $X$. A function $f : X \to [-\infty, +\infty]$ is a finite sequence and it belongs to $L^\infty$ if and only if none of the members of this sequence is infinite. Thus, $L^\infty(X)$ is isomorphic to $\mathbb{R}^n$ which is separable. (For a definition of an isomorphism of normed linear spaces see Definition 11.2.5 or mimic the proof that $\mathbb{R}^n$ is separable.) ♦

We next address the question under what conditions are continuous functions dense in an $L^p$ space. When the underlying space is a finite interval, such as $[0,1]$, even polynomials are dense. (See Exercise 10.4.19.) In a more general situation, we will focus on $\mathbb{R}^n$ and a subset $C_c(\mathbb{R}^n)$ of $C(\mathbb{R}^n)$, namely those functions that have a compact **support**. This means that each function in $C_c(\mathbb{R}^n)$ vanishes outside of some compact set. (Not necessarily the same one for each function.)

**Theorem 10.4.12.** *The set $C_c(\mathbb{R}^n)$ is dense in $L^p(\mathbb{R}^n)$, for $1 \leq p < \infty$.*

*Proof.* By Theorem 10.4.1 it suffices to show that $C_c(\mathbb{R}^n)$ is dense in $\mathcal{S} \cap L^p(\mathbb{R}^n)$. Further, each simple function in $L^p$ is of the form $\sum_{k=1}^n c_k \chi_{E_k}$, where $m(E_k) < \infty$ and we have suppressed the subscript $n$ in $m_n$. We will show that $\chi_{E_k}$ can be approximated by a function in $C_c(\mathbb{R}^n)$ and the result will then follow by an argument similar to the one at the end of the proof of Theorem 10.4.6.

So, let $E$ be a measurable set of finite measure and let $\varepsilon > 0$. Let us denote $\varepsilon_1 = \varepsilon^{1/p}$. By Exercise 5.3.9, there are an open set $G$ and a closed set $F$ such that $F \subset E \subset G$ and such that $m(G \setminus F) < \varepsilon_1/2$. For every $k \in \mathbb{N}$, let $R_k$ be the rectangle $[-k, k]^n$. Then $\{G \cap R_k\}$ is an increasing sequence of sets and $G = \cup_{k=1}^\infty (G \cap R_k)$, so it follows from the continuity of measure that $m(G) = \lim_k m(G \cap R_k)$. Consequently, there is $k \in \mathbb{N}$ for which $m(G \setminus (G \cap R_k)) < \varepsilon_1/2$. Let us fix that $k$ and denote $G' = G \cap R_k$, $E' = E \cap R_k$, and $F' = F \cap R_k$. Notice that $F'$ is compact and that

$$E \setminus E' = E \cap R_k^C \subset G \cap R_k^C = G \setminus G'.$$

From this we infer that $m(E \setminus E') \leq m(G \setminus G') < \varepsilon_1/2$. Now we define a function $f$ by the formula

$$f(x) = \frac{d(x, G'^C)}{d(x, F') + d(x, G'^C)}.$$

It is easy to see that $f$ is well-defined. Indeed, the denominator cannot be zero because it would imply that $x \in F'$, in which case it cannot be in $G'^C$ (see Exercise 2.3.16). The function $f$ is continuous because $d$ is (Exercise 2.3.38). Also, $f = 1$ on $F'$, $f = 0$ on $G'^C$, and $0 \leq f \leq 1$ on $\mathbb{R}^n$. Since $f$ vanishes on $G'^C$ its support is contained in the closure of $G'$, hence in a compact set $R_k$, whence $f \in C_c(\mathbb{R}^n)$. Finally,

$$\int_{\mathbb{R}^n} |f - \chi_E|^p = \int_{G^C} |f - \chi_E|^p + \int_{G \setminus G'} |f - \chi_E|^p + \int_{G' \setminus F'} |f - \chi_E|^p + \int_{F'} |f - \chi_E|^p.$$

Both $f$ and $\chi_E$ vanish on $G^C$ and they are both 1 on $F'$. On the remaining 2 sets we have that $|f - \chi_E| \leq 1$, so Exercise 1.1.15 implies that

$$\int_{\mathbb{R}^n} |f - \chi_E|^p \leq m(G \setminus G') + m(G' \setminus F') < \varepsilon_1/2 + m(G \setminus F) < \varepsilon_1.$$

It follows that $\|f - \chi_E\|_p < \varepsilon_1^{1/p} = \varepsilon$ and the theorem is proved. □

How far beyond $\mathbb{R}^n$ can we extend Theorem 10.4.12? First, the concept of continuity requires a topology on $X$. We will therefore assume that $X$ is a metric space, and consider functions $f : X \to \mathbb{R}$ that are continuous with respect to the topology induced by the metric. This allows us to talk about the class $C_c(X)$. Next, there has to be a relation between the metric and the measure or, more precisely, between the topology and the measure. Namely, we need to be able to approximate a measurable set by an open set (from above) and a closed set (from below). This property of a measure space is known as **regularity** and it was established for Euclidean spaces in Exercise 5.3.9. It will play a prominent role in Section 12.5. Also, we need the topology to be **$\sigma$-compact**, meaning that $X$ is a countable union of compact sets. Then we have the following generalization of Theorem 10.4.12.

**Theorem 10.4.13.** *Let $X$ be a metric space, let $\mathcal{B}(X)$ be the Borel $\sigma$-algebra and $\mu$ a measure on it, and let $(X, \mathcal{B}(X), \mu)$ be a $\sigma$-compact regular measure space. The set $C_c(X)$ is dense in $L^p(X, \mu)$, for $1 \leq p < \infty$.*

We leave the proof as Exercise 10.4.24. It is interesting to point out that conditions mentioned above cannot be omitted. (See Exercises 10.4.25 and 10.4.26.)

Our final remark concerns the proof of Theorem 10.4.12. It can be verbatim applied to the space $L^p[0,1]$, for $1 \leq p < \infty$ and it yields a result which we will often require.

**Corollary 10.4.14.** *The set $C[0,1]$ is dense in $L^p[0,1]$, for $1 \leq p < \infty$.*

ERHARD SCHMIDT (1876–1959) was a German mathematician, born in what is today Estonia. He studied in Berlin with Schwarz, and his doctoral advisor was Hilbert at the University of Göttingen. The topic of his dissertation was integral equations. After defending his Habilitation thesis at the University of Bonn, he taught in Zürich, Erlangen, and Breslau. Finally he settled at the University of Berlin in 1917. He assumed the leadership position reviving a somewhat stagnant situation. He was instrumental in founding the Institute of Applied Mathematics and served as its director. His administrative skills brought him the position of Dean in 1921 and Vice Chancelor in 1929. During WWII Schmidt had to carry out various Nazi resolutions against the Jews—a job that he apparently did not do well, since he was criticized at one point for not understanding the "Jewish question". After the war he served as Director of the Mathematics Research Institute of the German Academy of Science until 1958. He was a co-founder and the first editor of *Mathematische Nachrichten*. His most important research came between 1905 and 1908. In a series of papers he built on the ideas of Hilbert to create the concept of abstract Hilbert space and established many of its geometric properties. The orthogonalization technique that he used is now known as the Gram–Schmidt process, although it was introduced by Laplace. In the later part of his career, Schmidt became interested in topology, finding a new proof of the Jordan curve theorem.

## Exercises

10.4.1. Let $(X, \mathfrak{M}, \mu)$ be a measure space. Prove that $A \sim B \Leftrightarrow \mu(A \triangle B) = 0$ defines an equivalence relation on $\mathfrak{M}$.

10.4.2. Let $(X, \mathfrak{M}, \mu)$ be a measure space, $A \sim B \Leftrightarrow \mu(A \triangle B) = 0$, and let $d : \mathfrak{M}/\sim \times \mathfrak{M}/\sim \to \mathbb{R}$ be defined by $d([A],[B]) = \mu(A \triangle B)$. Prove that $(\mathfrak{M}/\sim, d)$ is a metric space.

10.4.3. Prove that $(\mathbb{R}^n, \mathcal{M}_n, m_n)$ is a separable measure space for every $n \in \mathbb{N}$.

10.4.4. Let $A$ denote the subset of $\ell^p$, $1 \leq p < \infty$, consisting of all sequences that have only finitely many nonzero terms. Prove that $A$ is dense in $\ell^p$. Conclude that $\ell^p$ is a separable Banach space for $1 \leq p < \infty$.

10.4.5. Let $(X, \mathfrak{M}, \mu)$ be a finite measure space and let $1 \leq p < r < \infty$. Consider the vector space $L^r(X, \mu)$ but with the norm $\|\cdot\|_p$. Is this normed space complete?

10.4.6. Let $(X, \mathfrak{M}, \mu)$ be a measure space. Prove that the span of the set $\{\chi_E : E \in \mathfrak{M}\}$ is dense in $L^\infty(X, \mu)$.

10.4.7. Let $f \in L^1(\mathbb{R})$. Prove that

$$\lim_{n\to\infty} \int_{\mathbb{R}} f(x) \sin nx \, dx = \lim_{n\to\infty} \int_{\mathbb{R}} f(x) \cos nx \, dx = 0.$$

10.4.8. Let $f \in L^1(\mathbb{R})$. Prove that

$$\lim_{\varepsilon\to 0} \int_{\mathbb{R}} |f(x+\varepsilon) - f(x)| \, dx = 0.$$

10.4.9.* Let $f, g, h$ be integrable functions over $(\mathbb{R})$ and let

$$f * g(x) = \int_{\mathbb{R}} f(t) g(x-t) \, dt.$$

Prove that:

(a) $f * g$ is integrable over $(\mathbb{R})$.

(b) If, in addition, one of the functions $f, g$ is bounded then $f * g$ is continuous.

(c) $f * g = g * f$.

(d) $(f * g) * h = f * (g * h)$.

10.4.10. Give an example of $L^p(X, \mu)$ that is separable even though $(X, \mathfrak{M}, \mu)$ is not separable.

10.4.11.* Let $(X, \mathfrak{M}, \mu)$ be a measure space and suppose that $L^p(X, \mu)$ is separable. Prove that there exists a countable collection $\mathcal{A} \subset \mathfrak{M}$ such that for every $E \in \mathfrak{M}$ with $\mu(E) < \infty$ and every $\varepsilon > 0$ there exists $A \in \mathcal{A}$ such that $\mu(A \triangle E) < \varepsilon$.

10.4.12.* Prove Corollary 10.4.8. Does it remain true in $\mathbb{R}^n$?

10.4.13. Let $A$ be the set of all $x = \{x_n\}$ in $\ell^2$ such that $\sum_{n=1}^{\infty} x_n = 0$. Prove or disprove: $A$ is dense in $\ell^2$.

10.4.14. Prove Corollary 10.4.10. Does it remain true in $\mathbb{R}^n$?

10.4.15. Let $(X, \mathfrak{M}, \mu)$ be a measure space and suppose that $X$ is an infinite set such that $\mu(X) \neq 0$. Prove or disprove: $L^\infty(X)$ is not separable.

10.4.16. Prove that $\ell^\infty$ is not separable.

In Exercises 10.4.17–10.4.21 prove that the given set is dense in $L^p[0,1]$, $1 \leq p < \infty$:

10.4.17. The set of step functions $\sum_{k=1}^{n} c_k \chi_{[x_{k-1},x_k)}$, where $0 = x_0 < x_1 < \cdots < x_n = 1$.

10.4.18. The set of continuous piecewise linear functions with finitely many corners.

10.4.19. The set of polynomials.

10.4.20. The set of trigonometric polynomials, i.e., functions of the form $T(x) = \sum_{k=1}^{n} a_k \cos kx + b_k \sin kx$.

10.4.21. The set of Lipschitz functions on $[0,1]$.

10.4.22. Let $1 \leq p < \infty$ and let $f$ be a function in $L^p[0,1]$ such that, for every $n \in \mathbb{N}$,

$$\int_0^1 x^n f(x)\, dx = 0.$$

Prove that $f = 0$ a.e. Does the result remain true if $p = \infty$?

10.4.23. Let $f : [0,1] \to \mathbb{R}$. Prove that $f$ is Lipschitz with $|f(x) - f(y)| \leq M|x-y|$ for all $x, y \in [0,1]$ if and only if there is a sequence $\{f_n\} \subset C^1[0,1]$ such that

(i) $|f_n'(x)| \leq M$ for all $x \in [0,1]$;

(ii) $\lim_n f_n(x) = f(x)$ for all $x \in [0,1]$.

10.4.24. Prove Theorem 10.4.13.

10.4.25.* Give an example of a measure space $(X, \mathfrak{M}, \mu)$ and a metric on $X$ such that $X$ is $\sigma$-compact but $C_c(X)$ is not dense in $L^p(X,\mu)$, for any $p \in [1, +\infty)$.

10.4.26.* Give an example of a regular measure space $(X, \mathfrak{M}, \mu)$ and a metric on $X$ such that $C_c(X)$ is not dense in $L^p(X,\mu)$, for any $p \in [1, +\infty)$.

10.4.27. Prove or disprove: The set of continuous function on $[0,1]$ is dense in $L^\infty[0,1]$.

10.4.28.* Let $S = \{f \in L^1(\mathbb{R}^3) : \int f = 0\}$. Prove that $S \cap L^2(\mathbb{R}^3)$ is a dense subset of $L^2(\mathbb{R}^3)$.

10.4.29. Let $AC[0,1]$ denote the normed space of absolutely continuous functions on $[0,1]$ with the norm $\|f\| := |f(0)| + \int_0^1 |f'(t)|\, dt$. Prove that $AC[0,1]$ is a separable Banach space.

10.4.30. Let $BV[0,1]$ denote the normed space of functions of bounded variation on $[0,1]$ with the norm $\|f\| := |f(0)| + V_0^1(f)$. Prove that $BV[0,1]$ is not a separable Banach space.

10.4.31. Let Lip$[0,1]$ denote the vector space of Lipschitz functions on $[0,1]$ with the norm

$$\|f\| := |f(0)| + \sup\{\frac{|f(x) - f(y)|}{|x-y|} : x, y \in [0,1], x \neq y\}.$$

Prove that Lip$[0,1]$ is not a separable Banach space.

10.4.32. Prove that the space $c_0$ is separable.

10.4.33. Let $e = (1, 1, 1, \dots)$ and for every $n \in \mathbb{N}$ let $\{e_n\}$ be the sequence with 1 in the $n$th spot and zeros elsewhere. Prove that the span of vectors $\{e, e_1, e_2, \dots\}$ is dense in $c$. Conclude that the space $c$ is separable.

10.4.34. Prove that the space $C[a, b]$ with $L^1$ norm is separable.

10.4.35. Let $f, g$ be two continuous functions on $(0, 1)$ and let

$$E(f, g) = \{x \in (0, 1) : f(x) \neq g(x)\}.$$

(a) Prove that $E(f, g)$ is a countable union of open intervals $\{I_n(f, g)\}$.

(b) Prove that

$$d(f, g) = \sum_{n=1}^{\infty} m\left(I_n(f, g)\right)$$

defines a distance on the set $C(0, 1)$ but the metric space $(C(0, 1), d)$ is not separable.

10.4.36. Let $B(X)$ be the vector space of all bounded, real-valued functions on a set $X$. For $f \in B(X)$ define

$$\|f\| = \sup\{|f(x)| : x \in X\}.$$

Prove that $B(X)$ is separable if and only if $X$ is a finite set.

10.4.37. Give an example of a measure space $(X, \mathfrak{M}, \mu)$ which is not separable even though $X$ is a separable normed space.

10.4.38. Let $X$ be a Banach space and let $Y$ be a dense subspace of $X$. Suppose that $Y$ is a Banach space with respect to the same norm. Prove that $X = Y$.

10.4.39.* Let $X$ be a normed linear space. Prove that $X$ is separable if and only if there exists a compact set $K \subset X$ for which the closed linear span equals $X$.

10.4.40.* Let $(X, \mathfrak{M}, \mu)$ be a measure space, let $p, q$ be conjugate exponents, let $S$ be a dense subset of $L^q(X, \mu)$, and suppose that $g \in L^p(X, \mu)$ satisfies $\int_X fg \, d\mu = 0$, for all $f \in S$. Prove that $g = 0$.

10.4.41.* Let $f \in L^1(\mathbb{R})$ and suppose that $\int_{\mathbb{R}} fg \, dm = 0$, for all $g \in C_c(\mathbb{R})$. Prove that $f = 0$ a.e.

10.4.42. Let $X$ be a normed linear space and suppose that $X$ is a closed linear span of $\{x_n\}_{n=1}^{\infty}$. Prove that $X$ is separable.

10.4.43. Prove that a proper closed vector subspace of a normed linear space is nowhere dense.

## 10.5 Hilbert Space

In applications the space $L^2$ stands out among all $L^p$ spaces. The reason is that much like Euclidean spaces it is endowed with an inner product. This will make understanding the geometry of $L^2$ much easier. In particular, it will allow us to define the concept of orthogonality and extend many results that hold in $\mathbb{R}^n$. Recall that if $x, y \in \mathbb{R}^n$, $x = (x_1, x_2, \ldots, x_n)$, $y = (y_1, y_2, \ldots, y_n)$ then the inner product is defined by $\langle x, y \rangle = \sum_{k=1}^{n} x_k y_k$. The following definition is inspired by this formula as well as the properties of this specific inner product.

**Definition 10.5.1.** Let $\mathcal{H}$ be a linear space. A function $\langle \cdot, \cdot \rangle : \mathcal{H} \times \mathcal{H} \to \mathbb{R}$ is called an **inner product** on $\mathcal{H}$ if for all $x, y, z \in \mathcal{H}$ and $\alpha, \beta \in \mathbb{R}$:

(i) $\langle \alpha x + \beta y, z \rangle = \alpha \langle x, z \rangle + \beta \langle y, z \rangle$;

(ii) $\langle x, y \rangle = \langle y, x \rangle$;

(iii) $\langle x, x \rangle > 0$ if $x \neq 0$.

A linear space $\mathcal{H}$ together with an inner product is called an **inner product space.**

The best-known example of an inner product space is $\mathbb{R}^n$. Here we present two additional examples and we leave it to the reader to verify the details (see Exercises 10.5.1–10.5.2).

**Example 10.5.2. An inner product space.**
Let $\mathcal{H} = \ell^2$ and $\langle x, y \rangle = \sum_{k=1}^{\infty} x_k y_k$. ♦

**Example 10.5.3. An inner product space.**
Let $(X, \mathfrak{M}, \mu)$ be a measure space and let $\mathcal{H} = L^2(X, \mu)$ with $\langle f, g \rangle = \int_X fg \, d\mu$. ♦

On every inner product space it is possible to define a function $\|\cdot\| : \mathcal{H} \to \mathbb{R}$ by

$$\|h\| = \sqrt{\langle h, h \rangle}. \tag{10.13}$$

We list some basic properties of this function (including the fact that it is a norm) and leave the verification to the reader (Exercises 10.5.3–10.5.6).

**Proposition 10.5.4.** ***Cauchy–Schwarz Inequality.*** *For any $x, y \in \mathcal{H}$,*

$$|\langle x, y \rangle| \leq \|x\| \|y\|.$$

**Proposition 10.5.5.** *The function $\| \, \|$ defined by (10.13) is a norm on $\mathcal{H}$.*

**Corollary 10.5.6.** *For a fixed $y \in \mathcal{H}$ the inner product $\langle x, y \rangle$ is a continuous function of $x$ in the norm topology. An analogous statement is true when we fix $x$.*

**Proposition 10.5.7.** ***The Parallelogram Identity.*** *For any* $x, y \in \mathcal{H}$,

$$\|x-y\|^2 + \|x+y\|^2 = 2\|x\|^2 + 2\|y\|^2.$$

*Remark* 10.5.8. In fact, more is true. If $X$ is a normed linear space in which the Parallelogram Identity holds, then there is an inner product on $X$ such that $\|x\|^2 = \langle x, x\rangle$. (See Exercise 10.5.11.)

**Definition 10.5.9.** An inner product space $\mathcal{H}$ is called a **Hilbert space** if it is a Banach space with respect to the norm defined by (10.13).

*Remark* 10.5.10. From now on, whenever we talk about Hilbert space we will assume that an inner product is given and that the norm is related to it by (10.13).

**Example 10.5.11.** $\ell^2$ is an inner product space (Example 10.5.2) that is complete (Example 10.2.14 and the Riesz–Fischer theorem). Therefore, it is a Hilbert space. ♦

**Example 10.5.12.** $L^2$ is an inner product space (Example 10.5.3) that is complete (the Riesz–Fischer theorem). Therefore, it is Hilbert space. ♦

**Example 10.5.13. An inner product space that is not Hilbert space.**

Let $\mathcal{H}$ be the set $C[-1,1]$ with the inner product $\langle f, g\rangle = \int_{-1}^{1} fg$. Then $\mathcal{H}$ is an inner product space that is not Hilbert space.

**Proof:** By Exercise 10.5.2, $\langle f, g\rangle$ is an inner product on $L^2(-1,1)$, and all the more on its subset $C[-1,1]$. The corresponding norm is given by

$$\|f\| = \left(\int_{-1}^{1} |f|^2\right)^{1/2},$$

and it remains to show that $\mathcal{H}$ is not complete with respect to this norm. We define a sequence of functions by

$$f_n(x) = \begin{cases} 0 & \text{if } -1 \le x < 0 \\ nx & \text{if } 0 \le x < \frac{1}{n} \\ 1 & \text{if } \frac{1}{n} \le x \le 1. \end{cases}$$

Then $\{f_n\}_{n=1}^{\infty}$ is a Cauchy sequence in $\mathcal{H}$ but it is not convergent. (See Exercise 10.5.7.) ♦

**Example 10.5.14. An inner product space that is not Hilbert space.**

The space $C[0,1]$ with $\| \; \|_{max}$ is not Hilbert space.

**Proof.** We need to show that there is no inner product on $C[0,1]$ such that $\| \; \|_{max}$ and this inner product are related as in (10.13). By Proposition 10.5.7, it suffices to exhibit 2 functions that do not satisfy the Parallelogram Identity. So, let

$$f(x) = \begin{cases} 1-2x & \text{if } 0 \le x < \frac{1}{2} \\ 0 & \text{if } \frac{1}{2} \le x \le 1, \end{cases} \qquad g(x) = \begin{cases} 0 & \text{if } 0 \le x < \frac{1}{2} \\ 2x-1 & \text{if } \frac{1}{2} \le x \le 1. \end{cases}$$

It is easy to verify that $f$ and $g$ do not satisfy the Parallelogram Identity, whence $C[0,1]$ with $\| \|_{max}$ is not Hilbert space. ♦

One of the nice properties of Hilbert space is that it can be decomposed into a direct sum of subspaces. We will now work on establishing the precise theorem. First, we need to articulate what type of subspaces we are interested in.

**Definition 10.5.15.** Let $\mathcal{H}$ be Hilbert space and $\mathcal{K} \subset \mathcal{H}$. If $\mathcal{K}$ is itself Hilbert space with respect to the algebraic and metric structure inherited from $\mathcal{H}$, we say that it is a **Hilbert subspace** of $\mathcal{H}$.

**Example 10.5.16. A Hilbert subspace.**

Let $\mathcal{H}$ be as in Example 10.5.3 and let $\mathcal{K}$ consists of those $L^2$ functions that vanish on $[0, 1/2]$. We leave the verification to the reader (Exercise 10.3.38).♦

Next, we make precise the concept of orthogonality.

**Definition 10.5.17.** Two vectors $x, y$ in the inner product space $\mathcal{H}$ are **orthogonal** if $\langle x, y \rangle = 0$. In that case we write $x \perp y$.

**Example 10.5.18. Orthogonal functions in $L^2(0, 2\pi)$.**

Let $\mathcal{H} = L^2([0, 2\pi], m)$ and let $f(x) = \cos x$, $g(x) = \sin x$. Then

$$\int_0^{2\pi} \cos x \sin x \, dx = 0$$

so $f \perp g$. ♦

We note that the orthogonality of these two functions has nothing to do with the angle at which their graphs intersect.

**Example 10.5.19. Orthogonal sets in $\mathbb{R}^2$.**

Let $\mathcal{K}_1 := \{(x, 0) : x \in \mathbb{R}\}$ and $\mathcal{K}_2 := \{(0, y) : y \in \mathbb{R}\}$. Then each vector in $\mathcal{K}_1$ is orthogonal to each vector in $\mathcal{K}_2$. ♦

Actually, much more is true: $\mathbb{R}^2$ is a direct sum of $\mathcal{K}_1 := \{(x, 0) : x \in \mathbb{R}\}$ and $\mathcal{K}_2 := \{(0, y) : y \in \mathbb{R}\}$, and they are both *subspaces* of $\mathbb{R}^2$. It turns out such a phenomenon is typical for Hilbert space.

**Definition 10.5.20.** If $\mathcal{K}$ is a subset of $\mathcal{H}$ we denote by $\mathcal{K}^\perp$ the set of vectors in $\mathcal{H}$ that are orthogonal to $\mathcal{K}$ and we call it the **orthogonal complement** of $\mathcal{K}$.

*Remark* 10.5.21. Although $\mathcal{K}$ can be any subset of $\mathcal{H}$, we will be mostly interested in the cased when it is a subspace of $\mathcal{H}$.

For Definition 10.5.20 to offer a generalization of Example 10.5.19 we need to establish that $\mathcal{K}^\perp$ is a subspace of $\mathcal{H}$.

**Proposition 10.5.22.** *Let $\mathcal{K}$ be a subset of $\mathcal{H}$. The set $\mathcal{K}^\perp$ is a Hilbert subspace of $\mathcal{H}$.*

*Proof.* Let $x, y \in \mathcal{K}^\perp$ and $\alpha, \beta \in \mathbb{R}$. If $z \in \mathcal{K}$, then

$$\langle \alpha x + \beta y, z \rangle = \alpha \langle x, z \rangle + \beta \langle y, z \rangle = 0 + 0 = 0,$$

so $\alpha x + \beta y \in \mathcal{K}^\perp$, hence $\mathcal{K}^\perp$ is a vector subspace of $\mathcal{H}$.

Let $x$ be a point in the closure of $\mathcal{K}^\perp$. Then there exists a sequence $\{x_n\}_{n=1}^\infty$ in $\mathcal{K}^\perp$ that converges to $x$. Let $y \in \mathcal{K}$. Then, taking into account the continuity of the inner product (Corollary 10.5.6),

$$\langle x, y \rangle = \langle \lim x_n, y \rangle = \lim \langle x_n, y \rangle = 0,$$

so $x \in \mathcal{K}^\perp$, hence $\mathcal{K}^\perp$ is closed. By Proposition 10.1.15, $\mathcal{K}^\perp$ is complete, and the proposition is proved. □

As we said earlier, the most interesting situation occurs when $\mathcal{K}$ is itself a Hilbert subspace of $\mathcal{H}$. Our ultimate goal is to show that in such a situation every vector can be written as a sum of two vectors, one from each of these subspaces. The next result is a big step in this direction.

**Proposition 10.5.23.** *Let $\mathcal{K}$ be a Hilbert subspace of $\mathcal{H}$, let $x \in \mathcal{H} \setminus \mathcal{K}$, and let*

$$d = d(x, \mathcal{K}) = \inf\{\|x - y\| : y \in \mathcal{K}\}.$$

*Then there is exactly one vector $y \in \mathcal{K}$ such that $d = \|x - y\|$.*

*Proof.* Let $\{y_n\}_{n=1}^\infty \subset \mathcal{K}$ be chosen so that

$$\lim_n \|x - y_n\| = d. \tag{10.14}$$

By the Parallelogram Identity, for any positive integrers $m$ and $n$,

$$\begin{aligned} 2\|y_m - x\|^2 + 2\|y_n - x\|^2 &= \|y_m - y_n\|^2 + \|(y_m - x) + (y_n - x)\|^2 \\ &= \|y_m - y_n\|^2 + 4\left\|\frac{y_m + y_n}{2} - x\right\|^2 \\ &\geq \|y_m - y_n\|^2 + 4d^2, \end{aligned}$$

because $(y_m + y_n)/2 \in \mathcal{K}$. It follows from (10.14) that the left side of this inequality has a limit $4d^2$, whence $\{y_n\}$ is a Cauchy sequence. Since $\mathcal{H}$ is complete, $\{y_n\}$ converges to a vector $y \in \mathcal{H}$. Moreover, $\{y_n\}$ is in the closed set $\mathcal{K}$ so $y \in K$. The continuity of the norm and (10.14) imply that $d = \|x - y\|$.

It remains to establish the uniqueness of $y$. Suppose that $z$ is another element of $\mathcal{K}$ satisfying $\|x - z\| = d$. By the Parallelogram Identity,

$$\begin{aligned} \|y - z\|^2 &= 2\|y - x\|^2 + 2\|x - z\|^2 - \|y + z - 2x\|^2 \\ &= 4d^2 - 4\left\|\frac{y + z}{2} - x\right\|^2 \leq 0, \end{aligned}$$

because $(y+z)/2 \in \mathcal{K}$ implies that

$$\left\| \frac{y+z}{2} - x \right\|^2 \geq d^2.$$

Thus, $\|y-z\|^2 = 0$, whence $z = y$. □

*Remark* 10.5.24. We call the vector $y$ the **orthogonal projection** of $x$ to a Hilbert subspace $\mathcal{K}$.

Now we are ready to establish the decomposition result announced earlier.

**Theorem 10.5.25.** *Let $\mathcal{K}$ be a Hilbert subspace of $\mathcal{H}$, let $x \in \mathcal{H} \setminus \mathcal{K}$, and let $y$ be the point in $\mathcal{K}$ closest to $x$. Then $z := x - y$ is orthogonal to $\mathcal{K}$. In other words, every vector $x \in \mathcal{H}$ can be decomposed in a unique way as $x = y + z$, where $y \in \mathcal{K}$ and $z \in \mathcal{K}^\perp$.*

*Proof.* Let $x \in \mathcal{H} \setminus \mathcal{K}$. By Proposition 10.5.23, there is a unique $y \in \mathcal{K}$ closest to $x$. Let $w$ be a nonzero vector in $\mathcal{K}$ and $t$ a nonzero real number. Then $y - tw \in \mathcal{K}$, so

$$\begin{aligned}\|x-y\|^2 \leq \|x-(y-tw)\|^2 &= \|(x-y)+tw\|^2 \\ &= \|x-y\|^2 + t^2\|w\|^2 + 2t\langle x-y, w\rangle.\end{aligned}$$

Simplifying this inequality, we obtain

$$\|w\|^2 + \frac{2}{t}\langle x-y, w\rangle \geq 0, \qquad \text{for each } t \in \mathbb{R}.$$

If $\langle x-y, w\rangle \neq 0$, taking

$$t = -\frac{\langle x-y, w\rangle}{\|w\|^2},$$

we obtain $-\|w\|^2 \geq 0$, which is impossible. It follows that $\langle x-y, w\rangle = 0$. Since $w \in \mathcal{K}$ is arbitrary, we conclude that $x-y$ is orthogonal to $\mathcal{K}$. Then $x = y+z$, where $z = x - y \in \mathcal{K}^\perp$.

The decomposition is unique. If $x = y_1 + z_1 = y_2 + z_2$ then $y_1 - y_2 = z_2 - z_1$ and this vector belongs to both subspaces $\mathcal{K}$ and $\mathcal{K}^\perp$, hence it must be 0 (Exercise 10.5.22). □

**Corollary 10.5.26.** *If $\mathcal{K}$ is a proper Hilbert subspace of $\mathcal{H}$, there exists a nonzero vector orthogonal to $\mathcal{K}$.*

The space that originated through the work of Hilbert is $\ell^2$. In a paper [108] from 1908, Schmidt introduced orthogonality, vector subspaces, and established the existence of a projection of a point on a closed subspace (all in $\ell^2$). The abstract Hilbert space appears in von Neumann's article [78] who coined the term Hilbert space. Before that, it was known that spaces $L^2$ and $\ell^2$ are isomorphic, which likely inspired Hilbert to create a unified theory of such spaces.

DAVID HILBERT (1862–1943) was a German mathematician. He was born and raised in Königsberg, Germany (Kaliningrad, Russia, from 1945). His mathematical talent was not noticed until the last year of high school. After receiving his doctorate from the University of Königsberg in 1884, he remained there until 1895 when he moved to the University of Göttingen, where he remained for the rest of his career. He is considered to have been one of the most influential and universal mathematicians of the nineteenth and early twentieth centuries. Before 1900 he was considered to be the world's leading expert in invariant theory (abstract algebra), he produced brilliant work in number theory, and his book *Grundlagen der Geometrie* (The Foundations of Geometry) contains the axiomatization of geometry. At the International Congress of Mathematicians in Paris in 1900 he presented 23 unsolved problems. To this day it is generally accepted as the most influential collection of open problems ever to be produced by an individual mathematician. It appears that it inspired Hilbert himself because his later work outshone the earlier. His work on integral equations led him to define what is now known as a Hilbert space. In mathematical physics, he developed *Einstein field equations* simultaneously with Einstein but inspired by their communication during that period. His work laid the foundation for the mathematical formulation of quantum mechanics. The epitaph on his tombstone in Göttingen consists of the famous lines he spoke at the conclusion of his retirement address: *We must know, we shall know.*

## Exercises

10.5.1. Prove that $\langle x, y\rangle = \sum_{k=1}^{\infty} x_k y_k$ is an inner product on $\ell^2$.

10.5.2. Prove that $\langle f, g\rangle = \int_X fg\, d\mu$ is an inner product on $L^2(X, \mu)$.

10.5.3. Prove Proposition 10.5.4. Use it to derive Corollary 10.5.6.

10.5.4. Prove that equality holds in Cauchy–Schwarz inequality if and only if $x = 0$ or $y = \lambda x$, for some $\lambda \in \mathbb{R}$.

10.5.5. Prove Proposition 10.5.5.

10.5.6. Prove Proposition 10.5.7.

10.5.7. Prove that the sequence $\{f_n\}$ in Example 10.5.13 is Cauchy but not convergent.

10.5.8. Prove that $L^p(0,1)$ is not Hilbert space if $p \neq 2$.

10.5.9. Prove that $\ell^p$ is not Hilbert space if $p \neq 2$.

10.5.10. Prove that $C[a,b]$ is not Hilbert space.

10.5.11.* Let $X$ be a normed linear space in which the Parallelogram Identity holds. Prove that there is an inner product on $X$ such that $\|x\|^2 = \langle x, x\rangle$, for all $\in X$.

10.5.12. Let $\mathcal{H}$ be the set of all absolutely continuous functions $f : [0,1] \to \mathbb{R}$ such that $f(0) = 0$ and $f' \in L^2[0,1]$. Let $\langle f, g\rangle = \int_0^1 f'(t)g'(t)\, dt$ for $f, g \in \mathcal{H}$. Prove that $\mathcal{H}$ is Hilbert space.

10.5.13. Prove the **Polarization Identity**: For any $x, y \in \mathcal{H}$,

$$\langle x, y\rangle = \frac{1}{4}\left(\|x+y\|^2 - \|x-y\|^2\right).$$

10.5.14. A vector $h$ in Hilbert space $\mathcal{H}$ is an *extreme point* of a set $A$ if

$$h = tx + (1-t)y, \quad 0 \le t \le 1, \quad x, y \in \mathcal{H} \quad \Rightarrow \quad x = y = h.$$

Prove that every unit vector in $\mathcal{H}$ is an extreme point of the unit sphere.

10.5.15. Let $\{x_n\}, \{y_n\}, x, y$ be vectors in Hilbert space such that $\lim x_n = x$ and $\lim y_n = y$. Prove that $\lim \langle x_n, y_n \rangle = \langle x, y \rangle$.

10.5.16. Let $(X, \mathfrak{M}, \mu)$ be a measure space and let $\{f_n\}, f \in L^2(X, \mu)$ such that

$$\lim_{n\to\infty} \int_X f_n f \, d\mu = \lim_{n\to\infty} \int_X f_n^2 \, d\mu = \int_X f^2 \, d\mu.$$

Prove that $\lim f_n = f$ in the norm of $L^2(X, \mu)$.

10.5.17. Let $\mathcal{H}$ be Hilbert space and let $X$ denote the set of all sequences $\mathbf{x} = \{x_n\}$, where for each $n \in \mathbb{N}$, $x_n \in \mathcal{H}$ and $\|\mathbf{x}\| := \sup\{\|x_n\| : n \in \mathbb{N}\} < \infty$. Prove that with the natural definition of addition and scalar multiplication $X$ is a normed linear space. Is it a Banach space? Is it Hilbert space?

In Exercises 10.5.18–10.5.21 a vector subspace $V$ of $L^2[0,1]$ is given consisting of some polynomials $p(x) = \sum_{k=0}^{\deg p} c_k x^k$. Find its orthogonal complement.

10.5.18. All polynomials.

10.5.19.* Polynomials of $x^2$, i.e., such that $c_{2k-1} = 0$, for all $k \in \mathbb{N}$.

10.5.20.* Polynomials with the property that $c_0 = 0$.

10.5.21.* Polynomials satisfying $\sum_{k=0}^{\deg p} c_k = 0$.

10.5.22. Let $\mathcal{K}$ be a subspace of Hilbert space $\mathcal{H}$. Prove that $\mathcal{K} \cap \mathcal{K}^\perp$ consists of zero vector alone.

10.5.23. Let $\mathcal{K}, \mathcal{M}$ be two subspaces of Hilbert space $\mathcal{H}$ such that $\mathcal{K} \subset \mathcal{M}$. Prove that $\mathcal{M}^\perp \subset \mathcal{K}^\perp$.

10.5.24.* Let $S$ be a subset of Hilbert space. Prove that $(S^\perp)^\perp$ equals the closed linear span of vectors in $S$.

10.5.25.* Let $V$ be a closed subspace of Hilbert space $\mathcal{H}$, and let $x_0 \in \mathcal{H}$. Show that

$$\min\{\|x - x_0\| : x \in V\} = \max\{|\langle x_0, y\rangle| : y \in V^\perp, \ \|y\| = 1\}.$$

10.5.26. Let $S$ be a subset of $\mathcal{H}$. Prove that $S^\perp = S^{\perp\perp\perp}$.

10.5.27. Let $\mathcal{K}, \mathcal{M}$ be two closed subspaces of Hilbert space $\mathcal{H}$ that are orthogonal to each other. Prove that $\mathcal{K} + \mathcal{M}$ is a closed subspace of $\mathcal{H}$.

In Exercises 10.5.28–10.5.30 $X$ is the vector space of all finite sequences equipped with the $\ell^2$ norm and the inner product. For the given subspace $\mathcal{M} \subset X$ determine $\overline{\mathcal{M}}$ and $\mathcal{M}^\perp$.

10.5.28. $\mathcal{M} = \{x = \{x_n\} \in X : \sum_n x_n = 0\}$.

10.5.29. $\mathcal{M} = \{x = \{x_n\} \in X : \sum_n x_n/n = 0\}$.

10.5.30. $\mathcal{M} = \{x = \{x_n\} \in X : \sum_n x_n/\sqrt{n} = 0\}$.

10.5.31. Let $\mathcal{K}, \mathcal{M}$ be two closed subspaces of Hilbert space $\mathcal{H}$. Prove that

(a) $(\mathcal{K} + \mathcal{M})^{\perp} = \mathcal{K}^{\perp} \cap \mathcal{M}^{\perp}$.

(b) $(\mathcal{K} \cap \mathcal{M})^{\perp} = \overline{\mathcal{K}^{\perp} + \mathcal{M}^{\perp}}$.

10.5.32. Let $\mathcal{M}$ be a subspace of Hilbert space $\mathcal{H}$, let $x \in \mathcal{H} \setminus \mathcal{M}$, let $d = d(x, \mathcal{M})$, and let $\{x_n\} \subset \mathcal{M}$ be such that $\lim_n \|x_n - x\| = d$. Prove that $\{x_n\}$ is a Cauchy sequence in $\mathcal{M}$.

## 10.6 Bessel's Inequality and Orthonormal Bases

In every Euclidean space there is a basis that consists of mutually orthogonal vectors. For example, in $\mathbb{R}^3$ such vectors are $\mathbf{i} = (1, 0, 0)$, $\mathbf{j} = (0, 1, 0)$, and $\mathbf{k} = (0, 0, 1)$. Our goal is to establish the existence of such a basis in general Hilbert space.

**Definition 10.6.1.** A subset $S$ of Hilbert space $\mathcal{H}$ is **orthogonal** if every two vectors in $S$ are orthogonal. If, in addition, each vector in $S$ has norm 1, then $S$ is **orthonormal**. We call $S$ an **orthonormal set** or an **orthonormal system**.

**Example 10.6.2. An orthonormal set.**

Let $e_n$ denote a sequence that has 1 in the $n$th position and zeros elsewhere. The set $S = \{e_1, e_2, e_3, \dots\}$ or any (finite or infinite) subset of $S$ is an orthonormal set in $\ell^2$. ♦

**Example 10.6.3. An orthonormal set.**

In $L^2([0, 2\pi])$ any (finite or infinite) subset of

$$\frac{1}{\sqrt{2\pi}}, \frac{\cos x}{\sqrt{\pi}}, \frac{\sin x}{\sqrt{\pi}}, \frac{\cos 2x}{\sqrt{\pi}}, \frac{\sin 2x}{\sqrt{\pi}}, \dots \tag{10.15}$$

is an orthonormal set.

This is a well-known fact from advanced calculus that can be verified easily. ♦

From the theory of Fourier series we know that if $f \in L^2([0, 2\pi])$ its Fourier coefficients can be calculated using the formulas

$$a_0 = \int_0^{2\pi} f(x) \frac{1}{\sqrt{2\pi}} \, dx, \quad a_n = \int_0^{2\pi} f(x) \frac{\cos nx}{\sqrt{\pi}} \, dx,$$
$$b_n = \int_0^{2\pi} f(x) \frac{\sin nx}{\sqrt{\pi}} \, dx.$$

This means that each of them equals $\langle f, g_n \rangle$, where $g_n$ is one of the functions in (10.15). This motivated the following definition.

**Definition 10.6.4.** Let $S = \{x_\alpha\}_{\alpha \in A}$ be an orthonormal system in $\mathcal{H}$ and let $x \in \mathcal{H}$. The **Fourier coefficients** of $x$ with respect to $S$ are defined by

$$\hat{x}_\alpha = \langle x, x_\alpha \rangle, \quad \alpha \in A.$$

The following result is left as Exercise 10.6.1.

**Theorem 10.6.5. *The General Pythagorean Identity.*** *Let $x_1, x_2, \ldots, x_n$ be $n$ orthonormal vectors in $\mathcal{H}$, let $\alpha_1, \ldots, \alpha_n \in \mathbb{R}$, and let $x = \alpha_1 x_1 + \cdots + \alpha_n x_n$. Then*

$$\alpha_k = \langle x, x_k \rangle, \quad 1 \le k \le n, \quad \text{and} \quad \|x\|^2 = \alpha_1^2 + \cdots + \alpha_n^2.$$

In Theorem 10.6.5 the vector $x$ belongs to the linear span of $x_1, x_2, \ldots, x_n$. If it does not we have the following inequality.

**Theorem 10.6.6.** *Let $x_1, x_2, \ldots, x_n$ be $n$ orthonormal vectors in $\mathcal{H}$ and $x \in \mathcal{H}$. Then*

$$\sum_{k=1}^{n} \langle x, x_k \rangle^2 \le \|x\|^2. \tag{10.16}$$

*Proof.* Let $s_n = \sum_{k=1}^{n} \langle x, x_k \rangle x_k$. Then, using the General Pythagorean Identity,

$$\begin{aligned}
0 &\le \|x - s_n\|^2 \\
&= \|x\|^2 + \|s_n\|^2 - 2\langle x, s_n \rangle \\
&= \|x\|^2 + \sum_{k=1}^{n} \hat{x}_k^2 - 2\left\langle x, \sum_{k=1}^{n} \hat{x}_k x_k \right\rangle \\
&= \|x\|^2 + \sum_{k=1}^{n} \hat{x}_k^2 - 2\sum_{k=1}^{n} \hat{x}_k \langle x, x_k \rangle \\
&= \|x\|^2 - \sum_{k=1}^{n} \hat{x}_k^2,
\end{aligned}$$

so inequality (10.16) is proved. □

Next, we would like to extend Theorem 10.6.6 to infinite orthonormal sets. In that case we need to define what we mean by

$$\sum_{\alpha \in A} \hat{x}_\alpha^2 \quad \text{and} \quad \sum_{\alpha \in A} \hat{x}_\alpha x_\alpha \tag{10.17}$$

because the set $A$ may be uncountable. For that we need the following theorem.

**Theorem 10.6.7.** *Let $\{x_\alpha\}_{\alpha \in A}$ be an orthonormal set in $\mathcal{H}$. Among all Fourier coefficients of any vector $x \in \mathcal{H}$ the nonzero coefficients form a countable set.*

*Proof.* For a fixed $n \in \mathbb{N}$ let

$$E_n = \{\hat{x}_\alpha : |\hat{x}_\alpha| \geq 1/n\}.$$

Suppose that $\hat{x}_{\alpha_1}, \hat{x}_{\alpha_2}, \ldots, \hat{x}_{\alpha_k}$ belong to $E_n$. By Theorem 10.6.6,

$$k \cdot \frac{1}{n^2} \leq \sum_{i=1}^{k} |\hat{x}_{\alpha_i}|^2 \leq \|x\|^2,$$

so $k \leq n^2\|x\|^2$, whence each set $E_n$ must be finite. Thus the set $\cup_{n=1}^{\infty} E_n$ is countable. □

From Theorem 10.6.7 we deduce that the sums in (10.17) are countable so if we denote by $\{\hat{x}_n\}_{n=1}^{\infty}$ the nonzero Fourier coefficients of $x$, these sums can be written as

$$\sum_{n=1}^{\infty} \hat{x}_n^2 \quad \text{and} \quad \sum_{n=1}^{\infty} \hat{x}_n x_n. \tag{10.18}$$

There remains the issue of convergence of these series and whether their sums depend on the order of terms.

It is not hard to see that the first one converges. Indeed, Theorem 10.6.6 applies to any of its partial sums, so they are all bounded and the series converges (absolutely, because all terms are positive).

Regarding the second series in (10.18), we need to show that it converges regardless of the ordering of its terms. We will prove a more general result.

**Proposition 10.6.8.** *Let $\{x_n\}_{n=1}^{\infty}$ be an orthonormal set in $\mathcal{H}$ and suppose that $\{\alpha_n\}_{n=1}^{\infty}$ are real numbers for which*

$$\sum_{n=1}^{\infty} \alpha_n^2 < \infty. \tag{10.19}$$

*Then the series*

$$\sum_{n=1}^{\infty} \alpha_n x_n \tag{10.20}$$

*converges unconditionally, i.e., regardless of the ordering of its terms, to an element of $\mathcal{H}$.*

*Proof.* Let us select an arbitrary ordering of the terms of (10.20) and denote the new series (with respect to this ordering) by

$$\sum_{n=1}^{\infty} \beta_n^2 \quad \text{and} \quad \sum_{n=1}^{\infty} \beta_n y_n. \tag{10.21}$$

If we denote partial sums of the series in (10.21) by $s_n$ and $t_n$, respectively then, for any $m, n \in \mathbb{N}$ such that $m \geq n$,

$$\|t_m - t_n\|^2 = \left\| \sum_{k=n+1}^{m} \beta_k y_k \right\|^2 = \left\langle \sum_{k=n+1}^{m} \beta_k y_k, \sum_{j=n+1}^{m} \beta_j y_j \right\rangle$$
$$= \sum_{k=n+1}^{m} |\beta_k|^2 = |s_m - s_n|^2 \to 0, \quad m, n \to \infty,$$

so $\{t_n\}_{n=1}^{\infty}$ is a Cauchy sequence in $\mathcal{H}$, hence it converges because $\mathcal{H}$ is complete. Thus, every rearrangement of (10.20) converges. That the sum is the same element of $\mathcal{H}$ is Exercise 10.6.9. □

Having established that the sums in (10.17) are well-defined and that they do not depend on the order of terms, we get an immediate consequence of Theorem 10.6.6.

**Theorem 10.6.9. *Bessel's Inequality.*** *Let $\{x_\alpha\}_{\alpha \in A}$ be an orthonormal set in $\mathcal{H}$ and let $x \in \mathcal{H}$. Then*

$$\sum_{\alpha \in A} |\hat{x}_\alpha|^2 \leq \|x\|^2, \tag{10.22}$$

*and the Fourier expansion of $x$,*

$$\sum_{\alpha \in A} \hat{x}_\alpha x_\alpha$$

*converges absolutely.*

A fundamental question is whether in every Hilbert space there exists an orthonormal set for which (10.22) would be an identity, and the Fourier expansion of every $x \in \mathcal{H}$ would converge to $x$. If the answer is affirmative, then a follow-up question would be how to characterize such orthonormal systems. As we will discover (Theorem 10.6.12 below), a part of the answer is contained in the following definition and the theorem.

**Definition 10.6.10.** An orthonormal set $\{x_\alpha\}_{\alpha \in A}$ is **complete (maximal)** in $\mathcal{H}$ if it is not properly contained in another orthonormal set.

**Theorem 10.6.11.** *Every Hilbert space different from $\{0\}$ contains a complete orthonormal system.*

*Proof.* First, there is always an orthonormal system in $\mathcal{H}$. Such is a set that consists of one unit vector. Let $S$ be an orthonormal system and let $\mathcal{S}$ denote the set of all orthonormal systems in $\mathcal{H}$ that contain $S$. This is a partially ordered set (by inclusion) and Hausdorff maximal principle (page 22) implies that there exists a maximal chain $\mathcal{C}$. Let $C$ denote the union of all sets in $\mathcal{C}$. We will show that $C$ is a complete orthonormal system.

If $C$ consists of a single element then it is trivially an orthonormal system. Otherwise, let $x_1, x_2 \in C$. Then there are orthonormal sets $A_1, A_2$ in $\mathcal{C}$ such that $x_1 \in A_1$ and $x_2 \in A_1$. Since $\mathcal{C}$ is a chain, we may assume that $A_1 \subset A_2$. Thus, both $x_1, x_2 \in A_2$ so

$$\langle x_1, x_2 \rangle = \begin{cases} 1 & \text{if } x_1 = x_2 \\ 0 & \text{if } x_1 \neq x_2. \end{cases}$$

Since $x_1, x_2$ were arbitrary, we conclude that $C$ is an orthonormal system.

To show that it is complete, suppose to the contrary that $C$ is contained in another orthonormal system $C'$. Then $C'$ is not contained in the chain $\mathcal{C}$, so if we add $C'$ to $\mathcal{C}$ we would get another chain, which would contradict the maximality of $\mathcal{C}$. □

The following theorem is the main result of this section.

**Theorem 10.6.12.** *Let $S = \{x_\alpha\}_{\alpha \in A}$ be an orthonormal set in $\mathcal{H}$. The following are equivalent:*

*(a) $\{x_\alpha\}_{\alpha \in A}$ is complete in $\mathcal{H}$.*

*(b) $x = \sum_{\alpha \in A} \hat{x}_\alpha x_\alpha$, for every $x \in \mathcal{H}$.*

*(c) The set of finite linear combinations of elements of $S$ is dense in $\mathcal{H}$.*

*(d) $\|x\|^2 = \sum_{\alpha \in A} |\hat{x}_\alpha|^2$, for every $x \in \mathcal{H}$.*

*(e) $\langle x, y \rangle = \sum_{\alpha \in A} \hat{x}_\alpha \hat{y}_\alpha$ for any $x, y \in \mathcal{H}$.*

*Proof.*

(a) $\Rightarrow$ (b) Let $x \in \mathcal{H}$. By Theorem 10.6.9, the series $\sum_{\alpha \in A} \hat{x}_\alpha x_\alpha$ is convergent. Let us denote its sum by $y$, and let $z = x - y$. For any $\beta \in A$, using the continuity of inner product,

$$\begin{aligned} \langle z, x_\beta \rangle &= \langle x, x_\beta \rangle - \langle y, x_\beta \rangle = \langle x, x_\beta \rangle - \left\langle \sum_{\alpha \in A} \hat{x}_\alpha x_\alpha, x_\beta \right\rangle \\ &= \langle x, x_\beta \rangle - \sum_{\alpha \in A} \hat{x}_\alpha \langle x_\alpha, x_\beta \rangle = \langle x, x_\beta \rangle - \hat{x}_\beta = 0. \end{aligned}$$

This shows that $z$ is orthogonal to every element of a complete orthonormal set $S$, so it must be 0, otherwise $S \cup \{z/\|z\|\}$ would be another orthonormal set.

(b) $\Rightarrow$ (c) Let $x \in \mathcal{H}$ and let $\varepsilon > 0$. By Theorem 10.6.9 and (b), there exists a vector

$$z = \sum_{k=1}^{n} \hat{x}_k x_k$$

such that $\|x - z\| < \varepsilon$. Since $z$ is a finite linear combination of elements of $S$, (c) is proved.

(c) ⇒ (a) Suppose that (a) is not true. Then $S$ is properly contained in an orthonormal set, so there exists $z \in \mathcal{H}$ that is orthogonal to every vector in $S$. It follows that $z$ is orthogonal to the closed linear span of $S$ which is $\mathcal{H}$. Thus, $z$ is orthogonal to itself, i.e., $\langle z, z\rangle = 0$, whence $z = 0$.

(b) ⇒ (d) By the continuity of the inner product,

$$\begin{aligned}\|x\|^2 = \langle x, x\rangle &= \left\langle \sum_{\alpha\in A} \hat{x}_\alpha x_\alpha, \sum_{\beta\in A} \hat{x}_\beta x_\beta \right\rangle \\ &= \sum_{\alpha\in A}\sum_{\beta\in A} \hat{x}_\alpha \hat{x}_\beta \langle x_\alpha, x_\beta\rangle = \sum_{\alpha\in A} |\hat{x}_\alpha|^2.\end{aligned}$$

(d) ⇒ (a) If (a) is not true then there is $z \neq 0$ that is orthogonal to every vector in $S$, which leads to a contradiction because

$$\|z\|^2 = \sum_{\alpha\in A} |\hat{z}_\alpha|^2 = \sum_{\alpha\in A} |\langle z, \hat{x}_\alpha\rangle|^2 = 0.$$

(d) ⇔ (e) Taking $x = y$ in (e) we obtain (d). In the other direction, applying (d) to $x + y$ we have

$$\|x + y\|^2 = \sum_{\alpha\in A} |\langle x + y, x_\alpha\rangle|^2,$$

which yields

$$\|x\|^2 + 2\langle x, y\rangle + \|y\|^2 = \sum_{\alpha\in A} \left(|\langle x, x_\alpha\rangle|^2 + 2\langle x, x_\alpha\rangle\langle y, y_\alpha\rangle + |\langle y, x_\alpha\rangle|^2\right)$$

Applying (d) to $x$ and $y$ separately and combining with the last equality, we obtain

$$\langle x, y\rangle = \sum_{\alpha\in A} \langle x, x_\alpha\rangle\langle y, y_\alpha\rangle = \sum_{\alpha\in A} \hat{x}_\alpha \hat{y}_\alpha. \qquad \square$$

*Remark* 10.6.13. In view of the equivalence (a) ⇔ (b), a complete orthonormal system is often called an **orthonormal basis** of $\mathcal{H}$.

*Remark* 10.6.14. The equality in (*d*) is known as **Parseval's Identity**.

**Example 10.6.15. An orthonormal basis for $\ell^2$.**

The set $\{e_n\}_{n=1}^{\infty}$ as defined in Example 10.6.2 is a complete orthonormal system in $\ell^2$.

**Proof:** By Exercise 10.4.4, $\{e_n\}$ satisfies condition (c) of Theorem 10.6.12. ◆

**Example 10.6.16. An orthonormal basis for $L^2[0, 2\pi]$.**

The set $S$ of functions listed in (10.15) is a complete orthonormal system in $L^2([0, 2\pi])$.

**Proof:** By Fejér's Theorem, if $f \in C[0, 2\pi]$ then the Cesàro means of its Fourier series converges uniformly to $f$. It follows that $C[0, 2\pi]$ coincides with the closure of the linear span of $S$ in the norm of $C[0, 2\pi]$:

$$C[0, 2\pi] = \overline{\bigvee S}^{C[a,b]}.$$

We know that if a sequence converges uniformly on $[0, 2\pi]$ then it converges in the norm of $L^2[0, 2\pi]$ (Exercise 10.3.1), so the Cesàro means converge in $L^2[0, 2\pi]$, i.e.,

$$C[0, 2\pi] \subset \overline{\bigvee S}^{L^2[0,2\pi]}. \tag{10.23}$$

If we take the closure of both sides of (10.23) in the norm of $L^2[0, 2\pi]$ and use the density of continuous functions in $L^2[0, 2\pi]$ (Theorem 10.4.13) we obtain that

$$L^2[0, 2\pi] \subset \overline{\bigvee S}^{L^2[0,2\pi]}.$$

Of course, the opposite inclusion is trivial, so it remains to apply the equivalence of (a) and (c) in Theorem 10.6.12. ♦

We close this section by considering applications of the presented material to the theory of the Fourier series. In classical analysis, most of the results for Fourier series are formulated by assuming that the functions are continuous or at worst piecewise continuous, and the convergence of the series is studied as pointwise, or in the mean (using Riemann integral). If instead we consider the space $L^2$, we get the same conclusions from weaker hypotheses. The reason is that these are just special cases of the general Hilbert space results.

- If $x \in L^2([0, 2\pi])$ the Fourier coefficients $\{a_k\}, \{b_k\}$ form sequences converging to 0. (Follows from Bessel's Inequality.) This is known as **Riemann–Lebesgue Lemma.**

- If $x \in L^2([0, 2\pi])$ partial sums $\{s_n\}$ of its Fourier series converge mean square to $x$, i.e.,

$$\lim_n \int_0^{2\pi} |x(t) - s_n(t)|^2 \, dt = 0.$$

(Theorem 10.6.12 (a) $\Leftrightarrow$ (b) and Example 10.6.16.)

- A very hard and deep question in the classical theory of the Fourier series is the following. Suppose that we are given a convergent trigonometric series, i.e., a series of the form

$$\frac{a_0}{2} + \sum_{k=1}^{\infty} (a_k \cos kt + b_k \sin kt), \tag{10.24}$$

that converges pointwise to a function $x(t)$. Does the Fourier series of $x$ coincide with the series (10.24)? When $x \in L^2([0, 2\pi])$, Parseval's Identity and Proposition 10.6.8 imply that the latter series converges in $L^2([0, 2\pi])$. For any $y \in S$, using the continuity of inner product,

$$\begin{aligned}\langle x, y\rangle &= \left\langle \frac{a_0}{2} + \sum_{k=1}^{\infty}(a_k \cos kx + b_k \sin kx), y\right\rangle \\ &= \frac{a_0}{2}\langle 1, y\rangle + \sum_{k=1}^{\infty}\left(a_k\langle \cos kt, y\rangle + b_k\langle \sin kt, y\rangle\right).\end{aligned}$$

Replacing $y$ with functions in (10.15) we get on the right side the coefficient corresponding to it in (10.24). On the other hand, $\langle x, y\rangle$ is by definition a Fourier coefficient of $x$. Thus, assuming that $x \in L^2([0, 2\pi])$, the answer to the question above is in the affirmative.

Notice that, if $x \in L^2([0, 2\pi])$ then its Fourier coefficients satisfy Parseval's Identity and

$$\frac{1}{2}|a_0|^2 + \sum_{k=1}^{\infty}(|a_k|^2 + |b_k|^2) < \infty. \tag{10.25}$$

The question of whether the converse is true was answered independently by Riesz and Fischer.

**Theorem 10.6.17.** ***Riesz–Fischer Theorem.*** *If sequences $\{a_n\}$ and $\{b_n\}$ satisfy (10.25) then there is a function $x \in L^2([0, 2\pi])$ so that $\{a_n\}$ and $\{b_n\}$ are its Fourier coefficients.*

*Proof.* By Proposition 10.6.8, the series

$$\frac{1}{2}a_0 + \sum_{n=1}^{\infty}(a_n \cos nt + b_n \sin nt) \tag{10.26}$$

converges unconditionally to a function $x \in L^2([0, 2\pi])$.

It remains to show that $\{a_n\}$ and $\{b_n\}$ are the Fourier coefficients of $x$. Notice that partial sums of the series (10.26) can be written as

$$s_n = \sum_{j=0}^{2n} c_j x_j,$$

where

$$c_j = \begin{cases} a_0\sqrt{\pi/2} & \text{if } j = 0 \\ a_k\sqrt{\pi} & \text{if } j = 2k \\ b_k\sqrt{\pi} & \text{if } j = 2k-1, \end{cases} \quad \text{and} \quad x_j = \begin{cases} 1/\sqrt{2\pi} & \text{if } j = 0 \\ \cos kt/\sqrt{\pi} & \text{if } j = 2k \\ \sin kt/\sqrt{\pi} & \text{if } j = 2k-1. \end{cases}$$

For every $n$, $\{x_1, x_2, \ldots, x_{2n}\}$ is an orthonormal system in $L^2([0, 2\pi])$. By the General Pythagorean Identity,

$$c_j = \langle s_n, x_j\rangle \quad \text{for all } 1 \le j \le 2n.$$

The result is now obtained by letting $n \to \infty$. □

Riesz [88] and Fischer [34] proved the same result independently and it appeared in the same French journal Comptes Rendus in 1907, although the former appeared in March and the latter in May and the proofs are different. Both results are set in $L^2$, but they can be generalized to any Hilbert space because every Hilbert space is isomorphic to some $L^2$ space. (For Hilbert space isomorphisms see Section 11.2.)

We hope that the reader is convinced in the advantages of having an orthonormal system in Hilbert space. More evidence is supplied throughout the exercises. Therefore, if we have an infinite collection of linearly independent vectors that is not an orthogonal set it would be very convenient, if possible, to replace it by an orthonormal system. Such a process is well-known in linear algebra under the name *Gram-Schmidt orthogonalization method.* It is possible to extend it to the case when the given collection of vectors is countable in a very natural manner—the procedure never stops. For collections that are not countable a little more is needed and we leave it as an exercise.

ERNST SIGISMUND FISCHER (1875–1954) was an Austrian mathematician. He came from a musical family. His father was a composer and professor at the world-famous Vienna Academy and his maternal grandfather was also a composer. Fischer was educated in Vienna, and he studied at the University of Vienna under Mertens and Leopold Gegenbauer (1849–1903). He was awarded his doctorate in 1899. He spent that year at the University of Berlin, then studied at Zürich and Göttingen with Minkowski. From 1902 he held a position at the German Technische Hochschule of Brünn (since 1945 Brno, in the Czech Republic). From 1911 to 1920, he was a professor at the University of Erlangen, where he collaborated with a German mathematician Emmy Noether (1882–1935), significantly impacting her mathematical perspectives. Renowned for his work in algebra, Fischer's most recognized achievement is the Riesz–Fischer Theorem. During World War I, Fischer served from 1915 to 1918. He married Ellis Strauss in 1917 and later worked at the University of Cologne until his retirement in 1938. The retirement was forced because of his Jewish roots. His contributions spanned various mathematical domains, exploring topics such as Hadamard determinants, Sylvester determinants, the Carathéodory Problem, and finite abelian groups. In 1931 he was the president of the German Mathematical Society.

## Exercises

10.6.1. Prove the General Pythagorean Identity (Theorem 10.6.5).

10.6.2. Let $S = \{x_\alpha\}_{\alpha\in A}$ be an orthonormal set in $\mathcal{H}$, let $\mathcal{K}$ be a Hilbert subspace of $\mathcal{H}$ generated by $S$, let $x \in \mathcal{H}$, and let $y$ be the orthogonal projection of $x$ to $\mathcal{K}$. Prove that:

(a) $y = \sum_{\alpha\in A} \hat{x}_\alpha x_\alpha$.

(b) The distance $d$ from $x$ to $\mathcal{K}$ is given by $d^2 = \|x\|^2 - \sum_{\alpha\in A} |\hat{x}_\alpha|^2$.

10.6.3. Let $V$ be the linear span of functions (10.15). Determine whether $V$ is dense in the space $L^2[a, b]$.

10.6.4.* For $m \in \mathbb{N}_0$ and $1 \le n \le 2^m$, let

$$\varphi_{m,n}(x) = \begin{cases} 2^{m/2} & \text{if } \frac{n-1}{2^m} \le x < \frac{n-1/2}{2^m}, \\ -2^{m/2} & \text{if } \frac{n-1/2}{2^m} \le x < \frac{n}{2^m}, \\ 0 & \text{otherwise.} \end{cases}$$

Also, let $\varphi_{0,0}(x) = 1$. Prove that the system $\{\varphi_{m,n}\}$ is an orthonormal basis of $L^2[0,1]$.

10.6.5.* For $m \in \mathbb{N}_0$, let

$$\varphi_m(x) = (-1)^{\lfloor 2^m x \rfloor}. \tag{10.27}$$

Prove that the system $\{\varphi_m\}$ is orthonormal but not complete in $L^2[0,1]$.

10.6.6.** Let $\{\varphi_m\}$ be as in (10.27) and, for $m_1 < m_2 < \cdots < m_n$, let

$$\varphi_{m_1,m_2,\ldots,m_n}(x) = \varphi_{m_1}(x)\varphi_{m_2}(x)\ldots\varphi_{m_n}(x).$$

Prove that the system $\{\varphi_{m_1,m_2,\ldots,m_n}\}$ is orthonormal and complete in $L^2[0,1]$.

10.6.7. Prove that Hilbert space $\mathcal{H}$ is separable if and only if it has a countable orthonormal basis.

10.6.8. Let $\mathcal{H}$ be Hilbert space and $S = \{x_n\}$ be an infinite orthonormal system in $\mathcal{H}$.

(a) Prove that $S$ is closed and bounded, but not totally bounded.

(b) If $\{a_n\} \subset \mathcal{H}$, prove that the set $\{a_n x_n\}$ is totally bounded if and only if $\lim a_n = 0$.

(c) If $A = \{x_n\} \subset \mathcal{H}$ is an orthogonal system, prove that it is totally bounded if and only if $\lim \|x_n\| = 0$.

10.6.9. Let $\mathcal{H}$ be Hilbert space, let $\{x_n\}$ be an orthonormal system in $\mathcal{H}$, let $\{\alpha_n\}_{n=1}^{\infty}$ be real numbers for which the series $\sum_n \alpha_n^2$ converges, and suppose that every rearrangement of the series $\sum_n \alpha_n x_n$ converges. Prove that it must converge to the same vector.

10.6.10. Prove that an orthonormal system $\{f_n\}$ in $L^2(a,b)$ is complete if and only if

$$\sum_{n=1}^{\infty} \left(\int_a^x f_n\right)^2 = x - a, \quad \text{for all } x \in [a,b].$$

10.6.11.* Let $\{f_n\}$ be an orthonormal system in $L^2(0,1)$ and suppose that there is $M > 0$ such that $|f_n(x)| \le M$ a.e., for all $n \in \mathbb{N}$. Let $\{c_n\}$ be a sequence of real numbers such that $\sum_{n=1}^{\infty} c_n f_n$ converges a.e. Prove that $\lim_n c_n = 0$.

10.6.12. Let $\mathcal{H}$ be Hilbert space, let $\{x_n\}$ be an orthonormal system in $\mathcal{H}$, and let $\{\alpha_n\}_{n=1}^{\infty} \in \ell^2$. Prove that $\sum_n \alpha_n x_n$ converges unconditionally. Does the series still converge if $\{\alpha_n\}_{n=1}^{\infty} \notin \ell^2$?

10.6.13. Let $\mathcal{H}$ be Hilbert space and let $\{x_n\}$ be an orthonormal system in $\mathcal{H}$. Find a sequence $\{\alpha_n\}$ of real numbers such that the series $\sum_n \alpha_n x_n$ converges unconditionally but not absolutely.

10.6.14. Let $\mathcal{H}$ be Hilbert space, let $\{x_n\}_{n=1}^{\infty}$ be an orthonormal basis of $\mathcal{H}$, and let $\{y_n\}_{n=1}^{\infty}$ be an orthonormal system in $\mathcal{H}$. Suppose that $\sum_n \|x_n - y_n\|^2 < 1$. Prove that $\{y_n\}_{n=1}^{\infty}$ must also be an orthonormal basis for $\mathcal{H}$. Is this is still true if we only assume that $\sum_n \|x_n - y_n\|^2 < \infty$?

10.6.15.* Let $\mathcal{M}$ be a dense subspace of a separable Hilbert space $\mathcal{H}$. Prove that $\mathcal{H}$ has an orthonormal basis consisting of elements of $\mathcal{M}$.

10.6.16. Let $\{e_n\}$ be an orthonormal basis of Hilbert space $\mathcal{H}$, let

$$x_n := e_{2n}, \quad y_n := \sqrt{1-4^{-n}}\, e_{2n} + 2^{-n}\, e_{2n-1}, \quad \text{for } n = 1, 2, 3, \dots,$$

and let $X$ and $Y$ be the closed linear span of $\{x_n\}$ and $\{y_n\}$, respectively. Prove that:

(a) $\{x_n\}$ and $\{y_n\}$ are orthonormal systems in $\mathcal{H}$, $X \cap Y = \{0\}$, and $X+Y$ is dense in $\mathcal{H}$;

(b) $v := \sum_{n=0}^{\infty} 2^{-n}\, e_{2n+1} \in \mathcal{H} \setminus (X+Y)$ and $X+Y$ is not closed in $\mathcal{H}$.

10.6.17. Let $\{e_n\}$ be an orthonormal basis of Hilbert space $\mathcal{H}$. Show that there exists $x, y \in \mathcal{H}$ such that, for all $n \in \mathbb{N}$, $\langle x, e_n \rangle = 1/n^{4/3}$ and $\langle y, e_n \rangle = 1/n^{2/3}$. Find $\langle x, y \rangle$.

10.6.18. Let $\{e_\alpha\}_{\alpha \in A}$ and $\{f_i\}_{i \in I}$ be two orthonormal bases in Hilbert space $\mathcal{H}$. Prove that sets $A$ and $I$ have the same cardinality.

10.6.19. Let $\mathcal{H}$ be Hilbert space. Prove that the closed unit ball $\overline{B(0,1)}$ is compact if and only if $\mathcal{H}$ is finite-dimensional.

10.6.20. Let $(X, \mathfrak{M}, \mu)$ be a measure space. If $E \in \mathfrak{M}$, we identify $L^2(E, \mu)$ with the subspace of $L^2(X, \mu)$ consisting of functions that vanish outside of $E$. Suppose that $\{E_n\}$ is a disjoint sequence in $\mathfrak{M}$ such that $X = \cup_{n=1}^{\infty} E_n$. Prove that:

(a) $\{L^2(E_n, \mu)\}_{n=1}^{\infty}$ is a sequence of mutually orthogonal subspaces of $L^2(X, \mu)$.

(b) Every $f \in L^2(X, \mu)$ can be written uniquely as $f = \sum_{n=1}^{\infty} f_n$, where $f_n \in L^2(E_n, \mu)$ and the series converges in norm).

(c) If $L^2(E_n, \mu)$ is separable for all $n \in \mathbb{N}$, then so is $L^2(X, \mu)$.

10.6.21. Let $(X, \mathfrak{M}, \mu)$ and $(Y, \mathfrak{N}, \nu)$ be $\sigma$-finite measure spaces such that $L^2(X, \mu)$ and $L^2(Y, \nu)$ are both separable, with orthonormal bases $\{f_n\}$ and $\{g_n\}$ respectively, and let $h_{mn}(x, y) = f_m(x) g_n(y)$. Prove that $\{h_{mn}\}_{m,n \in \mathbb{N}}$ is an orthonormal basis of $L^2(\mu \times \nu)$.

10.6.22. Let $\mathcal{H}$ be Hilbert space, let $\{x_n\}, x$ be vectors in $\mathcal{H}$ such that $\|x_n\| = \|x\| = 1$ for all $n \in \mathbb{N}$, and suppose that $|1 - \langle x_n, x\rangle| \leq (n-1)/n$. Prove that $\{x_n\}$ is not an orthonormal basis of $\mathcal{H}$.

10.6.23. Let $\{e_n\}$ be an orthonormal basis of $L^2([0,1])$. Extend each $e_n$ to $\mathbb{R}$ by making it 0 outside $[0,1]$ and for each $k \in \mathbb{Z}$ define $e_{n,k}(x) = e_n(x-k)$. Prove that $\{e_{n,k} : n \in \mathbb{N},\, k \in \mathbb{Z}\}$ is an orthonormal basis of $L^2(\mathbb{R})$.

10.6.24. Let $\{e_n\}$ be an orthonormal basis of Hilbert space $\mathcal{H}$, and for every $n \in \mathbb{N}$ let $x_n = \sum_{k=1}^n e_k + e_{n+1}$, $y_n = \sum_{k=1}^n e_k - e_{n+1}$, and $z_n = x_n - x_{n+1}$. Prove or disprove:

(a) The span of $\{x_n\}$ is dense in $\mathcal{H}$.

(b) The span of $\{y_n\}$ is dense in $\mathcal{H}$.

(c) The span of $\{z_n\}$ is dense in $\mathcal{H}$.

10.6.25. Let $\mathcal{M} = \{x \in \ell^2 : x_1 - x_2 = 0,\, x_2 + x_3 = 0\}$. Find the projection onto $\mathcal{M}^\perp$ of $\{1/n\}$.

10.6.26. Let $\mathcal{M} = \{f \in L^2([0,1]) : \int_0^1 f(t)\,dt = \int_0^1 tf(t)\,dt = \int_0^1 t^2 f(t)\,dt = 0\}$. Find the function in $\mathcal{M}$ that is closest to $g(t) = t^3$.

10.6.27. Determine the infimum of

$$\int_{-1}^{1} |p(x) - |x||^2\, dx,$$

over all choices of polynomials $p$ of degree at most three.

10.6.28. Let $\{f_n\}$ be an infinite orthonormal system in $L^2(\mathbb{R})$. Given $\varepsilon > 0$, define $E = \{x \in \mathbb{R} : f_n(x) \geq \varepsilon,\, n \in \mathbb{N}\}$. Prove that $E$ is measurable and find its measure.

10.6.29.* Let $\mu$ be a Borel measure on $[0, 2\pi]$, let $a_0 = \frac{1}{\pi}\int_0^{2\pi} d\mu$, $a_n = \frac{1}{2\pi}\int_0^{2\pi} \cos nx\, d\mu$, $b_n = \frac{1}{2\pi}\int_0^{2\pi} \sin nx\, d\mu$, for $n \in \mathbb{N}$, and suppose that $a_0^2 + \sum_{k=1}^\infty (a_k^2 + b_k^2) < \infty$. Prove that $\mu$ is absolutely continuous with respect to $\lambda = m|[0, 2\pi]$ and that $d\mu/d\lambda \in L^2([0, 2\pi])$.

10.6.30. Give an example of a nonseparable Hilbert space.

10.6.31. Let $F = \{x_n\}$ be a countable subset of Hilbert space $\mathcal{H}$ and suppose that every $x \in \mathcal{H}$ is a finite linear combination of some of the vectors in $F$. Prove that $\mathcal{H}$ is finite-dimensional.

10.6.32. Let $\{x_n\}$ be a sequence in Hilbert space $\mathcal{H}$ such that $\sum_{k=1}^\infty \|x_k\|^2 < \infty$ and let $s_n = \sum_{k=1}^n x_k$. Prove or disprove: $\{s_n\}$ is a Cauchy sequence in $\mathcal{H}$.

10.6.33. Let $X$ be a set of real-valued functions on $\mathbb{R}$ such that $f \in X$ if $f$ is zero except on a countable subset of $\mathbb{R}$, and

$$\|f\| := \left(\sum_{x \in \mathbb{R}} |f(x)|^2\right)^{1/2} < \infty.$$

Prove that $X$ is a Banach space with this norm.

10.6.34.* Let $\mathcal{H}$ be Hilbert space, let $\{w_j : j \in J\}$ be a linearly independent set of vectors in $\mathcal{H}$, and let $W = \overline{\bigvee(\{w_j : j \in J\})}$ be the space of all $\sum_{j \in J} c_j w_j$ with at most countably many $c_j \neq 0$ and $\sum \|c_j w_j\|^2 < \infty$. Prove that there is an orthonormal basis $\{b_i\}_{i \in I}$ of $W$ such that each $b_i$ belongs to $\overline{\bigvee(\{w_j : j \in J, j \leq i\})}$.

## 10.7 The Space $C(X)$

Let $X$ be a compact metric space. We define the space $C(X)$ as the set of all continuous functions $f : X \to \mathbb{R}$. It is obvious that $C(X)$ is a vector space with

$$(\alpha f + \beta g)(x) = \alpha f(x) + \beta g(x),$$

where $f, g \in C(X)$, $\alpha, \beta \in \mathbb{R}$, and $x \in X$. Moreover, since a continuous function is bounded on a compact metric space and it attains its extreme values, the formula

$$\|f\| := \max\{|f(x)| : x \in X\}$$

defines a mapping from $C(X)$ to $\mathbb{R}$, and it is not hard to verify that it has all properties of a norm (Exercise 10.7.1). Thus, $C(X)$ is a normed linear spaces and, just like $C[a, b]$, it is complete in the topology induced by the norm (Exercise 10.7.2). In other words, $C(X)$ is a Banach space. In this section we will explore the space $C(X)$. We will characterize its compact subsets (Theorem 10.7.10 below) and present an application of this result in the theory of differential equations.

*Remark* 10.7.1. It is possible to make a weaker assumption about $X$, namely that it is a compact Hausdorff space. In fact, in order to generalize $\mathbb{R}^n$ which is not compact, it is sometimes assumed that $X$ is only locally compact. However, such a presentation requires from a reader a strong background in topology, going beyond the level of this monograph.

It is natural to ask whether the complete metric space $C(X)$ is compact and we start there.

**Example 10.7.2. A bounded sequence in $C(X)$ without a convergent subsequence.**

Let $X = [0, 1]$ and a sequence $\{f_n\}$ defined by $f_n(x) = x^n$. This sequence is bounded in $C[0, 1]$, but has no convergent subsequence. Of course, convergence means convergence in the norm of $C(X)$, i.e., the uniform convergence. ♦

Since a metric space is compact if and only if it is sequentially compact (Theorem 2.3.7) it follows from Example 10.7.2 that $C(X)$ is not a compact metric space. This raises the question of what subsets of $C(X)$ are compact. In order to answer this question we start with a definition.

**Definition 10.7.3.** A collection $\mathcal{F}$ of real-valued functions on a metric space $X$ is **equicontinuous at $x_0 \in X$** if

$$\forall \varepsilon > 0 \quad \exists \delta > 0 \quad \forall f \in \mathcal{F} \quad \forall x \in X \quad d(x, x_0) < \delta \quad \Rightarrow \quad |f(x) - f(x_0)| < \varepsilon. \tag{10.28}$$

$\mathcal{F}$ is **equicontinuous on $X$** if it is equicontinuous at every point of $X$.

*Remark* 10.7.4. It is obvious that each function in an equicontinuous collection is itself continuous.

**Example 10.7.5. An equicontinuous subset of $C[0,1]$.**

The set $\mathcal{F} = \{f : [0,1] \to \mathbb{R} : |f'(x)| \leq 2, \forall x \in [0,1]\}$ is equicontinuous on $[0,1]$.

**Proof:** Let $\varepsilon > 0$ and set $\delta = \varepsilon/2$. If $f \in \mathcal{F}$ and $x, y \in [0,1]$ satisfy $|x - y| < \delta$ then, using Mean Value Theorem, there exists $c \in (x.y)$ such that

$$|f(x) - f(y)| = |f'(c)(x - y)| \leq 2|x - y| < 2\delta = \varepsilon,$$

so $\mathcal{F}$ is equicontinuous on $[0,1]$. ♦

**Example 10.7.6. A subset of $C[0,1]$ that is not equicontinuous.**

The set $\mathcal{F} = \{\sin nx : n \in \mathbb{N}\}$ is not equicontinuous at $x_0 = 0$.

**Proof.** Suppose to the contrary that $\mathcal{F}$ is equicontinuous at $x_0 = 0$. Let $\varepsilon = 1/2$ and $\delta = \delta(1/2)$. Choose

$$n = \left\lfloor \frac{\pi}{2\delta} \right\rfloor + 1, \quad x = \frac{\pi}{2n}.$$

Then $d(x, 0) = |\pi/(2n)| < \delta$ but $|f_n(x) - f_n(0)| = 1 > 1/2$. ♦

We know that in an abstract metric space, a set is sequentially compact if and only if it is complete and totally bounded (Theorem 2.3.7). This is certainly true in $C(X)$ as well but we hope to formulate a sufficient condition in more concrete terms. The critical issue is when it is possible, given a sequence of functions in $C(X)$, to extract a subsequence that converges in the norm of $C(X)$, i.e., uniformly on $X$. Our first step will be much more modest. We will establish only pointwise convergence on a countable dense set, albeit under very modest assumptions.

**Theorem 10.7.7.** *Let $X$ be a separable metric space, let $D$ be a countable, dense set of points in $X$, and let $\{f_n\}$ be real-valued functions such that for every $x \in D$, the sequence $\{f_n(x)\}$ is bounded. Then there exists a subsequence $\{f_{n_k}\}$ such that the sequence $\{f_{n_k}(x)\}$ is convergent for every $x \in D$.*

*Proof.* Let $D = \{x_k\}_{k=1}^{\infty}$. By assumption, the sequence $\{f_n(x_1)\}$ is bounded so Bolzano-Weierstrass Theorem implies that it has a subsequence $\{f_{1k}(x_1)\}$ that converges to a real number $a_1$. The sequence $\{f_{1k}(x_2)\}$ is bounded, so it has a convergent subsequence $\{f_{2k}(x_2)\}$ with limit $a_2$. Proceeding in this fashion we obtain a countable collection of subsequences of our original sequence:

$$\begin{array}{cccc} f_{11} & f_{12} & f_{13} & \cdots \\ f_{21} & f_{22} & f_{23} & \cdots \\ f_{31} & f_{32} & f_{33} & \cdots \\ . & . & . & \cdots \\ . & . & . & \cdots \\ . & . & . & \cdots \end{array}$$

where the sequence in the $n$-th row converges at the points $x_1, \ldots, x_n$, and each row is a subsequence of the one above it.

We define function $f$ on $D$ by $f(x_n) = a_n$ and consider the diagonal sequence $\{g_k\}$ defined by $g_k = f_{kk}$. Notice that for every $n \in \mathbb{N}$, $\lim_k g_k(x_n) = a_n$. Indeed,

$$g_n(x_n), g_{n+1}(x_n), g_{n+2}(x_n), \ldots$$

is a subsequence of $\{f_{nk}(x_n)\}_{k=1}^{\infty}$. Thus, $\lim_k g_k = f$ on $D$. □

Our next step is to expand the pointwise convergence from a countable dense set to the whole metric space. In order to ensure the convergence we need to strengthen the hypotheses.

**Theorem 10.7.8.** *Let $X$ be a separable metric space, let $D$ be a countable, dense set of points in $X$, and let $\{f_n\}$ be real-valued functions such that for every $x \in D$, the sequence $\{f_n(x)\}$ is Cauchy. Suppose also that $\{f_n\}$ is equicontinuous on $X$. Then $\{f_n(x)\}$ is a convergent sequence for all $x \in X$.*

*Proof.* Let $x_0 \in X$. We will show that $\{f_n(x_0)\}$ is a Cauchy sequence and it will follow that it is convergent from the completeness of $\mathbb{R}$.

Let $\varepsilon > 0$ and let $\delta = \delta(\varepsilon/3)$ coming from (10.28). In particular,

$$\forall n \in \mathbb{N} \quad \forall x \in X \quad d(x, x_0) < \delta \quad \Rightarrow \quad |f_n(x) - f_n(x_0)| < \varepsilon/3.$$

Since $D$ is dense, there exists $x \in D$ such that $d(x, x_0) < \delta$. Further, $x \in D$ implies that $\{f_n(x)\}$ is a Cauchy sequence, so there is $N \in \mathbb{N}$ such that

$$m, n \geq N \quad \Rightarrow \quad |f_n(x) - f_m(x)| < \varepsilon/3.$$

Suppose that $m, n \geq N$. Then

$$\begin{aligned} |f_n(x_0) - f_m(x_0)| &\leq |f_n(x_0) - f_n(x)| + |f_n(x) - f_m(x)| + |f_m(x) - f_m(x_0)| \\ &< \frac{\varepsilon}{3} + \frac{\varepsilon}{3} + \frac{\varepsilon}{3} = \varepsilon. \end{aligned}$$ □

Our final step is to show that with further strengthening of the hypotheses we obtain a *uniformly* convergent subsequence.

**Theorem 10.7.9.** ***The Arzelà–Ascoli Theorem.*** *Let $X$ be a compact metric space and $\{f_n\}$ a uniformly bounded equicontinuous sequence of real-valued functions on $X$. Then $\{f_n\}$ has a subsequence that converges uniformly on $X$ to a continuous function $f$ on $X$.*

*Proof.* First, we notice that, by Theorem 2.3.7 and Exercise 2.5.23, $X$ is separable. Thus we can apply Theorems 10.7.7–10.7.8 and we obtain a subsequence $\{g_n\} \subset \{f_n\}$ that converges pointwise on $X$ to a function $f$. Our goal is to prove that the convergence is uniform. Since $C(X)$ is complete it is sufficient to establish that $\{g_n\}$ is a Cauchy sequence in the uniform topology.

To that end, let $\varepsilon > 0$. Since $\{g_n\}$ is equicontinuous, there is $\delta > 0$ such that

$$\forall n \in \mathbb{N} \quad \forall u, v \in X \quad d(u,v) < \delta \quad \Rightarrow \quad |g_n(u) - g_n(v)| < \frac{\varepsilon}{3}. \tag{10.29}$$

Let $K$ be a positive integer such that $K > 1/\delta$. By Theorem 2.3.7, $X$ is totally bounded so there exists a finite set $\{x_i\}_{i=1}^k$ such that

$$X \subset \bigcup_{i=1}^{k} B(x_i, 1/K).$$

Since $\{g_n(x_i)\}_{n=1}^\infty$ converges for all $i$, it follows that there exists $N \in \mathbb{N}$ such that

$$m, n \geq N, \quad 1 \leq i \leq k \quad \Rightarrow \quad |g_n(x_i) - g_m(x_i)| < \frac{\varepsilon}{3}. \tag{10.30}$$

Finally, let $x \in X$. Choose $x_i$ such that $d(x_i, x) < 1/K$, which ensures that $d(x_i, x) < \delta$. If $m, n \geq N$ then (10.29) and (10.30) imply that

$$\begin{aligned} |g_n(x) - g_m(x)| &\leq |g_n(x) - g_n(x_i)| + |g_n(x_i) - g_m(x_i)| + |g_m(x_i) - g_m(x)| \\ &< \frac{\varepsilon}{3} + \frac{\varepsilon}{3} + \frac{\varepsilon}{3} = \varepsilon, \end{aligned}$$

and the proof is complete. □

The Arzelà-Ascoli Theorem allows us to answer the question: what subsets of $C(X)$ are compact? We leave the proof to the reader (Exercise 10.7.23).

**Theorem 10.7.10.** *Let $X$ be a compact metric space and $\mathcal{F} \subset C(X)$. Then $\mathcal{F}$ is compact if and only if it is closed, uniformly bounded, and equicontinuous.*

> A Jewish-Italian mathematician Giulio Ascoli (1843–1896) and Arzelà both worked on the case $X = [0,1]$. In 1883, Ascoli proved that the listed conditions are sufficient for the sequence $\mathcal{F}$ to have a uniformly convergent subsequence, and Arzelà established the opposite implication in 1889. While the results are there a reader might find it challenging to recognize them. A much clearer presentation can be found in Arzelà's paper from 1895. In 1906, Fréchet generalized these results to the case when $X$ is a metric space in [35]. Also, he connected the theorem to the compactness of the set $\mathcal{F}$.

Aside from its theoretical value, the Arzelà–Ascoli Theorem is a powerful tool in the applications of mathematics. Here we will present an existence theorem for a first-order initial value problem that is due to Peano.

**Theorem 10.7.11.** *Let $f(x, y)$ be a function continuous on a closed rectangle $R = [x_0 - a, x_0 + a] \times [y_0 - b, y_0 + b]$. Then there exists $h > 0$ so that the initial value problem*

$$\frac{dy}{dx} = f(x, y), \quad y(x_0) = y_0,$$

*has a solution on the interval $[x_0 - h, x_0 + h]$.*

*Proof.* The function $f$ is continuous on $R$ so it is bounded and there exists $M > 0$ such that $|f(x, y)| \leq M$, for all $(x, y) \in R$. We define $h = \min\{a, b/M\}$ and $I = [x_0 - h, x_0 + h]$.

Let $P = \{x_0 - h = t_0, t_1, \dots, t_{m-1}, t_m = x_0, t_{m+1}, \dots, t_{k-1}, t_k = x_0 + h\}$ be a partition of $I$. We will construct a polygonal line $\varphi$ that goes through $(x_0, y_0)$ devised by Euler as an approximation to a solution. On the interval $[t_m, t_{m+1}]$ it will go through $(x_0, y_0)$ with slope $f(x_0, y_0)$; on $[t_{m+1}, t_{m+2}]$ it will go through $(t_{m+1}, \varphi(t_{m+1}))$ with slope $f(t_{m+1}, \varphi(t_{m+1}))$; we continue this process until we reach $x_0 + h$. A symmetric construction is completed going from $x_0$ towards $x_0 - h$. Notice that the graph of $\varphi$ lies in the region bounded by lines that go through $(x_0, y_0)$ and have slopes $M$ and $-M$. (The definition of $h$ ensures that for $x \in I$, points in this angular region are within the rectangle $R$.)

For each partition of $I$ as above we get a function $\varphi$. Since their graphs are all in the region between two lines, the collection $\mathcal{F}$ of all functions $\varphi$ is uniformly bounded. Moreover, each of them satisfies the inequality $|\varphi(x_1) - \varphi(x_2)| \leq M|x_1 - x_2|$, so $\mathcal{F}$ is equicontinuous. Applying the Arzelà-Ascoli Theorem we obtain a sequence $\{\varphi_n\} \subset \mathcal{F}$ that converges to a function $\varphi$ uniformly on $I$. It is immediately obvious that $\varphi(x_0) = y_0$, so it remains to establish that $\varphi$ is a solution of the given differential equation. To accomplish this goal it suffices to demonstrate that

$$\varphi(x) = y_0 + \int_{x_0}^{x} f(x, \varphi(x))\, dx, \qquad \text{for all } x \in I. \tag{10.31}$$

We will prove this equality for $x \in (x_0, x_0 + h]$. The proof for the other part of $I$ is similar.

Let $x \in (x_0, x_0 + h]$. We will write $P_n = \{t_{0n}, t_{1n}, \dots, t_{m-1,n}, t_{mn} = x_0, t_{m+1,n}, \dots, t_{k-1,n}, t_{k_n,n}\}$ for the partition of $I$ whose associated polygonal line is $\varphi_n$. Then there exists $p$ such that $x \in (t_{pn}, t_{p+1,n}]$. (Of course, $p$ depends on $n$.) By the construction of $\varphi_n$, we have the following equalities:

$$\begin{aligned}
\varphi_n(t_{m+1,n}) - \varphi(t_{mn}) &= f(t_{mn}, \varphi_n(t_{mn}))(t_{m+1,n} - t_{mn}) \\
\varphi_n(t_{m+2,n}) - \varphi(t_{m+1,n}) &= f(t_{m+1,n}, \varphi_n(t_{m+1,n}))(t_{m+2,n} - t_{m+1,n}) \\
&\dots \\
\varphi_n(t_{pn}) - \varphi(t_{p-1,n}) &= f(t_{p-1,n}, \varphi_n(t_{p-1,n}))(t_{pn} - t_{p-1,n}) \\
\varphi_n(x) - \varphi(t_{pn}) &= f(t_{pn}, \varphi_n(t_{pn}))(x - t_{pn}).
\end{aligned}$$

Adding these equations yields

$$\varphi_n(x) - y_0 = \sum_{i=m}^{p-1} f(t_{i,n}, \varphi_n(t_{i,n}))(t_{i+1,n} - t_{i,n}) + f(t_{pn}, \varphi_n(t_{pn}))(x - t_{pn}). \tag{10.32}$$

Let us denote the right side in (10.32) by $s_n$, and let $\sigma_n$ denote the sum obtained from $s_n$ by replacing $\varphi_n$ with $\varphi$ throughout. Then

$$\sigma_n = \sum_{i=m}^{p-1} f(t_{i,n}, \varphi(t_{i,n}))(t_{i+1,n} - t_{i,n}) + f(t_{pn}, \varphi(t_{pn}))(x - t_{pn}).$$

Let $\varepsilon > 0$. The function $f$ is uniformly continuous on $R$ so there exists $\delta \in (0, \varepsilon)$ such that

$$(x, y_1), (x, y_2) \in R, \quad |y_1 - y_2| < \delta \quad \Rightarrow \quad |f(x, y_1) - f(x, y_2)| < \varepsilon. \tag{10.33}$$

The sequence $\{\varphi_n\}$ converges to $\varphi$ uniformly on $I$ so there exists $N \in \mathbb{N}$ such that

$$x \in I, \quad n \geq N \quad \Rightarrow \quad |\varphi_n(x) - \varphi(x)| < \delta. \tag{10.34}$$

For such $n$, using (10.33) and (10.34),

$$\begin{aligned}
|s_n - \sigma_n| &= \left| \sum_{i=m}^{p-1} \left[ f(t_{i,n}, \varphi_n(t_{i,n})) - f(t_{i,n}, \varphi(t_{i,n})) \right] (t_{i+1,n} - t_{i,n}) \right. \\
&\quad \left. + \left[ f(t_{pn}, \varphi_n(t_{pn})) - f(t_{pn}, \varphi(t_{pn})) \right] (x - t_{pn}) \right| \\
&\leq \sum_{i=m}^{p-1} \left| f(t_{i,n}, \varphi_n(t_{i,n})) - f(t_{i,n}, \varphi(t_{i,n})) \right| (t_{i+1,n} - t_{i,n}) \\
&\quad + \left| f(t_{pn}, \varphi_n(t_{pn})) - f(t_{pn}, \varphi(t_{pn})) \right| (x - t_{pn}) \\
&< \sum_{i=m}^{p-1} \varepsilon(t_{i+1,n} - t_{i,n}) + \varepsilon(x - t_{pn}) \\
&= \varepsilon(x - x_0).
\end{aligned}$$

It follows that, for such $n$,

$$\begin{aligned}
|\varphi(x) - y_0 - \sigma_n| &\leq |\varphi(x) - y_0 - s_n| + |s_n - \sigma_n| = |\varphi(x) - \varphi_n(x)| + |s_n - \sigma_n| \\
&< \delta + \varepsilon(x - x_0) < \varepsilon(1 + x - x_0) \leq \varepsilon(1 + h).
\end{aligned}$$

That way, we have proved that $\varphi(x) - y_0 = \lim \sigma_n$. On the other hand, $\sigma_n$ is just a Riemann sum for the function $f(x, \varphi(x))$ on $[x_0, x]$, so we get (10.31) and the theorem is proved. □

We close this section with a few words about the separability of $C(X)$. When $X$ is a closed interval on the real line, the Weierstrass Approximation

Theorem allows the countable dense set to be chosen among polynomials. The same is true when $X$ is a compact subset of $\mathbb{R}^n$. Beyond Euclidean spaces it is impossible to define a polynomial so the best one can say is contained in the following result. Recall that $C(X)$ is an algebra of functions (as in Definition 6.2.2). Also, we say that an algebra $\mathcal{A}$ of functions on $X$ **separates points** if for any $x, y \in X$ there is $f \in \mathcal{A}$ for which $f(x) \neq f(y)$.

**Theorem 10.7.12.** ***(Stone–Weierstrass Theorem)*** *Suppose $X$ is a compact Hausdorff space and $\mathcal{A}$ is a subalgebra of $C(X)$ which contains a nonzero constant function. Then $\mathcal{A}$ is dense in $C(X)$ if and only if it separates points.*

CESARE ARZELA (1847–1912) was an Italian mathematician born in Santo Stefano di Magra. He attended schools in Sarzana and Pisa, and studied mathematical and physical sciences at the Scuola Normale Superiore in Pisa on a scholarship. He completed his dissertation on potential theory under topologist Enrico Betti (1823–1892) in 1869. He could not afford to continue advanced studies abroad, so he embarked on a teaching career. He started in 1870 at the Lyceum of Macerata but kept an active research agenda and made significant contributions to elasticity theory and function series. He moved across various teaching positions before assuming a professorship at the University of Palermo in 1878. While at the Instituto Tecnico in Florence one of his students was Volterra. His tenure in Bologna, starting in 1880, proved influential. His research thrived, notably producing pioneering work in functional analysis and series convergence, including the Arzelà–Ascoli Theorem in 1895. His exceptional teaching extended to Galois theory, delivering the first such course in Italy. His textbooks on algebra and infinitesimal calculus were widely used, showing his significant impact on mathematical education in Italy. Among the students that Arzelà taught at Bologna were Tonelli and Vitali. Recognized for his contributions, Arzelà received honors from prestigious institutions and shared the Royal Prize for Mathematics from the Reale Accademia dei Lincei in 1907.

## Exercises

10.7.1. Prove that $\|f\| := \max\{|f(x)| : x \in X\}$ defines a norm on $C(X)$, if $X$ is a compact metric space.

10.7.2. Let $X$ be a compact metric space. Prove that $C(X)$ is complete in the topology induced by the norm.

In Exercises 10.7.3–10.7.10 determine whether the family of functions $\mathcal{F} = \{f_n : n \in \mathbb{N}\}$ is equicontinuous on $X$.

10.7.3. $f_n(x) = \dfrac{x^2}{x^2 + (1 - nx)^2}$, $X = [0, 1]$.

10.7.4. $f_n(x) = x^2 + \dfrac{1}{n}$, $X = \mathbb{R}$.

10.7.5. $f_n(x) = \arctan(nx)$, $X = [0, 1]$.

10.7.6. $f_n(x) = \dfrac{\sin nx}{n^2}$, $X = \mathbb{R}$.

10.7.7. $f_n(x) = n^2 x(1 - x^2)^n$, $X = [0, 1]$.

10.7.8. $f_n(x) = x + \dfrac{1}{nx}$, $X = [1, 2]$.

10.7.9. $f_n(x) = x^n$, $X = [0, 1/2]$.

10.7.10. $f_n(x) = \dfrac{nx}{1 + n^3x^2}$, $X = [0, 1]$.

10.7.11. Determine whether the family of functions $\mathcal{F} = \{\int_0^x f(t)\,dt : f \in C[0,1],\ \|f\|_{max} \leq 1\}$ is equicontinuous on $[0, 1]$.

10.7.12. Let $\mathcal{F}$ be an equicontinuous collection of functions in $C(X)$ and let $\overline{\mathcal{F}}$ denotes its closure. Prove that $\overline{\mathcal{F}}$ is equicontinuous.

10.7.13. Let $\mathcal{F}$ be a family of continuous functions on $[0, 1]$ having the property that there exists $M > 0$ such that $|f(x)| \leq M$ for all $f \in \mathcal{F}$ and all $x \in [0, 1]$. Let

$$\mathcal{F}^* = \{g \in C[0,1] : g(x) = f(0) + \int_0^x f(t)\,dt,\ f \in \mathcal{F}\}.$$

(a) Prove that the closure of $\mathcal{F}^*$ is a compact subset of $C[0, 1]$.

(b) Give an example to show that $\mathcal{F}^*$ need not be closed, even if $\mathcal{F}$ is closed.

10.7.14. Let $\mathcal{F}$ be a collection of functions in $C^1[0, 1]$, let $M > 0$, and suppose that $|f'(x)| \leq M$, for all $f \in \mathcal{F}$ and all $x \in [0, 1]$. Prove that $\mathcal{F}$ is equicontinuous.

10.7.15. Let $\mathcal{F}$ be a collection of functions in $C^1[0, 1]$, let $x_0 \in [0, 1]$, let $M > 0$, and suppose that $|f'(x_0)| \leq M$, for all $f \in \mathcal{F}$. Prove or disprove: $\mathcal{F}$ is equicontinuous at $x_0$.

10.7.16. A collection $\mathcal{F}$ of real-valued functions on a metric space $X$ is uniformly equicontinuous if

$$\forall \varepsilon > 0 \quad \exists \delta > 0 \quad \forall f \in \mathcal{F} \quad \forall x, y \in X \quad d(x, y) < \delta \quad \Rightarrow \quad |f(x) - f(y)| < \varepsilon.$$

Prove that if $X$ is compact every equicontinuous family of continuous functions is uniformly equicontinuous.

10.7.17.* Let $X$ be a compact metric space and $\mathcal{F}$ an equicontinuous sequence of real-valued functions. Prove that $\mathcal{F}$ is pointwise bounded if and only if it is uniformly bounded.

10.7.18.* Let $X$ be a metric space and $\{f_n\}$ a sequence in $C(X)$ that converges uniformly on $X$. Prove that $\{f_n\}$ is equicontinuous.

10.7.19. Let $\{f_n\}$ be a sequence in $C[a, b]$. Must $\{f_n\}$ have a uniformly convergent subsequence if:

(a) $\{f_n\}$ is equicontinuous?

(b) $\{f_n\}$ is uniformly bounded?

10.7.20.* For $1 \leq p < \infty$, a collection $\mathcal{S} \subset \ell^p$ is equisummable if

$$\forall \varepsilon > 0 \quad \exists n \in \mathbb{N} \quad \forall \{x_k\} \in \mathcal{S} \quad \sum_{k=n}^{\infty} |x_k|^p < \varepsilon.$$

Prove that $\mathcal{S}$ is compact if and only if it is closed, bounded, and equisummable.

10.7.21. Let $\{c_n\}$ be a sequence of nonnegative numbers in $\ell^2$ and let $\mathcal{S}$ be a subset of $\ell^2$ consisting of those $\{x_n\} \in \ell^2$ satisfying $|x_n| \leq c_n$, for all $n \in \mathbb{N}$. Prove that $\mathcal{S}$ is equisummable.

10.7.22.* Let $\{f_n\} \subset C(X)$ and suppose that $\lim_n f_n(x) = f(x)$ for all $x \in X$. Prove that there is a dense subset $D$ of $X$ such that $\{f_n\}$ is equicontinuous on $D$ and $f$ is continuous on $D$.

10.7.23. Prove Theorem 10.7.10.

10.7.24. Let $\{a_n\}$ be a sequence of real numbers in $[1, 2]$ and let the sequence $\{f_n\} \subset C[0, 1]$ be defined by

$$f_n(x) = \frac{1}{a_n} \sin(a_n x) + \cos(x + a_n).$$

Prove that it has a convergent subsequence.

10.7.25. Let $K$ be a continuous function on $[0, 1] \times [0, 1]$, and let $\{f_n\}$ be a bounded sequence in $C[0, 1]$, and define

$$g_n(x) = \int_0^1 K(x, y) f_n(y)\, dy.$$

Prove that $\{g_n\}$ is a sequence in $C[0, 1]$ that has a convergent subsequence.

10.7.26. Let $f \in L^1(0, 1)$, let $\{E_n\}$ be measurable subsets of $[0, 1]$, and let

$$g_n(x) = \int_0^x \chi_{E_n}(t) f(t)\, dt.$$

Prove that $\{g_n\}$ has a subsequence that converges uniformly to a continuous function on $[0, 1]$.

10.7.27. Let $\{f_n\}$ be an equicontinuous, uniformly bounded sequence of real-valued functions on $\mathbb{R}$. Prove that there is a subsequence $\{f_{n_k}\}$ that converges pointwise on $\mathbb{R}$ to a continuous function $f$ on $\mathbb{R}$ and that the convergence is uniform on bounded subsets of $\mathbb{R}$.

10.7.28.* Let $X$ be a compact metric space. Prove that $C(X)$ is separable.

10.7.29.* Let $\mu$ be a signed Borel measure on $[0, 1]$ such that

$$\int_0^1 \frac{1}{x + t}\, d\mu(t) = 0,$$

for all $x > 1$. Show that $\mu = 0$.

10.7.30. Let $g : \mathbb{R} \to [0, +\infty)$ and suppose that, for $f \in C^1[0, 1]$,

$$I(f) := \frac{1}{2} \int_0^1 (f'(x) + g(x))\, dx.$$

Finally, let $\mathcal{F} = \{f \in C^1[0, 1] : \|f\| \leq 1\}$. Prove that there is a function $h \in \mathcal{F}$ that minimizes $I(f)$, i.e., $I(h) = \inf\{I(f) : f \in \mathcal{F}\}$.

10.7.31. Prove that the collection of finite sums of the form $\sum_{i=1}^n g_i(x) h_i(y)$, where $n \in \mathbb{N}$, $g_i \in C[0, 1]$, and $h_i \in C[0, 1]$, is dense in $C([0, 1] \times [0, 1])$.

# 11

# *Continuous Linear Operators between Banach Spaces*

Some of the most important Banach spaces are those whose elements are functions. Many linear transformations (such as the derivative and integral) play a prominent role in these spaces. In this chapter we will focus on linear transformations between Banach spaces. Since Banach spaces are generalizations of Euclidean spaces, the linear transformations that we will study can be viewed as an analogue of those represented by matrices and studied in linear algebra.

## 11.1 Linear Operators

The most tractable transformations between normed spaces are linear transformations. In this section we will look at some of them and some of their fundamental properties. We will use the word *transformation* only when both normed spaces are finite-dimensional, in which case they can be identified with Euclidean spaces (Corollary 11.2.12). In the general case we will use the word *operator*.

**Definition 11.1.1.** Let $X$ and $Y$ be normed linear space and let $T$ be a function $T : X \to Y$. Such a function is called an **operator**. Further, $T$ is a **linear operator** if for all $g, h \in X$ and $\alpha, \beta \in \mathbb{R}$,

$$T(\alpha g + \beta h) = \alpha T(g) + \beta T(h). \tag{11.1}$$

*Remark* 11.1.2. Instead of writing $T(x)$ we will often abbreviate it to $Tx$. (Just like we do with matrices, i.e., linear operators between Euclidean spaces.)

We start with at few examples of linear operators and leave to the reader to verify that (11.1) holds.

**Example 11.1.3. A linear operator on $L^p$.**

Let $X = Y = L^p([0,1])$ and $g$ a continuous function on $[0,1]$. The operator $T$ defined by $Tf = gf$ is a linear operator which is called a *multiplication operator*. ♦

DOI: 10.1201/9781032716497-11

**Example 11.1.4. A linear operator on $\ell^p$.**
Let $X = Y = \ell^p$ and define $T$ by

$$T(x_1, x_2, x_3, \dots) = (0, x_1, x_2, x_3, \dots).$$

Then $T$ is a linear operator called the *shift operator*, or simply the *shift*. ♦

**Example 11.1.5. A linear operator on $C^1[0,1]$.**
Let $X = C^1[0,1]$ and $Y = C[0,1]$. The operator $T$ is defined by

$$Tf = f',$$

is a linear operator, called the *derivative operator*. ♦

If we define (in the usual way) the sum $T_1 + T_2$ of two linear operators $T_1$ and $T_2$ by

$$(T_1 + T_2)(x) = T_1(x) + T_2(x),$$

and the multiplication $\alpha T$ of a linear operator $T$ by a real number $\alpha$ by

$$(\alpha T)(x) = \alpha(T(x)),$$

then the set of all linear operators mapping one vector space to another is a vector space itself (Exercise 11.1.11). When $X$ and $Y$ are normed spaces, we can define an important property that an operator may have.

**Definition 11.1.6.** Let $X, Y$ be normed linear spaces. A linear operator $T : X \to Y$ is **bounded** if there is $M \geq 0$ such that

$$\|Tu\| \leq M\|u\|, \qquad \text{for all } u \in X.$$

The infimum of such numbers $M$ is the **operator norm** of $T$, denoted by $\|T\|$. If an operator is not bounded we say that it is **unbounded**. The collection of all bounded linear operators from $X$ to $Y$ is denoted by $\boldsymbol{\mathcal{L}(X, Y)}$. In the special case when $X = Y$, we frequently write $\boldsymbol{\mathcal{L}(X)}$ instead of $\boldsymbol{\mathcal{L}(X, X)}$.

Even though we use the word *bounded*, no linear operator is really a bounded function (except for the zero operator). Indeed, for every $M > 0$ there is a vector $y$ such that $\|Ty\| > M$. Namely, let $Tx \neq 0$ and $y = (M+1)x/\|Tx\|$. Then

$$Ty = T\left(\frac{(M+1)x}{\|T(x)\|}\right) = \frac{M+1}{\|T(x)\|}T(x),$$

so $\|Ty\| = M + 1 > M$. With this evidence available, why do we still use the word *bounded*? The answer is provided by Exercise 11.1.4: the boundedness of an operator $T$ really means that it maps bounded sets to bounded sets.

**Example 11.1.7. A bounded linear operator.**
The operator of multiplication by a continuous function presented in Example 11.1.3 is a bounded operator on $L^p(0,1)$.

**Proof:** Let $f \in L^p(0,1)$. Then

$$\|Tf\|_p = \left(\int_0^1 |gf|^p\right)^{1/p} \le \left(\int_0^1 \|g\|_{max}^p |f|^p\right)^{1/p}$$
$$= \|g\|_{max}\left(\int_0^1 |f|^p\right)^{1/p} = \|g\|_{max}\|f\|_p,$$

so $T$ is bounded and $\|T\| \le \|g\|_{max}$. We leave to the reader to prove the equality $\|T\| = \|g\|_{max}$ (Exercise 11.1.19). ♦

**Example 11.1.8. A bounded linear operator.**

The shift operator is bounded on $\ell^p$.

**Proof:** Let $x = (x_1, x_2, \dots) \in \ell^p$. Then

$$\|T(x_1, x_2, \dots)\|_p = \|(0, x_1, x_2, \dots)\|_p$$
$$= (0^p + x_1^p + x_2^p + \dots)^{1/p} = \|(x_1, x_2, \dots)\|_p,$$

so $T$ is bounded and $\|Tx\|_p \le \|x\|_p$, hence $\|T\| \le 1$. On the other hand, Exercise 11.1.3 implies that $\|T\| = 1$. Since $\|Tx\|_p = \|x\|_p$ for any $x \in \ell^p$, we say that $T$ is an *isometry*. ♦

**Example 11.1.9. An unbounded linear operator.**

The derivative operator defined in Example 11.1.5 is unbounded when both $C^1[0,1]$ and $C[0,1]$ are equipped with $\|\ \|_{max}$ norm.

**Proof:** Let $f_n(x) = x^n$, for $n \in \mathbb{N}$. Then $f_n \in C^1[0,1]$ and

$$\|f_n\| = \max\{|f_n(x)| : x \in [0,1]\} = \max\{|x^n| : x \in [0,1]\} = 1.$$

On the other hand,

$$\|Tf_n\| = \max\{|f_n'(x)| : x \in [0,1]\} = \max\{|nx^{n-1}| : x \in [0,1]\} = n.$$

If $T$ were bounded there would exist $M > 0$ satisfying $\|Tf_n\| \le M\|f_n\|$, for all $n \in \mathbb{N}$. Since this inequality is really $n \le M$, we see that it cannot be true for all $n$, so $T$ is unbounded. ♦

In a paper [41] from 1900 a Swedish mathematician Ivar Fredholm (1866–1927) initiated the study of infinite matrices, albeit using determinants. These were associated to integral equations that defined a mapping of $C[a,b]$ to itself. Hilbert then took over and wrote a series of papers dealing with the transformations of the space we now call $\ell^2$. He realized that infinite matrices need not come from integral equations and by considering them on their own he created functional analysis as a subject. In the words of a French mathematician Dieudonné, Hilbert's paper [61] from 1906 deserves to be considered as the first paper in functional analysis. While Hilbert based his study on quadratic forms associated to these infinite matrices, it was Riesz who focused on these transformations alone in 1913 paper [95]. Since Banach spaces were not invented yet, Riesz worked on operators acting on $C[a,b]$.

When $X$ and $Y$ are normed spaces, by using the words *operator norm*, we are creating an expectation that $\mathcal{L}(X,Y)$ is a normed space as well. The following proposition establishes that the operator norm indeed possesses the properties that we have come to expect from a norm, namely those listed in Definition 10.1.1. We leave the proof as Exercise 11.1.1.

**Proposition 11.1.10.** *Let $X, Y$ be normed linear spaces. The operator norm on $\mathcal{L}(X,Y)$ is a norm.*

We list some fundamental properties of the operator norm in addition to those established in Proposition 11.1.10 (see Exercise 11.1.2).

**Proposition 11.1.11.** *Let $X, Y$ be normed linear spaces and let $T \in \mathcal{L}(X,Y)$. Then*

*(a)* $\|T\| = \sup\{\|Tx\| : \|x\| \leq 1\} = \sup\{\|Tx\| : \|x\| = 1\}$.

*(b)* $\|Tx\| \leq \|T\|\|x\|$, *for all* $x \in X$.

*Remark* 11.1.12. Assertion (a) of Proposition 11.1.11 shows that a linear operator is bounded if and only if it is bounded when restricted to a unit ball of $X$, or more precisely to the unit sphere.

*Remark* 11.1.13. Notice that in the inequality in (b) all three norms have a different meaning. A pedantic way to write would be $\|Tx\|_Y \leq \|T\|\|x\|_X$, but such notation is used only for emphasis.

The following result brings forward a striking property of bounded linear operators: they are automatically continuous. In fact more is true (see also Exercise 11.1.5).

**Proposition 11.1.14.** *Let $X, Y$ be normed linear spaces, and let $T : X \to Y$ be a linear operator. Then $T$ is bounded if and only if it is continuous.*

*Proof.* If $T$ is bounded the result follows from Proposition 11.1.11 (b). Indeed, if $x, y \in X$ then

$$\|Tx - Ty\| = \|T(x-y)\| \leq \|T\|\|x-y\|,$$

so $T$ is in fact uniformly continuous.

In the other direction, suppose that $T$ is not bounded. By definition, there exists a sequence $\{x_n\}$ of nonzero vectors in $X$ such that

$$\|Tx_n\| \geq n\|x_n\|.$$

By dividing both sides by $n\|x_n\|$, we obtain a sequence $\{y_n\}$ for which

$$y_n := \frac{x_n}{n\|x_n\|} \to 0 \quad \text{and} \quad \|Ty_n\| \geq 1.$$

Thus, $T$ is not continuous at $x = 0$. In fact, $T$ is not continuous at any point. Indeed, for $x \neq 0$, the conclusion follows from Exercise 11.1.5. □

By Proposition 11.1.10, $\mathcal{L}(X,Y)$ is a normed space. In some cases, it is a Banach space. (See also Exercise 12.2.15.)

**Theorem 11.1.15.** *Let $X, Y$ be normed linear spaces. If $Y$ is a Banach space, then so is $\mathcal{L}(X,Y)$.*

*Proof.* It suffices to show that $\mathcal{L}(X,Y)$ is complete. To that end, let $\{T_n\}_{n=1}^{\infty}$ be a Cauchy sequence in $\mathcal{L}(X,Y)$. If $x \in X$ and $m, n \in \mathbb{N}$, then

$$\|T_n x - T_m x\| = \|(T_n - T_m)x\| \le \|T_n - T_m\|\|x\|,$$

so $\{T_n x\}_{n=1}^{\infty}$ is a Cauchy sequence in $Y$. By assumption, $Y$ is complete so $\{T_n x\}_{n=1}^{\infty}$ is a convergent sequence, converging to a vector $y \in Y$. We will define $Tx := y$. That way, we have a map

$$x \mapsto Tx = \lim_n T_n x.$$

We will show that $T \in \mathcal{L}(X,Y)$ and that $\lim T_n = T$ in $\mathcal{L}(X,Y)$.

First, $T$ is linear. Indeed, if $x, y \in X$ and $\alpha, \beta \in \mathbb{R}$ then, using linearity of each $T_n$,

$$T(\alpha x + \beta y) = \lim_n T_n(\alpha x + \beta y) = \lim_n (\alpha T_n x + \beta T_n y) = \alpha Tx + \beta Ty.$$

Next, we show that $T$ is bounded. Since $\mathcal{L}(X,Y)$ is a metric space, the Cauchy sequence $\{T_n\}$ is bounded, so there exists $M > 0$ such that $\|T_n\| \le M$, for all $n \in \mathbb{N}$. It follows from the continuity of the norm (Exercise 10.1.35) that for every $x \in X$,

$$\|Tx\| = \|\lim_n T_n x\| = \lim_n \|T_n x\| \le \limsup_n \|T_n\|\|x\| \le M\|x\|.$$

Thus, $T$ is bounded and $\|T\| \le M$, whence $T \in \mathcal{L}(X,Y)$.

Finally, we show that $\{T_n\}$ converges to $T$ in the operator norm. Let $\varepsilon > 0$. Then there exists $N \in \mathbb{N}$ such that

$$m \ge n \ge N \quad \Rightarrow \quad \|T_m - T_n\| < \varepsilon.$$

It follows that, for $m \ge n \ge N$ and $x \in X$,

$$\|T_m x - T_n x\| \le \|T_m - T_n\|\|x\| < \varepsilon\|x\|.$$

Taking the limit as $m \to \infty$ and using once again the continuity of the norm, we obtain that

$$\|(T - T_n)x\| = \|Tx - T_n x\| \le \varepsilon\|x\|,$$

which means that, for $n \ge N$, $\|T - T_n\| \le \varepsilon$, i.e., $\lim_n T_n = T$ in $\mathcal{L}(X,Y)$ and the theorem is proved. □

In the study of linear operators on a Banach space $X$, two subspaces of $X$ deserves special attention. One is the **kernel** of $T$, denoted by $\ker T$, is the set $\{x \in X : Tx = 0\}$. It is also known under the name **nullspace** of $T$, and the reader is likely to be familiar with this name from linear algebra where it is more common. The other subspace is the **range** of $T$, denoted by $\operatorname{ran} T$, which is $\{Tx : x \in X\}$. In algebra, such a set would be typically referred to as the **image** of $T$.

JOHN VON NEUMANN (1903–1957), born János von Neumann in Budapest, Hungary, was raised in a wealthy, non-observant Jewish family. As a child he demonstrated exceptional memory and intelligence by dividing eight-digit numbers in his head. He studied simultaneously mathematics in Budapest and chemistry in Berlin and Zürich and in 1926 got a degree in the latter and a Ph.D. in the former. He then spent a year at the University of Göttingen. After a few brief appointments in Berlin, Hamburg, and Princeton, he joined the Institute for Advanced Study in 1933 where he remained for the rest of his life. He made profound contributions to almost all mathematical fields. In quantum mechanics he discovered that quantities such as position or momentum can be represented as linear operators acting on Hilbert space thereby laying down the foundation for the mathematical study of quantum mechanics. Realizing that the most promising approach to nonlinear partial differential equations was using numerical methods, he made significant advances in hardware design, scientific computing, and the philosophy of computer science. During World War II, he served in advisory roles for the armed forces, contributing to the development of the hydrogen bomb. He received awards including the Medal for Merit, the Medal of Freedom, and the Enrico Fermi Award. He was a member of the American Academy of Arts and Sciences and the National Academy of Sciences. His incredible career was cut short by cancer in 1957.

## Exercises

11.1.1. Prove Proposition 11.1.10.

11.1.2. Prove Proposition 11.1.11.

11.1.3. Let $X, Y$ be normed linear spaces and $T \in \mathcal{L}(X, Y)$. Suppose that $M > 0$ and $\|T\| \leq M$.

(a) Prove that if there is a vector $x$ such that $\|Tx\| = M\|x\|$ then $\|T\| = M$;

(b) Prove that $\|T\| = M$ if and only if there exists a sequence $\{x_n\}_{n=1}^{\infty} \subset X$ such that $\|x_n\| = 1$ for all $n \in \mathbb{N}$ and $\lim_n \|Tx_n\| = M$;

(c) Prove that $\|T\| = M$ if and only if for every $\varepsilon > 0$ there exists $x \in X$ sich that $\|x\| = 1$ and $\|Tx\| > \|T\| - \varepsilon$.

(d) Give an example to show that the converse of (a) need not be true.

11.1.4. Let $X, Y$ be normed linear spaces, and let $T : X \to Y$ be a linear operator. Prove that $T$ is bounded if and only if it maps bounded sets to bounded sets.

11.1.5. Let $X$ be a normed linear space and $x_0 \in X$. Prove that a linear operator on a normed space is continuous if and only if it is continuous at $x_0$.

11.1.6. Let $X$ be a normed linear space and let $T, S \in \mathcal{L}(X, X)$. Prove that the composition $ST$ belongs to $\mathcal{L}(X, X)$ and that $\|ST\| \leq \|S\|\|T\|$. (We write $ST$ for $S \circ T$ just like when $S$ and $T$ are matrices.)

11.1.7. Let $X$ be a Banach space and $\mathcal{F}$ a family of bounded linear operators from $X$ to a normed linear space $Y$. Prove that $\mathcal{F}$ is equicontinuous if and only if it is uniformly bounded.

11.1.8. Let $\mathcal{H}$ be Hilbert space and $T \in \mathcal{L}(\mathcal{H})$. Prove that

$$\|T\| = \sup\{|\langle Tx, y\rangle| : \|x\| = \|y\| = 1\}.$$

11.1.9. Let $X, Y$ be normed linear spaces and $T : X \to Y$ a linear map such that $\{Tx_n\}$ is bounded in $Y$ whenever the sequence $\{x_n\}$ converges to 0 in $X$. Prove that $T \in \mathcal{L}(X, Y)$.

11.1.10. Let $X, Y$ be normed linear spaces, and let $T : X \to Y$ be a linear operator. Prove that $T$ is bounded if and only if it maps every Cauchy sequence in $X$ to a Cauchy sequence in $Y$.

11.1.11. Prove that the set of all linear transformations from a vector space $X$ to a vector space $Y$ is a vector space.

11.1.12.* Let $Z$ be a subspace of a normed linear space $X$, let $Y$ be a Banach space, and let $A : Z \to Y$ be a continuous linear operator. Prove that there exists a unique continuous linear operator $\overline{A} : \overline{Z} \to Y$ such that $\overline{A}|Z = A$ and $\|\overline{A}\| = \|A\|$.

11.1.13. Let $X, Y$ be normed spaces, let $T \in \mathcal{L}(X, Y)$, and let $\tilde{T} : X/\ker T \to Y$ be defined by $\tilde{T}([x]) = Tx$ (see Exercise 10.1.13). Prove that $\tilde{T}$ is a well-defined, injective, bounded, linear operator for which $\|\tilde{T}\| = \|T\|$.

11.1.14.* Let $X$ be a normed space and $V$ a closed subspace of $X$ (see Exercise 10.1.13). Prove that:

(a) The projection $\pi(x) := [x]$ from $X$ to $X/V$ has norm 1.

(b) If $Y$ is a normed linear space, $T \in \mathcal{L}(X, Y)$, and $V$ is a closed subspace of $\ker T$ then there is a unique $S \in \mathcal{L}(X/V, Y)$ such that $T = S \circ \pi$ and that $\|S\| = \|T\|$.

11.1.15. Let $\mathcal{H}$ be Hilbert space and let $T \in \mathcal{L}(\mathcal{H})$. Prove that $T = 0$ if and only if $\langle Tx, y\rangle = 0$ for all $x, y \in \mathcal{H}$.

11.1.16. Let $X$ be a Banach space and $T \in \mathcal{L}(X, X)$ with $\|T\| < 1$. Define $T^0 = I$. Prove that $\sum_{n=0}^{\infty} T^n$ converges in $\mathcal{L}(X, X)$.

11.1.17. Let $X, Y$ be normed linear spaces, let $\{T_n\}$ converge to $T$ in $\mathcal{L}(X, Y)$, and let $\{x_n\}$ converge to $x$ in $X$. Prove that $\{T_n x_n\}$ converges to $Tx$ in $Y$.

11.1.18. Let $X, Y$ be normed linear spaces and $\{T_n\}, T \in \mathcal{L}(X, Y)$. Prove that $\lim_n T_n = T$ if and only if $\{T_n x\}$ converges to $Tx$ uniformly on the unit ball of $X$.

11.1.19. Let $g$ be a continuous function on $[0, 1]$ and let $T : L^p([0, 1]) \to L^p([0, 1])$ be defined by $T(f) = gf$. Prove that $T$ is a bounded operator and $\|T\| = \|g\|_{max}$.

11.1.20. Let $p, q$ be real numbers such that $q > 1$ and $1/p + 1/q = 1$, and let $(a_{ij})$ be an infinite matrix of real numbers such that

$$\sum_{j=1}^{\infty}\sum_{i=1}^{\infty} |a_{ij}|^q < +\infty.$$

Prove that

$$y_j = \sum_{i=1}^{\infty} a_{ij} x_i, \quad j = 1, 2, \ldots$$

defines a bounded linear operator

$$y = Tx, \quad x = \{x_n\}, \quad y = \{y_n\}$$

from $\ell^p$ to $\ell^q$.

11.1.21. Let $K : [0,1] \times [0,1] \to \mathbb{R}$ be a measurable function such that

$$\int_0^1 |K(s,t)|\, dt \le M \qquad \text{for all } s \in [0,1].$$

Prove that

$$y(s) = \int_0^1 K(s,t) x(t)\, dt$$

defines an operator in $\mathcal{L}(L^\infty)$,

$$y = Ax, \quad x = x(t), \quad y = y(t).$$

11.1.22. Let $K : [0,1] \times [0,1] \to \mathbb{R}$ be a continuous function. Prove that

$$y(x) = \int_0^1 K(s,t) x(t)\, dt$$

defines an operator in $\mathcal{L}(C[0,1])$,

$$y = Ax, \quad x = x(t), \quad y = y(t).$$

11.1.23. Let $K : [0,1] \times [0,1] \to \mathbb{R}$ be a measurable function such that

$$\int_0^1 \|K(s,t)\|_{t,\infty}\, ds < +\infty,$$

where $\|K(s,t)\|_{t,\infty}$ means the $L^\infty$ norm of $K$ as a function of $t$ for a fixed $s$. Prove that

$$y(s) = \int_0^1 K(s,t) x(t)\, dt$$

defines an operator in $\mathcal{L}(L^1([0,1]))$,

$$y = Ax, \quad x = x(t), \quad y = y(t).$$

11.1.24. Let $q > 1$ and let $K : [0,1] \times [0,1] \to \mathbb{R}$ be a measurable function such that

$$\int_0^1 \int_0^1 |K(s,t)|^q\, ds dt < +\infty.$$

Prove that

$$y(s) = \int_0^1 K(s,t) x(t)\, dt$$

defines an operator in $\mathcal{L}(L^p([0,1]), L^q([0,1]))$, $(1/p + 1/q = 1)$

$$y = Ax, \quad x = x(t), \quad y = y(t).$$

11.1.25. Let $K : [0,1] \times [0,1] \to \mathbb{R}$ be a measurable function such that

$$\int_0^1 \int_0^1 |K(s,t)|^2 \, dsdt < +\infty,$$

Prove that

$$y(s) = x(s) - \int_0^1 K(s,t)x(t)\, dt$$

defines an operator in $\mathcal{L}(L^2([0,1]))$.

11.1.26.* Let $(X, \mathfrak{M}, \mu)$ be a measure space, let $h$ be a bounded measurable function on $X$, and let $T$ be the operator on $L^p(X, \mu)$ defined by $Tf = hf$. Prove that:

(a) $T$ is bounded and $\|T\| \le \|h\|_\infty$.

(b) $\|T\| = \|h\|_\infty$ if and only if each measurable set $E$ of infinite measure contains a measurable subset $A$ of finite positive measure, i.e., $0 < \mu(A) < \infty$.

11.1.27.* Let $(X, \mathfrak{M}, \mu)$ be a finite measure space and let $p \ge q$. Find the norm of the identity operator $I : L^p(X, \mu) \to L^q(X, \mu)$.

11.1.28.* Let $1 \le p \le q < \infty$. Determine for what functions $h$ on $[0,1]$ is the multiplication by $h$ a bounded operator from $L^p[0,1]$ to $L^q[0,1]$.

11.1.29. Let $T : C[0,1] \to C[0,1]$ be defined by $Tf(x) = x^3 f(x)$, for all $f \in C[0,1]$ and each $x \in [0,1]$.

(a) Prove that $T$ is a bounded linear operator.

(b) Let $I : C[0,1] \to C[0,1]$ denote the identity operator, i.e., $I(f) = f$ for each $f \in C[0,1]$. Prove that $\|I + T\| = 1 + \|T\|$.

11.1.30. Let $T : C[0,1] \to C[0,1]$ be defined by $Tf(x) = \int_0^x f(t)\, dt$, for all $f \in C[0,1]$ and each $x \in [0,1]$. Prove that $T$ is a bounded linear operator and find its norm.

11.1.31. Let $T : L^1[0,1] \to L^1[0,1]$ be defined by $Tf(x) = f(\sqrt{x})$, for all $f \in C[0,1]$ and each $x \in [0,1]$. Prove that $T$ is a bounded linear operator and find its norm.

11.1.32. For $f \in L^1[0,1]$ let $x_n = \int_0^1 f(t)t^n\, dt$, and let $T(f) = \{x_n\}$. Prove that the operator $T$ is a bounded linear operator mapping $L^1[0,1]$ to $c_0$ and determine its norm. (See Exercise 10.1.15.)

11.1.33. Let $(X, \mathfrak{M}, \mu)$ and $(Y, \mathfrak{N}, \nu)$ be $\sigma$-finite measure spaces, let $K$ be an $\mathfrak{M} \otimes \mathfrak{N}$ measurable function on $X \times Y$ and let $C > 0$. Suppose that

$$\int_X |K(x,y)|\, d\mu(x) \le C \text{ for a.e. } y \in Y, \text{ and}$$
$$\int_Y |K(x,y)|\, d\nu(y) \le C \text{ for a.e. } x \in X.$$

Prove that the operator $T$ defined on $L^p(X,\mu)$ by

$$Tf(x) = \int_Y K(x,y)f(y)\,d\nu(y)$$

is a bounded operator and that $\|Tf\|_p \leq C\|f\|_p$.

11.1.34.* Let $(X,\mathfrak{M},\mu)$ be a measure space and let $1 \leq p < r \leq \infty$. Prove that:

(a) $L^p(X,\mu) \cap L^r(X,\mu)$ is a Banach space with norm $\|f\| := \|f\|_p + \|f\|_r$.

(b) If $p < q < r$ the inclusion map $L^p(X,\mu) \cap L^r(X,\mu) \to L^q(X,\mu)$ is continuous.

11.1.35.* Let $(X,\mathfrak{M},\mu)$ be a measure space and let $1 \leq p < r \leq \infty$. Prove that:

(a) $L^p(X,\mu) + L^r(X,\mu)$ is a Banach space with norm $\|f\| := \inf\{\|g\|_p + \|h\|_r : f = g + h\}$.

(b) If $p < q < r$ the inclusion map $L^q(X,\mu) \to L^p(X,\mu) + L^r(X,\mu)$ is continuous.

## 11.2 Banach Space Isomorphisms

In this section we will define what it means for two Banach spaces to be isomorphic. Once we get the proper formulation, we will examine whether some well-known spaces are isomorphic.

Before we present a formal definition, let us consider what an isomorphism should be. For starters, every Banach space is a vector space, so an isomorphism between Banach spaces $X$ and $Y$ should be a vector space isomorphism, i.e., an invertible linear map from $X$ onto $Y$. In other words, it should be an invertible operator, so we start there.

**Definition 11.2.1.** Let $X, Y$ be normed spaces. A linear operator $T : X \to Y$ is **invertible** if there exists an operator $S : Y \to X$ such that

$$S(T(x)) = x, \quad T(S(y)) = y, \quad \text{for all } x \in X \text{ and } y \in Y.$$

We call $S$ the **inverse (operator)** of $T$ and denote it by $T^{-1}$.

*Remark* 11.2.2. The relations between $T$ and $T^{-1}$ are typically written as $T^{-1}Tx = x$ and $TT^{-1}y = y$, omitting parentheses.

Notice that we did not require the inverse operator in Definition 11.2.1 to be linear. That comes automatically.

**Proposition 11.2.3.** *If a linear operator $T : X \to Y$ has an inverse $T^{-1}$, then $T^{-1}$ is linear on $Y$.*

*Proof.* Let $y_1, y_2 \in Y$ and $\alpha, \beta \in \mathbb{R}$. Then

$$\begin{aligned} T^{-1}(\alpha y_1 + \beta y_2) &= T^{-1}(\alpha T T^{-1} y_1 + \beta T T^{-1} y_2) \\ &= T^{-1}(T(\alpha T^{-1} y_1 + \beta T^{-1} y_2)) \\ &= \alpha T^{-1} y_1 + \beta T^{-1} y_2, \end{aligned}$$

so $T^{-1}$ is a linear operator. □

An important question in operator theory is when is $T^{-1}$ bounded.

**Proposition 11.2.4.** *Let $T : X \to Y$ be a surjective operator. Then $T$ has a bounded inverse if and only if there exists $m > 0$ such that*

$$\|Tx\| \geq m\|x\|, \quad \textit{for all } x \in X. \tag{11.2}$$

*Proof.* ($\Leftarrow$) If (11.2) holds then $T$ must be injective. Indeed, if $Tx_1 = Tx_2$ then $T(x_1 - x_2) = 0$ so (11.2) implies that $x_1 = x_2$. Thus, $T$ is invertible so, using the notation $y = Tx$, (11.2) can be written as

$$\|y\| \geq m\|T^{-1}y\|, \quad \text{for all } y \in Y,$$

whence $\|T^{-1}\| \leq 1/m$ and $T^{-1}$ is bounded.

($\Rightarrow$) Suppose that $T$ is invertible and that $T^{-1}$ is bounded. In this case, there is $M > 0$ such that

$$\|T^{-1}(Tx)\| \leq M\|Tx\|, \quad \text{for all } x \in X,$$

which yields

$$\|Tx\| \geq \frac{1}{M}\|x\|,$$

and the result follows by taking $m = 1/M$. □

Now we can define an isomorphism of normed linear spaces.

**Definition 11.2.5.** An **isomorphism of normed spaces** $X$ and $Y$ is an invertible linear operator $T : X \to Y$ such that $T$ and its inverse $T^{-1}$ are continuous. If in addition $T$ is an isometry, i.e., $\|Tx\| = \|x\|$, for all $x \in X$, then $T$ is an **isometric isomorphism** and $X$ and $Y$ are **isometrically isomorphic**.

Definition 11.2.5 gives yet another reason to study linear operators between Banach spaces. Of course, linear operators play an important role in many applications such as calculus of variation and Fourier analysis.

We leave the proofs of the following easy corollaries as Exercises 11.2.12–11.2.13.

**Corollary 11.2.6.** *Two normed linear spaces are isomorphic if and only if there is a linear operator $T$ mapping $X$ onto $Y$ and positive constants $m, M$ such that*

$$m\|x\| \leq \|Tx\| \leq M\|x\|, \quad \textit{for all } x \in X.$$

**Corollary 11.2.7.** *Let $X$ be a vector space and let $\| \, \|_\alpha$ and $\| \, \|_\beta$ be two norms on $X$. Normed spaces $(X, \| \, \|_\alpha)$ and $(X, \| \, \|_\beta)$ are isomorphic if and only if there are positive constants $m, M$ such that*

$$m\|x\|_\alpha \leq \|x\|_\beta \leq M\|x\|_\alpha, \quad \textit{for all } x \in X. \tag{11.3}$$

*Remark* 11.2.8. When the norms $\| \, \|_\alpha$ and $\| \, \|_\beta$ on a vector space $X$ satisfy (11.3) we say that they are **equivalent norms.**

Equivalent norms on $X$ share several important properties (see Exercises 11.2.22–11.2.23).

**Example 11.2.9. Two equivalent norms on $\mathbb{R}^2$.**

Let $\|(x, y)\|_1 := |x| + |y|$ and $\|(x, y)\|_2 := \sqrt{x^2 + y^2}$. Then these are two equivalent norms on $\mathbb{R}^2$.

**Proof:** Each element $(x, y) \in \mathbb{R}^2$ can be identified with $(x, y, 0, 0, \dots) \in \ell^1$, so $\| \, \|_1$ is simply the $\ell^1$ norm. A similar argument shows that $\| \, \|_2$ is also a norm. Finally, it is an easy exercise in algebra to verify that

$$|x| + |y| \leq \sqrt{2}\sqrt{x^2 + y^2} \leq \sqrt{2}(|x| + |y|),$$

so two norms are equivalent and they yield the same topology. Figure 11.1 shows open balls $B_k = \{(x, y) : \|(x, y)\|_k < 1\}$, for $k = 1, 2$. (See also Example 2.1.10.) ♦

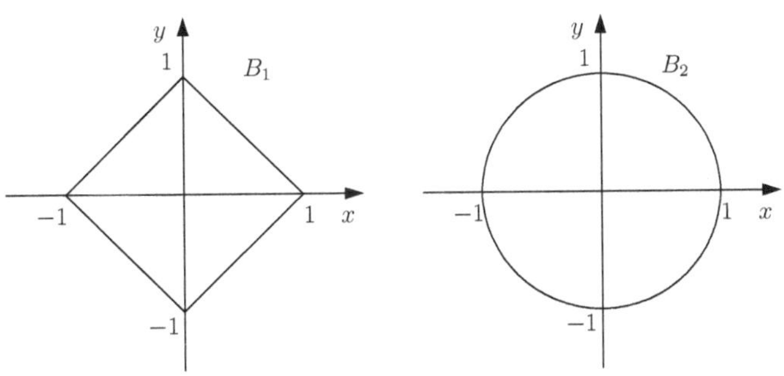


Figure 11.1: Open balls $B_k = \{(x, y) : \|(x, y)\|_k < 1\}$, for $k = 1, 2$

*Remark* 11.2.10. The equivalence of norms is an equivalence relation (Exercise 11.2.24).

Recall that if a vector space has a finite basis we say that it has finite dimension. By a basis we mean a linearly independent, spanning set.

**Theorem 11.2.11.** *Any two norms on a finite-dimensional linear space are equivalent.*

*Proof.* Let $\{e_1, \ldots, e_n\}$ be a basis for $X$. The Euclidean norm on $X$ is defined by

$$\|x\|_2 = \|x_1 e_1 + \cdots + x_n e_n\|_2 = \sqrt{x_1^2 + \cdots + x_n^2}. \tag{11.4}$$

Let $\| \, \|$ be any norm on $X$. We will show that it is equivalent to $\| \, \|_2$, i.e., that it satisfies

$$m\|x\|_2 \leq \|x\| \leq M\|x\|_2, \qquad \text{for all } x \in X. \tag{11.5}$$

The result will then follow from Remark 11.2.10.

First we consider the second inequality in (11.5). Let

$$M = \left( \sum_{k=1}^{n} \|e_k\|^2 \right)^{1/2}.$$

Then,

$$\begin{aligned} \|x\| = \| \sum_{k=1}^{n} x_k e_k \| &\leq \sum_{k=1}^{n} |x_k| \|e_k\| \\ &\leq \left( \sum_{k=1}^{n} |x_k|^2 \right)^{1/2} \left( \sum_{k=1}^{n} \|e_k\|^2 \right)^{1/2} = M\|x\|_2. \end{aligned}$$

Notice that $M > 0$ because each $e_k \neq 0$.

For the first inequality in (11.5) we consider the function $f : \mathbb{R}^n \to \mathbb{R}$ defined by

$$f(x_1, x_2, \ldots, x_n) = \|x_1 e_1 + x_2 e_2 + \cdots + x_n e_n\|.$$

The function $f$ is Lipschitz with constant $M$ because if we denote $x = (x_1, x_2, \ldots)$ and $y = (y_1, y_2, \ldots)$ then

$$\begin{aligned} |f(x) - f(y)| &= \left| \left\| \sum_{k=1}^{n} x_k e_k \right\| - \left\| \sum_{k=1}^{n} y_k e_k \right\| \right| \\ &\leq \left\| \sum_{k=1}^{n} x_k e_k - \sum_{k=1}^{n} y_k e_k \right\| \\ &\leq M \left\| \sum_{k=1}^{n} x_k e_k - \sum_{k=1}^{n} y_k e_k \right\|_2 \quad \text{[by the second inequality in (11.3)]} \\ &= M \sqrt{\sum_{k=1}^{n} (x_k - y_k)^2} \\ &= M\|x - y\|_{\mathbb{R}^n}. \end{aligned}$$

In particular, $f$ is continuous and it takes only strictly positive values on the unit sphere $S = \{x \in \mathbb{R}^n : \|x\|_2 = 1\}$. Since $S$ is a compact set, $f$ attains a minimum value $m > 0$, so

$$|f(x_1, x_2, \ldots, x_n)| \geq m \quad \text{for all } x = (x_1, x_2, \ldots, x_n) \in S.$$

It follows that

$$\|x\| \geq m \quad \text{for all } x \in X \text{ for which } \|x\|_2 = 1. \tag{11.6}$$

If $x$ is a nonzero vector in $X$, then $x/\|x\|_2$ satisfies (11.6) and we obtain that

$$\|x\| \geq m\|x\|_2, \quad \text{for all } x \in X \text{ such that } x \neq 0. \tag{11.7}$$

Clearly, (11.7) holds if $x = 0$. Thus the first inequality in (11.5) is proved. □

From Theorem 11.2.11 we deduce an important corollary.

**Corollary 11.2.12.** *If a finite-dimensional normed linear space has dimension $n$ then it is isomorphic to $\mathbb{R}^n$. Consequently, any two normed linear spaces of the same finite dimension are isomorphic.*

*Proof.* It suffices to show that if the dimension of $(X, \|\cdot\|)$ is $n$, then it is isomorphic to $\mathbb{R}^n$. To that end, let $\{e_1, \ldots, e_n\}$ be a basis of $X$ and let $T : \mathbb{R}^n \to (X, \|\cdot\|)$ be defined by

$$T(x) = T(x_1, \ldots, x_n) := \sum_{k=1}^{n} x_k e_k.$$

We leave to the reader (Exercise 11.2.11) to verify that $T$ is an invertible linear operator from $\mathbb{R}^n$ onto $X$. It follows from Theorem 11.2.11 that the norm on $X$ and the one defined by (11.4) are equivalent, so $T$ is bounded (hence continuous). Since $T$ is bijective it is invertible and its inverse transformation $T^{-1} : (X, \|\cdot\|) \to \mathbb{R}^n$ is also linear (Proposition 11.2.3). Moreover, by applying again Theorem 11.2.11 we have that

$$\|Tx\| = \|x_1e_1 + x_2e_2 + \cdots + x_ne_n\| \geq m\|x\|_2,$$

whence Proposition 11.2.4 ensures that $T^{-1}$ is bounded (hence continuous, by Proposition 11.1.14) and the proof is complete. □

In view of the last result, a finite-dimensional normed linear space shares many important properties with $\mathbb{R}^n$. We single out two of them.

**Corollary 11.2.13.** *Any finite-dimensional normed linear space is complete. Any finite-dimensional subspace of a normed linear space is closed.*

*Proof.* If $X$ is finite-dimensional it is isomorphic to $\mathbb{R}^n$, hence complete. Also, if $Y \subset X$ is finite-dimensional, then $Y$ is complete and hence closed. □

**Corollary 11.2.14.** *The closed unit ball in a finite-dimensional normed linear space is compact.*

*Proof.* Let $B = \{x \in X : \|x\| \leq 1\}$ and let $T : X \to \mathbb{R}^n$ be an isomorphism. If $x \in B$ then

$$\|Tx\| \leq \|T\|\|x\| \leq \|T\|,$$

so $T(B)$ is a bounded set. Also, $T(B) = (T^{-1})^{-1}(B)$. Since $B$ is closed (Exercise 10.1.36) and $T^{-1}$ is continuous, it follows that $T(B)$ is closed. Being a closed and bounded subset of $\mathbb{R}^n$, $T(B)$ is compact. Finally, $B$ is a continuous image of a compact set, $B = (T^{-1})(T(B))$, so $B$ is compact. □

*Remark* 11.2.15. In infinite-dimensional spaces, Corollary 11.2.14 need not be true. For example, the closed unit ball in $L^p[0,1]$ is not compact (Exercise 10.3.40).

HERMANN MINKOWSKI (1864–1909) was a German mathematician, born in Russia, in the territory of today's Lithuania. His family moved to Königsberg when he was eight. He showed a mathematical talent in high school, reading Dedekind, Dirichlet, and Gauss. At the University of Königsberg he struck a friendship with Hilbert. While still a student, he shared the Grand Prix of the French Academy of Sciences with Henry Smith, on a problem about quadratic forms. At the age of 21 he received his doctorate with a dissertation that continued the prize winning work. After several academic positions at the University of Bonn, Königsberg, and ETF Zürich (where Einstein was enrolled in several courses that Minkowski taught), in 1902 he moved to the University of Göttingen but died at the age of 44 from appendicitis. His main achievement was the *geometry of numbers*, a study of convex bodies and integer vectors in a space of $n$ dimensions that proved a valuable tool in number theory. He is best known for his pioneering work in the geometry of four dimensions. He astutely realized that Einstein's special theory of relativity could be best understood in a four-dimensional space, using time as the fourth dimension. It is less known that he suggested to Hilbert the topic of his historic 1900 address in Paris.

## Exercises

11.2.1. Let $X, Y$ be Banach spaces and let $S, T \in \mathcal{L}(X, Y)$. Suppose that the operator $ST$ is invertible. Prove or disprove: $T$ is invertible.

11.2.2. Let $X$ be a Banach space and let $T \in \mathcal{L}(X)$ satisfy $\|T\| \leq \eta < 1$. Prove that the operator $I - T$ has a bounded inverse $(I - T)^{-1}$ and that $\|(I - T)^{-1}\| \leq 1/(1 - \eta)$.

11.2.3. Let $X, Y$ be normed spaces and $T : X \to Y$ a linear map that has a unique left inverse $S$. Prove that $T$ is invertible.

11.2.4. Give an example of normed spaces $X, Y$ and an operator $T : X \to Y$ that is:

(a) left invertible but not right invertible;

(b) right invertible but not left invertible.

11.2.5.* Let $X$ be a linear space and let $S, T : X \to X$ be linear maps. Prove:

(a) $I - ST$ is injective if and only if $I - TS$ is injective;

(b) $I - ST$ is surjective if and only if $I - TS$ is surjective.

11.2.6. Let $X$ be a linear space and let $T : X \to X$ be a linear map that satisfies equation $p(T) = 0$ for some polynomial $p$ with $p(0) \neq 0$. Prove that $T$ is invertible.

11.2.7. Let $X$ be a linear space, let $T : X \to X$ be a bounded map that satisfies equation $T^2 = T$, and let $0 < |\lambda| < 1$. Find the inverse of $\lambda I - T$.

11.2.8. Let $X$ be a linear space, let $T : X \to X$ be a bounded operator with $\|T\| < 1$. Prove or disprove: the equation $x - Tx = y$ has a solution $x$ for every $y \in X$.

11.2.9.* Let $X$ be a Banach space. Prove that the set of invertible operators in $\mathcal{L}(X, X)$ is open.

11.2.10.* Let $\mathcal{H}$ be Hilbert space, let $\{x_n\}_{n=1}^{\infty}$ be an orthonormal basis of $\mathcal{H}$, let $0 \leq C < 1$, and let $\{y_n\}_{n=1}^{\infty}$ be a sequence in $\mathcal{H}$ such that for any scalars $\{\alpha_n\}$,

$$\|\sum_{k=1}^{n} \alpha_k (x_k - y_k)\| \leq C \|\sum_{k=1}^{n} \alpha_k x_k\|.$$

Prove that for each $x \in \mathcal{H}$ there are scalars $\{\beta_n\}$ such that $x = \sum_{k=1}^{\infty} \beta_k y_k$, with the series converging in the norm of $\mathcal{H}$.

11.2.11. Let $X$ be a normed linear space of dimension $n$, let $\{e_1, \ldots, e_n\}$ be a basis of $X$, and let $T : \mathbb{R}^n \to (X, \|\cdot\|)$ be defined by $T(x_1, \ldots, x_n) = \sum_{k=1}^{n} x_k e_k$. Prove that $T$ is an invertible linear operator.

11.2.12. Prove Corollary 11.2.6.

11.2.13. Prove Corollary 11.2.7.

11.2.14. Prove that any two infinite-dimensional separable Hilbert spaces are isometrically isomorphic and that any such isomorphism preserves the inner product.

11.2.15.* Prove that the space $C[0, 1]$ is isometrically isomorphic to its subspace $A = \{f \in C[0, 1] : f(t) = f(1 - t)\}$.

11.2.16.* Let $B(\ell^1)$ denote the vector space of bounded sequences $\{a_n\}$, where for each $n \in \mathbb{N}$, $a_n \in \ell^1$. The norm on $B(\ell^1)$ is defined by

$$\|\{a_n\}\| = \sup_n \|a_n\|_1.$$

Prove that the normed spaces $B(\ell^1)$ and $\mathcal{L}(\ell^1)$ are isometrically isomorphic.

11.2.17. Let $X, Y$ be Banach spaces and let $T \in \mathcal{L}(X, Y)$ be an isometry. Prove that the range of $T$ is a closed subspace of Y.

11.2.18. Let $X, Y, Z$ be normed linear spaces, let $F : X \to Y$ be an isometric isomorphism, and let $G \in \mathcal{L}(Y, Z)$. Prove that $\|G\| = \|G \circ F\|$.

11.2.19. Let $\mathcal{H}$ be Hilbert space. Prove that there is a measure space $(X, \mathfrak{M}, \mu)$ for which $\mathcal{H}$ is isometrically isomorphic to $L^2(X, \mu)$.

11.2.20. Let $\mathcal{H}$ be an infinite dimensional Hilbert space. Prove that there exists a continuous injective mapping $\gamma : [0, 1] \to \mathcal{H}$ such that $\gamma(b) - \gamma(a)$ is orthogonal to $\gamma(d) - \gamma(c)$ whenever $0 \le a \le b \le c \le d \le 1$.

11.2.21. Prove that if $T \in \mathcal{L}(\mathcal{H})$ then $T$ is an isometry on Hilbert space $\mathcal{H}$ if and only if $\langle Tu, Tv\rangle = \langle u, v\rangle$ for all $u, v \in \mathcal{H}$.

11.2.22. Let $X$ be a vector space and let $\| \|_\alpha$ and $\| \|_\beta$ be two equivalent norms on $X$. Prove that a sequence $\{x_n\}_{n=1}^\infty$ in $X$ is Cauchy with respect to $\| \|_\alpha$ if and only if it is Cauchy with respect to $\| \|_\beta$. Conclude that $(X, \| \|_\alpha)$ is complete if and only if $(X, \| \|_\beta)$ is.

11.2.23. Let $X$ be a vector space and let $\| \|_\alpha$ and $\| \|_\beta$ be two equivalent norms on $X$. Let $\mathcal{T}_\alpha$ and $\mathcal{T}_\beta$ be the corresponding topologies. Prove that $\mathcal{T}_\alpha = \mathcal{T}_\beta$.

11.2.24. Prove that the equivalence of norms is an equivalence relation.

11.2.25. Let $X$ be a normed linear space and let $T$ be a linear invertible the operator on $X$ such that

$$\|T^n x\| \le M\|x\|, \qquad \text{for all } x \in X,\ n \in \mathbb{Z}.$$

Prove that there exists an equivalent norm on $X$ which makes $T$ an isometry.

11.2.26. Prove that $\|f\| = |f(0)| + \|f'\|_\infty$ defines a norm on $C^1[0, 1]$ that is equivalent to the norm defined by $\|f\|_\alpha := \|f\|_\infty + \|f'\|_\infty$.

11.2.27.** Let $X$ be a linear space equipped with two norms $\| \|_1$ and $\| \|_2$ such that $(X, \| \|_1)$ is complete and the identity mapping $I : (X, \| \|_1) \to (X, \| \|_2)$ is continuous. Suppose that there exists $r > 0$ such that $B_2(0, 1) \subset B_1(0, r) + B_2(0, 1/2)$. Prove that these norms are equivalent.

11.2.28. Prove that there is $M > 0$ such that if $p(t) = a_0 + a_1 t + \cdots + a_n t^n$ then

$$\int_{-1}^{1} |p(t)|\, dt \le M \max\{|p(t)| : t = 0, 1, \ldots, n\}.$$

11.2.29.* Let

$$X = L^1(\mathbb{R}) + L^2(\mathbb{R}) = \{g + h : g \in L^1(\mathbb{R}) \text{ and } h \in L^2(\mathbb{R})\}.$$

Given $p \in [1, \infty)$, define

$$\|f\|_p^{inf} := \inf\{(\|g\|_1^p + \|h\|_2^p)^{1/p} : f = g + h, g \in L^1(\mathbb{R}) \text{ and } h \in L^2(\mathbb{R})\}.$$

Show that $\| \cdot \|_p^{inf}$ is a norm for each $p$, and all of these norms are equivalent.

11.2.30. Give an example of an open cover of the closed unit ball in $\ell^2$ that has no finite subcover.

11.2.31. Give an example of an open cover of the closed unit ball in $C[0,1]$ that has no finite subcover.

11.2.32.* The Hilbert cube is a subset of $\ell^2$ consisting of all sequences $\{x_n\}$ such that $|x_n| \leq 1/n$, for all $n \in \mathbb{N}$. Prove that it is a compact set.

11.2.33. Let $\{e_n\}$ be an orthonormal basis of Hilbert space $\mathcal{H}$, let $\{a_n\} \subset (0,+\infty)$, and let

$$A = \left\{ \sum_{n=1}^{\infty} c_n e_n : \{c_n\} \subset \mathbb{R},\ |c_n| \leq a_n,\ n \in \mathbb{N} \right\}.$$

Prove that $A$ is compact if and only if $\sum_n a_n^2$ is convergent.

11.2.34.* Let $X$ be an infinite dimensional normed linear space. Prove that there exists a sequence $\{x_n\} \subset X$ such that $\|x_n\| = 1$ for all $n \in \mathbb{N}$ and $\|x_m - x_n\| \geq 1/2$ for $m \neq n$.

11.2.35. Prove that the vector space of all polynomials in one variable is not a Banach space in any norm.

11.2.36. Let $X, Y$ be normed linear spaces. Prove that:

(a) $\|(x,y)\| = \|x\| + \|y\|$ defines a norm on $X \times Y$.

(b) $\|(x,y)\|_p := (\|x\|^p + \|y\|^p)^{1/p}$, for $p > 1$, defines a norm on $X \times Y$.

(c) $\|(x,y)\|_{max} := \max\{\|x\|, \|y\|\}$ defines a norm on $X \times Y$.

(d) All these norms are equivalent.

(e) If $X, Y$ are Banach spaces then so is $X \times Y$ (with respect to any of these norms).

11.2.37. Let $\mathcal{F}$ be a closed subspace of $L^p(\mathbb{R}, \mathcal{M}, m)$, $1 \leq p < \infty$. Prove that $\mathcal{F}$ is compact if and only if all of the following conditions hold:

(a) There is $M > 0$ such that $\|f\|_p \leq M$, for all $f \in \mathcal{F}$;

(b) for any $\varepsilon > 0$ there exists $R > 0$ for which

$$\int_{|x|>R} |f|^p < \varepsilon^p,$$

for all $f \in \mathcal{F}$;

(c) for any $\varepsilon > 0$ there exists $\delta > 0$ such that

$$|t| < \delta \quad \Rightarrow \quad \int_{\mathbb{R}} |f(x+t) - f(x)|^p < \varepsilon^p,$$

for all $f \in \mathcal{F}$.

## 11.3 The Uniform Boundedness Principle

In this section we will use the completeness of a Banach space to prove one of the most important results in the theory of linear operators, the so-called Uniform Boundedness Principle. The proof relies heavily on the Baire Category Theorem (Theorem 2.5.1).

**Theorem 11.3.1.** ***The Uniform Boundedness Principle.*** *Let $X$ be a Banach space and $\mathcal{F}$ a family of bounded linear operators from $X$ to a normed linear space $Y$. If, for every $x \in X$,*

$$\sup\{\|Tx\| : T \in \mathcal{F}\} < \infty$$

*then*

$$\sup\{\|T\| : T \in \mathcal{F}\} < \infty. \tag{11.8}$$

*Proof.* Let

$$A_n = \{x \in X : \|Tx\| \leq n \text{ for all } T \in \mathcal{F}\}.$$

Notice that each $A_n$ is closed. Indeed, if $\{x_k\}_{k=1}^{\infty} \subset A_n$ and $\lim x_k = x$, then for any $T \in \mathcal{F}$,

$$\|Tx\| = \|T(\lim_k x_k)\| = \lim_k \|Tx_k\| \leq n,$$

so $x \in A_n$, whence $A_n$ is closed.

By assumption, each $x \in X$ belongs to some $A_n$, so

$$X = \cup_{n=1}^{\infty} A_n.$$

Since $X$ is complete, Theorem 2.5.1 guarantees that, for some $n \in \mathbb{N}$, the set $\overline{A_n}$ has a nonempty interior. In fact, since $A_n$ is closed, the last statement is that, for some $n \in \mathbb{N}$, the set $A_n$ has a nonempty interior. Equivalently, there is a ball

$$B(x_0, r) \subset A_n.$$

If we fix such $n$ and $r$, and if $\|x\| \leq r$, then $x = (x + x_0) + (-x_0)$, $\|T(-x_0)\| = \|Tx_0\|$, and both $x + x_0$ and $x_0$ are in $B(x_0, r)$. Thus, for any $T \in \mathcal{F}$

$$\|Tx\| \leq \|T(x + x_0)\| + \|Tx_0\| \leq n + n.$$

It follows that for any unit vector $u$,

$$\|Tu\| = \frac{1}{r}\|T(ru)\| \leq \frac{1}{r}2n,$$

and hence,

$$\sup\{\|T\| : T \in \mathcal{F}\} \leq \frac{2n}{r}.$$

□

The next result is an important tool in analysis. It allows us to construct bounded linear operators as pointwise limits on a dense set, provided one has a uniform bound.

**Proposition 11.3.2.** *Let $\{T_n\}$ be a sequence of bounded linear operators from a normed linear space $X$ to a Banach space $Y$ and suppose that:*

*(a) $\|T_n\| \leq M$ for all $n \in \mathbb{N}$;*

*(b) $\lim_n T_n x$ exists for all $x$ in a set $D$ that is dense in a closed ball $B \subset X$.*

*Then there exists $\lim_n T_n x = Tx$ for all $x \in X$ and $T$ is a bounded linear operator satisfying $\|T\| \leq M$. If $D = X$, these assertions are also true if $Y$ is merely a normed vector space.*

*Proof.* First we will establish that if $x \in B$ then $\{T_n x\}$ is a convergent sequence. Let $\{x_n\} \subset D$ be chosen so that $\lim_n x_n = x$. By (a),

$$\begin{aligned}\|T_m x - T_n x\| &\leq \|T_m x - T_m x_k\| + \|T_m x_k - T_n x_k\| + \|T_n x_k - T_n x\| \\ &\leq 2M\|x_k - x\| + \|T_m x_k - T_n x_k\|,\end{aligned}$$

so (b) implies that

$$\limsup_{m,n} \|T_m x - T_n x\| \leq 2M\|x_k - x\|.$$

Letting $k \to \infty$ we get that $\{T_n x\}$ is a Cauchy sequence in $Y$. In view of the completeness of $Y$, we obtain that this sequence converges. This is the only place in the proof where completeness of $Y$ is needed. Of course, if $D = X$ then the conclusion that we just obtained is a part of hypothesis.

Next, we will show that $\{T_n x\}$ converges for all $x \in X$. This is clearly true for $x = 0$, so we will assume that $x \neq 0$. Let $B = B(x_0, r)$ and $x \in X \setminus \{0\}$. If $0 < |\alpha| \leq r/\|x\|$, then

$$x_0 + \alpha x \in B$$

so, as we have demonstrated above, the sequence $\{T_n(x_0 + \alpha x)\}$ converges. On the other hand,

$$T_n(x_0 + \alpha x) = T_n x_0 + \alpha T_n x.$$

The left side as well as the first term on the right side have limits when $n \to \infty$, so the same is true of the second term, i.e., there exists

$$\lim_n T_n x \qquad \text{for all } x \in X.$$

Finally, we will show that $Tx := \lim_n T_n x$ is a bounded linear operator satisfying $\|T\| \leq M$. Notice that, for all $n \in \mathbb{N}$,

$$T_n(\alpha x + \beta y) = \alpha T_n x + \beta T_n y \quad \text{and} \quad \|T_n x\| \leq M\|x\|,$$

so the result follows by letting $n \to \infty$. □

Combining Theorem 11.3.1 and Proposition 11.3.2 we get the following result.

**Theorem 11.3.3.** *Let* $X, Y$ *be Banach spaces and let* $\{T_n\}_{n=1}^{\infty}$ *and* $T$ *be operators in* $\mathcal{L}(X,Y)$. *Then* $\lim_n T_n x = Tx$, *for all* $x \in X$, *if and only if*

*(a) the sequence* $\{\|T_n\|\}$ *is bounded;*

*(b)* $\lim_n T_n x$ *exists on a dense subset of* $X$.

Next, we present a nice application of the Uniform Boundedness Principle.

**Proposition 11.3.4.** *The series* $\sum_{n=1}^{\infty} a_n b_n$ *converges absolutely for every convergent sequence* $\{b_n\}$ *if and only if* $\sum_{n=1}^{\infty} |a_n|$ *converges.*

*Proof.* The condition is sufficient because every convergent sequence $\{b_n\}$ is bounded and

$$\sum_{n=1}^{\infty} |a_n b_n| \leq \sup\{|b_n| : n \in \mathbb{N}\} \sum_{n=1}^{\infty} |a_n|. \tag{11.9}$$

To prove that it is necessary, even a weaker assumption that $\sum_{n=1}^{\infty} a_n b_n$ converges conditionally will do. Consider the sequence $\{x_n\}$, where

$$x_n = (b_1, b_2, \dots, b_n, 0, 0, \dots), \qquad \text{for } n \in \mathbb{N},$$

and define

$$T_n(x) = \sum_{k=1}^{n} a_k b_k, \qquad \text{for } x = \{b_n\}_{n=1}^{\infty}.$$

Thus, $T_n$ is a linear operator from the Banach space $c$ (Example 10.1.9) to $\mathbb{R}$. By assumption, for every $x \in c$, the limit $\lim_n T_n(x)$ exists. Consequently, for every $x \in c$, the sequence $\{T_n x\}$ is bounded. Applying the Uniform Boundedness Principle we obtain that the sequence $\{\|T_n\|\}$ is bounded. Further, the same argument as in (11.9) shows that $\|T_n(x)\| \leq \|x\| \sum_{k=1}^{n} |a_k|$, whence

$$\|T_n\| \leq \sum_{k=1}^{n} |a_k|. \tag{11.10}$$

In fact, choosing $x = (\operatorname{sgn} a_1, \operatorname{sgn} a_2, \dots, \operatorname{sgn} a_n, 0, 0, 0, \dots)$, Exercise 11.1.3 shows that the equality holds in (11.10) whence the series $\sum_{n=1}^{\infty} |a_n|$ converges. □

Recall that a sequence $\{a_n\}$ is *Cesàro convergent* if the sequence $\{\sigma_n\}$, defined by $\sigma_n = (a_1 + a_2 + \cdots + a_n)/n$ converges. Notice that $\{\sigma_n\} = C\{a_n\}$, where $C$ is the infinite matrix

$$C = \begin{bmatrix} 1 & 0 & 0 & 0 & \dots \\ \frac{1}{2} & \frac{1}{2} & 0 & 0 & \dots \\ \frac{1}{3} & \frac{1}{3} & \frac{1}{3} & 0 & \dots \\ \vdots & \vdots & \vdots & \vdots & \ddots \end{bmatrix}.$$

In view of the matrix $C$, the Cesàro convergence is frequently referred to as *$C$-convergence.* This procedure can be generalized to any matrix $A$, as long as every convergent sequence is $A$-convergent and $A\text{-}\lim x_n = \lim x_n$. Notice that, given a matrix $A = [a_{ij}]$ we need that each series

$$y_i = \sum_{k=1}^{\infty} a_{ik} x_k \tag{11.11}$$

converges, and that $\lim y_n = \lim x_n$. The following theorem that is due to Toepliz gives a necessary and sufficient condition for a matrix $A$ to fulfill these requirements.

**Theorem 11.3.5.** *Let $A = [a_{ij}]$ be a matrix. A necessary and sufficient condition that for every convergent sequence $\{x_n\}$ every series (11.11) converges and that $\lim y_n = \lim x_n$ is that the following 3 conditions hold:*

*(i) There exists $M > 0$ such that $\sum_{k=1}^{\infty} |a_{ik}| \leq M$, for all $i = 1, 2, \ldots$.*

*(ii) $\lim_i a_{ik} = 0$, for all $k = 1, 2, \ldots$.*

*(iii) $\lim_i \sum_{k=1}^{\infty} a_{ik} = 1$.*

*Proof.* Let us introduce the following notation. For a convergent sequence $x = \{x_n\}$ let $T_i(x) = \sum_{k=1}^{\infty} a_{ik} x_k$ and $T(x) = \lim x_n$. Then each $T_n$ and $T$ are linear operators mapping $c$ to $\mathbb{R}^*$. (We do not know apriori whether any of the operators $T_n$ is bounded.) Let $e = (1, 1, 1, \ldots) \in c$ and let $\{e_n\} \in c$ as usual be the sequence with 1 in the $n$th spot and zeros elsewhere. Then operators $\{T_i\}_{i \in \mathbb{N}}$ satisfy the following conditions:

(a) $\|T_i\| = \sum_{k=1}^{\infty} |a_{ik}|$, for all $i = 1, 2, \ldots$.

(b) $T_i(e_k) = a_{ik}$ for all $k = 1, 2, \ldots$.

(c) $T_i(e) = \sum_{k=1}^{\infty} a_{ik}$.

We leave the proof as Exercise 11.3.11.

Conditions (i)–(iii) now become:

(i') There exists $M > 0$ such that $\|T_i\| \leq M$, for all $i = 1, 2, \ldots$.

(ii') $\lim_i T_i(e_k) = T(e_k)$ for all $k = 1, 2, \ldots$.

(iii') $\lim_i T_i(e) = T(e)$.

Thus, the theorem can be rephrased as: *$\{T_i\}$ is a sequence of bounded linear operators from $c$ to $\mathbb{R}$ that converges pointwise to $T$ if and only if (i')–(iii') hold.* Now (i') follows directly from the Uniform Boundedness Principle, and (ii'), (iii') state that $\{T_i\}$ converges pointwise at $x = e_k$ and $x = e$. In the other direction, by Exercise 10.4.33, conditions (a)–(b) of Theorem 11.3.3 are satisfied so $T_i$ converges pointwise to $T$. □

Numerous special cases of the Uniform Boundedness Principle appeared in the second decade of the 20th century, such as Lebesgue in 1910 [74], German mathematicians Ernst Hellinger (1883–1950) and Otto Toeplitz (1881–1940) in 1910 [58], Helly in 1912 [59], and Hahn in 1916 [46]. They all used the so-called *sliding of the gliding hump* that Lebesgue originated. After a plethora of specific results and some generalizations, Banach proved the first theorem dealing with linear operators (Exercise 11.3.1) in his dissertation in 1922 (again using the sliding hump method). In 1927, Banach and Steinhaus [7] worked on a question about the Fourier series and needed a version of the Uniform Boundedness Principle. Saks was a referee and suggested that they replace the sliding hump proof by the one that uses the Baire Category theorem. Today, the Uniform Boundedness Principle is often referred to as the Banach–Steinhaus theorem or, less frequently, Banach–Saks–Steinhaus theorem. Hellinger was a professor at the University of Frankfurt until he was let go as a Jew. He was arrested in 1938 and sent to Dachau concentration camp. However, his friends were able to arrange a temporary job for him at Northwestern University. He was released from the Dachau camp after six weeks, on condition that he emigrate immediately. Toeplitz was an important figure in the development of functional analysis. He was dismissed from his post at Bonn University and he emigrated to Mandatory Palestine, where he was scientific advisor to the rector of the Hebrew University of Jerusalem. He studied matrices with constant diagonals which are now known as Toeplitz matrices.

HUGO STEINHAUS (1887–1972) was a Polish mathematician, born in Jasło, a part of Austria until 1918, in a family with Jewish roots. He studied mathematics at the University of Göttingen under influential mathematicians like Toeplitz, Carathéodory, and Klein, completing his doctoral dissertation with Hilbert as an advisor in 1911. After a brief military service he worked in Kraków in the government of the short-lived Kingdom of Poland. With Banach he founded the Mathematical Society of Kraków, which later became the Polish Mathematical Society. In 1920, he joined the University of Lwów. Occasionally, he took part in the informal gatherings at Scottish Café and contributed to the renowned "Scottish Book" of open mathematical problems. His mathematical accomplishments spanned functional analysis, trigonometric series, probability theory, and real functions. In 1923 he published the first rigorous account of the theory of tossing coins based on measure theory. In 1925 he was the first to define and discuss the concept of strategy in game theory. He initiated the modern research of the so-called "fair cake-cutting". His book *Mathematical Snapshots*, published in 1937, depicted mathematics as a reflection of reality, akin to poetry. *One Hundred Problems in Elementary Mathematics* is an instructive, stimulating collection of problems. In 1945 he moved to the University of Wrocław and worked on rebuilding Polish mathematics.

## Exercises

11.3.1. Let $X$ be a Banach space and $Y$ a normed linear space, and let $T_n : X \to Y$, $n \in \mathbb{N}$, be bounded linear operators. Suppose that, for every $x \in X$, the sequence $T_n x$ is convergent, and let $T$ be an operator defined by $Tx = \lim T_n x$. Prove that $T$ is a bounded linear operator.

11.3.2. Let $X$ be a Banach space and $\mathcal{F}$ an unbounded family of bounded linear operators from $X$ to a normed linear space $Y$. Prove that the set of $x \in X$ for which $\sup\{\|Tx\| : T \in \mathcal{F}\} = \infty$ is dense in $X$.

11.3.3. Let $X, Y, Z$ be Banach spaces and let $F : X \times Y \to Z$ be a separately continuous bilinear map, i.e., $F_x \in \mathcal{L}(Y, Z)$ for each $x \in X$ and $F^y \in \mathcal{L}(X, Z)$ for each $y \in Y$. Prove that $F$ is continuous.

11.3.4. Give an example to show that the Uniform Boundedness Principle (Theorem 11.3.1) does not hold if $X$ is not a Banach space.

11.3.5.* Let $X, Y$ be Banach spaces and $\{T_{j,k}\} \subset \mathcal{L}(X, Y)$. Suppose that for each $k \in \mathbb{N}$ there is $x \in X$ such that $\sup_j \|T_{j,k}x\| = \infty$. Prove that there exists $x \in X$ such that $\sup_j \|T_{j,k}x\| = \infty$ for all $k \in \mathbb{N}$.

If $X$ is a Banach space the sequence $\{T_n\} \subset \mathcal{L}(X)$ converges *strongly* to a linear operator $T : X \to X$ if $\lim T_n x = Tx$ for all $x \in X$. If $\mathcal{H}$ is Hilbert space the sequence $\{T_n\} \subset \mathcal{L}(\mathcal{H})$ converges *weakly* to a linear operator $T : \mathcal{H} \to \mathcal{H}$ if $\lim\langle T_n x, y\rangle = \langle Tx, y\rangle$ for all $x, y \in \mathcal{H}$.

11.3.6. Let $X$ be a Banach space. Prove that if $\{T_n\}$ converges strongly to $T$ and $\lim x_n = x$ then $\lim T_n x_n = Tx$.

11.3.7. Let $X$ be a Banach space. Prove that if $\{T_n\}$ converges strongly to $T$ and $\{S_n\}$ converges strongly to $S$ then $\{T_n S_n\}$ converges strongly to $TS$.

11.3.8. Let $\mathcal{H}$ be Hilbert space and let $A, B : \mathcal{H} \to \mathcal{H}$ be linear operators satisfying $\langle Bx, y\rangle = \langle x, Ay\rangle$ for all $x, y \in \mathcal{H}$. Prove that $A$ and $B$ are both bounded.

11.3.9. Let $\mathcal{H}$ be Hilbert space with orthonormal basis $\{e_n\}$ and let $\{T_n\}, \{S_n\}$ be sequences in $\mathcal{L}(\mathcal{H})$, defined by $T_n(\sum_k a_k e_k) = a_n e_n$ and $S_n(\sum_k a_k e_k) = \sum_k a_k e_{n+k}$. Prove that:

(a) $\{T_n\}$ converges to 0 strongly but not in the norm.

(b) $\{S_n\}$ converges to 0 weakly but not strongly.

11.3.10. Let $X, Y$ be Banach spaces, let $T \in \mathcal{L}(X, Y)$, let $\{T_n\}$ be a bounded sequence in $\mathcal{L}(X, Y)$, and suppose that $\lim_n \|T_n x - Tx\| = 0$ for all $x$ in a dense subset of $X$. Prove or disprove: $\lim_n \|T_n - T\| = 0$.

11.3.11. Prove assertions (a)–(c) made in the proof of Theorem 11.3.5.

11.3.12. Let $X, Y$ be Banach spaces and let $\{T_n\} \subset \mathcal{L}(X, Y)$ be operators such that for all $x \in X$ the series $\sum\limits_n T_n(x)$ converges. Prove that $\sum_{n=1}^{\infty} \|T_n\|/2^n$ converges.

11.3.13.* Let $p, q$ be real numbers such that $p > 1$ and $1/p + 1/q = 1$. Prove that the function $fg \in L^1(a, b)$ for every $g \in L^p$ if and only if $f \in L^q$.

11.3.14.* Prove that the function $fg \in L^1(a, b)$ for every $g \in L^1$ if and only if $f \in L^\infty$.

11.3.15.* Let $p, q$ be real numbers such that $p > 1$ and $1/p + 1/q = 1$. Prove that the series $\sum_{n=1}^{\infty} a_n b_n$ converges absolutely for every sequence $\{b_n\} \in \ell^p$ if and only if $\{a_n\} \in \ell^q$.

11.3.16.* Prove that the series $\sum_{n=1}^{\infty} a_n b_n$ converges absolutely for every sequence $\{b_n\} \in \ell^1$ if and only if $\{a_n\} \in \ell^\infty$.

11.3.17.* Prove that if the series $\sum_{n=1}^{\infty} a_n b_n$ converges for every sequence $\{b_n\} \in c_0$ then $\{a_n\} \in \ell^1$.

11.3.18.** Prove that $S_k := \sum_{n=1}^{\infty} a_n^{(k)} b_n$ converges for every sequence $\{b_n\} \in c_0$ and $\lim_k S_k = 0$ if and only if $\sup_k \|a^{(k)}\|_1 < \infty$ and $\lim_k a_n^{(k)} = 0$ for all $n \in \mathbb{N}$.

11.3.19.* Let $1 \leq q < p < \infty$. Determine for what functions $h$ on $[0, 1]$ is the multiplication by $h$ a bounded operator from $L^p[0, 1]$ to $L^q[0, 1]$.

11.3.20.** Let $s_n$ denote the $n$th partial sum of the Fourier series of a function $f \in C[0, 2\pi]$, let $D_n$ denote the Dirichlet kernel, and let $\| \ \|_1$ be the norm in $L^1(0, 2\pi)$. Prove that:

(a) there exist constants $m, M > 0$ such that $m \ln(n + 1) \leq \|D_n\|_1 \leq M \ln(n + 1)$;

(b) $\lim_n \|s_n\|_{max} / \ln(n + 1) = 0$;

(c) if $\{\lambda_n\}$ is a sequence of real numbers such that $\lim \lambda_n / \ln(n + 1) = 0$ then there exists $f \in C[0, 2\pi]$ for which the sequence $\{s_n(0)/\lambda_n\}$ is unbounded.

11.3.21. Let $X$ be a Banach space and $Y$ a normed linear space, and let $\{T_n\} \subset \mathcal{L}(X, Y)$. Suppose that whenever $\{x_n\} \subset X$ and $\lim_n \|x_n\| = 0$ we get $\lim_n \|T_n x_n\| = 0$. Prove that $\sup_n \|T_n\| < \infty$.

11.3.22. Let $X$ be a Banach space and $Y$ a normed linear space, and let $\{T_n\} \subset \mathcal{L}(X, Y)$. Prove that the following assertions are equivalent:

(a) For each norm covergent series $\sum_n x_n$ it follows that $T_n x_n \to 0$ in norm.

(b) $\sup_n \|T_n\| < \infty$.

11.3.23. Let $\{a_n\} \subset \mathbb{R}$ have the property that if $\{x_n\} \in c_0$ then $\{a_n x_n\} \in c_0$. Prove that $\{a_n\} \in \ell^\infty$.

11.3.24. Let $X$ be a Banach space and $\{T_n\} \subset \mathcal{L}(X, \mathbb{R})$. Prove that the following assertions are equivalent:

(a) If $\lim_n \|x_n\| = 0$ then the series $\sum_{n=1}^{\infty} T_n(x_n)$ converges.

(b) The series $\sum_{n=1}^{\infty} \|T_n\|$ converges.

11.3.25. Give an example of the implication (a) $\Rightarrow$ (b) in Exercise 11.3.24 is not true if $\{T_n\} \subset \mathcal{L}(X, \mathbb{Y})$, with $Y$ a Banach space.

11.3.26. Let $X$ be a normed linear space and let $\{x_n\} \subset X$ with the property that if $T \in \mathcal{L}(X, \mathbb{R})$ then $\sum_{k=1}^{\infty} |T(x_k)|$ converges. Prove that

$$\sup_{\|T\| \leq 1} \sum_{k=1}^{\infty} |T(x_k)| < \infty.$$

11.3.27. Let $X$ and $Y$ be Banach spaces and let $\{T_n\} \subset \mathcal{L}(X, Y)$. Prove that the following assertions are equivalent:

(a) If $\sum_n \|x_n\|$ converges then the sequence $\{\sum_{k=1}^n T_k(x_k)\}$ is bounded.

(b) The sequence $\{T_k\}$ is bounded.

11.3.28. Let $X$ be a Banach space and $\{T_n\} \subset \mathcal{L}(X, \mathbb{R})$. Suppose that $1 < p < \infty$ and $1/p + 1/q = 1$. Prove that the following assertions are equivalent:

(a) If $\sum_n \|x_n\|^p$ converges then the series $\sum_{n=1}^\infty T_n(x_n)$ converges.

(b) The series $\sum_{n=1}^\infty \|T_n\|^q$ converges.

11.3.29. Give an example of the implication (a) $\Rightarrow$ (b) in Exercise 11.3.28 is not true if $\{T_n\} \subset \mathcal{L}(X, Y)$, with $Y$ a Banach space.

11.3.30.* Let $1 \leq p < \infty$, let $[a_{ij}]_{i,j=1}^\infty$ be an infinite matrix of real numbers, and suppose that, for any $\mathbf{x} \in \ell^p$, the sequence $A\mathbf{x}$ belongs to $\ell^p$. Use Uniform Boundedness Principle to prove that the operator $T$, defined by $T(\mathbf{x}) = A\mathbf{x}$, is a bounded operator on $\ell^p$.

## 11.4 The Open Mapping and Closed Graph Theorems

Recall that a function is continuous if and only if the inverse image of an open set is open (Theorem 2.3.11). Here, we are interested in functions for which the *direct* image of an open set is open.

**Definition 11.4.1.** Let $X, Y$ be normed linear spaces. The map $f : X \to Y$ is **open** if $f(G)$ is open whenever $G$ is an open subset of $X$.

In this section we will look into the question of when a linear operator is an open map, and we will consider some applications of such criteria.

*Remark* 11.4.2. We stated the definition for $X, Y$ normed linear spaces because this is the framework of interest for us in this chapter. The concept of an open map can be naturally defined when $X, Y$ are any topological spaces.

Examples of maps that are not open are easy to find. One such is $f : \mathbb{R} \to \mathbb{R}$ defined by $f(x) = x^2$, which maps the open interval $(-1, 1)$ onto $[0, 1)$. It is harder to formulate sufficient conditions for a map to be open. In the case of Euclidean spaces, say $X = \mathbb{R}^n$ and $Y = \mathbb{R}^m$, $f$ is open if it has continuous partial derivatives and its derivative is surjective. (See [86, Theorem 12.3.1].) In the special case when $f$ is linear, the first condition is satisfied and, since $f$ equals its derivative, the second condition requires that $f$ be surjective. To summarize, if a linear map between Euclidean spaces is surjective then it is open. It turns out that the same result holds if Euclidean spaces are replaced by arbitrary Banach spaces. In fact, the proof that we will present follows a similar idea as in the finite-dimensional case.

**Theorem 11.4.3.** ***The Open Mapping Theorem.*** *Let $X, Y$ be Banach spaces and let $T \in \mathcal{L}(X, Y)$ be surjective. Then $T$ is an open map.*

*Proof.* Let $B_X, B_Y$ be the open unit balls in $X, Y$.

**Step I.** There exists $r > 0$ such that

$$\overline{B_Y} \subset r \cdot \overline{T(B_X)}. \tag{11.12}$$

Since we are assuming that $T$ is surjective, we have that

$$Y = \bigcup_{n=1}^{\infty} n \cdot T(B_X) = \bigcup_{n=1}^{\infty} n \cdot \overline{T(B_X)}.$$

Baire Theorem guarantees that $n \cdot \overline{T(B_X)}$ has a nonempty interior for some $n \in \mathbb{N}$. For such $n$, there exists $y_0 \in n \cdot \overline{T(B_X)}$ and $r_1 > 0$ for which

$$y_0 + r_1 B_Y \subset n \cdot \overline{T(B_X)}.$$

The last inclusion, when combined with $y_0 \in n \cdot \overline{T(B_X)}$, implies that

$$r_1 B_Y \subset n \cdot \overline{T(B_X)} - y_0 \subset 2n \cdot \overline{T(B_X)}.$$

If we denote $r = 2n/r_1$, then $B_Y \subset r \cdot \overline{T(B_X)}$, and (11.12) follows by taking the closure of both sides.

**Step II.** With $r$ as in Step I, for each $y \in Y$ and $\varepsilon > 0$ there exists $x \in X$ so that

$$\|y - Tx\| < \varepsilon \quad \text{and} \quad \|x\| \le r\|y\|. \tag{11.13}$$

Let $y \in Y$ and $\varepsilon > 0$ be given. If $y = 0$ we take $x = 0$. Otherwise, $y/\|y\| \in \overline{B_Y}$ and (11.12) implies that

$$\frac{y}{\|y\|} \in r \cdot \overline{T(B_X)},$$

whence $y/(r\|y\|) \in \overline{T(B_X)}$. It follows that there exists $x_0 \in B_X$ such that

$$\left\| \frac{y}{r\|y\|} - Tx_0 \right\| < \frac{\varepsilon}{r\|y\|}.$$

Multiplying through by $r\|y\|$ and setting $x = r\|y\|\, x_0$, we obtain (11.13).

**Step III.** With $r$ as before, for each $y \in Y$ there exists $x \in X$ such that

$$T(x) = y \quad \text{and} \quad \|x\| \le 2r\|y\|. \tag{11.14}$$

Let $y \in Y$. If $y = 0$, (11.14) holds with $x = 0$, so we may assume that $y \neq 0$. By Step II, taking $\varepsilon = \|y\|/2$, there exists $u_1 \in X$ such that

$$\|y - Tu_1\| < \frac{\|y\|}{2} \quad \text{and} \quad \|u_1\| \le r\|y\|.$$

Applying again Step II, this time to $y - Tu_1$ and $\|y\|/2^2$, we obtain $u_2 \in X$ such that

$$\|y - Tu_1 - Tu_2\| < \frac{\|y\|}{2^2} \quad \text{and} \quad \|u_2\| \le r\|y - Tu_1\| < \frac{r}{2}\,\|y\|.$$

Continuing this process we obtain a sequence $\{u_k\} \subset X$ such that, for every $k \in \mathbb{N}$,

$$\|y - Tu_1 - Tu_2 - \cdots - Tu_k\| < \frac{\|y\|}{2^k} \quad \text{and} \quad \|u_k\| < \frac{r}{2^{k-1}}\,\|y\|. \tag{11.15}$$

Let $x_n = \sum_{k=1}^{n} u_k$. Then the second inequality in (11.15) implies that

$$\|x_{n+m} - x_n\| = \left\| \sum_{k=n+1}^{n+m} u_k \right\| \le \sum_{k=n+1}^{n+m} \|u_k\| \le \sum_{k=n+1}^{\infty} \frac{r}{2^{k-1}}\,\|y\| = \frac{r}{2^{n-1}}\,\|y\|.$$

It follows that $\{x_n\}$ is a Cauchy sequence in $X$, hence it converges to $x \in X$. Since $T$ is continuous we have that

$$\lim Tx_n = Tx.$$

On the other hand, the first inequality in (11.15) is $\|y - Tx_k\| < \|y\|/2^k$ so

$$\lim Tx_n = y,$$

and we conclude that $y = Tx$.

Finally, using once again the second inequality in (11.15), we have that

$$\|x_n\| = \left\| \sum_{k=1}^{n} u_k \right\| \le \sum_{k=1}^{n} \|u_k\| < \sum_{k=1}^{n} \frac{r}{2^{k-1}}\,\|y\| \le 2r\|y\|,$$

and the inequality in (11.14) follows by letting $n \to \infty$.

**Step IV.** There is $\gamma > 0$ such that $\gamma B_Y \subset T(B_X)$.

Let $\gamma = 1/(2r)$ and suppose that $y \in \gamma B_Y$. By Step III, there is $x \in X$ such that $y = T(x)$ and $\|x\| \le \|y\|/\gamma < 1$. Thus, $y \in T(B_X)$.

**Step V.** $T$ is an open map.

Let $G \subset X$ be open and let $y_0 \in T(G)$. We will show that $y_0$ is an interior point of $T(G)$. First, there is $x_0 \in G$ such that $y_0 = Tx_0$. Second, $x_0$ is an interior point of $G$ so there is $R > 0$ such that $x_0 + RB_X \subset G$ which implies that

$$T(x_0 + RB_X) \subset T(G).$$

It follows that, with $\gamma$ as in Step IV,

$$y_0 + R\gamma B_Y \subset y_0 + RT(B_X) = T(x_0) + RT(B_X) = T(x_0 + RB_X) \subset T(G),$$

so $y_0$ is an interior point of $T(G)$ and the theorem is proved. □

We have seen in Proposition 11.2.3 that when $T$ is an invertible operator in $\mathcal{L}(X, Y)$, its inverse $T^{-1}$ is also a linear operator. Further, Proposition 11.2.4 gives a necessary and sufficient condition for $T^{-1}$ to be bounded. Perhaps unexpectedly, if $X$ and $Y$ are Banach spaces the boundedness is automatic.

**Corollary 11.4.4.** *Let $X, Y$ be Banach spaces and $T \in \mathcal{L}(X, Y)$ injective and surjective. Then $T^{-1}$ is continuous.*

*Proof.* Let $G$ be an open set in $X$ and let $A$ be the inverse of $T$. It suffices to show that the inverse image of $G$ under $A$ is an open set. However,

$$A^{-1}(G) = (T^{-1})^{-1}(G) = T(G),$$

which is open by the Open Mapping Theorem. Thus, $A$ is continuous. □

*Remark* 11.4.5. It is easy to see that Corollary 11.4.4 is equivalent to the Open Mapping Theorem. It is sometimes referred to as the Bounded Inverse Theorem.

> The first to state and prove the Open Mapping Theorem was a Polish mathematician Julius Schauder (1899–1943) in 1930 [105]. He also derived Corollary 11.4.4. A year earlier Banach had proved Corollary 11.4.4 [4] without relying on the Open Mapping Theorem. Both were popularized by Banach in his book [5]. Perhaps because of that, the Open Mapping Theorem is also known as Banach–Schauder Theorem.

We have defined the equivalence of two norms by inequalities (11.3). When the underlying vector space is complete in both topologies induced by these norms, one inequality suffices.

**Corollary 11.4.6.** *Let $\|\cdot\|_\alpha$ and $\|\cdot\|_\beta$ be two norms on a linear space $X$ for which both $(X, \|\cdot\|_\alpha)$ and $(X, \|\cdot\|_\beta)$ are Banach spaces. Suppose that there is $M \geq 0$ such that*

$$\|x\|_\beta \leq M\|x\|_\alpha, \qquad \text{for all} \quad x \in X. \tag{11.16}$$

*Then these norms are equivalent.*

*Proof.* Let $I : (X, \|\cdot\|_\alpha) \to (X, \|\cdot\|_\beta)$ be defined by $I(x) = x$. Clearly, $I$ is bijective, and (11.16) shows that it is bounded. By Corollary 11.4.4, $I^{-1} : (X, \|\cdot\|_\beta) \to (X, \|\cdot\|_\alpha)$ is continuous, hence bounded, whence there is $C > 0$ such that $\|x\|_\alpha \leq C\|x\|_\beta$, for all $x \in X$. □

Next, we present an application of the results obtained in this section. In the theory of Fourier series the Riemann-Lebesgue Lemma (page 347) asserts that the Fourier coefficients of an integrable function form a sequence that belongs to $c_0$, i.e., a sequence that converges to 0. Thus, it is natural to ask whether for every sequence $\{a_n\} \in c_0$ there is an integrable function for which these are its Fourier coefficients. We will demonstrate that the answer is in the negative.

It suffices to establish our claim in the case when each coefficient $b_n$ in (10.6) equals 0. If a function $f$ has such Fourier coefficients then it is an even function, and we denote by $X$ the space consisting of even functions in $L^1([0, 2\pi])$ or, more precisely, functions in $L^1([0, 2\pi])$ whose periodic extension to $\mathbb{R}$ are even. Now, the first two equations in (10.6) define an operator $T$ which, by the Riemann-Lebesgue Lemma, maps $X$ to $c_0$. The question that we are asking is: is $T$ surjective?

It is easy to verify that $T$ is linear, and that $X$ is a Banach space. Also, by Exercise 10.1.15, $c_0$ is a Banach space with respect to the norm

$$\|(x_1, x_2, x_3, \dots)\| = \sup\{|x_n| : n \in \mathbb{N}\}.$$

Next, $T$ is bounded because

$$\begin{aligned}\|Tf\|_{c_0} &= \sup_{n\in\mathbb{N}} \left| \int_0^{2\pi} f(x) \frac{\cos nx}{\sqrt{\pi}} \, dx \right| \\ &\leq \frac{1}{\sqrt{\pi}} \int_0^{2\pi} |f(x)| \, dx = \frac{1}{\sqrt{\pi}} \|f\|_X.\end{aligned}$$

Moreover, $T$ is injective. Indeed, it suffices to show that if all Fourier coefficients of $f$ are 0, then $f = 0$. To that end, using integration by parts it is easy to show that the function

$$F(x) = \int_0^x f(t) \, dt$$

has Fourier coefficients equal to 0. Since $F$ is continuous, its Fourier series (which is identically zero) converges to $F$, so $F$ is the zero function. That $f = 0$ a.e. now follows from Lemma 9.4.12.

Finally, if $T$ were surjective, it would be invertible, and Corollary 11.4.4 would imply that $T^{-1}$ is bounded. By Proposition 11.2.4, there would exist $m > 0$ such that

$$\|Tf\|_{c_0} \geq m\|f\|_X, \quad \text{for all } f \in X. \tag{11.17}$$

Let $D_n$ denote the Dirichlet kernel

$$D_n(x) = \frac{1}{2} + \sum_{k=1}^{n} \cos kx = \frac{\sin(n + 1/2)x}{2\sin(x/2)}.$$

A calculation shows that

$$T(D_n) = e_n = (0, 0, \dots, 0, 1, 0, 0, \dots),$$

with 1 in the $n$th spot, so $\|TD_n\|_{c_0} = 1$. On the other hand, using substitution $u = (n + 1/2)x$,

$$\|D_n\|_X \geq \int_0^{2\pi} \left| \frac{\sin(n+1/2)x}{x} \right| dx = \int_0^{(2n+1)\pi} \frac{|\sin u|}{u} du$$
$$= \sum_{k=0}^{2n} \int_{k\pi}^{(k+1)\pi} \frac{|\sin u|}{u} du \geq \sum_{k=0}^{2n} \int_{k\pi}^{(k+1)\pi} \frac{|\sin u|}{(k+1)\pi} du = \sum_{k=0}^{2n} \frac{2}{(k+1)\pi},$$

so $\|D_n\|_X \to \infty$, as $n \to \infty$. This contradicts (11.17), and we conclude that $T$ cannot be surjective. Therefore, there are sequences that converge to 0 but are not the Fourier coefficients of an integrable function.

## Closed Graph Theorem

Here, we will present an important consequence of the Open Mapping Theorem. So far, we worked exclusively with bounded operators. However, we have seen in Example 11.1.9 that the derivative is an unbounded operator so studying unbounded operators is a significant part of functional analysis. Such (unbounded) operators sometimes have another useful property.

**Definition 11.4.7.** A linear operator $T : X \to Y$ between normed linear spaces $X, Y$ is **closed** if for every $\{x_n\} \subset X$

$$x_n \to x_0 \quad \text{and} \quad Tx_n \to y_0 \quad \Rightarrow \quad Tx_0 = y_0. \tag{11.18}$$

*Remark* 11.4.8. It is easy to see that condition (11.18) can be replaced by

$$x_n \to 0 \quad \text{and} \quad Tx_n \to y_0 \quad \Rightarrow \quad y_0 = 0.$$

It is easy to verify that if $T$ is continuous then it is closed. However, the converse is not true.

**Example 11.4.9. A closed operator that is not continuous.**
Let $X = C^1[0,1]$, $Y = C[0,1]$, both with $\| \|_{max}$ norm, and let $T : X \to Y$ be defined by $Tf = f'$. The operator $T$ is not continuous (see Example 11.1.9). However, $T$ is closed. If $f_n \to f$ and $f_n' \to g$, then $\{f_n\}$ converges uniformly to $f$ and $\{f_n'\}$ converges uniformly to $g$, so $g = f'$ [86, Theorem 8.2.5]. ♦

There is an equivalent definition of a closed operator that uses the concept of a graph.

**Definition 11.4.10.** Let $T : X \to Y$ be a linear map between normed linear spaces $X, Y$. The **graph** of $T$ is the set $\{(x, Tx) : x \in X\}$.

Since the graph of an operator $T : X \to Y$ is a subset of $X \times Y$, a word or two is in order about the latter. By defining

$$(x_1, y_1) + (x_2, y_2) = (x_1 + x_2, y_1 + y_2) \quad \text{and} \quad \alpha(x, y) = (\alpha x, \alpha y),$$

$X \times Y$ becomes a vector space. The norm can then be defined by

$$\|(x,y)\| = \|x\| + \|y\|, \tag{11.19}$$

and it is easy to verify that it is indeed a norm. Moreover, when $X$ and $Y$ are Banach spaces, so is $X \times Y$. (See Exercise 11.2.36.)

The choice of the adjective *closed* to describe an operator is motivated by the following proposition.

**Proposition 11.4.11.** *Let $T : X \to Y$ be a linear map between normed linear spaces $X, Y$. Then $T$ is closed if and only if its graph is closed.*

*Proof.* ($\Rightarrow$) Let $T$ be a closed operator and denote its graph by $G$. Suppose that $(x,y) \in \overline{G}$. Then there is a sequence $\{(x_n, Tx_n)\}$ that converges to $(x,y)$. By definition of the norm (11.19), this implies that

$$\lim x_n = x \quad \text{and} \quad \lim Tx_n = y. \tag{11.20}$$

Since $T$ is closed, we see from (11.18) that $y = Tx$ so $(x,y) \in G$.

($\Leftarrow$) Let $\{x_n\} \subset X$ such that $x_n \to x_0$ and $Tx_n \to y_0$. Then $(x_n, Tx_n) \to (x_0, y_0)$ in $X \times Y$. Since the graph is closed, $(x_0, y_0) \in G$, i.e., $Tx_0 = y_0$. $\square$

Although far from obvious, Example 11.4.9 takes advantage of the fact that $C^1[0,1]$ with $\| \|_{max}$ norm is not complete (see Exercise 10.1.14). An attempt to find an example where both $X$ and $Y$ are Banach spaces would be doomed to fail because in such a situation every closed operator is continuous. Together with the Uniform Bounded Principle and the Open Mapping Theorem this is one of the three fundamental results in the theory of linear operators.

**Theorem 11.4.12.** ***The Closed Graph Theorem.*** *Let $X, Y$ be Banach spaces and let $T : X \to Y$ be linear. Then $T$ is continuous if and only if its graph is closed.*

*Proof.* ($\Rightarrow$) Let $G$ be the graph of a continuous operator $T$ and let $(x,y) \in \overline{G}$. Then there is a sequence $\{(x_n, Tx_n)\}$ that converges to $(x,y)$ which means that (11.20) holds. Since $T$ is continuous we obtain that $Tx = y$, so $(x,y) \in G$ and the graph of $T$ is closed.

($\Leftarrow$) Let

$$\|x\|_G = \|x\| + \|Tx\|.$$

We will show that $(X, \|\cdot\|_G)$ is complete. Once this is accomplished, the obvious inequality $\|x\| \le \|x\|_G$ and Corollary 11.4.6 will imply that there exists $M > 0$ such that $\|x\|_G \le M\|x\|$, whence $\|Tx\| \le M\|x\|$.

To prove that $(X, \|\cdot\|_G)$ is complete, let $\{x_n\}$ be a Cauchy sequence in the topology induced by $\|\cdot\|_G$. Since

$$\|x_m - x_n\|_G = \|x_m - x_n\| + \|Tx_m - Tx_n\|, \tag{11.21}$$

we see that both $\{x_n\}$ and $\{Tx_n\}$ are Cauchy sequences in Banach spaces $X, Y$ respectively, hence convergent. Let $x_n \to x$ and $Tx_n \to y$. The fact that $T$ is closed implies that $y = Tx$, so $Tx_n \to Tx$. Therefore,

$$\|x - x_n\|_G = \|x - x_n\| + \|Tx - Tx_n\| \to 0,$$

and the theorem is proved. □

*Remark* 11.4.13. We used the completeness of $X$ and $Y$ only in one direction. The fact that every bounded operator has a closed graph holds in any normed linear spaces.

We make a remark here that when checking the continuity of an operator between Banach spaces, we often try to prove the implication $x_n \to x \Rightarrow Tx_n \to Tx$. The Closed Graph Theorem allows us to assume that $\{Tx_n\}$ is convergent, which is frequently very helpful. (See Exercises 11.4.44 and 11.4.45.)

We close this section with an application of the Closed Graph Theorem. We will see more of them in Chapter 12.

**Proposition 11.4.14.** *Let* $T : (L^2(0,1), \| \|_2) \to (L^2(0,1), \| \|_2)$ *be a bounded operator such that if* $f \in L^2(0,1)$ *is a continuous function then so is* $Tf$*. Then the restriction* $T : (C[0,1], \| \|_{max}) \to (C[0,1], \| \|_{max})$ *is a bounded operator.*

*Proof.* It is clear that $T$ maps $C[0,1]$ to itself and that the restriction $T : (C[0,1], \| \|_2) \to (C[0,1], \| \|_2)$ is a bounded operator. Let us denote the norm of this restriction as $\|T\|_2$. It is not hard to show that for any $f \in C[0,1]$, $\|f\|_2 \leq \|f\|_{max}$. It follows that $\|Tf\|_2 \leq \|T\|_2\|f\|_2 \leq \|T\|_2\|f\|_{max}$ whence the operator $T : (C[0,1], \| \|_{max}) \to (C[0,1], \| \|_2)$ is bounded. Let $G$ denote the graph of the restriction of $T$ to $C[0,1]$. Since every bounded operator has a closed graph (Remark 11.4.13), we see that $G$ is closed in the topology induced by the norm $\|(f,g)\|_{max,2} := \|f\|_{max} + \|g\|_2$. This says that if $\{(f_n, g_n)\} \subset G$ and

$$\|(f_n, g_n) - (f, g)\|_{max,2} \to 0 \tag{11.22}$$

then $(f, g) \in G$.

Further, let us denote $\|(f,g)\|_{max,max} := \|f\|_{max} + \|g\|_{max}$. We will show that $G$ is closed in the topology induced by this norm. To that end, suppose that $\{(f_n, g_n)\} \subset G$ and $\|(f_n, g_n) - (f, g)\|_{max,max} \to 0$. Since $\|(f,g)\|_{max,2} \leq \|(f,g)\|_{max,max}$ for any $f, g \in C[0,1]$, we see that (11.22) holds, hence $(f, g) \in G$, so $G$ is closed in the topology induced by the norm $\|(f,g)\|_{max,max}$.

Finally, $(C[0,1], \| \|_{max})$ is a Banach space so the Closed Graph Theorem allows us to conclude that $T : (C([0,1]), \| \|_{max}) \to (C([0,1]), \| \|_{max})$ is a bounded operator. □

Closed Graph Theorem comes from Banach's 1932 book [5].

## Exercises

11.4.1. Let $X, Y$ be normed linear spaces, let $T \in \mathcal{L}(X,Y)$ be an open map, and let $W$ be a dense subset of $Y$. Prove that $T^{-1}(W)$ is dense in $X$.

11.4.2. Let $T$ be an operator on $\ell^2$ defined by $T(\{x_n\}) = \{x_n/n\}$. Prove that:

(a) $T$ is a bounded operator on $\ell^2$.

(b) The range of $T$ is not closed.

(c) $T$ is not open.

11.4.3. Let $X, Y$ be normed linear spaces and suppose that $X$ is finite dimensional. Prove that every linear operator $T : X \to Y$ is continuous and open.

11.4.4. Let $X, Y$ be Banach spaces and let $T \in \mathcal{L}(X,Y)$. Prove that $T$ is open if and only if 0 is an interior point of $\overline{T(B_X)}$.

11.4.5.* Let $X, Y$ be Banach spaces and let $T \in \mathcal{L}(X,Y)$. Prove that either $T$ is surjective or $T(X)$ is a set of first category.

11.4.6.* Prove or disprove: there exists a sequence $a = \{a_n\} \in \ell_1$ with $a_n > 0$ for all $n \in \mathbb{N}$ such that

$$\sum_{n=1}^{\infty} a_n |x_n| < \infty \quad \text{if and only if} \quad x = \{x_n\} \in \ell^\infty.$$

11.4.7. Let $X$ be a Banach space, and let $P \in \mathcal{L}(X)$ be a projection (i.e., a linear operator satisfying $P^2 = P$). Prove that $P$ is open, i.e., that the image of every open subset of $X$ is a relatively open set in $P(X)$.

11.4.8. Give an example of a Banach space $X$, a closed subspace $V \subset X$, and a linear operator $T \in \mathcal{L}(X)$ that is open but the restriction $T|V$ is not.

11.4.9. Let $X, Y$ be Banach spaces, let $T \in \mathcal{L}(X,Y)$ be surjective, and let $\tilde{T} : X/\ker T \to Y$ be defined by $\tilde{T}([x]) = Tx$. Prove that $\tilde{T}$ is an isomorphism onto $Y$.

11.4.10. Let $X$ be a normed linear space. Prove that $X$ is a Banach space if and only if there is a Banach space $Y$ and a continuous, linear, open mapping of $Y$ onto $X$.

11.4.11. Prove that the identity operator $I : (C[0,1], \|\cdot\|_\infty) \to (C[0,1], \|\cdot\|_1)$ is continuous and surjective, but not open.

11.4.12. Let $p \geq 2$ and let $X$ be a subspace of $L^p(0,1)$ that is closed in $L^1(0,1)$. Prove that $(X, \| \, \|_p)$ is isomorphic to a Hilbert space.

11.4.13.* Let $X$ be a subspace of $L^\infty(0,1)$ that is closed in $L^2(0,1)$. Prove that $(X, \| \, \|_\infty)$ is finite dimensional.

11.4.14. Let $X$ be a linear space such that $(X, \| \, \|_1)$ and $(X, \| \, \|_2)$ are both complete. Prove that these norms are equivalent if and only if $\|x_n\|_1 \to 0$ implies $\|x_n\|_2 \to 0$.

11.4.15. Let $\| \; \|$ be a norm on $C[a,b]$ so that $(C[a,b], \| \; \|)$ is complete and suppose that whenever $\|f_n - f\| \to 0$ there exists a subsequence $\{f_{n_k}\}$ that converges pointwise to $f$ on $[a,b]$. Prove that $\| \; \|$ is equivalent to $\| \; \|_{max}$.

11.4.16.* Let $X, Y$ be Banach spaces and let $T \in \mathcal{L}(X,Y)$. Prove that $X/\ker T$ is isomorphic to $\operatorname{ran} T$ if and only if $\operatorname{ran} T$ is closed.

11.4.17.** Let $X$ be a separable Banach space, let $\{x_n\}$ be a countable dense subset of the unit ball of $X$, and let $T : \ell^1 \to X$ be defined by $T(\{a_n\}) = \sum_n a_n x_n$. Prove that:

(a) $T$ is bounded.

(b) $T$ is surjective.

(c) $X$ is isomorphic to a quotient space of $\ell^1$.

11.4.18. Let $X, Y$ be normed spaces and let $T : X \to Y$ be a closed linear operator. Prove that $\ker T$ is closed.

11.4.19. Let $\mathcal{D}$ be a subset of $L^2(0,1)$ defined in the following way. A function $f$ belongs to $\mathcal{D}$ if and only if $f$ is equal almost everywhere to a function $g$ that is absolutely continuous on $[0,1]$, $g(0) = g(1)$, and $g' \in L^2(0,1)$. Let $T$ be the operator defined on $\mathcal{D}$ by $T(f) = g'$. Prove that $T$ is a well-defined linear operator on $\mathcal{D}$, and that $T$ is closed.

11.4.20. Let $X$ be a normed linear space and let $T : X \to X$ be a linear mapping such that $I - T$ is invertible. Prove that $T$ is closed.

11.4.21. Let $X$ be a subset of $\ell^1$ consisting of sequences $\{a_n\}$ for which $\sum n|a_n| < \infty$ and let $T : X \to \ell^1$ be defined by $T(\{x_n\}) = \{nx_n\}$. (See Exercise 10.1.28.) Prove that:

(a) $T$ is closed but not bounded.

(b) $T$ is invertible and $T^{-1} : \ell^1 \to X$ is bounded and surjective but not open.

11.4.22. Let $X$ be a normed linear space, let $D$ be a subspace of $X$, and let $T : D \to Y$ be a closed linear operator. Suppose that $A \in \mathcal{L}(X,Y)$. Prove that $T + A$ is closed.

11.4.23. Let $X, Y$ be normed linear spaces, let $T : X \to Y$ be a closed linear operator, and let $F$ be a compact set in $X$. Prove that $T(F)$ is a closed subset of $Y$. Does the assertion remain true if $F$ is only assumed to be a closed set in $X$?

11.4.24. Prove that the operator in Example 11.4.9 has closed graph.

11.4.25. Let $X, Y$ be Banach spaces, let $D$ be a subspace of $X$ and $T : D \to Y$ a closed linear operator. Prove that if $T$ is invertible then its range must be closed in $Y$.

11.4.26. Let $X, Y$ be Banach spaces, let $T : X \to Y$ be a linear operator, and suppose that $T$ maps closed sets onto closed sets. Prove that $T$ is a closed operator.

11.4.27. Let $X, Y$ be normed linear spaces, let $D$ be a subspace of $X$, and let $T : D \to Y$ be a closed, invertible linear operator. Prove that $T^{-1}$ is closed.

11.4.28.* Let $X$ be a normed linear space, let $D_A, D_B$ be subspaces of $X$, and let $A : D_A \to Y$ and $B : D_B \to Y$ be closed linear operators. Prove or disprove: the operator $A + B : D_A \cap D_B \to Y$ is closed.

11.4.29. Let $X, Y$ be Banach spaces, let $D$ be a subspace of $X$, and let $A : D \to Y$ be a closed linear operator. If $D$ is a closed subspace of $X$, prove that $A$ is bounded.

11.4.30. Let $X$ and $Y$ be Banach spaces and let $T : X \to Y$ be bounded and linear. Show that there is a constant $c > 0$ such that $\|Tx\| \geq c\|x\|$ for all $x \in X$ if and only if $T$ has trivial kernel and closed range.

11.4.31. Let $X, Y, Z$ be Banach spaces, let $T : X \to Y$ is a linear operator, and let $S$ be an injective operator in $\mathcal{L}(Y, Z)$. Prove that $S \circ T$ is bounded if and only if $T$ is bounded.

11.4.32. Let $X$ be a Banach space, and let $P : X \to X$ be a projection (i.e., a linear operator satisfying $P^2 = P$). Prove that $P$ is bounded if and only if its range and kernel are closed subspaces of $X$.

11.4.33. Let $\{X_k\}_{k=1}^N$ be closed subspaces of a Banach space $X$, let $X$ be a direct sum $X = X_1 + X_2 + \cdots + X_N$, and let $P_k : X \to X_k$ be projections (i.e., linear operators satisfying $P_k^2 = P_k$) for which $P_k(X) = X_k$ and $P_k|X_j = 0$, for each $j \neq k$. Prove that each $P_k$ is a bounded operator.

11.4.34.* Let $(X, \mathfrak{M}, \mu)$ be a measure space, let $1 \leq p, q \leq \infty$, and let $Y$ be a subspace of $L^p(X, \mu) \cap L^q(X, \mu)$ that is closed in both $L^p$ and $L^q$. Prove that the $L^p$ and $L^q$ norms are equivalent on $Y$.

11.4.35. Let $X, Y$ be normed spaces and $T : X \to Y$ a closed linear operator such that $T^{-1} : T(X) \to X$ is well defined. Prove that:

(a) $T^{-1}$ is closed;

(b) if $X, Y$ are Banach spaces and $T(X) = Y$ then $T^{-1}$ is bounded.

11.4.36. Give an example to show that the Closed Graph Theorem need not hold if $X$ is not complete.

11.4.37. Let $X, Y, Z$ be Banach spaces and let $\{A_i\}_{i \in I} \subset \mathcal{L}(Y, Z)$ with the property that if $A_i x = 0$ for all $i \in I$, then $x = 0$. Suppose that $A : X \to Y$ is a linear operator such that for all $i \in I$, the operator $A_i A$ is continuous. Prove that $A$ is continuous.

11.4.38. Let $(X, \mathfrak{M}, \mu)$ be a measure space, $p, q \in [1, \infty]$, $p \neq q$, and let $L^p(X, \mu) \subset L^q(X, \mu)$. Prove that there exists $C > 0$ such that $\|f\|_q \leq C\|f\|_p$, for all $f \in L^p(X, \mu)$.

11.4.39. Let $X, Y$ be Banach spaces, let $S : X \to Y$ be an unbounded linear operator, and let $G$ be the graph of $S$. Prove that:

(a) $G$ is not complete.

(b) If $T : X \to G$ is defined by $Tx = (x, Sx)$ then $T$ is closed but not bounded.

(c) $T$ is invertible and $T^{-1} : G \to X$ is bounded and surjective but not open.

11.4.40. Prove that the identity operator $I : C^1[0,1] \to AC[0,1]$ is continuous, where the norm on $AC[0,1]$ is $\|f\| = \|f\|_1 + \|f'\|_1$ (see Exercise 10.3.36) and the norm on $C^1[0,1]$ is $\|f\| := \max_{x\in[0,1]} |f(x)| + \max_{x\in[0,1]} |f'(x)|$ (see Exercise 10.1.23).

11.4.41.* Let $1 \leq p \leq \infty$, let $[a_{ij}]_{i,j=1}^{\infty}$ be an infinite matrix of real numbers, and suppose that, for any $\mathbf{x} \in \ell^p$, the sequence $A\mathbf{x}$ belongs to $\ell^p$. Use Closed Graph Theorem to prove that the operator $T$, defined by $T(\mathbf{x}) = A\mathbf{x}$, is a bounded operator on $\ell^p$.

11.4.42. Let $X, Y$ be Banach spaces and let $T \in \mathcal{L}(X, Y)$. Prove that the range of $T$ is closed in $Y$ if and only if there exists $C > 0$ such that $\inf\{\|x - y\| : Ty = 0\} \leq C\|Tx\|$ for all $x \in X$.

11.4.43. Let $1 \leq p < \infty$, let $k : [0,1] \times [0,1] \to \mathbb{R}$, and suppose that, for almost every $x \in [0,1]$, the function $k(x,y)f(y)$ is integrable (as a function of $y$) over $[0,1]$. Prove that the operator $T$, defined by $Tf(x) = \int_0^1 k(x,y)f(y)\,dy$, is a bounded operator on $L^p([0,1])$.

11.4.44.* Let $\mathcal{H}$ be Hilbert space, let $T : \mathcal{H} \to \mathcal{H}$ be a linear operator that is symmetric, i.e., $\langle Tx, y\rangle = \langle x, Ty\rangle$, for all $x, y \in \mathcal{H}$. Prove that $T$ is bounded. Give an example to show that the assertion is not true if $D$ is a dense subspace of $\mathcal{H}$ and $T : D \to \mathcal{H}$ is symmetric.

11.4.45.* Let $\mathcal{H}$ be Hilbert space, let $T : \mathcal{H} \to \mathcal{H}$ be a linear operator such that $\langle Tx, x\rangle \geq 0$, for all $x \in \mathcal{H}$. Prove that $T$ is bounded.

# 12

# *Duality*

In the previous chapter we studied linear operators between Banach spaces and we presented some of their fundamental properties. In the special case when the range is one-dimensional, i.e., when $T : X \to \mathbb{R}$, much more can be said about such linear transformations which we now call *functionals.* They are the focal point throughout this chapter, and we will see how they can lead to a better understanding of the underlying function space $X$.

## 12.1 Linear Functionals

In this section we will consider linear operators $T : X \to \mathbb{R}$ and some of their basic properties. Plus, we will take a first stab at the question of whether they can be found on every space $X$.

**Definition 12.1.1.** A **linear functional** on a linear space $X$ is a real-valued function $T$ on $X$ such that, if $x, y \in X$ and $\alpha, \beta \in \mathbb{R}$, then $T(\alpha x + \beta y) = \alpha T(x) + \beta T(y)$.

Of course, $T$ is just a linear operator, so we will continue to use the abbreviated notation $Tx$ for $T(x)$.

In the next two examples we will present some linear functionals. It is straightforward to verify that they are both linear and we leave the task to the reader.

**Example 12.1.2. A linear functional.**
Let $X = \{f : f \text{ is differentiable in } [0, 1]\}$ and $x_0 \in [0, 1]$. If we define $Tf = f'(x_0)$, then $T$ is a linear functional on $X$. ♦

**Example 12.1.3. A linear functional.**
Let $X = C[0, 1]$ and define $Tf = \int_0^1 f(x)\, dx$.Then $T$ is a linear functional on $X$. ♦

Definition 12.1.1 does not require $X$ to be a normed space but in most cases it will be. This is essential in order to define what it means for a linear functional to be *bounded.*

DOI: 10.1201/9781032716497-12

**Definition 12.1.4.** A linear functional $T$ on a normed linear space $X$ is **bounded** if there is $M \geq 0$ such that

$$|Tx| \leq M\|x\|, \qquad \text{for all } x \in X.$$

The infimum of all such numbers $M$ is the **norm** of $T$ and it is denoted by $\|T\|$.

*Remark* 12.1.5. Obviously, Definition 12.1.4 is just a special case of Definition 11.1.6.

**Example 12.1.6. A bounded linear functional.**

Let $X = C[0,1]$ with $\|\cdot\|_{\max}$ norm and $Tf = \int_0^1 f(x)\,dx$. Then $T$ is a bounded linear functional and $\|T\| = 1$.

**Proof:** For any $f \in C[0,1]$,

$$|Tf| = \left|\int_0^1 f(x)\,dx\right| \leq \int_0^1 |f(x)|\,dx \leq \int_0^1 \|f\|_{\max}\,dx = \|f\|_{\max},$$

so $T$ is bounded and $\|T\| \leq 1$. If $f = 1$ then $|Tf| = 1 = \|f\|_{\max}$, so $\|T\| = 1$ by Exercise 11.1.3. ♦

**Example 12.1.7. An unbounded linear functional.**

Let $X = \{f : f \text{ is differentiable in } [0,1]\}$ with $\|\|_{max}$ and $Tf = f'(1)$. Then $T$ is not bounded.

Reason: if $f(x) = x^n$ then $\|f\|_{max} = 1$, but $f'(1) = n$. ♦

From Proposition 11.1.14 we deduce an immediate corollary.

**Proposition 12.1.8.** *A linear functional on a normed linear space $X$ is continuous if and only if it is bounded.*

> The word *functional* was originally used as an adjective by a French mathematician Jacques Hadamard (1865–1963) 1903 [45]. By "functional operation", he meant a numerically valued function defined on a class of functions. He advised his protégé Fréchet to use it as a noun which he did in one section of his 1906 dissertation [35].

We know that $\mathbb{R}^n$ is a normed linear space that is finite-dimensional. Our next result shows that every linear functional on $\mathbb{R}^n$ is bounded.

**Theorem 12.1.9.** *Let $X$ be a normed linear space of finite dimension. Then every linear functional on $X$ is bounded.*

*Proof.* Let $T$ be a linear functional on $X$ and let $\{e_1, e_2, \ldots, e_n\}$ be a basis of $X$. For any $x \in X$ there are real numbers $x_1, x_2, \ldots x_n$ such that $x = x_1e_1 + x_2e_2 + \ldots x_ne_n$. We define $\|x\|_\infty := \max\{|x_k| : 1 \leq k \leq n\}$. (See Exercise 10.1.7.) Then

$$|Tx| = \left|T\left(\sum_{k=1}^n x_ke_k\right)\right| \leq \sum_{k=1}^n |x_k||Te_k| \leq \|x\|_\infty \sum_{k=1}^n |Te_k|.$$

By Theorem 11.2.11, all norms on $X$ are equivalent so $T$ is bounded. □

We have not yet looked at the question of whether on a given normed linear space there are any continuous functionals. We start with the case when $X$ is a vector space. In the absence of topology we can only ask about the existence of linear functionals and the answer to that question is in the affirmative.

**Proposition 12.1.10.** *Let $Y$ be a linear subspace of a linear space $X$. Then each linear functional on $Y$ has an extension to a linear functional on $X$. In particular, for each $x \in X \setminus \{0\}$, there is a linear functional $T$ such that $Tx \neq 0$.*

*Proof.* Let $Z$ be a subspace of $X$ such that $X = Y \oplus Z$. (See Exercise 12.1.1.) Let $T$ be a linear functional on $Y$. Then $T$ can be extended to $X$ by defining $Tx = Ty$, where $x = y + z$ is the decomposition as above. It is easy to verify that $T$ is a linear functional on $X$ (Exercise 12.1.3).

For the second assertion, let $x \in X \setminus \{0\}$. We define $S : \vee\{x\} \to \mathbb{R}$ by

$$S(\lambda x) = \lambda.$$

Then $S$ is linear and, by the first assertion, it can be extended to the whole $X$, while $S(x) = 1 \neq 0$. □

Proposition 12.1.10 shows that there is always a good supply of linear functionals on a linear space. What happens when we introduce topology? If it comes from a metric the resulting space may not support any continuous functionals.

**Example 12.1.11. A metric linear space without nonzero continuous functionals.**

Let $0 < p < 1$ and consider $L^p(0,1)$ with a metric defined by

$$d(f,g) = \int_0^1 |f(t) - g(t)|^p \, dt. \tag{12.1}$$

If $T$ is a nonzero linear functional on this metric space then $T$ is not continuous.

**Proof:** By Exercise 12.1.2, $d$ is a metric on $L^p(0,1)$. Let $T$ be a nonzero linear functional on $(L^p(0,1), d)$. There exists a nonzero function $f \in L^p(0,1)$ such that $|Tf| \geq 1$. Let

$$g(x) = \int_0^x |f(t)|^p \, dt, \qquad \text{for } x \in [0,1].$$

Since $|f|^p$ is integrable we have that $g$ is continuous on $[0,1]$. Further, $g(0) = 0$, $g(1) = d(f,0) > 0$, so there exists $s \in [0,1]$ such that $g(s) = d(f,0)/2$. Let

$$g_1 = f\chi_{[0,s]}, \quad g_2 = f\chi_{(s,1]}.$$

Then $f = g_1 + g_2$, $|f|^p = |g_1|^p + |g_2|^p$, and

$$\int_0^1 |g_1|^p = \int_0^s |f|^p = g(s) = \frac{1}{2}\int_0^1 |f|^p, \quad \int_0^1 |g_2|^p$$
$$= \int_0^1 |f|^p - \int_0^1 |g_1|^p = \frac{1}{2}\int_0^1 |f|^p.$$

Next, $|Tg_1 + Tg_2| = |Tf| \geq 1$, so there exists $k \in \{1,2\}$ such that $|Tg_k| \geq 1/2$. For that $k$ define $f_1 = 2g_k$. Then

$$|Tf_1| \geq 1 \quad \text{and} \quad \int_0^1 |f_1|^p = 2^{p-1}\int_0^1 |f|^p. \tag{12.2}$$

We conclude that given $f$ such that $|Tf| \geq 1$, there exists $f_1 \in L^p(0,1)$ satisfying (12.2). If we apply this conclusion to $f_1$, we obtain a function $f_2 \in L^p(0,1)$ such that

$$|Tf_2| \geq 1 \quad \text{and} \quad \int_0^1 |f_2|^p = 2^{p-1}\int_0^1 |f_1|^p = \left(2^{p-1}\right)^2\int_0^1 |f|^p.$$

Continuing, we get a sequence $\{f_n\} \subset L^p(0,1)$ such that

$$|Tf_n| \geq 1 \quad \text{and} \quad \int_0^1 |f_n|^p = \left(2^{p-1}\right)^n\int_0^1 |f|^p. \tag{12.3}$$

Since $0 < p < 1$, (12.3) implies that

$$\lim_n d(f_n, 0) = 0 \quad \text{but } \lim_n |Tf_n| \neq 0,$$

whence $T$ is not continuous. ♦

*Remark* 12.1.12. While it is possible, as Example 12.1.11 shows, that there are no nonzero bounded linear functionals on a metric space, the situation is different in a normed linear space. First of all, if a space is finite-dimensional there are plenty of bounded linear functionals. Indeed, if $X$ is such a normed linear space, the second assertion in Proposition 12.1.10 provides an abundance of linear functionals which are all bounded by Theorem 12.1.9. If $X$ is infinite-dimensional and $Y$ is its finite-dimensional subspace then there are bounded linear functionals on $Y$, and the question is when can they be extended to the whole space. Example 12.1.11 shows that in a metric space that may not be possible. In the next section we will see what the situation is in a normed linear space.

Before that, let us explain why these questions matter. The success of linear algebra in the second half of the 19th century led to the interest in solving infinite systems of linear equations, i.e., systems of infinitely many equations of the form

$$\sum_{k=1}^{\infty} a_{ik}x_k = b_i, \quad i = 1, 2, \ldots. \tag{12.4}$$

Riesz attacked a more general problem around 1910, replacing equations (12.4) by

$$\int_a^b f g_n = c_n, \quad n = 1, 2, \ldots. \tag{12.5}$$

In this context the question is as follows. If $p > 1$ and $q$ are conjugate numbers, and if we are given functions $\{g_n\}_{n=1}^\infty$ in $L^q(a, b)$ and real numbers $\{c_n\}_{n=1}^\infty$, is there a function $f \in L^p(a, b)$ for which (12.5) holds? Assuming that such a function exists, Riesz was able to prove that for any sequence $\{a_n\}$ of real numbers there exists $M > 0$ so that

$$\left| \sum_{n=1}^N a_n c_n \right| \le M \left( \int_a^b \left| \sum_{n=1}^N a_n g_n \right|^q \right)^{1/q}, \quad \text{for all } N \in \mathbb{N}. \tag{12.6}$$

(See Exercise 12.1.25.) Notice that we can define a linear functional $T$ on the closed linear span $\mathcal{M}$ of $\{g_n\}$ (in $L^q$) by

$$T g_n := \int_a^b f g_n, \quad n = 1, 2, \ldots.$$

Condition (12.6) ensures that $T$ is well-defined. The issue here is that if $g \in \mathcal{M}$ is a limit of two different sequences from $\mathcal{M}$,

$$g = \sum_{n=1}^\infty a_n g_n \quad \text{and} \quad g = \sum_{n=1}^\infty b_n g_n,$$

then $Tg$ has to be the same for both representations. That this is true is because (12.6) ensures that $\sum_{n=1}^\infty (a_n - b_n) g_n = 0$ implies $\sum_{n=1}^\infty (a_n - b_n) c_n = 0$, whence

$$\begin{aligned} T\left( \sum_{n=1}^\infty a_n g_n \right) &= \sum_{n=1}^\infty a_n T g_n = \sum_{n=1}^\infty a_n c_n = \sum_{n=1}^\infty b_n c_n \\ &= \sum_{n=1}^\infty b_n T g_n = T\left( \sum_{n=1}^\infty b_n g_n \right). \end{aligned}$$

Moreover, (12.6) shows that if $g$ belongs to $\mathcal{M}$, then $|Tg| \le M \|g\|_q$, i.e., $T$ is bounded on $\mathcal{M}$. Now comes the crucial point. If we could extend $T$ to the whole space $L^q(a, b)$ then it would imply that there is $f \in L^p(a, b)$ satisfying equations (12.5) *for all* $\{g_n\} \subset L^q(a, b)$. (This last implication is known as Riesz representation theorem and we will prove it in Section 12.3.)

Thus, it became of interest to pursue extensions of linear functionals on normed spaces. We will focus on this issue in the next section.

The fact that there are no bounded linear functionals on $L^p$, for $0 < p < 1$, was discovered by an American mathematician Mahlon Marsh Day (1913–1992) in 1940 [26], Hahn (1927) [49], and Banach (1929) [4]. Day was a professor at the University of Illinois at Urbana-Champaign, described as "the first American mathematician to study normed spaces from a geometric standpoint". His monograph [27] is considered to be classical in the field.

## Exercises

12.1.1. If $Y$ is a subspace of a vector space $X$, show there is a subspace $Z$ such that $Y \cap Z = \{0\}$ and $Y \oplus Z = X$. (This means that every $x \in X$ can be written in a unique way as $x = y + z$, with $y \in Y$ and $z \in Z$.

12.1.2. Let $0 < p < 1$. Prove that (12.1) defines a metric on $L^p(0,1)$.

12.1.3. Prove that the extension of $T$ to $X$ in the proof of Proposition 12.1.10 is a linear functional.

12.1.4. Find the norm of the linear functional on $\mathbb{R}^2$ defined by $T(x,y) = x - 7y$ if $\mathbb{R}^2$ is given the $\ell^p$ norm, $1 \le p \le \infty$.

12.1.5. Let $\delta$ be Dirac functional on $C[0,1]$ defined by $\delta(f) = f(0)$.

(a) Prove that $\delta$ is bounded when $C[0,1]$ is equipped with $\| \ \|_{max}$ and find its norm.

(b) Prove that $\delta$ is unbounded when $C[0,1]$ is equipped with $\| \ \|_1$ norm.

12.1.6. For what values of the constant $\alpha$ does

$$Tf = \int_0^1 x^\alpha f(x)\,dx,$$

define a bounded linear functional on $C[0,1]$? For such $\alpha$ determine the norm of $T$.

12.1.7.* For what values of the constant $\alpha$ does

$$Tf = \int_0^1 x^\alpha f(x)\,dx,$$

define a bounded linear functional on $L^p(0,1)$? For such $\alpha$ determine the norm of $T$.

12.1.8. Let $X = C[0,1]$ with $\| \cdot \|_{\max}$ norm, let $g \in C[0,1]$, and define $Tf = \int_0^1 f(t)g(t)\,dt$. Prove that $T$ is a bounded linear functional on $C[0,1]$ and find its norm.

12.1.9. Let $X = C^1[0,1]$ with the norm $\|f\|_{(1)} := \|f\|_\infty + \|f'\|_\infty$, and define $Tf = \int_0^1 tf(t)\,dt$. Prove that $T$ is a bounded linear functional on $C^1[0,1]$ and find its norm.

12.1.10. Let $X = C[0,1]$ with $\| \cdot \|_{\max}$ norm, let $n \in \mathbb{N}$, let $\{\alpha_k\}_{k=1}^n$ be real numbers, let $\{x_k\}_{k=1}^n \subset [0,1]$, and define

$$Tf = \sum_{k=1}^n \alpha_k f(x_k).$$

Prove that $T$ is a bounded linear functional on $C[0,1]$ and find its norm. What if the sequence $\{\alpha_k\}$ is infinite?

12.1.11. If $n \in \mathbb{N}$ let

$$T_n f := \int_0^\pi f(t) \sin nt \, dt, \qquad \text{for } f \in C[0,\pi].$$

Prove that $\|T_n\| = 2$ for all $n \in \mathbb{N}$.

12.1.12. For a continuous function $f$ on $[0, 2\pi]$, let $s_{n,x}(f) = s_n(f)(x)$, the $n$th partial sum of its Fourier series evaluated at $x \in [0, 2\pi]$. Prove that $s_{n,x}$ is a bounded linear functional on $C[0, 2\pi]$ and determine its norm.

12.1.13.* Let $X = C[0,1]$ with $\|\cdot\|_{\max}$ norm, let $g \in L^1[0,1]$, and define $Tf = \int_0^1 f(t)g(t)\,dt$. Prove that $T$ is a bounded linear functional on $C[0,1]$ and find its norm.

12.1.14. Let $X$ be the space $\ell^1$ with the $\ell^\infty$ norm and let $T$ be a linear functional on $X$ defined by $Tx = \sum_k x_k$. Prove or disprove: $T$ is continuous on $X$.

12.1.15. For $f \in C[0,1]$ let

$$Tf = \int_0^{1/2} f - \int_{1/2}^1 f.$$

(a) Prove that $\|T\| = 1$.

(b) Prove that there does not exist any $f \in C[0,1]$ such that $\|f\| = 1$ and $|Tf| = 1$.

12.1.16. For $x := \{x_n\} \in c_0$ let

$$Tx = \sum_{n=1}^\infty 2^{-n} x_n.$$

(a) Find $\|T\|$.

(b) Prove that there does not exist any $x \in c_0$ such that $\|x\| = 1$ and $Tx = \|T\|$.

12.1.17. For $x := \{x_n\} \in c$ let

$$Tx = \sum_{n=1}^\infty 2^{-n} x_n - \lim x_n.$$

(a) Find $\|T\|$.

(b) Prove that there does not exist any $x \in c$ such that $\|x\| = 1$ and $Tx = \|T\|$.

12.1.18. For $x := \{x_n\} \in \ell^1$ let

$$Tx = \sum_{n=1}^{\infty}(1 - 1/n)x_n.$$

(a) Find $\|T\|$.

(b) Prove that there does not exist any $x \in \ell^1$ such that $\|x\| = 1$ and $Tx = \|T\|$.

12.1.19. Let $T$ and $T'$ be two functionals on a vector space $X$ such that the kernel of $T$ is contained in the kernel of $T'$. Prove that $T' = cT$, for some real number $c$.

12.1.20. Let $X$ be a linear space, let $n \in \mathbb{N}$, and let $T_1, \ldots, T_n$ and $T$ be linear functionals on $X$ such that $\cap_{k=1}^{n} \ker T_k \subset \ker T$. Prove that $T$ is a linear combination of $T_1, \ldots, T_n$.

12.1.21. Prove that a linear functional on a normed space is bounded if and only if it maps bounded sets to bounded sets.

12.1.22. Let $X$ be a normed linear space and $x_0 \in X$. Prove that a linear functional on a normed space is continuous if and only it is continuous at $x_0$.

12.1.23. Let $X$ be a normed linear space and $T : X \to \mathbb{R}$ a continuous additive map, i.e., $T(x+y) = Tx+Ty$, for all $x, y \in X$. Prove that $T$ is homogeneous, i.e., $T(\alpha x) = \alpha Tx$, for all $\alpha \in \mathbb{R}$ and all $x \in X$.

12.1.24. Let $X$ be a normed linear space, and let $T$ be a bounded linear functional on $X$, and let $S$ be a set that is dense in $X$. Prove that if $Tx = 0$ for all $x \in S$ then $T = 0$ on all of $X$.

12.1.25. Prove (12.6).

12.1.26. Let $X$ be a normed space and let $T$ be a nonzero bounded linear functional on $X$. Prove that for all $x \in X$,

$$\min\{\|x - y\| : Ty = 0\} = \frac{|Tx|}{\|T\|}.$$

12.1.27. Let $X$ be a normed linear space and $T$ a nonzero linear functional on $X$. Prove that there exists $y \in X$ such that every vector $x \in X$ can be written uniquely in the form $x = \lambda y + z$, where $\lambda \in \mathbb{R}$ and $z \in \ker T$.

12.1.28. Let $Y$ be a subspace of a vector space $X$ and let $T$ be a functional on $X$ such that $T(Y)$ is a proper subset of $\mathbb{R}$. Show that $Tx = 0$ for all $x \in Y$.

12.1.29.* Let $X$ be a normed linear space and $T$ a nonzero linear functional on $X$. Prove that $T$ is continuous if and only if $\ker T^C$ has a nonempty interior.

12.1.30.* Let $X$ be a normed linear space and $T$ a linear functional on $X$. Prove that $T$ is continuous if and only if there is an open set $G \subset X$ and a real number $\alpha$ such that $\alpha \notin T(G)$.

12.1.31.* Let $X$ be a normed linear space and $T$ a linear functional on $X$. Prove that $T$ is continuous if and only if its kernel $\{x \in X : T(x) = 0\}$ is closed set.

12.1.32.* Let $X$ be a normed linear space that is not finite-dimensional. Prove that there exists an unbounded linear functional on $X$.

12.1.33.* Let $X$ be a normed linear space and $T$ a linear functional on $X$. Prove that the norm of $T$ is reciprocal to the distance in $X$ from 0 to the set $\{x \in X : Tx = 1\}$.

12.1.34.* Let $X$ be a normed linear space and $T$ a bounded linear functional on $X$. Suppose there exists a unit vector $x_0 \in X$ such that $\|x_0 - y\| \geq 1$ for every $y$ in the nullspace of $T$.

(a) Prove that $\|T\| = \|Tx_0\|$.

(b) Moreover, let $\mathcal{A} = \{f \in C[0,1] : \int_0^{1/2} f = \int_{1/2}^1 f\}$. Prove that if $g \in C[0,1]$ and $\|f - g\| \geq 1$ for all $f \in \mathcal{A}$, then $\|g\| \geq 1$.

## 12.2 The Hahn–Banach Theorem

In this section we will accomplish the goal articulated in the previous section. Namely, we will prove that, if $X$ is a normed linear space, there are many nonzero bounded linear functionals on it. This is the content of the Hahn–Banach theorem which is the main result of this section. We will then see some of its consequences. By Remark 12.1.12, there is a bounded linear functional $T$ is defined on a finite-dimensional subspace $Y$ of $X$, and the task at hand is to extend it to $X$. In addition, we would like the extension $\hat{T}$ to have the same norm as $T$. Aside from the existence of $\hat{T}$, some natural questions are whether we can demand that it has specific values, and to what extent. In particular, is there a bounded linear functional that vanishes on a proper subspace of a Banach space without vanishing on the whole space?

In order to prove that the asserted extension exists we propose the following plan. We will first construct an extension from $Y$ to a subspace whose dimension is one larger, without increasing its norm. (Since by extending a functional the norm cannot get smaller, this means that the norm remains the same.) Then we will use these one-dimensional extensions to obtain an extension from $Y$ to all of $X$. Our first step is known as the Hahn–Banach Lemma.

**Theorem 12.2.1.** ***The Hahn–Banach Lemma.*** *Let $X$ be a normed linear space, let $Y \subset X$ be a linear subspace, and let $T$ be a bounded linear functional on $Y$. Let $z \in X \setminus Y$ and $V = \{y + \lambda z : y \in Y, \lambda \in \mathbb{R}\}$. Then there exists a bounded linear functional $T_1$ on $V$ such that*

$$T_1 x = Tx, \quad \text{for } x \in Y, \quad \text{and} \quad \|T_1\| = \|T\|.$$

*Proof.* Without loss of generality we will assume that $\|T\| = 1$. (If $\|T\| = 0$, we define $T_1 = 0$ on $V$.) For $T_1$ to be linear on $V$, it has to satisfy

$$T_1(y + \lambda z) = T_1 y + \lambda T_1 z. \tag{12.7}$$

Thus, the question is how to define $T_1 z$. Notice that, if $y_1, y_2 \in Y$ then

$$Ty_1 + Ty_2 = T(y_1 + y_2) \leq \|y_1 + y_2\| \leq \|y_1 - z\| + \|z + y_2\|.$$

It follows that

$$Ty_1 - \|y_1 - z\| \leq -Ty_2 + \|z + y_2\|. \tag{12.8}$$

This shows that if $y_2$ is fixed, the set $\{Ty_1 - \|y_1 - z\| : y_1 \in Y\}$ is bounded and we define

$$T_1 z = \sup\{Ty_1 - \|y_1 - z\| : y_1 \in Y\}. \tag{12.9}$$

With this choice of $T_1 z$, we have that $T_1$, as defined by (12.7), is linear on $V$. Thus, it remains to show that $\|T_1\| \leq 1$, i.e., that

$$T_1 y + \lambda T_1 z \leq \|y + \lambda z\|, \qquad \text{for all } y \in Y,\ \lambda \in \mathbb{R}. \tag{12.10}$$

It follows from (12.8) and (12.9) that

$$\begin{aligned} Ty_1 - \|y_1 - z\| &\leq \sup\{Ty_1 - \|y_1 - z\| : y_1 \in Y\} && (12.11)\\ &= T_1 z \leq -Ty_2 + \|z + y_2\|, \quad \forall y_2 \in Y. && (12.12)\end{aligned}$$

If $\lambda = 0$, (12.10) follows from our assumption that $\|T\| = 1$. If $\lambda \neq 0$, let $y \in Y$, $y_2 = y/\lambda$, and $y_1 = -y/\lambda$. Then, using the fact that $T = T_1$ on $Y$, (12.11) becomes

$$-T_1\left(\frac{y}{\lambda}\right) - \left\|-\frac{y}{\lambda} - z\right\| \leq T_1 z \leq -T_1\left(\frac{y}{\lambda}\right) + \left\|\frac{y}{\lambda} + z\right\|,$$

which yields

$$-\frac{1}{|\lambda|}\|y + \lambda z\| \leq T_1\left(\frac{y + \lambda z}{\lambda}\right) \leq \frac{1}{|\lambda|}\|y + \lambda z\|.$$

Now we obtain (12.10) from the first inequality if $\lambda < 0$, and from the second if $\lambda > 0$. □

If $X$ is finite-dimensional, or at least if $Y$ has finite codimension in $X$, then we simply need to repeat the one-dimensional extension step a finite number of times. Otherwise, we use Zorn's Lemma.

**Theorem 12.2.2.** ***The Hahn–Banach Theorem.*** *Let $X$ be a normed linear space, let $Y \subset X$ be a linear subspace, and let $T$ be a bounded linear functional on $Y$. Then there exists a linear functional $\tilde{T}$ on $X$ such that*

$$\tilde{T}x = Tx, \qquad \textit{for } x \in Y, \qquad \textit{and} \qquad \|\tilde{T}\| = \|T\|.$$

*Proof.* Let $T'$ denote an extension of $T$ to a subspace $Y' \supset Y$ that does not increase the norm. Let $\mathcal{A}$ denote the collection of all such pairs $(T', Y')$. By the Hahn–Banach Lemma, $\mathcal{A}$ is nonempty. In this set we can introduce a partial ordering $\prec$ by

$$(T', Y') \prec (T'', Y'') \quad \text{if} \quad Y' \subset Y'' \quad \text{and} \quad T'' = T' \text{ on } Y'.$$

Let $\mathcal{A}_0$ be a totally ordered subfamily of $\mathcal{A}$ and let

$$Z = \bigcup\{Y' : (T', Y') \in \mathcal{A}_0\}.$$

Then $Z$ is a linear subspace of $X$. Indeed, it is obvious that if $z \in Z$ and $\alpha \in \mathbb{R}$, then $\alpha z \in Z$. If $x, y \in Z$ then $x \in Y'$ and $y \in Y''$ for some $(T', Y'), (T'', Y'') \in \mathcal{A}_0$. However, $\mathcal{A}_0$ is totally ordered so $x, y$ are both in $Y'$ or in $Y''$, and the same is true for their sum.

Also, $\mathcal{A}_0$ has an upper bound in $\mathcal{A}$. Indeed, let $z \in Z$. Then there exists $(T', Y') \in \mathcal{A}_0$ such that $z \in Y'$. Define $T_0$ on $Z$ by $T_0(z) = T'(z)$. The map $T_0$ is well-defined, linear, and $\|T_0\| \leq \|T\|$. (See Exercise 12.2.4.) Thus, $(T_0, Z)$ is an upper bound in $\mathcal{A}_0$. By Zorn's Lemma, $\mathcal{A}$ has a maximal element $(\tilde{T}, \tilde{Y})$. If $\tilde{Y} \neq X$ then $(\tilde{T}, \tilde{Y})$ could be extended using the Hahn–Banach Lemma, thereby contradicting its maximality. It follows that $\tilde{Y} = X$ and the theorem is proved. □

In view of the Hahn–Banach theorem, the collection of bounded linear functionals on $X$ is nontrivial (i.e., it consists of more than the zero functional). Moreover, by Exercise 11.1.11, it is a vector space and, by Proposition 11.1.10, it is a normed space.

**Definition 12.2.3.** The vector space of all bounded linear functionals on $X$, equipped with the operator norm, is called the **dual space** of $X$ and it is denoted by $X^*$.

By Theorem 11.1.15, the dual space is a Banach space. Notice that this is true regardless of whether $X$ is a complete normed space or not.

> The concept of a dual space was introduced independently by Hahn (1927) [49], and Banach (1929) [4]. The name *dual space* and its symbol $X^*$ were introduced by Bourbaki—a group of French top mathematicians formed in 1934 with the goal of standardizing calculus instruction in French universities. Initial members were Henri Cartan, Claude Chevalley, Jean Delsarte, Jean Dieudonné, René de Possel, and André Weil.

The next result is a very important consequence of the Hahn–Banach Theorem.

**Theorem 12.2.4.** *Let $X$ be a normed linear space and $x_0$ a nonzero vector in $X$. Then there is $T \in X^*$ such that $Tx_0 = \|x_0\|$ and $\|T\| = 1$.*

*Proof.* Let $Y$ be a one dimensional subspace of $X$ spanned by $x_0$, and let $T_0$ be a defined on $Y$ by $T_0(\lambda x_0) = \lambda \|x_0\|$. It is easy to verify that $T_0$ is a bounded linear functional of norm 1. The result now follows from the Hahn–Banach Theorem. □

Another application of the Hahn–Banach Theorem concerns the following question. If $Y$ is a vector subspace of $X$ is there a nonzero functional in $X^*$ that vanishes on $Y$?

**Theorem 12.2.5.** *Let $X$ be a normed linear space, and let $Y$ be a proper closed linear subspace of $X$, and let $x_0 \in X \setminus Y$ such that $d := d(x_0, Y) > 0$. Then there exists $T \in X^*$ such that*

*(a) $T(x) = 0$ for $x \in Y$,*

*(b) $T(x_0) = 1$,*

*(c) $\|T\| = 1/d$.*

*Proof.* Consider the linear subspace

$$Y' = \{x + tx_0 : x \in Y,\, t \in \mathbb{R}\}.$$

For every vector $y \in Y'$, the vector $x \in Y$ and the real number $t$ are uniquely determined, so we can define a linear functional $T'$ on $Y'$ by

$$T'(x + tx_0) = t.$$

It is not hard to verify that $T'$ is a linear functional that vanishes on $Y$, and that $T'(x_0) = 1$.

Regarding the norm of $T'$, we notice that

$$\|y\| \geq |t|d, \qquad \text{for all } y \in Y'. \tag{12.13}$$

Indeed, if $y \in Y$ then $t = 0$, so (12.13) is automatically satisfied. If $y \in Y' \setminus Y$, then $y = x + tx_0$ for some $t \neq 0$, so

$$\|y\| = \|x + tx_0\| = |t| \left\| \frac{1}{t}\, x + x_0 \right\| \geq |t|\, d,$$

because $x/t \in Y$ so $\|x/t + x_0\|$ cannot be less than the distance from $x_0$ to $Y$. In view of the definition of $T'$, (12.13) implies that

$$|T'y| = |t| \leq \frac{1}{d}\, \|y\|, \qquad \text{for all } y \in Y',$$

whence $\|T'\| \leq 1/d$.

On the other hand, let $\{x_n\} \subset Y$ be such that $\lim_n \|x_n - x_0\| = d$. Since $x_n \in Y$, $T'x_n = 0$, so

$$1 = |T'x_0| = |T'(x_n - x_0)| \leq \|T'\| \|x_n - x_0\| \to \|T'\|\, d.$$

Thus $\|T'\| \geq 1/d$, and we conclude that $\|T'\| = 1/d$. We can now invoke the Hahn–Banach theorem to extend $T'$ to a linear functional $T$ on all of $X$ that satisfies conditions (a)–(c). □

Using Theorem 12.2.5 we get some useful results.

**Theorem 12.2.6.** *Let $X$ be a normed linear space. If $X^*$ is separable, then so is $X$.*

*Proof.* Suppose that $X^*$ is separable. By Exercise 2.5.22, the same is true of the unit sphere of $X^*$ which therefore contains a countable dense set $\{T_n\}_{n=1}^{\infty}$. Since $\|T_n\| = 1$, for each $n$ there exists $x_n \in X$ such that

$$\|x_n\| = 1 \quad \text{and} \quad |T_n(x_n)| \geq \frac{1}{2}.$$

Now let $Y$ be the closed linear span of $\{x_n\}_{n=1}^{\infty}$. Then $Y$ is separable (Exercise 10.4.42) so it suffices to prove that $Y = X$. Suppose to the contrary that $Y$ is a proper closed linear subspace of $X$. By Theorem 12.2.5, there exists $T \in X^*$ such that $\|T\| = 1$ and $T$ vanishes on $Y$. In particular, $Tx_n = 0$ for all $n \in \mathbb{N}$ and it follows that

$$\frac{1}{2} \leq |T_n x_n| = |T_n x_n - Tx_n| \leq \|T_n - T\|\|x_n\| = \|T_n - T\|,$$

for all $n \in \mathbb{N}$. But this contradicts the hypothesis that the set $\{T_n\}$ is dense in the unit sphere of $X^*$. □

The Hahn–Banach theorem was first proved in 1912 [59] by Helly. The result was obtained in the framework of $C[a, b]$ as a part of the simplified proof that condition (12.6) is necessary and sufficient for the solution of (12.5) to exist. (The equivalence was originally proved by Riesz [94] in 1911.) The Hahn–Banach theorem was rediscovered independently in 1927 by Hahn [49] and Banach in 1929 [4]. Both Hahn and Banach used Helly's argument in their proofs. The attribution of the result to them appeared for the first time in [13].

EDUARD HELLY (1884–1943) was an Austrian mathematician. He wrote his doctoral thesis on Fredholm equations in 1907 under Wirtinger and Mertens at the University of Vienna. He then spent a year in Göttingen, where he studied under Hilbert, Klein, Minkowski, and a German mathematician and physicist Carl Runge (1856–1927). When he returned to Vienna he worked as a high school teacher, a tutor, and he wrote solution manuals. During this period, he proved the Hahn–Banach theorem for the space $C[a, b]$, the uniform boundedness principle for linear functionals, the Helly's selection principle (Exercise 9.2.28), and Helly's theorem about convex subsets in Euclidean space. When World War I began, he enlisted, sustained severe injuries, was captured by the Russians, and spent years in hospitals and prisoner-of-war camps in Siberia. In the post-war Vienna he worked for a bank and an insurance company. In 1938 he fled from the Nazis to the United States where he worked at Paterson Junior College (at the recommendation by Einstein), at Monmouth Junior College, and for the Signal Corps in Chicago. He eventually secured a position at the Illinois Institute of Technology but had two heart attacks, succumbing to the second one. He was unlucky not to receive due recognition and the opportunity to further extend his mathematical contributions.

The following theorem is another important consequence of Hahn–Banach theorem.

**Theorem 12.2.7.** *Let $X$ be a normed linear space and $Y$ a linear subspace of $X$. Then a point $x \in X$ belongs to the closure of $Y$ if and only if every functional $T \in X^*$ that vanishes on $Y$ must vanish at $x$.*

*Proof.* If $x$ lies in the closure of $Y$ and $T$ vanishes on $Y$, the continuity of $T$ ensures that $Tx = 0$.

In the other direction, we will use the contrapositive. Suppose that $x \in X \setminus \overline{Y}$. Since this set is open, there is a ball $B(x,r) \subset X \setminus \overline{Y}$, hence the distance $d(x, \overline{Y}) > r > 0$. By Theorem 12.2.5, there is a functional $T \in X^*$ that vanishes on $\overline{Y}$ but not at $x$. □

From Theorem 12.2.7 we get the following corollary. It gives a test whether a subspace of a Banach space is dense. (See Exercise 12.2.17).

**Corollary 12.2.8.** *A subspace $Y$ of a Banach space $X$ is dense in $X$ if and only if every bounded linear functional that vanishes on $Y$ must vanish on $X$.*

Since the dual $X^*$ of a normed space $X$ is a normed space itself, it has its own dual space, denoted by $(X^*)^*$, or more often $X^{**}$, and called the **second dual** space or **bidual** of $X$. There is a natural mapping $J : X \to X^{**}$ defined by $x \mapsto J_x$, where

$$J_x(T) = Tx, \qquad \text{for all } T \in X^*. \tag{12.14}$$

The following theorem introduces several fundamental properties of the second dual.

**Theorem 12.2.9.** *Let $X$ be a normed linear space and $J$ a map defined by (12.14). Then $J$ is an isometric isomorphism of $X$ onto its image which is a subspace of $X^{**}$. Moreover, if $X$ is complete, then the range of $J$ is closed.*

*Proof.* First, the image of $J$ is contained in $X^{**}$, i.e., $J_x \in X^{**}$ for all $x \in X$. Indeed, if $x \in X$, $T_1, T_2 \in X^*$, and $\alpha, \beta \in \mathbb{R}$, then

$$J_x(\alpha T_1 + \beta T_2) = (\alpha T_1 + \beta T_2)x = \alpha T_1 x + \beta T_2 x = \alpha J_x T_1 + \beta J_x T_2.$$

Thus, $J_x$ is a linear functional on $X^*$ and it is bounded because

$$|J_x(T)| = |Tx| \le \|T\|\|x\|, \qquad \text{for any } T \in X^*. \tag{12.15}$$

Consequently, $J_x \in X^{**}$. This also shows that $\|J_x\| \le \|x\|$.

In fact, as we will show, these norms are equal. To that end, let $x \in X$. By Theorem 12.2.4, there exists $T \in X^*$ such that $\|T\| = 1$ and $Tx = \|x\|$. Then

$$\|x\| = Tx = J_x(T) \le \|J_x\|\|T\| = \|J_x\|,$$

which together with (12.15) shows that $\|J_x\| = \|x\|$, i.e., $J$ is an isometry.

Next, if $x, y \in X$ and $\alpha, \beta \in \mathbb{R}$, then

$$J_{\alpha x + \beta y}(T) = T(\alpha x + \beta y) = \alpha Tx + \beta Ty = (\alpha J_x + \beta J_y)(T),$$

so $J$ is linear which implies that $J(X)$ is a vector subspace of $X^{**}$. Since $J$ is an isometry it is injective, so it is an isometric isomorphism. In other words, $X$ and $J(X)$ are isomorphic as normed linear spaces. Therefore, if $X$ is complete, then so is $J(X)$ (Exercise 11.2.22), and it follows that it is closed (Exercise 2.2.18). □

The map $J$ is called the **natural embedding** of $X$ into $X^{**}$. Theorem 12.2.9 shows that $X$ is isomorphic with a subspace of $X^{**}$. It is then natural to ask whether this subspace is proper or not.

**Definition 12.2.10.** A normed linear space $X$ is said to be **reflexive** if the natural embedding maps $X$ onto $X^{**}$.

*Remark* 12.2.11. Notice that, for $X$ to be reflexive, it is not sufficient that $X$ and $X^{**}$ are isometrically isomorphic. The isometric isomorphism must be realized by the natural embedding $J$.

In the next section we will look at examples of normed spaces, some reflexive and some not.

> Examples of reflexive spaces appear in the work of Hahn in 1927 [49] where he calls them *regular*. The word *reflexive* is due to an American mathematician and early civil rights activist Lee Alexander Lorch (1915–2014) in 1939 [75]. Nonreflexive sequence spaces were discovered by Helly in 1921 [60].

## Exercises

12.2.1. Let $\mathcal{M} = \{(x, y) \in \mathbb{R}^2 : 2x - y = 0\}$ and let $T : \mathcal{M} \to \mathbb{R}$ be defined by $f(x, y) = x$. Show that $T' : \mathbb{R}^2 \to \mathbb{R}$ defined by $T'(x, y) = x/5 + 2y/5$ is the unique Hahn–Banach extension of $T$.

12.2.2. Let $\mathcal{M} = \{\{x_n\} : x_1 = x_3 = x_5 = \cdots = 0\} \subset \ell^1$. Show that every non-zero functional $T : \mathcal{M} \to \mathbb{R}$ has infinitely many Hahn–Banach extensions to $\ell^1$.

12.2.3. Let $Y$ be the subspace of $C[a, b]$ consisting of all constant function and let $T \in Y^*$ be defined by $Tf = f(a)$, for all $f \in Y$. Define

$$T'f = \frac{1}{b-a}\int_a^b f, \qquad \text{for } f \in C[a, b].$$

Prove that $T'$ is a Hahn–Banach extension of $T$.

12.2.4. Prove that the function $T_0$ defined in the proof of the Hahn–Banach theorem is a well-defined, linear functional on the set $Z$ satisfying $\|T_0\|_{Z^*} \leq \|T\|_{Y^*}$.

In Exercises 12.2.5–12.2.8 $\mathcal{P}$ is the vector space of polynomials defined on $[0, 1]$ and $p(x) = \sum_{k=0}^{\deg p} a_k x^k$. Determine which of the following linear functionals on $\mathcal{P}$ can be extended to a bounded linear functional on $C[0, 1]$:

12.2.5. $Tp = a_0$.

12.2.6. $Tp = \sum_{k=0}^{\deg p} a_k$.

12.2.7.* $Tp = \sum_{k=0}^{\deg p} (-1)^k a_k$.

12.2.8.* $Tp = \sum_{k=0}^{N} c_k a_k$, where $N$ is a fixed positive integer and $\{c_k\}_{k=1}^{N}$ are fixed real numbers.

12.2.9. Let $X$ be a normed linear space and let $Y \subset X$ be a finite-dimensional linear subspace. Prove that there exists a closed subspace $Z \subset X$ such that $X = Y + Z$ and $Y \cap Z = \{0\}$.

12.2.10.** Let $X$ be a normed linear space and let $Y \subset X$ be a finite-dimensional linear subspace. Prove that there exists a vector $x_0 \in X$ such that $d(x_0, Y) = 1$.

12.2.11.* Let $X$ be a normed linear space and $x, y$ two distinct points in $X$. Prove that there exists $T \in X^*$ for which $Tx \neq Ty$.

12.2.12. Let $V$ be the subspace of $L^\infty(0,1)$ defined by

$$V = \{f \in L^\infty(0,1) : \lim_n n \int_{[0,1/n]} f(x)\, dx \text{ exists}\}.$$

Prove that there is $\varphi \in (L^\infty(0,1))^*$ such that $\varphi(f) = \lim_n n \int_{[0,1/n]} f(x)\, dx$ for every $f \in V$.

12.2.13. Prove that there exists a bounded linear functional LIM on $\ell^\infty$ with the following properties:

(a) $\inf x_n \leq \text{LIM}\{x_n\} \leq \sup x_n$.

(b) If $\lim x_n = L$ then $\text{LIM}\{x_n\} = L$.

(c) $\text{LIM}\{x_{n+1}\} = \text{LIM}\{x_n\}$.

12.2.14.** Let $X \neq 0$ and $Y$ be normed spaces for which $\mathcal{L}(X, Y)$ is a Banach space. Prove that $Y$ is also a Banach space.

12.2.15. Let $X$ and $Y$ be Banach spaces and $T : X \to Y$ a linear operator such that $\varphi \circ T \in X^*$ for every $\varphi \in Y^*$. Prove that $T$ is bounded.

12.2.16. Let $X$ and $Y$ be Banach spaces and $T : X \to Y$ an isometric isomorphism. Prove that there exists an isometric isomorphism $S : X^* \to Y^*$.

12.2.17. Prove Corollary 12.2.8.

12.2.18. Let $X$ be a normed linear space, let $Y \subset X$ be a closed linear subspace, let $z \in X \setminus Y$, and let $V = \{y + \lambda z : y \in Y, \lambda \in \mathbb{R}\}$. Prove that $V$ is a closed subspace of $X$.

12.2.19. Let $X$ be a normed linear space, let $Y \subset X$ be a linear subspace, and let $x_0 \in X \setminus Y$. Prove that

$$d(x_0, Y) = \sup\{Tx_0 : \|T\| = 1,\, T = 0 \text{ on } Y\}.$$

12.2.20. Let $X, Y$ be normed spaces, let $X_0$ a closed subspace of $X$, let $a \in X \setminus X_0$, and let $b \in Y$. Prove that there is $T \in \mathcal{L}(X, Y)$ that vanishes on $X_0$ and such that $Ta = b$.

12.2.21. Let $(X, \mathfrak{M}, \mu)$ be a finite measure space, let $Y$ be a Banach space, and let $A : X \to Y$ be such that $T \circ A$ is integrable for every $T \in Y^*$. Define $L : Y^* \to \mathbb{R}$ by $L(T) = \int_X T \circ A\, d\mu$. Prove that $L \in Y^{**}$.

12.2.22. Let $S$ be a subset of a normed linear space $X$. Prove that $S$ is bounded if and only if $T(S)$ is bounded for every $T \in X^*$.

12.2.23. Let $X$ be a normed linear space and $x_0 \in X$. Prove that

$$\|x_0\| = \sup\{|Tx_0| : T \in X^*,\ \|T\| = 1\}.$$

12.2.24. Prove that every finite-dimensional normed linear space is reflexive.

12.2.25. Prove that a Banach space $X$ is reflexive if and only if $X^*$ is reflexive.

12.2.26. Let $X$ be a reflexive normed space. Prove that if $Y$ is a closed subspace of $X$, then $Y$ is reflexive.

12.2.27.* Let $X$ be a reflexive normed space and let $Y$ be a Banach space such that $Y^*$ is isometrically isomorphic to $X^*$. Prove that $Y$ is isometrically isomorphic to $X$.

12.2.28.* Let $X$ be a reflexive normed space. Prove that every functional $T \in X^*$ attains its norm on $X$.

12.2.29. Let $X$ be a Banach space. If $\mathcal{M} \subset X$ and $\mathcal{N} \subset X^*$ are closed subspaces, let $\mathcal{M}^0 = \{T \in X^* : T|\mathcal{M} = 0\}$ and $\mathcal{N}^\perp = \{x \in X : Tx = 0, \text{ for all } T \in \mathcal{N}\}$. Prove that:

(a) $\mathcal{M}^0$ and $\mathcal{N}^\perp$ are closed subspaces of $X^*$ and $X$, respectively.

(b) $(\mathcal{M}^0)^\perp = \mathcal{M}$ and $(\mathcal{N}^\perp)^0 \supset \mathcal{N}$. If $X$ is reflexive $(\mathcal{N}^\perp)^0 = \mathcal{N}$.

12.2.30. Let $X$ be a normed space and $T, T_1, \ldots, T_n \in X^*$. Let $Y = \{x \in X : T_1(x) = \cdots = T_n(x) = 0\}$ and $S : Y \to \mathbb{R}$ be defined by $S(x) = T(x)$ on $Y$. If $\varphi$ is a bounded linear functional on $X$ that extends $S$ prove that there are $\alpha_1, \ldots, \alpha_n \in \mathbb{R}$ such that $\varphi = T + \alpha_1 T_1 + \cdots + \alpha_n T_n$.

12.2.31.* Let $\mathcal{A} = \{T \in \mathcal{L}(\ell^2) : \langle Te_1, e_1\rangle - 2\langle Te_1, e_2\rangle = 0\}$, and $\varphi$ a bounded linear functional on $\mathcal{A}$ defined by $\varphi(T) = \langle Te_1, e_1\rangle$. Prove that $\tilde{\varphi} : \mathcal{L}(\ell^2) \to \mathbb{R}$ defined by

$$\tilde{\varphi}(T) = \left\langle Te_1, \frac{4}{5}e_1 + \frac{2}{5}e_2 \right\rangle$$

is the unique Hahn–Banach extension of $\varphi$.

## 12.3 The Dual of $L^p(X, \mu)$, $1 \le p < \infty$

It is natural to look for concrete examples of dual spaces, and we will do that next. We will start with the space $L^p(X, \mu)$ and its dual space. Our goal is to shed more light on the latter.

**Example 12.3.1. A bounded linear functional on $L^p(X, \mu)$.**

Let $1 \leq p < \infty$, $1/p + 1/q = 1$ and $f \in L^q(X, \mu)$. Define $T_f : L^p(X, \mu) \to \mathbb{R}$ by $T_f(g) = \int_X fg\, d\mu$. Then,

$$|T_f(g)| = \left|\int_X fg\, d\mu\right| \leq \int_X |fg|\, d\mu \leq \|g\|_p \|f\|_q, \tag{12.16}$$

so $T_f \in L^p(X, \mu)^*$ and

$$\|T_f\| \leq \|f\|_q. \tag{12.17}$$

♦

The moral of this example is that if $f \in L^q$ then $T_f$ is bounded on $L^p$ and (12.17) holds. We will now prove the converse.

**Proposition 12.3.2.** *Let $(X, \mathfrak{M}, \mu)$ be a $\sigma$-finite measure space, let $1 \leq p < \infty$, $1/p + 1/q = 1$, and let $f$ be an integrable function over $X$. Define a functional $T_f$ on $L^\infty(X, \mu)$ by $T_f(g) = \int_X fg\, d\mu$. If $T_f$ extends to a bounded linear functional on $L^p(X, \mu)$ then $f \in L^q(X, \mu)$ and $\|f\|_q \leq \|T_f\|$.*

*Proof.* We will consider separately cases $p = 1$ and $p > 1$.

**Case I.** $p = 1$

Now $q = \infty$. Suppose to the contrary that $\|f\|_\infty > \|T_f\|$. (For all we know, we could have $\|f\|_\infty = \infty$, but $\|T_f\|$ is finite.) Then there exists $\varepsilon > 0$ and a set $E$ of positive measure such that $|f(x)| > \|T_f\| + \varepsilon$ on $E$ (Exercise 11.1.3). By Exercise 5.1.13, we may assume that $\mu(E) < \infty$. Let $g = \operatorname{sgn}(f)\chi_E$. Then $g$ is a bounded function and $\|g\|_1 = \mu(E)$. Thus,

$$\begin{aligned}\|T_f\|\|g\|_1 \geq |T_f g| &= \left|\int_X fg\, d\mu\right| \\ &= \int_E |f|\, d\mu > (\|T_f\| + \varepsilon)\, \mu(E) = (\|T_f\| + \varepsilon)\, \|g\|_1\end{aligned}$$

which is impossible. It follows that $\|f\|_\infty \leq \|T_f\|$ which also ensures that $f \in L^\infty$.

**Case II.** $p > 1$.

We will assume that $\mu$ is a finite measure on $X$ and leave the $\sigma$-finite case as Exercise 12.3.1. Let $E_n = \{x \in X : |f(x)| \leq n\}$ and $g = \chi_{E_n} \operatorname{sgn}(f)\, |f|^{q-1}$. Then $g$ is a bounded function,

$$\begin{aligned}\|g\|_p &= \left(\int_X \left|\chi_{E_n} \operatorname{sgn}(f)\, |f|^{q-1}\right|^p d\mu\right)^{1/p} = \left(\int_{E_n} |f|^{p(q-1)}\, d\mu\right)^{1/p} \\ &= \left(\int_{E_n} |f|^q\, d\mu\right)^{1/p} \quad \text{and} \\ T_f(g) &= \int_X f\chi_{E_n} \operatorname{sgn}(f)\, |f|^{q-1}\, d\mu = \int_{E_n} |f|^q\, d\mu \\ &= \left(\int_{E_n} |f|^q\, d\mu\right)^{1/q} \left(\int_{E_n} |f|^q\, d\mu\right)^{1/p} = \left(\int_{E_n} |f|^q\, d\mu\right)^{1/q} \|g\|_p.\end{aligned}$$

Since $T_f$ is bounded on $L^p(X,\mu)$ and $g \in L^p(X,\mu)$, we have that $|T_f g| \le \|T_f\| \|g\|_p$ and it follows that

$$\left(\int_{E_n} |f|^q \, d\mu\right)^{1/q} \le \|T_f\|.$$

Now the Monotone Convergence Theorem implies that $f \in L^q(X,\mu)$ and that $\|f\|_q \le \|T_f\|$. □

By Example 12.3.1, $\|T_f\| \le \|f\|_q$ and by Proposition 12.3.2 we have the opposite inequality, so the mapping $f \mapsto T_f$ from $L^q(X,\mu)$ to $L^p(X,\mu)^*$ is an isometry. It is automatically injective. Is it surjective? In other words, is every element of $L^p(X,\mu)^*$ of the form $T_f$, for some $f \in L^q(X,\mu)$? If so, how do we find $f$?

Suppose for a moment that the answer is affirmative, i.e., if $S$ is a given bounded linear functional then there is a function $f$ for which $S(g) = \int_X fg \, d\mu$. Then $S(\chi_E) = \int_E f \, d\mu$ for any measurable set $E$. By Example 7.3.7, $S(\chi_E)$ is a signed measure. If it is absolutely continuous with respect to $\mu$, then $f$ is the Radon–Nikodym derivative of that measure with respect to $\mu$ (see Remark 9.5.9). This is the plan, now let us get to details.

First, we will prove a weaker version of the anticipated result, namely when $\mu$ is a finite measure.

**Theorem 12.3.3.** *Let $(X,\mathfrak{M},\mu)$ be a finite measure space, $1 \le p < \infty$, and $q$ the conjugate of $p$. For $f \in L^q(X,\mu)$ define $T_f \in (L^p(X,\mu))^*$ by*

$$T_f(g) = \int_X fg \, d\mu, \qquad \text{for } g \in L^p(X,\mu).$$

*Then the mapping $f \mapsto T_f$ is an isometric isomorphism of $L^q(X,\mu)$ onto $(L^p(X,\mu))^*$.*

*Proof.* It is easy to see that the mapping is linear and Proposition 12.3.2 implies that it is an isometry as well, hence injective. It remains to show that it is surjective by taking an arbitrary $S \in (L^p(X,\mu))^*$ and constructing a function $f \in L^q(X,\mu)$ such that $S = T_f$.

If $E \in \mathfrak{M}$ then $\mu(E) \le \mu(X) < \infty$ so $\chi_E \in L^p(X,\mu)$ and we define

$$\nu(E) = S(\chi_E).$$

We will show that $\nu$ is a signed measure.

It is easy to see that $\nu$ does not take either of the values $+\infty, -\infty$ and that $\nu(\emptyset) = 0$. It remains to establish that it is $\sigma$-additive. To that end, let

$\{E_n\}_{n=1}^{\infty}$ be a disjoint sequence in $\mathfrak{M}$ and $E = \cup_{k=1}^{\infty} E_k$. We have that

$$\begin{aligned}
\left|\nu(E) - \sum_{k=1}^{n} \nu(E_k)\right| &= \left|S(\chi_E) - \sum_{k=1}^{n} S(\chi_{E_k})\right| = \left|S\left(\chi_E - \sum_{k=1}^{n} \chi_{E_k}\right)\right| \\
&\leq \|S\| \left\|\chi_E - \sum_{k=1}^{n} \chi_{E_k}\right\|_p = \|S\| \left\|\sum_{k=n+1}^{\infty} \chi_{E_k}\right\|_p \\
&= \|S\| \left(\int_X \sum_{k=n+1}^{\infty} |\chi_{E_k}|^p \, d\mu\right)^{1/p} = \|S\| \left(\sum_{k=n+1}^{\infty} \int_X |\chi_{E_k}|^p \, d\mu\right)^{1/p} \\
&= \|S\| \left(\sum_{k=n+1}^{\infty} \mu(E_k)\right)^{1/p} \to 0, \quad \text{when } n \to \infty,
\end{aligned}$$

as the tail of a convergent series. (Remember: $\mu(E) < \infty$.) That way, the series $\sum_{k=1}^{\infty} \nu(E_k)$ converges to $\nu(E)$ and $\nu$ is a signed measure. Moreover, it is absolutely continuous with respect to $\mu$. Indeed,

$$\mu(E) = 0 \quad \Rightarrow \quad \chi_E = 0 \quad \Rightarrow \quad S(\chi_E) = 0 \quad \Rightarrow \quad \nu(E) = 0.$$

Let $f = d\nu/d\mu$. Then

$$S(\chi_E) = \int_E f \, d\mu. \quad \text{for all } E \in \mathfrak{M}. \tag{12.18}$$

Notice that $\|\chi_X\|_p = \mu(X)^{1/p} < \infty$, so $\chi_X \in L^p$, whence $S(\chi_X)$ is finite. It follows that $f$ is integrable over $X$.

Equality (12.18) extends to all simple functions $s$

$$S(s) = \int_X fs \, d\mu. \tag{12.19}$$

Moreover, it extends to all functions $g \in L^{\infty}(X, \mu)$ because each such function is a uniform limit of a sequence of simple functions (except for a possible set of measure zero). By Proposition 12.3.2, $f \in L^q(X, \mu)$. Finally, if we define a functional $\Lambda$ on $L^p(X, \mu)$ by

$$\Lambda(g) = S(g) - \int_X fg \, d\mu,$$

then $\Lambda$ is bounded, hence continuous and by (12.19) it vanishes on all simple functions. However, simple functions are dense in $L^p(X, \mu)$ (Proposition 10.4.1), so $\Lambda = 0$ on $L^p(X, \mu)$ (Exercise 12.1.24), i.e, $S = T_f$. □

*Remark* 12.3.4. In view of the injectivity of the mapping $f \mapsto T_f$ we have that, given $S \in (L^p(X, \mu))^*$, the function $f$ satisfying $S = T_f$ is *unique*!

Now we will extend Theorem 12.3.3 to the case when the measure is $\sigma$-finite.

**Theorem 12.3.5.** ***The Riesz Representation Theorem for the Dual of $L^p(X, \mu)$.*** *Let $(X, \mathfrak{M}, \mu)$ be a $\sigma$-finite measure space, $1 \le p < \infty$, and $q$ the conjugate of $p$. For $f \in L^q(X, \mu)$ define $T_f \in (L^p(X, \mu))^*$ by*

$$T_f(g) = \int_X fg\, d\mu, \qquad \text{for } g \in L^p(X, \mu). \tag{12.20}$$

*Then the mapping $f \mapsto T_f$ is an isometric isomorphism of $L^q(X, \mu)$ onto $(L^p(X, \mu))^*$.*

*Proof.* Let $S \in (L^p(X, \mu))^*$. We will find a function $f \in L^q(X, \mu)$ for which $S = T_f$.

Since $X$ is $\sigma$-finite we can represent it as $X = \cup_{n=1}^{\infty} X_n$ where $\{X_n\}$ is an increasing sequence and $\mu(X_n) < \infty$, for all $n \in \mathbb{N}$. Notice that the space $L^p(X_n, \mu)$ can be embedded into $L^p(X, \mu)$ by extending each function on $X_n$ to be 0 on $X \setminus X_n$. That way, $S$ is also a bounded linear functional on $L^p(X_n, \mu)$ and this restriction $S_n$ of $S$ satisfies $\|S_n\| \le \|S\|$. It follows from Theorem 12.3.3 that for every $n$ there exists a unique function $f_n \in L^q(X, \mu)$ that is 0 on $X_n^C$,

$$\|f_n\|_q = \|T_{f_n}\| = \|S_n\| \le \|S\| \quad \text{and} \quad S(g) = \int_X f_n g\, d\mu, \tag{12.21}$$

for all $g \in L^p(X, \mu)$. Actually, Theorem 12.3.3 yields the last equality only for $g \in L^p(X_n, \mu)$, but the values of $g$ outside of $X_n$ do not affect the product $f_n g$, so we can extend it to the whole $L^p(X, \mu)$.

Since $X_n \subset X_{n+1}$, the restriction of $f_{n+1}$ to $X_n$ satisfies (12.21) as well, so the uniqueness part of Theorem 12.3.3 implies that $f_{n+1} = f_n$ on $X_n$. (See Remark 12.3.4.) This allows us to define a function $f$ by $f(x) = f_n(x)$ if $x \in X_n$. This function belongs to $L^q(X, \mu)$. Indeed, when $p > 1$ (so that $q$ is finite) we have that $|f_n|^q \to |f|^q$ a.e. on $X$. By Fatou's Lemma,

$$\int_X |f|^q\, d\mu \le \liminf \int_X |f_n|^q\, d\mu = \liminf \|f_n\|_q^q \le \|S\|^q,$$

which implies that $f \in L^q(X, \mu)$. For the case $p = 1$ see Exercise 10.2.32.

Thus, it remains to show that $S = T_f$. To that end, let $g \in L^p(X, \mu)$ and define $g_n(x)$ to be $g(x)$ on $X_n$ and 0 on $X_n^C$. Then $|fg_n| \le |fg|$ on $X$, and $fg$ is integrable, so Lebesgue Dominated Convergence Theorem implies that

$$\lim \int_X f g_n\, d\mu = \int_X fg\, d\mu. \tag{12.22}$$

On the other hand, $|g_n - g|^p \to 0$ a.e. on $X$ and $|g_n - g|^p \le |g|^p$ on $X$ so another application of Lebesgue Dominated Convergence Theorem yields

$$\int_X |g_n - g|^p\, d\mu \to 0,$$

i.e., $g_n \to g$ in $L^p(X, \mu)$. Since $S$ is continuous,

$$S(g) = \lim S(g_n). \tag{12.23}$$

Also, by (12.21) ,

$$S(g_n) = \int_{X_n} f_n g_n \, d\mu = \int_X f g_n \, d\mu.$$

If we now let $n \to \infty$, (12.23) and (12.22) imply that $S(g) = \int_X fg \, d\mu$, i.e., that $S = T_f$. □

*Remark* 12.3.6. Riesz Theorem covers the cases $1 \leq p < \infty$, but not $p = \infty$. The dual space of $L^\infty$ is not $L_1$. Indeed, under reasonable assumptions on the measure space, $L^1(X, \mu)$ is separable (see Theorem 10.4.6). If it were the dual space of $L^\infty(X, \mu)$, Theorem 12.2.6 would imply that $L^\infty(X, \mu)$ is also separable, which is not true except in some special cases (see Theorem 10.4.9). In particular, if $E$ is a Lebesgue measurable subset of $\mathbb{R}^n$, if $m$ is the Lebesgue measure on $\mathbb{R}^n$, and $m(E) \neq 0$, then $L^1(E, m)$ is not the dual space of $L^\infty(E, m)$. (See also Exercise 12.3.2.)

The special case of Riesz Theorem for $p = 2$ was announced simultaneously by Riesz and Fréchet in 1907, and proved by the latter in the same year in [37]. Riesz then extended in 1910 [92] the result by showing that the dual of $L^p(a, b)$ is $L^q(a, b)$, for $1 < p < \infty$. Nikodym generalized it in 1931 to finite measure spaces [80] and an American mathematician Edward McShane (1904–1989) in 1950 [76] to arbitrary measure spaces. The case $p = 1$ was demonstrated by Steinhaus in 1919 [115] for the interval $[0, 1]$. An American mathematician Nelson Dunford (1906–1986), extended this result to finite measure spaces in 1938 [29]. In 1951, his student Jacob T. Schwartz (1930–2009) reported an example due to T. Botts that the theorem fails in the absence of $\sigma$-finiteness. (See Exercise 12.3.4.) Dunford is known for his work in functional analysis, namely the integration of vector-valued functions, ergodic theory, and linear operators. He spent his whole career at Yale University. In 1981, he was awarded jointly with Schwartz the well-known Leroy P. Steele Prize of the American Mathematical Society for the three-volume work "Linear operators". Schwartz was a professor of computer science at the New York University Courant Institute of Mathematical Sciences. He founded the Department of Computer Science there, and he was a chair from 1964 to 1980.

Since $\ell^p$ spaces are special cases of $L^p$ spaces, we get an immediate corollary.

**Corollary 12.3.7.** *Every bounded linear functional $T$ on $\ell^p$, $1 \leq p < \infty$, is given by an $\ell^q$ sequence $\{c_n\}$ so that*

$$T(\{x_n\}) = \sum_{n=1}^{\infty} c_n x_n,$$

*for $\{x_n\} \in \ell^p$. Further, $\|T\| = \|\{c_n\}\|_q$ and for a given $T$ the sequence $\{c_n\}$ is uniquely determined.*

The Riesz Representation Theorem yields a very nice characterization of the dual space of $L^p(X, \mu)$, for $1 \leq p < \infty$. In the special case when $p = 2$, $L^2(X, \mu)$ is a Hilbert space, and we can say more. Notice that the the linear map $T_f$ is given by (12.20) which is also the definition of inner product on $L^2(X, \mu)$. (See Example 10.5.3.) Therefore, not only that every bounded linear functional $L^2(X, \mu)$ is of the form $T_f$, for some $f \in L^2(X, \mu)$, but it is given by the inner product

$$T_f g = \langle f, g \rangle, \quad \text{for all } g \in L^2(X, \mu).$$

Furthermore, as we will demonstrate, this is true in any Hilbert space, not just $L^2(X,\mu)$.

**Theorem 12.3.8.** ***The Riesz–Fréchet Representation Theorem.*** *Let $\mathcal{H}$ be Hilbert space. Define the map $T:\mathcal{H}\to\mathcal{H}^*$ by assigning to each $h\in\mathcal{H}$ the linear functional $T_h:\mathcal{H}\to\mathbb{R}$ defined by $T_h(x)=\langle h,x\rangle$. Then $T$ is a linear isometry of $\mathcal{H}$ onto $\mathcal{H}^*$.*

*Proof.* Let $h\in\mathcal{H}$. It is not hard to see that $T_h$ is linear and $|T_h(x)|=|\langle h,x\rangle|\leq\|h\|\|x\|$, so $\|T_h\|\leq\|h\|$. Thus, $T_h$ is a bounded linear functional on $\mathcal{H}$. On the other hand, if $h\neq 0$ then

$$T_h\left(\frac{h}{\|h\|}\right)=\left\langle h,\frac{h}{\|h\|}\right\rangle=\|h\|,$$

so $\|T_h\|=\|h\|$. In conclusion, $T$ is an isometry, which also implies that it is injective.

It follows from the linearity of the inner product that $T$ is linear, so it remains to show that it is surjective. To that end, let $\Lambda$ be an element of $\mathcal{H}^*$. Let $\mathcal{K}$ denote the set of vectors $z\in\mathcal{H}$ such that $\Lambda(z)=0$. The linearity of $\Lambda$ ensures that $\mathcal{K}$ is a vector subspace of $\mathcal{H}$ and it is closed because $\Lambda$ is continuous. If $\mathcal{K}=\mathcal{H}$, i.e., if $\Lambda(x)=0$ for all $x\in\mathcal{H}$, then $h=0$ yields $T_h=\Lambda$. Therefore, suppose that $\mathcal{K}$ is a proper subspace of $\mathcal{H}$. By Corollary 10.5.26, there is a nonzero vector $y_0\in\mathcal{H}$ that is orthogonal to $\mathcal{K}$. For a given vector $x\in\mathcal{H}$ consider

$$w=\Lambda(x)\,y_0-\Lambda(y_0)\,x.$$

Notice that

$$\Lambda(w)=\Lambda(\Lambda(x)\,y_0-\Lambda(y_0)\,x)=\Lambda(x)\Lambda(y_0)-\Lambda(y_0)\Lambda(x)=0,$$

so $w\in\mathcal{K}$. It follows that $\langle w,y_0\rangle=0$, i.e.,

$$\langle\Lambda(x)\,y_0-\Lambda(y_0)\,x,y_0\rangle=0.$$

Solving for $\Lambda(x)$ we obtain

$$\Lambda(x)=\frac{\Lambda(y_0)}{\|y_0\|^2}\langle x,y_0\rangle.$$

Thus, $\Lambda=T_h$, where

$$h=\frac{\Lambda(y_0)}{\|y_0\|^2}\,y_0,$$

and the theorem is proved. □

*Remark* 12.3.9. Every bounded linear functional on $\mathcal{H}$ is given by the inner product and $\mathcal{H}^*\cong\mathcal{H}$.

Riesz [89] and Fréchet [36] announced the representation theorem for $L^2(a,b)$ in 1907. The analogous result in an (axiomatically defined) abstract Hilbert space was established by Riesz in 1934 [101].

Exercise 12.3.9 asserts that, if $1 < p < \infty$ and $(X, \mathfrak{M}, \mu)$ is $\sigma$-finite, the space $L^p(X,\mu)$ is reflexive. In particular, it holds for Hilbert space $L^2(X,\mu)$. Once again, this carries over to every Hilbert space.

**Theorem 12.3.10.** *Every Hilbert space is reflexive.*

*Proof.* We will show that the map $J : \mathcal{H} \to \mathcal{H}^{**}$, defined by (12.14) is surjective. Let $\Psi \in \mathcal{H}^{**}$ and let $T : \mathcal{H} \to \mathcal{H}^*$ as in the Riesz-Fréchet Representation Theorem. Then $\Psi \circ T : \mathcal{H} \to \mathbb{R}$ is linear and bounded as a composition of two such maps. The Riesz- Fréchet Representation Theorem implies that there is $h_0 \in \mathcal{H}$ such that $\Psi \circ T = T(h_0)$. Then, for any $h \in \mathcal{H}$,

$$(\Psi \circ T)(h) = T(h_0)(h) = \langle h_0, h\rangle = \langle h, h_0\rangle = T(h)(h_0) = J(h_0)(T(h)),$$

so $\Psi = J(h_0)$ on $\{T(h) : h \in \mathcal{H}\} = \mathcal{H}^*$, and the theorem is proved. □

MAURICE FRECHET (1878–1973) was a French mathematician. His family ran an orphanage, a school, and finally, a boarding house to make ends meet. He attended high school in Paris, where his teacher was Jacques Hadamard who later became his doctoral advisor at the Ecole Normale Supérieure. During that time Fréchet wrote over 20 papers, and his groundbreaking doctoral dissertation introduced the concept of metric space and initiated the study of functionals on metric spaces. He started as a high school teacher in Besançon and Nantes, later becoming a professor of mechanics at the Faculty of Science in Poitiers. During World War I he spent two and a half years at the front, yet maintained his research output. From 1919, he was a professor and Director of the Mathematics Institute at the University of Strasbourg, where he organized the International Congress of Mathematicians in 1920. While in Strasbourg, he became interested in statistics which became his focus after he moved to Sorbonne in 1928, where he stayed until his retirement in 1948. Despite his significant contributions, he received more recognition abroad than in France. He was a member of the Polish Academy of Science and Arts, the Royal Netherlands Academy of Arts and Sciences, and the Royal Society of Edinburgh, but he was elected to the French Academy of Sciences at the age of 78. Notably, Fréchet supported Esperanto, publishing papers and articles in that constructed language.

## Exercises

12.3.1. In Proposition 12.3.2 prove the case when $p > 1$ and $\mu$ is $\sigma$-finite.

12.3.2. Use Hahn–Banach Theorem to prove that the dual space of $L^\infty(0,1)$ cannot be identified with $L^1(0,1)$.

12.3.3. Give an example of a measure space $(X, \mathfrak{M}, \mu)$ for which the Riesz Representation Theorem does extend to the case $p = \infty$.

12.3.4. The purpose of this exercise is to establish the following result: Theorem 12.3.5 need not hold for $p = 1$ if the measure space is not $\sigma$-finite. Let $X = (0,1)$ and let $\mathfrak{M}$ be the $\sigma$-algebra consisting of sets that are either finite or countable or their complements are such. Finally, let $\mu$ be the counting measure on $\mathfrak{M}$.

(a) Characterize the space $L^1(X, \mu)$;

(b) prove that $Tf := \int_X tf(t)\, d\mu$ defines a bounded linear functional on $L^1(X, \mu)$;

(c) prove that if $g \in L^\infty(X, \mu)$ then its range is countable;

(d) prove that, as $f$ ranges over the characteristic functions of singletons, $Tf$ takes on uncountably many values, whereas $\int_X gf\, d\mu$ takes on only countably many values;

(e) conclude that the mapping $g \mapsto T_g$ from $L^\infty(X, \mu)$ to $(L^1(X, \mu))^*$ is not surjective.

12.3.5. Prove or disprove each of the following statements: There is $T \in (L^q(0,1))^*$, $1 \le q < \infty$ such that $Tf = f'(0)$ for all

(a) polynomials in $\mathcal{P}_n$;

(b) polynomials;

(c) functions $f \in C^1[0,1]$.

12.3.6. Let $1 \le p < \infty$ and let $D \subset L^p(\mathbb{R})$ consists of all functions $f \in L^p(\mathbb{R})$ with compact support. Prove that $D$ is dense in $L^p(\mathbb{R})$. Prove that this is not true if $p = \infty$.

12.3.7. Let $p > 1$. Give an example of a subspace $X$ that is dense in $L^r(\mathbb{R})$, for all $1 < r < p$, but not in $L^p(\mathbb{R})$.

12.3.8. Let $p > 1$. Give an example of a subspace $X$ that is dense in $L^p(\mathbb{R})$, but not in $L^r(\mathbb{R})$ for any $1 < r < p$.

12.3.9. Let $(X, \mathfrak{M}, \mu)$ be a $\sigma$-finite measure space and $1 < p < \infty$. Prove that $L^p(X, \mu)$ is reflexive.

12.3.10. Let $Y$ be a subspace of $X$. Prove that:

(a) $L^p(X) \subset L^p(Y)$.

(b) $(L^p(Y))^* \subset (L^p(X))^*$.

(c) if $T \in (L^p(Y))^*$, then the functions $f_Y$ and $f_X$ provided by Theorem 12.3.3 coincide on $Y$ a.e.

12.3.11. Prove that in the case $1 < p < \infty$, Riesz Theorem extends to measures that are not $\sigma$-finite.

12.3.12. Let $X = \{a, b\}$ and define $\mu(\{a\}) = 1$, $\mu(\{b\}) = \mu(X) = \infty$, and $\mu(\emptyset) = 0$. What is the dual space of $L^p(X, \mu)$ if $1 \le p < \infty$? What if $p = \infty$?

In Exercises 12.3.13–12.3.16 prove that $T$ is a bounded linear functional on $L^p(X, \mu)$ and find the function $f \in L^q(X, \mu)$ for which $T = T_f$:

12.3.13. $X = [0,1]$, $\mu = m|[0,1]$, $Tg = \int_0^{1/2} x^2 g(x)\,dx$.

12.3.14. $X = [0,1]$, $\mu = m|[0,1]$, $Tg = \int_0^1 x^2 g(\sqrt{x})\,dx$.

12.3.15. $X = \mathbb{N}$, $\mu$ is the counting measure, $Tg = \sum_{n=1}^{\infty} g_{n+1}/n^2$.

12.3.16. $X = \mathbb{N}$, $\mu$ is the counting measure, $Tg = \sum_{n=1}^{\infty} g_{2n}/n^{3/2}$.

12.3.17. Let $(X, \mathfrak{M}, \mu)$ be a $\sigma$-finite measure space, let $1 < p < \infty$, and let $T$ be a bounded linear functional on $L^p(X,\mu)$. Prove that $T$ attains its norm.

12.3.18. Let $(X, \mathfrak{M}, \mu)$ be a measure space, let $1 < p < \infty$, and let $A$ be a bounded linear operator on $L^p(X,\mu)$. Suppose that

$$\int_X (Af)\,g = \int_X f\,(Ag), \qquad \text{for all } f, g \in L^p(X,\mu) \cap L^q(X,\mu).$$

Prove that $A$ extends in a unique way to a bounded operator on $L^q(X,\mu)$.

12.3.19. Let $1 < p < \infty$. Prove that $E = \{f \in L^p(\mathbb{R}) : \int_{\mathbb{R}} f = 0\}$ is a dense subspace of $L^p(\mathbb{R})$.

12.3.20. Let $(X, \mathfrak{M}, \mu)$ be a measure space, let $1 \le p < \infty$, let $x_0 \in X$, and let $T$ be defined by $Tf = f(x_0)$. Suppose that $\{x_0\} \in \mathfrak{M}$. Prove that $T$ is a bounded linear functional on $L^p(X,\mu)$ if and only if $\mu(\{x_0\}) > 0$.

12.3.21. Give an example of a measure space $(X, \mathfrak{M}, \mu)$ and a measurable function $f : X \to \mathbb{R}$ which does not belong $L^p(X,\mu)$ for any $1 \le p < \infty$ and yet has the property that $fg \in L^1(X,\mu)$ for any $g \in L^q(X,\mu)$, where $1 \le q < \infty$.

12.3.22. Give an example of a measure space $(X, \mathfrak{M}, \mu)$ for which the mapping $f \mapsto T_f$ from $L^\infty(X,\mu)$ to $(L^1(X,\mu))^*$ is not an isometry.

12.3.23. Let $(X, \mathfrak{M}, \mu)$ be a measure space, let $1 \le p < \infty$, let $1/p + 1/q = 1$, let $g \in L^1(X,\mu)$, and suppose that there is $M \ge 0$ for which $|\int_X gf\,d\mu| \le M\|f\|_p$, for every simple function $f$ in $L^p(X,\mu)$. Prove that $g$ belongs to $L^q(X,\mu)$ and that $\|g\|_q \le M$.

12.3.24. Let $(X, \mathfrak{M}, \mu)$ be a measure space, let $Y = L^2(X,\mu) \times L^2(X,\mu)$, and define a norm on $Y$ by $\|(f,g)\| = (\|f\|_2^3 + \|g\|_2^3)^{1/3}$. Prove that this is a norm that makes $Y$ a Banach space and describe $Y^*$.

12.3.25. Let $c$ be the normed linear space of all convergent real sequences with the norm $\|\{x_n\}\| = \sup_n |x_n|$. Prove that the dual space of $c$ is isometrically isomorphic to $\ell^1$.

12.3.26. Let $c_0$ be the normed linear space of all real sequences that converge to 0, with the norm $\|\{x_n\}\| = \sup_n |x_n|$. Prove that the dual space of $c_0$ is isometrically isomorphic to $\ell^1$.

12.3.27. Use Hahn–Banach Theorem to prove that the dual space of $\ell^\infty$ cannot be identified with $\ell^1$.

12.3.28. Let $\{x_n\}$ be an unbounded sequence in Hilbert space $\mathcal{H}$. Prove that there exists a vector $x \in \mathcal{H}$ such that the sequence $\{\langle x_n, x\rangle\}$ is unbounded.

12.3.29. Consider an inner product on $C[-1,1]$ given by

$$\langle f, g\rangle = \int_{-1}^{1} fg, \qquad \text{for } f, g \in C[-1,1].$$

Let $T$ be a linear functional on $C[-1,1]$ defined by $Tf = f(0)$. Show that there does not exist $g \in C[-1,1]$ such that $Tf = \langle f, g\rangle$ for every $f \in C[-1,1]$.

12.3.30. Let $X = \{f \in AC[0,1] : f(0) = 0,\ f' \in L^2(0,1)\}$.

(a) Prove that $X$ is Hilbert space with respect to the inner product

$$\langle f, g\rangle = \int_{0}^{1} f'g', \qquad \text{for } f, g \in X.$$

(b) Prove that the so-called *point evaluation functional* $T$ on $X$, defined by $Tf = f(x)$ is continuous.

(c) Find the function $h \in X$ for which $Tf = \langle f, h\rangle$.

12.3.31. Let $X$ be a Banach space, let $\mathcal{H}$ be Hilbert space, let $A : X \to \mathcal{H}$ be a linear operator, and let $\{e_k\}$ be an orthonormal basis of $\mathcal{H}$. Suppose that, for every $k \in \mathbb{N}$, the functional $\Lambda$ defined by $\Lambda x = \langle Ax, e_k\rangle$ is continuous. Prove that $A$ is continuous.

12.3.32. Give an example to show that the Riesz– Fréchet Representation Theorem need not hold if $\mathcal{H}$ is merely an inner product space that is not complete.

12.3.33. Let $X$ be an inner product space and suppose that the Riesz– Fréchet Representation Theorem holds, i.e., every continuous linear functional on $X$ can be represented as $Tx = \langle x, h\rangle$ for some $h \in X$. Prove that $X$ is complete.

12.3.34. Let $\mathcal{H}$ be Hilbert space with an orthonormal basis $\{e_n\}$. Prove or disprove: there exists a bounded linear functional $T$ on $\mathcal{H}$ such that

(a) $Te_n = 1/n$, for all $n \in \mathbb{N}$;

(b) $Te_n = 1/\sqrt{n}$, for all $n \in \mathbb{N}$.

12.3.35. Let $\mathcal{H}$ be Hilbert space and $T_1, T_2$ nonzero, bounded, linear functionals on $\mathcal{H}$ with the following property: if $x \in \mathcal{H}$ satisfies $|T_1x| = \|T_1\|\|x\|$ then $T_2x = 0$. Prove that if $x \in \mathcal{H}$ satisfies $|T_2x| = \|T_2\|\|x\|$ then $T_1x = 0$.

12.3.36. Let $\mathcal{H}$ be Hilbert space with an orthonormal basis $\{e_n\}$, and let

$$f_n = \frac{1}{\sqrt{n^2+n}}\left(\sum_{k=1}^{n} e_k - n e_{n+1}\right), \qquad n = 1, 2, 3, \ldots.$$

Prove that $\{f_n\}$ is an orthonormal basis for $\mathcal{H}$.

12.3.37. Let $\mathcal{H}$ be Hilbert space and $h \in \mathcal{H}$. Prove that $\|h\| = \sup\{\langle h, x\rangle : \|x\| \le 1\}$.

12.3.38. Let $\mathcal{H}_1, \mathcal{H}_2$ be Hilbert spaces and let $\mathcal{H} = \mathcal{H}_1 \oplus \mathcal{H}_2$, the direct orthogonal sum of $\mathcal{H}_1, \mathcal{H}_2$, equipped with an inner product $\langle (x_1, y_1), (x_2, y_2) \rangle := \langle x_1, y_1 \rangle_{H_1} + \langle x_2, y_2 \rangle_{H_2}$. Prove that:

(a) $\mathcal{H}$ is Hilbert space;

(b) the dual of $\mathcal{H}$ is isometrically isomorphic to $\mathcal{H}_1^* \oplus \mathcal{H}_2^*$.

12.3.39. Let $X$ be a Banach space and suppose that $X^*$ is isometrically isomorphic to $X$. Does it follow that $X$ is isometrically isomorphic to Hilbert space?

12.3.40. Let $X$ be an inner product space and suppose that every closed subspace $Y$ of $X$ is complemented, i.e., for every $x \in X$ there exists $y \in Y$ and $z \perp Y$ such that $x = y + z$. Prove that $X$ is complete.

12.3.41. Let $X$ be a separable inner product space with the property that every maximal orthonormal sequence in $X$ is an orthonormal basis of $X$. Prove that $X$ is complete.

12.3.42. Let $\mathcal{K}$ be a closed subspace of Hilbert space $\mathcal{H}$, and let $P$ denote the orthogonal projection on $\mathcal{K}$, i.e., the operator that maps each $x \in \mathcal{H}$ to its orthogonal projection on $\mathcal{K}$. Let $g \in \mathcal{H}$ and let $T$ denote the linear functional, $Tx := \langle g, x \rangle$. Find the vector $h \in \mathcal{H}$ for which $(T \circ P)x = \langle h, x \rangle$.

12.3.43. Let $\mathcal{H}$ be a separable Hilbert space, let $\{e_k\}$ be an orthonormal basis of $\mathcal{H}$, and let $S$ denote the shift operator, i.e., the linear operator that maps each $e_k$ to $e_{k+1}$. Let $g \in \mathcal{H}$ and let $T$ denote the linear functional, $Tx := \langle g, x \rangle$. Find the vector $h \in \mathcal{H}$ for which $(T \circ S)x = \langle h, x \rangle$.

12.3.44. Let $T$ be a bounded linear functional on a subspace $\mathcal{M}$ of Hilbert space $\mathcal{H}$. Prove that $T$ has a unique norm-preserving extension to a bounded linear functional on $\mathcal{H}$, and that this extension vanishes on $\mathcal{M}^\perp$.

12.3.45. Let $T$ be a bounded linear functional on Hilbert space $\mathcal{H}$, let $\mathcal{M}$ be a closed subspace of $\mathcal{H}$, and let $\varphi : \mathcal{M} \to \mathbb{R}$ be defined by $\varphi(x) = \|x\|^2 - Tx$. Prove that $\varphi$ is bounded below and that it attains its infimum at a unique point of $\mathcal{M}$.

12.3.46. Let $\mathcal{H}$ be Hilbert space, let $\{T_n\} \subset \mathcal{L}(\mathcal{H})$, and suppose that there is a set $A \subset \mathcal{H}$ of the second category that is dense in $\mathcal{H}$ and such that $\lim_n \langle T_n x, y \rangle = 0$ for each $x \in \mathcal{H}$ and $y \in A$. Prove that $\lim_n \langle T_n x, y \rangle = 0$ for all $x, y \in \mathcal{H}$.

12.3.47. Let $f$ be a function in $L^1(\mathbb{R})$ that does not belong to $L^2(\mathbb{R})$. Prove that there is an orthonormal basis $\{\varphi_n\}$ of $L^2(\mathbb{R})$ consisting of continuous functions such that $\int_{\mathbb{R}} f \varphi_n = 0$ for all $n \in \mathbb{N}$.

12.3.48. Let $\mathcal{H}$ be an inner product space and suppose that $\mathcal{M} = \mathcal{M}^{\perp\perp}$ for every closed subspace $\mathcal{M}$ of $\mathcal{H}$. Prove that $\mathcal{H}$ is complete.

12.3.49. Let $S$ be a subspace of $C[0,1]$ and suppose that there exists a constant $K \geq 0$ such that $\|f\|_{max} \leq K\|f\|_2$ for all $f \in S$. Prove that $S$ is finite-dimensional.

12.3.50. Let $T$ be a bounded linear functional on $\mathbb{R}^n$. Prove that $T$ has a unique norm-preserving extension to $\mathbb{R}^{n+1}$.

12.3.51. Let $X$ be a closed subspace of $L^2(0,1)$ which is contained in $C[0,1]$. Prove that:

(a) There exists $A > 0$ such that $\|f\|_{max} \leq A\|f\|_2$ for all $f \in X$.

(b) For each $x \in [0,1]$ there exists $h_x \in X$ such that $f(x) = \langle f, h_x \rangle$ for all $f \in X$, and $\|h_x\|_2 \leq A$.

(c) The dimension of $X$ is at most $A^2$.

## 12.4 The Dual Space of $L^\infty(X, \mu)$

In view of Riesz Representation Theorem, one might expect the dual space of $L^\infty$ to be $L^1$. Remark 12.3.6 points out that this is not true in general. In this section we will describe the dual space of $L^\infty(X, \mu)$ and shed more light on the failure of $L^1(X, \mu)$ to play this role.

We start with an example.

**Example 12.4.1. A bounded linear functional on $L^\infty(X, \mu)$.**

Let $f \in L^1(X, \mu)$. Define $T_f : L^\infty(X, \mu) \to \mathbb{R}$ by $T_f(g) = \int_X fg\, d\mu$. Then, $T_f$ is linear and

$$|T_f(g)| = \left|\int_X fg\, d\mu\right| \leq \int_X |fg|\, d\mu \leq \int_X \|g\|_\infty |f|\, d\mu = \|g\|_\infty \|f\|_1, \tag{12.24}$$

so $T_f \in L^\infty(X, \mu)^*$ and

$$\|T_f\| \leq \|f\|_1. \tag{12.25}$$

♦

Unfortunately, the form $T_f$, with $f \in L^1$, is not sufficient to describe *all* bounded linear functionals on $L^\infty$. There are deep reasons for that (see Remark 12.3.6), but we present a simple counterexample here.

**Example 12.4.2. A bounded linear functional on $L^\infty$ that is not given by an $L^1$ function.**

Consider the space $L^\infty(0,1)$ with respect to Lebesgue measure. There exists $F \in (L^\infty(0,1))^*$ that is not of the form $T_f$ for any $f \in L^1(0,1)$.

**Proof:** Notice that $C[0,1]$ is a closed subspace of $L^\infty(0,1)$. By Hahn–Banach theorem, there exists a nonzero functional $F$ on $L^\infty(0,1)$ that vanishes on $C[0,1]$. Suppose to the contrary that there is a function $f \in L^1(0,1)$ such that

$$F(g) = \int_0^1 fg, \quad \text{for all } g \in L^\infty(0,1).$$

In particular, we would have that

$$\int_0^1 fg = 0, \quad \text{for all } g \in C[0,1].$$

By Exercise 12.4.1, $f = 0$ a.e., so $F$ would be the zero functional. ♦

In order to present the missing linear functionals, we will sketch the theory of integration with respect to an additive signed measure and direct the reader for more information to [9]. We make a point that we will not need measures to be $\sigma$-additive but merely additive, meaning that if $A, B$ are disjoint sets in $\mathfrak{M}$ then $\nu(A \cup B) = \nu(A) + \nu(B)$. Furthermore, in what follows $\mathfrak{M}$ will be an algebra of subsets of $X$ (and not necessarily a $\sigma$-algebra). We start with a definition.

**Definition 12.4.3.** Let $(X, \mathfrak{M})$ be a measurable space and $\nu : \mathfrak{M} \to \mathbb{R}$ an additive signed measure. For $E \in \mathfrak{M}$, the **total variation** of $\nu$ over $E$, denoted by $|\nu|(E)$ is defined by

$$|\nu|(E) = \sup\{\sum_{k=1}^{n} |\nu(E_k)| : \{E_k\} \text{ disjoint sets in } \mathfrak{M},\ \cup_{k=1}^{n} E_k \subset E\}. \quad (12.26)$$

When $|\nu|(X) < \infty$ we say that $\nu$ is a **bounded additive signed measure** or shorter a **bounded additive measure.**

Before we proceed we present an example.

**Example 12.4.4. An additive measure that is not $\sigma$-additive.**

Let $\mathcal{A} = \{E \subset \mathbb{N} :$ is a finite set or $E^C$ is a finite set$\}$. (See Example 5.2.10.) For $E \in \mathcal{A}$ let $\nu(E) = 0$ if $E$ is a finite set and $\nu(E) = 1$ if $E^C$ is a finite set. Then $\nu$ is a finitely additive measure. It is not $\sigma$-additive because $\mathbb{N}$ is a countable union of singletons, each of measure 0, but $\nu(\mathbb{N}) = 1$. ♦

Several important properties of additive signed measures can be found in Exercise 12.4.3. One of them is an analogue of the Jordan Decomposition Theorem, namely that $\nu$ is a difference of two bounded additive nonnegative measures, $\nu = \nu^+ - \nu^-$. This allows us to define the integral of a measurable function $f$ by

$$\int_X f\,d\nu = \int_X f\,d\nu^+ - \int_X f\,d\nu^-. \quad (12.27)$$

For the integral to be defined, we require that $f$ be integrable with respect to both $\nu^+$ and $\nu^-$, or equivalently with respect to $|\nu|$. Thus, when considering the integral with respect to a bounded additive measure, we may assume that the measure is nonnegative. Just like in Chapter 7, we write $f = f^+ - f^-$ and define

$$\int_X f\,d\nu = \int_X f^+\,d\nu - \int_X f^-\,d\nu.$$

Therefore, in addition to the measure we may assume that the integrand is nonnegative as well. In the case when the latter is a simple function, countable additivity is not relevant and the definition of the integral remains as usual:

$$s = \sum_{k=1}^{n} c_k \chi_{E_k} \quad \Rightarrow \quad \int_X s\,d\nu = \sum_{k=1}^{n} c_k \nu(E_k).$$

When $f$ is a nonnegative measurable function, we define

$$\int_X f\,d\nu = \sup\left\{\int_X s\,d\nu : 0 \le s \le f,\ s \in \mathcal{S}\right\}. \tag{12.28}$$

We leave to the reader to prove that properties of the integral summarized in Proposition 7.2.11 remain true when $\mu$ is finitely additive (Exercise 12.4.5). However, the proof of Theorem 7.2.14 (linearity of integration) appeals to the Monotone Convergence Theorem which we do not have here. Nevertheless, the linearity is still true (Exercise 12.4.7).

Having introduced the integral with respect to a signed additive measure $\nu$, we can now look at functionals of the form

$$f \mapsto \int_X f\,d\nu, \tag{12.29}$$

defined on $L^\infty(X,\mu)$. Notice that we are dealing with two measures $\mu$ and $\nu$ and our next step is to clarify the relationship between them. A function $f \in L^\infty(X,\mu)$ is defined only a.e. $[\mu]$ and we identify two functions $f$ and $g$ if $f = g$ a.e. $[\mu]$. Obviously, an integral with respect to $\nu$ should yield the same result, i.e., $\int_X f\,d\nu = \int_X g\,d\nu$. In particular, if $E \in \mathfrak{M}$ and $\mu(E) = 0$ then $\chi_E = 0$ a.e. $[\mu]$ and we must have $\int_X \chi_E\,d\nu = 0$, hence $\nu(E) = 0$. The implication $\mu(E) = 0 \Rightarrow \nu(E) = 0$ thus forces $\nu$ to be absolutely continuous with respect to $\mu$. To summarize, we are interested in functionals as in (12.29) where $\nu$ is a bounded, additive, signed measure that is absolutely continuous with respect to $\mu$. We denote the class of such measures by $ba(X,\mathfrak{M},\mu)$.

**Example 12.4.5. A bounded linear functional on $L^\infty(X,\mu)$.**

Let $\nu \in ba(X,\mathfrak{M},\mu)$. Define $T_\nu : L^\infty(X,\mu) \to \mathbb{R}$ by $T_\nu(g) = \int_X g\,d\nu$. By Exercise 12.4.7 and remarks above, $T_\nu$ is linear. Moreover, using the estimate from Exercise 12.4.8, for every $g \in L^\infty(X,\mu)$,

$$|T_\nu(g)| = \left|\int_X g\,d\nu\right| \le \int_X |g|\,d|\nu| \le \int_X \|g\|_\infty\,d|\nu| = \|g\|_\infty |\nu|(X),$$

so $T_\nu \in L^\infty(X,\mu)^*$ and

$$\|T_\nu\| \le |\nu|(X).$$ ♦

It is easy to verify that $ba(X,\mathfrak{M},\mu)$ is a vector space. Moreover, if we define

$$\|\nu\| := |\nu|(X) \tag{12.30}$$

then $\|\nu\|$ is a norm on $ba(X, \mathfrak{M}, \mu)$. Exercise 12.4.9 delegates to the reader the responsibility of proving that not only (12.30) defines a norm on $ba(X, \mathfrak{M}, \mu)$ but that it makes it a Banach space.

Now we can state and prove the duality theorem.

**Theorem 12.4.6.** *Let $(X, \mathfrak{M}, \mu)$ be a measure space. For $\nu \in ba(X, \mathfrak{M}, \mu)$ define $T_\nu \in (L^\infty(X, \mu))^*$ by*

$$T_\nu(f) = \int_X f\, d\nu, \qquad \text{for } f \in L^\infty(X, \mu).$$

*Then the mapping $\nu \mapsto T_\nu$ is an isometric isomorphism of $ba(X, \mathfrak{M}, \mu)$ onto $(L^\infty(X, \mu))^*$.*

*Proof.* Example 12.4.5 shows that $T_\nu$ is indeed a bounded linear functional on $L^\infty(X, \mu)$ and $\|T_\nu\| \leq \|\nu\|$.

Next, we will show that the opposite inequality is true. To that end, let $\{E_k\}_{k=1}^n$ be disjoint sets in $\mathfrak{M}$, let $c_k = \operatorname{sgn} \nu(E_k)$, and let $s = \sum_{k=1}^n c_k \chi_{E_k}$. Then $\|s\|_\infty = 1$, so

$$\sum_{k=1}^n |\nu(E_k)| = \sum_{k=1}^n c_k \nu(E_k) = \int_X s\, d\nu = T_\nu(s) \leq \|T_\nu\| \|s\|_\infty = \|T_\nu\|.$$

Taking the supremum over all such collections $\{E_k\}$ we obtain that $\|\nu\| \leq \|T_\nu\|$. Consequently, the mapping $\nu \mapsto T_\nu$ is an isometry, hence injective, and it remains to prove that it is surjective.

Given a bounded linear functional $\Lambda$ on $L^\infty(X, \mu)$, we define a function $\nu$ on $\mathfrak{M}$ by

$$\nu(E) = \Lambda(\chi_E), \qquad \text{for all } E \in \mathfrak{M}. \tag{12.31}$$

We leave the verification that $\nu \in ba(X, \mathfrak{M}, \mu)$ as Exercise 12.4.17. If $s$ is a simple function, $s = \sum_{k=1}^n c_k \chi_{E_k}$, then

$$\Lambda(s) = \Lambda\left(\sum_{k=1}^n c_k \chi_{E_k}\right) = \sum_{k=1}^n c_k \Lambda(\chi_{E_k}) = \sum_{k=1}^n c_k \nu(E_k) = \int_X s\, d\nu.$$

If $f$ is a nonnegative function, by Exercise 12.4.19 there exists a sequence $\{s_n\}$ of simple functions such that $0 \leq s_n \leq f$, $\lim s_n = f$ a.e. $[\mu]$, and $\lim_n \int_X s_n\, d\nu = \int_X f\, d\nu$. Then

$$\int_X f\, d\nu = \lim_n \int_X s_n\, d\nu = \lim_n \Lambda(s_n) = \Lambda(\lim_n s_n) = \Lambda(f).$$

Finally, if $f = f^+ - f^-$ then, by definition

$$\int_X f\, d\nu = \int_X f^+\, d\nu - \int_X f^-\, d\nu = \Lambda(f^+) - \Lambda(f^-) = \Lambda(f).$$

Thus, $\Lambda = T_\nu$, and the theorem is proved. □

We encourage the reader to understand why $T_\nu$ does not necessarily define additional linear functionals on $L^p$, for $p \neq \infty$. (See Exercise 12.4.18.)

Theorem 12.4.6 was proved independently by an American mathematician Theophil Henry Hildebrandt (1888–1980) [62] and Soviet mathematicians Grigorii Mikhailovich Fichtenholz (1888–1959) and Kantorovich [33], both papers appearing in 1934. Hildebrandt was a professor at the University of Michigan, serving as chair of the mathematics department from 1934 until his retirement in 1957. He served two years, 1945 and 1946, as president of the American Mathematical Society. He earned a degree in music with a major in organ and he played the organ in his local church. Fichtenholz was one of the founders of the Leningrad School of Real Analysis. He authored a three-volume textbook titled "A Course of Differential and Integral Calculus" which is in many countries considered to be a classic in the mathematical literature.

LEONID KANTOROVICH (1912–1986) was a Soviet mathematician and economist who can be regarded as the founder of linear programming. He was a child prodigy, enrolling at Leningrad State University at the age of 14. By the time he was 22, he was a full professor. He started developing a research agenda in applied mathematics. In 1935, he introduced KK-spaces, providing the natural framework for developing the theory of linear inequalities. He made substantial contributions to economic theory starting in 1938, particularly in linear programming. Since much of the material was classified it is hard to assess the priority between him and the American George Dantzig (1914–2005). Kantorovich's 1942 publication, *The Best Use of Economic Resources* laid a foundation for the economics of technological innovation, demonstrating his dual mastery of mathematics and economics. His contributions to both fields were so rich that many contemporaries assumed it to be the work of two separate individuals—Kantorovich the mathematician and Kantorovich the economist. He is famous for integrating functional analysis into numerical mathematics. He also introduced large-block organization concepts that influenced computer system architecture. He wrote over 300 papers and books in mathematics, economics, and computer science. His numerous honors include the shared Nobel Prize in Economics in 1975, the Stalin State Prize and the Lenin Prize from the Soviet Government, two of the highest honors for a scientist in the USSR. He died from cancer in 1986.

## Exercises

12.4.1. Let $f \in L^1(0,1)$ and suppose that $\int_0^1 fg = 0$ for all $g \in C[0,1]$. Prove that $f = 0$ a.e.

12.4.2. Prove that the mapping defined by $Tf = f(0)$ is a bounded linear functional on $C[0,1]$. Show that its extension to $L^\infty(0,1)$ is not of the form $\int_0^1 g\,dm$ for any $g \in L^1(0,1)$.

12.4.3. Let $(X, \mathfrak{M})$ be a measurable space and $\nu : \mathfrak{M} \to \mathbb{R}$ an additive signed measure. For $A \in \mathfrak{M}$ define $\nu^+(A) = \sup\{\nu(E) : E \subset A,\, E \in \mathfrak{M}\}$ and $\nu^-(A) = \sup\{-\nu(E) : E \subset A,\, E \in \mathfrak{M}\}$. Prove that:

(a) both $\nu^+$ and $\nu^-$ are nonnegative additive measures;

(b) if there is $M > 0$ for which $\nu(A) \le M$, for all $A \in \mathfrak{M}$, then $\nu^+$ is a finite additive measure; similarly, if there is a constant $m$ for which $\nu(A) \ge m$, for all $A \in \mathfrak{M}$, then $\nu^-$ is a finite additive measure;

(c) if $\nu$ is bounded then $\nu(A) = \nu^+(A) - \nu^-(A)$, for all $A \in \mathfrak{M}$;

(d) if $\nu(A)$ is finite then $|\nu|(A) = \sup\{\nu(E_1) - \nu(E_2)\}$, where the supremum is taken over all $E_1, E_2 \subset A$ that belong to $\mathfrak{M}$;

(e) $|\nu|(A) = \nu^+(A) + \nu^-(A)$.

12.4.4. Let $(X, \mathfrak{M})$ be a measurable space, let $\nu : \mathfrak{M} \to \mathbb{R}$ be a bounded additive signed measure, and let $f$ be a measurable function on $X$. Prove that $f$ is integrable with respect to $|\nu|$ if and only if it is integrable with respect to both $\nu^+$ and $\nu^-$.

12.4.5. Let $(X, \mathfrak{M})$ be a measurable space, let $\mu : \mathfrak{M} \to \mathbb{R}$ be a bounded nonnegative additive measure. Prove that Proposition 7.2.11 remains true.

(a) Show that if $a \geq 0$ then $\int af = a \int f$.
(b) Show that $f \leq g$ a.e implies $\int f \leq \int g$.
(c) Show that $\mu(X \setminus X_0) = 0$ implies $\int_X f = \int_{X_0} f$.

12.4.6. Let $(X, \mathfrak{M})$ be a measurable space, let $\mu : \mathfrak{M} \to \mathbb{R}$ be a bounded nonnegative additive measure, let $A, B \in \mathfrak{M}$ be disjoint, and let $f$ be a non-negative measurable function on $X$. Prove that $f$ is integrable over $A \cup B$ if and only if it is integrable over both $A$ and $B$, in which case

$$\int_A f\, d\mu + \int_B f\, d\mu = \int_{A \cup B} f\, d\mu.$$

12.4.7. Let $(X, \mathfrak{M})$ be a measurable space, let $\mu : \mathfrak{M} \to \mathbb{R}$ be a bounded nonnegative additive signed measure. Prove that Theorem 7.2.14 remains true.

12.4.8. Let $(X, \mathfrak{M})$ be a measurable space, let $\nu : \mathfrak{M} \to \mathbb{R}$ be a bounded additive signed measure, and let $f$ be a measurable function on $X$. Prove that

$$\left| \int_E f\, d\nu \right| \leq \int_E |f|\, d|\nu|, \qquad \text{for all } E \in \mathfrak{M}.$$

12.4.9. Prove that the total variation $\|\nu\| := |\nu|(X)$ has all properties of a norm on $ba(X, \mathfrak{M}, \mu)$ which makes it a Banach space.

12.4.10. Let $(X, \mathfrak{M})$ be a measurable space and $\nu : \mathfrak{M} \to \mathbb{R}$ an additive signed measure. Prove that for any nonnegative function $f$ on $X$, $\int_X f\, d\nu^+ + \int_X f\, d\nu^- = \int_X f\, d|\nu|$.

12.4.11. Let $(X, \mathfrak{M})$ be a measurable space, let $\mu$ be a bounded additive measure on $X$, and let $f \in L^1(X, \mu)$. Prove that the expression

$$\nu(E) = \int_E f\, d\mu$$

defines a measure in $ba(X, \mathfrak{M}, \mu)$.

12.4.12. Let $(X, \mathfrak{M})$ be a measurable space and let $\nu : \mathfrak{M} \to \mathbb{R}$ be a bounded additive signed measure on $X$. Prove that:

(a) $\nu$ is $\sigma$-additive if and only if it is continuous from below, i.e., if (5.3) holds;

(b) assuming that $\nu$ is finite, $\nu$ is a measure if and only if it is continuous from above, i.e., if (5.4) holds.

12.4.13. Let $(X, \mathfrak{M})$ be a measurable space, let $\nu : \mathfrak{M} \to \mathbb{R}$ be a bounded additive signed measure, and let $\mu_1, \mu_2$ be countably additive bounded measures such that $\mu_1 \leq \nu \leq \mu_2$. Prove that $\nu$ is countably additive.

12.4.14. Let $(X, \mathfrak{M})$ be a measurable space, let $\mu$ and $\nu$ be bounded additive signed measures, and define

$$(\mu \wedge \nu)(E) := \inf\{\mu(T) + \nu(E \setminus T) : T \in \mathfrak{M},\ T \subset E\},$$
$$\mu \vee \nu := -\left((-\mu) \wedge (-\nu)\right).$$

Prove that $\mu \wedge \nu$ and $\mu \vee \nu$ are bounded additive signed measures on $\mathfrak{M}$.

12.4.15. Let $(X, \mathfrak{M})$ be a measurable space and let $\nu$ be bounded additive nonnegative measure. Prove that $\nu$ is $\sigma$-additive if and only if the monotone convergence theorem is true.

12.4.16. Let $(X, \mathfrak{M})$ be a measurable space and let $\nu$ be bounded additive nonnegative measure. Prove that $\nu$ is $\sigma$-additive if and only if the dominated convergence theorem is true.

12.4.17. Let $(X, \mathfrak{M}, \mu)$ be a measure space and $\Lambda$ a bounded linear functional on $L^\infty(X, \mu)$. Prove that (12.31) defines a bounded additive signed measure that is absolutely continuous with respect to $\mu$.

12.4.18. Give an example of a measure space $(X, \mathfrak{M}, \mu)$ and $\nu \in ba(X, \mathfrak{M}\mu)$ such that $T_\nu$ is not a nonzero bounded linear functional on $L^p(X, \mu)$, $1 \leq p < \infty$.

12.4.19. Let $(X, \mathfrak{M}, \mu)$ be a measure space, let $\nu \in ba(X, \mathfrak{M}, \mu)$, and let $f$ be a nonnegative function in $L^\infty(X, \mu)$. Prove that there exists a sequence $\{s_n\}$ of simple functions on $X$ such that $0 \leq s_n \leq f$, $\{s_n\}$ converges to $f$ in the norm of $L^\infty(X, \mu)$, and $\lim_n \int_X s_n \, d\nu = \int_X f \, d\nu$.

12.4.20. Give an example to show that the dual space of $\ell^\infty$ cannot be identified with $\ell^1$.

12.4.21. Let $(X, \mathfrak{M}, \mu)$ be a measure space, let $\{\nu_i\}_{i \in I} \subset ba(X, \mathfrak{M}, \mu)$, and suppose that there exists $\omega \in ba(X, \mathfrak{M}, \mu)$ that is an upper bound, i.e., $\nu_i \leq \omega$, for all $i \in I$. (This means that $\nu_i(E) \leq \omega(E)$ for each $E \in \mathfrak{M}$.) Prove that there is a least upper bound.

12.4.22. The purpose of this exercise is to establish the Vitali–Hahn–Saks theorem. Let $(X, \mathfrak{M}, \mu)$ be a finite measure space, let $\{\nu_n\}_{n=1}^\infty$ be a sequence of signed measures in $ba(X, \mathfrak{M}, \mu)$, and suppose that the limit $\nu(E) := \lim_n \nu_n(E)$ exists (as a finite number) for each $E \in \mathfrak{M}$. Prove that:

(a) the metric space $(\mathfrak{M}, d)$, where $d(A, B) := \mu(A \triangle B)$, is complete;

(b) for each $n \in \mathbb{N}$, $\nu_n : (\mathfrak{M}, d) \to \mathbb{R}$ is well-defined and continuous;

(c) operations $A \cup B$, $A \cap B$, and $A \setminus B$ are continuous maps from $(\mathfrak{M}, d) \times (\mathfrak{M}, d)$ to $(\mathfrak{M}, d)$, and $A^C$ is a continuous map from $(\mathfrak{M}, d)$ to $(\mathfrak{M}, d)$;

(d) for every $\varepsilon > 0$, there exists $N \in \mathbb{N}$, $A \in \mathfrak{M}$, and $r > 0$ such that

$$\mu(A \triangle E) < r \quad \text{and} \quad m, n \geq N \quad \Rightarrow \quad |\nu_m(E) - \nu_n(E)| \leq \varepsilon;$$

(e) there is $\delta > 0$ such that if $\mu(B) < \delta$ then $|\nu_n(B)| < \varepsilon$;

(f) $\nu$ is a signed measure. (This is the main conclusion of the theorem.)

12.4.23. Let $(X, \mathfrak{M}, \mu)$ be a measure space and $f \in L^1(X, \mu)$. Prove that

$$\forall \varepsilon > 0, \quad \exists \delta > 0, \quad \forall E \in \mathfrak{M}, \quad \mu(E) < \delta \quad \Rightarrow \quad \left| \int_E f \, d\mu \right| < \varepsilon.$$

12.4.24. Let $(X, \mathfrak{M})$ be a measurable space and let $\mu : \mathfrak{M} \to \mathbb{R}$ be an additive measure on $X$. Prove that $\mu$ is $\sigma$-additive if and only if it is $\sigma$-subadditive, i.e., if $E \subset \cup_{k=1}^{\infty} E_k$ then $\mu(E) \leq \sum_{k=1}^{\infty} \mu(E_k)$.

12.4.25. Let $X$ be the subspace of $L^\infty(\mathbb{R})$ consisting of the continuous functions that have a finite limit as $x \to +\infty$. For $f \in X$ let $Tf := \lim_{x \to +\infty} f(x)$. Prove that:

(a) $T$ extends to a bounded linear functional on $L^\infty(\mathbb{R})$.

(b) There is no function $g \in L^1(\mathbb{R})$ such that $Tf = \int_{\mathbb{R}} fg$ for all $f \in L^\infty(\mathbb{R})$.

12.4.26. Let $V$ be the subspace of $L^\infty(0,1)$ defined by

$$V = \{f \in L^\infty(0,1) : \lim_n n \int_{[0,1/n]} f(x) \, dx \text{ exists}\}.$$

Let $\varphi \in (L^\infty(0,1))^*$ defined by $\varphi(f) = \lim_n n \int_{[0,1/n]} f(x) \, dx$. (See Exercise 12.2.12.) Prove that there exists no $g \in L^1(0,1)$ such that $\varphi(f) = \int fg$ for all $f \in L^\infty(0,1)$.

12.4.27. Let $\varphi$ be a positive linear functional on $\ell^\infty$, meaning that $\varphi(x) \geq 0$ whenever $x = \{x_k\} \in \ell^\infty$ satisfies $x_k \geq 0$ for all $k \in \mathbb{N}$. Prove that $\varphi$ is bounded and that $\|\varphi\| = \varphi(e)$, where $e$ is the constant sequence $(1, 1, 1, \dots)$.

In Exercises 12.4.28–12.4.30, for each $n \in \mathbb{N}$, let $e_n$ denote the sequence $(0, 0, \dots, 0, 1, 0, 0, \dots)$ with 1 in the $n$th position. Is there a nonzero bounded linear functional $\varphi$ on $\ell^\infty$ such that:

12.4.28. $\varphi(e_n) = 0$ for all $n \in \mathbb{N}$?

12.4.29. $\varphi(x) = \lim_{n \to \infty} x_n$ for all $x = \{x_n\} \in c$?

12.4.30. $\varphi(x) = \sum_{n=1}^{\infty} x_n$ for all $x = \{x_n\} \in \ell^1$?

When $\{f_n\}$ is a sequence of measurable functions and $\nu_n(E) = \int_E f_n \, d\mu$, implication in part (e) of Exercise 12.4.22 is known as uniform integrability of $\{f_n\}$. Vitali proved in 1907 [122] that if a sequence of functions $\{f_n\}$ converges to $f$ a.e. on $[0,1]$ then $\lim_n \int_0^1 f_n = \int_0^1 f$ if and only if the functions $\{f_n\}$ are uniformly integrable. Hahn proved in 1922 [48] that if $\{f_n\}$ is a sequence of measurable functions for which $\lim_n \int_E f_n$ exists for any measurable set $E \subset [0,1]$ then $\{f_n\}$ is uniformly integrable and $\lim_n \int_E f_n$ is absolutely continuous with respect to Lebesgue measure. Saks proved the result above in 1933 [102]. Before this was published Nikodym proved in the same year part (f) of Exercise 12.4.22 [81] under the assumption that each $\nu_n$ is a countably additive.

## 12.5 The Dual Space of $C(X)$

In this section we will describe the dual space of $C(X)$ when $X$ is a compact metric space. To accomplish this goal we will need some properties that a measure on $X$ may have. Let $\mathcal{B}_F$ denote the smallest algebra that contains all closed sets. The reason that we consider an algebra rather than a $\sigma$-algebra is the same as in the previous section, namely we are interested in additive signed measures defined on algebras of sets. If such a measure is defined on $\mathcal{B}_F$ we will refer to it as a **Borel measure**, and to $\mathcal{B}_F$ as the **Borel algebra** on $X$.

**Definition 12.5.1.** Let $X$ be a topological space, let $\mathcal{B}_F$ be the Borel algebra and $\mu$ a positive Borel measure on $X$. Then $\mu$ is **regular** if for all $A \in \mathcal{B}_F$,

(a) $\mu(A) = \inf\{\mu(G) : A \subset G, \text{ and } G \text{ is open}\}$.

(b) $\mu(A) = \sup\{\mu(K) : K \subset A, \text{ and } K \text{ is compact}\}$.

Properties (a) and (b) are often referred to as **outer** and **inner** regularity.

*Remark* 12.5.2. Since we work under the assumption that $X$ is compact, when establishing that a measure is inner regular, it suffices to approximate $A$ from below by *closed* sets.

**Example 12.5.3. Lebesgue measure on $\mathbb{R}^n$ is regular.**
This is the content of Exercise 5.3.9. ♦

**Example 12.5.4. A Borel measure that is not regular.**
Let $X = \mathbb{R}$, $\mathfrak{M} = \mathbb{P}(\mathbb{R})$, and $\mu$ the counting measure. If $A = \{3\}$ then $\mu(A) = 1$, but any open set containing $A$ has infinite measure. ♦

Definition 12.5.1 can be extended to finitely additive, *signed* measures.

**Definition 12.5.5.** Let $X$ be a topological space, let $\mathcal{B}_F$ be the Borel algebra and $\nu$ a Borel, additive, signed measure on $X$. Then $\nu$ is **regular** if its total variation is regular.

The measures with this property will be important in what follows. If $X$ is a topological space and $\mathcal{B}_F$ is the Borel algebra on $X$, we will denote by

$rba(X, \mathcal{B}_F)$ the set of all Borel, bounded, additive, regular signed measures on $X$. This is a vector space (Exercise 12.5.5) that can be equipped with the norm that was introduced in Definition 12.4.3, namely the total variation of $\nu$ over $X$.

There are obvious connections between $rba(X, \mathcal{B}_F)$ and $ba(X, \mathfrak{M}, \mu)$, which was introduced in Section 12.4. The next result presents one that is not as evident.

**Theorem 12.5.6.** *Let $X$ be a compact metric space and let $\nu$ be a finite, additive signed measure on Borel algebra of subsets of $X$. There exists a measure $\lambda \in rba(X, \mathcal{B}_F)$ such that*

$$\int_X f\, d\lambda = \int_X f\, d\nu, \qquad \text{for all } f \in C(X). \tag{12.32}$$

*Proof.* Formula (12.27) defines the integral with respect to an additive measure. In view of remarks made there, we will assume without loss of generality that $\nu$ is a finite, additive measure, i.e., $\nu(E) \geq 0$ for all $E \in \mathcal{B}_F$.

Throughout this proof the letter $F$, with or without subscripts, will denote a closed set. Similarly, the letter $G$ will carry the meaning of an open set. We start by defining

$$\lambda_1(F) := \inf\{\nu(G) : F \subset G\}, \quad \lambda_2(E) := \sup\{\lambda_1(F) : F \subset E\}, \quad \lambda := \lambda_2|\mathcal{B}_F. \tag{12.33}$$

Since $\nu$ is bounded by assumption, then so are $\lambda_1$, $\lambda_2$, and $\lambda$.

Let $\mathfrak{M}$ denote the collection of all sets that are $\lambda$-measurable, i.e.,

$$\mathfrak{M} = \{A \subset X : \lambda(E) = \lambda(E \setminus A) + \lambda(E \cap A), \quad \text{for all } E \subset X\}.$$

**Step I:** $\mathfrak{M}$ contains all closed sets.

Let $F$ be a closed set and let $E \subset X$. We choose sets $F_1, G_1, G$ so that

$$F_1 \subset E, \quad G_1 \supset F \cap F_1, \quad \text{and} \quad G \supset F_1 \setminus G_1.$$

(See Figure 12.1.) Then, the closed set $F_1$ is contained in the open set $G \cup G_1$.

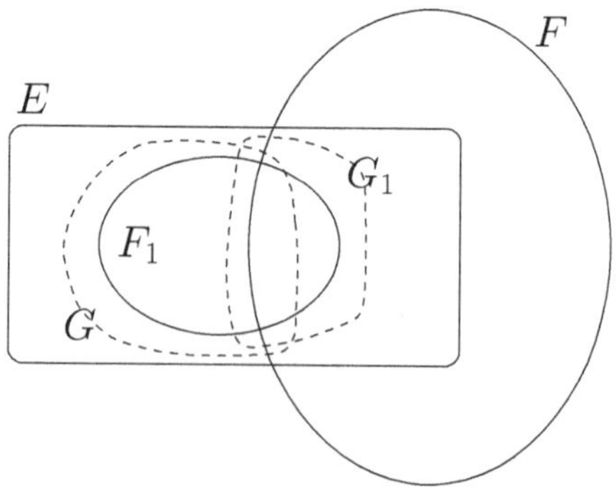


Figure 12.1: $F_1 \subset E$, $G_1 \supset F \cap F_1$, and $G \supset F_1 \setminus G_1$

Indeed, $F_1 = (F_1 \setminus G_1) \cup (F_1 \cap G_1) \subset G \cup G_1$, so

$$\lambda_1(F_1) \le \nu(G \cup G_1) \le \nu(G) + \nu(G_1).$$

Taking the infimum over all open sets $G$ that contain $F_1 \setminus G_1$, we get

$$\lambda_1(F_1) \le \lambda_1(F_1 \setminus G_1) + \nu(G_1).$$

Notice that $F_1 \setminus G_1$ is a closed set contained in $F_1 \setminus F$ (Exercise 1.1.13) so $\lambda_1(F_1 \setminus G_1) \le \lambda_2(F_1 \setminus F_1)$. Further, $G_1$ is an open set that contains $F \cap F_1$. Taking the infimum over all such sets $G_1$, we obtain

$$\lambda_1(F_1) \le \lambda_2(F_1 \setminus F) + \lambda_1(F \cap F_1).$$

Also, $F_1 \subset E$ implies that $\lambda_2(F_1 \setminus F) \le \lambda_2(E \setminus F)$. Taking supremum over all $F_1 \subset E$ yields

$$\lambda_2(E) \le \lambda_2(E \setminus F) + \lambda_2(E \cap F). \tag{12.34}$$

Next, we focus on the opposite inequality. We make a clean slate concerning the relationships between various $F$ and $G$ sets. Once again, let $F$ be a closed set and $E \subset X$. We choose sets $F_1, F_2, G$ so that

$$F_1 \subset E \setminus F, \quad F_2 \subset E \cap F, \quad \text{and} \quad G \supset F_1 \cup F_2.$$

Then $F_1$ and $F_2$ are disjoint sets (Figure 12.2). Since they are also compact,

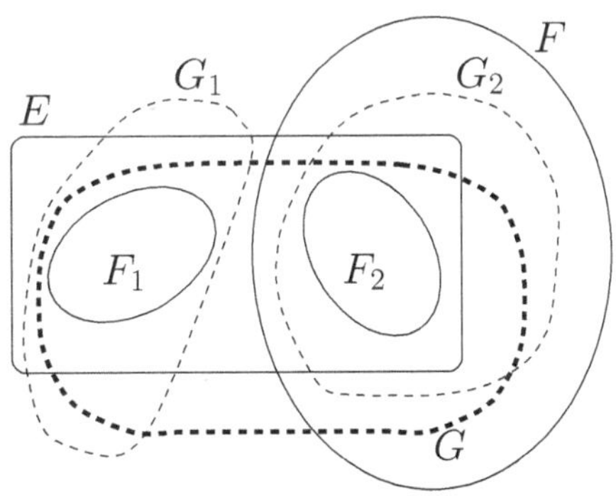


Figure 12.2: $F_1 \subset E \setminus F$, $F_2 \subset E \cap F$, and $G \supset F_1 \cup F_2$

there are open disjoint sets $G_1 \supset F_1$ and $G_2 \supset F_2$ (Exercise 2.3.17). The sets $G \cap G_1$ and $G \cap G_2$ are disjoint, the former contains $F_1$ and the latter $F_2$, and their union is contained in $G$, so

$$\nu(G) \ge \nu(G \cap G_1) + \nu(G \cap G_2).$$

Taking the infimum over all such sets $G_1$, $G_2$, and $G$ (in that order), we get

$$\lambda_1(F_1 \cup F_2) \ge \lambda_1(F_1) + \lambda_1(F_2).$$

Notice that $F_1 \cup F_2 \subset E$, so $\lambda_2(E) \geq \lambda_1(F_1 \cup F_2)$. Also, $F_1 \subset E \setminus F$, and $F_2 \subset E \cap F$, so taking supremum over all such $F_1, F_2$, we obtain

$$\lambda_2(E) \geq \lambda_2(E \cap F) + \lambda_2(E \setminus F).$$

Combining with (12.34) yields

$$\lambda_2(E) = \lambda_2(E \setminus F) + \lambda_2(E \cap F). \tag{12.35}$$

This shows that $\mathfrak{M}$ contains all closed sets.

**Step II:** $\mathfrak{M}$ is an algebra that contains $\mathcal{B}_F$.

We notice that $\mathfrak{M}$ contains $X$, $\emptyset$, and it is closed under taking complements. Therefore, it suffices to prove that it is closed under an intersection. To that end, let $A, B \in \mathfrak{M}$, and let $E \subset X$. Since $A, B \in \mathfrak{M}$,

$$\lambda(E \cap B) = \lambda(E \cap B \cap A) + \lambda(E \cap B \cap A^C) \quad \text{and} \quad \lambda(E) = \lambda(E \cap B) + \lambda(E \cap B^C).$$

From these equalities we conclude that

$$\lambda(E) = \lambda(E \cap B \cap A) + \lambda(E \cap B \cap A^C) + \lambda(E \cap B^C). \tag{12.36}$$

Further, using again the fact that $B \in \mathfrak{M}$,

$$\begin{aligned} \lambda(E \cap (A \cap B)^C) &= \lambda(E \cap (A \cap B)^C \cap B) + \lambda(E \cap (A \cap B)^C \cap B^C) \\ &= \lambda(E \cap B \cap A^C) + \lambda(E \cap B^C), \end{aligned} \tag{12.37}$$

because $(A \cap B)^C \supset B^C$. Combining (12.36) and (12.37), we obtain

$$\lambda(E) = \lambda(E \cap B \cap A) + \lambda(E \cap (A \cap B)^C),$$

whence $A \cap B \in \mathfrak{M}$. Thus, $\mathfrak{M}$ is an algebra and it contains $\mathcal{B}_F$ by Step I.

**Step III:** $\lambda \in rba(X, \mathcal{B}_F)$.

We have already established that $\lambda$ is bounded, and we leave additivity of $\lambda$ on $\mathfrak{M}$ as an exercise (Exercise 12.5.7).

Notice that $\lambda_1$ and $\lambda_2$ agree on closed sets, so the second formula in (12.33) implies the inner regularity of $\lambda$. By Exercise 12.5.1, $\lambda$ is outer regular.

**Step IV:** $\lambda$ satisfies (12.32).

Clearly, it suffices to establish (12.32) for nonnegative functions, and by scaling we may assume that $0 \leq f \leq 1$ on $X$ and that $f$ is not the zero function. Let $\varepsilon > 0$. By (12.28), there exists a simple function $s = \sum_{k=1}^{n} \alpha_k \chi_{E_k}$ for which $s \leq f$ and

$$\int_X f \, d\lambda - \varepsilon \leq \int_X s \, d\lambda.$$

This will be all the more true if, on each set $E_k$ where $s$ is constant, we replace its value by the largest possible, i.e., by $a_k = \inf\{f(x) : x \in E_k\}$. That way,

$$\int_X f \, d\lambda - \varepsilon \leq \sum_{k=1}^{n} a_k \lambda(E_k).$$

Since $\lambda$ is regular, there exist closed sets $F_k \subset E_k$ for which $\lambda(F_k) > \lambda(E_k) - \varepsilon / \sum_{k=1}^{n} a_k$. (The denominator is not zero, otherwise $f$ would be identically equal to 0.) It follows that

$$\int_X f\,d\lambda - 2\varepsilon \leq \sum_{k=1}^{n} a_k \lambda(F_k) \leq \sum_{k=1}^{n} c_k \lambda(F_k),$$

where $c_k = \inf\{f(x) : x \in F_k\}$. Next, we choose disjoint open sets $G_k \supset F_k$ for which

$$b_k := \inf\{f(x) : x \in G_k\} \geq c_k - \frac{\varepsilon}{n\|\nu\|}. \tag{12.38}$$

(See Exercise 12.5.6.) Moreover, if $G$ is an arbitrary open set and $F$ is closed subset, the first formula in (12.33) shows that $\lambda(F) \leq \nu(G)$. (Remember, we have established in Step III that $\lambda_1 = \lambda_2 = \lambda$ on $\mathcal{B}_F$.) It follows that

$$\begin{aligned}\int_X f\,d\lambda &\leq 2\varepsilon + \sum_{k=1}^{n} c_k \lambda(F_k) \leq 2\varepsilon + \sum_{k=1}^{n} c_k \nu(G_k)\\ &\leq 3\varepsilon + \sum_{k=1}^{n} b_k \nu(G_k) \leq 3\varepsilon + \int_X f\,d\nu,\end{aligned}$$

where the last inequality follows from the definition of $b_k$ in (12.38). Since $\varepsilon$ is arbitrary we conclude that $\int_X f\,d\lambda \leq \int_X f\,d\nu$. Replacing $f$ by $1 - f$, and taking advantage of the fact that $\lambda(X) = \nu(X)$ (Exercise 12.5.8) now yields the opposite inequality, whence (12.32) holds and the theorem is proved. $\square$

Next, we introduce a subclass of $rba(X, \mathcal{B}_F)$. It consists of all signed measures in $rba(X, \mathcal{B}_F)$ that are *countably* additive. We denote the space of such measures by $rca(X, \mathcal{B}_F)$.

There is no apriori reason to expect that every measure in $rba(X, \mathcal{B}_F)$ be $\sigma$-additive. Nevertheless, when $X$ is compact this is true.

**Theorem 12.5.7.** *Let $X$ be a compact metric space and let $\nu$ be a finite, additive, regular signed measure on Borel algebra of subsets of $X$. Then $\nu$ is $\sigma$-additive. Consequently, $\nu$ is a signed measure (not just an additive measure).*

*Proof.* First, we will show that $|\nu|$ is $\sigma$-additive on $\mathcal{B}_F$. Let $\varepsilon > 0$ and suppose that $E$ is a disjoint union of sets $\{E_k\}$, $E = \cup_{k=1}^{\infty} E_k$. By definition, $|\nu|$ is regular so there exists a closed set $F \subset E$ and open sets $\{G_k\}$ such that $E_k \subset G_k$ and

$$|\nu|(E \setminus F) < \varepsilon \quad \text{and} \quad |\nu|(G_k \setminus E_k) < \frac{\varepsilon}{2^k}.$$

The set $F$ is closed, hence compact and it is contained in $\cup_{k=1}^{\infty} G_k$, so there exists $m \in \mathbb{N}$ such that

$$F \subset \bigcup_{k=1}^{m} G_k.$$

Further,

$$\sum_{k=1}^{\infty}|\nu|(G_k) = \sum_{k=1}^{\infty}(|\nu|(E_k) + |\nu|(G_k \setminus E_k))$$
$$\leq \sum_{k=1}^{\infty}\left(|\nu|(E_k) + \frac{\varepsilon}{2^k}\right) = \sum_{k=1}^{\infty}|\nu|(E_k) + \varepsilon.$$

It follows that

$$\sum_{k=1}^{\infty}|\nu|(E_k) \geq \sum_{k=1}^{\infty}|\nu|(G_k) - \varepsilon \geq \sum_{k=1}^{m}|\nu|(G_k) - \varepsilon$$
$$\geq |\nu|\left(\bigcup_{k=1}^{m} G_k\right) - \varepsilon \geq |\nu|(F) - \varepsilon > |\nu|(E) - 2\varepsilon,$$

because $|\nu|(E) = |\nu|(F) + |\nu|(E \setminus F) < |\nu|(F) + \varepsilon$. Letting $\varepsilon \to 0$ we obtain that, for any $n \in \mathbb{N}$,

$$\sum_{k=1}^{\infty}|\nu|(E_k) \geq |\nu|(E) \geq |\nu|\left(\bigcup_{k=1}^{n} E_k\right) = \sum_{k=1}^{n}|\nu|(E_k).$$

Letting $n \to \infty$ now forces all these inequalities to be equalities, so $|\nu|$ is countably additive. Notice that

$$\sum_{k=1}^{\infty}|\nu|(E_k) = |\nu|(E) \leq \|\nu\| < \infty,$$

because $\nu$ is finite, so the series converges.

Now we can show that $\nu$ is also countably additive on $\mathcal{B}_F$. With $E$ and $\{E_k\}$ as before, and $n \in \mathbb{N}$,

$$\left|\nu(E) - \sum_{k=1}^{n-1}\nu(E_k)\right| = \left|\nu(E) - \nu\left(\bigcup_{k=1}^{n-1} E_k\right)\right|$$
$$= \left|\nu\left(\bigcup_{k=n}^{\infty} E_k\right)\right| \leq |\nu|\left(\bigcup_{k=n}^{\infty} E_k\right) = \left|\sum_{k=n}^{\infty}|\nu|(E_k)\right| \to 0,$$

as the tail of a convergent series. Consequently, $\nu$ is countably additive on $\mathcal{B}_F$, and the theorem is proved. □

Now we are ready for the representation theorem.

**Theorem 12.5.8.** *Let $X$ be a compact metric space. The map $T : \nu \mapsto T_\nu$ where*

$$T_\nu(f) = \int f\, d\nu,$$

*is an isometric isomorphism from the set $rca(X, \mathcal{B})$ onto the dual space of $C(X)$.*

*Proof.* By Exercise 12.5.9, $T_\nu$ is a bounded linear functional on $C(X)$, for every $\nu \in rca(X, \mathcal{B})$. Thus, $T$ is a well-defined map, and it is easy to see that it is linear. We will show that $T$ is a surjective isometry.

We start with surjectivity. Let $\Lambda$ be a bounded linear functional on $C(X)$. Let $\mu$ be the counting measure on $X$. If $f$ is a continuous function on $X$ then it is bounded (because $X$ is compact) and all the more $f \in L^\infty(X, \mu)$. Moreover, $\|f\|_{max} = \|f\|_\infty$. Thus, $(C(X), \| \|_{max})$ can be viewed as a closed subspace of $(L^\infty(X, \mu), \| \|_\infty)$. By Hahn–Banach Theorem, $\Lambda$ can be extended to a bounded linear functional $F$ on $L^\infty(X, \mu)$ in such a way that $\|\Lambda\| = \|F\|$. As we have seen in Theorem 12.4.6, there exists $\nu \in ba(X, \mathcal{B}_F, \mu)$ for which

$$F(f) = \int_X f\, d\nu, \qquad \text{for } f \in L^\infty(X, \mu), \qquad \text{and} \qquad \|F\| = \|\nu\|.$$

Restricting to $C(X)$ we get

$$\Lambda(f) = \int_X f\, d\nu, \qquad \text{for } f \in C(X), \qquad \text{and} \qquad \|\Lambda\| = \|\nu\|.$$

In other words, $\Lambda = T_\nu$ for some $\nu \in ba(X, \mathcal{B}_F, \mu)$. Notice that our choice of $\mu$ imposes no restrictions on $\nu$ because every measure is absolutely continuous with respect to the counting measure. In view of this, we will write $ba(X, \mathcal{B}_F)$, suppressing the measure $\mu$.

By Theorem 12.5.6, there is $\lambda \in rba(X, \mathcal{B}_F)$ for which $T_\lambda = T_\nu$. In fact, Theorem 12.5.7 implies that $\lambda \in rca(X, \mathcal{B}_F)$. Since $\mathcal{B}_F$ is an algebra, it is a ring so Carathéodory Extension Theorem, applied to $\lambda^+$ and $\lambda^-$ yields an extension of $\lambda$ to the Borel $\sigma$-algebra $\mathcal{B}$. If we denote this extension by $\lambda$ again, we get that $\lambda \in rca(X, \mathcal{B})$. Thus, $\Lambda = T_\lambda$ whence the mapping $T$ is surjective.

It remains to show that $T$ is an isometry. By Exercise 12.5.9, $\|T_\lambda\| \le \|\lambda\|$. In the other direction, it is a consequence of (12.26) that there exists a positive integer $n$ and disjoint Borel sets $\{E_k\}_{k=1}^n$ such that

$$\left| \sum_{k=1}^{n} |\lambda(E_k)| - \|\lambda\| \right| < \varepsilon/3.$$

By definition, the regularity of $\lambda$ implies the regularity of $|\lambda|$, so there are closed sets $\{F_k\}_{k=1}^n$ such that $F_k \subset E_k$ and

$$|\lambda|(E_k \setminus F_k) < \frac{\varepsilon}{3n}.$$

Furthermore, there are open sets $\{G_k\}_{k=1}^n$ such that $G_k \supset F_k$ and

$$|\lambda|(G_k \setminus F_k) < \frac{\varepsilon}{3n}.$$

Finally, by Exercise 2.3.38, there exist continuous functions $\{f_k\}_{k=1}^n$ such that

$$0 \le f_k \le 1, \quad f_k = 1 \text{ on } F_k, \quad \text{and} \quad f_k = 0 \text{ on } G_k^C.$$

Now we define $a_k = \operatorname{sgn} \lambda(E_k)$ and $f = \sum_{k=1}^{n} a_k f_k$. Then,

$$\begin{aligned}
|T_\lambda(f) - \|\lambda\|| &= \left|\int_X f\,d\lambda - \|\lambda\|\right| = \left|\sum_{k=1}^{n} a_k \int_{G_k} f_k\,d\lambda - \|\lambda\|\right| \\
&\le \left|\sum_{k=1}^{n} a_k \int_{G_k} f_k\,d\lambda - \sum_{k=1}^{n} a_k \int_{F_k} f_k\,d\lambda\right| \\
&\quad + \left|\sum_{k=1}^{n} a_k \int_{F_k} f_k\,d\lambda - \sum_{k=1}^{n} a_k\lambda(E_k)\right| + \left|\sum_{k=1}^{n} a_k\lambda(E_k) - \|\lambda\|\right|.
\end{aligned}$$

Notice that $G_k \supset F_k$, $|a_k| = 1$, and $|f_k| \le 1$, so

$$\begin{aligned}
\left|\sum_{k=1}^{n} a_k \int_{G_k} f_k\,d\lambda - \sum_{k=1}^{n} a_k \int_{F_k} f_k\,d\lambda\right| &= \left|\sum_{k=1}^{n} a_k \int_{G_k\setminus F_k} f_k\,d\lambda\right| \\
&\le \sum_{k=1}^{n} |a_k| \int_{G_k\setminus F_k} |f_k|\,d|\lambda| \le \sum_{k=1}^{n} \int_{G_k\setminus F_k} d|\lambda| = \sum_{k=1}^{n} |\lambda|(G_k \setminus F_k)
\end{aligned}$$

Therefore,

$$\begin{aligned}
|T_\lambda(f) - \|\lambda\|| &\le \sum_{k=1}^{n} |\lambda|(G_k \setminus F_k) + \left|\sum_{k=1}^{n} a_k\lambda(F_k) - \sum_{k=1}^{n} a_k\lambda(E_k)\right| \\
&\quad + \left|\sum_{k=1}^{n} |\lambda(E_k)| - \|\lambda\|\right| < \frac{\varepsilon}{3} + \sum_{k=1}^{n} |\lambda|(E_k \setminus F_k) + \frac{\varepsilon}{3} < \varepsilon.
\end{aligned}$$

By Exercise 11.1.3 (c), $\|T_\lambda\| = \|\lambda\|$ and the proof is complete. □

The case $X = [a, b]$ was resolved by Riesz in 1910 [93]. He represented a linear functional as the so-called Lebesgue–Stieltjes integral. Radon generalized it in 1913 [87] to compact sets in Euclidean spaces, representing the functional as an integral with respect to a regular measure. Banach extended it to any compact metric space, in a note that was published only as an appendix in a book of Saks in 1937 [103], who later produced a proof of his own in 1938 [104]. The result has received numerous further generalizations that go well beyond the scope of this text.

OTTON NIKODYM (1887–1974) was a Polish mathematician. After the early loss of both parents he was raised by maternal grandparents of Italian and French origin. He studied mathematics at the University of Lvóv and later began teaching mathematics and physics at a Kraków high school. In 1924, he married Stanisława Nikodym, the first Polish woman to obtain a PhD in mathematics. That same year, under strong pressure from Sierpiński, he agreed to take his doctoral examination at Warsaw University. With the degree in hand, he moved to the Jagiellonian University in Kraków and started to publish in 1925. The Nikodyms spent the academic year 1926/27 at the Sorbonne. From 1930 till 1945 they lived in Warsaw and until the beginning of World War II both lectured at the University. During this period Nikodym published 32 papers and four textbooks. After the war, they settled in Ohio in 1948, where he worked at the private Kenyon College. When he retired in 1965, the Nikodyms moved to Utica, New York. He continued his research, sponsored in part by the Atomic Energy Commission and National Science Foundation. In 1971 he got an electric shock and never regained consciousness before he died in 1974. Nikodym contributed significantly to measure theory, functional analysis, and the foundations of quantum mechanics. He was one of the founders of the Polish Mathematical Society. He authored several books on mathematics and delivered popular radio lectures.

## Exercises

12.5.1. Let $X$ be a compact metric space and let $\mu$ be a finite measure on Borel $\sigma$-algebra. Prove that a set $A$ has property (a) of Definition 12.5.1 if and only if $A^C$ has property (b).

12.5.2. Let $\mu$ be a finite measure on the $\sigma$-algebra of Borel subsets of a metric space. Prove that $\mu$ is regular.

12.5.3. Let $X$ be a compact metric space and $\mu \in rca(X, \mathcal{B})$. Prove that the same is true of $\mu^+$ and $\mu^-$.

12.5.4. Let $F$ be a closed subset of a compact metric space $X$, let $\mu \in rca(X, \mathcal{B})$, and let $\nu(A) = \mu(A \cap F)$ for all Borel sets $A$. Prove that $\nu \in rca(F, \mathcal{B}|F)$.

12.5.5. Prove that $rba(X, \mathcal{B}_F)$, the set of all Borel, bounded, additive, regular, signed measures on $X$, is a vector space.

12.5.6. Let $X$ be a compact metric space, let $f \in C(X)$, and let $\varepsilon > 0$. Suppose that $\{F_k\}_{k=1}^n$ are disjoint closed subsets of $X$. Prove that there exist disjoint open sets $\{G_k\}_{k=1}^n$ such that, for all $1 \leq k \leq n$, $F_k \subset G_k$ and $\inf\{f(x) : x \in G_k\} \geq \inf\{f(x) : x \in F_k\} - \varepsilon$.

12.5.7. In the proof of Theorem 12.5.6 prove that $\lambda$ is additive on $\mathfrak{M}$.

12.5.8. In the proof of Theorem 12.5.6 prove that $\lambda(X) = \nu(X)$.

12.5.9. Let $\nu \in rca(X, \mathcal{B})$. Prove that $T_\nu$, as defined in Theorem 12.5.8, is a bounded linear functional on $C(X)$ and that $\|T_\nu\| \leq \|\nu\|$.

12.5.10. Let $F$ be a closed subset of a compact metric space $X$, let $\mu \in rca(F, \mathcal{B}|F)$, and let $Tf = \int_F f|F \, d\mu$. Prove that:

(a) $T$ is a bounded linear functional on $C(X)$.

(b) If $\nu \in rca(X, \mathcal{B})$ is the measure for which $Tf = \int_X f \, d\nu$, then $\nu(A) = \mu(A \cap F)$ for all Borel set $A$.

In Exercises 12.5.11–12.5.16 find the norm of a functional $T$ on $C[-1, 1]$.

12.5.11. $Tf = \int_0^1 f$.

12.5.12. $Tf = \int_{-1}^1 \operatorname{sgn} x \, f(x) \, dx$.

12.5.13. $Tf = \int_{-1}^1 f - f(0)$.

12.5.14. $Tf = \frac{f(\varepsilon)+f(-\varepsilon)-2f(0)}{\varepsilon^2}$, for a fixed $\varepsilon \in [-1, 1] \setminus \{0\}$.

12.5.15. $Tf = \sum_{n=1}^{\infty} \frac{(-1)^n}{n^2} f\left(\frac{1}{n}\right)$.

12.5.16. $Tf = \int_{-1}^{1} f - \frac{1}{2n+1} \sum_{k=-n}^{n} f\left(\frac{k}{n}\right)$, for a fixed $n \in \mathbb{N}$.

12.5.17.* For $f \in C[0,1]$ let $T_n(f) = \int_0^1 nf(t)t^n\,dt$. Prove that each $T_n$ is a bounded linear functional on $C[0,1]$, that the sequence $\{T_n\}$ converges in $(C[0,1])^*$, and determine its limit.

12.5.18. Let $T$ be a bounded linear functional on $C[0,1]$ defined by $Tf = \int_0^1 tf(t)\,dt$

(a) Find $\|T\|$ and a function $f \in C[0,1]$ for which $Tf = \|T\|$;

(b) Let $Y = \{f \in C[0,1] : f(1) = 0\}$ and $T_1 = T|Y$. Prove that $\|T_1\| = \|T\|$ but that $\|T_1\|$ is not attained on $Y$.

12.5.19. Let $\{a_n\} \subset [0,1]$ and, for $f \in C[0,1]$ let

$$Tf = \sum_{n=1}^{\infty} \frac{f(a_n)}{2^n}.$$

Prove that there is no $g \in L^1([0,1])$ such that $Tf = \int_0^1 fg$ for all $f \in C[0,1]$.

12.5.20. Let $\{c_k\}_{k=1}^n$ be real numbers, let $\{x_k\}_{k=1}^n$ be distinct real numbers in $[0,1]$, and let $T$ be a bounded linear functional on $C[0,1]$ defined by $Tf = \sum_{k=1}^n c_k f(x_k)$. Find $\|T\|$.

In Exercises 12.5.21–12.5.26 find $\nu([1/3, 2/3])$ where $\nu \in rca([0,1])$ is such that $Tf = \int_0^1 f\,d\nu$.

12.5.21. $Tf = f(0)$.

12.5.22. $Tf = \int_0^{1/2} f(x)\,dx - 3\int_{1/2}^1 f(x)\,dx$.

12.5.23. $Tf = \int_0^1 f(x)\lfloor 10x \rfloor\,dx$.

12.5.24. $Tf = \int_0^1 f(x)\varphi(x)\,dx$, where $\varphi$ is the Cantor–Lebesgue function.

12.5.25. $Tf = \sum_{n=1}^{\infty} \frac{1}{n} f\left(\frac{1}{\sqrt{n}}\right)$.

12.5.26. $Tf = \int_0^1 f(x)\,dx \int_{3x-1}^1 e^{t^2}\,dt$.

12.5.27.* Let $x \in [0, 2\pi]$. Prove that the set of continuous functions whose Fourier series diverges at $x$ is dense in $C[0, 2\pi]$.

12.5.28.* Prove that $C[a,b]$ is not reflexive.

12.5.29. Give an example of a compact metric space $X$ for which the space $C(X)$ is reflexive.

12.5.30.** Let $X$ be a subset of $\mathbb{R}$. Prove or disprove: $C(X)$ is reflexive if and only if $X$ is a finite set.

12.5.31. Let $T$ be a positive linear functional on $C(X)$, meaning that $Tf \geq 0$ whenever $f$ is a nonnegative function in $C(X)$. Prove that $T$ is bounded and that $\|T\| = T(1)$, where 1 is the constant function 1.

12.5.32.** Prove that every bounded linear functional on $C(X)$ can be written as a difference between two positive linear functionals (see Exercise 12.5.31).

## 12.6 Weak Convergence

Let $\{x_n\}_{n=1}^{\infty}$ be a sequence in a normed linear space $X$. Until now, the statement $\lim_n x_n = x$ signified convergence in the norm. This type of convergence is sometimes overly restrictive. For example, Exercise 10.6.19 reveals that the closed unit ball in Hilbert space is not compact unless the space is finite-dimensional. (This is true in many Banach spaces but proofs are usually harder.) Therefore, it is of interest to introduce another "weaker" mode of convergence, meaning one less demanding from a sequence to converge. (Fewer open sets!)

**Definition 12.6.1.** Let $X$ be a normed linear space. We say that a sequence $\{x_n\} \subset X$ **converges weakly** to $x$, and we write $x_n \rightharpoonup x$, if

$$\lim_n T(x_n) = T(x)$$

for every $T \in X^*$.

*Remark* 12.6.2. When weak convergence is discussed we will always use the adjective *weak*. As for norm convergence, we will sometimes call it **strong convergence**. The word *convergence* without a qualifier will always mean the latter.

The inequality

$$|T(x_n) - T(x)| \leq \|T\| \|x_n - x\|$$

shows that convergence in the norm implies weak convergence. However, the converse is not generally true.

**Example 12.6.3. A sequence that converges weakly but not in the norm.**

Let $X = \ell^2$ and for each fixed $n \in \mathbb{N}$ let $e_n$ denote a sequence that has 1 in the $n$th position and zeros elsewhere. The sequence $\{e_n\}$ of vectors in $\ell^2$ converges weakly to 0, but it does not converge in the norm.

**Proof:** By Corollary 12.3.7, $X^*$ can be identified with $\ell^2$, and every bounded linear functional $T$ acts as as an $\ell^2$ sequence $\{c_n\}$ on $x = \{x_n\} \in \ell^2$ by

$$T(x) = \sum_{n=1}^{\infty} c_n x_n.$$

In particular, for any $k \in \mathbb{N}$, $T(e_k) = c_k$. Since $\{c_n\} \in \ell^2$ we have that $\lim c_n = 0$, hence $\lim T(e_n) = 0$. By definition, $\{e_n\}$ is a weakly convergent sequence. On the other hand, $\{e_n\}$ is not even a strong Cauchy sequence because, for any $m, n \in \mathbb{N}$, $\|e_n - e_m\|_2 = \sqrt{2}$. ♦

This example shows that in $\ell^2$ weak convergence does not imply strong convergence. In Euclidean spaces it does.

**Theorem 12.6.4.** *Every weakly convergent sequence in $\mathbb{R}^n$ is strongly convergent.*

*Proof.* Let $\{x_k\}_{k=1}^\infty$ be a weakly convergent sequence in $\mathbb{R}^n$ and suppose that $x_k \rightharpoonup x$. For $1 \le j \le n$, we consider a mapping $T_j$ defined on $\mathbb{R}^n$ by

$$T_j((x^{(1)}, x^{(2)}, \dots, x^{(n)})) = x^{(j)},$$

for $x = (x^{(1)}, x^{(2)}, \dots, x^{(n)}) \in \mathbb{R}^n$. It is not hard to verify that each $T_j$ is a bounded linear functional on $\mathbb{R}^n$, so by assumption, $\lim_k T_j(x_k) = T_j(x)$. This means that, for any $1 \le j \le n$,

$$\lim_k x_k^{(j)} = x^{(j)},$$

which is just another way of saying that $\{x_k\}_{k=1}^\infty$ converges to $x$ coordinatewise. It is a well known fact that this implies that $\{x_k\}_{k=1}^\infty$ converges to $x$ strongly. □

Theorem 12.6.4 remains true if $\mathbb{R}^n$ is replaced by any finite-dimensional normed linear space. (See Exercise 12.6.12.) When the space is infinite-dimensional, Example 12.6.3 shows that the statement need not be true. On the other hand, there are infinite-dimensional spaces where the two modes of convergence are equivalent. (See Exercise 12.6.44.)

> Riesz introduced the concept of weak convergence for the spaces $L^p[a, b]$, $1 < p < \infty$, in his 1910 paper [92]. He called it "schwache" (meaning "weak" in German).

In addition to weak convergence, we can now talk about **weak topology**. Namely, it is the smallest topology in which all bounded linear functionals are continuous. (By Proposition 12.1.8, they are already continuous in the bigger norm topology.) We will continue to use the adjective *weak* whenever we talk about the weak topology or any related concepts, such as *weak compactness*, *weak density*, etc. For the norm topology (and analogous properties) we will continue to use either *strong*, or *norm*, or nothing. (E.g., strongly compact, norm compact, and compact will mean the same thing.)

In general, the norm topology is too strong and many closed sets are not compact. We had previously mentioned that this is the case with the closed unit ball in Hilbert space and, by Exercise 10.3.40, the same phenomenon occurs in $L^p[0, 1]$. On the other hand, the following result that is due to Riesz [92], shows that it is sequentially compact in the weak topology. We leave the proof to the reader (Exercise 12.6.13).

**Theorem 12.6.5.** *A closed and bounded set in $L^p([0, 1])$ is weakly sequentially compact, if $1 < p < \infty$.*

*Remark* 12.6.6. Theorem 12.6.5 remains true in any reflexive Banach space but a proof would require tools that we did not develop. For Hilbert space even the converse is true (Exercise 12.6.35).

In a measure space one can, in addition to weak and strong convergence of a sequence of functions, also consider other modes of convergence and ask about their relationship with weak convergence.

**Example 12.6.7. A sequence in $L^p$ that converges in measure but not weakly.**

Let $f_n = n\chi_{[0,1/n]}$, for $n \in \mathbb{N}$. Then $\{f_n\}$ converges in measure on $[0,1]$ but it does not converge weakly in $L^p([0,1])$. We leave the proof to the reader (Exercise 12.6.14). ♦

**Example 12.6.8. A sequence in $L^p$ that converges uniformly but not weakly.**

Let

$$f_n = \frac{1}{n}\chi_{[1,e^n]}, \quad \text{for } n \in \mathbb{N}.$$

Then $\{f_n\}$ converges uniformly on $\mathbb{R}$, but it does not converge weakly in $L^p(\mathbb{R})$ for any $p \in [1,+\infty)$. We leave the proof to the reader (Exercise 12.6.15). ♦

Together with Example 12.6.3, these examples show that weak convergence is different from other modes that we have studied. We will now look at some of its properties.

**Theorem 12.6.9.** *Every weakly convergent sequence is bounded.*

*Proof.* Let $\{x_n\}_{n=1}^{\infty}$ be a weakly convergent sequence in $X$ and let $T \in X^*$. The equality (12.14) implies that $\{J_{x_n}(T)\}$ is a convergent sequence of real numbers, hence bounded. In other words, $\{J_{x_n}\}$ is a sequence of bounded linear functionals on $X^*$ that is pointwise bounded. By the Uniform Boundedness Principle, it is uniformly bounded, i.e., there is $M > 0$ such that

$$\|J_{x_n}\| \le M, \quad \text{for all } n \in \mathbb{N}.$$

Further, $J$ is an isometry, i.e., $\|J_{x_n}\| = \|x_n\|$, so the sequence $\{x_n\}$ is bounded. □

Another useful feature of weak convergence is that it can be verified only on a dense subset.

**Theorem 12.6.10.** *A sequence $\{x_n\}$ in a normed linear space $X$ converges weakly to $x_0$ if and only if both:*

*(a) There exists $M > 0$ such that $\|x_n\| \le M$, for all $n \in \mathbb{N}$.*

*(b) $\lim_n Tx_n = Tx_0$, for every $T$ in a set $D$ that is dense in $X^*$.*

*Proof.* If $x_n \rightharpoonup x_0$ then (a) follows from Theorem 12.6.9 and (b) is trivial.

Suppose that (a) and (b) hold and let $T \in X^*$. By (b), for every $\varepsilon > 0$ there exists a functional $T' \in D$ such that

$$\|T - T'\| < \frac{\varepsilon}{2M}.$$

It follows that

$$\begin{aligned}|Tx_n - Tx_0| &= |T(x_n - x_0)| \\ &\le |T'(x_n - x_0)| + |(T - T')(x_n - x_0)| \\ &\le |T'x_n - T'x_0| + \|T - T'\|\|x_n - x_0\| \\ &\le |T'x_n - T'x_0| + \frac{\varepsilon}{2M} 2M,\end{aligned}$$

so (b) implies that

$$\limsup_n |Tx_n - Tx_0| \le \varepsilon.$$

Since $\varepsilon$ and $T$ are arbitrary we conclude that $x_n \rightharpoonup x_0$. □

Next, we will use all these general results to examine the relation between weak and strong convergence in concrete spaces. The following result follows almost directly from Theorem 12.6.10 (Exercise 12.6.17).

**Theorem 12.6.11.** *Let $(X, \mathfrak{M}, \mu)$ be a measure space. A sequence $\{f_n\}$ in $L^p(X, \mu)$, $1 \le p < \infty$, converges weakly to $f_0$ if and only if both:*

*(a) There exists $M > 0$ such that $\|f_n\|_p \le M$, for all $n \in \mathbb{N}$.*

*(b) $\lim_n \int_E f_n \, d\mu = \int_E f_0 \, d\mu$, for every $E \in \mathfrak{M}$ that has finite measure.*

Since $\ell^p$ spaces are just a special case of $L^p$ spaces, we get an immediate corollary.

**Corollary 12.6.12.** *A sequence $\{x_n\}$ in $\ell^p$, $1 \le p < \infty$, converges weakly to $x_0$ if and only if both:*

*(a) There exists $M > 0$ such that $\|x_n\| \le M$, for all $n \in \mathbb{N}$.*

*(b) $\lim_n x_n^{(j)} = x_0^{(j)}$, for every $j \in \mathbb{N}$.*

Example 12.6.3 shows that in general weak convergence does not imply convergence in norm. Notice that the sequence $\{e_n\}$ used there converges weakly to 0, but $\|e_n\| = 1$, for all $n \in \mathbb{N}$, so the sequence $\{\|e_n\|\}$ does not converge to the norm of the limit function. Once this obstruction is removed, we get strong convergence.

**Theorem 12.6.13.** *Let $(X, \mathfrak{M}, \mu)$ be a measure space. A sequence $\{f_n\}$ in $L^p(X, \mu)$, $p > 1$, converges strongly to $f_0$ if and only if both:*

*(a) $f_n \rightharpoonup f_0$.*

*(b) $\lim_n \|f_n\| = \|f_0\|$.*

*Proof.* If $\{f_n\}$ converges strongly to $f_0$ then (a) is immediate, and (b) follows from the inequality

$$\big|\|f_n\| - \|f_0\|\big| \le \|f_n - f_0\|.$$

The proof of the converse is quite involved and we leave it to the reader as Exercise 12.6.18. □

The property of the space $L^p(X, \mu)$ space expressed in Theorem 12.6.13 is known as **Radon–Riesz property**. In addition to spaces of integrable functions, every Hilbert space has this trait. We leave the proof as Exercise 12.6.40.

**Theorem 12.6.14.** *Let* $\{x_n\}$ *converge weakly to* $x$ *in Hilbert space* $\mathcal{H}$. *Then* $\{x_n\}$ *converges to* $x$ *in the norm if and only if* $\lim \|x_n\| = \|x\|$.

> The Radon–Riesz property was first used by Radon in 1913 [87]. Several authors used it throughout 1920s, one of them being Riesz in 1928 [99].

The problem imposed by the fact that $\lim_n \|e_n\| \neq 0$ can be attacked in a different way. Namely, the vector $(e_1 + e_2 + \cdots + e_n)/n$ has the norm $1/\sqrt{n}$ which goes to 0. In other words, taking the averages seems to do the trick. This is indeed so.

**Theorem 12.6.15.** ***The Banach–Saks Theorem.*** *Let* $\{x_n\}_{n=1}^{\infty}$ *converge weakly to* $x$ *in Hilbert space* $\mathcal{H}$. *Then there is a subsequence* $\{x_{n_k}\}$ *of* $\{x_n\}$ *for which* $\{(x_{n_1} + \cdots + x_{n_k})/k\}$ *converges to* $x$ *in the norm of* $\mathcal{H}$.

*Proof.* Let $y_n = x_n - x$. Then $y_n \rightharpoonup 0$. Further, if we can extract a subsequence $\{y_{n_k}\}$ for which $\{(y_{n_1} + \cdots + y_{n_k})/k\}$ converges to 0, then $\{(x_{n_1} + \cdots + x_{n_k})/k\}$ converges to $x$. Thus, we may assume that $x = 0$.

Being weakly convergent the sequence $\{x_n\}$ is bounded, so there is $M > 0$ such that $\|x_n\|^2 \leq M$ for all $n \in \mathbb{N}$. We will construct the subsequence $\{x_{n_k}\}$ inductively. Let $x_{n_1} = x_1$. Since $\{x_n\}$ converge weakly to 0, $\lim \langle x_{n_1}, x_n \rangle = 0$. It follows that there is $n_2 > n_1$ such that $|\langle x_{n_1}, x_{n_2} \rangle| < 1$. Then

$$\|x_{n_1} + x_{n_2}\|^2 = \|x_{n_1}\|^2 + \|x_{n_2}\|^2 + 2\langle x_{n_1}, x_{n_2} \rangle \leq M + M + 2 < (2 + M)2.$$

Suppose that we have selected $x_{n_1}, x_{n_2}, \ldots, x_{n_k}$ satisfying

$$\|x_{n_1} + x_{n_2} + \cdots + x_{n_k}\|^2 \leq (2 + M)k. \tag{12.39}$$

We choose $n_{k+1} > n_1, \ldots, n_k$ so that $|\langle x_{n_1} + \cdots + x_{n_k}, x_{n_{k+1}} \rangle| < 1$. Then

$$\begin{aligned} \|x_{n_1} + x_{n_2} + \cdots + x_{n_{k+1}}\|^2 &= \|x_{n_1} + x_{n_2} + \cdots + x_{n_k}\|^2 + \|x_{k_{n+1}}\|^2 \\ &\quad + 2\langle x_{n_1} + \cdots + x_{n_k}, x_{n_{k+1}} \rangle \leq (2 + M)k + M + 2 \\ &= (2 + M)(k + 1). \end{aligned}$$

Thus, we have a subsequence $\{x_{n_k}\}$ of $\{x_n\}$ satisfying (12.39). It follows that

$$\left\| \frac{1}{k} (x_{n_1} + \cdots + x_{n_k}) \right\|^2 \leq \frac{2 + M}{k} \to 0,$$

and the theorem is proved. □

Banach and Saks proved in a joint paper in 1930 [6] that the spaces $L^p(0,1)$, $p > 1$, also have the property presented in Theorem 12.6.15. A Japanese–American mathematician Shizuo Kakutani (1911–2004) extended this result in 1938 to a large class of Banach spaces (that includes Hilbert space but not $L^1$). In 1965, a Polish mathematician Wiesław Szlenk (1935–1995) gave proof in the case $p = 1$.

STANISŁAW SAKS (1897–1942) was a Polish mathematician of Jewish origin. He attended high school and the University in Warsaw, where he excelled in mathematics. He served in the Polish Army during World War I and during the insurrections in Upper Silesia 1919–21, when the Polish population rebelled against the German administration. He was awarded the Cross of Valour for his patriotic actions. Saks' academic career flourished with influential mentors like Stefan Mazurkiewicz (1888–1945) and Sierpiński. He published papers on set theory, differentiability, and homomorphism of two-dimensional varieties. He began teaching at Warsaw Technical University and at the University of Warsaw in 1926. His collaborations with mathematicians like Banach and Antoni Zygmund (1900–1992) led to significant contributions, notably in functional analysis (the Banach-Saks property). He published groundbreaking monographs on the integral theory [103] and analytic functions. He spent the academic year 1931–32 in the United States, mostly at Brown University. Despite commendable academic achievements, he faced challenges getting a promotion, largely due to anti-Semitism. During World War II, he aided the Polish Army, worked with Banach in Lwów which was under Soviet control, and continued his mathematical pursuits in the Scottish Café. Tragically, Saks met his untimely demise in 1942. He was arrested by the Gestapo upon returning to Warsaw and perished in prison.

## Exercises

12.6.1. Let $X$ be a normed linear space and suppose that $x_n \rightharpoonup x_0 \in X$. Prove that $\|x_0\| \le \liminf \|x_n\|$.

12.6.2.* A sequence $\{x_n\}$ in a normed linear space $X$ is *weakly Cauchy* if $\Lambda(x_n)$ is a Cauchy sequence for every $\Lambda \in X^*$. The space $X$ is *weakly complete* if every weakly Cauchy sequence in $X$ is weakly convergent. Prove that every reflexive Banach space is weakly complete.

12.6.3. Prove that a subspace of a normed linear space is closed if and only if it is weakly closed.

12.6.4. Let $X, Y$ be Banach spaces and let $T : X \to Y$ be a linear operator. Prove that $T$ is bounded if and only if whenever $\{x_n\} \subset X$ converges weakly to $x$, then $\{Tx_n\}$ converges weakly to $Tx$.

12.6.5. Prove that a weakly convergent sequence can have only one limit.

12.6.6.* Let $X, Y$ be Banach spaces and let $\{T_n\}$ be a bounded sequence in $\mathcal{L}(X, Y)$. Let $\mathcal{M}$ be a set of all $x \in X$ for which every subsequence of $\{T_n x\}$ has a convergent subsequence. Prove that $\mathcal{M}$ is a closed subspace of $X$.

12.6.7. Let $X$ be a normed linear space. Prove that every weakly compact subset of $X$ is norm-bounded.

12.6.8. Let $X$ be a normed linear space. Prove that the weak topology is generated by the sets of the form $U_{a,r,T_1,\ldots,T_n} := \{x \in X : |T_k(x) - T_k(a)| < r,\ 1 \le k \le n,\ T_k \in X^*\}$.

12.6.9. Let $X$ be an infinite dimensional normed linear space. Prove that every weakly open subset of $X$ is unbounded.

12.6.10. Let $X$ be a normed linear space and $T$ a linear functional on $X$. Prove that $T$ is continuous in the weak topology if and only if it is continuous in the norm topology.

12.6.11. Let $\mathcal{H}$ be Hilbert space and $T, T_1, T_2, \ldots$ bounded linear operators on $\mathcal{H}$ such that $\lim_n \langle T_n x, y\rangle = \langle Tx, y\rangle$ for all $x, y \in \mathcal{H}$. Prove that there exists $C > 0$ s.t. $\|T_n\| \leq C$ for all $n \in \mathbb{N}$.

12.6.12. Prove that in a finite-dimensional normed linear space, every weakly convergent sequence is strongly convergent.

12.6.13.* Prove Theorem 12.6.5.

12.6.14. Prove the assertions made in Example 12.6.7.

12.6.15. Prove the assertions made in Example 12.6.8.

12.6.16. Give an example of a measure space and a sequence of functions in $L^p(X, \mu)$ that converges weakly but not in measure.

12.6.17. Prove Theorem 12.6.11.

12.6.18.* The purpose of this exercise is to establish that conditions (a)–(b) of Theorem 12.6.13 implies the strong convergence of $\{f_n\}$ to $f_0$. We will follow the original proof of Riesz.

(a) Show that there is no loss of generality in assuming that $\|f_0\| = 1$.

(b) Define $g_n = f_n |f_0|^{p-1} \operatorname{sgn} f_0$. Prove that

$$\lim_n \int_X g_n \, d\mu = 1 \quad \text{and} \quad \lim_n \int_X |g_n| \, d\mu = 1.$$

(c) Define

$$I_n(t) = \int_X |f_0|^{pt} |f_n|^{p(1-t)}, \quad \text{for } 0 \leq t \leq 1.$$

Show that

$$I_n(1) = 1, \quad \lim_n I_n(0) = 1, \quad \text{and} \quad \lim_n I_n(1 - 1/p) = 1.$$

(d) Prove that the function $\ln I_n(t)$ is convex on $[0, 1]$. Conclude that

$$\lim_n \int_X |f_0|^{p/2} |f_n|^{p/2} \, d\mu = 1.$$

(e) Use part (d) to prove that $\{|f_n|^{p/2}\}$ converges to $|f_0|^{p/2}$ in $L^2(X, \mu)$ and conclude that there is a subsequence of $\{f_n\}$ for which $\{|g_n|\}$ converges to $|f_0|^p$ a.e.

(f) Use part (b) to prove that there is a further subsequence for which $\{g_n\}$ converges to $|f_0|^p$ a.e. Conclude that, for that subsequence, $\{f_n\}$ converges to $f_0$ a.e.

(g) Use part (f) and the assumption (b) of the problem to prove that a subsequence of $\{f_n\}$ converges to $f_0$ in the norm of $L^p(X,\mu)$.

(h) Explain why the conclusion in (g) is sufficient to guarantee that the whole sequence converges strongly to $f_0$.

12.6.19. Prove or disprove: If a sequence $\{f_n\}$ converges weakly in $L^p([0,1])$ then it has a subsequence that converges a.e. in $[0,1]$.

12.6.20. Let $\{f_n\}$ be a sequence in $L^2[0,2\pi]$ defined by $f_n(x) = \sin nx$. Prove that $\{f_n\}$ converges weakly but not strongly.

12.6.21. Let $\{f_n\}$ be a sequence in $L^2[0,1]$ that converges in measure and suppose there is $M > 0$ such that $\|f_n\| \leq M$, for all $n \in \mathbb{N}$. Prove that $\{f_n\}$ is weakly convergent. Does the result hold for $L^p[0,1]$ for $1 < p < \infty$?

12.6.22. Let $\{f_n\}_{n=1}^{\infty}$ be a sequence of nonnegative functions in $L^2(0,1)$, and suppose that $\{f_n\}$ converges to a function $f$ in the norm of $L^2(0,1)$. Prove that $f \geq 0$. Does the statement remain true if $\{f_n\}$ converges *weakly* to $f$?

12.6.23. Give an example of a measure space $(X,\mathfrak{M},\mu)$ and a bounded sequence $\{f_n\}$ in $L^1(X,\mu)$ such that $\{f_n\}$ converges to $f$ a.e. but does not converge weakly to $f$ in $L^1(X,\mu)$.

12.6.24.* Let $\{f_n\}$ be a sequence in $L^p[0,1]$, $1 < p < \infty$. Prove that $\{f_n\}$ converges weakly to $f_0$ if and only if both:

(a) there exists $M > 0$ such that $\|f_n\|_p \leq M$, for all $n \in \mathbb{N}$;

(b) $\lim_n \int_0^x f_n\, dm = \int_0^x f_0\, dm$, for all $x \in [0,1]$.

Give an example to show that the result is false for $p = 1$.

12.6.25. Prove that the closed unit ball of $L^1[0,1]$ is not weakly sequentially compact.

12.6.26. Give an example of a measure space $(X,\mathfrak{M},\mu)$ for which the unit ball of $L^1(X,\mu)$ is weakly compact. Is it possible to take $X \subset \mathbb{R}$, $\mathfrak{M} = \mathcal{M}|X$, and $\mu = m|X$?

12.6.27. Let $(X,\mathfrak{M},\mu)$ be a measure space, let $1 \leq p < \infty$, let $1/p + 1/q = 1$, and suppose that $\{f_n\}$ converges weakly to $f$ in $L^p(X,\mu)$ and $\{g_n\}$ converges to $g$ in the norm of $L^q(X,\mu)$. Prove that

$$\lim_n \int_X f_n g_n\, d\mu = \int_X fg\, d\mu.$$

12.6.28.* Let $h \in C(\mathbb{R})$ be a periodic function with period $T$, let $\int_0^T h = 0$, let $1 \leq p < \infty$, and let $f_n : [a,b] \to \mathbb{R}$ be defined by $f_n(x) = h(nx)$ for $n \in \mathbb{N}$. Prove that $\{f_n\}$ converges weakly to 0 in $L^p([a,b])$.

12.6.29.* Let $1 < p < \infty$, let $h \in L^p(\mathbb{R})$, and let $f_n : \mathbb{R} \to \mathbb{R}$ be defined by $f_n(x) = h(x-n)$ for $n \in \mathbb{N}$. Prove that $\{f_n\}$ converges weakly to 0 in $L^p(\mathbb{R})$. Is this true for $p = 1$?

12.6.30.* Let $1 \leq p < r < \infty$, let $\{f_n\}, f \in L^r[0,1]$. Prove or disprove:

(a) If $\{f_n\}$ converges weakly to $f$ in $L^r[0,1]$ then it converges weakly to $f$ in $L^p[0,1]$.

(b) If $\{f_n\}$ converges weakly to $f$ in $L^p[0,1]$ then it converges weakly to $f$ in $L^r[0,1]$.

12.6.31. Let $(X, \mathfrak{M}, \mu)$ be a measure space, let $1 < p < \infty$, let $\{f_n\}, f \in L^p(X,\mu)$, and suppose that the sequence $\{f_n\}$ is bounded in $L^p(X,\mu)$. Consider the following four properties:

(i) $\lim_n f_n = f$ a.e.

(ii) $\{f_n\}$ converges weakly to $f$ in $L^p(X,\mu)$.

(iii) $\lim_n \|f_n\|_p = \|f\|_p$.

(iv) $\{f_n\}$ converges to $f$ in the norm of $L^p(X,\mu)$.

If $\{f_n\}$ possesses two of these properties, is there a subsequence that possesses all four properties?

12.6.32. Give an example to show that Theorem 12.6.13 is not true in $L^1[0,1]$.

12.6.33. Let $(X, \mathfrak{M}, \mu)$ be a measure space, let $1 < p < \infty$, let $\{f_n\}_{n=1}^{\infty}$ be a bounded sequence in $L^p(X,\mu)$, and suppose that $\{f_n\}_{n=1}^{\infty}$ converges to $f$ a.e. Prove that $\{f_n\}_{n=1}^{\infty}$ converges weakly to $f$ in $L^p(X,\mu)$.

12.6.34. Prove that every orthonormal sequence in Hilbert space converges weakly to 0.

12.6.35. Let $X$ be an inner product space. Prove that $X$ is complete if and only if every bounded sequence has a weakly convergent subsequence.

12.6.36.** Let $\alpha \in \mathbb{R}$, $\alpha \neq 1$, and let $\{x_n\}$ be a sequence in Hilbert space such that

$$\langle x_m, x_n \rangle = \begin{cases} \alpha & \text{if } m \neq n, \\ 1 & \text{if } m = n. \end{cases}$$

Prove that $\{x_n\}$ is weakly convergent.

12.6.37. Let $\{x_n\}$ be an orthogonal sequence in Hilbert space. Prove that the following are equivalent:

(a) $\sum_{n=1}^{\infty} x_n$ converges in the norm.

(b) $\sum_{n=1}^{\infty} x_n$ converges weakly.

(c) $\sum_{n=1}^{\infty} \|x_n\|^2$ converges.

12.6.38. Find the weak closure of the unit sphere in Hilbert space.

12.6.39. Let $\{e_n\}_{n=1}^{\infty}$ be an orthonormal basis of Hilbert space. Prove that $x_n$ converges weakly to $x$ if and only if $\{x_n\}$ is bounded and for each $k$, $\{\langle x_n, e_k\rangle\}$ converges to $\{\langle x, e_k\rangle\}$. Give an example to show the boundedness of $\{x_n\}$ cannot be omitted.

12.6.40. Prove Theorem 12.6.14.

12.6.41. Give an example of a (strongly) closed set that is not weakly closed

12.6.42. Let $\{f_n\}$ be a sequence in $C(X)$. Prove that $\{f_n\}$ converges weakly to some function $f$ if and only if $\{f_n\}$ is norm bounded and it converges pointwise on $X$.

12.6.43. Prove that the closed unit ball of $\ell^p$, $1 < p < \infty$, is weakly sequentially compact.

12.6.44.* The purpose of this exercise is to establish that every weakly convergent sequence in $\ell^1$ is strongly convergent.

(a) Supposing that the sequence $\{x_n\}$ is not strongly convergent, show that we may assume that the sequence $\{x_n\}$ converges weakly to 0 and that $\|x_n\|_1 = 1$.

(b) Denoting $x_n = (x_n^{(1)}, x_n^{(2)}, x_n^{(3)}, \ldots)$, construct inductively a sequence of integers $0 = k_0 < k_1 < k_2 < \ldots$ and a subsequence $\{a_n\} \subset \{x_n\}$ for which

$$\sum_{j=1}^{k_{i-1}} |a_n^{(j)}| < \frac{1}{4} \quad \text{and} \quad \sum_{j=k_{i-1}+1}^{k_i} |a_n^{(j)}| > \frac{3}{4}, \quad \text{for all } n \in \mathbb{N}.$$

(c) Define a mapping $T_y$ on $\ell^1$ by

$$T_y(x) = \sum_{j=1}^{\infty} x^{(j)} y^{(j)},$$

where $y^{(m)} = \operatorname{sgn} x_{n_k}^{(m)}$ for $k_{n-1}+1 \leq m \leq k_n$. Prove that $T_y$ is a bounded linear functional and that $|T_y(a_n)| > 1/2$ for all $n \in \mathbb{N}$.

(d) Conclude that $\{x_n\}$ does not converge weakly to 0.

12.6.45. Let $\{e_n\}$ be a sequence in $c_0$, such that for every $n \in \mathbb{N}$, $e_n$ has 1 in the $n$th spot and 0 elsewhere: $e_n = (0, 0, \ldots, 0, 1, 0, 0, \ldots)$. Prove that the sequence $\{e_n\}$ converges weakly to 0 in $c_0$ and in $\ell^p$, for $1 < p < \infty$, but not in $\ell_1$.

12.6.46. Prove that the closed unit ball of $\ell^1$ is not weakly sequentially compact.

12.6.47.** Prove that there is no Banach space whose dual is isomorphic to $c_0$.

12.6.48.* Let $\{a_k\}$ be a sequence of nonzero numbers and

$$E_a = \{x = (x_1, x_2, \ldots) \in \ell_2 : \sum_{k=1}^{\infty} \frac{(x_k)^2}{a_k^2} \leq 1\}$$

(a) Show that $E_a$ is not compact if $a_k = 1$, for all $k \in \mathbb{N}$.
(b) Show that $E_a$ is compact if

$$\sum_{k=1}^{\infty} a_k^2 < +\infty.$$

# *Epilogue*

In this book, we presented the development of measure theory which was a foundation to build a significant improvement of the concept of the integral, and all this led to the birth of functional analysis and its quick growth into one of the main research areas of mathematical analysis. In our closing remarks we will take a look at how all these topics continued to evolve.

**Measure Theory.** In this monograph we considered the abstract measure to be a nonnegative, $\sigma$-additive function $\mu$, defined on a $\sigma$-algebra $\mathfrak{M}$. We went beyond these constraints when we considered finitely additive measures, or when we allowed $\mu$ to take negative values, i.e., when we discussed signed measures. The next logical step is *complex measures*, namely functions $\mu : \mathfrak{M} \to \mathbb{C}$. We stayed away from such measures in order not to overburden the exposition and to remain faithful to the moniker "Real Analysis". Nevertheless, almost everything that we learned about signed measures remains true when $\mu$ is a complex measure.

In the 1930s, a further generalization appeared when $\mathbb{R}$ or $\mathbb{C}$ was replaced by an arbitrary vector space, $X$. Much progress was achieved in the case when $X$ is a Banach space and as of this writing it represents an important, highly specialized research area under the general heading of the Banach space theory.

**Integration Theory.** In more than one place we talked about the inadequacies of the Riemann integral and we illustrated the fact that the Lebesgue integral represents a significant leap forward. Yet, it is by no means perfect. For example, the function

$$f(x) = \begin{cases} x^2 \sin \frac{1}{x^2}, & \text{if } x \neq 0, \\ 0, & \text{if } x = 0, \end{cases}$$

is differentiable everywhere, including at $x = 0$, but its derivative is not Lebesgue integrable. Therefore, the search for an even "better" integral continued. Already in 1912, Denjoy introduced a generalization of the integral. In 1914, a German mathematician Oskar Perron (1880–1975) came up with a different notion of the integral that was later found to be equivalent to the Denjoy's. The third and considerably simpler equivalent definition was discovered in 1957 by Czech mathematician Jaroslav Kurzweil (1926–2022), and a similar one was independently introduced in 1961 by an English mathematician Ralph Henstock (1923–2007). Both contributed significantly to the "new" integration theory leading to the concept known today as the

Henstock–Kurzweil integral. This integral is superior to both the Lebesgue and the Riemann integrals because the class of integrable functions includes all Riemann and Lebesgue integrable functions and many more. At the same time, the main theorems remain true, such as the Monotone Convergence Theorem and the Dominated Convergence Theorem.

We also mention that it is possible to consider integrals where the integrand is an operator-valued function, i.e., $f : \mathbb{R} \to \mathcal{L}(\mathcal{H})$. For such a function it is possible to talk about integrability and define spaces analogous to the usual function spaces, e.g., $L^p$ spaces.

**Banach Spaces.** We encountered many Banach spaces of functions and sequences, such as $L^p$, $C[a, b]$, $\ell^p$, or $c$. These (and many others) are known as the *classical Banach spaces.* The study of these spaces flourished in the early 20th century, and it inspired the investigation of general Banach spaces and their geometry. For example, the Radon–Riesz property holds in every $L^p$ space (Theorem 12.6.13) but not in every Banach space, so it is interesting to characterize Banach spaces that possess this property. In the absence of such a result, it is useful to formulate a sufficient condition. A hypothesis of this type is that $X$ be *uniformly convex*, meaning that whenever $\|x_n\| = \|y_n\| = 1$, if $\|x_n + y_n\| \to 2$ then $\|x_n - y_n\| \to 0$.

As another illustration, we look at the notion of Schauder basis. A sequence $\{x_n\} \subset X$ is a Schauder basis for a Banach space $X$ if, for every $x \in X$, there is a unique sequence $\{c_n\}$ of scalars so that $x = \sum_{n=0}^{\infty} c_n x_n$. It is not hard to see that the sequence $\{e_n\}$, where each $e_n$ is itself a sequence with 1 in the $n$th spot and zeros elsewhere is a Schauder basis for $\ell^p$. In fact, every classical Banach space has a Schauder basis. In 1932, Banach asked whether this is universally true, i.e., whether every separable Banach space has a Schauder basis. The question remained open for decades until a Swedish mathematician Per Enflo gave a negative answer in 1972. Namely, he constructed a separable Banach space without a Schauder basis.

It would require a separate book to present all unsolved problems in Banach space theory. Actually, there are several such books, and many expository articles are devoted to this topic. Even though answers are lacking, these questions have inspired new techniques and methods and the research in this area of mathematics is thriving.

**Operator Theory.** We defined linear transformations between Banach spaces and established some of their fundamental properties. However, answers to some deeper questions are hard to come by, largely because there are numerous gaps in our understanding of the underlying spaces. (Recall that one of the first tasks in advanced calculus is to get a solid understanding of the real line.) A more rewarding strategy is to focus, for the time being, on linear operators on Hilbert space. This is not to say that this line of investigation is particularly easy, but rather that many questions that go beyond the Hilbert space is at a demoralizing level of difficulty.

Questions about Hilbert space operators fall roughly into 2 groups. In one, the object under scrutiny is a single operator of a special type and the task is to accumulate as much information about it. Some typical questions involve the structure of such an operator. Let us explain what that means. In the finite-dimensional Hilbert space, every linear transformation is given by a matrix, and in a suitable basis this matrix is represented in the Jordan canonical form. Thus, the Hilbert space is a direct sum of subspaces and on each of them, the linear transformation is given by a Jordan block, so understanding a linear transformation frequently comes down to answering a question about Jordan blocks. There is no such canonical form when the Hilbert space is infinite-dimensional, and we don't know even the answer to the yes-no question: does every operator $T$ have an invariant subspace, i.e., is there a closed subspace $\mathcal{M} \subset \mathcal{H}$ such that for every $x \in \mathcal{M}$, $Tx \in \mathcal{M}$? This question is known as the *invariant subspace problem* and the answer is known only for some classes of operators.

In the second group, questions deal with algebras of operators and their features. Some of the questions that arise when describing such algebras are whether they contain specific operators (e.g., projections), or what traits are shared by all operators in the algebra under scrutiny (e.g., do they have a common invariant subspace).

Once again, we refrain from further discussion—our goal is merely to give a glimpse into what is a lively research area.

**Conclusion.** An interested reader has acquired a good foundation to continue the study of mathematical analysis. We recommend the traditional approach: take a one or two-semester course in functional analysis. Typically, the aim of such a course is to serve as a gateway to mathematical research. It should help build a more comprehensive knowledge of Banach spaces, both the classical ones and the general theory, as well as of operators on Hilbert space, both specific types and operator algebras. We encourage the reader to continue along this path and wish them the best of luck on this arduous but rewarding journey.

# Solutions and Answers to Selected Exercises

**1.1.4.** Hint: use the identity $E = (E \cap F) \cup (E \cap F^C)$.
**1.1.6.** By Exercise 1.1.2 (a), $A \setminus (B \cap C) = (A \setminus B) \cup (A \setminus C) = (A \setminus B) \cup ((A \cap B) \setminus C) \cup ((A \cap B^C) \setminus C) \subset (A \setminus B) \cup ((A \cap B) \setminus C) \cup (A \cap B^C) = (A \setminus B) \cup ((A \cap B) \setminus C)$.
**1.1.9.** Hint: use Exercise 1.1.2 parts (b) and (c), and (1.1).
**1.1.13.** Hint: use the equivalence between $E \subset F$ and $F^C \subset E^C$.
**1.1.14.** By Exercise 1.1.2 (c), $(A \triangle B) \setminus B = ((A \setminus B) \cup (B \setminus A)) \setminus B = ((A \setminus B) \setminus B) \cup ((B \setminus A) \setminus B) = A \setminus B$. By Exercise 1.1.2 (b), $B \setminus (A \triangle B) = B \setminus ((A \setminus B) \cup (B \setminus A)) = (B \setminus (A \setminus B)) \cap (B \setminus (B \setminus A)) = B \cap (B \cap A) = B \cap A$. Thus, $(A \triangle B) \triangle B = (A \setminus B) \cup (B \cap A) = A$.
**1.1.18.** $[(A \cap S) \cup (B \cap S^C)]^C = (A \cap S)^C \cap (B \cap S^C)^C = (A^C \cup S^C) \cap (B^C \cup S) = (A^C \cap B^C) \cup (A^C \cap S) \cup (S^C \cap B^C) \cup (S^C \cap S) = (A^C \cap B^C \cap S) \cup (A^C \cap B^C \cap S^C) \cup (A^C \cap S) \cup (S^C \cap B^C) = ([(A^C \cap B^C) \cup A^C] \cap S) \cup ([(A^C \cap B^C) \cup B^C] \cap S^C) = (A^C \cap S) \cup (B^C \cap S^C)$.
**1.1.20.** By definition, $A \triangle B = (A \setminus B) \cup (B \setminus A)$ and $A_n \triangle B_n = (A_n \setminus B_n) \cup (B_n \setminus A_n)$, so it suffices to show $A \setminus B \subset \cup_{n=1}^{\infty}(A_n \setminus B_n)$. To that end, $A \setminus B = (\cup_{n=1}^{\infty} A_n) \setminus (\cup_{k=1}^{\infty} B_k) = (\cup_{n=1}^{\infty} A_n) \cap (\cup_{k=1}^{\infty} B_k)^C = \cup_{n=1}^{\infty}(A_n \cap (\cup_{k=1}^{\infty} B_k)^C)$. For each $n$, $B_n \subset \cup_{k=1}^{\infty} B_k$ so $(\cup_{k=1}^{\infty} B_k)^C \subset B_n^C$, hence $A \setminus B \subset \cup_{n=1}^{\infty}(A_n \cap B_n^C) = \cup_{n=1}^{\infty}(A_n \setminus B_n)$.
**1.1.22.** (a) $(\cap_{n=1}^{\infty} A_n) \cup (\cap_{m=1}^{\infty} B_m) = \cap_{n=1}^{\infty}(A_n \cup (\cap_{m=1}^{\infty} B_m)) = \cap_{n=1}^{\infty} \cap_{m=1}^{\infty} (A_n \cup B_m)$. When $n$ is fixed the sequence $\{A_n \cup B_m\}_{m=1}^{\infty}$ is decreasing so $\cap_{m=1}^{\infty}(A_n \cup B_m) = \cap_{m=n}^{\infty}(A_n \cup B_m)$. Thus, $(\cap_{n=1}^{\infty} A_n) \cup (\cap_{m=1}^{\infty} B_m) = \cap_{n=1}^{\infty} \cap_{m=n}^{\infty} (A_n \cup B_m) = \cap_{m=1}^{\infty} \cap_{n=1}^{m} (A_n \cup B_m)$. When $m$ is fixed the sequence $\{A_n \cup B_m\}_{n=1}^{m}$ is decreasing so $\cap_{n=1}^{m}(A_n \cup B_m) = A_m \cup B_m$. Thus, $(\cap_{n=1}^{\infty} A_n) \cup (\cap_{m=1}^{\infty} B_m) = \cap_{m=1}^{\infty}(A_m \cup B_m)$.
(b) Let $A_{2k} = B_{2k-1} = \emptyset$ and $A_{2k-1} = B_{2k} = X \neq \emptyset$. Then $(\cap_{n=1}^{\infty} A_n) \cup (\cap_{m=1}^{\infty} B_m) = \emptyset$ and $\cap_{m=1}^{\infty}(A_m \cup B_m) = X$.
**1.1.28.** By definition, $\limsup E_n = \cap_{n \in \mathbb{N}} \cup_{i \geq n} E_i$, hence $x \in \limsup E_n$ if and only if for every $n$ there is $i \geq n$ for which $x \in E_i$. This implies that the number of $E_i$ that contain $x$ cannot be finite. On the other hand, if $x \in E_i$ for infinitely many $i$, then surely for every $n$ there is one such $i$ bigger than $n$.
**1.1.30.** $\limsup E_n = \cap_{i=1}^{\infty} \cup_{n=i}^{\infty} (-1/n, 1 - 1/n] = \cap_{i=1}^{\infty} (-1/i, 1) = [0, 1)$, and $\liminf E_n = \cup_{i=1}^{\infty} \cap_{n=i}^{\infty} (-1/n, 1 - 1/n] = \cup_{i=1}^{\infty} [0, 1 - 1/i]) = [0, 1)$.
**1.2.2.** (a) ($\subset$) Let $y \in f(f^{-1}(B))$. There is $x \in f^{-1}(B)$ such that $y = f(x)$. Since $x \in f^{-1}(B)$ we have that $f(x) \in B$. Together with $y = f(x)$ we get $y \in B$. ($\supset$) Let $y \in B$. Since $f$ is surjective, there is $x \in X$ for which $y = f(x)$. Now, $y \in B$ implies $f(x) \in B$ hence $x \in f^{-1}(B)$. Finally, $y = f(x) \in f(f^{-1}(B))$.
(b) ($\supset$) Let $x \in A$. Then $f(x) \in f(A)$ so, by definition, $x \in f^{-1}(f(A))$. ($\subset$) Let $x \in f^{-1}(f(A))$. By definition, $f(x) \in f(A)$. Thus, there is $z \in A$ such that $f(x) = f(z)$, so the injectivity of $f$ implies that $x = z \in A$.
**1.2.6.** (a) $x \in f_0^{-1}(E)$ if and only if $f_0(x) \in E$ if and only if $x \in X_0$ and $f(x) \in E$ if and only if $x \in X_0 \cap f^{-1}(E)$.
(b) Using part (a), $f^{-1}(E) = [f^{-1}(E) \cap X_0] \cup [f^{-1}(E) \cap (X \setminus X_0)] = f_0^{-1}(E) \cup [f^{-1}(E) \cap (X \setminus X_0)]$.
**1.2.7.** (c) It suffices to prove that $f$ is injective if and only if $f(\cap_{i \in I} A_i) \supset \cap_{i \in I} f(A_i)$.
($\Rightarrow$) Let $y \in \cap_{i \in I} f(A_i)$. For all $i \in I$, $y \in f(A_i)$ so for all $i \in I$, there is $x_i \in A_i$ such that $y = f(x_i)$. Since $f$ is injective all $x_i$ must be equal, so $y \in f(A_i)$ for all $i$, hence $y \in f(\cap_{i \in I} A_i)$.
($\Leftarrow$) Suppose $f$ is not injective. There are $u, v \in X$ s.t. $f(u) = f(v)$. Let $A_1 = \{u\}$, $A_2 = \{v\}$, and $A_i = X$ for all other $i \in I$. Then $\cap_{i \in I} f(A_i) = \{f(u)\}$ but $\cap A_i = \emptyset$ so $f(\cap_{i \in I} A_i) = \emptyset$.
**2.1.3.** (a) $d(f, g) = \int_a^b |f(x) - g(x)|\, dx \geq 0$ and $d(f, g) = 0$ if and only if $f(x) = g(x)$ for all $a \leq x \leq b$, [86, Exercise 6.5.17]. (b) $d(f, g) = \int_a^b |f(x) - g(x)|\, dx = \int_a^b |g(x) - f(x)|\, dx = d(g, f)$. (c) $d(f, h) = \int_a^b |f(x) - h(x)|\, dx \leq \int_a^b (|f(x) - g(x)| + |g(x) - h(x)|)\, dx = \int_a^b |f(x) - g(x)|\, dx + \int_a^b |g(x) - h(x)|\, dx = d(f, g) + d(g, h)$.
**2.1.8.** Hint: prove that $f(x) = x/(1 + x)$ is increasing for $x \geq 0$.

**2.1.16.** A ball $B(f,r)$ is a strip bounded by the graphs of $f+r$ and $f-r$ and contains all functions with a graph between these two. Closed balls include functions whose graph has common points with either boundary graph, or Open does not.
**2.1.19.** Let $\mathbf{a}=(a,b)$ satisfy $\|\mathbf{a}\|<1$, i.e., $a^2+b^2<1$. The disk $B(\mathbf{a},r)$ is contained in $G$ if $r<1-\|\mathbf{a}\|$. Indeed, if $\mathbf{x}\in B(\mathbf{a},r)$ then $\|\mathbf{x}-\mathbf{a}\|<r$ so $\|\mathbf{x}\|\le\|\mathbf{a}\|+\|\mathbf{x}-\mathbf{a}\|<\|\mathbf{a}\|+1-\|\mathbf{a}\|=1$. Similarly, if $\|\mathbf{a}\|>1$ the disk $B(\mathbf{a},r)$ is contained in $G$ if $r<\|\mathbf{a}\|-1$.
**2.1.22.** (a) By Theorem 2.1.15, $\operatorname{Int}A$ is an open set. We show that $\operatorname{Int}G=G$ for any open set $G$. The inclusion $\operatorname{Int}G\subset G$ follows from the definition of $\operatorname{Int}G$. In the opposite direction, using again Theorem 2.1.15, $\operatorname{Int}G$ is the largest open set contained in $G$, and $G$ is one such set.

(b) If $x\in\operatorname{Int}A$ then $x$ is an interior point of $A$ so there is $B(x,r)\subset A$, hence $B(x,r)\subset B$, whence $x$ is an interior point of $B$.
**2.1.26.** ($\subset$) If $(x,y)\in\operatorname{Int}(A\times B)$ then there is $r>0$ s.t. $B((x,y),r)\subset A\times B$. In particular, if $|x-z|<r$ then $(z,y)\in A\times B$ hence $z\in A$, whence $B(x,r)\subset A$, i.e., $x\in\operatorname{Int}A$. Similarly, $y\in\operatorname{Int}B$. ($\supset$) If $(x,y)\in\operatorname{Int}A\times\operatorname{Int}B$ then there is $r>0$ s.t. $B(x,r)\subset A$ and $B(y,r)\subset B$. Then $B((x,y),r)\subset A\times B$. Indeed, if $\|(u,v)-(x,y)\|<r$ then $|x-u|<r$ and $|y-v|<r$, so $u\in A$ and $v\in B$. Thus, $(x,y)\in\operatorname{Int}(A\times B)$.
**2.1.28.** Its complement is $G=\mathbb{Q}\setminus[\sqrt{2},\pi]$ and it is an open set in $\mathbb{Q}$. If $x\in G$ and $x>\pi$ there is $\varepsilon>0$ s.t. $x-\pi>\varepsilon$, so $(x-\varepsilon,x+\varepsilon)\cap\mathbb{Q}\subset G$. Similar if $x<\sqrt{2}$.
**2.1.32.** Hint: prove that $\mathbb{Q}$ is an $F_\sigma$ set.
**2.1.33.** Hint: use Theorem 2.1.22.
**2.1.36.** (a) ($\Rightarrow$) (b) Let $x$ be a cluster point of $A$. By definition, every $B(x,r)$ contains at least one point of $A$, so it is not contained in $A^C$, hence $x$ is not an interior point of $A^C$. By Exercise 2.1.21, $x$ is not in $A^C$.

(b) ($\Rightarrow$) (c) By definition, if $F$ contains all of its cluster points then it contains $\overline{F}$. On the other hand, Theorem 2.1.22 implies that $F\subset\overline{F}$.

(c) ($\Rightarrow$) (a) By Theorem 2.1.22, $\overline{A}$ is a closed set, hence so is $A$.
**2.1.37.** ($\subset$) Let $\{F_i : i\in I\}$ be all closed sets that contain $A$ and $F=\cap_{i\in I}F_i$. By Exercise 2.1.27, $F$ is a closed set and it contains $A$ so, by Theorem 2.1.22, it contains $\overline{A}$. ($\supset$) By Theorem 2.1.22, $\overline{A}$ is a closed set and it contains $A$ so there is $i\in I$ s.t. $\overline{F}_i=\overline{A}$, whence $\overline{A}\supset F$.
**2.1.38.** (a) ($\subset$) Let $x\in(\overline{A})^C$. By Theorem 2.1.22, $(\overline{A})$ is closed so $(\overline{A})^C$ is open, hence there is $B(x,r)\subset(\overline{A})^C\subset A^C$, whence $x\in\operatorname{Int}(A^C)$. ($\supset$) Let $x\in\operatorname{Int}(A^C)$. By Theorem 2.1.15, $\operatorname{Int}(A^C)$ is open so there is $B(x,r)\subset\operatorname{Int}(A^C)\subset A^C$. Suppose $x\in\overline{A}$. Then either $x\in A$ or it is a cluster point of $A$. In either case, there is an element of $A$ in $B(x,r)$, a contradiction. Thus, $x\in(\overline{A})^C$.

(b) follows from (a) if we replace $A$ with $A^C$.
**2.1.43.** By definition, $\partial A=\overline{A}\setminus\operatorname{Int}A$. Now, $x\notin\operatorname{Int}A$ if and only if for every $r>0$, $B(x,r)\cap A^C\ne\emptyset$. Also, $x\in\overline{A}$ if and only if $x\in A$ or $x$ is a cluster point of $A$, which is equivalent to $B(x,r)\cap A\ne\emptyset$ for any $r>0$.
**2.1.44.** Hint: use Exercise 2.1.36 and Theorem 2.1.26.
**2.2.3.** Let $y\in G+x_0$. Then $y-x_0$ is in the open set $G$, so there is $r>0$ s.t. $y-x_0\in B(y-x_0,r)\subset G$. Since $\lim x_n=x_0$, there is $n$ s.t. $k\ge n\Rightarrow\|x_k-x_0\|<r$. For such $k$, $\|(y-x_k)-(y-x_0)\|<r$ so $y-x_k\in B(y-x_0,r)\subset G$, whence $y\in x_k+G$. By definition, $y\in\liminf(G+x_n)$ so $G+x_0\subset\liminf(G+x_n)$.
**2.2.7.** $\sum_{i=1}^{n}\sum_{j=1}^{\infty}a_{ij}=\sum_{j=1}^{\infty}\sum_{i=1}^{n}a_{ij}\le\sum_{j=1}^{\infty}\sum_{i=1}^{\infty}a_{ij}$. Now let $n\to\infty$. The opposite inequality can be proved in a similar way.
**2.2.10.** Let $F_1=F\cap(a,b)$ and let $x\in\overline{F_1}$. There is $\{x_n\}\subset F_1$ that converges to $x$. Since $x_n\in F$ and $F$ is closed we have that $x\in F$. Moreover, $x_n\in(a,b)$ implies $x\in[a,b]$ but $x\ne a$ and $x\ne b$ because $x\in F\subset(a,b)\cup(c,d)$. Thus $x\in(a,b)$ which, together with $x\in F$, implies that $x\in F_1$ whence $F_1$ is closed.
**2.2.12.** Hint: for a closed set $F$ consider $\cap_{n=1}^{\infty}\cup_{x\in F}B(x,1/n)$.
**2.2.16.** Hint: consider the Taylor series of a function that is not a polynomial.
**2.2.17.** Hint: consider continuous functions that are 1 on much of $[0,1/2]$ and $-1$ on much of $[1/2,1]$.
**2.2.18.** Let $(M,d)$ be a metric space, let $A\subset M$ be a subspace that is complete, and let $x$ be a cluster point of $A$. By Exercise 2.2.5, there is a sequence $\{x_n\}\subset A$ that converges to $x$. By Exercise 2.2.1, $\{x_n\}$ is a Cauchy sequence in $A$ so it must converge to an element of $A$ (because $A$ is complete). In view of Theorem 2.2.5, there can be only one limit, namely $x$, so $x\in A$. Thus, $A$ contains all of its cluster points so it is closed by Exercise 2.1.36.
**2.2.21.** Let $\lim x_{2n}=x$, $\lim x_{2n+1}=y$, and $\lim x_{3n}=z$. First, $\{x_{6n}\}$ is a subsequence of both $\{x_{2n}\}$ and $\{x_{3n}\}$. By Exercise 2.2.20, it has to converge to both $x$ and $z$, so Theorem 2.2.5 ensures that $x=z$. Considering $\{x_{6n+3}\}$ we obtain that $y=z$.
**2.2.24.** Hint: use Proposition 2.2.17.
**2.2.25.** Cauchy but not rapidly Cauchy. Indeed, $1/n-1/(n+1)=1/[n(n+1)]$ so if $1/[n(n+1)]\le\varepsilon_n^2$ then $\varepsilon_n\ge1/\sqrt{n(n+1)}>1/(n+1)$, so $\sum\varepsilon_n$ cannot converge.

**2.2.30.** Cauchy but not rapidly Cauchy. Indeed, it is easy to see that $\lim f_n(x) = 0$ for all $x \in [0,1]$. Moreover, $|f_n(x) - 0| \leq 1/n$, so the convergence is uniform, i.e., in $C[0,1]$. Consequently, it is a Cauchy sequence. On the other hand, the function $f_{n+1} - f_n$ is decreasing on $[0,1]$, so $\sup\{|f_{n+1}(x) - f_n(x)| : x \in [0,1]\} = 1/n - 1/(n+1) = 1/[n(n+1)]$. If $1/[n(n+1)] \leq \varepsilon_n^2$ then $\varepsilon_n \geq 1/\sqrt{n(n+1)} > 1/(n+1)$, so $\sum \varepsilon_n$ cannot converge.
**2.3.1.** Let $\{x_n\}$ be a Cauchy sequence and let $\varepsilon = 1$. There is $N \in \mathbb{N}$ s.t. $m \geq n \geq N \Rightarrow d(x_m, x_n) < 1$. Let $K = \max\{d(x_N, x_1), d(x_N, x_2), \dots, d(x_N, x_{N-1}), 1\}$. Then for any $i, j \in \mathbb{N}$, $d(x_i, x_j) \leq d(x_i, x_N) + d(x_N, x_j) \leq 2K$, so $\{x_n\}$ is bounded.
**2.3.9.** ($\Rightarrow$) Let $A$ be an infinite subset of $M$. By Example 1.4.1, $A$ has a countable subset $B = \{x_n\}$. Since $M$ is compact, there is a subsequence $\{x_{n_k}\}$ that converges to $x \in M$. Then $x$ is a cluster point of $A$.

($\Leftarrow$) Let $\{x_n\}$ be a sequence in $M$. If there is a constant subsequence $x_{n_k} = \alpha$ for $k = 1, 2, \dots$ then the said subsequence converges to $\alpha \in M$. Otherwise, the set $C = \{x_1, x_2, \dots\}$ is infinite, so it has a cluster point $\beta \in M$. By Exercise 2.2.5, there is a subsequence $\{y_{n_k}\} \subset \{x_n\}$ that converges to $\beta$.
**2.3.10.** Hint: use Theorem 2.3.7, Exercise 2.2.18, and Theorem 2.2.14.
**2.3.11.** (a) Combining Exercise 2.2.18 and Theorem 2.2.14, we see that in the complete metric space $\mathbb{R}$ a subset $A$ is complete if and only if it is closed. Therefore, a set $A \subset \mathbb{Q}$ is complete in $(\mathbb{Q}, d)$ if and only if it is a closed subset of $\mathbb{R}$.

(b) By Theorem 2.3.7, a set $A$ is compact if and only if it is totally bounded and complete. By part (a), a set $A \subset \mathbb{Q}$ is complete if and only if it is closed. In Euclidean spaces, a set is totally bounded if and only if it is bounded (Exercise 2.3.3) so $A$ is compact if and only if it is closed and bounded.
**2.3.14.** Hint: find a sequence $\{c_n\}$ such that $d(c_m, c_n) = 1$ when $m \neq n$.
**2.3.15.** Let $x \in \overline{A}$. Then there is $\{x_n\} \subset A$ that converges to $x$. The set $K = \{x_n\} \cup \{x\}$ is sequentially compact, hence compact. By assumption, $A \cap K$ is compact, hence sequentially compact. Therefore, $\{x_n\}$ must have a convergent subsequence that converges to an element of $A \cap K$, whence $x \in A$.
**2.3.22.** Hint: from a sequence extract a subsequence, then a further subsequence, etc.
**2.3.24.** No! If a sequence is convergent it must be eventually constant. If $\{x_n\}$ is not such, it cannot have a convergent subsequence.
**2.3.30.** Hint: prove the contrapositive.
**2.3.31.** ($\leq$) is obvious. ($\geq$) Let $\alpha = \sup\{|f(x)| : x \in \overline{E}\}$ and $\varepsilon > 0$. There is $x \in \overline{E}$ s.t. $|f(x)| > \alpha - \varepsilon/2$. Since $f$ is continuous at $x$, there is $\delta > 0$ s.t. $d(x,y) < \delta \Rightarrow |f(x) - f(y)| < \varepsilon/2$. Further, there is $y \in E$ s.t. $d(x,y) < \delta$. For this $y$, $|f(y)| \geq |f(x)| - |f(x) - f(y)| \geq |f(x)| - \varepsilon/2 > \alpha - \varepsilon/2 - \varepsilon/2 = \alpha - \varepsilon$. Thus, $\sup\{|f(x)| : x \in E\} \geq \alpha - \varepsilon$ whence $\sup\{|f(x)| : x \in E\} \geq \alpha$.
**2.3.32.** See [86, Theorem 3.6.9].
**2.3.37.** Let $a \in A$ satisfy $d(y,a) < d(y,A) + \varepsilon$. Then $h(x) - h(y) = d(x,A) - d(y,A) < d(x,A) - d(y,a) + \varepsilon < d(x,A) - d(x,a) + d(x,y) + \varepsilon \leq d(x,y) + \varepsilon$. Similarly, $h(y) - h(x) < d(x,y) + \varepsilon$. Thus $|h(x) - h(y)| < d(x,y) + \varepsilon$ so $|h(x) - h(y)| \leq d(x,y)$.
**2.3.41.** Yes, because $|F(f) - F(g)| = |f(a) - g(a)| \leq \sup\{|f(x) - g(x)| : x \in [0,1]\} = d(f,g)$. Therefore, if $\varepsilon > 0$ taking $\delta = \varepsilon$, we have that $d(f,g) < \delta$ implies $F(f) - F(g)| < \varepsilon$.
**2.3.44.** (b) $\Leftrightarrow$ (c) by switching to complements.

(a) $\Rightarrow$ (b) Let $y \in \mathbb{R}$ be fixed, let $A = \{x \in M : f(x) < y\}$, and let $x_0 \in A$. By assumption, there is $\delta > 0$ s.t. $d(x, x_0) < \delta \Rightarrow f(x) < y$, hence $B(x_0, \delta) \subset A$ whence $A$ is open.

(b) $\Rightarrow$ (a) Let $\varepsilon > 0$, let $x_0 \in M$, and let $A = \{x \in M : f(x) < f(x_0) + \varepsilon\}$. Since $A$ is open, there is $B(x_0, \delta) \subset A$. Then, for any $x \in B(x_0, \delta)$, $f(x) < f(x_0) + \varepsilon$, so $f$ is upper semicontinuous at $x_0$.
**2.3.46.** Hint: prove that $\Phi^{-1}(B(z,\delta)) \subset B(\Phi^{-1}(z), \varepsilon)$.
**2.4.2.** By considering $g - f$, and taking advantage of Exercise 2.4.1, we may assume that $f = 0$. Suppose $L := \lim_{x \to a} g(x) < 0$. Let $\varepsilon = -L/2$. There is $\delta > 0$ such that $0 < d(x,a) < \delta$ implies $|g(x) - L| < -L/2$. It follows that $g(x) - L < -L/2$, whence $g(x) < L/2 < 0$, a contradiction.
**2.4.3.** Let $\{f_n\}$ be a sequence in $C[0,1]$ that converges to $g$. This means that $\{f_n\}$ converges uniformly to $g$. Since $|F(f_n) - F(g)| = |f_n(0) - g(0)| \leq \sup\{|f_n(x) - g(x)| : x \in [0,1]\} \to 0$ we get $\lim_n F(f_n) = F(g)$, i.e., $\lim_{f \to g} F(f) = F(g) = g(0) = 0^2 = 0$.
**2.4.5.** Hint: prove that if $d(x,y) < \varepsilon < 1/2$ then $|x - y| < 2\varepsilon$.
**2.4.8.** By definition, $\limsup_{x \to 0} f(x) = \lim_{h \to 0+} \sup\{f(x) : 0 < |x| < h\} = \lim_{h \to 0+} \sup\{a_n : 0 < 1/n < h\} = \lim_{h \to 0+} \sup\{a_n : n > 1/h\}$. Using substitution $t = 1/h$ we obtain that $\lim_{t \to +\infty} \sup\{a_n : n > t\} = \lim_{k \to +\infty} \sup\{a_n : n > k\} = \inf_k \sup_n a_n$ and, by Exercise 2.4.6, this is exactly $\limsup a_n$.
**2.4.9.** ($\Rightarrow$) If the limit exists then $\forall \varepsilon > 0$, $\exists \delta > 0$ s.t. $a - \delta < x < a + \delta \Rightarrow f(a) - \varepsilon < f(x) < f(a) + \varepsilon$, so both limit superiors and both limit inferiors are between $f(a) - \varepsilon$ and $f(a) + \varepsilon$.

($\Leftarrow$) If all 4 are equal to $L$ and $\varepsilon > 0$, there is $\delta > 0$ s.t. $a < x < a + \delta \Rightarrow \sup f(x) - L < \varepsilon$ and $a - \delta < x < a \Rightarrow \sup f(x) - L < \varepsilon$, so $f(x) < L + \varepsilon$. Similarly, we obtain $f(x) > L - \varepsilon$, whence $\lim_{x \to a} f(x) = L$.

**2.4.13.** ($\leq$) For any $\{x_k\} \subset M$, we have $f_k(x_k) \leq \sup\{f_k(x) : x \in M\}$. Summing over $k \in \mathbb{N}$, $\sum_{k=1}^{\infty} f_k(x_k) \leq \sum_{k=1}^{\infty} \sup\{f_k(x) : x \in M\}$ and the inequality persists when we take supremum on the left side.

($\geq$) Let $\varepsilon > 0$. For each $k$ there is $x_k \in M$ s.t. $f_k(x_k) > \sup\{f_k(x) : x \in M\} - \varepsilon/2^k$. Summing, $\sum_{k=1}^{\infty} f_k(x_k) > \sum_{k=1}^{\infty} \sup\{f_k(x) : x \in M\} - \varepsilon$. Taking supremum on the left side yields $\sup\{\sum_{k=1}^{\infty} f_k(x_k) : \{x_k\} \subset M\} > \sum_{k=1}^{\infty} \sup\{f_k(x) : x \in M\} - \varepsilon$, so now let $\varepsilon \to 0$.

**2.4.14.** Inequality 1 (from left to right): taking infimums over $0 < |x - a| < h$, $\inf f(x) \leq f(x)$ and $\inf g(x) \leq g(x)$, for all $x$, so $\inf f(x) + \inf g(x) \leq f(x) + g(x)$, hence $\inf f(x) + \inf g(x) \leq \inf(f(x) + g(x))$. Now take the limit as $h \to 0^+$.

Inequality 2: apply Inequality 1 to $f(x) + g(x)$ and $-g(x)$ and use Exercise 2.4.10 (a).

Inequality 4 is proved just as Inequality 1, and Inequality 3 follows from Inequality 4 applied to $f(x) + g(x)$ and $-f(x)$, with the aid of Exercise 2.4.10 (a).

**2.4.17.** Hint: use Exercise 2.4.14.

**2.4.28.** Hint: try $f(x) = x^{\alpha}$.

**2.4.31.** Answer: $\alpha \geq 1/2$.

**2.4.32.** No! Take $f(x) = n + 1$ and $g(x) = n$. Then $d(f, g) = 1$ and $d(F(f), F(g)) = d(f^2, g^2) = 2n + 1$, so $d(F(f), F(g))/d(f, g) = 2n + 1 \to \infty$, as $n \to \infty$.

**2.5.2.** ($\Rightarrow$) Suppose $\operatorname{Int} A^C \neq \emptyset$. Then there is $B(x, r) \subset A^C$ so $A \cap B(x, r) = \emptyset$, contradicting Proposition 2.5.5. ($\Leftarrow$) Suppose $\overline{A} \neq M$. Then $(\overline{A})^C$ is a nonempty open set so there is $B(x, r) \subset (\overline{A})^C \subset A^C$, hence $\operatorname{Int} A^C \neq \emptyset$ because it contains $x$.

**2.5.5.** Hint: use Proposition 2.5.5.

**2.5.6.** (b) First, $\overline{[\cup \overline{\overline{A_n}}]^C} = \overline{[\cup \overline{A_n}]^C} = \overline{\cap (\overline{A_n})^C} = \cap \overline{(\overline{A_n})^C}$. By assumption, each $A_n$ is dense so Proposition 2.5.9 (c) implies that $\overline{[\cup \overline{\overline{A_n}}]^C} = M$. Using Proposition 2.5.9 (c) again, we get $\cup \overline{A_n}$ is nowhere dense, hence so is $\cup A_n$.

**2.5.7.** By definition, $\partial A = \overline{A} \setminus \operatorname{Int} A$. When $A$ is closed, $\partial A = A \setminus \operatorname{Int} A$. If $\partial A$ is not nowhere dense then there is $B(x, r) \subset \partial A \subset A$, so $x \in \operatorname{Int} A$, a contradiction. When $A$ is open, by Exercise 2.1.38, $\partial A = \partial(A^C)$, and $A^C$ is closed.

**2.5.10.** ($\Rightarrow$) Suppose $G$ is open and $A = \partial G$. By definition, $A = \overline{G} \setminus \operatorname{Int} G = \overline{G} \setminus G = \overline{G} \cap G^C$ so it is closed. The conclusion now comes from Exercise 2.5.7.

($\Leftarrow$) We will show that $A = \partial(A^C)$. Since $A$ is closed, $A^C$ is open so $\partial(A^C) = \overline{A^C} \setminus \operatorname{Int}(A^C) = \overline{A^C} \setminus A^C = \overline{A^C} \cap A$. By Proposition 2.5.9, $A^C$ is dense so $\partial(A^C) = A$.

**2.5.13.** Suppose both $E$ and $E^C$ are countable unions of closed sets. If $F$ is one of these sets then $F^C$ is dense so, by Proposition 2.5.9, $F$ is nowhere dense. It would follow that $M$ is a countable unions of nowhere dense sets, which would contradict Theorem 2.5.10.

Finally, if $\mathbb{Q} \cap [0, 1] = \cap G_k$ then $[0, 1] \setminus (\mathbb{Q} \cap [0, 1]) = [0, 1] \setminus (\cap G_k) = [0, 1] \cap (\cup G_k^C) = \cup([0, 1] \cap G_k^C)$, a countable union of closed sets. Since $Q \cap [0, 1]$ is a countable union of closed sets $\{r_k\}$, by the first part would have a contradiction.

**2.5.20.** Let $a \in \mathbb{R}$. Sets $A^C$ and $B := a - A^C$ both are complements of sets of first category so they must have nonempty intersection (otherwise $A \cup B^C = (A^C \cap B)^C = \mathbb{R}$). Let $x \in A^C \cap B$. Then $x \in B$ so $x = a - y$ for $y \in A^C$ hence $a = x + y \in A^C + A^C$.

**2.5.26.** Hint: for a fixed $\varepsilon$ apply Baire's Theorem to sets $K_n = \{x : |f(mx)| \leq \epsilon \text{ for all } m \geq n\}$.

**2.5.28.** Hint: conisder sets $F_{mn} = \{x : |f_m(x)| \leq n\}$.

**3.1.6.** Hint: use Exercise 3.1.4.

**3.1.8.** Hint: prove that $A$ and $B$ are closed sets.

**3.1.9.** First, $\cup \operatorname{Int} A_i$ is open so it equals $\cup_n (a_n, b_n)$. We will show that $\cup A_i \setminus \cup \operatorname{Int} A_i \subset \{a_n\} \cup \{b_n\}$. Let $x \in \cup A_i \setminus \cup \operatorname{Int} A_i$. Clearly, $x \notin \cup_n (a_n, b_n)$. On the other hand, $x \in A_j$, for some $j$. Since $A_j = \operatorname{Int} A_j \cup \partial A_j$ and $x \notin \operatorname{Int} A_j$, $x$ must be an endpoint of $A_j$. However, $\operatorname{Int}(A_j)$ is a connected set, hence it must be contained in one $(a_k, b_k)$ and it follows that $\partial A_j \subset [a_k, b_k]$. Therefore, $x = a_k$ or $x = b_k$.

**3.1.10.** Hint: find the cardinality of the collection of all open rectangles.

**3.1.11.** Let $G$ be an open set in $\mathbb{R}^n$ and let $x \in G$. There exists $r \in \mathbb{Q}$ s.t. $B(x, r) \subset G$. Also, there is $y \in \mathbb{Q}^n \cap B(x, r/2)$. Now, $x \in B(y, r/2) \subset G$. Indeed, if $z \in B(y, r/2)$ then $\|z - x\| \leq \|z - y\| + \|y - x\| < r$, so $z \in B(x, r)$, hence $z \in G$. Thus, $G = \cup\{B(y, r/2) : y \in \mathbb{Q}^n, r \in \mathbb{Q}\}$.

**3.1.12.** First, we make a commitment: if a number in ternary system ends with a 1, say $x = 0.a_1a_2 \ldots a_n1$ then we will write it as $x = 0.a_1a_2 \ldots a_n02222\ldots$. So, let $x \in F$. Then $x \in F_1$, hence $x \notin O_1 = (1/3, 2/3)$. In ternary system, $O_1 = (0.1, 0.2)$ so either the first ternary digit of $x$ is not 1 or $x = 0.1$. However, following our agreement, in ternary system $0.1 = 0.02222\ldots$, and we conclude that the first digit of $x$ is either 0 or 2. Next, $x \in F_2$ hence $x \notin (1/9, 2/9) \cup (7/9, 8/9)$ or, in ternary system, $x \notin (0.01, 0.02) \cup (0.21, 0.22)$. Thus, either the second ternary digit of $x$ is not 1 or $x = 0.01$ or $x = 0.21$. Since $0.01 = 0.002222\ldots$ and $0.21 = 0.202222\ldots$ we see that the second digit of $x$ is either 0 or 2. Continuing this way, we see that every digit of $x$ is either 0 or 2. Moreover, the argument shows that each digit is uniquely determined to be either 0 or 2.

**3.1.20.** Suppose $X$ is countable. Let $x_1 \in X$ and define $I_1 = [x_1 - 1, x_1 + 1]$. We discard all members of $X$ that are not in $I_1$. Since $X$ is still a perfect set, $I_1 \cap X$ is infinite so there is $x_2 \in I_1 \setminus \{x_1\}$ and a closed interval $I_2 \subset I_1 \setminus \{x_1\}$ that contains $x_2$. Discard all members of $X$ (except for $x_1$) that are not in $I_2$ and continue. We obtain a nested sequence of closed intervals $\{I_n\}_{n=1}^{\infty}$ that satisfies $x_k \notin I_{n+1}$ for $k = 1, \dots, n$, and $E_n := I_n \cap X \neq \emptyset$ for $n \in \mathbb{N}$. Then $\{E_n\}_{n=1}^{\infty}$ is a decreasing sequence of nonempty closed sets, and $E_1$ is bounded. Therefore, by the Nested Set Theorem, $\cap_{n=1}^{\infty} E_n$ is nonempty. Since $\cap_{n=1}^{\infty} E_n \subset X$, it must contain one of the elements of $x$ that was never discarded, i.e., $x_k \in \cap_{n=1}^{\infty} E_n$ for some $k$. But then $x_k \in I_{k+1}$, a contradiction.
**3.1.23.** Let $x \in F^C$. Since $F$ is closed, $F^C$ is open so $x$ is an interior point of $F^C$. The function $\chi_F$ is constant on $F^C$, hence continuous at $x$. Let $y \in F$. By Theorem 3.1.8, $y$ is an accumulation point of both $F$ and $F^C$, so there are sequences $\{x_n\} \subset F$ and $\{y_n\} \subset F^C$ such that $\lim x_n = \lim y_n = y$. Since $\lim \chi_F(x_n) = 1 \neq 0 = \lim \chi_F(y_n)$, $\chi_F$ is not continuous at $y$.
**3.1.27.** Let $\alpha \in (0, \log_3 2]$, let $C(\alpha) = 5 \cdot 3^{1-\alpha}$, and let $x, y \in [0,1]$. Choose $k \in \mathbb{N}$ s.t. $3^{-k} < |x - y| \leq 3^{-k+1}$. First, $|\phi(x) - \phi(y)| \leq |\phi(x) - \phi_k(x)| + |\phi_k(x) - \phi_k(y)| + |\phi_k(y) - \phi(y)|$. It can be shown, as in the proof of Proposition 3.1.6, that $\sup |\phi(x) - \phi_k(x)| \leq 2^{-k+1}$. Thus, $|\phi(x) - \phi(y)| \leq 2 \cdot 2^{-k+1} + (3/2)^k |x-y|$, because for $\phi_k$ the slope is either 0 or $(3/2)^k$. Continuing, $|\phi(x) - \phi(y)| \leq 2 \cdot 2^{-k+1}|x - y|3^k + (3/2)^k|x - y| = |x - y|^{\alpha}|x - y|^{1-\alpha}5(3/2)^k = C(\alpha)|x - y|^{\alpha}|x - y|^{1-\alpha}(3/2)^k 3^{\alpha-1}$. Since $1 - \alpha > 0$ we have that $|x - y|^{1-\alpha} \leq (3^{-k+1})^{1-\alpha}$ so $|x - y|^{1-\alpha}(3/2)^k 3^{\alpha-1} \leq 3^{(-k+1)(1-\alpha)}(3/2)^k 3^{\alpha-1} = 2^{-k}3^{k\alpha} \leq 2^{-k}3^{k \log_3 2} = 1$.
**3.1.28.** Hint: use induction to prove for finite sums.
**3.1.33.** Define $O = [0,1] \setminus F$ and fix any point $x \in (0,1) \cap F$. If $t \in O \cap [0,x)$ then $\varphi(t) = f(t) \leq f(x)$ whence the continuity of $\varphi$ yields $\varphi(x) \leq f(x)$. To prove the reverse inequality, fix $\varepsilon > 0$ and $x \in (0,1) \cap F$. Since $\varphi$ is continuous, there exists $\delta > 0$ such that $\varphi(t) < \varphi(x) + \varepsilon$ for all $t \in [0,1]$ satisfying $|x - t| < \delta$. Since $F$ is nowhere dense it cannot contain the open interval $(x, x + \delta)$. Therefore there exists $t \in O \cap (x, x + \delta)$ such that $f(x) \leq f(t) = \varphi(t) < \varphi(x) + \varepsilon$. Let $\varepsilon \to 0$.
**4.1.7.** Let $\varepsilon > 0$. By definition, there is $A \in \mathcal{E}$ that contains $E$ and $c(A) < c^*(E) + \varepsilon$. Let $A = \cup_{k=1}^{n} R_k$. By Exercise 4.1.1, there are open rectangles $\{R_k'\}_{k=1}^{\infty}$ such that $R_k \subset R_k'$, for all $k \in \mathbb{N}$, and such that $\sum_{k=1}^{\infty} \text{vol}\,(R_k') \leq \sum_{k=1}^{\infty} \text{vol}\,(R_k) + \varepsilon$. The set $A' = \cup_{k=1}^{n} R_k'$ is open, it contains $E$, and $c(A') = \sum_{k=1}^{n} \text{vol}\, R_k' \leq \sum_{k=1}^{n} \text{vol}\, R_k + \varepsilon = c(A) + \varepsilon < c^*(E) + 2\varepsilon$.
**4.1.9.** Hint: start with the case when $A$ is a rectangle.
**4.1.12.** Hint: cover $E$ by rectangles so that none intersect both $E_1$ and $E_2$.
**4.1.15.** Let $\alpha = \nu([0,1]^2)$. Then $\nu([0,1)^2) = n^2 \times \nu([0,1/n)^2)$, so $\nu([0,1/n)^2) = \alpha(1/n)^2$; for any $r \in \mathbb{Q}$, we have $\nu([0,r)^2) = \alpha r^2$. Thus $\nu = \alpha c$ on squares with the length of a side a rational number. Next we consider rectangles. Suppose that $R = I_1 \times I_2$ and that lengths of $I_1$ and $I_2$ are rational numbers $L_1 = p/q$ and $L_2 = r/s$. Then $L_1 = (ps)/(qs)$ and $L_2 = (qr)/qs)$ so $R$ is the union of $(ps)(qr)$ squares, each with a side of length $1/(qs)$. Therefore, $\nu(R) = (ps)(qr)\nu(([0,1/(qs))^2) = (ps)(qr)\alpha c(([0,1/(qs))^2) = \alpha c(R)$. Next, if lengths of $I_1$ and $I_2$ are real numbers $M_1$ and $M_2$, then there are $a,b,c,d \in \mathbb{Q}$ s.t. $a < M_1 < b$ and $c < M_2 < d$. Then there are rectangles $R_1, R_2$ with sides $a,c$ and $b,d$ so that $R_1 \subset R \subset R_2$. By the previous, $\nu(R_1) = \alpha c(R_1) = \alpha ac$ and $\nu(R_2) = \alpha c(R_2) = \alpha bd$. By letting $a,b \to M_1$ and $c,d \to M_2$ we get $\nu(R) = \alpha M_1 M_2 = \alpha c(R)$. If $E \in \mathcal{E}$ then $E$ is a finite disjoint union if rectangles, so $\nu = \alpha c$ carries over to $\mathcal{E}$. Finally, a Jordan measurable set $A$ has content $c(A) = c_*(A) = c^*(A)$, so there are sequences $\{A_k\}, \{B_k\} \subset \mathcal{E}$ such that $A_k \subset A \subset B_k$ and $\lim c(A_k) = \lim c(B_k) = c(A)$. Then $\alpha c(A_k) = \nu(A_k) \leq \nu(A) \leq \nu(B_k) = \alpha c(B_k)$, so when $k \to \infty$ we get $\nu(A) = \alpha c(A)$.
**4.1.16.** It suffices to show that the outer content is 0. Let $L$ be the length of segment $I$ and let $\varepsilon > 0$. First we consider the case when $I$ is not parallel to any of the coordinate axes. Choose $n \in \mathbb{N}$ s.t. $n > L^2/(2\varepsilon)$. Split $I$ in $n$ segments $I_k$ of the same length $d_k = L/n$ and build a rectangle $R_k$ so that $I_k$ is a diagonal of $R_k$. Then $\text{vol}\,(R_k) = a_k b_k \leq (a_k^2 + b_k^2)/2 = d_k^2/2 = (L/n)^2/2 = L^2/(2n^2)$. The sum of the volumes of these rectangles is $L^2/(2n) < \varepsilon$. Thus, $c^*(I) < \varepsilon$ whence $c^*(I) = 0$. When $I$ is parallel to one of the axes, we build squares $\{R_k\}$ so that $I_k$ is a side of $R_k$. Then $\text{vol}\,(R_k) = L^2/n^2$ so the sum of the volumes is $L^2/n < 2\varepsilon$.
**4.1.21.** Hint: use Exercises 2.1.35 and 1.1.2.
**4.1.24.** By Exercises 4.1.22 and 2.1.35, $\partial[A \cap B] = \overline{A \cap B} \cap \overline{[A \cap B]^C} \subset \overline{A} \cap \overline{B} \cap \overline{[A \cap B]^C} = \overline{A} \cap \overline{B} \cap \overline{A^C \cup B^C} = \overline{A} \cap \overline{B} \cap [\overline{A^C} \cup \overline{B^C}] = [\overline{A} \cap \overline{B} \cap \overline{A^C}] \cup [\overline{A} \cap \overline{B} \cap \overline{B^C}] \subset [\overline{A} \cap \overline{A^C}] \cup [\overline{B} \cap \overline{B^C}] = \partial(A) \cup \partial(B)$.
**4.1.27.** Yes! ($\Rightarrow$) Let $A \in \mathcal{J}$. By assumption, $c(A) = 0$ and, by Theorem 4.1.9, $c(\partial(A)) = 0$. Thus, $c(\overline{A}) = 0$. Now $\overline{A}$ must have nonempty interior otherwise it would contain an open disk, hence a rectangle of positive area, which would imply that $c_*(\overline{A}) > 0$, a contradiction.

($\Leftarrow$) Suppose $c(A) > 0$. Then $c_*(A) > 0$ so $A$ contains an elementary set of positive content, which contains a rectangle of positive area hence a disk. All the more, $\overline{A}$ contains a disk so $\overline{A}$ has nonempty interior whence $A$ is not nowhere dense.

**4.1.29.** Union: obvious. Difference: First for two rectangles $R$ and $S$. Use endpoints of all intervals (whose products are $R$ and $S$) to obtain rectangles $\{R_k\}_{k=1}^m$, arrangend in such order that $R = \cup_{i=1}^p R_i$ and $S = \cup_{i=q}^r R_i$, whence $R \setminus S$ is a union of some rectangles among $\{R_k\}_{k=1}^m$, hence an elementary set. When $A$ is a union of rectangles, $A = \cup_{k=1}^m R_k$ then $A \setminus S = \cup_{k=1}^m (R_k \setminus S)$, an elementary set by the first part. Finally, if $A = \cup_{k=1}^m R_k$ and $B = \cup_{i=1}^p S_i$ then $A \setminus B = (A \setminus S_1) \setminus \cup_{i=2}^p S_i$. By the previous, $A \setminus S_1$ is an elementary set $E_1$, and $E_1 \setminus \cup_{i=2}^p S_i = (E_1 \setminus S_2) \setminus \cup_{i=3}^p S_i$, etc.
**4.1.32.** Hint: consider $X$ and $Y$ to be finite sets.
**4.2.3.** Hint: use Heine–Borel Theorem to reduce one set of rectangles to a finite set, then Theorems 4.1.10 and 4.1.8.
**4.2.5.** Hint: define $R_1 = (a,b) \times (c,d)$ and $R_2 = (a, b+h) \times (c,d)$ and use Propositions 4.2.4 and 4.2.5 to conclude that $m(R_2 \setminus F) = h(d-c) + m(R_1 \setminus F)$.
**4.2.6.** By definition, $m(F) = m(R) - m(R \setminus F)$, for some open rectangle $R$. Since $R$ is a Jordan set and $F$ is assumed to be a Jordan set, by Theorem 4.1.15, so is $R \setminus F$. Further, Proposition 4.2.3 implies that $m(F) = c(R) - c(R \setminus F)$ whence additivity of $c$ (Theorem 4.1.10) yields $m(F) = c(F)$.
**4.2.17.** Hint: use Propositions 4.2.17 and 4.2.7.
**4.2.22.** Let $\{A_n\}$ be a Cauchy sequence in the metric space under investigation. By Proposition 2.2.22, there is a subsequence $\{B_n\} \subset \{A_n\}$ s.t. $d(B_n, B_k) < 1/2^n$, for $k \geq n$. Define $P_n = \cup_{j \geq n} B_j$ and $Q_n = \cap_{j \geq n} B_j$. Then $P_n \triangle B_n = P_n \setminus B_n \subset \cup_{j \geq n}(B_{j+1} \setminus B_n)$ so $d(P_n, B_n) < 1/2^{n-1}$ and similarly $d(B_n, Q_n) < 1/2^{n-1}$, whence $d(P_n, Q_n) < 1/2^n$. Notice that for every $n \in \mathbb{N}$, $Q_n \subset \liminf_k B_k \subset \limsup_k B_k \subset P_n$ so $d(\liminf_k B_k, \limsup_k B_k) < 1/2^n$, whence $\liminf_k B_k = \limsup_k B_k$, meaning that $\{B_k\}$ converges. The result now follows from Proposition 2.2.17.
**4.2.23.** Hint: consider the set $E = \{(x,y) \in [0,1] \times [0,1] : x \in \mathbb{Q}\}$.
**4.2.24.** By Theorem 4.2.18, $m^*(A \cup B) \leq m^*(A) + m^*(B)$, so we prove the opposite inequality. WLOG we assume that both $m^*(A)$ and $m^*(B)$ are finite. Let $G_1 = \cup_{a \in A} B(a, \alpha/3)$ and $G_2 = \cup_{b \in B} B(b, \alpha/3)$. Then $G_1$ and $G_2$ are open disjoint sets. Indeed, if $x \in G_1 \cap G_2$ then there are $a \in A$ and $b \in B$ for which $\|x - a\| < \alpha/3$ and $\|x - b\| < \alpha/3$, and it would follow that $\|a - b\| < 2\alpha/3$. Also, $A \subset G_1$ and $B \subset G_2$.

Let $\varepsilon > 0$. By definition, there exists an open set $G \supset A \cup B$ such that $m^*(A \cup B) + \varepsilon > m(G)$. Notice that $E := G \cap (G_1 \cup G_2)$ is also an open set and $A \cup B \subset E \subset G$, so $m^*(A \cup B) + \varepsilon > m(E)$. Therefore, using Propositions 4.2.17 and 4.2.5, $m^*(A) + m^*(B) \leq m(G_1 \cap G) + m(G_2 \cap G) = m(E) < m^*(A \cup B) + \varepsilon$. Now let $\varepsilon \to 0$.
**4.2.27.** (a) ($\leq$) is Theorem 4.2.19. ($\geq$) By definition, there is a compact set $F \subset E \cap (\cup_{k=1}^n R_k)$ s.t. $m(F) > m_*(E \cap (\cup_{k=1}^n R_k)) - \varepsilon$. Let $F_k = F \cap R_k$, for $k = 1, 2, \dots, n$. Notice that $F_k \subset E \cap R_k$ and $F = \cup_{k=1}^n F_k$. Therefore, using Proposition 4.2.17 and Corollary 4.2.20, $m_*(E \cap ((\cup_{k=1}^n R_k)) < m(F) + \varepsilon = \sum_{k=1}^n m(F_k) + \varepsilon \leq \sum_{k=1}^n m_*(E \cap R_k) + \varepsilon$. Now let $\varepsilon \to 0$.

(b) ($\geq$) is Theorem 4.2.18. ($\leq$) By definition, there is an open set $G \supset E \cap (\cup_{k=1}^n R_k)$ s.t. $m(G) < m_*(E \cap (\cup_{k=1}^n R_k)) + \varepsilon$. WLOG we assume that $G \subset \cup_{k=1}^n R_k$, otherwise replace it by $G \cap (\cup_{k=1}^n R_k)$. Let $G_k = G \cap R_k$, for $k = 1, 2, \dots, n$. Notice that $G_k \supset E \cap R_k$ and $G = \cup_{k=1}^n G_k$. Therefore, using Proposition 4.2.17 and Corollary 4.2.20, $m^*(E \cap (\cup_{k=1}^n R_k)) > m(G) - \varepsilon = \sum_{k=1}^n m(G_k) - \varepsilon \geq \sum_{k=1}^n m_*(E \cap R_k) - \varepsilon$. Now let $\varepsilon \to 0$.
**4.2.30.** Notice that, if $R$ is an open rectangle containing $E$ and $C = R \setminus E$ then, replacing $E$ with $C$ in Theorem 4.2.22, we obtain $m^*(C) + m_*(R \setminus C) = m(R)$, hence $m^*(R \setminus E) + m_*(E \cap R) = m(R)$.

Since $A \in \mathcal{E}$ it can be written as a finite disjoint union of rectangles, $A = \cup_{k=1}^n R_k$. Applying the presented modification of Theorem 4.2.22 to each $R_k$ we get $m^*(R_k \setminus E) + m_*(E \cap R_k) = m(R_k)$. Summing these equalities yields $m(A) = \sum_{k=1}^n m^*(R_k \setminus E) + \sum_{k=1}^n m_*(E \cap R_k)$. By Exercise 4.2.27, we get $c(A) = m^*(A \setminus E) + m_*(E \cap A)$.
**4.2.31.** Let $G$ be an open set that contains $E$ and $F = [0,1] \times [0,1] \setminus G$. Then $F$ is a closed set, hence measurable by Proposition 4.2.17, yet it does not intersect $E$, so it must have measure 0. Clearly, $G \cup F \supset [0,1] \times [0,1]$ so, applying Theorem 4.2.18, $1 = m([0,1] \times [0,1]) \leq m(G) + m(F) = m(G)$. Thus, $m^*(E) = \inf\{m(G) : G \supset E,\, G \text{ open}\} \geq 1$. On the other hand, $E \subset [0,1] \times [0,1]$ so $m^*(E) \leq m^*([0,1] \times [0,1]) = 1$.
**4.3.4.** By Corollary 4.3.9, there are $G_\delta$ sets $G_A \supset A$ and $G_B \supset B$ s.t. $m(G_A) = m^*(A)$ and $m(G_B) = m^*(B)$. First, $m(G_A \cup G_B) \leq m(G_A) + m(G_B) = m^*(A) + m^*(B) = m(A \cup B) \leq m(G_A \cup G_B)$. Thus, $m(G_A \cup G_B) = m(G_A) + m(G_B)$ hence $m(G_A \cap G_B) = 0$. Since $A \cap G_B \subset G_A \cap G_B$ we have $m(A \cap G_B) = 0$. Finally, $A = [(A \cup B) \setminus G_B] \cup (A \cap G_B)$ whence $A$ is measurable. Analogous for $B$.
**4.3.12.** By definition of $\mathcal{A}$, if $A \in \mathcal{A}$ then $A^C \in \mathcal{A}$. Let $\{A_k\} \subset \mathcal{A}$. If all $A_k$ are countable then so is $\cup A_k$ hence $\cup A_k \in \mathcal{A}$. Otherwise, there is $m \in \mathbb{N}$ for which $A_m^C$ is countable. Then $(\cup A_k)^C \subset A_m^C$ so $(\cup A_k)^C$ is countable whence $\cup A_k \in \mathcal{A}$.
**4.3.15.** Hint: consider $A_n = [-n, n] \times [-n, n]$ and $\mathcal{A}_n = \{B \subset R^2 : B \subset A_n \text{ or } B^C \subset A_n\}$.
**4.3.18.** (a) By definition, $\mathcal{T} \subset \sigma(\mathcal{T})$. It follows that $\mathcal{S}$ is contained in the $\sigma$-algebra $\sigma(\mathcal{T})$. Since $\sigma(\mathcal{S})$ is the smallest $\sigma$-algebra that contains $\mathcal{S}$, we conclude that $\sigma(\mathcal{S}) \subset \sigma(\mathcal{T})$. (b) It is obvious that $\mathcal{S} \cap \mathcal{T} \subset \mathcal{S}$. By (a), $\sigma(\mathcal{S} \cap \mathcal{T}) \subset \sigma(\mathcal{S})$. Similarly, $\sigma(\mathcal{S} \cap \mathcal{T}) \subset \sigma(\mathcal{T})$. It follows that $\sigma(\mathcal{S} \cap \mathcal{T}) \subset \sigma(\mathcal{S}) \cap \sigma(\mathcal{T})$. (c) Let $A$ be a set. Define $\mathcal{S}$ to have only one element, namely $A$,

and that $\mathcal{T}$ contains only $A^C$. Then $\mathcal{S} \cap \mathcal{T} = \emptyset$, so $\sigma(\mathcal{S} \cap \mathcal{T}) = \{\emptyset, X\}$. On the other hand, $\sigma(\mathcal{S}) = \sigma(\mathcal{T}) = \{\emptyset, A, A^C, X\}$.
**4.3.19.** Hint: consider the intersection of all $\sigma$-algebras that contain $\mathcal{A}$.
**4.3.22.** Hint: prove that the right-hand side is a $\sigma$-algebra that contains $S$.
**4.3.23.** ($\supset$) if $\mathcal{F} \subset \mathcal{A}$ then $\sigma(\mathcal{F}) \subset \sigma(\mathcal{A})$ so $\cup\{\sigma(\mathcal{F})\} \subset \sigma(\mathcal{A})$. ($\subset$) Let $\mathcal{C}$ denote the right-hand side of the equality. If $Z \in \mathcal{A}$ then $Z \in \sigma(\{Z\})$ so $Z \in \mathcal{C}$ hence $\mathcal{A} \subset \mathcal{C}$. Thus, it remains to show that $\mathcal{C}$ is a $\sigma$-algebra because then $\sigma(\mathcal{A}) \subset \mathcal{C}$. Complement: If $Z \in \mathcal{C}$ then there is $\mathcal{F}$ s.t. $Z \in \sigma(\mathcal{F})$ so $Z^C \in \sigma(\mathcal{F})$. Union: If $\{Z_n\} \subset \mathcal{C}$ then there are $\{\mathcal{F}_n\}$ s.t. $Z_n \in \sigma(\mathcal{F}_n)$ so $Z_n \in \sigma(\mathcal{F}_1 \cup \mathcal{F}_2 \cup \dots)$ hence $\cup Z_n \in \sigma(\mathcal{F}_1 \cup \mathcal{F}_2 \cup \dots)$ whence $\cup Z_n \in \mathcal{C}$.
**4.3.24.** Let $\mathcal{A} = \{A_1, A_2, \dots\}$ and $\mathfrak{M} = \{\cup_{j\in J} A_j : J \subset \mathbb{N}\}$. We prove that $\mathfrak{M}$ is a $\sigma$-algebra. It contains $X$: take $J = \mathbb{N}$. Complement: $[\cup_{j\in J} A_j]^C = \cup_{j\in J^C} A_j$. Union: $\cup_k[\cup_{j\in J_k} A_j] = \cup_{j\in\cup J_k} A_j$. It is easy to see that $\mathcal{A} \subset \mathfrak{M} \subset \sigma(\mathcal{A})$. But $\sigma(\mathcal{A})$ is the smallest $\sigma$-algebra that contains $\mathcal{A}$ so $\mathfrak{M} = \sigma(\mathcal{A})$.
**4.3.29.** Hint: consider $\mathcal{A} = \{A_k\}_{k=1}^n \subset P(X)$, $\mathcal{B} = \mathcal{A} \cup \{A_k^C\}_{k=1}^n$, $\mathcal{C}$ the set of all intersections $B_{i_1} \cap B_{i_2} \cap \dots \cap B_{i_m}$ of elements of $\mathcal{B}$, and $\mathcal{D}$ the set of all unions $C_{i_1} \cup C_{i_2} \cup \dots \cup C_{i_m}$ of elements of $\mathcal{C}$.
**5.1.8.** Let $\{A_n\} \subset \mathfrak{M}$. Define $E_1 = A_1$, $E_2 = A_2 \setminus A_1$, $E_3 = A_3 \setminus (A_1 \cup A_2), \dots$. Then, for every $k$, $E_k \subset A_k$ and $E_k \in \mathfrak{M}$. Moreover, the sets $\{E_n\}$ are disjoint and $\cup A_n = \cup E_n$. Thus, $\mu(\cup A_n) = \mu(\cup E_n) = \sum \mu(E_n) \le \sum \mu(A_n)$.
**5.1.9.** $\mu(\emptyset) = 0$ and $\mu \ge 0$ are trivial. If $\{E_n\}$ are disjoint sets and $E = \cup E_k$ then $\mu(E) = \lim \mu_n(E) = \lim \mu_n(\cup E_k) = \lim \sum \mu_n(E_k) \le \lim \sum \nu(E_k) = \sum \nu(E_k) = \nu(E) < \infty$, so the series $\sum \mu_n(E_k)$ converges uniformly. Thus, $m(E) = \sum \lim \mu_n(E_k) = \sum \mu(E_k)$.

Without $\nu$, $\mu$ need not be a measure. Example: let $\mu_n$ be Lebesgue measure on $E_n = [n, n+1)$, and 0 elsewhere. Then $\mu(\cup_k E_k) = \lim_n \mu_n(\cup_k E_k) = \lim 1 = 1$, but $\mu(E_k) = \lim_n \mu_n(E_k) = 0$.
**5.1.11.** Yes, if increasing! $\mu(\emptyset) = 0$ and $\mu \ge 0$ are trivial. If $\{E_n\}$ are disjoint sets and $E = \cup E_k$ then $\mu(E) = \lim_k \mu_k(E) = \lim_k \mu_k(\cup_n E_n) = \sup_k \sum_n \mu_k(E_n) = \sup_k \sup_n \sum_{i=1}^n \mu_k(E_i)$. By Exercise 2.2.8, $\mu(E) = \sup_n \sup_k \sum_{i=1}^n \mu_k(E_i) = \sup_n \lim_k \sum_{i=1}^n \mu_k(E_i) = \sup_n \sum_{i=1}^n \mu(E_i) = \sum_{i=1}^\infty \mu(E_i)$.

No, if decreasing! $X = \mathbb{N}$, $\mathfrak{M} = \mathbb{P}(X)$, $\mu_n(\{k\}) = 1$ if $k > n$ and 0 if $k \le n$. Then $\mu_n \ge \mu_{n+1}$ and $\mu_n(\{k\}) \to 0$ so $\mu(\{k\}) = 0$. But $\mu_n(\mathbb{N}) = \infty$ so $\mu(\mathbb{N}) = \infty$.
**5.1.12.** First, $\emptyset \in \mathfrak{M}'$ (take $A = B = \emptyset$). Second, if $A \subset E \subset B$ then $B^C \subset E^C \subset A^C$ and $A^C \setminus B^C = B \setminus A$. Finally, if $E = \cup E_n$ and $A_n \subset E_n \subset B_n$ then define $A := \cup A_n$ and $B := \cup B_n$ so that $A \subset E \subset B$ and $\mu(B \setminus A) = \mu[(\cup B_n) \setminus A] \le \sum \mu(B_n \setminus A) \le \sum \mu(B_n \setminus A_n) = 0$. Conclusion: $\mathfrak{M}'$ is a $\sigma$-algebra.

The extension of $\mu$ is well defined: if $A' \subset E \subset B'$ then $\mu(A') = \mu(A)$. Indeed, if either one is infinite, say $\mu(A) = \infty$ then $\mu(B) = \mu(B') = \infty$, so $\mu(A') + \mu(B' \setminus A') = \infty$ implies that $\mu(A') = \infty$. If both $\mu(A)$ and $\mu(A')$ are finite then $\mu(A) - \mu(A') \le \mu(B') - \mu(A') = \mu(B' \setminus A') = 0$, hence $\mu(A) \le \mu(A')$, and similarly $\mu(A') \le \mu(A)$. Finally, if $E = \cup E_n$ and $A_n \subset E_n \subset B_n$ we define $A := \cup A_n$ and $B := \cup B_n$. Then $\mu(B \setminus A) = 0$ and $\{A_n\}$ are disjoint so $\mu(E) = \mu(A) = \mu(\cup A_n) = \sum \mu(A_n) = \sum \mu(E_n)$, so $\mu$ is a measure on $\mathfrak{M}'$.
**5.1.14.** Hint: use the Borel–Cantelli lemma.
**5.1.19.** $\mu(\emptyset) = 0$ and $\mu \ge 0$ hold because $\mu$ is a finitely additive measure. Let $\{E_n\}$ be disjoint sets and $E = \cup E_k$. Then $\mu(\cup_i^\infty E_n) = \mu[(\cup_1^{n-1} E_k) \cup (\cup_n^\infty E_k)] = \sum_1^{n-1} \mu(E_k) + \mu(\cup_n^\infty E_k)$; now let $n \to \infty$. The second term goes to 0 because $\{\cup_n^\infty E_k\}_{n=1}^\infty$ is a decreasing sequence with empty intersection.
**5.1.22.** Denote $A_n = \cup_{k=n}^\infty E_k$ and notice that the sequence $\{A_n\}$ is decreasing, so continuity of measure implies that $\mu(\limsup E_n) = \mu(\cap A_n) = \lim \mu(A_n) \ge \limsup \mu(E_n)$.
**5.1.25.** Hint: try a finite set for $X$.
**5.1.27.** Hint: start by replacing $A_1$ with $A_1^C$.
**5.1.28.** Hint: use induction.
**5.2.3.** (c) Assume all $\mu^*(E_n) < \infty$. Let $\varepsilon > 0$. For each $n$, select $\{E_{nk}\}$ covering $E_n$ so that $\sum_k F(E_{nk}) < \mu^*(E_n) + \varepsilon 2^{-n}$. Then $\cup E_n \subset \cup_{n,k} E_{nk}$ so $\mu^*(\cup_n E_n) \le \sum_{n,k} F(E_{nk}) \le \sum_n \mu^*(E_n) + \varepsilon$.
**5.2.5.** $\{\emptyset, X\}$ because for any other set $A$, $\mu^*(A) + \mu^*(A^C) = 2$.
**5.2.10.** First, $\mu^*(A) = \inf\{\sum \mu(E_k) : A \subset \cup E_k\} = 1$, because $X$ is the only cover, unless $A = \emptyset$. Second, $E \in \mathfrak{M}$ if and only if $\mu^*(A) = \mu^*(A \cap E) + \mu^*(A \setminus E)$ for all $A$. Taking $A = X$ we get $1 = \mu^*(E) + \mu^*(E^C)$ so either $E = \emptyset$ or $E^C = \emptyset$ hence $E = \emptyset$ or $E = X$. Thus $\mathfrak{M} = \{\emptyset, X\}$.
**5.2.14.** For $n = 2$: $\mu^*(A_1 \cup A_2) = \mu^*[(A_1 \cup A_2) \cap E_1] + \mu^*[(A_1 \cup A_2) \setminus E_1] = \mu^*(A_1) + \mu^*(A_2)$. For finite sums use induction. For infinite sums: ($\le$) is subadditivity; ($\ge$) $\mu^*(\cup_1^\infty A_k) \ge \mu^*(\cup_1^n A_k) = \sum_1^n \mu^*(A_k)$. Let $n \to \infty$.
**5.2.17.** ($\Rightarrow$) Take $E = A$. ($\Leftarrow$) let $B \subset X$. Claim $\mu^*(B) = \mu^*(B \cap A) + \mu^*(B \setminus A)$. ($\le$) is subadditivity. ($\ge$) $B \cap A = (B \cap E) \cup [(B \cap A) \setminus E] \subset (B \cap E) \cup (A \setminus E)$ so $\mu^*(B \cap A) \le \mu^*(B \cap E) + \varepsilon$. Also $B \setminus E \supset B \setminus A$ so $\mu^*(B \setminus E) \ge \mu^*(B \setminus A)$. Now, $\mu^*(B) = \mu^*(B \cap E) + \mu^*(B \setminus E) \ge \mu^*(B \cap A) - \varepsilon + \mu^*(B \setminus A)$. Let $\varepsilon \to 0$.

**5.2.22.** (a) If $\mu^*(E) = \infty$ take $A = X$. Otherwise, there are sets $\{E_k : k \in \mathbb{N}\}$ s.t. $E \subset \cup_{k=1}^{\infty} E_k$ and $\sum_{k=1}^{\infty} \mu(E_k) \leq \mu^*(E) + \varepsilon$. Thus, take $A = \cup_{k=1}^{\infty} E_k$.

(b) ($\Rightarrow$) By (a), for each $n \in \mathbb{N}$ there exists $B_n \in A_\sigma$ such that $E \subset B_n$ and $\mu^*(B_n) \leq \mu^*(E) + 1/n$. Define $B := \cap_n B_n$, so that $B \in A_{\sigma\delta}$ and $E \subset B$. Since $\mu^*(E) < \infty$, it follows that $\mu^*(B \setminus E) = \mu^*(B_n) - \mu^*(B_n \cap E) = \mu^*(B_n) - \mu^*(E) \leq 1/n$ for all $n \in \mathbb{N}$. This implies that $\mu^*(B \setminus E) = 0$.

($\Leftarrow$) If $F \subset X$, then $\mu^*(F \cap E) + \mu^*(F \cap E^C) \leq \mu^*(F \cap E) + \mu^*(F \cap E^C \cap B^C) + \mu^*(F \cap E^C \cap B) \leq \mu^*(F \cap B) + \mu^*(F \cap B^C) + \mu^*(F \cap (B \setminus E)) = \mu^*(F \cap B) + \mu^*(F \cap B^C) = \mu^*(F)$ because $B \in A_{\sigma\delta}$ is measurable.

**5.2.23.** Hint: prove that $\mu(B \setminus A) = 0$ and $\mu(A \setminus B) = 0$.

**5.2.25.** Hint: consider the ring of all finite unions of intervals $[x, y)$ and the restriction of $\mu$ to it.

**5.2.27.** Hint: approximate a set $B$ from above and prove $\mu^*(B \cap E) + \mu^*(B \cap E^C) \leq \mu^*(B)$.

**5.2.28.** Let $\mu$ be a Carathéodory extension of $\mu'$ to $\mathfrak{M}$, and let $\mu_1$ be another extension of $\mu'$ to $\mathfrak{M}$. To show that $\mu = \mu_1$ on $\mathfrak{M}$, it suffices to write $X = \cup X_n$, where $\mu'(X_n) < \infty$, and to prove that $\mu = \mu_1$ on $\mathfrak{M}|X_n$. So, without loss of generality, assume that $\mu'$ is finite. Let $E \in \mathfrak{M}$. Choose $\{E_n\} \subset \mathfrak{E}$ s.t. $E \subset \cup E_n$. Then $\mu_1(E) \leq \mu_1(\cup E_n) \leq \sum \mu_1(E_n) = \sum \mu(E_n)$. Taking infimum over all selections $\{E_n\}$, we get $\mu_1(E) \leq \mu^*(E) = \mu(E)$. Similarly, if $F \in \mathfrak{E}$ contains $E$, then $\mu_1(F \setminus E) \leq \mu(F \setminus E)$. Finally, $\mu_1(E) + \mu_1(F \setminus E) = \mu_1(F) = \mu(F) = \mu(E) + \mu(F \setminus E) \leq \mu_1(E) + \mu_1(F \setminus E)$, so $\mu(E) - \mu_1(E) = \mu_1(F \setminus E) - \mu(F \setminus E) \leq 0$, hence $\mu(E) \leq \mu_1(E)$.

**5.2.32.** (a) By definition, $\mathcal{A}$ is closed under union. Moreover, the complement of $(a, b] \cap \mathbb{Q}$ is $((-\infty, a] \cap \mathbb{Q}) \cup ((b, \infty] \cap \mathbb{Q}) \in \mathcal{A}$. If $B = \cup_{k=1}^{n} (a_k, b_k] \cap \mathbb{Q}$ then $B^C = \cap_{k=1}^{n} ((-\infty, a_k] \cup (b_k, \infty]) \cap \mathbb{Q}) = (\cap_{k=1}^{n} (-\infty, a_k]) \cap \mathbb{Q}) \cup (\cap_{k=1}^{n} (b_k, \infty]) \cap \mathbb{Q}) = (-\infty, \min\{a_k : 1 \leq k \leq n\}] \cap \mathbb{Q}) \cup (\max\{b_k : 1 \leq k \leq n\}, +\infty] \cap \mathbb{Q}) \in \mathcal{A}$.

(b) Let $\mathfrak{M}$ be the $\sigma$-algebra generated by $\mathcal{A}$. Since $\mathbb{Q}$ is countable and $\mathfrak{M}$ is closed under countable unions (and every set in $\mathfrak{M}$ is a subset of $\mathbb{Q}$), it suffices to show that $\{x\} \in \mathfrak{M}$ for all $x \in \mathbb{Q}$. Given $x \in \mathbb{Q}$ and $n \in \mathbb{N}$, it is clear that $(x - 1/n, x] \cap \mathbb{Q} \in \mathfrak{M}$. Therefore $\{x\} = \cap_n ((x - 1/n, x] \cap \mathbb{Q}) \in \mathfrak{M}$, as required.

(c) By definition $\mu(\emptyset) = 0$. Let $\{E_n\}_{n \in \mathbb{N}}$ be a sequence of disjoint members of $\mathcal{A}$ whose union lies in $\mathcal{A}$. If $E_n = \emptyset$ for all $n \in \mathbb{N}$, then $\cup_n E_n = \emptyset$ and hence $\mu(\cup_n E_n) = 0 = \sum_n \mu(E_n)$. Otherwise $\cup_n E_n \neq \emptyset$, and $E_m \neq \emptyset$ for some $m \in \mathbb{N}$, so $\mu(\cup_n) = \infty$ and $\sum_n \mu(E_n) \geq \mu(E_m) = \infty$. Therefore $\mu$ is a measure on $\mathcal{A}$. Define $\mu_1, \mu_2 : \mathbb{P}(\mathbb{Q}) \to [0, \infty]$ by $\mu_1(E) = \infty$, if $E \neq \emptyset$, and $\mu(\emptyset) = 0$; and $\mu_2(E) = \infty$, if $E$ contains $2^{-n}m$ for some $m \in \mathbb{Z}$ and $n \in \mathbb{N}$, 0, otherwise. Clearly $\mu_1|\mathcal{A} = \mu = \mu_2|\mathcal{A}$, because every nonempty interval contains a rational of the form $2^{-n}m$ for some $m \in \mathbb{Z}$ and $n \in \mathbb{N}$. The proof that $\mu_1$ and $\mu_2$ are measures is similar to the one for $\mu$. But $\mu_1 \neq \mu_2$ because $\mu_1(\{1/3\}) = \infty$ whereas $\mu_2(\{1/3\}) = 0$.

**5.2.34.** (a) By definition, $\mu^+(E) = \inf\{\sum \mu(E_k)\} = \inf\{\sum \mu^*(E_k)\}$. For each choice $\{E_k\}$, $\sum \mu^*(E_k) \geq \mu^*(E)$ so it holds for the infimum.

(b) ($\Leftarrow$) $\mu^+(E) = \inf\{\sum \mu^*(E_k)\} \leq \mu^*(A) + \mu^*(\emptyset) + \mu^*(\emptyset) + \cdots = \mu^*(A) = \mu^*(E)$. The reverse inequality is (a). ($\Rightarrow$) $\mu^*(E) = \mu^+(E) = \inf\{\sum \mu^*(E_k)\}$ so for each $n$ there are sets $\{E_{kn}\}_k$ s.t. $E \subset \cup_k E_{kn}$ and $\sum_k \mu^*(E_{kn}) < \mu^*(E) + 1/n$. Let $E_n = \cup_k E_{kn}$ so $E_n \in \mathfrak{M}$, $E_n \supset E$. Let $A = \cap_n E_n \in \mathfrak{M}$. Then $A \supset E$ so $\mu^*(A) \geq \mu^*(E)$. In the opposite direction, $\mu^*(A) = \mu^*(\cap_n E_n) \leq \mu^*(E_m)$ for any $m$ so $\mu^*(A) \leq \mu^*(\cup_k E_{km}) \leq \sum_k \mu^*(E_{km}) < \mu^*(E) + 1/m$. This holds for all $m \in \mathbb{N}$ so $\mu^*(A) \leq \mu^*(E)$.

**5.2.36.** Hint: from the cover $E \subset \cup A_k$ define $A = \cup_{k=1}^{N} A_k$.

**5.3.2.** Hint: consider $\limsup(E - x_n)$.

**5.3.4.** Hint: start with $E = [a, b]$.

**5.3.5.** (a) Let $\{R_k\}$ be rectangles in $\mathbb{R}^n$ and suppose that $E \subset \cup R_k$. Without loss of generality we may assume that the rectangles have disjoint interiors. Notice that $E \subset \cup R_k$ if and only if $\alpha E + \beta \subset \cup(\alpha R_k + \beta)$, and the rectangles $\{\alpha R_k + \beta\}$ have disjoint interiors. Then $m^*(\alpha E + \beta) \leq m[\cup(\alpha R_k + \beta)] = |\alpha|^n m(\cup R_k)$. Taking infimum over all covers $\{R_k\}$ of the set $E$, we get $m^*(\alpha E + \beta) = |\alpha|^n m^*(E)$.

(b) For any $A$, $A \cap (E + \beta) = \beta + (A - \beta) \cap E$, and $A \cap (E + \beta)^C = \beta + (A - \beta) \cap E^C$. Therefore, $m^*[A \cap (E + \beta)] + m^*[A \cap (E + \beta)^C] = m^*[(A - \beta) \cap E] + m^*[(A - \beta) \cap E^C]$. Thus, $E + \beta$ is measurable if and only if $E$ is. Similarly, $A \cap \alpha E = \alpha[(1/\alpha A) \cap E]$ and $A \cap (\alpha E)^C = \alpha[(1/\alpha A) \cap E^C]$. Therefore, $m^*[A \cap \alpha E] + m^*[A \cap (\alpha E)^C] = |\alpha|^n m^*[(1/\alpha A) \cap E]] + |\alpha|^n m^*[(1/\alpha A) \cap E^C]$. Thus, $\alpha E$ is measurable if and only if $E$ is.

**5.3.11.** Since $m(E) < (1/\alpha) m(E)$ there are intervals $\{I_n\}$ covering $E$ such that $\sum m(I_n) < (1/\alpha) m(E)$. Next, $E \subset \cup(E \cap I_n)$ so $m(E) \leq \sum m(E \cap I_n)$. Thus $0 < \sum[m(E \cap I_n) - \alpha m(I_n)] < \infty$ hence there is $n$ s.t. $m(E \cap I_n) - \alpha m(I_n) > 0$.

**5.3.14.** Hint: spit a cover of $A \cup B$ into disjoint covers of $A$ and $B$.

**5.3.16.** Denote $F_n := E \cap (\cup_{k=1}^{n} G_k)$. By Exercise 5.2.21, there are measurable sets $A_k \supset F_k$ s.t. $m(A_k) = m^*(F_k)$. Notice that $F_k \subset F_{k+n} \subset A_{k+n}$ so if we denote $B_k = \cap_{i=k}^{\infty} A_k$ then $B_k$ is measurable subset of $A_k$, contains $F_k$ and the sequence $\{B_k\}$ is increasing to a measurable set $B$ that contains $E$ (because $F_k$ increases to $E$). Thus, $m^*(E) \leq m^*(B) = m(B) = \lim m(B_k) \leq \lim m(A_k) = \lim m^*(F_k) \leq m^*(E)$.

**5.3.17.** Hint: prove that $m^*(A \cup B) + m^*(A \cap B) \leq m^*(A) + m^*(B)$.
**5.3.19.** Hint: use Exercises 5.3.4 and 5.3.9.
**5.3.21.** ($\Rightarrow$) RHS $= m^*(R) = \text{vol}(R)$. ($\Leftarrow$) Let $G$ be an open set. By Theorem 3.1.3, every open set in $\mathbb{R}^n$ is a countable union of rectangles with disjoint interiors, so $G = \cup R_k$. Since the boundaries of these rectangles form a set of measure 0, we may assume that rectangles $\{R_k\}$ are disjoint. Then $m(G) = m(\cup R_k) = \sum m^*(R_k \cap A) + m^*(R_k \setminus A) \geq m^*(\cup R_k \cap A) + m^*(\cup R_k \setminus A) = m^*(G \cap A) + m^*(G \setminus A)$, and the opposite inequality holds because $m^*$ is subadditive. If $E \subset \mathbb{R}^n$, given $\varepsilon > 0$, by Theorem 5.3.8, there is an open set $G \supset E$ such that $m(G) \leq m^*(E) + \varepsilon$. Then $m^*(E) \geq m(G) - \varepsilon = m^*(G \cap A) + m^*(G \setminus A) - \varepsilon \geq m^*(E \cap A) + m^*(E \setminus A) - \varepsilon$. Letting $\varepsilon \to 0$ we get an inequality and the opposite one follows from subadditivity of $m^*$.
**5.3.24.** Hint: consider $X_\alpha = X + \alpha x$.
**5.3.31.** First, $G := \{(x, f(x) : x \in \mathbb{R}\}$ is closed. Indeed, if $\lim(x_n, f(x_n)) = (x, y)$, then $x = \lim x_n$ and $y = \lim f(x_n) = f(x)$, so $(x, y) \in G$. By Theorem 5.3.7, $G$ is measurable. Next, $f$ is continuous on $[-n, n]$ hence uniformly continuous, so there is a partition $-n = x_0 < x_1 < \cdots < x_{k_n} = n$ s.t. $\sup\{f(x) : x \in [x_i, x_{i+1}]\} - \inf\{f(x) : x \in [x_i, x_{i+1}]\} < \varepsilon/(2n)$. Let $A_n = \{(x, f(x)) : x \in [-n, n]\}$. Then $A_n \subset \cup_{i=1}^{k_n} [x_i, x_{i+1}] \times [\inf\{f(x) : x \in [x_i, x_{i+1}]\}, \sup\{f(x) : x \in [x_i, x_{i+1}]\}]$, so $m(A_n) \leq \sum_i (x_{i+1} - x_i)\varepsilon/(2n) = \varepsilon$. Since $\varepsilon$ is arbitrary, $m(A_n) = 0$. But $\{A_n\}$ is an increasing sequence that converges to $G$, so $m(G) = \lim m(A_n) = 0$.
**5.3.35.** Hint: find $m(F^C)$.
**5.3.42.** Hint: Use Fat Cantor set.
**5.3.45.** For $k \in N$, let $U_k = \cup_{n=1}^{\infty}(r_n - 2^{-(n+k)}, r_n + 2^{-(n+k)})$. $U_k$ is a dense open set with Lebesgue measure $\leq 2^{2-k}$, thus $\cap_{k=1}^{\infty} U_k$ is a set of the second category (Baire), hence uncountable, and has Lebesgue measure 0.
**5.3.46.** Hint: consider sets $A_n := \{x + r_n : x \in A\}$.
**5.4.4.** Hint: extend the Cantor–Lebesgue function to $\mathbb{R}$.
**5.4.5.** Suppose that $B \subset A$ but $m(B) > 0$. Let $\Lambda$ and $C_E$ be as in the proof of Theorem 5.4.2 and take $A = C_E$. It was shown that the sets $\{C_E + \lambda : \lambda \in \Lambda\}$ are pairwise disjoint, so all the more this is true of $\{B + \lambda : \lambda \in \Lambda\}$. By Lemma 5.4.1, $m(B) = 0$.
**5.4.9.** Since $f$ is continuous and increasing it has a continuous inverse $g$ ([86, Theorem 3.6.9]). Thus, $\{E : f(E) \in \mathcal{B}\} = \{E : g^{-1}(E) \in \mathcal{B}\}$ and by Exercise 4.3.9, it is a $\sigma$-algebra. Moreover, it contains all open sets because $g$ is continuous. It follows that it contains $\mathcal{B}$, so for every $E \in \mathcal{B}$, the set $f(E) = g^{-1}(E) \in \mathcal{B}$.
**5.4.15.** Since $E \times F = (E \times \mathbb{R}^m) \cap (\mathbb{R}^n \times F)$, it suffices to prove when $F = \mathbb{R}^m$. Let $\mathcal{A} = \{E \subset R^n : E \times \mathbb{R}^m \in \mathcal{B}_{m+n}\}$. Then $\mathcal{A}$ is a $\sigma$-algebra that contains all open sets [straightforward] in $\mathbb{R}^n$, hence $\mathcal{B}_n \subset \mathcal{A}$ so if $E \in \mathcal{B}_n$ then $E \times \mathbb{R}^m \in \mathcal{B}_{m+n}$.
**5.4.18.** Hint: define $\mu_x(A) := \chi_A(x)$ and $\mu(A) = \sum_{x \in \mathbb{Q}^n} \chi_A(x)$.
**5.4.20.** Hint: use Example 1.1.2 and Exercise 1.1.14.
**5.4.21.** Let $x_0 \in \mathbb{R}$ and $(x_n)_{n \in \mathbb{N}}$ a sequence that converges to $x_0$. By Exercise 2.2.4, $F + x_0 \supset \limsup(F + x_n)$ so continuity of measure implies that $\mu(F + x_0) \geq \mu(\limsup(F + x_n)) = \lim_n \mu\left(\cup_{k=n}^{\infty}(F + x_k)\right) \geq \limsup_n \mu(F + x_n)$. We apply [86, Theorem 2.8.3] to conclude that for every $\varepsilon > 0$ there is $n \in \mathbb{N}$ s.t. $k \geq n \Rightarrow \mu(F + x_k) < \limsup_n \mu(F + x_n) + \varepsilon \leq \mu(F + x_0) + \varepsilon$, whence $f$ is upper semicontinuous at $x_0$ and therefore on $\mathbb{R}$.
**5.4.22.** Let $\mathcal{D} = \{B \in \mathcal{B} : \mu(B) = \sup\{\mu(F) : F \subset B \text{ closed}\} = \inf\{\mu(G) : G \supset B \text{ open}\}\}$. We will show that $\mathcal{D}$ is a $\sigma$-algebra that contains every open set, which will imply that it contains $\mathcal{B}$ . Easy for complements: if $F \subset E \subset G$ then $G^C \subset E^C \subset F^C$ and use the identity $V^C \setminus W^C = W \setminus V$. Harder for countable unions: $F_k \subset E_k \subset G_k \Rightarrow \cup F_k \subset \cup E_k \subset \cup G_k$. Since $\cup F_k$ need not be closed, we have to replace it by a finite union. For each $i \in \mathbb{N}$ we have $F_i$ such that $\mu(E_i \setminus F_i) < \varepsilon/2^{i+1}$ so by Exercise 1.2.8, $\mu(\cup_{i \in \mathbb{N}} E_i \setminus \cup_{i \in \mathbb{N}} F_i) \leq \sum_{i \in \mathbb{N}} \varepsilon/2^{i+1} = \varepsilon/2$. Thus we know: $\mu(\cup_{i \in \mathbb{N}} F_i) \geq \mu(\cup_{i \in \mathbb{N}} E_i) - \varepsilon/2$. Finally, we use continuity of measure to conclude that there exists $n$ such that $\mu(\cup_{i<n} F_i) > \mu(\cup_{i \in \mathbb{N}} E_i) - \varepsilon$. Thus, $\mathcal{D}$ is a $\sigma$-algebra. Finally, every open set is in $\mathcal{D}$: the approximation from above is trivial, and from below follows from Exercise 2.2.12.
**5.4.26.** First, $A = [(A \cup B) \setminus B] \cup (A \cap B)$. Second, $(A \cup B) \setminus B \in \mathfrak{M}$ because both $A \cup B$ and $B$ are in $\mathfrak{M}$. Finally, $A \cap B \in \mathfrak{M}$ because $A \cap B \subset B$ and $\mathfrak{M}$ is complete.
**5.4.28.** Let $E = A \cup B$, where $A \subset N$, $\mu(N) = 0$, and $B \in \mathfrak{M}$.
($\leq$) $\mu^*(E) \leq \mu^*(A) + \mu^*(B) = \mu^*(B) = \mu_0(E)$.
($\geq$) $\mu_0(E) = \mu_0(A) + \mu_0(B) = \mu_0(B) = \mu(B) \leq \mu^*(A \cup B) = \mu^*(E)$.
**5.5.6.** Hint: consider $A = E \cap P$.
**5.5.8.** By hypothesis, for every $n \in \mathbb{N}$, there exists $E_n \in \mathfrak{M}$ such that $\mu(E_n) < 1/n^2$ and $\nu(E_n^C) < 1/n^2$. Hence, $\sum_{n \in \mathbb{N}} \mu(E_n) \leq \sum_{n \in \mathbb{N}} 1/n^2 < \infty$ and $\sum_{n \in \mathbb{N}} \nu(E_n^C) \leq \sum_{n \in \mathbb{N}} 1/n^2 < \infty$. Let $A = \limsup_n E_n$. By Borel–Cantelli's lemma we get $\mu(A) = 0$ and $\nu(\limsup_n E_n^C) = 0$. By Exercise 1.1.25, $A^C = \liminf_n E_n^C$. Thus, $\nu(A^C) \leq \nu(\limsup_n E_n^C) = 0$. From $\mu(A) = 0$ and $\nu(A^C) = 0$ we obtain $\mu \perp \nu$.
**5.5.10.** Let $n \in \mathbb{N}$ and $(P_n, N_n)$ Hahn decomposition for $\nu - \mu/n$. Let $P = \cup P_n$, $N = \cap N_n = P^C$. Now, $N$ is a negative set for all $\nu - \mu/n$ so $0 \leq \nu(N) \leq \mu(N)/n \to 0$ hence $\nu(N) = 0$. If $\mu(P) = 0$ then $\nu \perp \mu$. If $\mu(P) > 0$ then there is $n$ s.t. $\mu(P_n) > 0$ so take $E = P_n$ and $\varepsilon = 1/n$.

**5.5.13.** Denote the supremum by $L(E)$. Claim $|\nu|(E) = L(E)$. ($\leq$) Let $(P, N)$ be a Hahn decomposition. Then $|\nu(E)| = \nu^+(E)+\nu^-(E) = \nu(E\cap P)-\nu(E\cap N) \leq |\nu(E\cap P)|+|\nu(E\cap N)| \leq L(E)$ by taking $P = F_1$, $N = F_2$. ($\geq$) Let $\{F_i\}$ be as above. By Exercise 5.5.12, $\sum |\nu(F_i)| = \sum |\nu^+(F_i) - \nu^-(F_i)| \leq \sum[\nu^+(F_i) + \nu^-(F_i)] = \sum |\nu|(F_i) = |\nu|(\cup F_i) \leq |\nu|(E)$. Now take supremum.
**5.5.15.** It is a signed measure by the paragraph at the beginning of Section 5.5. Let $L = \mu \wedge \nu$. Then $L = (\mu - \nu - |\mu - \nu|)/2 + \nu \leq \nu$ and $L = (\nu - \mu - |\mu - \nu|)/2 + \mu \leq \mu$. If $\omega \leq \mu, \nu$, denote $\mu_1 = \mu - \omega \geq 0$, and $\nu_1 = \nu - \omega \geq 0$. Then $L = (\mu_1 + \nu_1 - |\mu_1 - \nu_1|)/2 + \omega \geq \omega$, where the last inequality follows from Exercise 5.5.14. Indeed, it shows that $\lambda + \mu \geq |\nu|$, so apply it to $\lambda = \mu_1$, $\mu = \nu_1$, and $\nu = \mu_1 - \nu_1$.
**5.5.17.** Hint: in one direction use Exercise 5.5.15, in the other apply Exercise 5.5.14 to $\lambda = \mu - \nu$.
**5.5.23.** A Hahn decomposition $(A, B)$ for $\nu$ is $A = \mathbb{R} \setminus \{0\}$, $B = \{0\}$, so $\nu^+ = \nu(E \cap A) = m(E \cap A) - \mu(E \cap A) = m(E \cap A) = m(E \setminus \{0\}) = m(E)$, and $\nu^-(E) = \nu(E \cap B) = m(E \cap B) - \mu(E \cap B) = -\mu(E \cap B) = -\delta(E)$. Now, $|\nu| = m + \mu$ and $|\nu|(E) = m(E) + \delta(E)$. Thus, $|\nu|([0, 1]) = 1 + 1 = 2$.
**5.5.26.** Hint: prove for $(\mu + \nu)^+$ and $(\mu + \nu)^-$.
**5.5.28.** Hint: find Hahn decomposition for $\alpha\nu$.
**5.5.29.** ($\Rightarrow$) $\nu = \nu^+ - \nu^- \leq \nu^+ + \nu^- = |\nu|$, and $-\nu = -\nu^+ + \nu^- \leq \nu^+ + \nu^- = |\nu|$;

($\Leftarrow$) If $(P, N)$ is a Hahn decomposition for $\nu$ then $\nu^+(E) = \nu(E \cap P)$, $\nu^-(E) = -\nu(E \cap N)$ so $|\nu|(E) = \nu(E \cap P) - \nu(E \cap N) \leq \mu(E \cap P) + \mu(E \cap N) = \mu(E)$.
**6.1.5.** Hint: look for an example with $X$ a finite set.
**6.1.6.** ($\Leftarrow$) By Exercise 1.2.6, for any open set $G$, $(f|C)^{-1}(G) = f^{-1}(G) \cap C = [f^{-1}(G) \cap A] \cup [f^{-1}(G) \cap B] = (f|A)^{-1}(G) \cup (f|B)^{-1}(G) \in \mathfrak{M}$, because $\mathfrak{M}$ is closed under unions.

($\Rightarrow$) By Exercise 1.2.6, for any open set $G$, $(f|A)^{-1}(G) = f^{-1}(G) \cap A = f^{-1}(G) \cap A \cap C = (f|C)^{-1}(G) \cap A \in \mathfrak{M}$, because $\mathfrak{M}$ is closed under intersections.
**6.1.9.** Hint: consider Fat Cantor set.
**6.1.12.** Hint: use Luzin's Theorem and Borel–Cantelli Lemma.
**6.1.21.** The set $E := f^{-1}(\{0\}) \cap A$ is measurable. It is contained in $A$, so $\mu(E) \leq \mu(A)$ and it contains $B$, so $\mu(A) = \mu^*(B) \leq \mu^*(E) = \mu(E)$. Thus, $\mu(E) = \mu(A)$ and $f = 0$ on $E$ whence $f = 0$ a.e. on $A$.
**6.1.23.** Write $x \in [0, 1]$ as $x = \sum a_n/2^n$ where $a_n = 0$ or 1. Define $h(x) = \sum 2a_n/3^n$. Then $h$ is strictly increasing, hence invertible and Borel, by Exercise 6.1.19. Also, $h$ maps $[0, 1]$ onto a set $B$ which is a subset of the Cantor set, hence measurable and of measure 0. Define $g = f \circ h^{-1}$ if $x \in B$ and 0 otherwise. Then $g$ is measurable because $g^{-1}(G) = h[f^{-1}(G)] \subset h([0, 1]) = B$. Finally, $f = g \circ h$.
**6.1.26.** Hint: consider what it takes for $f^{-1}(\{x\})$ to be measurable.
**6.1.27.** (a) This is Exercise 4.3.20.

(b) ($\Rightarrow$) If $f$ is $\mathcal{F}$-measurable then, for each Borel set $G$, $f^{-1}(G) \in \mathcal{F} \subset \mathcal{A}$, whence $f$ is $\mathcal{A}$-measurable. Suppose $f \neq f \circ T$. Then there is $x \in X$ s.t. $f(x) = a \neq b = f(T(x))$ so $x \in f^{-1}(a)$ and $T(x) \in f^{-1}(b)$ Since $f$ is $\mathcal{F}$-measurable, $f^{-1}(b) \in \mathcal{F}$ so $f^{-1}(b) = Tf^{-1}(b)$ hence $x \in f^{-1}(a)$ and $x \in f^{-1}(b)$, a contradiction.

($\Leftarrow$) If $f$ is $\mathcal{A}$-measurable then $f^{-1}(G) \in \mathcal{A}$ for each Borel set $G$. Since $f = f \circ T$ we have $f^{-1}(G) = (f \circ T)^{-1}(G) = T^{-1} \circ f^{-1}(G)$ so $T[f^{-1}(G)] = f^{-1}(G)$ whence $f^{-1}(G) \in \mathcal{F}$.
**6.1.28.** Hint: prove that $f(\mathbb{R})$ is countable.
**6.1.30.** (a) Straightforward from Exercise 1.2.7 and Example 1.2.4.

(b) By Exercise 4.3.10, the family $\mathcal{A}' = \{f^{-1}(B) : B \in \sigma(\tau')\}$ is a $\sigma$-algebra on $X$. It contains $\tau$ and hence also $\sigma(\tau) = \mathfrak{M}$. Notice that $f^{-1}$ is measurable which means that if $E \in \mathfrak{M}$ then $f(E) \in \mathfrak{N}$. Therefore, $\{f(A) : A \in \mathcal{A}'\} \supseteq \mathfrak{N}$. But by construction, $\{f(A) : A \in \mathcal{A}'\} = \sigma(\tau')$ and therefore $\mathfrak{N} \subseteq \sigma(\tau')$. The opposite inclusion is obvious because $\tau' \subset \mathfrak{N}$ implies $\sigma(\tau') \subset \mathfrak{N}$.
**6.2.7.** For $k \geq 1$ partition the interval $[a, b]$ into $2k$ intervals $I_1^{(k)}, \dots, I_{2^k}^{(k)}$ of length $2^{-k}(b - a)$. For $t \in \mathbb{R}$ and $1 \leq n \leq 2^k$ let $L_n^{(k)}(t) = \chi_{f(I_n^{(k)})}(t)$. Note that $t \mapsto L_n^{(k)}(t)$ is measurable because it has at most two points of discontinuity, therefore the function $N_k(t) = L_1^{(k)}(t) + \cdots + L_{2^k}^{(k)}(t)$ is also measurable. Note that $N_k(t)$ simply counts on how many of the intervals $I_1^{(k)}, \dots, I_{2^k}^{(k)}$ the function $f$ attains the value $t$ at least once. Thus, the sequence $\{N_k\}_{k=1}^{\infty}$ of measurable functions is increasing and therefore the pointwise limit $F(t) = \lim_k N_k(t)$ exists and is a measurable function of $t$. Observe that by definition $N(t) \geq N_k(t)$ for all $k$, so $N(t) \geq F(t)$. Let us argue that $F(t) \geq N(t)$. Let $m$ be an integer such that $N(t) \geq m$. Then there exist points $a \leq x_1 < \cdots < x_m \leq b$ such that $f(x_i) = t$. If $k$ is so large that $(b - a)/2^k < \min\{x_{i+1} - x_i : i = 1, \dots, m - 1\}$ then $N_k(t) \geq m$. This shows $F(t) \geq N(t)$ and thus $F(t) = N(t)$, establishing measurability. Remark: $N$ is known as Banach indicatrix.
**6.2.9.** Let us use notation $f_i(x) = n_i$. Then $\{x : f_i(x) = n\}$ is the union of intervals $[0.n_1 \dots n_{k-1}n, 0.n_1 \dots n_{k-1}(n + 1))$ and $\{x : f_i(x) \leq n\} = \cup_{k \leq n}\{x : f_i(x) = k\}$. Therefore, each

$f_i$ is measurable and by Proposition 6.2.6, so is $f$. Finally, we will show that the set of numbers that have a 9 in its decimal representation has measure 1. For those in the first decimal, the measure is $\alpha_1 = 1/10$. Among the first two, $\alpha_2 = \alpha_1 + (1/10)(1-\alpha_1) = (1/10)+(9/10)(1/10)$. Among the first 3, $\alpha_3 = \alpha_2 + (1/10)(1-\alpha_2) = (1/10)+(9/10)\alpha_2 = (1/10)+(9/10)(1/10)+(9/10)^2(1/10)$. Continuing, we obtain the sum $(1/10)\sum_{k=0}^{\infty}(9/10)^k = 1$.
**6.2.13.** Hint: show that $\limsup_{y\to\infty} f(x,y) = \inf_n \sup_{y\in\mathbb{Q},y\geq n} f(x,y)$.
**6.2.14.** For every $\varepsilon > 0$ there exists a continuous $g_\varepsilon : [a,b] \to \mathbb{R}$ such that the set $\{x \in [a,b] : f(x) \neq g_\varepsilon(x)\}$ is measurable and its measure is less than $\varepsilon$. Define $A_n := \cup_{k\geq n}\{x \in [a,b] : f(x) \neq g_{2^{-k}}(x)\}$. Now $A_n \supset A_{n+1} \supset A_{n+2} \dots$ and $\mu(A_n) \leq \sum_{k=n}^{\infty} \mu(\{x \in [a,b] : f(x) \neq g_{2^{-k}}(x)\}) \leq \sum_{k=1}^{\infty} 1/2^k = 1$. Also, $\lim_n \mu(A_n) \leq \lim_n \sum_{k=n}^{\infty} 1/2^k = 0$. Now by the continuity of measure $\mu(\cap_{n\in\mathbb{N}} A_n) = \lim_n \mu(A_n) = 0$. Consider now $x \notin \cap_{n\in\mathbb{N}} A_n$. Then there exists $j$ s.t. $x \notin A_j$. So $x \notin \cup_{k\geq j}\{x \in [a,b] : f(x) \neq g_{2^{-k}}(x)\}$. Thus, $g_{2^{-k}}(x) = f(x)$ for all $k \geq j$ hence $\lim g_{2^{-n}} = f$ a.e. Therefore, $f$ is a pointwise limit of continuous functions a.e. so $f$ is measurable.
**6.2.16.** It suffices to prove the case when $g = 0$, because $h = h_1 + h_2$, where $h_1 = f$ on $A$ and 0 on $A^C$, and $h_2 = g$ on $A^C$ and 0 on $A$. Let $B := \{x : f > \alpha\}$. If $\alpha \geq 0$ then $\{x : h > \alpha\} = \{x : f > \alpha\} \in \mathfrak{M}$. If $\alpha < 0$ then $\{x : h > \alpha\} = \{x : f > \alpha\} \cup A^C \in \mathfrak{M}$.
**6.2.17.** Hint: use Theorem 5.4.7.
**6.2.24.** Suppose first that $f = \chi_E$ and $E \in \mathfrak{M}_0$. By Theorem 5.4.13, $E = A \cup B$, where $A \in \mathfrak{M}$ and $B \subset Z$ where $Z \in \mathfrak{M}$ and $\mu(Z) = 0$. Let $g := \chi_A$, so that $g$ is $\mathfrak{M}$-measurable, and $W := \{x \in X : f(x) \neq g(x)\}$. Note that $W \subset B \subset Z$. Since $Z \in \mathfrak{M}$ and $\mu(Z) = 0$, we have that $W \in \mathfrak{M}_0$ and $\mu_0(W) = 0$ whence $f = g$ a.e. $[\mu_0]$. In fact, $f = g$ on $Z^C$ and $\mu(Z) = 0$, so $f = g$ a.e. $[\mu]$. The case when $f$ is a simple function follows easily. When $f$ is a nonnegative measurable function, Theorem 6.2.13 provides a sequence $\{s_n\}$ of simple functions s.t. $s_n \uparrow f$. By the previous, for each $n$ there is a simple function $t_n$ such that $t_n$ is $\mathfrak{M}$-measurable, $s_n = t_n$ on $Z_n^C$, and $\mu(Z_n) = 0$. It follows that $t_n = s_n \uparrow f$ on $\cap_n Z_n^C = (\cup_n Z_n)^C$ which has $\mu$ measure 0. Since each $t_n$ is $\mathfrak{M}$-measurable, Exercise 6.2.4 ensures that its limit $h$ is equal a.e. $[\mu]$ to a $\mathfrak{M}$-measurable function $g$, whence $f = g$ except on a set of $\mu$ measure 0. The general case is reduced to the previous since $f = f^+ - f^-$.
**6.2.25.** Hint: use Corollary 6.2.4, Exercise 6.1.29), and Proposition 6.2.6). Also, adopt the proof of Theorem 6.2.13.
**6.3.3.** $m\{x \in [0,1] : x^n > \eta\} = 1 - \eta^{1/n} \to 0$.
**6.3.9.** Let $E_k = \{x \in E : |f(x) - g(x)| > 1/k\}$. For $x \in E_k$ we have $1/k \leq |f(x) - f_n(x)| + |f_n(x) - g(x)|$ so either $|f(x) - f_n(x)| \geq 1/(2k)$ or $|g(x) - f_n(x)| \geq 1/(2k)$, which means that $m(E_k) \leq m\{x \in E : |f_n(x) - f(x)| > 1/(2k)\} + m\{x \in E : |f_n(x) - g(x)| > 1/(2k)\} \to 0$ as $n \to \infty$. Therefore $m(E_k) = 0$ for all $k$, so $m(\cup_{k=1}^{\infty} E_k) = 0$. Since $f = g$ on $E \setminus \cup_{k=1}^{\infty} E_k$, $f = g$ a.e. on $E$.
**6.3.13.** Hint: consider $f_n = \chi_{(n,\infty)}$.
**6.3.17.** First we prove for $f = \chi_E$. By Theorem 5.3.8, for every $n$ there is a closed set $F_n$ and an open set $G_n$ s.t. $F_n \subset E \subset G_n$ and $m(G_n \setminus F_n) < 1/n$. Let $h_n(x) = d(x, G_n^C)/[d(x, G_n^C) + d(x, F_n)]$. By Exercise 2.3.37, $h_n$ is continuous. It equals $f$, except on $G_n \setminus F_n$, so $h_n \xrightarrow{m} f$ and Theorem 6.3.10 implies that there is a subsequence of $\{h_n\}$ that converges a.e. The case when $f$ is a simple function follows easily. If $f \geq 0$, by Theorem 6.2.13, $f = \lim s_n$. As before, there are continuous functions $\{h_n\}$ s.t. $m\{x : h_n(x) \neq s_n(x)\} < 1/[n(n+1)]$. Let $A_n = \{x : h_n(x) \neq s_n(x)\}$ and $A = \limsup A_n$. Then $m(A) \leq m(\cup_{i\geq n} A_i) \leq \sum_{i\geq n} m(A_i) < \sum_{i\geq n} 1/[i(i+1)] = 1/n \to 0$, so $m(A) = 0$. By Exercise 1.1.28, if $x \in A^C$ then $h_n(x) = s_n(x)$ for $n \geq N = N(x)$, so $h_n \to f$ pointwise on $A^C$, hence a.e. Finally, $f = f^+ - f^-$.
**6.3.19.** Hint: use Exercise 6.3.12.
**6.3.23.** Let $\mathcal{A}$ be the smallest such $\sigma$-algebra. Since $\mathcal{B}$ is one such, we have that $\mathcal{A} \subset \mathcal{B}$. Let $F$ be a closed set in the given metric space. By Exercise 2.2.12, $F = \cap G_n$, where $\{G_n\}$ are open sets. For every $n$, $F$ and $G_n^C$ are closed disjoint sets, so $f_n(x) := d(x, G_n^C)/[d(x, G_n^C) + d(x, F)]$ is continuous by Exercise 2.3.37, and $0 \leq f_n \leq 1$. Define $f = \sum f_n/2^n$. The series converges uniformly, so $f$ is also continuous and bounded. Notice that $f(x) = 1$ if and only if $x \in F$ so $F = f^{-1}(\{1\})$. In order for the continuous function $f$ to be measurable, the set $F$ must be in $\mathcal{A}$. Thus $\mathcal{A}$ contains every closed set, hence the whole $\mathcal{B}$.
**6.3.24.** Let $\phi(x) = 2x$ on $[0,1/2]$, $\phi(x) = 2-2x$ on $[1/2,1]$ and 0 elsewhere. Then $\lim_n \phi(2^n x) = 0$ for all $x \in \mathbb{R}$. Let $\{r_n\}$ be an enumeration of $\mathbb{Q}$ and $f_n(x) := \sum_k 2^{-k}\phi(2^n(x - r_k))$. The series converges uniformly, so $f_n$ is continuous and $\lim_n f_n(x) = 0$ for all $x \in \mathbb{R}$. For any $(a,b)$, there is $r_k \in (a,b)$ so if $n$ is large enough then $x = r_k + 1/2^{n+1} \in (a,b)$ and $2^n(x - r_k) = 1/2$ hence $\phi(2^n(x - r_k)) = 1$ so $f_n(x) \geq 1/2^k$ so $\sup\{f_n(x) : a < x < b\} \geq 1/2^k$ whence $\{f_n\}$ does not converge uniformly on $(a,b)$. If $E$ is any set on which $\{f_n\}$ converges uniformly, then $\sup\{f_n(x) : x \in E\} = \alpha_n \to 0$. Since $f_n$ is continuous, the same result is obtained when supremum is taken over clos $(E)$ (Exercise 2.3.31), so $\{f_n\}$ converges uniformly on clos $(E)$. By the previous clos $(E)$ cannot contain an interval so it is nowhere dense.

**6.3.25.** ($\Rightarrow$) Suppose there is $\varepsilon > 0$ s.t. $\mu(\limsup_n E_{n,\varepsilon}) \neq 0$. Then, the set of all $x$ s.t. $x \in E_{n,\varepsilon}$ for infinitely many $n$ is a set of positive measure. For each such $x$, there is a subsequence $\{f_{n_k}\}$ s.t. $f_{n_k}(x) \geq \varepsilon$, so for such $x$, $\lim f_n(x) \neq 0$ hence $\{f_n\}$ does not converge to 0 a.e.

($\Leftarrow$) Suppose that $\{f_n\}$ does not converge to 0 a.e. Then there is a set $A$ of positive measure such that for $x \in A$, $\lim f_n(x) \neq 0$. Let $\varepsilon > 0$. For $x \in A$ there is a subsequence $\{f_{n_k}\}$ such that $f_{n_k}(x) \geq \varepsilon$, i.e. $x \in E_{n_k,\varepsilon}$ so $x \in \limsup_n E_{n,\varepsilon}$ hence $A \subset \limsup_n E_{n,\varepsilon}$ whence $\mu(\limsup_n E_{n,\varepsilon}) \geq \mu(A) > 0$.

**6.3.27.** Hint: use Exercise 6.3.12 and $f_n(x) = x + (1/n)\chi_{(0,n)}$.

**6.3.30.** Hint: write $X = \cup_{n=1}^{\infty} Y_n$ and apply Egorov's Theorem to each $Y_n$.

**7.2.2.** By definition, $\int_E f\,d\mu = \sup\{\int_E s : s \in \mathcal{S}(X),\ 0 \leq s \leq f \text{ on } E\} = \sup\{\int_X s\chi_E : s \in \mathcal{S}(X),\ 0 \leq s \leq f \text{ on } E\} = \sup\{\int_X s\chi_E : s\chi_E \in \mathcal{S}(X),\ 0 \leq s \leq f \text{ on } E\} = \sup\{\int_X t : t \in \mathcal{S}(X),\ 0 \leq t \leq f\chi_E \text{ on } E\} = \int_X f\chi_E$.

**7.2.6.** Let $M = \sup f$. Then, using Definition 7.2.9, $\int M - \int f = \int(M - f) = \sup\{\int t : t \in \mathcal{S}(X),\ 0 \leq t \leq M - f\} = \sup\{\int M - \int(M - t) : t \in \mathcal{S}(X),\ 0 \leq f \leq M - t\} = \int M + \sup\{-\int(M - t) : t \in \mathcal{S}(X),\ 0 \leq f \leq M - t\} = \int M - \inf\{\int(M - t) : t \in \mathcal{S}(X),\ 0 \leq f \leq M - t\}$. Subtracting $\int M$ from both sides and multiplying by $-1$, we get $\int f = \inf\{\int(M - t) : t \in \mathcal{S}(X),\ 0 \leq f \leq M - t\} = \inf\{\int s : s \in \mathcal{S}(X),\ 0 \leq f \leq s\}$.

No, if $\mu(X) = \infty$! Take $f > 0$ that is integrable (e.g. $1/(1 + x^2)$). Then any simple function $s \geq f$ must also be positive everywhere, so if $s = \sum c_k\chi_{A_k}$ then $c_k > 0$ for all $k = 1, \ldots, n$. Also, some $A_j$ must have infinite measure, i.e., $m(A_j) = \infty$. Therefore $\int s = \sum c_k m(A_k) \geq c_j m(A_j) = \infty$.

**7.2.10.** Hint: consider the sets $E_n = \{x \in E : f(x) > 1/n\}$ and $E = \{x \in E : f(x) > 0\}$.

**7.2.12.** ($\Rightarrow$) If $A$ is as above, then $\int_A f = \int_X f\chi_A \leq \int_X f < \infty$ so $\sup\{\int_A f\} \leq \int_X f$.

($\Leftarrow$) Suppose $f$ not integrable. Let $C > 0$. We show that there is $A \in \mathfrak{M}$ such that $\mu(A) < \infty$ and $f|A$ is bounded but $\int_A f > C$. First, there is simple function $s$ s.t. $0 \leq s \leq f$ and $\int_X s \geq C+1$. Let $E_n = \{x : f(x) \in [n - 1, n)\}$. By Corollary 7.2.15, $\int_X s = \int_X \sum \chi_{E_n} s = \sum \int_{E_n} s$, so there is $N$ s.t. $\sum_{k=1}^{N} \int_{E_k} s \geq C$. Define $A = \cup_{k=1}^{N} E_k$ if all sets $E_k$ have finite measure. If any of the sets $E_k$ has infinite measure, we can replace it by a set $A_k \subset E_k$ of finite measure such that $\mu(A_k) > C/(k - 1)$. Indeed, in that case $\int_{A_k} f \geq (k - 1)\mu(A_k) > C$.

**7.2.13.** Hint: replace $f$ with $\sum_{k=0}^{\infty} c_k\chi_{E_k}$.

**7.2.16.** Since $\{f_n\}$ are nonnegative, we can apply Fatou's Lemma to obtain $F(x) = \int_0^x f = \int_0^x \liminf f_n\,dm \leq \liminf \int_0^x f_n\,dm = \liminf F_n(x)$. Using Fatou's Lemma once again, $\int_0^{10}(f + F)\,dm \leq \int_0^{10}(f + \liminf F_n)\,dm = \int_0^{10} \liminf(f_n + F_n)\,dm \leq \liminf \int_0^{10}(f_n + F_n)\,dm$.

**7.2.18.** ($\Rightarrow$) $\{f_1 - f_n\}$ is increasing, nonnegative so, by Monotone Convergence Theorem, $\lim \int(f_1 - f_n) = \int \lim(f_1 - f_n) = \int f_1$ hence $\int f_1 = \int f_1 - \lim \int f_n$ whence $\lim \int f_n = 0$.

($\Leftarrow$) Since $\{f_n\}$ is monotone, there is $f := \lim f_n$. By Monotone Convergence Theorem, $\lim \int(f_1 - f_n) = \int \lim(f_1 - f_n) = \int(f_1 - f)$. By linearity of integration (Theorem 7.2.14), the left side equals $\int f_1 - \lim \int f_n = \int f_1$ and the right side equals $\int f_1 - \int f$ so $\int f = 0$. By Proposition 7.2.20, $f = 0$.

**7.2.20.** Hint: start with $f = \chi_A$.

**7.2.25.** Hint: consider sets $A = \{x : f(x) > 1\}$ and $B = \{x : f(x) < 1\}$.

**7.2.29.** Hint: consider $f_k(x) = 1/\sqrt{x - r_k}$ and $f = \sum f_k/2^k$.

**7.2.29.** By Exercise 6.2.23, there is an increasing sequence $\{s_n\}$ of simple functions with finite support $A$ that converges to $f$, and by Monotone Convergence Theorem, $\int_X s_n \to \int_X f$. Thus, there is $n \in \mathbb{N}$ for which $\int_X s_n > \int_X f - \varepsilon$ hence $\int_X |f - s_n| = \int_X(f - s_n) < \varepsilon$.

**7.2.35.** Suppose $f_n \to f$ in measure on $X$. Fix $\varepsilon > 0$ and let $\eta = \varepsilon/m(X)$. Let $X' := \{x \in X : |f_n(x) - f(x)| > \eta\}$. Then $d(f_n, f) = \int_X |f_n - f|/(1 + |f_n - f|) = \int_{X'} |f_n - f|/(1 + |f_n - f|) + \int_{X \setminus X'} |f_n - f|/(1+|f_n - f|) \leq m(X') + \eta m(X) < m(\{x \in X : |f_n(x) - f(x)| > \eta\} + \varepsilon$, where the first inequality follows because $|f_n - f|/(1 + |f_n - f|) \leq 1$ on $X'$ and $|f_n - f|/(1 + |f_n - f|) \leq \eta$ on $X \setminus X'$. Since $f_n \to f$ in measure, $\limsup_n d(f_n, f) \leq \varepsilon$ whence $\lim_n d(f_n, f) = 0$.

Conversely, suppose $\lim_n d(f_n, f) = 0$. Fix $\eta > 0$. The function $x/(1+x)$ is strictly increasing on $[0, \infty)$ so, using Chebyshev's inequality, $m(|f_n - f| > \eta) \leq m(|f_n - f|/(1 + |f_n - f|) > \eta/(1 + \eta)) \leq (1 + \eta)/\eta \int_X |f_n - f|/(1 + |f_n - f|) = [(1 + \eta)/\eta]\,d(f_n, f) \to 0$ as $n \to \infty$.

**7.2.38.** Suppose that there is $F \neq 0$ and let $f \in M$ satisfy $F(f) \neq 0$. Since $F$ is continuous there is $\delta > 0$ such that if $g \in M$ and $d(g, 0) < \delta$ then $|F(g)| < |F(f)|$. Choose $N > 1/\delta$. Let $\chi_k$ be the characteristic function of $((k - 1)/N, k/N)$ and $f_k = Nf\chi_k$. Then $d(f_k, 0) = \int_0^1 |f_k|/(1 + |f_k|) = \int_{(k-1)/N}^{k/N} N|f|/(1 + N|f|) \leq \int_{(k-1)/N}^{k/N} 1 = 1/N < \delta$, so $|F(f_k)| < |F(f)|$. However, $f = (f_1 + \cdots + f_N)/N$, so $|F(f)| = |F(f_1) + \cdots + F(f_N)|/N \leq (|F(f_1)| + \cdots + |F(f_N)|)/N < F(f)$, a contradiction.

**7.2.41.** ($\Rightarrow$) Suppose that $X = \cup X_n$ and $\mu(X_n) < \infty$. Without loss of generality, $\{X_n\}$ are disjoint and $\mu(X_n) > 0$. Define $f = \sum 1/[n^2\mu(X_n)]\chi_{X_n}$.

($\Leftarrow$) Let $X_n = \{x : f(x) \geq 1/n\}$. By Chebyshev's Inequality, $\mu(X_n) < n \int f < \infty$ and $X = \cup X_n$.

**7.3.2.** Hint: reduce to $f \geq 0$ and consider $E_m = \{x \in X : x_m \leq f(x) \leq x_{m+1}\}$ and $g_P = \sum x_k \chi_{E_k}$, $h_P = \sum x_{k+1}\chi_{E_k}$.
**7.3.4.** Hint: apply summation by parts to $E_n := \{x : M/n \geq |f(x)| \geq M/(n+1)\}$.
**7.3.10.** First, $f$ is integrable if and only if $f^+$ and $f^-$ are integrable over $X$. For a nonnegative $f$, denote $f_n = f\chi_{X_n}$. Then $f = \sum f_n$ and $\int f = \sum \int f_n$ so $f$ is integrable if and only if each $f_n$ is integrable and the series $\sum \int f_n$ converges. For a general $f$, $f_n = f\chi_{X_n}$, $f_n^+ = f^+\chi_{X_n}$ and $f_n^- = f^-\chi_{X_n}$, so $f$ is integrable if and only if each $f_n$ is integrable and both series $\sum \int f_n^+$ and $\sum \int f_n^-$ converge, which is equivalent to $\sum \int |f_n| < \infty$ i.e., $\sum \int_{X_n} |f| < \infty$.
**7.3.12.** There is $\{f_{n_k}\}$ such that $\lim_k \int_X f_{n_k}\, d\mu = \liminf_n \int_X f_n\, d\mu$. Since $\{f_{n_k}\}$ also converges in measure to $f$, there is $\{g_n\} \subset \{f_{n_k}\}$, which converges pointwise to $f$ a.e. Since $g_n + g \geq 0$, Fatou's lemma yields $\int_X \lim_n (g_n + g) \leq \lim_n \int_X (g_n + g)$ hence, using the linearity of the Lebesgue integral, $\int_X f = \int_X \lim_n g_n \leq \lim_n \int_X g_n = \lim_k \int_X f_{n_k} = \liminf_n \int_X f_n$.
**7.3.16.** Hint: apply Fatou's Lemma to $g_n + |f| - |f_n - f|$.
**7.3.19.** (a) Let $f_n = n(|1 + f/n| - 1)$. Since $f$ is integrable it is finite a.e., and when it is finite $f/n \to 0$, so for $n$ large, $f/n > -1$ hence $f_n = f$. Thus, $\lim f_n = f$ a.e. Also $|f_n| = n(|1 + f/n| - 1) \leq n(1 + |f|/n - 1| = |f|$. Now apply Dominated Convergence Theorem to obtain that the limit is $\int f$.

(b)($\Rightarrow$) $|1 + \lambda f| - 1 \geq \lambda f$ so $\int(|1 + \lambda f| - 1) \geq \int \lambda f = 0$ hence $\int |1 + \lambda f| \geq m(E)$.

($\Leftarrow$) Take $\lambda = 1/n$. Then $0 \leq n\int(|1 + f/n| - 1) \to \int f$. Take $\lambda = -1/n$. Then $0 \leq n\int(|1 - f/n| - 1) \to \int(-f)$. From these 2 inequalities, we get $\int f = 0$.
**7.3.20.** Hint: use $\ln(1 + t) \leq 2\sqrt{t}$.
**7.3.22.** The function $g(t) := \ln(1+e^t) - t$ is decreasing for $t > 0$, so it attains its maximum at $t = 0$ and $g(0) = \ln 2$. Therefore, $\ln(1 + e^t) \leq t + \ln 2$, when $t \geq 0$. For $t < 0$ we have $\ln(1 + e^t) \leq \ln 2$. Thus, $\ln(1 + e^{nf(x)}) \leq nf(x) + \ln 2$ when $f(x) \geq 0$, and $\ln(1 + e^{nf(x)}) \leq \ln 2$ when $f(x) < 0$. We conclude that $\ln(1 + e^{nf(x)})/n \leq f^+(x) + (\ln 2)/n$, so we cann apply Dominated Convergence Theorem to obtain $\int_0^1 \lim_n \ln(1 + e^{nf(x)})/n = \int_0^1 f^+$.
**7.3.24.** By Mean Value theorem, there is $0 < s < 1$ s.t. $\phi(tg(x)) = tg(x)\phi'(stg(x))$. $\phi$' is bounded and $g$ integrable so $f$ is well defined. Next, $|[\phi(sg(x)) - \phi(tg(x)]/[s-t] - \phi'(tg(x))g(x)| = |\phi'(rg(x))g(x) - \phi'(tg(x))g(x)| \leq 2\sup|\phi'||g|$ which is integrable. Apply Dominated Convergence Theorem, when $s \to t$.
**7.3.28.** Hint: use continuity of measure.
**7.3.35.** Suppose that $\mu\{x \in X : f(x) \notin F\} > 0$. Since $F^C$ is open, it is a countable union of open intervals $I_n := (a_n - r_n, a_n + r_n)$. Then there is $k$ s.t. $E := f^{-1}(I_k)$ has positive measure and $f(E) \subset I_k$. Therefore, $|(1/\mu(E))\int_E f\, d\mu - a_k| = |(1/\mu(E))\int_E (f - a_k)\, d\mu \leq (1/\mu(E))\int_E |f - a_k|\, d\mu < (1/\mu(E))\int_E r_k\, d\mu = r_k$, whence $(1/\mu(E))\int_E f\, d\mu \in I_k \subset F^C$, contradicting the assumption that it should be in $F$.
**7.3.39.** Let $A = \{x : f(x) \geq g(x)\}$. Then $2\omega(E) = 2\int_A f\, d\mu + 2\int_{E\setminus A} g\, d\mu = \int_E f\, d\mu + \int_E g\, d\mu + \int_A (f - g)\, d\mu + \int_{E\setminus A}(g - f)\, d\mu = \lambda(E) + \nu(E) + (\lambda - \nu)^+(E) + (\lambda - \nu)^-(E) = \lambda(E) + \nu(E) + |\lambda - \nu|(E) = 2(\lambda \vee \nu)(E)$. For the other, replace $f$ and $g$ by $-f$ and $-g$. Obtain $2\int_E \max\{-f, -g\}\, d\mu = -\lambda(E) - \nu(E) + |-\lambda + \nu|(E)$ so $-2\int_E \min\{f, g\}\, d\mu = -\lambda(E) - \nu(E) + |-\lambda + \nu|(E)$ hence $2\rho = \lambda + \nu - |\lambda - \nu| = 2\lambda \wedge \nu$.
**7.3.41.** ($f \geq 0$): Let $g := \sup\{s : s \in \mathcal{S}(X),\ s \leq f \text{ on } X\}$, $h := \inf\{t : t \in \mathcal{S}(X),\ t \geq f \text{ on } X\}$. Then $\int_X g = \int_X h$. Indeed, $\int_X g = \sup\left\{\int_X s : s \in \mathcal{S}(X),\ s \leq f \text{ on } X\right\}$ $= \inf\left\{\int_X t : t \in \mathcal{S}(X),\ t \geq f \text{ on } X\right\} = \int_X h$. (See Exercise 7.2.6.) Also, it is obvious that $g \leq f \leq h$ so $g \leq h$ hence $g = h$ a.e. The set $\{x : g(x) < f(x)\} \subset \{x : g(x) < h(x)\}$ and the latter has measure 0 so the former has measure 0 with respect to the completion. Thus $f = g$ a.e. hence $f$ is measurable.

If $f$ is not nonnegative, we use boundedness. There is $M$ s.t. $|f| \leq M$ so $f + M \geq 0$. Given that $f$ vanishes outside of the set of finite measure so do all simple functions involved hence $s + M$ and $t + M$ are also simple functions, and $s + M \leq f + M \leq t + M$. We obtain by the first part that $f + M$ is measurable whence $f$ is measurable.
**7.4.4.** Hint: take $f = \chi_G$, where $G$ is an open subset of $[0, 1]$ that contains all rationals.
**7.4.11.** Since $|f|$ is continuous on $(0, 1]$ and $|f|$ is Lebesgue integrable over $[0, 1]$, we may assume that $f \geq 0$. Now, L $-\int_{a_n}^b f =$ R $-\int_{a_n}^b f$. When $n \to \infty$, the left side converges, by Monotone Convergence Theorem, to L$-\int_a^b f$, which is finite by assumption. Thus, the right side converges to a finite number. By definition, that means that the improper R-integral of $f$ over $[a, b]$ converges.
**7.4.15.** Yes, if $f \geq 0$. In that case $\int_0^1 f = \lim \int_{1/n}^1 f$ by Monotone Convergence Theorem.

No, without $f \geq 0$. On each interval $[1/(n+1), 1/n]$ define $f$ to be the function whose graph consists of two triangles of the same area, one above the other below the $x$-axis, both with height $n$. Then $f$ is continuous on $(0, 1]$, $\int_{1/n}^1 f = 0$ but $f$ is not integrable over $[0, 1]$, because $|f|$ is not integrable. Reason: $\int_{1/(n+1)}^{1/n} |f| = 1/(n + 1)$ and the sum of the areas goes to infinity.

**7.4.18.** Riemann proper: Yes! By Theorem 7.4.3, $f, g$ are continuous a.e. hence, by Exercise 2.3.29, so is $h$.

Riemann improper: No! Take $a_k = 1/2^k$ and define $f = 2^k/k$ on $[a_{2k}, a_{2k+1})$ and $f = -2^k/k$ on $[a_{2k+1}, a_{2k+2})$. Then the Rieman improper integral converges as an alternating harmonic series. If $g = -f$, then $h = |f|$ which is not integrable because now we have harmonic series.

Lebesgue: Yes! Notice that $\max\{f, g\} = (f+g)/2 + |f(x) - g(x)|/2$ and the result follows from Theorem 7.3.6 and the fact that a function $h$ is Lebesgue integrable if and only if $|h|$ is.

**7.4.19.** The function $f(x) = xe^{-x^2}$ has a convergent improper Riemann integral over $[0, \infty)$ so by Theorem 7.4.8, $f$ is Lebesgue integrable. Since $\sin(x/n) \le x/n$, we get $\chi_{[0,n^2]} e^{-x^2} n \sin \frac{x}{n} \le f(x)$ on $[0, \infty)$, so by Dominated Convergence Theorem we have $\int_0^\infty \lim_n \chi_{[0,n^2]} e^{-x^2} n \sin \frac{x}{n} = \int_0^\infty xe^{-x^2}$. Appealing again to Theorem 7.4.8, we see that this Lebesgue integral equals the improper Riemann integral which is 1/2.

**7.4.25.** Hint: consider $f(x) = x^\alpha \sin(x^\beta)$.

**7.4.27.** The function $f(x) = x^{-3/4}$ has a convergent improper Riemann integral over $[0, 1]$ so by Theorem 7.4.4, $f$ is Lebesgue integrable. Further, $n^3 x^{3/4}/(1 + n^4 x^2) \le f(x)$. Indeed, we need to establish the inequality $n^3 x^{3/2}/(1 + n^4 x^2) \le 1$ or, equivalently, $u^3/(1 + u^4) \le 1$. If $u \in [0, 1]$ then $u^3/(1 + u^4) \le u^3 \le 1$. If $u \ge 1$ then $u^3/(1 + u^4) \le u^3/u^4 \le 1$. By Dominated Convergence Theorem, we obtain $\int_0^1 \lim_n n^3 x^{3/4}/(1 + n^4 x^2) = 0$.

**7.4.28.** Hint: prove that $\sum |a_k|/\sqrt{|x - r_k|}$ is integrable.

**7.4.29.** (a) The function $f$ has a convergent improper Riemann integral over $\mathbb{R}$ so by Theorem 7.4.4, $f$ is Lebesgue integrable and $\int_{\mathbb{R}} f = 2$. By Corollary 7.2.15, $\int_{\mathbb{R}} |g| = \sum_{n=1}^\infty 2^{-n} \int_{\mathbb{R}} f(x - r_n)\, dx = \sum_{n=1}^\infty 2^{-n} \int_{\mathbb{R}} f = 2$, so $g$ is integrable.

(b) Let $I \subset \mathbb{R}$ be a nondegenerate interval. There exists $n \in \mathbb{N}$ such that $r_n$ is an interior point of $I$. Let $\{x_k\} \subset I$ be a sequence that converges to $r_n$. Then $\lim_k g(x_k) \ge \lim_k f(x_k - r_n)/2^n = \infty$.

(c) Follow from (b) and Exercise 2.3.29.

(d) Let $E \subset \mathbb{R}$ be a set of measure zero and suppose that $h$ is integrable and is equal to $g$ on $E^C$. For each $k \in \mathbb{N}$, the set $(r_n, r_n + 1/k) \cap I$ has positive measure, so the sequence $\{x_k\}$ can be selected in $((r_n, r_n + 1/k) \cap I) \setminus E$. Then $\lim_k h(x_k) = \lim_k g(x_k) = \infty$ so $h$ has property (b) and (c) follows from (b) and Exercise 2.3.29.

(e) By Proposition 7.3.4, $g < \infty$ a.e. so $g^2 < \infty$ a.e. Let $I \subset \mathbb{R}$ be a nondegenerate interval. There exists $n \in \mathbb{N}$ such that $r_n$ is an interior point of $I$ and $\delta > 0$ such that $(r_n, r_n + \delta) \subset I$. Then, $\int_I g^2 \ge \int_{r_n}^{r_n+\delta} g^2 \ge \int_{r_n}^{r_n+\delta} 2^{-2n} f(x - r_n)^2\, dx = 2^{-2n} \int_0^\delta f^2 = 2^{-2n} \int_0^\delta (1/x)\, dx = \infty$.

**7.4.31.** Let $n \in \mathbb{N}$. If $x \in [(n + 1/6)\pi, (n + 5/6)\pi]$ then $|\sin(x)| \ge 1/2$, and $1/x \ge 1/[(n+1)\pi]$. Therefore $\int_0^\infty |f| \ge \int \sum_{n=1}^\infty 1/[2(n+1)\pi] \chi_{[(n+1/6)\pi,(n+5/6)\pi]}$
$= \sum_{n=1}^\infty \int 1/[2(n+1)\pi] \chi_{[(n+1/6)\pi,(n+5/6)\pi]} = \sum_{n=1}^\infty 4\pi/6/[2(n+1)\pi] = \infty$.

On the other hand, $\int_0^\infty \sin x/x = \sum_n \int_{n\pi}^{(n+1)\pi} \sin x/x$ and it is clear that this is an alternating series. Moreover, $|\int_{n\pi}^{(n+1)\pi} \sin x/x| \le \int_{n\pi}^{(n+1)\pi} (1/x) = \ln([n+1]/n) \to 0$, so the series converges by Alternating Series Test.

**7.4.32.** Suppose that $f$ is Lebesgue integrable over $[a, b]$. Then so is $|f|$, and we cann apply Dominated Convergence Theorem to obtain $\lim_n \int_{[a_n,b]} f = \lim_n \int_{[a,b]} f\chi_{E_n} = \int_{[a_n,b]} f$. Thus, it would follow that $\lim_n \int_{a_n}^b f(x)\, dx$ exists, a contradiction.

**7.4.36.** Answer: Lebesgue integrable for $\beta \ge 0$ and $\alpha + \beta > -1$, or $\beta < 0$ and $\alpha > -1$. Riemann integral converges for $\beta \ge 0$ and $\alpha + \beta > -1$, or $\beta < 0$ and $\alpha > \beta - 1$.

**7.4.43.** Hint: see Exercise 7.2.17.

**8.1.4.** Hint: let $F \notin \mathcal{B}$ and $E := \{(x, x) : x \in F\}$.

**8.1.5.** Hint: use induction on the number of rectangles in $Q$.

**8.1.7.** Let $P_1(x, y) = x$; let $F(E) := P_1(E \cap D)$, where $D = \{(x, x) : x \in X\}$. Let $\mathcal{S} = \{E \subset X \times X : F(E) \in \mathfrak{M}\}$. Claim $\mathcal{S}$ is a $\sigma$-algebra that contains measurable rectangles. Then it must contain $\mathfrak{M} \otimes \mathfrak{M}$. Proof of the claim. Let $A \in \mathcal{S}$. Then $F(A^C) = [F(A)]^C \in \mathfrak{M}$. Let $\{A_k\} \subset \mathcal{S}$. Then $F(\cup A_k) = \cup F(A_k) \in \mathfrak{M}$. Finaly, if $A, B \in \mathcal{S}$, then $F(A \times B) = A \cap B \in \mathfrak{M}$.

**8.1.9.** Let $\mathfrak{M} = \mathfrak{M}_1 \otimes \mathfrak{M}_2 \otimes \mathfrak{M}_3$ be the $\sigma$-algebra generated by sets $A_1 \times A_2 \times A_3$. We will prove that $\mathfrak{M} = \mathfrak{M}_1 \otimes (\mathfrak{M}_2 \otimes \mathfrak{M}_3)$. The other equality can be proved in a similar fashion.

Since $\mathfrak{M}_1 \times \mathfrak{M}_2 \times \mathfrak{M}_3 \subset \mathfrak{M}_1 \times (\mathfrak{M}_2 \otimes \mathfrak{M}_3) \subset \mathfrak{M}_1 \otimes (\mathfrak{M}_2 \otimes \mathfrak{M}_3)$, and the last one is a $\sigma$-algebra, we have that $\mathfrak{M} \subset \mathfrak{M}_1 \otimes (\mathfrak{M}_2 \otimes \mathfrak{M}_3)$. In the other direction, let $\mathcal{A} = \{W \in \mathfrak{M} : W = X_1 \times E\}$. Then $\mathcal{A} = \{X_1 \times E \in \mathfrak{M}\} = X_1 \times \{E : X_1 \times E \in \mathfrak{M}\}$. By Exercise 8.1.8, $\mathcal{A} \supset X_1 \times (\mathfrak{M}_2 \otimes \mathfrak{M}_3)$, hence $X_1 \times (\mathfrak{M}_2 \otimes \mathfrak{M}_3) \subset \mathfrak{M}$. Finally, let $A \in \mathfrak{M}_1$ and $E \in \mathfrak{M}_2 \otimes \mathfrak{M}_3$. Then $A \times E = (A \times X_2 \times X_3) \cap (X_1 \times E)$ and $A \times X_2 \times X_3 \in \mathfrak{M}$ (obvious) and $X_1 \times E \in \mathfrak{M}$ (by above), so $A \times E \in \mathfrak{M}$. Since $\mathfrak{M}_1 \otimes (\mathfrak{M}_2 \otimes \mathfrak{M}_3)$ is the smallest $\sigma$-algebra that contains all rectangles $A \times E$, we get $\mathfrak{M}_1 \otimes (\mathfrak{M}_2 \otimes \mathfrak{M}_3) \subset \mathfrak{M}$.

**8.1.16.** Let $\{A_k\}$ be an increasing sequence in $\mathcal{A}$. By definition, $A_k \in \mathcal{G}$ and $A_k \cap C \in \mathcal{G}$, for all $C$. $\mathcal{G}$ is a monotone class so $\cup A_k \in \mathcal{G}$ and $(\cup A_k) \cap C = \cup(A_k \cap C) \in \mathcal{G}$, because $\{A_k \cap C\}$ is an increasing sequence in $\mathcal{G}$. Simlar proof in case $\{A_k\}$ is decreasing.
**8.1.19.** Let $\mathcal{F} = \{A \in \mathcal{G} : A \cup B \in \mathcal{G}$, for all $B \in \mathcal{S}\}$. Since $\mathcal{S}$ is an algebra we get that $\mathcal{S} \subset \mathcal{F}$. Also, $\mathcal{F}$ is a monotone class, Indeed, if $\{A_k\} \subset \mathcal{F}$ is an increasing sequence then each $A_n \in \mathcal{G}$ and $A_n \cup B \in \mathcal{G}$ for any $B \in \mathcal{S}$. Therefore, $\cup A_n \in \mathcal{G}$ (it is a monotone class) and $(\cup A_n) \cup B = \cup(A_n \cup B) \in \mathcal{G}$, for the same reason. Thus $\mathcal{F}$ is a monotone class that contains $\mathcal{S}$.

A similar argument shows that $\mathcal{T}$ is a monotone class, so it remains to prove that it contains $\mathcal{S}$. Let $B \in \mathcal{S}$. We need to show that $B \in \mathcal{T}$, meaning that $B \in \mathcal{G}$ (obvious), and that $B \cup C \in \mathcal{G}$ for any $C \in \mathcal{G}$. However, if $C \in \mathcal{G}$ then $C \in \mathcal{F}$ (because $\mathcal{G}$ is the smallest monotone class that contains $\mathcal{S}$), and by definition of $\mathcal{F}$, $C \cup B \in \mathcal{G}$.
**8.1.20.** Hint: use Exercises 8.1.18, 8.1.15, and 8.1.17.
**8.1.21.** (a) and (b) Use Exercise 5.2.28.

(c) No! Let $\mathcal{A}$ be the algebra consisting of all finite unions of intervals $I_{j,k} = [k/2^j, (k+1)/2^j)$, with $j \in \mathbb{N}$ and $0 \le k < 2^j$. Then $\sigma(\mathcal{A}) = \mathcal{B}[0,1]$, the $\sigma$-algebra of all Borel subsets of $[0,1]$. Take $\mu$ to be the counting measure, and $\nu(E) = \infty$ for all $E \ne \emptyset$ and $\nu(\emptyset) = 0$. Then $\mu(A) = \nu(A) = \infty$ for all $A \in \mathcal{A}$, but $\mu(\{1/2\}) = 1$ while $\nu(\{1/2\}) = \infty$.
**8.1.27.** Let $\{r_n\}$ be an ordering of rational numbers. Let $A_{1,m} = (r_1 - 1/m, r_1 + 1/m)$ and $A_{k,m} = (r_k - 1/m, r_k + 1/m) \setminus \cup_1^{k-1}(r_i - 1/m, r_i + 1/m)$. Define $f_m : [a,b] \times \mathbb{R} \to \mathbb{R}$ as follows. Given $(x,y)$, choose $k$ s.t. $y \in A_{k,m}$ and $f_m(x,y) := f(x, r_k)$. Since $f$ is a continuous function of $y$, we have that $\lim_m f_m = f$. Thus, need to show that $f_m$ is measurable.

Let $m$ be fixed and let $G \subset \mathbb{R}$ be open. Then $f_m^{-1}(G) = \{(x,y) : f_m(x,y) \in G\} = \cup_k\{(x,y) : y \in A_{k,m}$ and $f_m(x,y) \in G\} = \cup_k\{(x,y) : y \in A_{k,m}$ and $f(x,r_k) \in G\} = \cup_k[\{x : f(x,r_k) \in G\} \times A_{k,m}]$. Since all $A_{k,m}$ are measurable, it suffices to show that $\{x : f(x,r_k) \in G\}$ is measurable for each $k$. However, this is just $g^{-1}(G)$, where $g = f^{r_k}$, which is measurable.
**8.1.31.** Let $\{y_n\}$ be a dense set in the interval $[0,\delta)$ and $E_n = \{x : |f(x,y_n) - f(x,0)| \ge \varepsilon\}$. Since each $E_n$ is measurable, we will show that $E_{\varepsilon,\delta} = \cap_{n=1}^{\infty} E_n$. In fact, $(\subset)$ is trivial so we will prove $(\supset)$. Suppose $x \notin E_{\varepsilon,\delta}$. Then there is $y' < \delta$ s.t. $|f(x,y') - f(x,0)| < \varepsilon$. Since $f_x$ is continuous at $y'$, there is $\eta > 0$ s.t. $|y' - z| < \eta \Rightarrow |f(x,y') - f(x,z)| < \varepsilon - |f(x,y') - f(x,0)|$. Choose $n$ s.t $|y' - y_n| < \eta$. Then $|f(x,y_n) - f(x,0)| \le |f(x,y_n) - f(x,y')| + |f(x,y') - f(x,0)| < \varepsilon$.
**8.2.4.** Suppose $\mu(A) > 0$ and $\lambda(E_x) < 1$ for all $x \in A$. Then $1 = (\mu \times \lambda)(E) = \int_X \lambda(E_x)\,d\mu(x) = \int_A \lambda(E_x)\,d\mu(x) + \int_{A^C} \lambda(E_x)\,d\mu(x) < \mu(A) + \mu(A^C) = 1$, a contradiction.
**8.2.9.** $\eta^2 = (\mu \times \lambda)(E) = \int_X \lambda(E_x)\,d\mu(x) \ge \int_A \lambda(E_x)\,d\mu(x) \ge \eta\mu(A)$ so $\mu(A) \le \eta$. Sharp: take $E = [0,1/2] \times [0,1/2]$, $\lambda = \mu = m$, and $\eta = 1/2$. If $0 \le x \le 1/2$ then $A = [0,1/2]$ and if $x > 1/2$ then $A = \emptyset$. Thus, if $\eta = 1/2$, then $\mu(A) = \eta$.
**8.2.10.** Hint: find $\int \chi_E \, d(\mu \times \lambda)$ in two ways.
**8.2.17.** (a) $\lambda(\emptyset) = 0$ is easy. If $\{E_k\}$ are disjoint and $E = \cup E_k$, denoting by $\mathcal{F}$ the collection of finite subsets of $[0,1]$, we have $\lambda(E) = \sup\{\sum_{y\in A} m_y(([0,1] \times \{y\}) \cap (\cup E_k)) : A \in \mathcal{F}\} = \sup\{\sum_{y\in A}\sum_k m_y(([0,1] \times \{y\}) \cap E_k) : A \in \mathcal{F}\} \le \sum_k \sup\{\sum_{y\in A} m_y(([0,1] \times \{y\}) \cap E_k) : A \in \mathcal{F}\} = \sum_k \lambda(E_k)$. In the other direction, let $\varepsilon > 0$ and for every $k$ choose $A_k$ s.t. $\sum_{y\in A_k} m_y(([0,1] \times \{y\}) \cap E_k) \ge \lambda(E_k) - \varepsilon/2^k$. Let $N \in \mathbb{N}$ and $A = \cup_{k=1}^N A_k$. Then, $\sum_{y\in A} m_y(([0,1] \times \{y\}) \cap E_k) \ge \lambda(E_k) - \varepsilon/2^k$, and it follows that $\sum_{k=1}^N \lambda(E_k) - \sum_{k=1}^N \varepsilon/2^k \le \sum_{k=1}^N \sum_{y\in A} m_y(([0,1] \times \{y\}) \cap E_k) = \sum_{y\in A} m_y(([0,1] \times \{y\}) \cap (\cup_1^N E_k)) \le \sum_{y\in A} m_y(([0,1] \times \{y\}) \cap E) \le \lambda(E)$. Letting $N \to \infty$ we get $\sum_{k=1}^{\infty} \lambda(E_k) - \varepsilon \le \lambda(E)$, so take $\varepsilon \to 0$.

(b) $\lambda(E \times F) = \sup\{\sum_{y\in A} m_y(([0,1] \times \{y\}) \cap (E \times F)) : A$ is a finite subset of $[0,1]\} = \sup\{\sum_{y\in A} m_y(E \times (\{y\} \cap F)) : A$ is a finite subset of $[0,1]\}$. Notice that $m_y(E \times (\{y\} \cap F))$ is 0 if $y \notin F$ and $m(E)$ if $y \in F$. Therefore, $\sum_{y\in A} m_y(E \times (\{y\} \cap F))$ is 0 if $A \cap F = \emptyset$ and $m(E)\nu(A \cap F)$ if $A \cap F \ne \emptyset$, whence $\sum_{y\in A} m_y(E \times (\{y\} \cap F)) = m(E)\nu(A \cap F)$. Taking supremum over finite subsets of $[0,1]$, we obtain $\lambda(E \times F) = m(E)\nu(F)$.

(c) We use the outer measure $\omega^*$ as defined in Example 5.2.5. Notice that $\Delta = \cap_n \cup_{k<2^n} [k/2^n, (k+1)/2^n] \times [k/2^n, (k+1)/2^n]$ so it is measurable. Suppose $\Delta \subset \cup_{n\ge 0} E_n \times F_n$, where $E_n$ is Borel and $F_n$ is arbitrary. Let $\Delta_n = \Delta \cap (E_n \times F_n)$. Denote $P_1(x,y) = x$ and $P_2(x,y) = y$. Since $[0,1] \subset P_1(\cup_n(\Delta_n)) = \cup_n P_1(\Delta_n) = \cup_n E_n \cap F_n$, we must have $m(E_n \cap F_n) > 0$ for some $n$. For that $n$, $m(E_n) \ge m(E_n \cap F_n) > 0$ and $m(F_n) \ge m(E_n \cap F_n) > 0$ so $F_n$ must be uncountable, hence $\nu(F_n) = \infty$. Thus, $m(E_n)\nu(F_n) = \infty$, and so $(m \times \nu)(\Delta) = (m \times \nu)^*(\Delta) = \infty$.

On the other hand, $\lambda(\Delta) = 0$. Indeed, $\lambda(\Delta) = \sup\{\sum_{y\in A} m_y(([0,1] \times \{y\}) \cap \Delta) : A$ is a finite subset of $[0,1]\}$, and for every $y \in [0,1]$, $m_y(([0,1] \times \{y\}) \cap \Delta) = m_y(\{y\} \times \{y\}) = 0$, so the sum is zero, and taking supremum still gives 0.
**8.2.18.** We associate to $\mu \times \lambda$ an outer measure $\mu^*$ defined as in Example 5.2.5, with the ring $\mathcal{E}$ of all elementary sets, i.e., finite unions of measurable rectangles. If $\mu^*(E) = \infty$ then $(\mu \times \lambda)(E) = \infty$ and we can take $F = X \times Y$. Suppose that $\mu^*(E) < \infty$. By definition of $\mu^*$, there are sets $\{E_n\} \subset \mathcal{E}$ for which $\sum_n (\mu \times \lambda)(E_n) < \mu^*(E) + \varepsilon$. Let $A = \cup_n E_n$. Then $A \in \mathcal{E}_\sigma$ and $(\mu \times \lambda)(A) \le \sum_n (\mu \times \lambda)(E_n) < \mu^*(E) + \varepsilon$. We conclude that for every $k$, there is a set $A_k \in \mathcal{E}_\sigma$

that contains $E$ and satisfies $(\mu\times\lambda)(A_k) < \mu^*(E)+1/k$. If we define $H_n = \cap_{k=1}^n A_k$, then $\{H_n\}$ is a decreasing sequence, and for every $n$, $H_n \supset E$ and $(\mu\times\lambda)(H_n) < \mu^*(E)+1/n$. If $F := \cap_n H_n$ then $F \in \mathcal{E}_{\sigma\delta}$, $E \subset F$ and the continuity of measure implies that $(\mu\times\lambda)(E) \le (\mu\times\lambda)(F) = \lim_n(\mu\times\lambda)(H_n) \le \lim_n \mu^*(E)+1/n = \mu^*(E)$.

**8.2.24.** Let $g(x,y) = x-y$. Then $F^{-1}([\alpha,\infty)) = g^{-1}(f^{-1}([\alpha,\infty)))$. Since $f$ is measurable, so is the set $f^{-1}([\alpha,\infty))$, so it suffices to prove that if $E$ is a measurable subset of $\mathbb{R}$ then $g^{-1}(E)$ is a measurable subset of $\mathbb{R}^2$. If $E$ is open, it is true because $g$ is measurable. If $E$ is a $G_\delta$ set, then $E = \cap G_n$ so $g^{-1}(E) = g^{-1}(\cap G_n) = \cap g^{-1}(G_n)$ hence it is measurable as an intersection of measurable sets. Since every measurable set is a difference between a $G_\delta$ set and a set of measure 0, it remains to show that if $m(E) = 0$ then $m_2(g^{-1}(E)) = 0$. Let $B_k = \{y : |y| \le k\}$. Since $\mathbb{R} = \cup B_k$ we have that $g^{-1}(E) = \cup_k[g^{-1}(E)\cap B_k]$ so it suffices to show that $m_2(g^{-1}(E)\cap B_k) = 0$, for all $k$. So fix $k\in\mathbb{N}$. Let $\varepsilon > 0$. There is an open set $G \supset E$ s.t. $m(G) < \varepsilon$. Now $m_2(g^{-1}(E)\cap B_k) \le m_2(g^{-1}(G)\cap B_k) = \iint \chi_{g^{-1}(G)\cap B_k}\,dxdy = \iint \chi_{g^{-1}(G)}\chi_{B_k}\,dxdy$. By Theorem 8.2.1, this equals $\int_{B_k} dy \int_{\{x:x-y\in G\}} dx = \int_{B_k} m\{x : x-y \in G\}\,dy$. Using the translation invariance of Lebesgue measure, we get $\int_{B_k} m(G)\,dy = \pi k^2 m(G) < \varepsilon\pi k^2$. Since $\varepsilon$ is arbitrary we obtain $m_2(g^{-1}(E)\cap B_k) = 0$.

**8.2.25.** Hint: use Theorem 5.4.13 and Exercise 8.2.19.

**8.2.27.** For any $\alpha$, $\chi_{A_\alpha} = \chi_A\chi_{[0,\alpha]}$ so by Exercise 8.1.25, $\chi_{A_\alpha}$ is measurable whence $A_\alpha$ is measurable. Further, $\mu\times m(A_\alpha) = \int \chi_{A_\alpha}\,d\mu\times m = \int \chi_A\,d\mu \int \chi_{[0,\alpha]}\,dm = \alpha\mu(A)$. As for $A_\infty$, by continuity of measure, $\mu\times m(A_\infty) = \lim_{\alpha\to\infty}\mu\times m(A_\alpha) = \infty$ unless $\mu(A) = 0$ in which case $\mu\times m(A_\alpha) = 0$.

**8.2.28.** Hint: write $E = \{(u,v) : u = x+f(x),\ v = x-f(x),\ x\in\mathbb{R}\}$ and find $m(E_u)$.

**8.2.29.** Hint: define $H = \{(x,t) : f(x) > t \ge g(x) \text{ or } g(x) > t \ge f(x)\}$ and apply Theorem 8.2.1 to $\chi_H$.

**8.2.30.** For each $x$, $Q_x$ is the set of all successors of $x$, hence countable. For each $y$, $Q^y$ is the set of all predecessors of $y$, hence at most countable. By Exercise 8.1.33, $f_x = \chi_{Q_x} = 1$ a.e., and $f^y = \chi_{Q^y} = 0$ a.e., so they are both measurable. Thus $\int dx \int f\,dy = 1$ but $\int dy \int f\,dx = 0$. The issue is that $Q \notin \mathcal{M}\otimes\mathcal{M}$ (so that Theorem 8.2.1 does not apply).

**8.2.37.** Since $\mathcal{B}_r$ is the smallest $\sigma$-algebra that contains all rectangles, and each rectangle lies in $\mathcal{B}_p\times\mathcal{B}_q$ then it lies in $\mathcal{B}_p\otimes\mathcal{B}_q$, which is also a $\sigma$-algebra. Thus, $\mathcal{B}_r \subset \mathcal{B}_p\otimes\mathcal{B}_q$. In the other direction, let $A\in\mathcal{B}_p$ and $B\in\mathcal{B}_q$. It suffices to show that $A\times B\in\mathcal{B}_r$. This follows from $A\times B = A\times\mathbb{R}^q\cap\mathbb{R}^p\times B$ and Exercise 8.1.11.

Let $A\in\mathcal{M}_p\setminus\mathcal{B}_p$ and let $B\in\mathcal{B}_q$. Then $A\times B\in\mathcal{M}_p\otimes\mathcal{M}_q$. On the other hand, $A\times B\notin\mathcal{B}_p\otimes\mathcal{B}_q$. Indeed, Theorem 8.1.4 would then imply that $A\in\mathcal{B}_p$.

**8.3.2.** Hint: define $f(x) = \int_{\mathbb{R}} \chi_{[0,f(x)]}(t)\,dt$ and use Tonelli's Theorem.

**8.3.3.** By Exercise 6.1.19, $\varphi$ and $\psi$ are measurable. Further, $F(x,t) := \chi_{(0,\infty)}(|f(x)|-t)$ is measurable. By Tonelli's theorem, the iterated integrals of $F$ are equal. First, $\int_0^\infty dt\int_X F\,d\mu = \int_0^\infty dt \int_X \chi_{\{x:|f(x)|>t\}}\,d\mu = \int_0^\infty \mu(\{x : |f(x)| > t\})\,dt = \int_0^\infty(\varphi(t)+\psi(t))\,dt$. On the other hand, we have that $\int_X d\mu\int_0^\infty F(x,t)\,dt = \int_X d\mu\int_0^{|f(x)|} dt = \int_X |f(x)|\,d\mu$.

**8.3.9.** The function $g(x,y) := ye^{-x}$ is integrable over $E$. Indeed, using Tonelli's theorem, $\iint_E g\,d(m\times m) = \int_0^\infty\left(\int_0^{\sqrt{x}} ye^{-x}\,dy\right)dx = \int_0^\infty (x/2)e^{-x}\,dx = \frac{1}{2} < \infty$. Let $f(x,y) = (y/x)e^{-x}\sin x\,\chi_E(x,y)$. Since $|f| \le g$, $f$ is integrable. By Fubini's Theorem, $\int_{\mathbb{R}^2} f\,d(m\times m) = \int_0^\infty\left(\int_0^\infty f\,dy\right)dx = \int_0^\infty\left(\int_0^{\sqrt{x}}(y/x)e^{-x}\sin x\,dy\right)dx = (1/2)\int_0^\infty e^{-x}\sin x\,dx = 1/4$.

**8.3.14.** Hint: define $f(x,y) := e^{-sx}\sin(2xy)$ and apply Fubini's Theorem.

**8.3.15.** Define a measurable function $h : (0,a)\times(0,a)\to\mathbb{R}$ by $h(t,x) := t^{-1}f(t)\chi_E(t,x)$, where $E := \{(t,x)\in(0,a)\times(0,a) : x < t\}$ is an open set, hence measurable. Then $g(x) = \int_0^a t^{-1}f(t)\chi_{(x,a)}(t)\,dt = \int h_x$ for all $x\in(0,a)$. By Tonelli's theorem, $\int_{(0,a)\times(0,a)}|h| = \int_0^a\int_0^a t^{-1}|f(t)|\chi_E(t,x)\,dxdt = \int_0^a\int_0^t t^{-1}|f(t)|\,dxdt = \int_0^a |f(t)|\,dt < \infty$. Thus, $h$ is integrable over $(0,a)\times(0,a)$. Since $g = \int h_x$, Fubini's theorem ensures that $g$ is also integrable. Moreover, $\int_0^a g = \int_0^a\int_0^a h_x(t)\,dtdx = \int_{(0,a)\times(0,a)} h = \int_0^a\int_0^a t^{-1}|f(t)|\chi_E(t,x)\,dxdt = \int_0^a\int_0^t t^{-1}f(t)\,dxdt = \int_0^a f(t)\,dt$.

**8.3.16.** Hint: use Exercise 8.2.30.

**8.3.17.** Hint: use Exercise 8.2.22.

**8.3.22.** No! Let $S = \{(x,y) : x-1\le y\le x+1\}$, let $f(x,y) = 0$ if $(x,y)\notin S$, and for $(x,y)\in S$, $f(x,y) = 1$ if $x\le y$ and $f(x,y) = -1$ if $x > y$. Then $\int_{\mathbb{R}} f_x = \int_{\mathbb{R}} f^y = 0$, but $f$ is not integrable because $|f| = \chi_S$ and $m(S) = \infty$.

**9.1.1.** The expression $[f(x+h)-f(x)]/h$ takes different values depending on whether $x\in\mathbb{Q}$ or $x\in\mathbb{Q}^C$. If $x\in\mathbb{Q}$, then it equals $[f(x+t)-1]/t$. We get, $D^+f(x) = 0$, $D_+f(x) = -\infty$, $D^-f(x) = +\infty$, and $D_-f(x) = 0$. If $x\in\mathbb{Q}^C$, then we have $f(x+t)/t$. We get $D^+f(x) = +\infty$, $D_+f(x) = 0$, $D^-f(x) = 0$, and $D_-f(x) = -\infty$.

**9.1.9.** The inequality comes from properties of limit superior and limit inferior (Exercise 2.4.16). As an example use $f(x) = x\chi_Q - x\chi Q^C$, $g(x) = -f(x)$ and Exercises 9.1.3 and 9.1.8.

**9.1.13.** ($\Leftarrow$) By Exercise 9.1.8, $D_+(-f(x)) \geq 0$ so all the more $D^+(-f(x)) \geq 0$, whence Theorem 9.1.5 implies that $-f$ is nondecreasing, i.e., $f$ is nonincreasing.

($\Rightarrow$) Suppose that $D^+f(x_0) > 0$ for some $x_0 \in (a,b)$. By definition of $D^+$, there is $\delta > 0$ s.t. $0 < h < \delta \Rightarrow \sup\{[f(x) - f(x_0)]/[x - x_0] : x_0 < x < x + h\} \geq C > 0$. Therefore, there is $h > 0$ and $x \in (x_0, x_0 + h)$ for which $f(x) - f(x_0) \geq C(x - x_0) > 0$, contradicting the assumption that $f$ is nonincreasing on $(a,b)$.

**9.1.16.** (a) By Exercise 2.4.17 (c), $D^+(fg)(c) = \limsup_{h\to 0^+}[(fg)(c+h) - (fg)(c)]/h = \limsup_{h\to 0^+}[(f(c+h) - f(c))g(c+h) + f(c)(g(c+h) - g(c))]/h \leq g(c)\limsup_{h\to 0^+}[f(c+h) - f(c)]/h + f(c)\limsup_{h\to 0^+}[g(c+h) - g(c)]/h = f(c)D^+g(c) + g(c)D^+f(c)$.

(b) Let $\varphi$ be as in Example 9.1.2 and let $f(x) = 2 + \varphi(x)$, $g(x) = 1/f(x)$. Then $D^+(fg) = 0$. On the other hand, $D^+f(0) = 1$, $f(0) = 2$, $g(0) = 1/2$, and $D^+g(0) = 1/4$. (Take $t_n = 1/[3\pi/2 + 2n\pi]$.)

(c) Let $f(x) = g(x) = x\chi_Q - x\chi_{Q^C}$. Then $(fg)(x) = x^2$ so $D^+f(\sqrt{2}) = 2\sqrt{2}$. On the other hand, $D^+f(\sqrt{2}) = D^+g(\sqrt{2}) = +\infty$ and $f(\sqrt{2}) = g(\sqrt{2}) = -\sqrt{2}$.

**9.1.21.** Suppose $D^+f(x) \geq M$. By Exercise 9.1.15, $g(x) := f(x) - Mx$ has $D^+g = D^+f - M \geq 0$ so $g$ is increasing by Theorem 9.1.5. Thus, $g(x+h) - g(x) \geq 0$, for $h > 0$. Substituting, we get $[f(x+h) - f(x)]/h \geq M$. Also, $g(x) - g(x-h) \geq 0$ for $h > 0$, so $[f(x-h) - f(x)]/h \geq M$. Similarly for other derivatives.

In the other direction, if $[f(x+h) - f(x)]/h \geq M$ then $D^+f \geq M$ and $D_+f \geq M$, and similar for $D^-$ and $D_-$. Suppose $D^+f$ is continuous at $x_0$. Then, for each $\varepsilon > 0$ there is $\delta > 0$ s.t. $|x - x_0| < \delta \Rightarrow D^+f(x_0) - \varepsilon < D^+f(x) < D^+f(x_0) + \varepsilon$. By the previous, all four Dini derivatives are within the same bounds, hence continuous at $x_0$, and all of them are equal at $x_0$, i.e. $f$ is differentiable at $x_0$.

**9.1.23.** Hint: define $E_n := \{x \in [a,b] : [f(x+h) - f(x)]/h \geq M - \varepsilon$, for some $h \geq 1/n\}$ and prove that there is $n$ s.t. $m(E_n) > m(E) - \varepsilon$.

**9.1.28.** Hint: show that if $g$ is continuous then $\sup\{g(t) : 0 < t < h\} = \sup\{g(t) : t \in (0,h) \cap \mathbb{Q}\}$.

**9.1.29.** Hint: use Exercises 2.4.11, 6.1.20, and 9.1.8.

**9.2.3.** Hint: extend the collection of intervals to a Vitali covering of $E$.

**9.2.6.** Hint: approximate $E$ by small open sets.

**9.2.8.** By Exercise 9.2.7, $g_n$ is Riemann integrable. It follows that $\int_a^b g_n(x)\,dx$ $= 1/(1/n)\int_a^b[f(x+1/n) - f(x)]\,dx = n(\int_{a+1/n}^{b+1/n} f(x)\,dx - \int_a^b f(x)\,dx) = n(\int_b^{b+1/n} f - \int_a^{a+1/n} f)$ $= n((1/n)f(b) - \int_a^{a+1/n} f)$. Finally, $f$ is increasing so $\int_a^{a+1/n} f \geq (1/n)f(a)$, whence $\int_a^b g_n(x)\,dx \leq f(b) - f(a)$.

**9.2.9.** Hint: consider intervals of the form $[\sup\{f(x) : x < x_0\}, \inf\{f(x) : x < x_0\}]$.

**9.2.14.** For every $n$, there is an open set $G_n \supset E$, such that $m(G_n) < 1/2^n$. Let $f_n := m(G_n \cap [a,x])$. Then each $f_n$ is nonnegative, $< 1/2^n$, increasing, and continuous because $|f_n(x) - f_n(y)| \leq m([G_n \cap [a,x]]\triangle[G_n \cap [a,y]]$, and $(G_n \cap [a,x]) \setminus (G_n \cap [a,y]) = G_n \cap ([a,x] \setminus [a,y])$ with measure $\leq |x-y|$. Let $f := \sum f_n$. Uniform convergence (Weierstrass M-Test) implies that $f$ is nonnegative, increasing and continuous. Let $x \in E$ and $n \in \mathbb{N}$. Then $x \in G_n$ and there is $h_n > 0$ s.t. $[x, x+h_n] \subset G_n$ so $f_n(x+h_n) = m(G_n \cap [a, x+h_n]) = m(G_n \cap [a,x]) + m(G_n \cap [x, x+h_n]) = f_n(x) + h_n$ hence $[f_n(x+h_n) - f_n(x)]/h_n = 1$. Let $N \in \mathbb{N}$, let $\{t_n\}$ converge to 0. Then $f'(x) = \lim_n[f(x+t_n) - f(x)]/t_n = \lim_n \sum_k [f_k(x+t_n) - f_k(x)]/t_n \geq \lim_n \sum_{k\leq N}[f_k(x+t_n) - f_k(x)]/t_n$. Choose $n_0$ s.t. $n \geq n_0 \Rightarrow t_n \leq h_1, h_2, \ldots, h_N$. Then $\sum_{k\leq N}[f_k(x+t_n) - f_k(x)]/t_n = N$, so $f'(x) \geq N$.

**9.2.18.** By definition of $D^+$ (see also Exercise 9.1.21), there is $\delta > 0$ s.t. $x_0 < x < x_0 + \delta \Rightarrow |(f(x) - f(x_0)|/|x - x_0| \leq M$. Let $\varepsilon > 0$. There are intervals $\{I_n\}_{n=1}^{\infty}$ for which $E \subset \cup I_n$ and $m^*(E) + \varepsilon > \sum \ell(I_n)$. We will make a few assumptions about these intervals without loss of generality. First, $\ell(I_n) < \delta$; otherwise split $I_n := (a_n, b_n) = (a_n, a_n + \delta] \cup [a_n + \delta, a_n + 2\delta] \cup \cdots \cup [a_n + m\delta, b_n)$. Next, $\inf\{x \in E : x > a_n\} = a_n$; otherwise replace $I_n$ by $[\inf\{x \in E : x > a_n\}, b_n]$. Finally, each $a_n \in E$. Otherwise, consider a decreasing sequence $\{\beta_{nk}\}_{k=1}^{\infty} \subset E$ that converges to $a_n$ and replace $I_n$ by $[\beta_{n1}, b_n] \cup [\beta_{n,k-1}, \beta_{nk}] \cup \ldots$. Notice that these changes add to $\cup I_n$ or remove a countable number of points, so $E \subset (\cup I_n) \cup E_0$, where $E_0$ is a countable set, hence of measure 0, whence $f(E) \subset (\cup f(I_n)) \cup f(E_0)$ and $m^*(f(E)) \leq m^*(\cup f(I_n)) + m^*(f(E_0)) \leq \sum m^*(f(I_n)) \leq \sum[f(b_n) - f(a_n)] \leq \sum M(b_n - a_n) = M\sum \ell(I_n) < M[m^*(E) + 2\varepsilon]$. Since $\varepsilon$ is arbitrary, we obtain that $m^*f(E) \leq Mm^*(E)$.

**9.2.24.** There are $\{a_n\}$ and $\{b_n\}$ s.t. $0 < a_n < b_n < 1$, $b_n \to 1$, $a_n \to 0$, and $\lim_k f_k(a_n) = 1$, $\lim_k f_k(b_n) = 1$, for all $n$. By Fatou's Lemma, $\int_{a_n}^{b_n} \liminf_k f_k' \leq \liminf_k \int_{a_n}^{b_n} f_k' \leq \liminf_k[f_k(b_n) - f_k(a_n)] = 0$. The function $\liminf_k f_k'$ is nonnegative, integrable, so is 0 a.e. on $[a_n, b_n]$. Since $a_n, b_n$ are arbitrary, $\liminf_k f_k' = 0$ a.e. in $(0,1)$.

**9.2.29.** Suppose that every function $f \in C[0,1]$ has an interval $(a,b)$ on which it is monotone. Then there is some $I_n \subset (a,b)$ on which it is monotone so $f \in E_n$. In that case $C[0,1] = \cup E_n$. We will obtain a contradiction by showing that each $E_n$ is closed and nowhere dense. Fix $n$. Closed: let $\{f_k\} \in E_n$ and $\lim f_k = f$. Then for any $x < y$ in $I_n$ and any $k$, $f_k(x) < f_k(y)$. [Similar for decreasing.] Letting $k \to \infty$ we get $f(x) \le f(y)$. Nowhere dense: Let $r > 0$ and $f \in E_n$. Pick $x < y$ in $I_n$ s.t. $f(x) < f(y)$ and $f(y) - f(x) < r$. Construct a function $g$ that is equal to $f$ outside of $[x,y]$ and its graph connects $(x, f(x))$ and $(y, f(y))$ by a polygonal line that increases in $[x, x_1]$ and $[x_2, y]$, decreases in $[x_1, x_2]$, and never goes below $f(x)$ or above $f(y)$. Then $g \notin E_n$ but $\|g - f\| \le f(y) - f(x) < r$, so $g \in B(f,r)$.
**9.3.4.** Let $P = \{a, x, b\}$. Then $|f(x) - f(a)| + |f(b) - f(x)| \le V_a^b(f)$ so $|f(x) - f(a)| \le V_a^b(f)$ hence $|f(x)| \le |f(a)| + V_a^b(f)$.
**9.3.6.** If $P = \{x_0, \dots, x_n\}$ is a partition of $[a,b]$, then $||f|(x_k) - |f|(x_{k-1})| = ||f(x_k)| - |f(x_{k-1})|| \le |f(x_k) - f(x_{k-1})|$. Thus, $V(|f|, P) \le V(f,P) \le V_a^v(f)$.
**9.3.9.** Hint: consider $P = \{0, 2/(4n+3), 2/(4n+1), 2/(4n-1), \dots, 2/5, 2/3, 1\}$.
**9.3.14.** $f(x) = x^2 \sin \pi/(2x^2)$, for $x \ne 0$, and $f(0) = 0$. It is differentiable, but not BV. Take $P = \{0, 1/\sqrt{2n+1}$,
$1/\sqrt{2n}, \dots, 1/\sqrt{2}, 1\}$. Then $V(f,P) = 1/(2n+1) + 2[1/(2n-1) + 1/(2n-3) + \cdots + 1]$.
**9.3.19.** Hint: prove that $V_a^b(f) \le L(f) \le b - a + V_a^b(f)$.
**9.3.20.** If $f'$ is integrable then so is $|f'|$ and $|f(x_k) - f(x_{k-1}| = |f'(\xi)|(x_k - x_{k-1})$, for some $\xi$ between $x_{k-1}$ and $x_k$. Thus, $V(f,P)$ is a Riemann sum for $|f'|$. When $\|P\| \to 0$, we get the equality $V_a^b(f) = \int_a^b |f'|$. If $|f'|$ has a convergent improper integral, then the equality $V_{a+\varepsilon}^b(f) = \int_{a+\varepsilon}^b |f'|$ holds for every $\varepsilon > 0$. Taking the limit and using the right continuity of $V$ ($f$ continuous + Exercise 9.3.18) we get the result.
**9.3.21.** Hint: use Exercises 9.1.21 and 9.3.18.
**9.3.23.** $e^{50} - 1$.
**9.3.27.** Suppose that $\{f(x), g(x)\}$ is the square. For every $n$, select $n^2$ points $(i/n, j/n)$, $1 \le i,j \le n$. Let $\{x_k\}_{k=1}^{n^2}$ be their preimages, i.e. each $x_k$ satisfies $(f(x_k), g(x_k)) = (i/n, j/n)$ for a different pair $(i,j)$. Now $V_0^1 f + V_0^1 g \ge \sum_{k=1}^{n^2-1} [|f(x_{k+1}) - f(x_k| + |g(x_{k+1}) - g(x_k| \ge \sum_{k=1}^{n^2-1}(1/n) = (n^2-1)/n \to \infty$, a contradiction. If they are continuous, there is Peano's curve.
**9.3.28.** Suppose $V_a^b(f) < \infty$. Fix $\varepsilon > 0$ and choose a partition $P = \{x_0, \dots, x_k\}$ of $[a,b]$ such that $V(f,P) \ge V_a^b(f) - \varepsilon/2$. By pointwise convergence, there exists an index $N$ such that $|f_n(x_i) - f(x_i)| < \varepsilon/(4k)$, $i = 0, 1, \dots, k$, for all $n \ge N$. We therefore have $V_a^b(f) \le V(f,P) + \varepsilon/2 = \sum_{i=1}^k |f(x_i) - f(x_{i-1})| + \varepsilon/2 \le \sum_{i=1}^k |f(x_i) - f_n(x_i)| + \sum_{i=1}^k |f_n(x_i) - f_n(x_{i-1})| + \sum_{i=1}^k |f(x_{i-1}) - f_n(x_{i-1})| + \varepsilon/2 < V(f_n, P) + \varepsilon \le V_a^b(f_n) + \varepsilon$ for all $n \ge N$. But this implies $V_a^b(f) \le \liminf V_a^b(f_n)$.

Now suppose $V_a^b(f) = \infty$. Fix a real number $M > 0$ and choose a partition $P = \{x_0, \dots, x_k\}$ of $[a,b]$ such that $V(f,P) \ge M$. By pointwise convergence, for any $\varepsilon > 0$ there exists an index $N$ such that $|f_n(x_i) - f(x_i)| < \varepsilon/(2k)$, $i = 0, 1, \dots, k$ for all $n \ge N$. Using the same steps as before, we can conclude $M \le V(f,P) \le V_a^b(f_n)$ for all $n \ge N$. Since $M$ can be arbitrarily large, we must have $\liminf V_a^b(f_n) = \infty$.
**9.3.33.** The total variation of $f$ on $[0,x]$ is given by $V_0^x f = \sin x$ if $0 \le x < \pi/2$, $2 - \sin x$ if $\pi/2 < x \le 3\pi/2$, and $4 + \sin x$, if $3\pi/2 < x \le 2\pi$. We can write $f(x) = h(x) - g(x)$ where $h(x) = f(x) + V_0^x f = 2\sin x$ if $0 \le x < \pi/2$, $2\pi$ if $2 < x \le 3\pi/2$, and $4 + 2\sin x$, if $3\pi/2 < x \le 2\pi$, and $g(x) = V_0^x f$.
**9.3.37.** Hint: try $f_1 = [V + f]/2$ and $f_2 = [V - f]/2$.
**9.4.3.** There is $M$ s.t. $|f|, |g| \le M$ on $[a,b]$. Then $|(fg)(x_k) - (fg)(x_{k-1})| = |f(x_k)g(x_k) - f(x_{k-1})g(x_{k-1})| \le |f(x_k) - f(x_{k-1})||g(x_k)| + |f(x_{k-1})||g(x_k) - g(x_{k-1})| \le M|f(x_k) - f(x_{k-1})| + M|g(x_k) - g(x_{k-1})|$.
**9.4.7.** (a) Let $\varepsilon > 0$. Choose $n$ s.t. $\sum_{n+1}^\infty 1/2^k < \varepsilon/2$. Define $\delta = \varepsilon/2n$. If $\sum(b_i - a_i) < \delta$ then $\sum |f(b_i) - f(a_i)| = \sum_k \sum_i m(G_k \cap [a_i, b_i]) \le \sum_{k=n+1}^\infty m(G_k) + \sum_{k=1}^n \sum_i m(G_k \cap [a_i, b_i]) < \varepsilon/2 + n\sum_i (b_i - a_i) < \varepsilon$.

(b) ($\Leftarrow$) $|f(x) - f(y)| = \sum_{n=1}^N m(G_n \cap [x,y]) \le N|y - x|$. ($\Rightarrow$) Suppose there is no such $N$. Since $\{G_n\}$ is decreasing all $G_n \ne \emptyset$. Let $x \in G_n$, and $\delta > 0$ s.t. $(x - \delta, x + \delta) \subset G_n$. If $a, b \in (x - \delta, x + \delta)$ then $|f(b) - f(a)| = \sum_{k=1}^\infty m(G_k \cap [a,b]) \ge \sum_{k=1}^n m(G_n \cap [a,b]) = \sum_{k=1}^n m([a,b]) = n(b-a)$, so $f$ cannot be Lipschitz.
**9.4.9.** $f$ is continuous so for any interval $[c,d]$, it attains its minimum and maximum at $x_{min}$ and $x_{max}$. Then $m(f([c,d])) = f(x_{max}) - f(x_{min}) \le M|x_{max} - x_{min}| \le Mm([c,d])$. If $G$ is an open set, then $G = \cup I_k$ so $f(G) = \cup f(I_k)$ hence $m[f(G)] \le \sum m[f(I_k)] \le \sum Mm(I_k) = Mm(G)$. If $E$ is measurable, there is an open set $G \supset E$ s.t. $m(E) \le m(G) < m(E) + \varepsilon$. Then $m[f(E)] \le m[f(G)] \le Mm(G) < M[m(E) + \varepsilon]$. Let $\varepsilon \to 0$.

If $m(E) = 0$, $m[f(E)] \le Mm(E) = 0$.

The converse is not true: Take any simple function that is not continuous.

**9.4.10.** Hint: start with closed sets.

**9.4.12.** ($\Leftarrow$) First we show that if $[c,d] \subset [a,b]$ then $|f(d)-f(c)| \leq \int_c^d |f'|$. Indeed, $|f(d)-f(c)| \leq m[f([c,d])]$. Let $E$ be the set where $f'$ is defined, and $F = [c,d]\setminus E$. By Corollary 9.3.12, $m(F) = 0$ so by the $N$ property, $m[f(F)] = 0$ whence $|f(d) - f(c)| \leq m^*(f(E))$. Write $E_n = \{x \in E : (n-1)\varepsilon \leq |f'(x)| \leq n\varepsilon\}$. Then $f$ is Lipschitz on each $E_n$ and $E = \cup E_n$. Therefore, using Exercise 9.4.9, $|f(d) - f(c)| \leq \sum m^*[f(E_n)] \leq \sum n\varepsilon m(E_n) = \sum(n-1)\varepsilon m(E_n) + \sum \varepsilon m(E_n) \leq \sum \int_{E_n} |f'| + \varepsilon m(E) = \int_E |f'| + \varepsilon m(E)$. Now let $\varepsilon \to 0$.

Next, let $\varepsilon > 0$. By Corollary 9.3.12, $f'$ is integrable so there is $\delta > 0$ s.t. $m(A) < \delta \Rightarrow \int_A |f'| < \varepsilon$. Take $\{[a_k, b_k]\}$ s.t. $\sum(b_k - a_k) < \delta$. Then, by the previous, $\sum |f(b_k) - f(a_k)| \leq \sum \int_{a_k}^{b_k} |f'| < \varepsilon$.

($\Rightarrow$) Continuity is obvious and $f$ is of bounded variation, by Theorem 9.4.6. Finally, let $\varepsilon > 0$. Then there is $\delta > 0$ from the definition of absolute continuity. Let $m(E) = 0$. There is open set $G \supset E$, s.t. $m(G) < \delta$. Write $G = \cup(a_k, b_k)$. Then $f(E) \subset f(G) \subset f(\cup[a_k, b_k]) = \cup f([a_k, b_k]) = \cup[f(c_k), f(d_k)]$, where $f$ attains minimum and maximum on $c_k$ and $d_k$. Then $\sum |c_k - d_k| \leq \sum(b_k - a_k) < \delta$ so $m^*[f(E)] \leq \sum |f(c_k) - f(d_k)| < \varepsilon$.

**9.4.15.** Let $\varepsilon > 0$. Since $f$ is continuous at 0 there is $\delta_1 = \delta_1(\varepsilon/2)$. Also, $f$ is absolutely continuous on $[\delta_1, 1]$ so there is $\delta_2 = \delta_2(\varepsilon/2)$. Define $\delta = \min\{\delta_1, \delta_2\}$. Let $\{[a_k, b_k]\}_{k=1}^n$ be a finite, nonoverlapping collection of closed intervals contained in $[0,1]$ that satisfy $\sum_{k=1}^n (b_k - a_k) < \delta$. Assume the intervals are ordered so that $a_1 < b_1 \leq a_2 < b_2 \leq \cdots \leq a_n < b_n$. Also assume there exists an index $m$ such that $\{(a_k, b_k)\}_{k=1}^m$ is contained in $[0, \delta_1)$ and $\{(a_k, b_k)\}_{k=m+1}^n$ is contained in $[\delta_1, 1]$. (If $\delta_1 \in [a_k, b_k]$ for some $k$, replace $[a_k, b_k]$ with $[a_k, \delta_1]$ and $[\delta_1, b_k]$.) Since $b_m < \delta_1$ and $f$ is increasing, we have $\sum_{k=1}^m |f(b_k) - f(a_k)| = f(b_m) - f(a_1) - \sum_{k=2}^m (f(a_k) - f(b_{k-1})) \leq f(b_m) - f(0) < \varepsilon/2$ where the last estimate follows from continuity at 0. Since $\sum_{m+1}^n (b_k - a_k) < \delta_2$, we also have $\sum_{k=m+1}^n |f(b_k) - f(a_k)| < \varepsilon/2$ by absolute continuity of $f$ on $[\delta_1, 1]$. Combining, we obtain $\sum_{k=1}^n |f(b_k) - f(a_k)| < \varepsilon$.

**9.4.20.** Hint: in one direction use Exercises 9.3.38 and 9.4.17; in the other Propositon 9.3.10, Exercise 9.3.38, and Corollary 9.2.9.

**9.4.21.** Hint: start with $E$ an interval.

**9.4.32.** Let $x \in \mathbb{R}$ and let $F(z) = \int_0^z |f(t) - f(x)|\, dm(t)$. Since $f - f(x)$ is integrable, $F \in AC[x-1, x+1]$ so $F$ is differentiable a.e. Choose $x$ for which $F$ is differentiable. Then $\int_{(x-h,x+h)} |f - f(x)|\, dm = F(x+h) - F(x-h)$ and $\lim_{h\to 0^+} [F(x+h) - F(x-h)]/h = 2F'(x)$. (See [86, Exercise 4.2.16].) But $F'(z) = |f(z) - f(x)|$ so $F'(x) = 0$.

**9.4.34.** Hint: define $f(x) = \int_0^x \chi_A$ and prove that $f' \geq \alpha$ a.e.

**9.4.37.** (a) If $x, y \in [0,1]$ and $x < y$ then $F(y) = m([0,y] \cap A) = m([0,x] \cap A) + m((x,y] \cap A) > F(x)$, so $F$ is strictly increasing on $[0,1]$. Let $\varepsilon > 0$. If $\{(a_1, b_1), (a_2, b_2), \ldots, (a_n, b_n)\}$ is a finite collection of disjoint intervals such that $\sum_{k=1}^n (b_k - a_k) < \varepsilon$ then $\sum_{k=1}^n |F(b_k) - F(a_k)| = \sum_{k=1}^n |m([0,b_k] \cap A) - m([0,a_k] \cap A)| = \sum_{k=1}^n m((a_k, b_k] \cap A) \leq \sum_{k=1}^n m((a_k, b_k]) < \varepsilon$, so $F$ is absolutely continuous on $\mathbb{R}$. Moreover $F(b) - F(a) = m([0,b] \cap A) - m([0,a] \cap A) = m((a,b] \cap A) = \int_a^b \chi_A\, dm$ for all $a, b \in \mathbb{R}$ with $a < b$. This implies that $F' = \chi_A$ a.e. so $F' = 0$ on $[0,1] \setminus A$, which has positive measure because $m([0,1] \cap A) < m([0,1])$.

(b) Let $\varepsilon > 0$. If $\{(a_1, b_1), (a_2, b_2), \ldots, (a_n, b_n)\}$ is a finite collection of disjoint intervals such that $\sum_{k=1}^n (b_k - a_k) < \varepsilon$ than $\sum_{k=1}^n |G(b_k) - G(a_k)| = \sum_{k=1}^n |m([0,b_k] \cap A) - m([0,b_k] \setminus A) - m([0,a_k] \cap A) + m([0,a_k] \setminus A)| = \sum_{k=1}^n |m((a_k, b_k] \cap A) - m((a_k, b_k] \setminus A)| \leq \sum_{k=1}^n (m((a_k, b_k] \cap A) + m((a_k, b_k] \setminus A)) = \sum_{k=1}^n m((a_k, b_k]) < \varepsilon$, so $G$ is absolutely continuous on $\mathbb{R}$. In particular $G$ is differentiable on some $E \subset \mathbb{R}$ with $m(E^C) = 0$. Moreover $\int_a^b G'\, dm = G(b) - G(a) = m([0,b] \cap A) - m([0,b] \setminus A) - m([0,a] \cap A) + m([0,a] \setminus A) = m((a,b] \cap A) - m((a,b] \setminus A) = \int_a^b \chi_A\, dm - \int_a^b \chi_{A^C}\, dm = \int_a^b (\chi_A - \chi_{A^C})\, dm$ for all $a, b \in \mathbb{R}$ such that $a < b$. By Lemma 9.4.12, $G' = (\chi_A - \chi_{A^C})$ a.e. If $a, b \in [0,1]$, $a < b$, and $G$ is monotone on $(a,b)$, then $G' \geq 0$ or $G' \leq 0$ on $(a,b) \cap E$ (because the difference quotient at each point is either non-negative or non-positive). This is a contradiction because $\chi_A - \chi_{A^C}$ takes on the values 1 and $-1$ on $A \cap (a,b)$ and $A^C \cap (a,b)$ respectively, both of which have positive measure and hence $G'(x) = 1$ and $G'(y) = -1$ for some $x, y \in (a,b) \cap E$.

**9.4.39.** ($\Rightarrow$) Fix $\varepsilon > 0$ and $\delta > 0$. Define the collection of closed, bounded, nondegenerate intervals $\mathcal{F} = \{[c,d] \subset (a,b) : c < d \text{ and } f(d) - f(c) < \varepsilon(d-c)/(b-a)\}$. Since $f$ is singular, there exists a set $E_0$ with $m(E_0) = 0$ such that $f'(x) = 0$ for all $x \in (a,b) \setminus E_0$. Let $E = (a,b) \setminus E_0$. Because $f$ is increasing and $\lim_{h\to 0}[f(x+h) - f(x)]/h = 0$ for all $x \in E$, $\mathcal{F}$ covers $E$ in the sense of Vitali. By the Vitali Covering Lemma, there exists a finite, disjoint subcollection $\{(c_k, d_k)\}_{k=1}^N$ of $\mathcal{F}$ for which $m(E \setminus \cup_{k=1}^N [c_k, d_k]) < \delta$. Without loss, assume the subcollection is ordered so that $c_1 < d_1 < c_2 < d_2 < \cdots < c_N < d_N$. Then $(a,b) \setminus \cup_{k=1}^N [c_k, d_k] = \cup_{k=1}^n (a_k, b_k)$, where $n = N+1$, $a_1 = a$, $a_k = d_{k-1}$ if $k \geq 2$, $b_k = c_k$ if $k \leq n-1$, and $b_n = b$. Then $m(\cup_{k=1}^n (a_k, b_k)) = m(E \setminus \cup_{k=1}^N [c_k, d_k] \setminus E_0) + m(E \setminus \cup_{k=1}^N [c_k, d_k] \cap E_0) \leq m(E \setminus \cup_{k=1}^N [c_k, d_k]) + m(E_0) < \delta$. Since $[c_k, d_k] \in \mathcal{F}$ for $k = 1, \ldots, N$, we also have $\sum_{k=1}^n [f(b_k) - f(a_k)] = f(b) - f(a) - \sum_{k=1}^N [f(d_k) - f(c_k)] > f(b) - f(a) - \varepsilon[\sum_{k=1}^n (d_k - c_k)]/(b-a) \geq f(b) - f(a) - \varepsilon$.

($\Leftarrow$) Fix $\varepsilon > 0$. By Corollary 9.3.12, $f'$ is integrable so there exists $\delta$ s.t. $m(A) < \delta \Rightarrow \int_A f' < \varepsilon/2$ (Theorem 7.3.12). Choose a partition $a = b_0 \le a_1 < b_1 < a_2 < b_2 < \cdots < a_n < b_n \le a_{n+1} = b$ for which $\sum_{k=1}^n [b_k - a_k] < \delta$ and $f(b) - f(a) - \sum_{k=1}^n [f(b_k) - f(a_k)] < \varepsilon/2$. The former implies that $\int_{\cup_{k=1}^n (a_k,b_k)} f' < \varepsilon/2$. We then have $f(b) - f(a) - \sum_{k=1}^n [f(b_k) - f(a_k)] = \sum_{k=1}^{n+1} [f(a_k) - f(b_{k-1})] \ge \sum_{k=1}^{n+1} \int_{b_{k-1}}^{a_k} f' = \int_{[a,b]\setminus\cup_{k=1}^n (a_k,b_k)} f'$, where the inequality follows from Corollary 9.2.9. Thus $\int_a^b f' = \int_{\cup_{k=1}^n (a_k,b_k)} f' + \int_{[a,b]\setminus\cup_{k=1}^n (a_k,b_k)} f' < \varepsilon$. It follows that $\int_a^b f' = 0$. But $f' \ge 0$ because $f$ is increasing, so $f' = 0$ a.e.

**9.5.1.** (a) $\Rightarrow$ (b) $\nu(E) = 0 \Rightarrow \nu^+(E) - \nu^-(E) = 0 \Rightarrow 0 \le \nu^+(E) = \nu^-(E) \le 0 \Rightarrow \nu^+(E) = \nu^-(E) = 0$;

(b) $\Rightarrow$ (c) $\nu^+(E) = \nu^-(E) = 0 \Rightarrow |\nu|(E) = \nu^+(E) + \nu^-(E) = 0$;

(c) $\Rightarrow$ (a) $|\nu|(E) = 0 \Rightarrow \nu^+(E) = \nu^-(E) = 0 \Rightarrow \nu(E) = \nu^+(E) - \nu^-(E) = 0$.

**9.5.3.** Suppose that $\nu_n \perp \mu$ for all $n \in \mathbb{N}$. For each $n \in \mathbb{N}$ there exists a measurable set $E_n$ such that $\nu_n(E_n) = 0$ and $\mu(E_n^C) = 0$. Define $E := \cap_{n=1}^\infty E_n$ and note that $E^C = \cup_{n=1}^\infty E_n^C$; in particular $\mu(E^C) = 0$. It is also clear that $\sum_{n=1}^\infty a_n \nu_n(E) = 0$. Therefore $\sum_{n=1}^\infty a_n \nu_n \perp \mu$.

**9.5.8.** We prove (a) $\Leftrightarrow$ (b). Then (a) $\Leftrightarrow$ (c) follows by interchanging $\mu$ and $\nu$.

($\Rightarrow$) There are sets $A, B$ s.t. $\nu(B) = 0 = \mu(A)$. Then, if $E \subset A$ we get $(\mu-\nu)(E) = -\nu(E) \le 0$ and if $E \subset B$ we get $(\mu - \nu)(E) = \mu(E) \ge 0$. Thus $(B, A)$ is a Hahn decomposition for $(\mu - \nu)$. Now, for any set $E$, $(\mu - \nu)^+(E) = (\mu - \nu)(E \cap B) = \mu(E \cap B) = \mu(E)$.

($\Leftarrow$) If $(P, N)$ is a Hahn decomposition for $(\mu-\nu)$ then $\mu(E) = (\mu-\nu)^+(E) = (\mu-\nu)(E\cap P) = \mu(E \cap P) - \nu(E \cap P)$ so $\nu(E \cap P) = \mu(E \cap P) - \mu(E) \le 0$ hence $\nu(E \cap P) = 0$ whence $\nu(P) = 0$. Also $\mu(E \cap P) - \mu(E) = 0$ so $\mu(P^C) = 0$. Thus $\mu \perp \nu$.

**9.5.16.** (a) $\mu_1 \times \mu_2(E) = 0 \Rightarrow \int_X \mu_2(E_x)\, d\mu_1 = 0 \Rightarrow \mu_2(E_x) = 0$ except on a set $A \subset X$ for which $\mu_1(A) = 0 \Rightarrow \nu_2(E_x) = 0$ except on a set $A$ for which $\nu_1(A) = 0 \Rightarrow \int_X \nu_2(E_x)\, d\nu_1 = 0 \Rightarrow \nu_1 \times \nu_2(E) = 0$.

(b) $\nu_1 \times \nu_2(E) =$ [ by definition ] $\int_X \nu_2(E_x)\, d\nu_1 = \int_X d\nu_1 \int_{E_x} f_2(y) d\mu_2(y)$ $= \int_Y f_2(y) d\mu_2(y) \int_X \chi_E d\nu_1 = \int_Y f_2(y) d\mu_2(y) \nu_1(E^y) = \int_Y f_2(y) d\mu_2(y) \int_{E^y} f_1(x) d\mu_1(x) =$ [ Fubini's theorem ] $\int_E f_1(x) f_2(y) d(\mu_1 \times \mu_2) \Rightarrow d(\nu_1 \times \nu_2)/d(\mu_1 \times \mu_2) = [d\nu_1/d\mu_1]\,[d\nu_2/d\mu_2]$.

**9.5.17.** For each $n \in \mathbb{N}$ let $E_n := \{x \in X : f(x) < -1/n\}$, so that $-\lambda(E_n)/n = \int_{E_n} -d\lambda/n \ge \int_{E_n} f d\lambda = \nu(E_n) \ge 0$ and hence $\mu(E_n) \le \lambda(E_n) = 0$. It follows that $\mu(\cup_{n=1}^\infty E_n) = 0$, so $f \ge 0$ $[\mu]$-a.e. because $\cup_{n=1}^\infty E_n = \{x \in X : f(x) < 0\}$.

Now set $F := \{x \in X : f(x) \ge 1\}$. Since $\nu$ is $\sigma$-finite, there is a sequence $\{F_n\}_{n=1}^\infty$ of subsets of $F$ which cover $F$ such that $\nu(F_n) < \infty$ for all $n \in \mathbb{N}$. Since $\nu(F_n) = \int_{F_n} f\, d\lambda \ge \int_{F_n} 1\, d\lambda = \lambda(F_n) = \mu(F_n) + \nu(F_n)$ it is clear that $\mu(F_n) = 0$ for all $n \in \mathbb{N}$, in which case $\mu(F) = 0$ and hence $f < 1$ $[\mu]$-a.e.

Let $E \in \mathfrak{M}$ with $\nu(E) < \infty$. Then $\int_E (1 - f)\, d\lambda + \nu(E) = \int_E 1\, d\lambda = \lambda(E) = \mu(E) + \nu(E)$. It follows that $\int_E (1 - f)\, d\lambda = \mu(E)$. The equality extends to all $E \in \mathfrak{M}$ because $\nu$ is $\sigma$-finite, using Corollary 7.2.15 and the additivity of $\mu$. Therefore $d\mu/d\lambda = (1 - f)$. Since $\mu \ll \lambda$ (in fact $\mu \le \lambda$) and $\lambda \ll \mu$ (as $\nu \ll \mu$), by Exercise 9.5.15 (d), $d\lambda/d\mu = 1/(1 - f)$. By Exercise 9.5.15 (c), $d\nu/d\mu = [d\nu/d\lambda][d\lambda/d\mu] = f/(1 - f)$.

**9.5.22.** The existence of a measurable $f$ that satisfies $\nu(E) = \int_E f\, d|\nu|$ is a consequence of Radon–Nikodym theorem for signed measures. Let $(P, N)$ be a Hahn decomposition for $\nu$. Then $\nu(P) = \int_P f\, d|\nu|$, but $\nu$ is positive on $P$, so we have $\nu(A) = \int_A \chi_P\, d\nu = \int_A \chi_P\, d|\nu|$, for any $A \subset P$. By uniqueness in Radon–Nikodym theorem, $f = \chi_P$ on $P$, i.e, $f = 1$ on $P$ a.e. Similarly, we find that $f = -1$ a.e. on $N$.

**9.5.23.** $f = d\mu/d\nu$ must be constant on $A$ otherwise it would not be measurable. Indeed, let $I_k = [k, k + 1]$, for $k \in \mathbb{Z}$. Only one of these sets satisfies $\nu(f^{-1}(B) \cap A) > 0$. (If 2 sets do then their intersection must have also a positive measure, and $f(A)$ equals that number.) Next we bisect the obtained interval and once again only one of the 2 sets has the inverse image of positive measure. Continuing, we get a nested collection of closed sets whose diameter goes to 0, so there is a unique number $C$ in all of them, and thus $f = C$. Further, $\mu(E) = \int_E C d\nu = C\nu(E)$. Thus $C$ can be an arbitrary constant.

**9.5.26.** ($\ge$) If $f$ is a measurable function on $X$ such that $|f| \le 1$ then $|\int_E f\, d\nu| \le \int_E |f|\, d|\nu| \le \int_E 1\, d|\nu| = |\nu|(E)$.

($\le$) Case when $\nu(E)$ is finite: if $f := d\nu/d|\nu|$ (which exists because $\nu \ll |\nu|$) then, by Exercise 9.5.22, $|f| = 1$ $|\nu|$-almost everywhere. Therefore, the right side equals $|\int_E f[d\nu/d|\nu|]\, d|\nu|| = |\int_E f^2\, d|\nu|| = |\nu|(E)$.

Case when $\nu(E)$ is infinite: let $(A, B)$ be Hahn decomposition for $\nu$ and $g := \chi_B - \chi_A$. Then $|g| \le 1$ and $\int_E g\, d\nu = \int(\chi_B - \chi_A)\chi_E\, d\nu = \int \chi_B \chi_E\, d\nu^+ - \int -\chi_A \chi_E\, d\nu^- = \int \chi_{B\cap E} d\nu^+ + \int \chi_{A\cap E} d\nu^- = \nu^+(B\cap E) + \nu^-(A\cap E) = \nu^+(E) + \nu^-(E) = |\nu|(E) \ge 0$. Thus, $|\nu|(E) = |\int_E g\, d\nu| \le \sup\{|\int_E f\, d\nu| : |f| \le 1\}$.

**9.5.27.** Hint: consider $\lambda = \mu + \nu$ and $A = \{x \in X : d\mu/d\lambda(x) d\nu/d\lambda(x) > 0\}$.

**9.5.31.** Hint: the counting measure.
**9.5.34.** Hint: use Exercise 5.5.17.
**10.1.5.** By definition, $V_0^1(f) = \sup V(f,P)$, taken over all partitions $P$ of $[0,1]$. Clearly, if $f = 0$ then $V_0^1(f) = 0$. If $V_0^1(f) = 0$, then $V(f,P) = 0$ for any partition $P$, so $|f(x) - f(y)| = 0$ for any $x, y \in [0,1]$. Since $f(0) = 0$, we have that $f = 0$. The rest follows from Exercises 9.3.15 and 9.3.16.
**10.1.10.** ($\Rightarrow$) Let $\sum \|x_k\| < \infty$. Let $s_n = \sum_{k=1}^n x_k$. Then $\|s_{n+p} - s_n\| = \|\sum_{k=n+1}^{n+p} x_k\| \leq \sum_{k=n+1}^{n+p} \|x_k\| < \varepsilon$. ($\Leftarrow$) Let $\{x_n\}$ be Cauchy. By Proposition 2.2.21, it has a rapidly Cauchy subsequence $\{y_n\}$ so $\sum \|y_{n+1} - y_n\| < \infty$. By assumption, $\sum(y_{n+1} - y_n) < \infty$. Thus, $\lim_n \sum_{k=1}^n (y_{k+1} - y_k)$ exists hence $\lim_n y_n$ exists. By Proposition 2.2.17, $\lim x_n$ exists.
**10.1.11.** Hint: if $\{x_n\}$ is a Cauchy sequence consider $\|x_n/\|x_n\| - x_m/\|x_m\|\|$.
**10.1.13.** (a) Straightforward.

(b) Known from linear algebra.

(c) First, independence of the representative. If $x \sim y$ then $\inf\{\|x + z\| : z \in V\} = \inf\{\|y + w\| : w \in V\}$. Indeed, $y = x+v$, for some $v \in V$. Then $\inf\{\|y+w\| : w \in V\} = \inf\{\|x+v+w\| : w \in V\} = \inf\{\|x+(v+w)\| : v+w \in v+V\} = \inf\{\|x+(v+w)\| : v+w \in V\} = \inf\{\|x+z\| : z \in V\}$. Scalar: $\|[\alpha x]\| = \inf\{\|\alpha x + y\| : y \in V\} = \inf\{\|\alpha x + \alpha z\| : z \in V\} = |\alpha| \inf\{\|x + z\| : z \in V\} = |\alpha|\|[x]\|$, when $\alpha \neq 0$, and trivial when $\alpha = 0$. Triangle: $\|[x]\| = \inf\{\|x + z\| : z \in V\}$ and there is a sequence $\{z_n\} \subset V$ s.t. $\|[x]\| = \lim_n \|x + z_n\|$. Similarly, there is a sequence $\{w_n\} \subset V$ s.t. $\|[y]\| = \lim_n \|y + w_n\|$. Then $\|[x] + [y]\| = \|[x + y]\| = \inf\{\|x + y + z\| : z \in V\} \leq \|x + y + z_n + w_n\| \leq \|x + z_n\| + \|y + w_n\|$. Letting $n \to \infty$, $\|[x] + [y]\| \leq \|[x]\| + \|[y]\|$. Finally, if $x = 0$ then $\|[x]\| = 0$ because in the infimum we can take $y = 0$. If $\|[x]\| = 0$, then there is a sequence $\{z_n\} \subset V$ s.t. $\|[x]\| = \lim_n \|x + z_n\|$. Thus $\lim_n(x + z_n) = 0$ so $\lim_n z_n = -x$. But $z_n \in V$ which is closed so $-x \in V$ hence $x \in V$ whence $[x] = 0$.

(d) Will use Exercise 10.1.10. Consider $\{[x_n] : n \in \mathbb{N}\}$ such that the series $\sum_{n=1}^\infty \|[x_n]\|$ converges. From the definition of the norm in $X/V$ we have that for each $n \in \mathbb{N}$ there exists $y_n$ s.t. $\|x_n + y_n\| \leq \|[x_n]\| + 1/2^n$. It follows that $\sum_{n=1}^\infty \|x_n + y_n\|$ converges. But $X$ is Banach $\sum_{n=1}^\infty (x_n + y_n)$ converges to some $x \in X$. Then from the definiton of the norm in $X/V$ it follows that $\sum_{n=1}^\infty [x_n]$ converges to $[x]$ in $X/V$. Indeed, $\lim_n \sum_1^n [x_k] = \lim_n \sum_1^n [x_k + y_k] = \lim_n [\sum_1^n (x_k + y_k)] = \lim_n [x - t_n] = \lim_n ([x] - [t_n]) = [x] - \lim_n [t_n]$. Since $\lim t_n = 0$, in view of $\|[t_n]\| \leq \|t_n\|$ (take $y = 0$), we get that $\lim_n \sum_1^n [x_k] = [x]$.

(e) Since $V \neq X$, there exists $x \in X \setminus V$. If $\varepsilon \geq 1$ we are done, otherwise $1 < 1/(1 - \varepsilon)$, so $\|[x]\| < \|[x]\|/(1 - \varepsilon)$ and there exists $y \in V$ such that $\|x - y\| < \|[x]\|/(1 - \varepsilon)$. The idea is that $y$ is close to $x$, so $x - y$ is close to being "normal" to $V$. Thus we define $z := \|x - y\|^{-1}(x - y)$, and note that $\|z\| = 1$ while $\|[z]\| = \|[x - y]\|/\|x - y\| = \|[x]\|/\|x - y\| > 1 - \varepsilon$.
**10.1.16.** Vector space is straightforward. Norm is straightforward. Let $\{f_n\}$ be Cauchy in $B(X)$. For every $\varepsilon > 0$ there is $N$ s.t. $m \geq n \geq N \Rightarrow \|f_m - f_n\| < \varepsilon$ i.e. for each $x$, $|f_m(x) - f_n(x)| < \varepsilon$ so for each $x$, $\{f_n(x)\}$ is Cauchy, hence convergent. Let $f(x) := \lim f_n(x)$. For each $x$, $|f_m(x) - f_n(x)| < \varepsilon$ implies, by letting $m \to \infty$, $|f(x) - f_n(x)| \leq \varepsilon$, so $\{f_n\}$ converges uniformly to $f$. Finally, $f$ is bounded because there is $n$ s.t. for all $x$, $|f(x) - f_n(x)| \leq \varepsilon$ so $|f(x)| \leq |f_n(x)| + \varepsilon$ and $\{f_n\}$ is bounded.
**10.1.18.** $\|x_m - x_n\| = |x_m(1) - x_n(1)| + \sum_{k=1}^\infty |x_m(k) - x_n(k) - x_m(k+1) + x_n(k+1)| \geq |x_m(1)-x_n(1)|+\sum_{k=1}^N |x_m(k)-x_n(k)-x_m(k+1)+x_n(k+1)| \geq |[x_m(1)-x_n(1)]+\sum_{k=1}^N [x_n(k) - x_m(k) - x_n(k+1) + x_m(k+1)]| = |x_m(N+1) - x_n(N+1)|$. Thus, if $\{x_n\}$ is Cauchy in $X$ then $\{x_n(k)\}$ is Cauchy, hence convergent. Let $x(k) := \lim_n x_n(k)$. Let $\varepsilon > 0$. There is $K$ s.t. $m \geq n \geq K \Rightarrow \|x_m - x_n\| < \varepsilon$ so $|x_m(1)-x_n(1)|+\sum_{k=1}^N |x_m(k)-x_n(k)-x_m(k+1)+x_n(k+1)| < \varepsilon$. Letting $m \to \infty$, we get $|x(1)-x_n(1)|+\sum_{k=1}^N |x(k)-x_n(k)-x(k+1)+x_n(k+1)| \leq \varepsilon$. Now letting $N \to \infty$ we get $\|x - x_n\| \leq \varepsilon$, so $\lim_n x_n = x$. Finally, $x \in X$ because $\|x\| \leq \|x_n\| + \|x - x_n\| \leq \|x_n\| + \varepsilon$.
**10.1.21.** Same norm as in Exercise 10.1.20, but a subspace $Y$ that needs to be closed. If $\{y_n\}$ is a sequence in $Y$ that converges to $y$, need to show that $y \in Y$, i.e., that $t_n := \sum_{i=1}^n y(i)$ converges. Let $\varepsilon > 0$. Choose $k$ s.t. $\|y - y_k\| < \varepsilon$ and fix that $k$. Since $y_k \in Y$, the series $\sum_i y_k(i)$ converges so there is $N$ s.t. if $m \geq n \geq N$ then $|\sum_{i=n}^m y_k(i)| < \varepsilon$. For such $m, n$, we have $|\sum_{i=n}^m y(i)| = |\sum_{i=1}^m y(i) - \sum_{i=1}^{n-1} y(i)| \leq |\sum_{i=1}^m y(i) - \sum_{i=1}^m y_k(i)| + |\sum_{i=1}^m y_k(i) - \sum_{i=1}^{n-1} y_k(i)| + |\sum_{i=1}^{n-1} y_k(i) - \sum_{i=1}^{n-1} y(i)| \leq \|y - y_k\| + |\sum_{i=n}^m y_k(i)| + \|y - y_k\| < 3\varepsilon$.
**10.1.22.** Hint: define $\mu$ as $\mu(A) := \lim \mu_n(A)$ and use Exercise 5.5.13.
**10.1.26.** Subspace of $C(\mathbb{R})$: straightforward. Complete: Let $\{f_n\}$ be a Cauchy sequence in $C_b(\mathbb{R})$. For each $x$, $\{f_n(x)\}$ is a Cauchy sequence hence convergent. Let $f(x) := \lim_{n\to\infty} f_n(x)$. Let $N$ be such that $n, m \geq N \Rightarrow \|f_n - f_m\| < \varepsilon$. Then for all $n \geq N$, $|f(x) - f_n(x)| = \lim_{m\to\infty} |f_m(x) - f_n(x)| \leq \varepsilon$, for all $x$, and hence $\|f - f_n\| \leq \varepsilon$. Thus, $\{f_n\}$ converges to $f$ in the norm, hence uniformly, whence $f$ is continuous. Finally, $f$ is bounded. Indeed, for all $x$, $|f(x)| \leq |f(x) - f_N(x)| + |f_N(x)| \leq \|f - f_N\| + \|f_N\| \leq \varepsilon + \|f_N\| < \infty$. Thus, $f \in C_b(\mathbb{R})$.
**10.1.27.** Hint: prove that $f \in C_0(\mathbb{R})$ if and only if $\lim_{x\to\pm\infty} |f(x)| = 0$.
**10.1.33.** Hint: approximate $x = (x_1, x_2, x_3, \dots)$ by $y_n = (x_1, x_2, \dots, x_n, \lim x_n, \lim x_n, \dots)$.
**10.2.5.** Hint: use Theorem 7.4.4 and Exercise 7.4.32.
**10.2.7.** Hint: use Dominated Convergence Theorem.

**10.2.9.** Since $f$ is absolutely continuous, $f(x) = f(0) + \int_0^x f(t)\,dt = \int_0^x f(t)\,dt$. It follows that $|f(x)| \le \left(\int_0^x f^2(t)\,dt\right)^{1/2} \left(\int_0^x 1\,dt\right)^{1/2} = \sqrt{x}\left(\int_0^x f^2(t)\,dt\right)^{1/2}$. It remains to show that $g(x) := \int_0^x f^2(t)\,dt \to 0$, as $x \to 0+$. Let $g_n = \int_0^{1/n} f^2 = \int_0^1 f^2 \chi_{[0,1/n]}$. Since $f^2$ is continuous it is in $L^2$, so Dominated Convergence Theorem implies that $\lim_n g_n = 0$.
**10.2.11.** Hint: such a function cannot have a polynomial in the denominator.
**10.2.14.** ($\Leftarrow$) Straightforward. ($\Rightarrow$) Examine (10.8). The first inequality $|f+g| \le |f| + |g|$ is equality when $\operatorname{sgn} f = \operatorname{sgn} g$. The second inequality is Hölder's Inequality. By Exercise 10.2.13, $|f|^p \| |f+g|^{p-1} \|_q^q = |f+g|^{(p-1)q} \|f\|_p^p$ so $|f|^p \|f+g\|_p^p = |f+g|^p \|f\|_p^p$ hence $|f| \|f+g\|_p =$ [because of $\operatorname{sgn} f = \operatorname{sgn} g$ ] $= (|f| + |g|)\|f\|_p$ so $|f|(\|f+g\|_p - \|f\|_p) = |g| \|f\|_p$. Thus, [because of $\operatorname{sgn} f = \operatorname{sgn} g$], $f(\|f+g\|_p - \|f\|_p) = g\|f\|_p$. It is clear that both $a := (\|f+g\|_p - \|f\|_p)$ and $b := \|f\|_p$ are nonnegative and if both were 0, we would have $f - g = 0$.
**10.2.16.** Hint: define $A = \{x : |f(x)| > t\}$, $g = f\chi_A$, $h = f\chi_{A^C}$.
**10.2.19.** Hint: consider $f(x) = 1/[x^{1/\alpha}(1 + \ln^2 x)]$.
**10.2.23.** For any $p \ge 1$, $\int |f|^p \le \|f\|_\infty^p \mu(X)$, so $\|f\|_p \le \|f\|_\infty \mu(X)^{1/p}$. Letting $p \to \infty$ gives $\limsup \|f\|_p \le \|f\|_\infty$. On the other hand, let $E = \{x \in X : f(x) \ge \|f\|_\infty - \varepsilon\}$. Therefore, $\int |f|^p \ge \int_E |f|^p \ge \int_E (\|f\|_\infty - \varepsilon)^p$, so $\|f\|_p \ge (\|f\|_\infty - \varepsilon)\mu(E)^{1/p}$. Taking the limit $p \to \infty$ gives $\liminf \|f\|_p \ge \|f\|_\infty - \varepsilon$. Since $\varepsilon$ is arbitrary, we obtain $\lim_{p\to\infty} \|f\|_p = \|f\|_\infty$.

In one direction, $\|f\|_{p+1}^{p+1} = \int_X |f|^{p+1} d\mu \le \int_X \|f\|_\infty |f|^p d\mu \le \|f\|_\infty \|f\|_p^p$. In the other direction, by Proposition 10.2.15, $\|f\|_p \le \mu(X)^{1/(p(p+1))} \|f\|_{p+1}$, so $\|f\|_p^{p+1} \le \mu(X)^{1/p} \|f\|_{p+1}^{p+1}$, hence $\|f\|_p \le \mu(X)^{1/p} \|f\|_{p+1}^{p+1} / \|f\|_p^p$.
**10.2.25.** By Hölder's inequality, $|\int_0^\varepsilon f| \le (\int_0^\varepsilon |f|^p)^{1/p} \varepsilon^{1/q}$ so $1/\varepsilon^\lambda |\int_0^\varepsilon f| \le (\int_0^\varepsilon |f|^p)^{1/p} \varepsilon^{1/q - \lambda}$. Since $|f|^p$ is integrable, Dominated Convergence Theorem implies that $\int_0^\varepsilon |f|^p \to 0$, so the given limit is 0 whenever $1/q - \lambda \ge 0$. When $p = 1$, $1/q = 0$, and the same proof shows that if $-\lambda \ge 0$, the given limit is 0. When $1/q - \lambda < 0$ the limit need not be 0. Proof: consider $f(x) = x^{\lambda - 1}$. Then $|f|^p = x^{p(\lambda-1)}$ and $p(\lambda - 1) > p(1/q - 1) = -1$, so $f \in L^p$. [Works for $p = 1$ the same.] However, $\int_0^\varepsilon f = \varepsilon^{(\lambda-1)+1}/[(\lambda - 1) + 1] = \varepsilon^\lambda/\lambda$, so the given limit $= 1/\lambda$.
**10.2.28.** Let $\alpha = \int g$. Then $|\int fg| = |\int(fg - \alpha f)| \le \|f\|_2 \|g - \alpha\|_2$ and $\int (g-\alpha)^2 = \int g^2 - 2\alpha \int g + \alpha^2 = \int g^2 - 2(\int g)(\int g) + (\int g)^2 = \int g^2 - (\int g)^2$.
**10.2.29.** Hint: for $0 < \varepsilon < 1$ define $\delta_p := 2[(1-\varepsilon)/2]^q$.
**10.2.33.** Hint: define $f = \sum_1^\infty \chi_{E_k}/[m(E_k)]^{1/q}$, where $\{E_n\}$ are some disjoint measurable subsets of $E$.
**10.2.36.** No! Let $g(x) = \chi_{(0,1)}(x)/\sqrt{x}$, let $\mathbb{Q} \cap (0,\infty) = \{r_k\}$, let $\{a_k\} \subset \mathbb{R}$ s.t. $\sum |a_k| < \infty$, and let $f(x) = \sum |a_k| g(x - r_k)$. Since $\int |g| = 2$ we have $f \in L^1$. Let $a < b$ and choose $q = r_m \in (a,b)$. Then $\int_a^b f^2 > \int_q^b f^2 > \int_q^b a_m^2 [g(x - r_m)]^2 = a_m^2 \int_0^{b-q} 1/x = \infty$.
**10.2.39.** We will prove the first equality. The second one can be proved similarly.

($q < \infty$) We will prove that $\lim_n (\int |f|^{q-1/n})^{1/(q-1/n)} = (\int |f|^q)^{1/q}$ and for that it suffices to prove that $\lim_n \int |f|^{q-1/n} = \int |f|^q$. Let $A = \{x : |f| > 1\}$ and $B = A^C$. On $A$, we have that $\{|f|^{q-1/n}\}$ is increasing, so by Monotone Convergence Theorem, $\lim_n \int_A |f|^{q-1/n} = \int_A |f|^q$. On $B$, the strategy depends on whether $\int_B |f|^q$ is finite or infinite. If finite, apply Dominated Convergence Theorem. [For $n$ large, $q - 1/n > p$, so we can apply Dominated Convergence Theorem.] If infinite, then so is $\int_B |f|^{q-1/n} \ge \int_B |f|^q$.

($q = \infty$): Case I. $\|f\|_\infty < \infty$. Let $A = \{x : |f(x)| > \|f\|_\infty - \varepsilon\}$. Then $\|f\|_\infty \ge \|f\|_n/[\mu(X)]^{1/n} = (\int_X |f|^n/\mu(X))^{1/n} \ge (\int_A |f|^n/\mu(X))^{1/n} > (\int_A (\|f\|_\infty - \varepsilon)^n/\mu(X))^{1/n} = (\|f\|_\infty - \varepsilon)[\mu(A)/\mu(X)]^{1/n} \to \|f\|_\infty - \varepsilon$. Case II. $\|f\|_\infty = \infty$. Let $B = \{x : |f(x)| \ge M\}$. Then $(\int_X |f|^n)^{1/n} \ge (\int_B |f|^n)^{1/n} \ge M\mu(B)^{1/n}$ so $\liminf \|f\|_n \ge M$.
**10.2.42.** No! Let $\{A_n\}$ be disjoint sets s.t. $m(A_n) = 1/(2n^2)$ and let $f = \sum m(A_n)^{-1/(2p)} \chi_{A_n}$.
**10.2.47.** By assumption, there exists a real number $M$ such that $\|f_n\|_{p+\theta} \le M$ for all $n$. By Fatou's Lemma, we have $\int_X |f|^{p+\theta} \le \liminf_{n\to\infty} \int_X |f_n|^{p+\theta} \le M^{p+\theta} < \infty$. Therefore $f \in L^{p+\theta}(X)$ so $f$ and $\{f_n\}$ belong to $L^p(X)$. It remains to show that $\|f_n - f\|_p \to 0$. By Hölder's Inequality, we have $\int_A |f_n - f|^p \le \|f_n - f\|_{p+\theta}^p \mu(A)^{\theta/(p+\theta)}$ for any measurable set $A \subset X$. By Minkowski's Inequality, $\|f_n - f\|_{p+\theta} \le \|f_n\|_{p+\theta} + \|f\|_{p+\theta} \le 2M$. Therefore, $\int_A |f_n - f|^p \le 2^p M^p \mu(A)^{\theta/(p+\theta)}$. Let $\varepsilon > 0$. By Egorov's Theorem, there is a set $F$ s.t. $\{f_n\}$ converges uniformly to $f$ on $F$ and $\mu(X \setminus F) < \varepsilon$. Choose $N$ s.t. $n \ge N \Rightarrow |f_n - f| < [\varepsilon/\mu(F)]^{1/p}$ on $F$. Now, $\int_X |f_n - f|^p = \int_F + \int_{X\setminus F}$. Clearly, $\int_F |f_n - f|^p \le (\varepsilon/\mu(F))\mu(F) = \varepsilon$. OTOH, as we have shown, $\int_{X\setminus F} |f_n - f|^p \le \|f_n - f\|_{p+\theta}^p \mu(X \setminus F)^{\theta/(p+\theta)} \le 2M\varepsilon^{\theta/(p+\theta)}$. Thus, $\|f_n - f\|_p \le (\varepsilon + 2M\varepsilon^{\theta/(p+\theta)})^{1/p}$, for $n \ge N$ so $\|f_n - f\|_p \to 0$.

**10.3.1.** Let $\varepsilon > 0$. There is $N$ s.t. $n \geq N \Rightarrow |f_n(x) - f(x)| < \varepsilon/\mu(X)$. For such $n$, $\int_X |f_n(x) - f(x)| < [\varepsilon/\mu(X)]\mu(X) = \varepsilon$. If $\mu(X) = \infty$ not true. Example: $X = (1, +\infty)$, $\mu = m$, $f_n(x) = (1/x)\chi(1, n)$, converges to $f(x) = 1/x$, and $\sup|f_n - f| = \sup|f\chi(n, \infty)| = 1/n \to 0$. But $1/x$ is not in $L^1$.

**10.3.3.** Let $\{f_n\}$ be Cauchy in $L^p$. By Proposition 2.2.21, it has a rapidly Cauchy subsequence $\{g_n\}$. By Theorem 10.3.2, it converges both in the norm of $L^p(X, \mu)$ and pointwise a.e. to a function $f$ in $L^p(X, \mu)$. By Proposition 2.2.17, $\{f_n\}$ converges to $f$ in $L^p$.

**10.3.5.** Hint: use Exercise 2.2.23.

**10.3.11.** Norm: Straighthorward except for $\|f\| = 0$ and triange inequality.

If $\|f\| = 0$ then for each $n$ there are $g_n$ and $h_n$ s.t. $f = g_n + h_n$ and $(\|g_n\|_1^p + \|h_n\|_2^p)^{1/p} < 1/n$ so $\|g_n\|_1, \|h_n\|_2 < 1/n$ hence $g_n \to 0$, $h_n \to 0$ whence $f = 0$. Triangle: Consider the function $H(x) = (1 + x^p)^{1/p}$. It is convex for $x > 0$, so if $a, b, c, d > 0$ then $H(a/(a+b)x + b/(a+b)y) \leq a/(a+b)H(x) + b/(a+b)H(y)$. For $x = c/a$ and $y = d/b$ we get $H((c+d)/(a+b)) \leq a/(a+b)H(c/a) + b/(a+b)H(d/b)$. Substituting the formula for $H$, we get $[1 + [(c+d)/(a+b)]^p]^{1/p} \leq a/(a+b)(1 + (c/a)^p)^{1/p} + b/(a+b)(1 + (d/b)^p)^{1/p}$ so $(a+b)[1 + [(c+d)/(a+b)]^p]^{1/p} \leq a(1 + (c/a)^p)^{1/p} + b(1 + (d/b)^p)^{1/p}$ hence $[(a+b)^p + (c+d)^p]^{1/p} \leq (a^p + c^p)^{1/p} + (b^p + d^p)^{1/p}$. Substituting $a = \|g_1\|_1$, $b = \|g_2\|_1$, $c = \|h_1\|_2$, $d = \|h_2\|_2$, we get $[(\|g_1\|_1 + \|g_2\|_1)^p + (\|h_1\|_2 + \|h_2\|_2)^p]^{1/p} \leq (\|g_1\|_1^p + \|h_1\|_2^p)^{1/p} + (\|g_2\|_1^p + \|h_2\|_2^p)^{1/p}$. Applying the triangle inequality that holds for $\|\cdot\|_1$ and $\|\cdot\|_2$, we get $[(\|g_1 + g_2\|_1)^p + (\|h_1 + h_2\|_2)^p]^{1/p} \leq (\|g_1\|_1^p + \|h_1\|_2^p)^{1/p} + (\|g_2\|_1^p + \|h_2\|_2^p)^{1/p}$. When $f_k = g_k + h_k$, we have $\|f_1 + f_2\|_p^{inf} = \|(g_1 + g_2) + (h_1 + h_2)\|_p^{inf} \leq (\|g_1 + g_2\|_1^p + \|h_1 + h_2\|_2^p)^{1/p} \leq$ [what we proved ] $\leq (\|g_1\|_1^p + \|h_1\|_2^p)^{1/p} + (\|g_2\|_1^p + \|h_2\|_2^p)^{1/p}$. Taking infimum over all $g_k$ and $h_k$ we get $\|f_1 + f_2\|_p^{inf} \leq \|f_1\|_p^{inf} + \|f_2\|_p^{inf}$.

Banach: If $\|f_m - f_n\|_p^{inf} < \varepsilon$ then there are decompositions $f_k = g_k + h_k$ s.t. $(\|g_m - g_n\|_1^p + \|h_m - h_n\|_2^p)^{1/p} < 2\varepsilon$ hence $\|g_m - g_n\|_1 < 2\varepsilon$ and $\|h_m - h_n\|_2 < 2\varepsilon$ whence $\{g_k\}$ is Cauchy in $L^1$ and $\{h_k\}$ is Cauchy in $L^2$. Thus, $\lim g_k = g$ in $L^1$, $\lim h_k = h$ in $L^2$ so $\|g - g_n\|_1 \leq 2\varepsilon$ and $\|h - h_n\|_2 \leq 2\varepsilon$. Let $f = g + h$. Then $\|f - f_n\|_p^{inf} \leq (\|g - g_n\|_1^p + \|h - h_n\|_2^p)^{1/p} \leq [(2\varepsilon)^p + (2\varepsilon)^p]^{1/p} = 2^{1/p}(2\varepsilon)$, so $\{f_n\}$ converges to $f$ in the norm of $X$.

**10.3.14.** Let $f \in L^r$, and define $f_n := f\chi_{0 \leq |f| \leq n}$. We have $f_n \in L^p \cap L^q$ for each $n$, and by monotone convergence, $\|f - f_n\|_r \to 0$ as $n$ goes to infinity. If $q = \infty$ the proof is valid unless $r = \infty$. In that case, the question becomes whether $L^p \cap L^\infty$ is dense in $L^\infty$ which is trivial because $L^p \cap L^\infty = L^\infty$.

**10.3.16.** $\|f_n\|_p \to 0$ so $\|f_n - 0\|_p \to 0$ hence $\{f_n\}$ converges to 0 in $L^p$ whence there is a subsequence $\{f_{n_k}\}$ that converges a.e. Let us denote it by $\{g_n\}$. We will extract a subsequence $\{g_{n_k}\}$ for which there is $h$. Ignoring the set of measure 0, we may assume that $g_n \to 0$ everywhere on a set $A$. That means that for each $x \in A$, $\{|g_n(x)|\}$ has a maximal element, i.e. there is $n = n(x)$ s.t. $|g_n(x)| \geq |g_i(x)|$, for all $i$. Let $E_n := \{x \in A : |g_n(x)| \geq |g_i(x)|$, for all $i\}$ and $A_n := E_n \setminus (E_1 \cup \cdots \cup E_{n-1})$. Then $A$ is a disjoint union of $A_n$ and $\lim_n \int_{A_n} |g_n|^p \leq \lim_n \int_A |g_n|^p = \lim_n \|g_n\|^p \to 0$ (because $\{g_n\} \subset \{f_n\}$). Thus, there is a subsequence $\{g_{n_k}\}$ of $\{g_n\}$ s.t. $\sum_k \int_{A_{n_k}} |g_{n_k}|^p < \infty$. Now define $h = |g_{n_k}|$ on $A_{n_k}$ and 0 elsewhere.

**10.3.18.** $p > 1$: Define $\{f_n\}$ by $f_n = \chi_{[0,1]} - \frac{1}{n}\chi_{[1,n+1]}$. Then $f_n \in E_p$, but $f_n \to \chi_{[0,1]}$ in $L^p$, and $\chi_{[0,1]} \notin E_p$.

$p = 1$: Let $f_n \in E_p$ converge to $f \in L^p$. Then $f_n^+ \to f^+$ and $f_n^- \to f^-$ in $L^1$ so $|\int f_n^+ - \int f^+| \leq \int |f_n^+ - f^+| \to 0$ and $\int f_n^- \to \int f^-$. Thus, $0 = \int f_n^+ - \int f_n^- \to \int f$.

**10.3.21.** Since $f_n \in AC[0,1]$, we have that $f_n(x) = \int_0^x f_n'$. Then, $\sup_{x \in [0,1]} |f_n(x) - f_m(x)| \leq \sup_{x \in [0,1]} |\int_0^x [f_m'(t) - f_n'(t)]\,dt| \leq \sup_{x \in [0,1]} \int_0^x |f_m'(t) - f_n'(t)|\,dt = \int_0^1 |f_m'(t) - f_n'(t)|\,dt \to 0$. Thus, $\{f_n\}$ converges uniformly to a function $f$ that is continuous. Also, there is $g \in L^1$ s.t. $\{f_n'\}$ converges to $g$ in $L^1$ norm. Then $|\int_0^x [f_n' - g]| \leq \int_0^1 |f_n' - g| \to 0$ so $f(x) = \lim f_n(x) = \lim \int_0^x f_n' = \int_0^x g$, whence $f \in AC[0,1]$.

**10.3.23.** Hint: use a partition $P = \{0, x, 1\}$.

**10.3.27.** (a) No! $L^q$ is a proper subset of $L^p$ so there are functions in the closed unit ball of $L^p$ that are not in $L^q$ at all.

(b) Yes! Let $B$ be the closed unit ball of $L^q$, and let $\{f_n\} \subset B$ that converges to a function $f$ in $L^p$ norm. By Riesz–Fischer theorem, there is a subsequence that converges a.e. WLOG, assume that $f_n \to f$ a.e. Then $|f_n|^q \to |f|^q$ a.e. By Fatou's Lemma, $\int |f|^q \leq \liminf \int |f_n|^q \leq 1$, so $f \in B$.

**10.3.29.** Hint: prove that $\mathcal{F} = \partial(\mathcal{F})$.

**10.3.34.** It is a metric, as concavity of $t \mapsto t^p$ for $t \geq 0$ implies that $(a+b)^p \leq a^p + b^p$ when $a$ and $b$ are nonnegative real numbers. Let $\{f_n\}$ be a Cauchy sequence. We extract a subsequence $\{g_k\} := \{f_{n_k}\}$ such that $d(g_{k+1}, g_k) \leqslant 2^{-k}$. Let $A_k := \{x : |g_{k+1}(x) - g_k(x)| \geqslant 2^{-k}\}$. Then $2^{-kp}\mu(A_k) \leqslant \int |g_{k+1} - g_k|^p d\mu \leqslant 2^{-k}$, so $\mu(A_k) \leqslant 2^{-k(1-p)}$. As $1 - p > 0$, a Borel–Cancelli's like argument shows that $\mu(\limsup_{k \to +\infty} A_k) = 0$. So, for almost all $x$, we can find $K(x)$ such

that for $k \geq K(x)$, $|g_{k+1}(x) - g_k(x)| \leqslant 2^{-k}$. For such $x$, $\{g_n(x)\}$ is as a Cauchy sequence of real numbers that converges to $g(x)$. The inequality $d(g_{k+1}, g_k) \leqslant 2^{-k}$ shows that $d(g_n, g_k) \leqslant 2^{-k+2}$. Letting $n \to \infty$ and using Fatou's lemma, we obtain that $d(g, g_k) \leqslant 2^{-k+2}$ whence $d(g, 0) \leq d(g, g_n) + d(g_n, 0) < \infty$, so $g \in L^p$.
**10.3.37.** Hint: apply Baire theorem to $E_{n,m} := \{f \in V : \|f\|_{L^{q_n}} \leq m\}$.
**10.4.7.** For $f = \chi_{(a,b)}$, see [86, Theorem 9.2.3]. Thus, it holds for simple functions. Let $\varepsilon > 0$. By Proposition 10.4.1, for $f \in L^1(\mathbb{R})$ there is a simple function $s$ s.t. $\int |f - s| < \epsilon/2$. Since $\int s \cos nx \to 0$, there is $N$ s.t. $n > N \Rightarrow |\int s \cos nx| < \varepsilon/2$. Now $|\int f \cos nx| \leq |\int (f - s) \cos nx| + |\int s \cos nx| \leq \int |f - s| + \varepsilon/2 < \varepsilon$.
**10.4.9.** (a) The function $g(x-t)$ is measurable as a composition of a continuous function (translation by $t$) and a measurable function (Exercise 6.2.1). Then $f(t)g(x-t)$ is measurable as a product of 2 measurable functions (Corollary 6.2.4). By Tonelli's theorem, $\int_{\mathbb{R}^2} |f(t)g(x-t)| = \int_{\mathbb{R}} |f * g| = \int_{\mathbb{R}} |f(t)|\,dt \int_{\mathbb{R}} |g(x-t)|\,dx$. By Exercise 7.3.36, this equals $\int_{\mathbb{R}} |f(t)|\,dt \int_{\mathbb{R}} |g(x)|\,dx < \infty$.

(b) Suppose $f$ is bounded. Then $|f * g(x+h) - f * g(x)| \leq \sup |f| \int_{\mathbb{R}} |g(x+h-t) - g(x-t)|dt$, so the result follows from Exercise 10.4.8.

(c) Use Exercise 7.3.36 for the change of variables $t = x - u$.

(d) $[(f * g) * h](x) = \int (f * g)(y)h(x-y)\,dy = \int dy \int f(t)g(y-t)h(x-y)dt$. Applying Fubini's theorem and substitution $u = y - t$, we get $\int dt \int f(t)g(y-t)h(x-y)dy = \int f(t)\,dt \int g(u)h(x-u-t)\,du = \int f(t)\,dt[(g * h)(x-t)] = [f * (g * h)](x)$.
**10.4.10.** Hint: use the counting measure.
**10.4.17.** Every continuous function $f$ is Riemann integrable, so by definition of Darboux sums, there is a sequence of step functions $\{s_n\}$ that converges uniformly to $f$, and all the more in $L^p$. Since $C[0,1]$ is dense in $L^p[0,1]$ we get that step functions are as well.
**10.4.23.** ($\Leftarrow$) $|f(x) - f(y)| = \lim |f_n(x) - f_n(y)| = \lim |f_n'(\xi)||x - y| \leq M|x - y|$.

($\Rightarrow$) $f$ is Lipschitz hence absolutely continuous, so $f(x) = f(0) + \int_0^x f'$ and $f' \in L^1$. By Exercise 10.4.19, there is a sequence of continuous functions $\{h_n\}$ that converges to $f'$ in the norm of $L^1$. To ensure that $|g_n| \leq M$, define $g_n(x) = h_n(x)$ if $|h_n(x)| \leq M$, and $\pm M$ if $h_n(x) > M$ or $h_n(x) < -M$. Since $|f'| \leq M$ we have that $\{g_n\}$ that converges to $f'$ in the norm of $L^1$. Finally, define $f_n(x) = f(0) + \int_0^x g_n$. Then (i) is obvious and $|f_n(x) - f(x)| = |\int_0^x (f'(x) - g_n(t))\,dt| \leq \|f' - g_n\|_1 \to 0$.
**10.4.24.** Hint: see the proof of Theorem 10.4.12 and Exercise 5.4.22.
**10.4.25.** $X = \mathbb{R}$, $\mathfrak{M} = \mathbb{P}(\mathbb{R})$, $\mu$ is the counting measure, $d(x,y) = |x - y|$. Let $f = \chi_{\{0\}} \in L^p$, and let $\varepsilon = 1/3$. If $g \in C_c(\mathbb{R})$ is such that $\|f - g\|_1 < 1/3$ then $g(0) \geq 2/3$, and the continuity of $g$ implies that there is $x \in \mathbb{R}$ for which $g(x) > 1/2$. The $\int_{\mathbb{R}} |f - g| \geq \int_{\{x\}} |f - g| = |f(x) - g(x)| = g(x) > 1/2 > \varepsilon$.
**10.4.26.** Hint: see Example 2.5.17.
**10.4.29.** Hint: consider $\mathcal{F} = \{s = \sum c_k \chi_{(\alpha_k, \beta_k)} : c_k, \alpha_k, \beta_k \in \mathbb{Q}\}$.
**10.4.31.** Hint: consider $f_a(x) = \int_0^x \chi_{(0,a)}$ for $0 < a < 1$.
**10.4.32.** The set $A = \{\text{finite sequences of rational numbers}\}$ is countable. Let $x = \{x_n\} \in c_0$ and let $\varepsilon > 0$. There is $N$ s.t. $n \geq N \Rightarrow |x_n| < \varepsilon$. For $1 \leq k \leq N$, choose $r_k \in \mathbb{Q}$ s.t. $|r_k - x_k| < \varepsilon$. Then $y := (r_1, r_2, \ldots, r_N, 0, 0, \ldots) \in A$ and $\|y - x\| = \sup |x_k - y_k| < \varepsilon$.
**10.4.35.** (a) $E(f,g)^C$ is closed, because it $= h^{-1}(0)$, where $h = f - g$. (b) distance is straightforward except triangle: if $f = g$ and $g = h$ then $f = h$, so if $f \neq h$ then $f \neq g$ or $g \neq h$ so $E(f,h) \subset E(f,g) \cup E(g,h)$ hence $\cup I_n(f,h) \subset (\cup I_n(f,g)) \cup (\cup I_n(g,h))$. Non-separable: let $F = \{f(x) = a : a \in (0,1)\}$. Then $F$ is uncountable, and $d(f,g) = 1$, for any $f, g \in F$.
**10.4.40.** Let $A := \{g(x) > 0\}$, $B := \{g(x) < 0\}$. If $g \neq 0$, then $\mu(A) \neq 0$ or $\mu(B) \neq 0$. Say $\mu(A) \neq 0$. There is $E \subset A$ s.t. $0 < \mu(E) < \infty$. Take $h = \chi_E$ and $\alpha := \int hg > 0$. Choose $f \in S$ s.t. $\|f - h\|_q < \alpha/\|g\|_p$. Then $|\int fg| = |\int hg + \int (f-h)g| \geq |\int hg| - |\int (f-h)g| \geq \alpha - \|f-h\|_q\|g\|_p > \alpha - (\alpha/\|g\|_p)\|g\|_p = 0$, a contradiction.
**10.5.3.** $0 \leq \|x + ty\|^2 = \langle x, x\rangle + 2t\langle x, y\rangle + t^2\langle y, y\rangle$ so $4\langle x, y\rangle^2 - 4\langle x, x\rangle\langle y, y\rangle \leq 0$. Next, let $f(y) = \langle x, y\rangle$. Then $|f(y_1) - f(y_2)| = |\langle x, y_1 - y_2\rangle| \leq \|x\|\|y_1 - y_2\|$.
**10.5.4.** ($\Leftarrow$) Easy. ($\Rightarrow$) If $x \neq 0$, let $\lambda = \|y\|/\|x\|$. Assuming that $\langle x, y\rangle = \|x\|\|y\|$, we have $\|y - \lambda x\|^2 = \|y\|^2 - 2\lambda\langle x, y\rangle + \lambda^2\|x\|^2 = \|y\|^2 - 2\lambda\|x\|\|y\| + \lambda^2\|x\|^2 = 0$.
**10.5.6.** Hint: use (10.13).
**10.5.8.** Hint: use the Parallelogram Identity.
**10.5.11.** Hint: define $\langle x, y\rangle = (1/2)[\|x + y\|^2 - \|x\|^2 - \|y\|^2]$.
**10.5.14.** Hint: argue by contradiction and apply Cauchy–Schwarz Inequality to $x$ and $y$.
**10.5.18.** By Exercise 10.4.19, polynomials are dense in $L^2[0,1]$ so $V^\perp = \{0\}$.
**10.5.19.** $V^\perp = \{0\}$. Proof: if $f \in L^2$ there is $g \in C[0,1]$ s.t. $\|f - g\|_2 < \varepsilon$. Now $g$ is continuous so $h(x) := g(\sqrt{x})$ is continuous hence there is a polynomial $p$ s.t. $\|h - p\|_{max} < \varepsilon$. It follows that $|h(x^2) - p(x^2)| < \varepsilon$, for all $x \in [0,1]$ whence $|g(x) - q(x)| < \varepsilon$, for all $x \in [0,1]$, where $q(x) = p(x^2)$. Now $\|f - q\|_2 < 2\varepsilon$.

**10.5.24.** Let $V$ be the closed linear span of vectors in $S$. Claim: $V^\perp = S^\perp$. ($\subset$) By Exercise 10.5.23, $S \subset V \Rightarrow V^\perp \subset S^\perp$. ($\supset$) Let $x \in S^\perp$. Then $\langle x, y\rangle = 0$ for all $y \in S$ hence for all $y$ in the span of $S$ whence for all $y$ in the closed span of $S$ which is $V$. Thus, $x \in V^\perp$. Since $V$ is a closed subspace, $\mathcal{H} = V \oplus V^\perp = V \oplus S^\perp$ so $V = (S^\perp)^\perp$.
**10.5.28.** $\overline{\mathcal{M}} = X$. Indeed, let $y \in X$, $y = \{y_1, \dots, y_N, 0, 0, \dots\}$. Let $\sum_1^N y_n = \alpha$. Define $\{x_n\}$ by $x_k = y_k$ for $1 \le k \le N$, $x_k = -\alpha/K$ for $N+1 \le k \le N+K$, and $x_n = 0$ for $k \ge N+K+1$. It is easy that $\{x_n\} \in \mathcal{M}$ and $\|x - y\|^2 = \sum_{k=N+1}^{N+K} (\alpha/K)^2 = \alpha^2/K \to 0$ as $K \to \infty$.
**10.6.2.** (a) Mimic the proof of Theorem 10.6.11 to extend $\{x_\alpha\}_{\alpha\in A}$ to a complete o.n.s. $\{x_\alpha\}_{\alpha\in A} \cup \{x_\beta\}_{\beta\in B}$. Let $\beta \in B$. For each $\alpha \in A$, $x_\beta \perp x_\alpha$ so $x_\beta \perp S$ hence $x_\beta \perp \mathcal{K}$ whence $x_\beta \in \mathcal{K}^\perp$. Since $x = \sum \hat{x}_\alpha x_\alpha + \sum \hat{x}_\beta x_\beta$ and $\sum \hat{x}_\alpha x_\alpha \in \mathcal{K}$, $\sum \hat{x}_\beta x_\beta \in \mathcal{K}^\perp$ we have the unique decomposition by Theorem 10.5.25, i.e., $y = \sum_{\alpha\in A} \hat{x}_\alpha x_\alpha$.

(b) By Theorem 10.5.25, $d(x, \mathcal{K}) = \|x - y\|$ and from $x = \sum \hat{x}_\alpha x_\alpha + \sum \hat{x}_\beta x_\beta$ we get that $x - y = \sum \hat{x}_\beta x_\beta$ so [ by Theorem 10.6.12 ] $d^2 = \|x - y\|^2 = \|z\|^2$. Because of orthogonality we have $\|x\|^2 = \|y\|^2 + \|z\|^2$ so $\|z\|^2 = \|x\|^2 - \|y\|^2 = \|x\|^2 - \sum |\hat{x}_\alpha|^2$ [again using Theorem 10.6.12].
**10.6.3.** The answer depends on whether $b - a \le 2\pi$ or not. Case 1: $b - a \le 2\pi$. Then $V$ is dense. Proof: $[a, b] \subset [a, a + 2\pi]$, so suffices to prove that $V$ is dense in $L^2[a, a + 2\pi]$, because every $f \in L^2[a, b]$ can be extended to $g \in L^2[a, a + 2\pi]$, and if $h \in V$ approximates $g$, then $\|h - f\|_{L^2[a,b]} \le \|h - g\|_{L^2[a,a+2\pi]}$. Finally, $V$ dense in $L^2[a, a + 2\pi]$ iff the functions (10.15) translated by $a$ have a dense span in $L^2[0, 2\pi]$ if the functions (10.15) have a dense span in $L^2[0, 2\pi]$ (which we know is true). The last equivalence holds by Parsevals' Identity because if $A_n = \int_0^{2\pi} \cos n(x+a) f(x)\,dx$ and $B_n = \int_0^{2\pi} \sin n(x+a) f(x)\,dx$, then $A_n^2 + B_n^2 = a_n^2 + b_n^2$. Indeed, $A_n = \cos naa_n - \sin nab_n$ and $B_n = \cos nab_n + \sin naa_n$.

Case 2: $b - a > 2\pi$. Then $V$ is not dense. Proof: $[a, b] \supset [a, a + 2\pi]$, and functions in (10.15) are periodic with period $2\pi$. If we consider a function in $L^2[a, b]$ that is not periodic, it cannot be approximated.
**10.6.5.** Orthonormal is straightforward. When computing $\int_0^1 \varphi_m \varphi_n$, where $m > n$, it suffices to notice that on each of the intervals where $\varphi_n$ is constant, the integral of $\varphi_m$ is zero. Not complete: Split $[0, 1]$ into 4 parts and define $f$ as $1, -1, -1, 1$ on these intervals. Since $f = \varphi_1\varphi_2$, for any $n > 2$, $\int_0^1 f\varphi_n = \int_0^{1/2} \varphi_2\varphi_n - \int_{1/2}^1 \varphi_2\varphi_n = \int_0^{1/4} \varphi_n - \int_{1/4}^{1/2} \varphi_n + \int_{1/2}^{3/4} \varphi_n - \int_{3/4}^1 \varphi_n = 0 - 0 + 0 - 0 = 0$. Also, $\langle f, \varphi_1\rangle = \int_0^1 \varphi_1^2\varphi_2 = \int_0^1 \varphi_2 = 0$ and $\langle f, \varphi_2\rangle = \int_0^1 \varphi_2^2\varphi_1 = \int_0^1 \varphi_1 = 0$. Finally, $\langle f, \varphi_0\rangle = \int_0^1 \varphi_1\varphi_2 = \langle \varphi_1, \varphi_2\rangle = 0$.
**10.6.9.** Let $\sum_n \beta_n y_n$ be a rearrangement of $\sum_n \alpha_n x_n$, denote their partial sums by $B_n$ and $A_n$ and suppose that they converge to vectors $B$ and $A$. For large $m, n$, $\|A - A_m\| < \varepsilon$ and $\|B - B_n\| < \varepsilon$, so to show that $A = B$, we must show that for large $m, n$, $\|A_m - B_n\| < \varepsilon$. For every $k$ there is $n \ge k$ and $m \ge n$ s.t. $\{\alpha_i x_i\}_{i=1}^k \subset \{\beta_i y_i\}_{i=1}^n \subset \{\alpha_i x_i\}_{i=1}^m$. Now $A_m - B_n = (\alpha_1 x_1 + \cdots + \alpha_m x_m) - (\beta_1 y_1 + \cdots + \beta_n y_n)$ so after canceling there are only some $\alpha_i x_i$ left, but necessarily for $i > k$ (the earlier ones were among the canceled $\beta_i y_i$). Thus, $\|A_m - B_n\|^2 \le \sum_{i=k+1}^m |\alpha_i|^2 < \varepsilon$ if $k$ is large enough.
**10.6.10.** Hint: write the equality as $\sum |\langle g, f_n\rangle|^2 = \|g\|^2$.
**10.6.11.** Hint: use Egorov's Theorem.
**10.6.15.** Let $\{x_n\}$ be a countable dense set in $\mathcal{H}$. Since $\mathcal{M}$ is dense, for each $n$ there is $\{x_{mn}\} \subset \mathcal{M}$ s.t. $\lim_m x_{mn} = x_n$. Let $D = \{x_{mn}\}$: it is countable and dense in $\mathcal{H}$ (because it is dense in $\mathcal{M}$). Let $B$ be a maximal linearly independent subset of $D$. Then the span of $B$ contains $D$, so it is dense in $\mathcal{H}$. Let $B = \{y_n\}$ and let $\{e_n\}$ be the sequence obtained from $\{y_n\}$ using Gramm–Schmidt. Since $\{y_n\} \subset D \subset \mathcal{M}$ we have that $\{e_n\} \subset \mathcal{M}$. Also, the span of $\{e_n\}$ is the same as the span of $\{y_n\}$ which is dense in $\mathcal{H}$.
**10.6.18.** Hint: consider $B_{e_\alpha} := \{f_i : \langle e_\alpha, f_i\rangle \ne 0\}$.
**10.6.25.** If $x \in \mathcal{M}$ then $x = (x_1, x_1, -x_1, x_4, x_5, \dots)$. If $y \in \mathcal{M}^\perp$ then $y_1x_1 + y_2x_1 - y_3x_1 + y_4x_4 + \cdots = 0$ for any $x \in \mathcal{M}$. It follows that $y_1 + y_2 - y_3 = 0$ and $y_k = 0$ for $k \ge 4$. Thus, $\mathcal{M}^\perp$ is 2-dimensional space. One o.n.b. for $\mathcal{M}^\perp$ is $\{(e_1 - e_2)/\sqrt{2}, (e_1 + e_2 + 2e_3)/\sqrt{6}\}$. If $w = \{1/n\}$ then $\langle w, (e_1 - e_2)/\sqrt{2}\rangle = 1/(2\sqrt{2})$ and $\langle w, (e_1 + e_2 + 2e_3)/\sqrt{6}\rangle = 13/(6\sqrt{6})$ so the projection is $1/(2\sqrt{2})(e_1 - e_2)/\sqrt{2} + 13/(6\sqrt{6})(e_1 + e_2 + 2e_3)/\sqrt{6} = (29/12)e_1 + (23/12)e_2 + (13/3)e_3$.
**10.6.28.** $E = \cap_n \{x : f_n(x) \ge \varepsilon\} \in \mathcal{M}$ because each $f_n$ is measurable. We show $m(E) = 0$. Suppose not. Then there is $A \subset E$ such that $0 < m(A) < \infty$. The function $\chi_A$ is in $L^2$ so Bessel's inequality gives $\sum_n |\int_{\mathbb{R}} \chi_A f_n|^2 \le \|\chi_A\|^2$ so $\sum_n |\int_A f_n|^2 \le \int_{\mathbb{R}} |\chi_A|^2 = m(A)$ hence $m(A) \ge \sum_n |\int_A f_n|^2 \ge \sum_n |\int_A \varepsilon|^2 = \sum_n \varepsilon^2 m(A)^2 = \infty$, a contradiction.
**10.6.29.** Hint: define $f = a_0 + \sum_{k=1}^\infty (a_k \cos kx + b_k \sin kx)$ and $\nu(E) = \int_E f\,dx$, then show that $\mu = \nu$.
**10.7.3.** Not at $x = 0$! Let $\varepsilon = 1/2$. Given $\delta > 0$ choose $n > 1/\delta$. Then $x = 1/n < \delta$ but $f_n(1/n) = 1 > \varepsilon$.
**10.7.4.** Yes! For any $x_0$, given $\varepsilon > 0$ choose $\delta$ s.t. $\delta(\delta + 2|x_0|) < \varepsilon$. Then $|f_n(x) - f_n(x_0)| = |x - x_0||x + x_0| \le \delta(|x - x_0| + 2|x_0|) < \varepsilon$.
**10.7.14.** Hint: apply Mean Value Theorem.

**10.7.17.** ($\Leftarrow$) is obvious. ($\Rightarrow$) Suppose not. Then for each $n$ there is $f_n \in \mathcal{F}$ and $x_n \in X$ s.t. $|f_n(x_n)| \geq n$. Since $X$ is compact, $\{x_n\}$ has a convergent subsequence $\{x_{n_k}\}$ that converges to $x \in X$. $\mathcal{F}$ is pointwise bounded so there is $M_x$ s.t. $|f(x)| \leq M_x$, for all $f \in \mathcal{F}$. $\mathcal{F}$ is equicontinuous at $x$, so there is $\delta$ s.t. $d(x_{n_k}, x) < \delta \Rightarrow |f_{n_k}(x_{n_k}) - f_{n_k}(x)| < 1$. It follows that $n_k \leq |f_{n_k}(x_{n_k})| \leq |f_{n_k}(x_{n_k}) - f_{n_k}(x)| + |f_{n_k}(x)| \leq 1 + M_x$, a contradiction.
**10.7.20.** ($\Leftarrow$) Given $\varepsilon > 0$, pick $n$ as in the display, and define a mapping $\Phi : \mathcal{S} \to \mathbb{R}^n$ by $\Phi(x) = (x_1, \dots, x_n)$. The boundedness of $\mathcal{S}$ implies that $\Phi(\mathcal{S})$ is bounded, hence totally bounded. If $x, y \in \mathcal{S}$ with $|\Phi(x) - \Phi(y)|_p = (\sum_{k=1}^{n} |x_k - y_k|^p)^{1/p} < \varepsilon$, then $\|x - y\|_p \leq (\sum_{k=1}^{n} |x_k - y_k|^p)^{1/p} + (\sum_{k>n} |x_k - y_k|^p)^{1/p} < \varepsilon + 2\varepsilon = 3\varepsilon$. By Exercise 2.3.46, $\mathcal{S}$ is totally bounded.

($\Rightarrow$) Since $\mathcal{S}$ is compact it is closed and bounded. Equisummable: Let $\varepsilon > 0$. $\mathcal{S}$ is also totally bounded so there are balls $\{B(f_k, \varepsilon)\}_{k=1}^{n}$ that cover $\mathcal{S}$. Each $f_k = \{f_{k,i}\}_{i=1}^{\infty} \in \ell^p$, so there is $N_k$ s.t. $\sum_{i \geq N_k} |f_{k,i}|^p < \varepsilon$. Let $N := \max\{N_k : 1 \leq k \leq n\}$. Suppose now that $g \in \mathcal{S}$. Then $g$ must belong to one of the $\varepsilon$-balls, say one with center at $f_j$. Then $[\sum_{k \geq N} |g_k|^p]^{1/p} \leq [\sum_{k \geq N} |g_k - f_{j,k}|^p]^{1/p} + [\sum_{k \geq N} |f_{j,k}|^p]^{1/p} \leq \|g - f_j\| + \varepsilon^{1/p} < \varepsilon + \varepsilon^{1/p}$.
**10.7.23.** ($\Leftarrow$) By Arzelà–Ascoli theorem, every sequence in $\mathcal{F}$ has a convergent subsequence which converges uniformly to a function $f$, so $f \in \overline{\mathcal{F}} = \mathcal{F}$. ($\Rightarrow$) Since $C(X)$ is a metric space, a compact set $\mathcal{F}$ is closed and bounded. Equicontinuous: Let $x_0 \in X$. Let $\varepsilon > 0$. $\mathcal{F}$ is also totally bounded so there is an $\varepsilon$-net $\{f_1, \dots, f_n\}$. Each $f_k$ is continuous, so there is $\delta_k$ s.t. $d(x, x_0) < \delta_k \Rightarrow |f_k(x) - f_k(x_0)| < \varepsilon$. Let $\delta := \min\{\delta_k : 1 \leq k \leq n\}$. Suppose now that $f \in \mathcal{F}$ and $x \in X$ satisfies $d(x, x_0) < \delta$. Then $f$ must belong to one of the $\varepsilon$-balls, say one with center at $f_i$. Then $|f(x) - f(x_0)| \leq |f(x) - f_i(x)| + |f_i(x) - f_i(x_0)| + |f_i(x_0) - f(x_0)| \leq \|f - f_i\| + \varepsilon + \|f - f_i\| < 3\varepsilon$.
**10.7.26.** $|g_n| \leq \|f\|_1$, so $\{g_n\}$ is bounded. Let $\varepsilon > 0$. There is $\delta > 0$ s.t. $m(E) < \delta$ implies $\int_E |f| < \varepsilon$. If $0 \leq x \leq y \leq 1$ and $y - x < \delta$ then $|g_n(y) - g_n(x)| = |\int_x^y \chi_{E_n} f| \leq \int_x^y |f| < \varepsilon$. Thus, $\{g_n\}$ is equicontinuous. Apply Arzelà–Ascoli theorem.
**10.7.27.** Hint: select in succession subsequences that converge to $f_k$, then use the diagonal subsequence.
**10.7.28.** Hint: consider $d_n := d(x, x_n)$ where $\{x_n\}$ is a dense set in $X$.
**11.1.2.** Hint: prove that $\sup\{\|Tx\| : \|x\| = 1\} \leq \sup\{\|Tx\| : \|x\| \leq 1\} \leq \|T\| \leq \sup\{\|Tx\| : \|x\| = 1\}$.
**11.1.3.** (a) Suppose that $\|T\| = \alpha < M$. Then, for this specific $x$, by Proposition 11.1.11, $M\|x\| = \|Tx\| \leq \|T\|\|x\| = \alpha\|x\| < M\|x\|$, a contradiction.

(b) ($\Leftarrow$) Suppose that $\|T\| = \alpha < M$. By assumption, for every $n$ there is $x_n$ of norm 1, such that $\|Tx_n\| > \alpha$. By Proposition 11.1.11, $\alpha < \|Tx_n\| \leq \|T\|\|x_n\| = \|T\|$, a contradiction.

($\Rightarrow$) If $\|T\| = M$ then, by Proposition 11.1.11, $M = \sup\{\|Tx\| : \|x\| = 1\}$, so $M - 1/n$ is not an upper bound of $\{\|Tx\| : \|x\| = 1\}$. Therefore, there is $x_n \in X$ such that $\|x_n\| = 1$ and $\|Tx_n\| \geq M - 1/n$. Since, by assumption, $\|Tx_n\| \leq \|T\|\|x_n\| = \|T\| \leq M$, using Squeeze Theorem we get the desired conclusion.

(c) Equivalent to (b).

(d) Let $X = \{f \in C[0,1] : f(0) = 0\}$ and $Tf = \int_0^1 f(x)\,dx$. Clearly, $T$ is linear and $|Tf| = |\int_0^1 f(x)\,dx| \leq \|f\|_{max} \int_0^1 dx = \|f\|_{max}$, so $\|T\| \leq 1$. Let $f_n = 1$ on $(1/n, 1)$, and $f(x) = x/n$ on $(0, 1/n)$. Then $Tf_n = \int_0^{1/n} x/n\,dx + \int_{1/n}^1 dx = 1/(2n^3) + (1 - 1/n) \to 1$. By (b), $\|T\| = 1$. Suppose that there is $f \in X$ for which $Tf = \|f\|_{max}$. Since $f(0) = 0$ and $f$ is continuous, there is $\varepsilon > 0$ such that $f(x) < \|f\|_{max}/2$ on $(0, \varepsilon)$. Then $Tf = \int_0^1 f(x)\,dx = \int_0^\varepsilon f(x)\,dx + \int_\varepsilon^1 f(x)\,dx \leq \|f\|_{max}/2 \int_0^\varepsilon dx + \|f\|_{max} \int_\varepsilon^1 dx = \|f\|_{max}[\varepsilon/2 + (1 - \varepsilon)] = \|f\|_{max}[1 - \varepsilon/2] < \|f\|_{max}$, a contradiction.
**11.1.8.** Hint: use Exercise 11.1.3 (c).
**11.1.10.** ($\Rightarrow$) If $\{x_n\}$ is a Cauchy sequence in $X$ and $y_n = Tx_n$ then $\|y_m - y_n\| = \|Tx_m - Tx_n\| = \|T(x_m - x_n)\| \leq \|T\|\|x_m - x_n\|$. ($\Leftarrow$) Suppose $T$ is not bounded. By definition, there is a sequence $\{z_n\}$ in $X$ s.t. $\|z_n\| = 1$ and $\|Tz_n\| \to \infty$. Passing to a subsequence we get that $\|Tz_n\| \geq n^2$. Now, let $x_n = z_n/n$. Then $\|x_n\| = 1/n \to 0$, so $\{x_n\}$ is a Cauchy sequence in $X$ (because it is convergent). On the other hand, $\|Tx_n\| = \|Tz_n/n\| \geq n \to \infty$, so $\{Tx_n\}$ is not a Cauchy sequence in $Y$ (Exercise 2.3.1), a contradiction.
**11.1.13.** Well-defined: If $[x] = [y]$ then there is $z \in \ker T$ s.t. $y = x + z$. Then $\tilde{T}([y]) = T(y) = T(x + z) = Tx = \tilde{T}([x])$. Linear: $\tilde{T}(a[x] + b[y]) = \tilde{T}([ax + by]) = T(ax + by) = aTx + bTy = a\tilde{T}([x]) + b\tilde{T}([y])$. Bounded: by definition, for every $\varepsilon > 0$ there is $y \in \ker T$ s.t. $\|x + y\| \leq \|[x]\| + \varepsilon$. Now, $\|\tilde{T}([x])\| = \|Tx\| = \|T(x + y)\| \leq \|T\|\|x + y\| \leq \|T\|(\|[x]\| + \varepsilon)$. Letting $\varepsilon \to 0$, we get $\|\tilde{T}([x])\| \leq \|T\|\|[x]\|$, so $\tilde{T}$ is bounded and $\|\tilde{T}\| \leq \|T\|$. In the opposite direction, $\|Tx\| = \|\tilde{T}[x]\| \leq \|\tilde{T}\|\|[x]\| \leq \|\tilde{T}\|\|x\|$. Injective: If $\|\tilde{T}([x])\| = 0$ then $\|Tx\| = 0$ so $x \in \ker T$ hence $[x] = 0$.
**11.1.19.** $\|Tf\|_p = \|gf\|_p = [\int |gf|^p]^{1/p} \leq [\int (\|g\|_{max}|f|)^p]^{1/p} = \|g\|_{max}\|f\|_p$ so $\|T\| \leq \|g\|_{max}$. In the other direction: $g$ is continuous so it attains its maximum at $x_0 \in [0,1]$. Let $\varepsilon > 0$. There is an interval $I$ containing $x_0$ s.t. $|g(x)| \geq \|g\|_{max} - \varepsilon$, for all $x \in I$. Define $f = \chi_I/[m(I)]^{1/p}$. It

is clear that $f \in L^p$, $\|f\|_p = 1$, and $\|gf\|_p = [\int_0^1 |gf|^p]^{1/p} = [\int_I |g|^p/m(I)]^{1/p} \geq [\int_I |(\|g\|_{max} - \varepsilon)|^p/m(I)]^{1/p} = \|g\|_{max} - \varepsilon$ so $\|g\|_{max} - \varepsilon \leq \|gf\|_p = \|Tf\| \leq \|T\|\|f\|_p = \|T\|$. Letting $\varepsilon \to 0$ we get $\|T\| \geq \|g\|_{max}$.

**11.1.20.** Applying Hölder's inequality with conjugate exponents $q$ and $p$, $\|Tx\|_q = \|\{\sum_{i=1}^\infty a_{ij}x_i\}\|_q = [\sum_j |\sum_{i=1}^\infty a_{ij}x_i|^q]^{1/q} \leq [\sum_j[\sum_{i=1}^\infty |a_{ij}|^q[\sum_{i=1}^\infty |x_i|^p]^{q/p}]]^{1/q} = [\sum_{i=1}^\infty |x_i|^p]^{1/p}]\sum_j[\sum_{i=1}^\infty |a_{ij}|^q]^{1/q} = C\|x\|_p$.

**11.1.27.** Applying Hölder's inequality with conjugate exponents $p/q$ and $p/(p-q)$, we obtain $\|If\|_q = \|f\|_q = (\int |f|^q)^{1/q} \leq [(\int |f|^{q(p/q)})^{q/p}(\int 1)^{(p-q)/p}]^{1/q} = \|f\|_p[\mu(X)]^{(p-q)/(pq)}$ so $\|I\| \leq [\mu(X)]^{(p-q)/(pq)}$. To prove the equality take $f = \mu(X)^{-1/p}$.

**11.1.28.** Hint: if $p < q$ then $h = 0$. If $p = q$ then $h \in L^\infty$.

**11.1.31.** Linear: easy. Bounded: for $f \in C[0,1]$, $\|Tf\| = \int_0^1 |f(\sqrt{x})|\,dx$ and this is a Riemann integral, so substituting $t = \sqrt{x}$ yields $\int_0^1 |f(t)|2t\,dt \leq 2\|f\|$. But $C[0,1]$ is dense in $L^1[0,1]$ (see Exercise 10.4.19), so Exercise 11.1.12 shows that $T$ is a bounded operator on $L^1[0,1]$ with the same norm, i.e., $\|T\| \leq 2$. Equality: use $f_n(x) = n\chi_{[1-1/n,1]}$. Then $\|f_n\| = 1$ and $\|Tf_n\| = 2 - 1/n$.

**11.1.33.** Hint: consider separately cases $1 < p < \infty$, $p = 1$, and $p = \infty$. Use Hölder's inequality and Tonelli's Theorem.

**11.1.34.** (a) Since $L^p$ and $L^r$ are vector spaces so is their intersection. Also, $\|\cdot\|$ is a norm by Exercise 10.1.9. Let $\{f_n\}_{n=1}^\infty$ be a Cauchy sequence in $L^p \cap L^r$. The inequalities $\|f_m - f_n\|_p \leq \|f_m - f_n\|$ and $\|f_m - f_n\|_r \leq \|f_m - f_n\|$ imply that $\{f_n\}_{n=1}^\infty$ is a Cauchy sequence in both $L^p$ and $L^r$. Let $g_p \in L^p$ and $g_r \in L^r$ be the respective limits of this sequence. By Riesz–Fischer Theorem, there is a subsequence of $\{f_n\}_{n=1}^\infty$ that converges a.e. to both $g_p$ and $g_r$, so $g_p = g_r \in L^p \cap L^r$, whence $\{f_n\}_{n=1}^\infty$ converges to $g_p$ in the norm of $L^p \cap L^r$. Thus, $L^p \cap L^r$ is a Banach space.

(b) Let $p < q < r$. By Exercise 10.2.22, there exists $\lambda \in (0,1)$ such that, for all $f \in L^p \cap L^r$, $\|f\|_q \leq \|f\|_p^\lambda\|f\|_r^{1-\lambda} \leq \|f\|^\lambda\|f\|^{1-\lambda} = \|f\|$. Therefore, the inclusion $L^p \cap L^r \to L^q$ is continuous.

**11.2.1.** Hint: consider the shift operator.

**11.2.3.** Hint: consider $TS - I + S$.

**11.2.9.** Hint: for $T$ invertible, consider $I - T^{-1}S$ and use Exercise 11.2.2.

**11.2.10.** We define an operator $T : \sum_{k=1}^n \alpha_k x_k \mapsto \sum_{k=1}^n \alpha_k(x_k - y_k)$, and we can extend it to the whole $\mathcal{H}$ because if $\sum_{k=1}^\infty \alpha_k x_k$ converges then so does $\sum_{k=1}^\infty \alpha_k(x_k - y_k)$. Then $T \in \mathcal{L}(\mathcal{H})$ and $\|T\| \leq C$. Moreover, by Exercise 11.2.2, $T - I$ is invertible, hence onto, and is given by $(T-I)\sum_{k=1}^\infty \alpha_k x_k = \sum_{k=1}^\infty -\alpha_k y_k$. Thus, for every $y \in \mathcal{H}$ there is $x$ s.t. $y = (T-I)x$ and if $x = \sum_{k=1}^\infty \alpha_k x_k$ then $y = \sum_{k=1}^\infty -\alpha_k y_k$.

**11.2.12.** ($\Rightarrow$) Since $T$ is continuous we can take $M = \|T\|$; since $T^{-1}$ is continuous we can take $m = 1/\|T^{-1}\|$. ($\Leftarrow$) 1-1: $Tx = Ty$ implies $\|T(x-y)\| = 0$ so [use $\|Tx\| \geq m\|x\|$], $\|x - y\| = 0$ hence $x = y$. Thus, $T$ is a bijection. Bounded: from $\|Tx\| \leq M\|x\|$. Bounded inverse: let $y \in Y$ and let $y = Tx$. Then $\|T^{-1}y\| = \|T^{-1}Tx\| = \|x\| \leq (1/m)\|Tx\| = (1/m)\|y\|$.

**11.2.14.** By Exercise 10.6.7, $\mathcal{H}$ has a countable o.n. basis $\{f_n\}$. Define $F : \mathcal{H} \to \ell^2$ by $F(x) = \{(\hat{x})_n\}$. By Bessel's inequality, $\{(\hat{x})_n\} \in \ell^2$ so $F$ is well defined. (1 to 1) If two vectors in $\mathcal{H}$ have the same Fourier coefficients, by Theorem 10.6.12 (b) they are equal, so $F$ is 1 to 1. (Onto): If $\{a_n\} \in \ell^2$ defined $x := \sum a_n f_n$. (Converges by Bessel.) Now $(\hat{x})_n = \langle x, f_n\rangle = \langle \sum a_i f_i, f_n\rangle = a_n$, so $F(x) = \{a_n\}$. Linear: $F(ax+by) = \{\langle ax+by, f_n\rangle\} = \{a\langle x, f_n\rangle + b\langle y, f_n\rangle\} = \{a(\hat{x})_n + b(\hat{y})_n\} = aF(x) + bF(y)$. (Isometry) $\langle F(x), F(y)\rangle = \langle\{(\hat{x})_n\}, \{(\hat{y})_n\}\rangle = \sum_n (\hat{x})_n(\hat{y})_n = \langle x, y\rangle$ by Theorem 10.6.12 (e).

**11.2.15.** Hint: define $T : C[0,1] \to C[0,1/2]$ by $Tf(x) = f(2x)$, and $L : C[0,1/2] \to A$ by $Lg(x) = g(x)$ when $0 \leq x \leq 1/2$ and $g(1-x)$ when $1/2 < x \leq 1$.

**11.2.19.** Hint: use the counting measure and $F(x) = \sum_{i \in I} \hat{x}_i \chi_{\{i\}}$.

**11.2.25.** $\|x\|_\alpha := \sup_n \|T^n x\|$. It is a norm. Nonnegative is obvious. If $x = 0$ then $T^n x = 0$ so $\|x\|_\alpha = 0$; If $\|x\|_\alpha = 0$, taking $n = 0$ obtain $\|x\| = 0$ so $x = 0$. $\|ax\|_\alpha = \sup\|T^n(ax)\| = |a|\sup\|T^n x\| = |a|\|x\|_\alpha$. Triangle: $\|x+y\|_\alpha = \sup\|T^n(x+y)\| \leq \sup\{\|T^n x\| + \|T^n y\|\} \leq \|x\|_\alpha + \|y\|_\alpha$. Equivalent: $\|x\|_\alpha \geq \|x\|$ [taking $n = 0$]. By the given condition, $\|x\|_\alpha \leq M\|x\|$. Isometry: $\|Tx\|_\alpha = \sup_n \|T^{n+1}x\| = \|x\|_\alpha$, because $\{n+1 : n \in \mathbb{Z}\} = \{n : n \in \mathbb{Z}\}$.

**11.2.27.** Since $I$ is continuous, hence bounded, we get that $\|x\|_2 \leq C\|x\|_1$ for some $C > 0$. Need the other inequality. We will show that $B_2(0,1) \subset B_1(0,2r)$. If so, then if $x \neq 0$ then $x/(2\|x\|_2) \in B_2(0,1)$ so $x/(2\|x\|_2) \in B_1(0,2r)$ hence $\|x\|_1 \leq 4r\|x\|_2$. Prove $B_2(0,1) \subset B_1(0,2r)$. Let $x \in B_2(0,1)$. Define a sequence $\{x_n\}$ as follows. Let $x_1 = x$. Notice that by scaling, the assumption $B_2(0,1) \subset B_1(0,r) + B_2(0,1/2)$ yields $B_2(0,1/2^{n-1}) \subset B_1(0,r/2^{n-1}) + B_2(0,1/2^n)$, for all $n \in \mathbb{N}$. Thus, if $\|x_n\|_2 < 1/2^{n-1}$ there are $\|y_n\|_1 < r/2^{n-1}$ and $\|z_n\|_2 < 1/2^n$ such that $x_n = y_n + z_n$, and we define $x_{n+1} := z_n$. The sequence $\{s_n\}$ of partial sums of $\sum y_k$ is 1-Cauchy because $\|s_m - s_n\|_1 \leq \sum_{k=n+1}^m \|y_k\|_1 \leq \sum_{k=n+1}^m r/2^{k-1} \leq r/2^{n-1}$. Since $X$ is 1-complete, there is $y \in X$, $y = \sum_1^\infty y_k$. In view of continuity of $I$, the series converges in 2-norm as well. Notice that $x = x_1 = y_1 + x_2 = y_1 + y_2 + x_3 \cdots = \sum_1^n y_k + x_{n+1}$. Since $\|x_{n+1}\|_2 < 1/2^n$, we have that $\lim x_n = 0$ in 2-norm, so $\sum y_k$ converges in 2-norm to $x$ whence

$x = y$. Thus, $\|x\|_1 = \|y\|_1 = \|\sum y_k\|_1 \le \sum \|y_k\|_1 < \sum r/2^{k-1} = 2r$. Thus, $x \in B_1(0, 2r)$ hence $B_2(0,1) \subset B_1(0, 2r)$.

**11.2.31.** Let $a_0 = 0$ and let $\{a_n\}$ be an increasing sequence converging to 1. Let $f_n$ be 0 except on $(a_{n-1}, a_n)$ where its graph forms a triangle of height 1. Let $\mathcal{G} = \{B(f, 1/3) : f \in B(0,1)\}$. If there were a finite subcover of $\mathcal{G}$, there would be two functions from $\{f_n\}$ in the same ball. But $\|f_m - f_n\| = 1$ so they cannot be in the same ball.

**11.2.32.** Closed: If $\{x^{(n)}\}$ is a sequence in Hilbert cube $H$ and $\lim x^{(n)} = x$, then $\lim_n x_k^{(n)} = x_k$ for all $k$. Therefore, $|x(k)| \le 1/k$, whence $x \in H$.

Totally bounded: Let $\varepsilon > 0$. There is $n$ s.t. $\sum_{k \ge n+1} 1/k^2 < \varepsilon/2$. Let $A$ denote the subset of $H$ consisting of sequences supported in the first $n$ coordinates. Then $A$ is isomorphic to a subset of $\mathbb{R}^n$. It is obvious that it is bounded and using the same argument as above we see that it is closed, whence it is compact. The set $\{B(x, \varepsilon/2) : x \in A\}$ covers $A$, so there are $x^{(1)}, \dots, x^{(m)} \in A$ s.t. $A \subset \cup_{i=1}^m B(x^{(i)}, \varepsilon/2)$. Finally, let $x \in H$. If we denote $x_A := (x_1, \dots, x_n, 0, 0, \dots)$ then $\|x - x_A\| < \varepsilon/2$. Further, there is $k$ for which $\|x_A - x^{(k)}\| < \varepsilon/2$, hence $\|x - x^{(k)}\| < \varepsilon$.

**11.3.4.** Hint: consider finite linear combinations of $\{e_n\}$ and $T_n(x) = nx_n$.

**11.3.6.** Let $\{x_n\}_{n=1}^\infty$ be a sequence in $X$ which converges to $x \in X$. By assumption, $\{T_n x\}_{n=1}^\infty$ converges to $Tx$, so it is bounded and by UBP there exists $M > 0$ such that $\|T_n\| \le M$ for all $n \in \mathbb{N}$. Given $\varepsilon > 0$, there exists $N \in \mathbb{N}$ such that $\|x_n - x\| < \varepsilon/(2M)$ and $\|T_n x - Tx\| < \varepsilon/2$ for all $n \in \mathbb{N}$ with $n \ge N$. It follows that $\|T_n x_n - Tx\| \le \|T_n x_n - T_n x\| + \|T_n x - Tx\| < \|T_n(x_n - x)\| + \varepsilon/2 \le \|T_n\|\|x_n - x\| + \varepsilon/2 < M\varepsilon/(2M) + \varepsilon/2 = \varepsilon$ for all $n \in \mathbb{N}$ with $n \ge N$, which implies that $\{T_n x_n\}_{n=1}^\infty$ converges to $Tx$.

**11.3.10.** Hint: consider the operator $T(x_1, x_2, \dots) = (x_2, x_3, \dots)$.

**11.3.13.** ($\Leftarrow$) Hölder's Inequality. ($\Rightarrow$) Consider the sequence $\{f_n\}$, defined by (7.8), and define $F_n(g) = \int_a^b f_n g$. Since $f_n \in L^q$, we get that $F_n$ is a linear operator from the Banach space $L^p$ to $\mathbb{R}$ and $\|F_n(g)\| \le \|g\|_p \|f_n\|_q$, whence $\|F_n\| \le \|f_n\|_q$. In fact, choosing $g = |f_n|^{q/p} \operatorname{sgn}(f_n)$, we get $\|g\|_p = [\int |f_n|^q]^{1/p} = \|f_n\|_q^{q/p}$ so $|\int f_n g| = \int |f_n|^{1+q/p} = \int |f_n|^q = \|f_n\|_q^q = \|g_n\|_p \|f_n\|_q^{q-q/p} = \|g_n\|_p \|f_n\|_q$, whence $\|F_n\| = \|f_n\|_q$. The sequence $\{F_n\}$ is pointwise convergent hence pointwise bounded. Indeed, for any $g \in L^p$, $\lim_n F_n(g) = \lim_n \int f_n g = \int fg$, by Dominated Convergence Theorem. By UBP, $\{F_n\}$ is uniformly bounded, i.e. there is $M$ s.t. $\|F_n\| \le M$, for all $n$ so $\|f_n\|_q \le M$, for all $n$ so $\lim_n \|f_n\|_q \le M$ hence [ by Monotone Convergence Theorem], $\|f\|_q \le M$.

**11.3.19.** Hint: let $1/p + 1/r = 1/q$ and $1/r + 1/r' = 1$, and show that $h \in L^{r'}$.

**11.3.20.** (a) Let $a_j = \pi j/(2n+1)$. Then $\|D_n\|_1 = \int_0^{2\pi} |D_n| = 2\int_0^\pi |D_n| = 2\sum_{j=0}^{2n} \int_{a_j}^{a_{j+1}} |D_n|$. For $j = 0$ we have $\int_0^{a_1} |D_n| \le \int_0^{a_1} (2n+1) = \pi$. For $j \ge 1$ we use the inequality $|\sin(x/2)| \ge x/\pi$ which implies $|D_n| \le (\pi/x)|\sin(n+1/2)x|$ so $\int_{a_j}^{a_{j+1}} |D_n| \le \int_{a_j}^{a_{j+1}} (\pi/x)|\sin(n+1/2)x| \le (\pi/a_j)|\int_{a_j}^{a_{j+1}} \sin(n+1/2)x| = (\pi/a_j)|\cos(n+1/2)x/(n+1/2)|_{a_j}^{a_{j+1}}| \le (\pi/a_j)[2/(n+1/2)] = 4/j$ so $\|D_n\|_1 \le 2\pi + 2\sum_{j=1}^{2n} 4/j$ and it is well known that the latter is of the same rate of growth as $\ln(n+1)$. Thus, $\|D_n\|_1 \le M\ln(n+1)$. The other inequality: $\|D_n\|_1 = 4\int_0^{\pi/2} |D_n|$. Using the inequality $\sin x \le x$, we get $|D_n| \ge |\sin(n+1/2)x|2/x$, so $\|D_n\|_1 \ge 4\int_0^{\pi/2} |\sin(n+1/2)x|2/x$. Using substitution $u = (2n+1)x/2$ the integral becomes $\int_0^{(2n+1)\pi/4} |\sin u|/u \ge \int_0^{n\pi/2} |\sin u|/u = \sum_{j=1}^n \int_{(j-1)\pi/2}^{j\pi/2} |\sin u|/u \ge \sum_{j=1}^n \int_{(j-1)\pi/2}^{j\pi/2} |\sin u|/(j\pi/2)$
$= (2/\pi)\sum_{j=1}^n (1/j) \int_{(j-1)\pi/2}^{j\pi/2} |\sin u| = (2/\pi)\sum_{j=1}^n (1/j)$. It follows that $\|D_n\|_1$
$\ge 8/\pi \sum_{j=1}^n (1/j)$ and consequently $\|D_n\|_1 \ge m\ln(n+1)$.

(b) Recall that $s_n(x) = (2/\pi)\int_0^\pi [f(x+t) + f(x-t)]/2D_n(t)dt$. It follows that $\|s_n\|_{max} \le (2/\pi)\|f\|_{max}\|D_n\|_1 \le M\ln(n+1)\|f\|_{max}$. If $F_n(f) := s_n(f)/\ln(n+1)$ then $\{F_n\}$ is a bounded sequence of linear operators from $C[0, 2\pi]$ to $\mathbb{R}$. Moreover, if $p$ is a trigonometric polynomial of degree $N$, then for $n \ge N$, $s_n(p) = p$, so $s_n(p)/\ln(n+1)$ converges uniformly to 0, i.e. in the norm of $C[0, 2\pi]$. By Fejer's theorem, trigonometric polynomials are dense in $C[0, 2\pi]$, so the result follows from Theorem 11.3.3.

(c) Define $T_n(f) = [s_n(f)](0)$, linear operators from $C[0, 2\pi]$ to $\mathbb{R}$. We will show that $\|T_n\| \ge (1/\pi)\|D_n\|_1$. Well known: $[s_n(f)](0) = (1/\pi)\int_0^{2\pi} f(t)D_n(t)dt$. The function $f = \operatorname{sgn} D_n$ would give the result but it is not continuous. But it is in $L^1$, so there is a sequence of continuous functions $\{f_k\}$ s.t. for $k$ large, $\|\operatorname{sgn} D_n - f_k\|_1 < \varepsilon/n$ ($n$ is fixed). In fact, we can construct $f_k$ s.t. $\|f_k\|_{max} = 1$. Then $\|T_n\| \ge |T_n(f_k)| = |[s_n(f_k)](0)| = |(1/\pi)\int_0^{2\pi} f_k(t)D_n(t)dt| \ge |(1/\pi)\int_0^{2\pi} \operatorname{sgn} D_n(t)D_n(t)dt| - |(1/\pi)\int_0^{2\pi} [f_k(t) - \operatorname{sgn} D_n(t)]D_n(t)dt|$
$= (1/\pi)\|D_n\|_1 - |(1/\pi)\int_0^{2\pi} [f_k(t) - \operatorname{sgn} D_n(t)]D_n(t)dt| \ge (1/\pi)\|D_n\|_1 - (1/\pi)\|f_k$
$- \operatorname{sgn} D_n\|_1 \|D_n\|_{max} = \|D_n\|_1/\pi - (1/\pi)\varepsilon$ so $\|T_n\| \ge (1/\pi)\|D_n\|_1 \ge C\ln(n+1)$. Now we have a sequence of operators $\{T_n/\lambda_n\}$ s.t. $\lim_n \|T_n/\lambda_n\| \ge \lim_n C\ln(n+1)/|\lambda_n| = \infty$, i.e., it is not uniformly bounded. By UBP, it cannot be pointwise bounded, so there is a function $f \in C[0, 2\pi]$ s.t. $\{T_n(f)/\lambda_n\}$ is unbounded.

**11.4.4.** Hint: ($\Leftarrow$) Mimic the proof of Open Mapping Theorem. ($\Rightarrow$) $T(B_X)$ is an open set

**11.4.6.** No! If there were such a sequence then we would have an operator $T : \ell^\infty \to \ell^1$ defined by $T(\{x_n\}) = \{a_n x_n\}$. It is linear and bounded, because $\|T(\{x_n\})\|_1 = \|\{a_n x_n\}\|_1 = \sum a_n|x_n| \leq \|\{x_n\}\|_\infty \|\{a_n\}\|_1$ so $\|T\| \leq \|a\|_1$. Moroever, $T$ is surjective. Indeed, if $\{y_n\} \in \ell^1$ then $\sum |y_n| < \infty$ so $\sum a_n|y_n/a_n| < \infty$ hence [by assumption] $\{y_n/a_n\} \in \ell^\infty$ and $T(\{y_n/a_n\}) = \{y_n\}$. By Open Mapping Theorem, $T$ would be open. The set $S$ of finite sequences is dense in $\ell^1$ so, by Exercise 11.4.1, $T^{-1}(S)$ is dense in $\ell^\infty$. It is easy to see that $T^{-1}(S) \subset S$, and $S$ is not dense in $\ell^\infty$, a contradiction.

**11.4.8.** Hint: take $A$ an operator that is not open and $T = \left(\begin{smallmatrix} A & I \\ 0 & 0 \end{smallmatrix}\right)$.

**11.4.12.** It is easy to see that $\|x\|_2 \leq \|x\|_p$ for all $x \in X$. We show that $X$ is closed in $L^p$ and that $\|x\|_p \leq C\|x\|_2$ for all $x \in X$ and some $C > 0$. Closed: Let $\{x_n\} \subset X$, $\|x_n - x\|_p \to 0$. Then $\|x_n - x\|_1 \to 0$, but $X$ is closed in 1-norm so $x \in X$. Inequality: Since $\|x\|_1 \leq \|x\|_p$, by Corollary 11.4.6, there is $C > 0$ s.t. $\|x\|_p \leq C\|x\|_1$, for all $x \in X$. Thus, for all $x \in X$, $\|x\|_2 \leq \|x\|_p \leq C\|x\|_1 \leq C\|x\|_2$.

**11.4.14.** ($\Rightarrow$) Trivial; ($\Leftarrow$) Claim: there is $M > 0$ s.t. $\|x\|_2 \leq M\|x\|_1$, for all $x$. If this is true then the result follows from Corollary 11.4.6. Proof of the claim: Suppose not. Then for each $n$, there is $y_n$ s.t. $\|y_n\|_2 > n\|y_n\|_1$. Let $x_n := y_n/\|y_n\|_2$. Then $\|x_n\|_2 = 1 > n\|x_n\|_1$, so $\|x_n\|_1 < 1/n \to 0$ and $\|x_n\|_2 = 1$ does not go to 0, a contradiction.

**11.4.17.** (a) $\|T(\{a_n\})\| = \|\sum a_n x_n\| \leq \sum |a_n|\|x_n\| \leq \sum |a_n| = \|\{a_n\}\|_1$.

(b) Let $x \in X$. Then $x/\|x\|$ is in the unit ball, so there is $x_1$ s.t. $\|x/\|x\| - x_1\| < 1/2$ hence $\|x - \|x\|x_1\| < \|x\|/2$. We denote $a_1 = \|x\|$. Next, $[x - \|x\|x_1]/\|x - \|x\|x_1\|$ is in the unit ball, so there is $x_2$ s.t. $\|[x - \|x\|x_1]/\|x - \|x\|x_1\| - x_2\| < 1/2$ hence $\|x - \|x\|x_1 - x_2\|x - \|x\|x_1\|\| < (1/2)\|x - \|x\|x_1\| < \|x\|/2^2$. We denote $a_2 = \|x - \|x\|x_1\| < \|x\|/2$. Continuing, we get a sequence $\{a_n\}$ s.t. $\|x - a_1x_2 - \cdots - a_nx_n\| < \|x\|/2^n$. It follows that $x = \sum a_n x_n = T(\{a_n\})$. Finally, $\{a_n\} \in \ell^1$ because $\sum |a_n| \leq \sum \|x\|/2^n < \infty$.

(c) By Exercise 11.4.16, $X$ is isomorphic to $\ell^1/\ker T$.

**11.4.21.** (a) Let $\{f_n\}_{n=1}^\infty$ be a sequence in $X$ which converges to $f \in X$ such that $\{Tf_n\}_{n=1}^\infty$ converges to $g \in \ell^1$. Given $\varepsilon > 0$, there exists $N \in \mathbb{N}$ such that $\sum_{n=N}^\infty n|f(n)| < \varepsilon/4$ and $\sum_{n=N}^\infty |g(n)| < \varepsilon/4$. Moreover, there exists $M \in \mathbb{N}$ such that $\|g - Tf_m\| < \varepsilon/4$ and $\|f - f_m\| < \varepsilon/(4N)$ for all $m \in \mathbb{N}$ with $m \geq M$. It follows that $\sum_{n=1}^\infty |Tf(n) - Tf_m(n)| = \sum_{n=1}^{N-1} |nf(n) - nf_m(n)| + \sum_{n=N}^\infty |nf(n) - Tf_m(n)| \leq \sum_{n=1}^{N-1} |nf(n) - nf_m(n)| + \sum_{n=N}^\infty |nf(n)| + \sum_{n=N}^\infty |Tf_m(n)| < \sum_{n=1}^{N-1} N|f(n) - f_m(n)| + \varepsilon/4 + \sum_{n=N}^\infty |Tf_m(n) - g(n)| + \sum_{n=N}^\infty |g(n)| < N\|f - f_m\| + \varepsilon/4 + \|Tf_m - g\| + \varepsilon/4 < \varepsilon$ for all $m \in \mathbb{N}$ with $m \geq M$. Therefore $\{Tf_n\}_{n=1}^\infty$ converges to $Tf$, so $T$ is closed.

For each $m \in \mathbb{N}$ define $f_m \in X$ by $f_m(n) := 1$ if $m = n$ and 0 if $m \neq n$. Then $\|f_m\| = 1$ but $\|Tf_m\| = \sum_{n=1}^\infty |Tf_m(n)| = \sum_{n=1}^\infty |nf_m(n)| = \sum_{n=1}^\infty nf_m(n) = m$ for all $m \in \mathbb{N}$. This shows that $T$ is not bounded.

(b) Clearly $T^{-1}f(n) = f(n)/n$ for all $f \in \ell^1$ and $n \in \mathbb{N}$. It follows that $\|T^{-1}f\| = \sum_{n=1}^\infty |T^{-1}f(n)| = \sum_{n=1}^\infty |f(n)/n| = \sum_{n=1}^\infty |f(n)|/n \leq \sum_{n=1}^\infty |f(n)| = \|f\|$ for all $f \in \ell^1$, so $T^{-1}$ is bounded. It is obvious that $T^{-1}$ is surjective. If $T^{-1}$ were open then $T$ would be continuous, contradicting part (a).

**11.4.23.** Hint: $T(\{x_n\}) = \{x_n/n\}$ on $\ell^2$.

**11.4.26.** Suppose that $T$ is not closed. By Proposition 11.4.11, its graph is not closed so there is $x_n \to x$ and $Tx_n \to y \neq Tx$. Replace $x_n - x$ by $x_n$, so $x_n \to 0$ but $Tx_n \to y \neq 0$. If some $Tx_n = y$, replace $x_n$ by $(1 + 1/n)x_n$. The set $A = \{x_n\} \cup \{0\}$ is closed set so the same holds for $T(A) = \{Tx_n\} \cup \{0\}$ which does not contain $y$. Contradiction.

**11.4.28.** Hint: $(Ax)_n = -(Bx)_n = n^2 x_n$ for $n > 1$, $(Ax)_1 = \sum_{n=1}^\infty nx_n$ and $(Bx)_1 = 0$, with $D(A) = D(B) = \{x : \sum_{n=1}^\infty n^4|x_n|^2 < \infty\}$.

**11.4.32.** ($\Rightarrow$) $P$ is continuous so $P^{-1}(0)$ is closed. If $y \in \overline{P(X)}$ then there is $\{x_n\}$ s.t. $Px_n \to y$. Then $Px_n = P^2x_n \to Py$, so $y = Py \in P(X)$. ($\Leftarrow$) Suppose $x_n \to x$ and $Px_n \to y$. $Px_n \in P(X)$ which is closed so $y \in P(x)$. Thus, $y = Pz$ implies $Py = P^2z = Pz = y$. Also, $Px_n - x_n \in \ker P$ which is closed so $y - x \in \ker P$ so $Px = Py = y$. Now Closed Graph Theorem.

**11.4.34.** Hint: use Closed Graph Theorem to show that the identity map on $Y$ is continuous.

**11.4.38.** Need to show that the embedding operator $I : L^p(X,\mu) \to L^q(X,\mu)$ is bounded. Suppose that $\{f_n\} \subset L^p$ s.t. $f_n \to f$ and $If_n \to g$. Then $\|f_n - f\|_p \to 0$ and $\|f_n - g\|_q \to 0$. By Riesz–Fischer, there is a subsequence $\{f_{n_k}\}$ that converges to $f$ a.e. and $\|f_{n_k} - g\|_q \to 0$. Then there is a further subsequence $\{h_n\}$ that converges to $f$ a.e. and to $g$ a.e. so $f = g$ a.e. By Closed Graph theorem, $I$ is bounded.

**11.4.41.** Hint: First show that the matrix $R_k$ with the same $k$th row as $A$ and the rest zeros is bounded, then use Closed Graph theorem.

**11.4.44.** Let $\{x_n\} \subset \mathcal{H}$ and $\lim x_n = x$, $\lim Tx_n = y$. Denote $z_n = x_n - x$ and $z = y - x$. The assumptions now say $\lim z_n = 0$ and $\lim Tz_n = z$, so $\langle z, z\rangle = \lim\langle Tz_n, z\rangle = \lim\langle z_n, Tz\rangle = 0$ whence $z = 0$. By Closed Graph Theorem, $T$ is bounded.

Let $\varphi(x) = x^{-1/3}$, let $D = \{f : \varphi f \in L^2[0,1]\}$. Then $D$ is dense in $L^2[0,1]$ because it contains all $\sin 2n\pi x$ and $\cos 2n\pi x$. Let $T : D \to L^2[0,1]$ be defined by $Tf = \varphi f$. Then $T$ is

symmetric, but not bounded because $f_n(x) = x^{-1/6+1/n} \in D$ and $\|f_n\| = [3n/(2n+6)]^{1/2} \leq 2$, while $\|Tf_n\| = \sqrt{n/2} \to \infty$.
**12.1.5.** (a) $|\delta(f)| = |f(0)| \leq \|f\|_{max}$ so $\|\delta\| \leq 1$. Equality for $f = 1$ so $\|T\| = 1$. (b) $f_n = 2n - 2xn^2$ for $0 \leq x \leq 1/n$, 0 otherwise. Then $\|f_n\|_1 = 1$ but $\delta(f_n) = 2n$.
**12.1.8.** $|Tf| = |\int_0^1 fg| \leq \|f\|_{max}\|g\|_1$ so $\|T\| \leq \|g\|_1$. Equality is attained with $f = \operatorname{sgn} g$, but $\operatorname{sgn} g$ is not continuous. Approximate $\operatorname{sgn} g$ by continuous functions $\{h_n\}$ that converge to $\operatorname{sgn} g$ a.e. and that satisfy $|h_n| \leq 1$. Then $\int_0^1 (h_n - \operatorname{sgn} g)g \to 0$ (by Dominated Convergence Theorem: $|h_n g| \leq |g|$) so $|Th_n| = \int_0^1 h_n g| = \int_0^1 (h_n - \operatorname{sgn} g)g + \int_0^1 |g|) \to \|g\|_1$.
**12.1.13.** Hint: Use Luzin's theorem.
**12.1.16.** (a) $|Tx| \leq \sum_1^\infty |x_n|/2^n \leq \|x\| \sum_1^\infty 1/2^n = \|x\|$, so $\|T\| \leq 1$. We show that $\|T\| = 1$. Let $x^{(k)} = (1, 1, \ldots, 1, 0, 0, \ldots)$ with $k$ 1s. Then $\|x^{(k)}\| = 1$ and $|Tx^{(k)}| = \sum_1^k 1/2^k = 1 - 1/2^k \to 1$. (b) Suppose there is such $x$. Then there is $N$ s.t. $n \geq N \Rightarrow |x_n| < 1/2$. Now $|Tx| \leq \sum_1^{N-1} |x_n|/2^n + \sum_N^\infty |x_n|/2^n \leq \sum_1^{N-1} 1/2^n + \sum_N^\infty (1/2)/2^n = \sum_1^\infty 1/2^n - 1/2^N = 1 - 1/2^N < 1$, a contradiction.
**12.1.29.** ($\Rightarrow$) $\ker T = T^{-1}(0)$ is closed, so its complement is open. ($\Leftarrow$) $0 \notin T[B(a,r)] = Ta + T[B(0,r)]$ so $\pm Ta \notin T[B(0,r)]$. Thus, if $\|x\| < r$ then $Tx \neq \pm Ta$. We will show that if $\|x\| < r$ the $|Tx| \leq |Ta|$. Indeed, if $|Tx| > |Ta|$ then $|Ta|x/|Tx|$ has norm less than $r$ but $|T[|Ta|x/|Tx|]| = Ta$, which is impossible. Thus, if $\|x\| < r$ then $|Tx| \leq |Ta|$, so $|T(x/r)| \leq |T(a/r)|$, whence $\|T\| \leq |Ta|/r$.
**12.1.30.** Hint: show that $(\ker T)^C$ contains an open ball.
**12.1.33.** Hint: write $x = ax_0 + z$ with $a \in \mathbb{R}$, $Tx_0 = 1$, and $z \in \ker T$.
**12.2.1.** The norm of $T|\mathcal{M}$ is $\sup\{|T(x,y)| : (x,y) \in \mathcal{M}$ and $\|(x,y)\| = 1\} = \sup\{|x| : \|(x,2x)\| = 1\} = 1/\sqrt{5}$. Write every $(u,v) \in \mathbb{R}^2$ as $(x,2x) + \lambda(1,0) = (x+\lambda, 2x)$. Then $T'(u,v) = T'(x,2x) + \lambda T'(1,0) = x + \lambda p$. We need to determine $p = T'(1,0)$ s.t. $\|T'\| \leq 1/\sqrt{5}$. Equivalently, $|x + \lambda p| \leq \|(x+\lambda, 2x)\|/\sqrt{5} = \sqrt{[(x+\lambda)^2 + 4x^2]/5}$ so $x^2 + 2\lambda xp + \lambda^2 p^2 \leq x^2 + 2x\lambda/5 + \lambda^2/5$ hence $2\lambda x(p - 1/5) \leq \lambda^2(1/5 - p^2)$. This inequality must fail when $x \to \pm\infty$ unless $p = 1/5$, i.e., $T(1,0) = 1/5$. It remains to find $\lambda$ as a function of $u$ and $v$. Since $(u,v) = (x+\lambda, 2x)$, we get $\lambda = u - x = u - v/2$. Thus, $T'(u,v) = x + \lambda p = v/2 + (u - v/2)/5 = u/5 + 2v/5$.
**12.2.7.** No! Take $f \in C[0,1]$ such that $f(1) \neq 0$, and extend $f$ to $[-1,1]$ in two ways: $f_1$ is even, $f_2$ is odd. Let $\{p_n^{(k)}\}$ converge uniformly to $f_k$. On $[0,1]$ both sequences converge to $f$. If $T$ could be extended to $C[0,1]$, we would have $Tp_n^{(k)} \to Tf$. But $Tp_n^{(k)} = p_n^{(k)}(-1)$ converges to $f_k(-1)$, and $f_1(-1) = f(1)$, $f_2(-1) = -f(1)$. So $\lim Tp_n^{(1)} \neq \lim Tp_n^{(2)}$.
**12.2.9.** Let $\{x_1, \ldots, x_m\}$ be a basis for $Y$ and $T : \mathbb{R}^m \to Y$ the corresponding isomorphism. Clearly $\|T(\cdot)\|$ is a norm on $\mathbb{R}^m$, and it is equivalent to $\|\cdot\|_1$ by Theorem 11.2.11. In other words $T$ and $T^{-1}$ are bounded. The projections $P_1, \ldots, P_m : \mathbb{R}^m \to \mathbb{R}$ induce linear functionals $f_1, \ldots, f_m : Y \to \mathbb{R}$ such that $|f_i(x)| = |P_i(T^{-1}(x))| \leq \|T^{-1}(x)\|_1 \leq \|T^{-1}\|\|x\|$ for all $i \in \{1, \ldots, m\}$ and $x \in Y$. By Hahn–Banach, there exist linear functionals $F_1, \ldots, F_m : X \to \mathbb{R}$ extending $f_1, \ldots, f_m$ such that each is bounded with norm at most $\|T^{-1}\|$. Let $Z$ be the closed subspace $\cap_{i=1}^m \ker(F_i)$. Note that $Y \cap Z = \cap_{i=1}^m \ker(f_i) = T(\cap_{i=1}^m \ker(P_i)) = T(\{0\}) = \{0\}$. Moreover, if $x \in X$ then $\sum_{i=1}^m F_i(x)x_i \in Y$ and $x - \sum_{i=1}^m F_i(x)x_i \in Z$, because $F_j(x - \sum_{i=1}^m F_i(x)x_i) = 0$ for all $j \in \{1, \ldots, m\}$. This shows that $X = Y + Z$.
**12.2.13.** Hint: Consider the subspace $\{Sx - x : x \in \ell^\infty\}$, where $S(x_1, x_2, x_3, \ldots) = (x_2, x_3, x_4, \ldots)$, and a linear functional $\Phi_0$ on $W \oplus \mathbb{R} \cdot 1$, $\Phi_0(y + c \cdot 1) = c$.
**12.2.14.** Let $\{y_n\}$ be a Cauchy sequence and let $x_0 \in X$ be a unit vector. There exists a linear functional $\varphi$ on $X$ such that $\|\varphi\| = 1 = \varphi(x_0)$. Let $T_n x = \varphi(x)y_n$ and note that $T_n x_0 = y_n$. Then $\|(T_n - T_m)x\| = |\varphi(x)|\|y_n - y_m\| \leq \|y_n - y_m\|\|x\|$, for all $x \in X$ and so $\|T_n - T_m\| \leq \|y_n - y_m\|$, which shows that $\{T_n\}_{n=1}^\infty$ is a Cauchy sequence in $\mathcal{L}(X,Y)$. Let $T$ be the limit of this sequence and conclude by noting that $\lim_{n\to\infty} y_n = \lim_{n\to\infty} T_n x_0 = Tx_0$.
**12.2.18.** There exists $T \in X^*$ such that $T(z) \neq 0$ and $T(Y) = \{0\}$. Let $\{y_n + \lambda_n z\}_{n=1}^\infty$ be a sequence in $V$ that converges to $y \in X$. Then $T(y) = \lim_{n\to\infty} T(y_n + \lambda_n z) = \lim_{n\to\infty} \lambda_n T(z)$, since $T$ is continuous, so $\{\lambda_n\}_{n=1}^\infty$ converges to $a := T(y)/T(z)$, which implies that $\{\lambda_n z\}_{n=1}^\infty$ converges to $az$. Therefore $\{y_n\}_{n=1}^\infty = \{(y_n + \lambda_n z) - \lambda_n z\}_{n=1}^\infty$ converges to $y - az$, which lies in $Y$ because $Y$ is closed. It follows that $y \in V$, which shows that $V$ is closed.
**12.2.22.** ($\Rightarrow$) Assume $\|x\| \leq M$, for all $x \in S$. Then, for any $T \in X^*$, $\sup_{x\in S} |Tx| \leq \sup_{x\in S} \|T\|\|x\| \leq \|T\|M$.

($\Leftarrow$) Let $T \in X^*$. Then there is $M_T$ s.t. $|Tx| \leq M_T$ for all $x \in S$. Then $|Tx| = |J_x(T)| \leq M_T$ for all $x \in S$, i.e., $\{J_x\}$ is pointwise bounded on $X^*$ so by UBP $\{J_x\}$ is uniformly bounded hence there is $M$ s.t. $\|J_x\| \leq M$ for all $x \in S$ whence $\|x\| \leq M$ for all $x \in S$.
**12.2.24.** Hint: show that $X$, $X^*$, and $X^{**}$ are isomorphic to the same $\mathbb{R}^n$.
**12.2.26.** Denote by $P$ the restriction of $X^*$ to $Y^*$, $P(x^*)y = x^*y$, for $y \in Y$. $P$ is bounded because $\|Px^*\| \leq \|x^*\|$. Denote by $Q$ the extension from $Y^{**}$ to $X^{**}$, $Q(y^{**})x^* = y^{**}(Px^*)$, for $x^* \in X^*$. $Q$ is bounded, because $|Q(y^{**})x^*| = |y^{**}(Px^*)| \leq \|y^{**}\|\|Px^*\| \leq \|y^{**}\|\|x^*\|$ so $\|Qy^{**}\| \leq \|y^{**}\|$. Let us show that $J^{-1}Q(Y^{**}) \subset Y$. Suppose not. There is $y^{**} \in Y^{**}$ s.t. $x := J^{-1}Q(y^{**}) \notin Y$. Since $Y$ is closed, by Theorem 12.2.5, there is $x^*$ s.t. $x^*(x) \neq 0$ but $x^*$ vanishes

on $Y$. Thus, $Px^* = 0$. But then $x^*(x) = Jx(x^*) = Qy^{**}(x^*) = y^{**}(Px^*) = 0$, a contradiction. Therefore, $J^{-1}Q(Y^{**}) \subset Y$. Now we can show that $Y$ is reflexive. Let $\tilde{J}$ denote the canonical embedding from $Y$ to $Y^{**}$. We will show that it is onto. Let $y_0^{**} \in Y^{**}$. Define $x_0 = J^{-1}Qy_0^{**}$, which is in $Y$ as we proved. We will show that $\tilde{J}x_0 = y_0^{**}$. Let $y^* \in Y^*$. Denote by $x^*$ an extension s.t. $Px^* = y^*$. Then $(\tilde{J}x_0)y^* = y^*(x_0) = x^*(x_0) = (Jx_0)x^* = (Qy_0^{**})x^* = y_0^{**}(Px^*) = y_0^{**}(y^*)$.

**12.2.27.** Hint: define $H : Y \to X$ by $H = J^{-1} \circ G^{-1} \circ J'$, where $J : X \to X^{**}$ and $J' : Y \to Y^{**}$ are canonical embeddings, and $G : X^{**} \to Y^{**}$ by $G(x^{**}) = x^{**} \circ F$, where $F : Y^* \to X^*$ is an isometric isomorphism.

**12.3.6.** Let $T \in (L^p)^*$ and suppose that $T$ vanishes on $D$. We will show that $T = 0$. There is $g \in L^q$ s.t. $Tf = \int_{\mathbb{R}} fg$. Let $I$ and $J$ be any 2 finite intervals and $f = \chi_I/m(I) - \chi_J/m(J)$. Then $f \in D$ so $Tf = 0$ hence $\int_I g/m(I) - \int_J g/m(J) = 0$. Since this is true for any intervals $I, J$, the quantity $\int_I g/m(I)$ is independent of $I$, hence constant. In particular, if $I = [0, x]$, we get $(1/x)\int_0^x g = C$ so $\int_0^x g = Cx$ hence (taking derivative) $g = C$ a.e. But $g \in L^q(\mathbb{R})$ so, if $p > 1$, $C = 0$ hence $g = 0$ a.e. whence $T = 0$. If $p = 1$, $C$ must be 0, otherwise $0 = T(\chi_{[0,1]}) = C\int_0^1 1 = C$, a contradiction. For $p = \infty$, take $g = 1 \in L^\infty$. If $f$ is compactly supported then there is $M > 0$ s.t. $f = 0$ outside $[-M, M]$ so $\|f - g\|_\infty \geq \sup\{|f - g| : x > M\} = 1$.

**12.3.7.** Hint: let $g(x) = x^{-1/q}[\ln(x/2)]^{-2/q}\chi_{[0,1]}$ and $X$ the set of all $f : \mathbb{R} \to \mathbb{R}$ that are measurable, bounded, compactly supported, and $\int_{\mathbb{R}} fg = 0$.

**12.3.13.** Linearity is obvious. Boundedness: $|Tg| \leq \|g\|_p(\int_0^{1/2} x^{2q})^{1/q} = \|g\|_p 1/[2^{(2q+1)/q}(2q+1)^{1/q}]$. Take $f(x) = x^2\chi_{(0,1/2)}$.

**12.3.19.** Let $T \in (L^p)^*$ and suppose that $T$ vanishes on $E$. We will show that $T = 0$. There is $g \in L^q$ s.t. $Tf = \int_{\mathbb{R}} gf$, for all $f \in L^p$, and $\|T\| = \|g\|_q$. Let $h > 0$, $x, y \in \mathbb{R}$. Define $f := \chi(y, y + h) - \chi(x, x + h)$. Then $f \in E$ so $Tf = 0$ hence $\int_x^{x+h} g = \int_y^{y+h} g$. It follows that $\frac{1}{h}\int_y^{y+h}[g(t) - g(y)]\,dt = \frac{1}{h}\int_x^{x+h}[g(t) - g(x)]\,dt + g(x) - g(y)$. Denoting by $G$ an antiderivative of $g$, we get $\int_x^{x+h}[g(t) - g(x)]\,dt = G(x + h) - G(x) - hg(x)$, so $\lim_{h\to 0}\frac{1}{h}\int_x^{x+h}[g(t) - g(x)] = G'(x) - g(x) = 0$ a.e. When $h \to 0+$, we are left with $0 = g(x) - g(y)$ so $g$ =const. Since $g \in L^q(\mathbb{R})$, must have $g = 0$ hence $T = 0$, and the result follows from Corollary 12.2.8.

**12.3.24.** Hint: prove that $Y^*$ is isometrically isomorphic to $L^2(X, \mu) \times L^2(X, \mu)$ with norm $\|(u, v)\| = (\|u\|_2^{3/2} + \|v\|_2^{3/2})^{2/3}$.

**12.3.25.** To each $y \in \ell^1$ we associate $\phi_y : c \to \mathbb{R}$, defined by $\phi_y(x) = y_1 \lim x_n + \sum x_n y_{n+1}$. The map $y \mapsto \phi_y$ is linear and bounded because $|\phi_y(x)| \leq \|x\|\|y\|_1$. We have equality here: given $\varepsilon > 0$ choose $N$ s.t. $\sum_{N+1}^\infty |y_k| < \varepsilon/2$. Let $x = (\operatorname{sgn} y_2, \ldots, \operatorname{sgn} y_N, \operatorname{sgn} y_1, \operatorname{sgn} y_1, \ldots)$. Then $\|x\| = 1$ and $\phi_y(x) = |y_1| + \cdots + |y_N| + \sum_{N+1}^\infty(\operatorname{sgn} y_1)y_k = \|y\|_1 - \sum_{N+1}^\infty(\operatorname{sgn} y_k - \operatorname{sgn} y_1)y_k$, and $|\sum_{N+1}^\infty(\operatorname{sgn} y_k - \operatorname{sgn} y_1)y_k| \leq 2\sum_{N+1}^\infty |y_k| < \varepsilon$. It remains to show that the mapping $y \mapsto \phi_y$ is surjective. Let $T \in c^*$. Let $e = (1, 1, \ldots)$. Define $y := (Te - \sum_1^\infty Te_k, Te_1, Te_2, \ldots)$. Then $y \in \ell^1$ because $\sum |Te_k| < \infty$. Indeed, for any positive integer $n$, $\sum_1^n |Te_k| = \sum_1^n(\operatorname{sgn} Te_k)Te_k = T((\operatorname{sgn} Te_1), (\operatorname{sgn} Te_2), \ldots, (\operatorname{sgn} Te_n), 0, 0, \ldots) \leq \|T\|\|(\operatorname{sgn} Te_1), \ldots, (\operatorname{sgn} Te_n), 0, 0, \ldots)\|$ $= \|T\|$. Finally, $T = \phi_y$. It suffices to verify on a set which has a dense span in $c$, such as $\{e_k\} \cup \{e\}$, which is routine.

**12.3.31.** Hint: define $\phi_k x := \langle x, e_k\rangle$ and use Exercise 11.4.37.

**12.3.36.** Hint: suppose that $T$ is a bounded linear functional that vanishes on the span of $\{f_n\}$ and show that it vanishes on $\mathcal{H}$.

**12.3.39.** No! (Unless $p = q = 2$.) Let $X = L^p + L^q$, with $\|(f, g)\| = (\|f\|_p^2 + \|g\|_q^2)^{1/2}$. We will prove that if $X \cong Y \oplus Z$ then $T : X^* \to Y^* \oplus Z^*$, defined by $Tx^* = (x^*|Y, x^*|Z)$, is an isometric isomorphism. Linear: straightforward. Bounded: $\|Tx^*\| = \|(x^*|Y, x^*|Z)\| = (\|x^*|Y\|^2 + \|x^*|Z\|^2)^{1/2} = \sup\{(\|x^*f\|^2 + \|x^*g\|^2)^{1/2} : \|f\|^2 + \|g\|^2 = 1\} \leq \|x^*\|$. In fact we have equality: choose $f, g$ s.t. $\|f\|_p = 1$, $\|g\|_q = 1$ and $\|x^*f\| \geq \|x^*\| - \varepsilon$, $\|x^*g\| \geq \|x^*\| - \varepsilon$. Then $\|Tx^*\| = \sup\{(|x^*f\|^2 + \|x^*|g\|^2)^{1/2} : \|f\|^2 + \|g\|^2 = 1\} = \sup\{(\|x^*f\|^2/(\|f\|^2 + \|g\|^2) + \|x^*g\|^2/(\|f\|^2 + \|g\|^2))^{1/2}\} \geq [(\|x^*\| - \varepsilon)^2 + (\|x^*\| - \varepsilon)^2]^{1/2}/\sqrt{2} = \|x^*\| - \varepsilon$. Letting $\varepsilon \to 0$ we get the opposite inequality, so $T$ is an isometry. Onto: Let $y^* \in Y$, $z^* \in Z$. Define $x^* : X \cong Y \oplus Z \to \mathbb{R}$ by $x^*(y + z) = y^*(y) + z^*(z)$, then $x^*$ is bounded. Indeed, $|x^*(y + z)| = |y^*(y) + z^*(z)| \leq \|y^*\|\|y\| + \|z^*\|\|z\| \leq [\|y^*\|^2 + \|z^*\|^2]^{1/2}[\|y\|^2 + \|z\|^2]^{1/2} = [\|y^*\|^2 + \|z^*\|^2]^{1/2}\|(y, z)\|$ as $y^*$ and $z^*$ are, so $x^* \in X^*$ and obviously $Tx^* = (y^*, z^*)$. Conclusion: $X^*$ is isometrically isomorphic to $L^q + L^p$, which is isometrically isomorphic to $L^p + L^q = X$, via $F(x, y) = (y, x)$. Finally, $X$ is not Hilbert space because its closed subspace $L^p$ is not.

**12.3.40.** Hint: use Exercise 12.3.33.

**12.3.47.** Hint: use Exercises 12.1.29, 2.5.2, and 10.6.15.

**12.3.48.** We will show that the embedding $i : \mathcal{H} \hookrightarrow \overline{\mathcal{H}}$ of $\mathcal{H}$ into its completion is onto. Given $z \in \overline{\mathcal{H}} \setminus \{0\}$, the subspace $\mathcal{M} = \{x \in \mathcal{H} : \langle i(x), z\rangle = 0\}$ is closed in $\mathcal{H}$ because it is the kernel of a bounded linear functional. Further, $\mathcal{M}^\perp \neq \{0\}$. Otherwise, $\mathcal{M} = \mathcal{M}^{\perp\perp} = \{0\}^\perp = \mathcal{H}$ which implies $z = 0$. For fixed $y \in \mathcal{M}^\perp \setminus \{0\}$ and the linear functionals $\varphi(x) = \langle x, y\rangle$ and

$\psi(x) = \langle i(x), z\rangle$ we have $\ker(\psi) = \mathcal{M} \subset \ker(\varphi)$. Indeed, $\langle x, z\rangle = 0$ implies $x \in \mathcal{M} = \mathcal{M}^{\perp\perp}$ so $\langle x, y\rangle = 0$. By Exercise 12.1.19, $\varphi = a\psi$ for some scalar $a$. Now, the functionals on $\overline{\mathcal{H}}$ given by $\langle \cdot, i(y)\rangle$ and $\langle \cdot, az\rangle$ [i.e., $\varphi$ and $a\psi$] coincide on the dense subspace $i(\mathcal{H})$ and hence on $\overline{\mathcal{H}}$ which implies $i(y) = az$. Since $a \neq 0$ this gives $z \in i(\mathcal{H})$.

**12.3.49.** Let $\overline{S}$ denote the closure of $S$ with respect to the $L^2$ norm. It obviously suffices to show that $\overline{S}$ is finite-dimensional. First we show that $\overline{S}$ is still contained in $C[0,1]$. Let $f \in \overline{S}$. There is a sequence $\{f_n\} \in S$ with $\|f_n - f\|_2 \to 0$ as $n \to \infty$. For any $n, m$, we have $\|f_n - f_m\|_{max} \leq K\|f_n - f_m\|_2$, and since $\{f_n\}$ converges in $L^2$, it is also Cauchy in $L^2$, so by the above inequality it is also a Cauchy sequence in $C[0,1]$. Since $C[0,1]$ is complete, there is some $g \in C[0,1]$ with $\|f_n - g\|_{max} \to 0$ as $n \to \infty$. Note that since $\|h\|_2 \leq \|h\|_{max}$ for any $h \in C[0,1]$, we have $\|g - f\|_2 \leq \|g - f_n\|_2 + \|f_n - f\|_2 \leq \|g - f_n\|_{max} + \|f_n - f\|_2 \to 0$ as $n \to \infty$. Thus, $f = g$ in $L^2$, hence $f$ is continuous, whence $\overline{S} \subset C[0,1]$. (More precisely, every $f \in \overline{S}$ is equal a.e. to a unique function $g \in C[0,1]$.)

For each $x \in [0,1]$, let $\varphi_x : \overline{S} \to \mathbb{R}$ be defined by $\varphi_x(f) = f(x)$. This is clearly a linear functional on $\overline{S}$ and it is bounded because $|\varphi_x(f)| = |f(x)| \leq \|f\|_{max} \leq K\|f\|_2$. Since $\overline{S}$ is a closed subspace of the Hilbert space $L^2$, it is also a Hilbert space and, by Riesz–Fréchet theorem, for each $x$ there exists some $g_x \in \overline{S}$ such that $f(x) = \varphi_x(f) = \langle f, g_x\rangle$ for all $f$ in $\overline{S}$. Note also that for each $x$, $\|g_x\|_2^2 = |\langle g_x, g_x\rangle| = |\varphi_x(g_x)| \leq \|g_x\|_{max} \leq K\|g_x\|_2$, so $\|g_x\|_2 \leq K$. Now let $\{f_1, \ldots, f_N\}$ be any linearly independent set in $\overline{S}$. By applying the Gram–Schmidt process if necessary we may assume that it is an orthonormal set. Then by Bessel's inequality, we have for each $x$ that $\sum_{j=1}^{N} |f_j(x)|^2 = \sum_{j=1}^{N} \langle f_j, g_x\rangle^2 \leq \|g_x\|_2^2 \leq K^2$. Then integrating both sides from 0 to 1 we get $K^2 \geq \sum_{j=1}^{N} \int_0^1 |f_j(x)|^2 = \sum_{j=1}^{N} \|f_j\|_2^2 = N$. This shows that a linearly independent set in $\overline{S}$ can have at most $K^2$ elements and thus $\dim(\overline{S}) \leq K^2 < \infty$.

**12.4.4.** We will show that for any measurable $f$, $\int f d|\nu| = \int f d\nu^+ + \int f d\nu^-$. From here the assertion follows: if $f$ is $|\nu|$-integrable then, by definition, so is $|f|$, and the proved formula then implies that both $\int |f| d\nu^+$ and $\int |f| d\nu^-$ are finite. Also, if $f$ is $\nu^+$-integrable and $\nu^-$-integrable, the same holds for $|f|$, hence $\int |f| d|\nu|$ is finite. Thus, we focus on the equality $\int f d|\nu| = \int f d\nu^+ + \int f d\nu^-$. When $f$ is a characteristic function, the assertion follows from the equality $|\nu|(A) = \nu^+(A) + \nu^-(A)$, proved in Exercise 12.4.3 (e). The transition to simple functions is immediate. The equality for nonnegative functions can be established as two opposite inequalities. In one direction, by definition, $\int f\, d|\nu| = \sup\{\int_X s\, d|\nu| : 0 \leq s \leq f,\, s \in \mathcal{S}\} = \sup\{\int_X s\, d\nu^+ + \int_X s\, d\nu^- : 0 \leq s \leq f,\, s \in \mathcal{S}\} \leq \sup\{\int_X s\, d\nu^+\} + \sup\{\int_X s\, d\nu^- :\} = \int f d\nu^+ + \int f d\nu^-$. In the opposite direction, given $\varepsilon > 0$ we choose simple functions $s_1, s_2$ such that $0 \leq s_1, s_2 \leq f$ and $\int_X s_1\, d\nu^+ > \int_X f\, d\nu^+ - \varepsilon/2$ and $\int_X s_2\, d\nu^- > \int_X f\, d\nu^- - \varepsilon/2$. Then, we define $s = \max\{s_1, s_2\}$ which is again a simple function satisfying $0 \leq s \leq f$ and the estimates $\int_X s\, d\nu^+ > \int_X f\, d\nu^+ - \varepsilon/2$ and $\int_X s\, d\nu^- > \int_X f\, d\nu^- - \varepsilon/2$. This implies that $\int_X f\, d|\nu| \geq \int_X s\, d|\nu| = \int_X s\, d\nu^+ + \int_X s\, d\nu^- > \int_X f\, d\nu^+ + \int_X f\, d\nu^- - \varepsilon$, so it remains to take the limit as $\varepsilon \to 0$. Finally, for general functions, it follows from the decomposition $f = f^+ - f^-$.

**12.4.7.** In view of Proposition 7.2.11 (a), we focus on $\int f + \int g = \int (f+g)$. It is straightforward to verify the equality when $f, g$ are simple functions. For $f, g \geq 0$ choose $s, t \in \mathcal{S}$ such that $0 \leq s \leq f$, $0 \leq t \leq g$, $\int f < \int s + \varepsilon/2$, and $\int g < \int t + \varepsilon/2$. Then $\int f + \int g < \int s + \int t + \varepsilon = \int (s+t) + \varepsilon \leq \int (f+g) + \varepsilon$, so $\int f + \int g \leq \int (f+g)$. For the reverse inequality, let $h \in \mathcal{S}$ satisfy $0 \leq h \leq f+g$. It suffices to show that $\int h\, d\nu \leq \int f\, d\nu + \int g\, d\nu$. Let $E = \{x \in X : f(x) \leq h(x),\ \text{and}\ g(x) \leq h(x)\}$. First, on $E^C$, $f(x) > h(x)$ or $g(x) > h(x)$ so $\int_{E^C} f \geq \int_{E_C} h$ or $\int_{E^C} g \geq \int_{E_C} h$, hence in either case $\int_{E^C} f + \int_{E^C} g \geq \int_{E_C} h$. In view of Exercise 12.4.6, it remains to show the inequality on $E$. Since $h$ is bounded by definition, then $f$ and $g$ are bounded on $E$. By Theorem 6.2.13, there are simple functions $s$ and $t$ satisfying $0 \leq s \leq f \leq s+\varepsilon/2$ and $0 \leq t \leq g \leq t+\varepsilon/2$. Now $\int_E f + \int_E g \geq \int_E s + \int_E t = \int_E (s+t) \geq \int_E (f+g-\varepsilon) \geq \int_E (h-\varepsilon) = \int_E h - \int_E \varepsilon = \int_E h - \varepsilon\, \mu(E) \to \int_E h$, when $\varepsilon \to 0$.

**12.4.10.** By Exercise 12.4.3 (e), $|\nu| = \nu^+ + \nu^-$, so we will prove that if $\lambda = \mu + \nu$ are additive measures and $f \geq 0$, then $\int f d\lambda = \int f d\mu + \int f d\nu$. It is straightforward to verify that it works when $f = \chi_E$, hence for simple functions. ($\leq$) $\int f d\lambda = \sup\{\int s d\lambda : 0 \leq s \leq f\} = \sup\{\int s d\mu + \int s d\nu : 0 \leq s \leq f\} \leq \sup\{\int s d\mu : 0 \leq s \leq f\} + \sup\{\int s d\nu : 0 \leq s \leq f\} = \int f d\mu + \int f d\nu$. ($\geq$) Choose simple functions $s_1$ and $s_2$ s.t. $0 \leq s_1, s_2 \leq f$ and $\int s_1 d\mu > \int f d\mu - \varepsilon/2$ and $\int s_2 d\nu > \int f d\nu - \varepsilon/2$. Let $s = \max\{s_1, s_2\}$. Then $s$ is a simple function s.t. $0 \leq s \leq f$ and we have that $\int f d\lambda \geq \int s d\lambda = \int s d\mu + \int s d\nu \geq \int s_1 d\mu + \int s_2 d\nu \geq \int f d\mu + \int f d\nu - \varepsilon$. Now let $\varepsilon \to 0$.

**12.4.15.** ($\Rightarrow$) Theorem 7.2.12. ($\Leftarrow$) If Monotone Convergence Theorem holds, pick a disjoint family of sets $\{A_k\}$. Then $f_n := \sum_{k=1}^{n} \chi_{A_k}$ is a monotone sequence of measurable functions with pointwise limit $f = \chi_{\cup_{k=1}^{\infty} A_k}$. Therefore, $\sum_{k=1}^{n} \mu(A_k) = \int f_n d\mu \to \int f d\mu = \mu(\cup_{k=1}^{\infty} A_k)$, so apply Exercise 12.4.12.

**12.4.20.** Hint: Consider Banach limit (Exercise 12.2.13).

**12.4.21.** Hint: define $\nu(E) := \inf\{\nu_1(E_1) + \cdots + \nu_n(E_n) : E = \cup E_i, n \in \mathbb{N}\}$.

**12.5.2.** Hint: show that the collection of all Borel sets that are regular is a $\sigma$-algebra that contains all closed sets.
**12.5.3.** Bounded, absolutely continuous, and $\sigma$-additive is straightforward; regular: by definition, $|\mu|$ is regular and $\mu^+(A) = |\mu|(A \cap P)$, $\mu^-(A) = |\mu|(A \cap N)$, so it suffices to prove for one of them, say for $\mu^+$. If $\{G_n\}$ decreases towards $A$ take $\{G'_n\}$ decreasing towards $P$, then $\{G_n \cap G'_n\}$ decreases towards $A \cap P$ so $\mu^+(A) = |\mu|(A \cap P) \geq |\mu|(G_n \cap G'_n) - \varepsilon \geq |\mu|(G_n \cap P) - \varepsilon = \mu^+(G_n) - \varepsilon$; similar for inner regularity.
**12.5.10.** (a) Straightforward. (b) By Exercise 12.5.4, $\lambda(A) := \mu(A \cap F)$ belongs to $rca(X, \mathcal{B})$ so it can be identified with a bounded linear functional $L$ defined by $Lf = \int f d\lambda$. But $\lambda(F^C) = 0$ so $Lf = \int_F f d\lambda$. Further, $\lambda = \mu$ on measurable subsets of $F$ and it follows that $Lf = \int_F f d\mu = \int_F f|F d\mu = Tf$. Thus, $T = L$ whence $\nu = \lambda$ by uniqueness in Theorem 12.5.8.
**12.5.12.** $|Tf| = |\int_{-1}^1 \operatorname{sgn} x f(x)\, dx| \leq \|f\|_{max} \int_{-1}^1 1\, dx = 2\|f\|_{max}$, so $\|T\| \leq 2$. Let $f_n(x) = -1$ for $-1 \leq x \leq -1/n$, $f_n(x) = nx$, for $-1/n \leq x \leq 1/n$, and $f_n(x) = 1$ for $1/n \leq x \leq 1$. Then $Tf_n = \int_{-1}^{-1/n} dx + \int_{-1/n}^{1/n} |nx|\, dx + \int_{1/n}^1 dx = 2(1 - 1/n) + 1/n \to 2$. Thus, $\|T\| = 2$.
**12.5.18.** (a) $|Tf| = |\int_0^1 tf(t)\, dt| \leq \|f\| \int_0^1 t\, dt = \|f\|/2$ so $\|T\| \leq 1/2$. For $f = 1$ we get $\|T\| = 1/2$.

(b) Take $f = 1$ on $[0, 1 - 1/n]$ and $f(1) = 0$. Then $Tf \geq \int_0^{1-1/n} tdt = (1 - 1/n)^2/2 \to 1/2$, so $\|T_1\| = 1/2$. Not attained: if there was $f \in Y$ s.t. $|Tf| = 1/2\|f\|$. WLOG, $\|f\| = 1$, so $1/2 = |Tf| = |\int_0^1 tf(t)|$. Since $f(1) = 0$ there is $a < 1$ s.t. $|f(t)| < 1/3$ for $a < t < 1$. Now, $1/2 \leq \int_0^1 t|f(t)| = \int_0^a t|f| + \int_a^1 t|f| < \int_0^a tdt + \int_a^1 t/3dt = a^2/2 + (1/2 - a^2/2)/3 = a^2/3 + 1/6 < 1/3 + 1/6 = 1/2$. Contradiction.
**12.5.22.** $\nu = m|[0, 1/2] - 3m|[1/2, 1]$ so $\nu([1/3, 2/3]) = m([1/3, 1/2]) - 3m([1/2, 2/3]) = 1/6 - 3(1/6) = -1/3$.
**12.5.27.** Hint: show that the norm of $s_{n,x}$ is $\|D_n\|_1/\pi$ and use Exercises 11.3.20 and 11.3.2.
**12.5.29.** Hint: take $X = \{x_1, \ldots, x_n\}$.
**12.5.32.** Hint: for $f \geq 0$ define $\Lambda^+(f) := \sup\{\Lambda(h) : 0 \leq h \leq f\}$. For general $f$, $\Lambda^+(f) := \Lambda^+(f^+) - \Lambda^+(f^-)$, and $\Lambda^- = \Lambda^+ - \Lambda$.
**12.6.1.** Hint: xonsider using Hahn–Banach theorem.
**12.6.4.** ($\Rightarrow$) Let $\varphi \in Y^*$. Then $\psi := \varphi \circ T \in X^*$. If $\{x_n\}$ converges weakly to $x$, then $\lim \psi(x_n) = \psi(x)$ so $\lim \varphi \circ T(x_n) = \varphi \circ T(x)$. Thus, $\lim \varphi(Tx_n) = \varphi(Tx)$ whence $\{Tx_n\}$ converges weakly to $Tx$. ($\Leftarrow$) Let $\lim x_n = x$ and $\lim Tx_n = y$ in the norms of $X$ and $Y$. Then $\{x_n\}$ converges weakly to $x$ and $\{Tx_n\}$ converges weakly to $y$. By assumption $\{Tx_n\}$ converges weakly to $Tx$ so $y = Tx$. Now, $T$ is bounded by Closed Graph Theorem.
**12.6.6.** It is obvious that $\mathcal{M}$ is a subspace. Let $\{x_n\} \subset \mathcal{M}$, $\lim x_n = x$. Let $\{n_i\} \subset \mathbb{N}$. Will show that is has a subsequence $\{m_i\}$ for which $\{T_{m_i}x\}$ converges. Select a subsequence $\{n_{1,i}\}$ s.t. $\{T_{n_{1,i}}x_1\}$ cvgs weakly to $y_1$. From this one extract a subsequence $\{n_{2,i}\}$ s.t. $\{T_{n_{2,i}}x_2\}$ cvgs weakly to $y_2$. Continue. Now, $m_i := n_{i,i}$ defines a subsequence of $\{n_i\}$ and $\{T_{m_i}x_k\}$ cvgs weakly to $y_k$, for all $k \in \mathbb{N}$. Next, let $y^* \in Y^*$. Then $\|(J_{y_k} - J_{y_j})(y^*)\| = \|y^*y_k - y^*y_j\| = \lim_i \|y^*(T_{m_i}x_k) - y^*(T_{m_i}x_j)\| \leq \liminf_i \|y^*\|\|T_{m_i}x_k - T_{m_i}x_j\|$. It follows that $\|y_k - y_j\| = \|J_{y_k} - J_{y_j}\| \leq \liminf_i \|T_{m_i}x_k - T_{m_i}x_j\| \leq \liminf_i \|T_{m_i}\|\|x_k - x_j\| \leq M\|x_k - x_j\|$. Since $\{x_k\}$ is Cauchy, so is $\{y_k\}$, hence cvgt, $\lim y_k = y$. For any $y^* \in Y^*$, $|y^*(T_{m_i}x) - y^*y| \leq |y^*(T_{m_i}x) - y^*(T_{m_i}x_k)| + |y^*(T_{m_i}x_k) - y^*(y_k)| + |y^*(y_k) - y^*y|$. Therefore, $|y^*(T_{m_i}x) - y^*y| \leq \|y^*\|M\|x - x_k\| + \|y^*\|\|T_{m_i}x_k - y_k\| + \|y^*\|\|y_k - y\| \to 0$, as $k \to \infty$. Thus, $\{T_{m_i}x\}$ converges weakly to $y$, whence $x \in \mathcal{M}$.
**12.6.10.** ($\Rightarrow$) Since weak topology is contained in the norm topology, every weakly open set is open in the norm topology. If T is continuous in the weak topology, then $T^{-1}(G)$ is weakly open hence open in the norm topology.

($\Leftarrow$) If $T$ is continuous in the norm topology then $T \in X^*$. By definition, weak topology is smallest one where each $T \in X^*$ is continuous. Thus, $T$ is continuous in the weak topology.
**12.6.13.** Hint: Let $K$ be a bounded set in $L^p([0, 1])$ and apply the Arzelà-Ascoli Theorem to $\mathcal{F} = \{F : F(x) = \int_0^x f(t)\, dt,\ f \in K\}$.
**12.6.21.** Let $E$ be a measurable subset of $[0, 1]$, let $\{f_n\}$ converge to $f$ in measure, and let $E_n = \{x \in E : |f_n(x) - f(x)| > 1/n\}$. Then $\int_E (f_n - f) = \int_{E_n} + \int_{E-E_n}$. First, $|\int_{E_n}(f - f_n)| \leq 2Mm(E_n)^{1/2} \to 0$. Second, $|\int_{E-E_n}(f - f_n)| \leq (1/n)m(E - E_n) \leq 1/n \to 0$. Now apply Theorem 12.6.11. So, the result hold for $1 < p < \infty$.
**12.6.22.** By Riesz–Fischer Theorem, there exists a subsequence $\{f_{n_k}\}$ that converges to $f$ except on a set $E$ of measure 0. For each $k$, $\exists E_k$ of measure zero such that $f_{n_k}(x) \geq 0$ for $x \in (0, 1) \setminus E_k$. Let $F = (\cup E_k) \cup E$. Then $m(F) = 0$ and outside $F$, $f(x) = \lim f_{n_k}(x) \geq 0$. For weak convergence, argue by contradiction. Suppose that $A = \{x : f(x) < 0\}$ and that $m(A) > 0$. Then $\int f_n\chi_A \to \int f\chi_A < 0$. On the other hand, $f_n\chi_A > 0 \Rightarrow \int f_n\chi_A > 0 \Rightarrow \int f\chi_A \geq 0$.
**12.6.24.** Hint: $f_n(x) = 1$ for $k/2^n + 1/2^{2n+1} \leq x < (k+1)/2^n$ and $0 \leq k \leq 2^n - 1$, and $f_n(x) = 1 - 2^{n+1}$ elsewhere on $[0, 1]$.

**12.6.28.** Since $h$ is continuous, it is integrable over $[0,T]$. Define $M = \int_0^T |h|$. Since $h$ is periodic, $|\int_c^d h| \leq M$ for any real numbers $c$ and $d$. Therefore for any $x \in [a,b]$, we have $\int_a^x f_n = 1/n \int_{na}^{nx} h \leq M/n$ which implies

$$\lim_{n\to\infty} \int_a^x f_n = 0 \tag{12.40}$$

Let $H = \max_{x\in[0,T]} |h(x)|$. We then have $\int_a^b |f_n|^p = 1/n \int_{na}^{nb} |h|^p \leq (b-a)H^p$. We can conclude that $\{f_n\}$ is bounded in $L^p[a,b]$, so $\{f_n\}$ converges weakly to 0 by Exercise 12.6.24 if $1 < p < \infty$.

It remains to show the result holds for $p = 1$. If $H = 0$, then $f_n = 0$ and the result follows trivially. So assume $H > 0$. Fix $g \in L^\infty[a,b]$. If $g$ is a step function, (1) can be used to conclude $\lim_{n\to\infty} \int_a^b gf_n = 0$. Now suppose $g$ is not a step function. Since $[a,b]$ is bounded, $g \in L^1[a,b]$. By Exercise 10.4.17, for any $\varepsilon > 0$ there exists a step function $s \in L^1[a,b]$ such that $\int_a^b |g-s| < \varepsilon/H$. We then have $|\int_a^b gf_n| \leq |\int_a^b sf_n| + \int_a^b |g-s||f_n| < |\int_a^b sf_n| + \varepsilon$ which implies $\lim_{n\to\infty} |\int_a^b gf_n| \leq \varepsilon$. Since $\varepsilon$ can be made arbitrarily small, we can conclude $\lim_{n\to\infty} \int_a^b gf_n = 0$. Therefore $\{f_n\}$ converges weakly to 0 in $L^p([a,b])$.

**12.6.30.** (a) Yes! For any $T \in (L^p[0,1])^*$ and $f \in L^r[0,1]$, we have $|T(f)| \leq \|T\|\|f\|_p \leq \|T\|\|f\|_r$. Thus $T \in (L^r[0,1])^*$, so $(L^p[0,1])^* \subset (L^r[0,1])^*$. We conclude that $\{f_n\}$ converges weakly to $f \in L^r[0,1]$ implies $\{f_n\}$ converges weakly to $f$ in $L^p[0,1]$.

(b) No! Suppose $p > 1$. Define $f_n = n^{1/p}\chi_{[0,1/n)}$ and $f = 0$. Then $\|f_n\|_p = 1$, $\|f_n\|_r = n^{1/p-1/r}$ so $\{f_n\}$ is in $L^r[0,1]$ and $L^p[0,1]$. For any measurable subset $A$ of $[0,1]$, we have $|\int_A f_n| \leq n^{1/p-1}$ which implies $\lim_{n\to\infty} \int_A f_n = 0 = \int_A f$. By Theorem 12.6.11, $\{f_n\}$ converges weakly to $f \in L^p[0,1]$. However, $\lim_{n\to\infty} \|f_n\|_r = \infty$, so $\{f_n\}$ does not converge weakly to $f$ in $L^r[0,1]$ by Theorem 12.6.9.

**12.6.33.** Hint: consider using Egorov's Theorem.

**12.6.35.** ($\Rightarrow$) Let $Y$ be the closed span of the sequence $\{x_n\}$. Then $Y$ is separable Hilbert space with onb $\{e_k\}$. Take a subsequence s.t. $\langle x_{n_k}, e_1\rangle$ converges, then the next one so that it converges for both $e_1, e_2$, etc. Then take the diagonal subsequence which now has $\langle x_{n_k}, h\rangle$ converge for every $h$ in the span of $\{e_k\}$. For $y$ in the closed span, approximate by $h$, $\|y-h\| < \varepsilon$. Then $\langle x_{n_k}, y\rangle = \langle x_{n_k}, h\rangle + \langle x_{n_k}, y-h\rangle$, the first one converges, the second is $\leq M\varepsilon$.

($\Leftarrow$) Let $K$ be the completion of $X$. Let $\{x_n\}$ be a Cauchy sequence in $X$. Then $\{x_n\}$ tends to some element $y \in K$. Let $x_{n_k} \to x$ weakly in $X$. Let $F$ be a continuous linear functional on $K$. Then $\lim F(x_n) = F(y)$ . By weak convergence of the subsequence we must have $\lim F(x_n) = F(x)$ (since the restriction of $F$ to $X$ is a continuous linear functional on $X$). Thus $F(y) = F(x)$ for all $F \in K^*$ which implies $y = x \in X$.

**12.6.36.** Hint: show that $y_n = (x_1 + \cdots + x_n)/n$ is weakly convergent and that $\{x_n\}$ converges weakly to the same limit.

**12.6.43.** Hint: extract a diagonal subsequence and apply Corollary 12.6.12.

**12.6.47.** Suppose that $X^* = c_0$. By Theorem 12.2.9, $X$ is a closed subspace of $X^{**} = \ell^1$. Let us fix $n \in \mathbb{N}$. Since $e_n \in c_0 = X^*$ and $\|e_n\| = \sup|(e_n)_k| = 1$, by Exercise 11.1.3, there is $\{v_k\} \subset X$, $\|v_k\| = 1$, s.t. $\lim_k \langle e_n, v_k\rangle = 1$. Since $\|v_k\| = 1$ and $X \subset \ell^1$, we have that $\|v_k\|_1 = 1$. Notice that $\langle e_n, v_k\rangle = (v_k)_n$. Thus, given $\varepsilon > 0$, there is $k_0$ s.t. $k \geq k_0$ implies $|(v_k)_n - 1| < \varepsilon$, hence $\sum_{j\neq n} |(v_k)_j| < \varepsilon$. Therefore, for $j \neq n$, $(v_k)_j \to 0$, as $k \to \infty$. It follows that $\{v_k\}$ converges weakly to $e_n$. Indeed, If $\{\gamma_n\} \in \ell^\infty = (\ell^1)^*$, then $\sum_j (v_k)_j\gamma_j \to \sum_j (e_n)_j\gamma_j = \gamma_n$, as $k \to \infty$. The limit is correct because $|\sum_j (v_k)_j\gamma_j - \gamma_n| \leq |\sum_{j\neq n}(v_k)_j\gamma_j| + |(v_k)_n - 1||\gamma_n| \leq 2\varepsilon\|\gamma\|_\infty$. By Exercise 12.6.44, $\{v_k\}$ converges strongly to $e_n$. Since $X$ is closed in $\ell^1$, we get that $e_n \in X$. This can be done for every $n$, so the span of $\{e_n\}$ is in $X$. But the span is dense in $\ell^1$ so $X = \ell^1$, hence $X^* = \ell^\infty$, which is a contradiction.

# Bibliography


[1] René-Louis Baire, *Sur les fonctions de variables réelles*, Annali di Mat. (3), Vol. 3, (1899), 1–123.

[2] Stefan Banach, *Sur les opérations dans les ensembles abstraits et leur application aux équations intégrales*, Fundam. Math., 3, (1922), 133–181.

[3] Stefan Banach, *Sur le théorème de M. Vitali*, Fund. math. 5, (1924), 130–136.

[4] Stefan Banach, *Sur les fonctionelles linéaires. II*, [On linear functionals, II], Stud. Math., 1, (1929), 223–239.

[5] Stefan Banach, *Théorie des opérations linéaires*, Monogr. Mat., Warszawa, 1, PWN - Panstwowe Wydawnictwo Naukowe, Warszawa, 1932.

[6] Stefan Banach and Stanislaw Saks, *Sur la convergence forte dans les champs $L^p$*, Stud. Math., 2, (1930), 51–57.

[7] Stefan Banach and Hugo Steinhaus, *Sur le principe de la condensation de singularités*, Fundam. Math., 9, (1927), 50–61.

[8] Stefan Banach and Alfred Tarski, *Sur la décomposition des ensembles de points en parties respectivement congruentes*, Fund. math., 6, (1924), 244–277.

[9] K. P. S. Bhaskara Rao and M. Bhaskara Rao, *Theory of charges*, Pure Appl. Math., 109, Academic Press, Inc., New York, 1983.

[10] Irénée-Jules Bienaymé, *Considérations à l'appui de la découverte de Laplace*, C. R. Acad. Sci., Paris 37, (1853), 309–324.

[11] Ralph P. Boas, *Can we make mathematics intelligible?*, American Mathematical Monthly 88 (1981), 727–731.

[12] Salomon Bochner, *Eine Bemerkung zum Satz von Fubini*, Fundam. Math., 20 (1933), 277–280.

[13] Frederic Bohnenblust and Andrew Sobczyk, *Extensions of functionals on complex linear spaces*, Bull. Am. Math. Soc., 44, (1938), 91–93.

[14] Émile Borel, *Leçons sur la théorie des fonctions*, Paris: Gauthier-Villars et Fils., 1898.

[15] Émile Borel, *Les probabilités denombrables et leurs applications arithmétiques*, Rend. Circ. Mat. Palermo, 27, (1909), 247–271

[16] Emile Borel, *Mécanique Statistique et Irréversibilité*, [Statistical mechanics and irreversibility]. J. Phys. 5e série 3, (1913), 189–196.

[17] Viktor Bunyakovskiĭ, *Sur quelques inégalités concernant les intégrales ordinaires et les intégrales aux différences finies*, [On some inequalities about integrals and integrals of finite differences]. Mémoires de l'Acad. de St. Pétersbourg (VII) 1, No. 9, (1859).

[18] Francesco Paolo Cantelli, *Sulla probabilità come limite della frequenza*, Rom. Acc. L. Rend. (5), 26, No. 1, (1917), 39–45.

[19] Georg Cantor, *Über unendliche lineare Punktmannigfaltigkeiten. No. 6.*, Math. Ann., 23, (1883), 453–488.

[20] Georg Cantor, *De la puissance des ensembles parfaits de points*, Acta Math., 4, (1884), 381–392.

[21] Constantin Carathéodory, *Über das lineare Maß von Punktmengen – eine Verallgemeinerung des Längenbegriffs*, Nachr. Ges. Wiss. Göttingen, Math.-Phys. Kl., 1914, (1914), 404–426.

[22] Constantin Carathéodory, *Bemerkungen zum Riesz-Fixcherschen Satz und zur Ergodentheorie*, Abh. Math. Sem. Hansischen Univ., 14, (1941), 351–389.

[23] Constantin Carathéodory, *Vorlesungen über reelle Funktionen*, Leipzig und Berlin: B. G. Teubner, 1918.

[24] Pafnuty Chebyshev, *Des valeurs moyennes*, J. Math. Pures Appl. 2., 12, (1867), 177–184.

[25] John B. Conway, *Subnormal operators*, Research Notes in Math., 51, Pitman Advanced Publishing Program, Boston-London-Melbourne, 1981.

[26] Mahlon M. Day, *The spaces $L^p$ with* $0 < p < 1$, Bull. Am. Math. Soc., 46, (1940), 816–823.

[27] Mahlon M. Day, *Normed Linear Spaces*, Ergebnisse der mathematik und Ihrer Grenzgebiete, Springer-Verlag, Berlin 1958.

[28] Charles de la Vallée-Poussin, *Sur l'intégrale de Lebesgue*, Trans. Am. Math. Soc., 16, (1915), 435–501.

[29] Nelson Dunford, *Uniformity in linear spaces*, Trans. Amer. Math. Soc., 44, (1938), No. 2, 305–356.

[30] Dmitrii Egorov, *Sur les suites de fonctions mesurables*, C. R. Acad. Sci., Paris, **152**, (1911), 244–246.

[31] Georg Faber, *Über stetige Funktionen. (Zweite Abhandlung)*, Math. Ann., 69, (1910), 372–443.

[32] Pierre Fatou, *Séries trigonométriques et séries de Taylor*, Acta Math., 30, (1906), 335–400.

[33] Grigorii M. Fichtenholz and Leonid V. Kantorovich, *Sur les operations linéaires dans l'espace des fonctions bornées*, Studia Math. 5 (1934), 69–98.

[34] Ernst Sigismund Fischer, *Sur la convergence en moyenne*, C. R. Acad. Sci., Paris, 144, (1907), 1022–1024.

[35] Maurice Fréchet, *Sur quelques points du calcul fonctionnel*, Rend. Circ. Mat. Palermo, Vol. 22, (1906), 1–74.

[36] Maurice Fréchet, *Sur les ensembles de fonctions et les opérations linéaires*, C. R. Acad. Sci., Paris, 144, (1907), 1414–1416.

[37] Maurice Fréchet, *Sur les opérations linéaires*, III. [On linear operations, III] Trans. Amer. Math. Soc., 8, (1907), no. 4, 433–446.

[38] Maurice Fréchet, *Sur l'intégrale d'une fonctionnelle étendue à un ensemble abstrait*, Bull. Soc. Math. Fr., 43, (1915), 248–265.

[39] Maurice Fréchet, *Des familles et fonctions additives d'ensembles abstraits*, Fund. math. 4, No. 1, (1924) 329–365.

[40] Maurice Fréchet, *Les espaces abstraits et leur théorie considérée comme introduction à l'analyse générale*, Collection de monographies sur la théorie des fonctions. Paris: Gauthier-Villars, 1928.

[41] Ivar Fredholm, *Sur une nouvelle méthode pour la résolution du problème de Dirichlet*, [On a new method for solving the Dirichlet problem], Stockh. Öfv. 57, (1900), 39–46.

[42] Guido Fubini, *Sugli integrali multipli*, Rom. Acc. L. Rend. (5), 16 (1), (1907), 608–614.

[43] Bernard R. Gelbaum and John M. H. Olmsted, *Counterexamples in Analysis*, Mineola, NY: Dover Publications, Inc., 2003.

[44] Hermann Grassmann, *Die Lineale Ausdehnungslehre, ein neuer Zweig der Mathematik*, [The Theory of Linear Extension, a New Branch of Mathematics], Leipzig 1844.

[45] Jacques Hadamard, *Sur les opérations fonctionelles.*, C. R. Acad. Sci., Paris, 136, (1903), 351–354.

[46] Hans Hahn, *Über die Darstellung gegebener Funktionen durch singuläre Integrale. 1. u. 2. Mitteilung.*, Wien. Denkschr., 93, (1916), 656–692.

[47] Hans Hahn, *Theorie der reellen Funktionen*, Berlin: J. Springer, 1921.

[48] Hans Hahn, *Über Folgen linearer Operationen*, Monatsh. Math. Phys., 32, No. 1, (1922), 3–88.

[49] Hans Hahn, *Über lineare Gleichungssysteme in linearen Räumen*, [About linear systems of equations in linear spaces], J. Reine Angew. Math., 157, (1927), 214–229.

[50] Hans Hahn, *Über die Multiplikation total-additiver Mengenfunktionen*, Ann. Sc. Norm. Super. Pisa, II. Ser., 2, (1933), 429–452.

[51] Paul R. Halmos, *Measure theory*, (University Series in Higher Mathematics) New York: D. Van Nostrand Co., Inc.; London: Macmillan & Co., Ltd., 1950.

[52] Paul R. Halmos, *Naive Set Theory*, (Undergraduate Texts in Mathematics) New York: Springer-Verlag, 1974.

[53] Hermann Hankel, *Untersuchungen über die unendlich oft oscillirenden und unstetigen Functionen*, Math. Ann., 20, (1882), 63–112.

[54] Axel Harnack, *Ueber den Inhalt von Punktmengen.*, Math. Ann., 25, (1884), 241–250.

[55] Axel Harnack, *Die allgemeinen Sätze über den Zusammenhang der Functionen einer reellen Variabeln mit ihren Ableitungen. II. Theil*, Math. Ann., 24, (1884), 217–252.

[56] Felix Hausdorff, *Grundzüge der Mengenlehre*, Leipzig: Veit & Comp., 1914.

[57] Felix Hausdorff, *Zur Theorie der linearen metrischen Räume*, J. Reine Angew. Math., 167, (1932), 294-311.

[58] Ernst Hellinger and Otto Toeplitz, *Grundlagen für eine Theorie der unendlichen Matrizen*, Math. Ann., 69, (1910), 289–330.

[59] Eduard Helly, *Über lineare Funktionaloperationen*, Wien. Ber., 121, (1912), 265–297.

[60] Eduard Helly, *Über Systeme linearer Gleichungen mit unendlich vielen Unbekannten*, Monatsh. Math. Phys., 31, (1921), 60–91.

[61] David Hilbert, *Grundzüge einer allgemeinen Theorie der linearen Integralgleichungen. Vierte Mitteilung*, Nachr. Ges. Wiss. Göttingen, Math.-Phys. Kl., 1906, (1906), 157–227.

[62] Theophil Henry Hildebrandt, *On bounded linear functional operations*, Trans. Amer. Math. Soc., 36 (1934), no. 4, 868–875.

[63] Ernest William Hobson, *Generalization of a theorem of F. Riesz*, J. Lond. Math. Soc., 1, (1926), 211–218.

[64] O. Hölder, *Ueber einen Mittelwertsatz.*, Gött. Nachr., 1889, (1889), 38–47.

[65] Camille Jordan, *Sur la série de Fourier*, C. R. Acad. Sci., Paris, 92, (1881), 228–230.

[66] Camille Jordan, *Remarques sur les intégrales définies*, Journ. de Math. (4), 8, (1892), 69–99.

[67] Casimir Kuratowski, *Sur l'opération $\overline{A}$ de l'analysis situs*, Fundam. Math., Vol. 3, (1922), 182–199.

[68] Henri Lebesgue, *Intégrale, longueur, aire.*, Annali di Mat. (3), Vol. 7, (1902), 231–359.

[69] Henri Lebesgue, *Leçons sur l'intégration et la recherche de fonctions primitives*, Paris: Gauthier-Villars, 1904.

[70] Henri Lebesgue, *Sur une propriété des fonctions*, C. R. Acad. Sci., Paris, 137, (1904), 1228–1230.

[71] Henri Lebesgue, *Leçons sur les séries trigonométriques professées au Collège de France*, Paris: Gauthier-Villars, 1906.

[72] Henri Lebesgue, *Sur la recherche des fonctions primitives par l'intégration*, Rom. Acc. L. Rend. (5), 16, No. 1, (1907), 283–290.

[73] Henri Lebesgue, *Sur l'intégration des fonctions discontinues*, Ann. Sci. Éc. Norm. Supér. (3), 27, (1910), 361–450.

[74] Henri Lebesgue, *Sur les intégrales singulières*, Ann. Fac. Sci. Univ. Toulouse, III. Ser., 1, (1910), 25–117.

[75] Edgar Lorch, *On a calculus of operators in reflexive vector spaces*, Trans. Am. Math. Soc., 45, (1939), 217–234.

[76] Edward McShane, E. J., *Linear functionals on certain Banach spaces*, Proc. Amer. Math. Soc., 1, (1950), 402–408.

[77] Hermann Minkowski, *Geometrie der Zahlen. In 2 Lieferungen. Lfg. 1*, Leipzig: B. G. Teubner, 1896.,

[78] John von Neumann, *Allgemeine Eigenwerttheorie Hermitescher Funktionaloperatoren*, Math. Ann., 102, (1929), 49–131.

[79] Otton Marcin Nikodym, *Sur une généralisation des intégrales de M. J. Radon*, Fundam. Math., 15, (1930), 131–179.

[80] Otton Marcin Nikodym, *Contribution à la théorie des fonctionnelles linéaires en connexion avec la théorie de la mesure des ensembles abstraits*, Mathematica, Cluj, 5, (1931), 130–141.

[81] Otton Marcin Nikodym, *Sur les suites convergentes de fonctions parfaitement additives d'ensemble abstrait*, Monatsh. Math. Phys., 40, No. 1, (1933), 427–432.

[82] William F. Osgood, *Non-Uniform Convergence and the Integration of Series Term by Term*, Am. J. Math., Vol. 19, (1897), 155–190.

[83] Giuseppe Peano, *Applicazioni geometriche del calcolo infinitesimale*, Torino. Bocca. (1887).

[84] Giuseppe Peano, *Calcolo geometrico secondo l'Ausdehnungslehre di H. Grassmann preceduto dalle operazioni della logica deduttiva*, [Geometric calculations according to the *Extensions theory* of H. Grassmann preceded by the operations of deductive logic.] Fratelli Bocca Editori, Torino, 1888.

[85] Giuseppe Peano, *Arithmetices principia, nova methodo exposita*, [The principles of arithmetic, presented in a new manner]. Bocca, Turin, 1889.

[86] John Srdjan Petrovic, *Advanced Calculus: Theory and Practice*, Chapman and Hall/CRC; 2nd edition (2020).

[87] Johann Radon, *Theorie und Anwendungen der absolut additiven Mengenfunktionen*, Wien. Ber., 122, (1913), 1295–1438.

[88] Frigyes Riesz, *Sur les systèmes orthogonaux de fonctions*, C. R. Acad. Sci., Paris, 144, (1907), 615–619.

[89] Frigyes Riesz, *Sur une espèce de géométrie analytique des systèmes de fonctions sommables*, C. R. Acad. Sci., Paris, 144, (1907), 1409–1411.

[90] Frigyes Riesz, *Sur les suites de fonctions mesurables*, C. R. Acad. Sci., Paris, 148, (1909), 1303–1305.

[91] Frigyes Riesz, *Über integrierbare Funktionensysteme*, Mat. Fiz. Lapok, 19, (1910), 1–56.

[92] Frigyes Riesz, *Untersuchungen über Systeme integrierbarer Funktionen*, [A study of systems of integrable functions], Math. Ann., 69, (1910), 449–497.

[93] Frigyes Riesz, *Sur les opérations fonctionnelles linéaires*, C. R. Acad. Sci., Paris, 149, (1910), 974–977.

[94] Frigyes Riesz, *Sur certains systèmes singuliers d'équations intégrales*, Ann. Sci. Éc. Norm. Supér. (3), 28, (1911), 33–62.

[95] Frigyes Riesz, *Les systèmes d'équations linéaires à une infinite d'inconnues*, Paris: Gauthier-Villars, (Collection Borel), 1913.,

[96] Friedrich Riesz, *Über lineare Funktionalgleichungen*, Acta Math., 41, (1916), 71–98.

[97] Frigyes Riesz, *Sur le théorème de M. Egoroff et sur les opérations fonctionnelles linéaires*, Acta Litt. Sci. Szeged, 1, (1922), 18–26.

[98] Frigyes Riesz, *Elementarer Beweis des Egoroffschen Satzes*, Monatsh. Math. Phys., 35, (1928), 243–248.

[99] Frigyes Riesz, *Sur la convergence en moyenne.*, Acta Litt. Sci. Szeged, 4, (1928), 58–64, 182–185.

[100] Frigyes Riesz, *Sur l'existence de la dérivée des fonctions monotones et sur quelques problèmes qui s'y rattachent*, Acta Litt. Sci. Szeged, 5, (1932), 208–221.

[101] Frigyes Riesz, *Zur Theorie des Hilbertschen Raumes*, [On the theory of Hilbert spaces], Acta Litt. Sci. Szeged, 7, (1934), 34–38.

[102] Stanislaw Saks, *Addition to the note: "On some functionals" [Trans. Amer. Math. Soc.* **35** *(1933), no. 2, 549–556; 1501701]*, Trans. Amer. Math. Soc., 35, No. 4, (1933), 965–970.

[103] Stanislaw Saks, *Theory of the integral.* 2. ed. English translation by L. C. Young. With two additional notes by S. Banach. Dover Publications, Inc., New York, 1964.

[104] Stanislaw Saks, *Integration in abstract metric spaces*, Duke Math. J., 4, No. 2, (1938), 408–411.

[105] Juliusz Schauder, *Über die Umkehrung linearer, stetiger Funktionaloperationen*, [About the inversion of linear, continuous functional operations], Stud. Math., 2, (1930), 1–6.

[106] Ludwig Scheeffer, *Allgemeine Untersuchungen über Rectification der Curven*, Acta Math., 5, (1884), 49–82.

[107] Ludwig Scheeffer, *Zur Theorie der Functionen einer reellen Veränderlichen.*, Acta Math., 5, (1884), 183–194, 279–296.

[108] Erhard Schmidt, *Über die Auflösung linearer Gleichungen mit unendlich vielen Unbekannten*, [On the solution of linear equations with infinitely many variables], Rend. Circ. Mat. Palermo, 25, (1908), 53–77.

[109] Hermann Amandus Schwarz, *Ueber ein die Flächen kleinsten Flächeninhalts betreffendes Problem der Variationsrechnung*, [A problem of variational calculus: the surfaces with a minimum surface area]. Acta soc. scient. Fenn. 15 (1885), 315–362.

[110] Carlo Severini, *Sulle successioni di funzioni ortogonali*, Atti Acc. Gioenia (5), 3, (1910).

[111] Wacław Sierpiński, *Sur une propriété générale des ensembles de points*, C. R. Acad. Sci., Paris, 162, (1916), 716–717.

[112] Wacław Sierpiński, *Démonstration d'un théorème sur les fonctions additives d'ensemble*, Fund. math, 5, (1924), 262–264.

[113] Wacław Sierpiński, *Remarque sur le théorème de M. Egoroff*, C. R. Soc. Sci. Varsovie, Cl. III, 21, (1928), 84–87.

[114] Henry J. Stephen Smith, *On the Integration of Discontinuous Functions*, Proc. Lond. Math. Soc., Vol. 6, (1874/75), 140–153.

[115] Hugo Steinhaus, *Additive und stetige Funktionaloperationen*, Math. Z., 5, (1919), No. 3–4, 186–221.

[116] Otto Stolz, *Über einen zu einer unendlichen Punktmenge gehörigen Grenzwert*, Math. Ann., 23, (1883), 152–156.

[117] Karl Stromberg, *The Banach–Tarski paradox*, Am. Math. Mon., 86, (1979), 151–161.

[118] Leonida Tonelli, *Sull'integrazione per parti*, Atti della Accademia Nazionale dei Lincei. (5). 18 (2), (1909), 246–253.

[119] Giuseppe Vitali, *Sul problema della misura dei gruppi di punti di una retta*, Bologna: Gamberini e Parmeggiani, 1905.

[120] Giuseppe Vitali, *Una proprietà delle funzioni misurabili*, Ist. Lombardo, Rend., II. Ser., 38, (1905), 599–603.

[121] Giuseppe Vitali, *Sulle funzioni integrali*, Torino Atti, 40, (1905), 1021–1034.

[122] Giuseppe Vitali, *Sull' integrazione per serie*, Rend. Circ. Mat. Palermo, 23, (1907), 137–155.

[123] Giuseppe Vitali, *Sui gruppi di punti e sulle funzioni di variabili reali*, Torino Atti, 43, (1908), 229–246.

[124] Vito Volterra, *Alcune osservazioni sulle funzioni punteggiate discontinue*, Battaglini G., Vol. 19, (1881), 76–87.

[125] Norbert Wiener, *Limit in terms of continuous transformations*, Bull. Soc. Math. France, 50, (1924), 119–134.

[126] William Henry Young and Grace Chisholm Young, *On the existence of a differential coefficient*, Proc. Lond. Math. Soc. (2), 9, (1911), 325–335.

# Subject Index

# Author Index

Kolibri 360 GmbH
Ferdinand-Jühlke-Straße 7
99095 Erfurt, Deutschland
produktsicherheit@kolibri360.de